Advanced Polymer Simulation and Processing- Volume I

Advanced Polymer Simulation and Processing- Volume I

Editors

Célio Bruno Pinto Fernandes
Salah Aldin Faroughi
Luís L. Ferrás
Alexandre M. Afonso

MDPI • Basel • Beijing • Wuhan • Barcelona • Belgrade • Manchester • Tokyo • Cluj • Tianjin

Editors
Célio Bruno Pinto Fernandes
University of Minho
Portugal

Salah Aldin Faroughi
Texas State University
USA

Luís L. Ferrás
University of Porto
Portugal

Alexandre M. Afonso
University of Porto
Portugal

Editorial Office
MDPI
St. Alban-Anlage 66
4052 Basel, Switzerland

This is a reprint of articles from the Special Issue published online in the open access journal *Polymers* (ISSN 2073-4360) (available at: https://www.mdpi.com/journal/polymers/special_issues/Polym_Simul_Process).

For citation purposes, cite each article independently as indicated on the article page online and as indicated below:

LastName, A.A.; LastName, B.B.; LastName, C.C. Article Title. *Journal Name* **Year**, *Volume Number*, Page Range.

Volume 1
ISBN 978-3-0365-6664-1 (Hbk)
ISBN 978-3-0365-6665-8 (PDF)

Volume 1-2
ISBN 978-3-0365-6662-7 (Hbk)
ISBN 978-3-0365-6663-4 (PDF)

Contents

About the Editors

Célio Bruno Pinto Fernandes

Dr. Célio Fernandes obtained his Education of Mathematics degree from the University of Minho (Portugal) in 2005. After graduating, C. Fernandes obtained his Applied Mathematics MSc degree at the University of Porto (Portugal) in 2007. During this period, C. Fernandes employed spectral methods to describe the melt flow that occurs in the polymer extrusion process. Afterwards, C. Fernandes joined the Department of Polymer Engineering at the University of Minho (Portugal) where he completed his Ph.D. degree in Science and Engineering of Polymers and Composites in 2012. During this period, C. Fernandes made important contributions to the field of the Injection Molding Process by applying multi-objective evolutionary algorithms to solve the inverse problem of finding the best injection molding parameters to achieve predefined criteria. C. Fernandes was a visiting Post-Doctoral researcher at MIT, USA in 2017. C. Fernandes has been working with the open-source computational fluid dynamics library OpenFOAM. He has established new numerical methods for the solution of viscoelastic matrix-based fluids using the finite volume method, such as an immersed boundary method able to fully-resolve particle-laden viscoelastic flows and developed a fully implicit log-conformation tensor coupled algorithm for the solution of incompressible non-isothermal viscoelastic flows.

Salah Aldin Faroughi

Dr. Salah A. Faroughi is an Assistant Professor in the Ingram School of Engineering at Texas State University. He holds a B.Sc. and M.Sc. in Mechanical Engineering. He obtained his Ph.D. in Civil and Environmental Engineering at Georgia Institute of Technology, where he investigated the effect of microstructures (e.g., particle shape, size, type, orientation, and rearrangement) on the rheophysics and thermophysics of complex particulate matter (e.g., fluids and composites). He did his postdoctoral study at the Massachusetts Institute of Technology in the department of Mechanical Engineering. His postdoc research focused on the development of high-performance computing algorithms augmented using physics-based deep learning models to explore the dynamics of particle-laden viscoelastic materials. His studies provided answers to multiple long-lived fundamental questions and resulted in several well-cited papers published in prestigious journals such as *Nature*, *Journal of Applied Physics*, *Physical Review E*, and *JNNFM*, among many others.

Luís L. Ferrás

L. L. Ferrás is an Assistant Professor at the Department of Mechanical Engineering, Faculty of Engineering, University of Porto (FEUP) and a researcher at the Centre for Mathematics, University of Minho, Portugal. He received his Ph.D. in Science and Engineering of Polymers and Composites from the University of Minho in 2012, a Ph.D. in Mathematics from the University of Chester in 2019, and was a visiting researcher at MIT in 2016. His current research interests are numerical analysis, applied mathematics, partial and fractional differential equations, mathematical modeling, computational mechanics, computational fluid dynamics, complex viscoelastic flows, rheology, anomalous diffusion, and machine learning.

Alexandre M. Afonso

Afonso graduated in Chemical Engineering from Faculty of Engineering of the University of Porto (FEUP) in 2000, with a final-year Research Project at the Universidad Politecnica de Catalunya graded with an Honor Grade (10/10). In 2005, Afonso completed an MSc in Heat and Fluid mechanics and, in 2010, completed a PhD degree in Biological and Chemical Engineering from FEUP. Currently, Afonso is an Assistant Professor at the Department of Mechanical Engineering at FEUP.

Preface to "Advanced Polymer Simulation and Processing- Volume I"

Polymer-processing techniques are of the utmost importance for producing polymeric parts. They must produce parts with the desired qualities, which are usually related to mechanical performance, dimensional conformity, and appearance. Aiming to maximize the overall efficiency of the polymer-processing techniques, advanced modeling codes along with experimental measurements are needed to simulate and optimize the processes. Our objective with this reprint is to provide a text that exploits the digital transformation of the plastics industry, both through the creation of more robust and accurate modeling tools and the development of cutting-edge experimental techniques.

We would like to thank all those who have supported us in completing this work.

Célio Bruno Pinto Fernandes, Salah Aldin Faroughi, Luís L. Ferrás, and Alexandre M. Afonso
Editors

polymers

MDPI

Editorial

Advanced Polymer Simulation and Processing

Célio Fernandes [1,*], Salah A. Faroughi [2], Luís L. Ferrás [3,4] and Alexandre M. Afonso [4]

1 LASI-Associate Laboratory of Intelligent Systems, Institute for Polymers and Composites, Polymer Engineering Department, School of Engineering of the University of Minho, Campus of Azurém, 4800-058 Guimarães, Portugal
2 Geo-Intelligence Laboratory, Ingram School of Engineering, Texas State University, San Marcos, TX 78666, USA; salah.faroughi@txstate.edu
3 Center of Mathematics (CMAT), University of Minho, Campus de Azurém, 4800-058 Guimarães, Portugal; lferras@fe.up.pt
4 CEFT, Department of Mechanical Engineering, University of Porto, 4200-465 Porto, Portugal; aafonso@fe.up.pt
* Correspondence: cbpf@dep.uminho.pt

Citation: Fernandes, C.; Faroughi, S.A.; Ferrás, L.L.; Afonso, A.M. Advanced Polymer Simulation and Processing. *Polymers* **2022**, *14*, 2480. https://doi.org/10.3390/polym14122480

Received: 27 May 2022
Accepted: 16 June 2022
Published: 18 June 2022

Polymer processing techniques are of paramount importance in the manufacture of polymer parts. The key focus is on producing parts with the desired quality, which usually refers to mechanical performance, dimensional conformity, and appearance. To maximize the overall efficiency of polymer processing techniques, advanced modeling codes are needed along with experimental setups to simulate and optimize the processes.

To improve modeling codes for polymer processing techniques, Fernandes et al. [1] developed an incompressible, non-isothermal finite volume method based on the arbitrary Lagrangian–Eulerian formulation (ALE) to calculate the viscous flow of polymer melts obeying the Herschel–Bulkley constitutive equation. The new method is employed to compute the extrudate swell ratio for Bingham and Herschel–Bulkley flows (shear thinning and shear thickening). Spanjaards et al. [2] numerically investigated the effect of thixotropy on the swelling of a 2D planar extrudate for constant and fluctuating flow rates, and the effect of thixotropy on the swelling behavior of a 3D rectangular extrudate for a constant flow rate. It was concluded that the presence of a low-viscosity outer layer and a high-viscosity core in the die has a pronounced effect on the swelling ratio for thixotropic fluids. Marschik et al. [3] proposed new leakage–flow models that allow the effect of flight clearance to be included in the analysis of the melt-conveying zones of the extrusion process. They derived regression models to locally predict the shear-thinning flow through the flight clearance. Castelo et al. [4] proposed a moving least squares meshless interpolation technique to simulate Newtonian, generalized Newtonian, and viscoelastic fluid flows. The code verification and testing were performed using numerical stabilizers for the Oldroyd-B flow solution in a 2D cavity and for a Phan–Thien–Tanner fluid in a complex 3D geometry. Faroughi et al. [5] developed a meta-model using a stacking technique to accelerate the calculation of the drag coefficient of a spherical particle moving through viscoelastic fluids. The meta-model combines random forest (RF), extreme gradient boosting (XGBoost), and deep neural network (DNN) models and outputs a prediction based on the individual learner's predictions and a DNN meta-regulator. The meta-model consistently outperformed the individual models in predicting the drag ground truth, and it provided accurate prediction in just a fraction of time compared with the conventional drag calculation. Huang et al. [6] proposed an artificial backpropagation neural network (BPNN) to render result predictions for the injection molding process. By inputting the plastic temperature, mold temperature, injection speed, holding pressure, and holding time in the molding parameters, the end of filling pressure, maximum cooling time, warpage, and shrinkage were accurately predicted by the BPNN. Baral et al. [7] proposed an accurate ab initio molecular dynamics and density functional theory calculations to investigate the structure and properties of the arginine–glycine–aspartate (RGD) sequence. The microscopic parameters determined from the quantum mechanical calculations proved useful in defining the range and strength of the complex molecular interactions between the RGD peptide and the integrin receptor. The study is also important in the context of conditions prevailing in the human body and relevant to health issues.

With a view to improving experimental setups for polymer processing techniques, Drozdov et al. [8] developed constitutive equations for the thermo-mechanical behavior of poly(ether ether ketone) under uniaxial deformation based on experimental observations made for uniaxial tensile tests, relaxation tests, and creep tests at various stresses in a wide temperature range. The activation energies for the elastoplastic, viscoelastic, and viscoelastoplastic responses adopt similar values at temperatures above the glass transition point. Castéran et al. [9] used the extrusion process for the controlled thermo-mechanical degradation of polyethylene in recycling applications. A Carreau–Yasuda model was developed to predict the rheological behavior based on in-line measurements of the die pressure of a reactive extrusion process. The linear viscoelastic behaviors were also used to predict the molecular weight distributions of the final products by an inverse rheological method. In addition, support vector machine regression (SVR) and sparse proper generalized decomposition (sPGD) techniques were chosen to predict the process outputs. Hirsch et al. [10] experimentally and numerically investigated the anisotropic mechanical behavior of a hybrid injection molding process using a continuous fiber-reinforced thermoplastic. The prediction of the mechanical behavior of the hybrid test structure under flexural loading by numerical simulation was significantly improved, resulting in a reduction in the deviation between the numerically predicted and experimentally measured flexural strength. Torres-Alba et al. [11] developed a hybrid cooling model based on the use of newly designed fluted conformal cooling channels combined with inserts of Fastcool material. The obtained results are in line with the sustainability criteria for green molds, which focus on reducing the cycle time and improving the quality of the complex molded parts. Li et al. [12] proposed a method to predict the warpage of crystalline parts molded by the rapid heat cycle molding process. Multi-layer models were created to predict the warpage with the same thicknesses as the skin–core structures in the molded parts. The numerical prediction results were compared with the experimental results, which showed that the average errors between the predicted warpage and the average experimental warpage were less than 10%. Lozano et al. [13] reviewed additive manufacturing (AM) technology to produce customized products with more complex geometries and short life cycles (flexibility) to keep up with the new variables imposed by the manufacturing environment. They address specific issues related to the characterization of the injection molding materials and molds most commonly used in this type of technology, their mechanical properties (part and mold), designs for all types of geometries, and costs. In addition, they highlight the advantages of this alternative manufacturing process, which is considered a desirable technology worldwide.

The editors are confident that this book will help researchers further understand the principles of polymer processing techniques, such as extrusion and injection molding, from both numerical and experimental perspectives. This book is also a useful resource for readers interested in the latest technologies, such as additive manufacturing. It will further advance the development of and improvements in data-driven algorithm research and promote the overlap and integration with polymer processing techniques.

Author Contributions: Conceptualization, C.F.; writing, C.F.; review and editing, C.F., S.A.F., L.L.F., and A.M.A. All authors have read and agreed to the published version of the manuscript.

Funding: This research was funded by FEDER through the COMPETE 2020 Programme and National Funds through FCT—Portuguese Foundation for Science and Technology under the projects UIDB/05256/2020 and UIDP/05256/2020. It was also funded by FCT through CMAT (Centre of Mathematics of the University of Minho) through projects UIDB/00013/2020 and UIDP/00013/2020.

Conflicts of Interest: The authors declare no conflict of interest.

References

1. Fernandes, C.; Fakhari, A.; Tukovic, Ž. Non-isothermal free-surface viscous flow of polymer melts in pipe extrusion using an open-source interface tracking finite volume method. *Polymers* **2021**, *13*, 4454. [CrossRef]
2. Spanjaards, M.; Peters, G.; Hulsen, M.; Anderson, P. Numerical study of the effect of thixotropy on extrudate swell. *Polymers* **2021**, *13*, 4383. [CrossRef] [PubMed]
3. Marschik, C.; Roland, W.; Dörner, M.; Steinbichler, G.; Schöppner, V. Leakage-flow models for screw extruders. *Polymers* **2021**, *13*, 1919. [CrossRef] [PubMed]
4. Castelo, A.; Afonso, A.M.; Bezerra, W.D.S. A Hierarchical grid solver for simulation of flows of complex fluids. *Polymers* **2021**, *13*, 3168. [CrossRef] [PubMed]
5. Faroughi, S.A.; Roriz, A.I.; Fernandes, C. A Meta-model to predict the drag coefficient of a particle translating in viscoelastic fluids: A machine learning approach. *Polymers* **2022**, *14*, 430. [CrossRef] [PubMed]
6. Huang, Y.-M.; Jong, W.-R.; Chen, S.-C. Transfer learning applied to characteristic prediction of injection molded products. *Polymers* **2021**, *13*, 3874. [CrossRef] [PubMed]
7. Baral, K.; Adhikari, P.; Jawad, B.; Podgornik, R.; Ching, W.-Y. Solvent effect on the structure and properties of RGD peptide (1FUV) at body temperature (310 K) using ab initio molecular dynamics. *Polymers* **2021**, *13*, 3434. [CrossRef] [PubMed]
8. Drozdov, A.D.; Christiansen, J. Thermo-mechanical behavior of poly(ether ether ketone): Experiments and modeling. *Polymers* **2021**, *13*, 1779. [CrossRef] [PubMed]
9. Castéran, F.; Delage, K.; Hascoët, N.; Ammar, A.; Chinesta, F.; Cassagnau, P. Data-driven modelling of polyethylene recycling under high-temperature extrusion. *Polymers* **2022**, *14*, 800. [CrossRef] [PubMed]
10. Hirsch, P.; John, M.; Leipold, D.; Henkel, A.; Gipser, S.; Schlimper, R.; Zscheyge, M. Numerical simulation and experimental validation of hybrid injection molded short and continuous fiber-reinforced thermoplastic composites. *Polymers* **2021**, *13*, 3846. [CrossRef] [PubMed]
11. Torres-Alba, A.; Mercado-Colmenero, J.M.; Caballero-Garcia, J.D.D.; Martin-Doñate, C. A Hybrid cooling model based on the use of newly designed fluted conformal cooling channels and fastcool inserts for green molds. *Polymers* **2021**, *13*, 3115. [CrossRef] [PubMed]
12. Li, J.; Bei, J.; Liu, W.; Xia, X.; Zhou, B.; Peng, X.; Jiang, S. Warpage prediction of RHCM crystalline parts based on multi-layers. *Polymers* **2021**, *13*, 1814. [CrossRef] [PubMed]
13. Lozano, A.B.; Álvarez, S.H.; Isaza, C.V.; Montealegre-Rubio, W. Analysis and advances in additive manufacturing as a new technology to make polymer injection molds for world-class production systems. *Polymers* **2022**, *14*, 1646. [CrossRef] [PubMed]

polymers

MDPI

Article

A Hybrid Cooling Model Based on the Use of Newly Designed Fluted Conformal Cooling Channels and Fastcool Inserts for Green Molds

Abelardo Torres-Alba, Jorge Manuel Mercado-Colmenero, Juan De Dios Caballero-Garcia and Cristina Martin-Doñate *

Department of Engineering Graphics Design and Projects, University of Jaen, 23071 Jaen, Spain; ata00001@red.ujaen.es (A.T.-A.); jmercado@ujaen.es (J.M.M.-C.); jdcg0004@red.ujaen.es (J.D.D.C.-G.)
* Correspondence: cdonate@ujaen.es; Tel.: +34-953-212-821; Fax: +34-953-212-334

Abstract: The paper presents a hybrid cooling model based on the use of newly designed fluted conformal cooling channels in combination with inserts manufactured with Fastcool material. The hybrid cooling design was applied to an industrial part with complex geometry, high rates of thickness, and deep internal concavities. The geometry of the industrial part, besides the ejection system requirements of the mold, makes it impossible to cool it adequately using traditional or conformal standard methods. The addition of helical flutes in the circular conformal cooling channel surfaces generates a high number of vortexes and turbulences in the coolant flow, fostering the thermal exchange between the flow and the plastic part. The use of a Fastcool insert allows an optimal transfer of the heat flow in the slender core of the plastic part. An additional conformal cooling channel layout was required, not for the cooling of the plastic part, but for cooling the Fastcool insert, improving the thermal exchange between the Fastcool insert and the coolant flow. In this way, it is possible to maintain a constant heat exchange throughout the manufacturing cycle of the plastic part. A transient numerical analysis validated the improvements of the hybrid design presented, obtaining reductions in cycle time for the analyzed part by 27.442% in comparison with traditional cooling systems. The design of the 1 mm helical fluted conformal cooling channels and the use of the Fastcool insert cooled by a conformal cooling channel improves by 4334.9% the thermal exchange between the cooling elements and the plastic part. Additionally, it improves by 51.666% the uniformity and the gradient of the temperature map in comparison with the traditional cooling solution. The results obtained in this paper are in line with the sustainability criteria of green molds, centered on reducing the cycle time and improving the quality of the complex molded parts.

Keywords: injection molding; conformal cooling; industrial design; sustainability; green channels; numerical simulation; temperature maps

Citation: Torres-Alba, A.; Mercado-Colmenero, J.M.; Caballero-Garcia, J.D.D.; Martin-Doñate, C. A Hybrid Cooling Model Based on the Use of Newly Designed Fluted Conformal Cooling Channels and Fastcool Inserts for Green Molds. *Polymers* **2021**, *13*, 3115. https://doi.org/10.3390/polym13183115

Academic Editor: Célio Bruno Pinto Fernandes

Received: 19 August 2021
Accepted: 13 September 2021
Published: 15 September 2021

Publisher's Note: MDPI stays neutral with regard to jurisdictional claims in published maps and institutional affiliations.

1. Introduction

The plastic injection molding process is today one of the most widespread manufacturing processes due in large part to its efficiency and its ability to manufacture parts with a complex geometry at a competitive production cost [1]. The production process of parts by molding is a cyclical process that consists of four well-differentiated phases: the injection of the melt flow into the cavity, the packing of the plastic material until the gate freezes, the cooling of the plastic part until it solidifies, and finally, the ejection of the mold part [2]. Among these four phases, the cooling phase is the most influential in the production cost of the part as well as in the geometric and dimensional quality. The cycle time in the production of a mold is a measure of its productive efficiency [3]. A small decrease in the industrial cycle time leads to a large decrease in the energy expenditure of the injection molding process, making it more efficient and sustainable. However, achieving decreases in cycle time while maintaining the quality and requirements specified by the client is a highly complex process that involves studying in detail the heat-exchange process between surface layers, internal areas of the piece, and the fluid that circulates through the cooling channels.

The factors with the greatest influence on the heat exchange between molten plastic and coolant are the geometry, the layout, and the dimensions of the cooling channels. Injection molding cooling channels are traditionally manufactured using subtractive technologies such as CNC. The use of these technologies requires the use of dimensional technical specifications regarding the geometric design of the layout of the channels. In compliance with these specifications, it is necessary to guarantee the structural integrity of the mold against the high pressures and stresses to which it is subjected in its productive life.

Subtractive manufacturing technologies present high levels of material waste; therefore, unfortunately, they are not in line with current sustainability requirements [4,5]. The SLM additive manufacturing process allows for the manufacture of conformal channels adapted to the free shape of the geometric surface of the plastic part. In this way, it is possible to eliminate or reduce the existence of hotspots caused by differences in thickness, accumulations of material, or deep areas that are difficult to cool. The lack of uniformity in the cooling and, therefore, on the surface and internal temperatures of the piece can cause visible problems during its ejection or at a later time such as residual stresses, differential shrinkage, warpages, etc. In light of this, several authors have proposed different solutions regarding the geometry of the conformal channels with the aim of decreasing the cycle time, seeking energy savings in the process, and an improvement in the line of sustainability [6–10].

The layout of the cooling channels, topology, and sizing has a great influence on the variables of the heat-exchange process, such as pressure drop, cooling efficiency, coolant flow speed, etc. In the case of conformal cooling channels, their design allows them to be perfectly adapted to the requirements established by the geometry of the plastic part.

The spiral geometry is one of the most used topologies in the design of the layout in conformal cooling channels [11]. Linear zigzag geometry [12,13] is used in cases where spiral channels are difficult to implement. It should be noted that spiral geometry, in comparison to zigzag topology, presents sharp turns—increasing pressure drops, slowing the flow rate, and thus weakening the cooling efficiency. The application of the spiral-shaped and zigzag conformal channels decreases as the difficulty of the geometric surface of the plastic part to be manufactured increases. In these cases, mesh-topological conformal channels [14] and vascularized conformal systems [15] inspired by the design of blood vessels with complex topology and non-uniform diameter can be applied for complex pieces. Unfortunately, the design of a lattice topology conformal cooling system requires a detailed study regarding flow distribution and pressure drops since the usual design rules are useful for channel geometries with uniform diameters and shapes. In line with the admissible pressure drop in the layout, there is a minimum channel diameter below which the channel cannot be divided into sub-branches [16]. In addition, from a functional point of view, in the event of obstruction in the channels due to foreign bodies, it would be difficult to remove it due to the communication of the channels with each other.

When comparing the conformal cooling channels with respect to the geometry of their cross-section, it has been observed that the surface area of the conformal channels is an important parameter in the reduction of the cooling time and the improvement of the quality of the part [17]. Along these lines, although the circular cross-section is the most common in the design of conformal channels, research has been carried out to develop conformal channels with non-circular cross-sections such as square, rectangular, rhomboid, elliptical, water drop, etc., [18,19]. Several authors have made use of a square section for the cooling channels by making cooling slots in the mold [20,21]. The implementation of this solution is viable through traditional methods; however, it is difficult to produce with additive manufacturing since it requires supports that prevent the collapse of the material in the upper zone of the channel. In order to avoid deviations with the circular surface of the canal or even collapse in the upper zone, Kamat et al. [22] modified the circular section to a triangular self-supporting teardrop profile. The semicircular-shaped cooling channels consist of two parts, a semicircular part and another straight part parallel to the contour of the cavity. Unfortunately, although the semicircular conformal channel allows better tracking of the surface of the piece compared to the circular conformal channel and an improvement of heat dissipation [23,24], the sharp corner at the junction of the semicircular part and the straight part can cause stress concentration and crack propagation. Xi et al. [25], Wang et al. [26], and Jiang et al. [27] designed grooves on the inner surface of square channels in general cooling applications outside of injection molding. The ribs can enlarge the contact between the coolant and the channel surface. Along these lines, Freitas et al. [28] proposed a finned design on the circular or square channels to further expand the surface of the conformal channels; however, manufacturing the complex fin shape poses difficulties in additive manufacturing.

The steel used in the manufacture of molds and dies has disadvantages due to its low thermal conductivity regarding heat-transfer efficiency. Although increasing the coolant flow rate can increase the cooling efficiency, this option may be limited by the design of the mold, also leading to higher pumping costs [29]. Core cooling in molds can be performed with straight or perforated baffles in the mold core inserts. This results in a hollow, and consequently structurally weaker, core.

Another way to cool slender cores is to use inserts made of materials with high thermal conductivity, including copper, beryllium-copper, or high-strength sintered copper tungsten materials [30]. The high thermal conductivity of these materials increases the production speed while maintaining the strength to corrosion and oxidation, which is key for the lifetime of the mold [31]. Fastcool 50 is a hardened and tempered hot-work tool steel that has been recently developed with the aim of providing a tool steel with high thermal conductivity and high wear resistance at a reduced cost. Fastcool 50 material can go up to 54 Hrc while its thermal conductivity can reach almost twice that of conventional hot-work tool steels with the same level of hardness. It is especially indicated in applications subject to a high level of wear and in configurations in which the tool requires high thermal conductivity at high hardness [32]. Additionally, Fastcool 50 material allows for very high polish levels (mirror and higher), which are required in certain plastic injection molding applications. This tool steel allows one to considerably increase the productivity of the molding process as well as production problems related to the generation of hot/cold spots and poor distribution of the surface temperature.

Cycle time is currently the most important parameter in measuring the efficiency of the injection molding process [33–35]. The molding process is characterized by its high productivity in such a way that any reduction in cycle time brings great economic benefits to the company. The reduction of the cycle time depends to a great extent on the adaptability of the cooling channels to the complexity of the part. For a system of conformal channels located in problematic areas of the part, it is possible to reduce the cycle time by up to 30% compared to the use of conventional channels [36,37].

For a complete conformal channel system for the mold, the cycle time reduction can be more than 50% for parts with complex shapes and structures [38] or even 70% for some specific cases [39]. Several authors have investigated the influence of the use of conformal channels to reduce the cycle time. Shaifullah et al. [40] managed to reduce the cycle time by 35% with shaped cooling channels compared to straight cooling channels. In [41], they decreased the cycle time by 20% with a square section compared to conventional straight cooling channels. Xu et al. [42] applied a model based on unit cells for the dimensioning of conformal cooling, the results obtained by Xu indicated a 15% reduction of the cycle time. Colmenero et al. [43] performed a method to maximize the efficiency of shaped cooling channels manufactured by additive manufacturing. Research results showed that cooling and cycle time can be reduced by more than 50%. Additive manufacturing can create three-dimensional lattices and porous structures with specific mechanical and thermal properties. Mercado et al. developed a method based on the use of lattice to reduce the cycle time and improve the uniformity of the cooled part [44]. Using the design of porous structures, Brooks et al. [45] presented a study based on the use of shaped cooling layers designed with unsupported unit cells. These results showed a 26% decrease in cooling time compared to traditional cooling systems. Unfortunately, there are still major drawbacks in using uniform porous structures to achieve shaped cooling because the pressure drop in the porous structure is usually much greater than that of shaped cooling channels.

The design of a sustainable injection mold involves planning and optimizing its design by working on several lines simultaneously. First, the mold must be designed with the aim of minimizing the cycle time while maintaining uniformity in the surface temperatures of the part. In this way, it is possible to reduce the energy cost of production and rejections due to lack of quality. On the other hand, the mold must be designed and manufactured using technologies that reduce production wastes. Finally, the plastic part must use recycled plastic materials capable of meeting the requirements and specifications of the designed part.

In light of the problems posed in the state of the art, this paper presents a hybrid cooling model formed by a new design of fluted conformal channels in combination with the use of inserts designed with Fastcool material. The hybrid cooling design presented in the paper is shown as an alternative to traditional cooling design solutions with the aim of achieving high energy savings while maintaining a high level of quality in the manufacture of the part in line with the requirements of industrial sustainability. The research presented analyzes the thermal influence of the geometry of the conformal layout section for cases of complex geometry where the real space limitations given by the design of the manufacturing tooling and the topology of the part complicate the layout of the channels and the compliance with industry standards.

The hybrid cooling design developed in the paper was applied to an industrial piece with complex geometry with high rates of variation in its thickness and deep concavities in its internal area. Additionally, in the part presented, the design of the ejection system of the mold complicating the placement of the cooling channels, making it impossible to cool the part adequately by traditional methods. The results obtained in the research indicate that the introduction of the design with helical flutes in the conformal cooling channels improves the cycle time with respect to traditional cooling, above the geometries in circular, spiral, and water drop. Likewise, the transient numerical analysis carried out in this paper validates the combined use of fluted conformal cooling channels together with the use of Fastcool inserts for the areas with greater thickness, improving the temperature profile. The results obtained by the research are in line with the sustainability criteria in the design of green molds, capable of reducing the cycle time and manufacturing quality pieces with recycled plastic materials capable of meeting the high levels of quality and requirements established by the industrial customer.

2. Materials and Methods

2.1. Geometrical Design and Analysis for the Plastic Parts Manufactured through the Injection Molding Process

In this section, the geometrical, functional, and manufacturing features associated with the plastic part under study are described. Technical details regarding the selection of the plastic material for the injection molding manufacturing process, boundary conditions, and geometrical features are also specified. The piece under study is characterized by presenting a highly complex geometry, with a great influence on the performance of the molding manufacturing process. The piece presented in the paper corresponds to a real plastic piece of complex geometry manufactured industrially, belonging to a consumer product with a high annual production ratio. In industrial parts manufactured by the injection molding process, the parameters of cycle time and uniformity in cooling are indicators of the efficiency of the production process and the sustainability of the process. The injection molding process is a cyclical process; therefore, any decrease in cycle time implies high economic savings for the company, as well as an improvement in sustainability and energy savings.

Figure 1 shows the geometry of the industrial part. The dimensions of the bounding box of the part are 103 × 81 × 68 mm^3. The piece has an interior area, characterized by the inclusion of a small hexagonal section blind hole. The blind hole crosses the part almost entirely; therefore, a reinforcement area is required in the upper area to avoid possible warpages and fractures in the part. Unfortunately, the reinforcing geometry creates a largely localized area over the thickness in the part due to material accumulation in that problem region. The maximum thickness in the reinforcement zone is 7 mm, while the rest of the piece has a general thickness of 2 mm (see Figure 1). Thus, the thickness ratio obtained for this case study is 3.5:1. Manufacturing a part by molding with this thickness gradient is a challenge since, as the thickness of the wall increases, so do the stresses and displacements associated with the effects of shrinkage. Additionally, the material accumulation zone has a greater heat accumulation due to a slower cooling process. This variation in the temperature of the part results in a prolonged cooling time and an uneven shrinkage, causing the part to warp heterogeneously. This accumulation of heat in the part also affects the productivity of the manufacturing process, as it causes a considerable increase in the total manufacturing cycle time. In addition, and as indicated in Figure 1, the geometry of the part has another added complication, which is the difficulty of access of the cooling to the problematic area. The reduced dimensions of the central hole of the piece do not allow the inclusion of traditional cooling channels while maintaining the structural integrity of the mold.

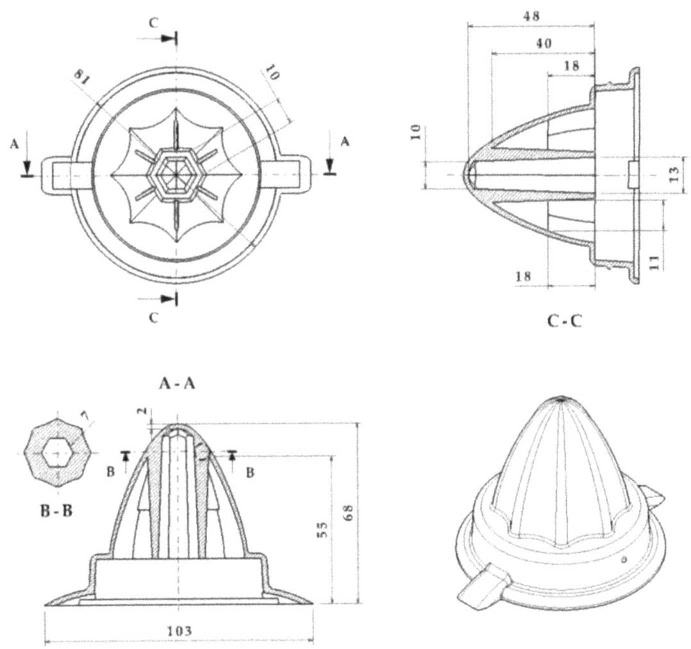

Figure 1. Description of the geometry of the plastic part. A-A section defined by A-A cutting plane, B-B section defined by B-B cutting plane and C-C section defined by C-C cutting plane.

The geometric design of the injection mold cooling system for the analyzed case study presents manufacturing difficulties for its main elements: perforated straight channels and baffles. The design of a traditional mold requires that the channels meet dimensional requirements that guarantee the structural integrity of the mold. These geometric requirements are related to the distance between channels, the distance to the ejectors, and the distance to the surface of the mold. The drills in charge of housing baffle-type cooling devices in the core must, in turn, meet a set of dimensional requirements related to the distance between channels, the surface of the part to be molded, and the ejectors of the system. Figure 2 shows the current design of the two-cavity mold following traditional manufacturing methods to manufacture the part under study. The mold cavity plate is cooled by a set of 8 mm diameter drilled straight holes which are located unevenly with respect to the surface of the mold, due to the requirements of traditional manufacturing using computerized numerical control (CNC) (see Figure 2). Likewise, the core plate is cooled using 6 mm diameter perforated straight channels and a 12 mm diameter and 40 mm deep baffle, located in the central part of the core of the plastic part (see Figure 2). The separation distance between channels themselves and the channels and the piece meets the criteria for sizing injection molds established by the industry, always being greater than the minimum safety distance that ensures the structural integrity of the mold. Figure 3 shows the assembly drawing of the mold, where the location of the cooling elements next to the ejectors is indicated. Figures 2 and 3 show how the need to include four ejectors to extract the part from the mold greatly reduces the space available for cooling.

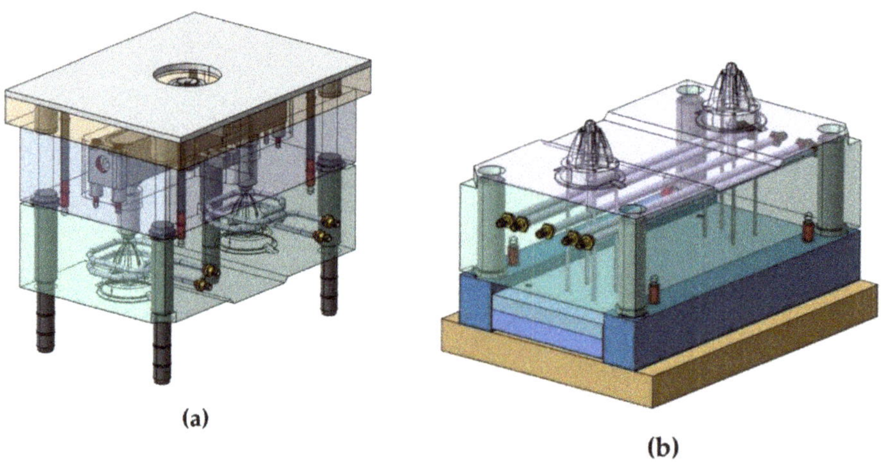

Figure 2. Mold of the plastic part injection area (**a**) and ejection area (**b**).

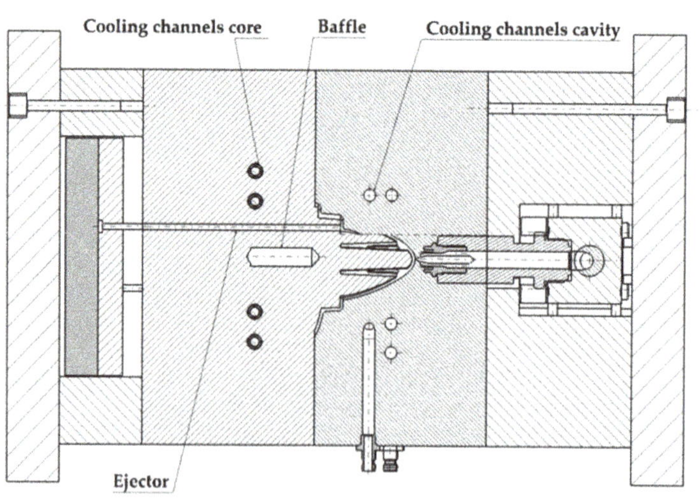

Figure 3. Mold cooling design using traditional straight channels and baffles.

2.2. Green Conformal Cooling Channels Design

The cooling of concave cores in injection molds poses great difficulties in complex parts due, in large part, to the number of geometric and technological requirements caused by the manufacturing process. These requirements are sometimes difficult or impossible to solve using traditional cooling elements or systems.

Conformal cooling channels solve these problems, taking advantage of the free space between elements of the mold, reducing the distance to the surface of the plastic part in areas of difficult access and proposing new geometric designs and sections in the cooling channels.

Likewise, this manuscript analyzes in detail the influence that the geometry of the cooling channel section has on the heat-exchange process in industrial parts with highly complex geometric requirements. This manuscript presents a new conformal channel design, based on the introduction of helical flutes along the circular surface of conformal-type cooling channels. The objective of the flutes is to increase the turbulence of the coolant flow inside the cooling channel, thus improving the heat exchange between the plastic part and the coolant flow.

To evaluate the thermal performance of the fluted sections presented in this manuscript, three additional configurations of conformal channels were designed based on the use of circular, elliptical, and water drop geometries adapted to a spiral axis geometric sweep topology with which the geometric CAD modeling of the conformal cooling channels was carried out.

The dimensioning of the sections, the separation of the channels to the study piece, and the distance between channels were carried out in accordance with the requirements established in [46] and according to the topological requirements of the plastic piece. Although the geometric contour of the sections is variable, the proposed sections were designed in such a way that their hydraulic diameter D_h (see Equation (1) remains constant, maintaining the same number of turns in all the proposed designs. The hydraulic diameter D_h was calculated according to Equation [1], where A_s [m^2] is the area of the cooling channel section and P [m] the perimeter of the channel section.

$$D_h = 4 \cdot \frac{A_s}{P} \quad (1)$$

In this way, the thermal performance between the different geometric designs proposed is compared. The sections presented in this manuscript meet all the necessary requirements for their manufacture using additive technology. The dimensions and angles in the geometric contours of the channels presented meet the sustainability criteria regarding the design of green channels, avoiding the use of supports inside, as well as the elastic and structural collapse of the metal material of the injection mold. Table 1 indicates the values of the set of geometric and dimensional specifications used in the design of the layout of the different cooling channels applied in the cavity area of the mold.

Table 1. Geometric and dimensional specifications used in the design of the layout of the different channels in the cavity plate of the mold.

Nomenclature	Units	Circular	Elliptical	Water Drop	Fluted Circular 1 mm	Fluted Circular 0.6 mm
A_s	mm^2	28.27	27.49	35.93	39.27	39.27
P	mm	18.84	18.98	22.77	30.25	27.76
D_h	mm	6	5.79	6.31	5.19	5.66
s	mm	7	7	7	7	7
p	mm	11	11	11	11	11

where A_s [mm^2] represents the area of the cooling channel section, P [mm] represents the perimeter of the cooling channel, D_h [mm] represents the hydraulic diameter of the cooling channel section, s [mm] represents the minimum distance between the center of the cooling channel section and the surface of the plastic part, and p [mm] represents the distance between centers of the cooling channel sections.

Figure 4 presents the design of the fluted conformal cooling channels located in the cavity plate of the mold cavity. Figure 5 shows the design of the conformal channels with circular, elliptical, and water drop sections, respectively.

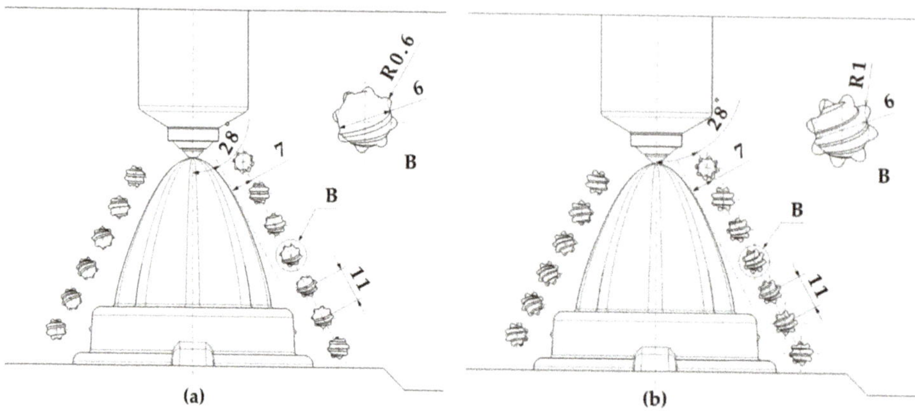

Figure 4. Design of the fluted conformal cooling channels (**a**) 0.6 mm and (**b**) 1 mm.

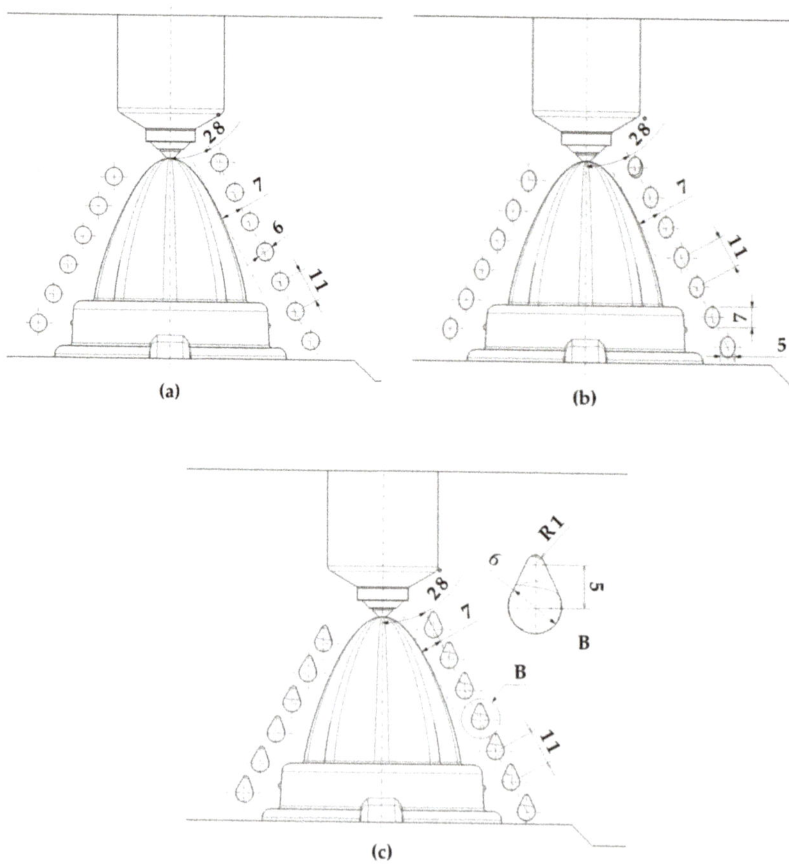

Figure 5. Conformal cooling channel design (**a**) circular, (**b**) elliptical, and (**c**) water drop.

Although conformal cooling is suitable for cooling complex areas, sometimes the geometry of the plastic part has space limitations that make it difficult to place conformal channels. This problem usually occurs when cooling thin areas and geometric regions with deep cores. The plastic piece under study is framed within this last group, presenting great geometric limitations characterized by internal areas of difficult access together with a central hole of great depth and width 10 mm.

This geometry prevents the use of conformal green channels in the upper area with a higher thickness ratio. Since, firstly, there is not enough space in the core to access the channels, and secondly, the diameter of the channels would be too small to provide functionality in cooling. Figure 6 shows the layout corresponding to the conformal channels designed in the core area. Unfortunately, as seen in Figure 6 for geometric and dimensional reasons, it is not possible to cool the internal area of the core using conformal type channels.

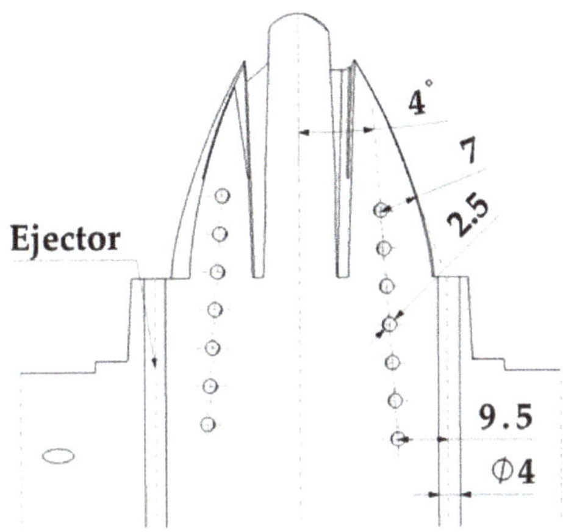

Figure 6. Design of conformal cooling channels in the core area.

To solve the problem posed, this manuscript presents a hybrid cooling design, based on the use of Fastcool inserts in combination with conformal-type cooling systems. Fastcool inserts are characterized by their high thermal transmission coefficient. This allows the heat exchange in high-temperature regions in the plastic part to be greatly accelerated and improved. However, the area of the mold in contact with the insert accumulates a large amount of heat flow that the insert is not able to dissipate, generating local points with a great accumulation of heat and high temperatures. The use of conformal cooling channels, in these cases, can be very useful, not so much for cooling the plastic part directly but for cooling the Fastcool insert itself, above all, in those plastic pieces in which the use of conformal channels is not enough. In this way, it is possible to minimize the cooling time of the part making the molding process sustainable, while maintaining the surface quality of the molded part. Table 2 shows the geometric data of the conformal channel used for core cooling and the Fastcool insert. Figure 7 presents the application of the hybrid Fastcool-conformal cooling system presented in this manuscript. Figure 8 shows the final design of the mold.

Table 2. Geometric parameters used in the design of conformal cooling channels.

Nomenclature	Units	Description	Circular
A_s	mm^2	Section area	4.9
P	mm	Perimeter	7.85
D_h	mm	Hydraulic diameter	2.5
s	mm	Distance channel center—mold surface	7
p	mm	Pitch between channels	7

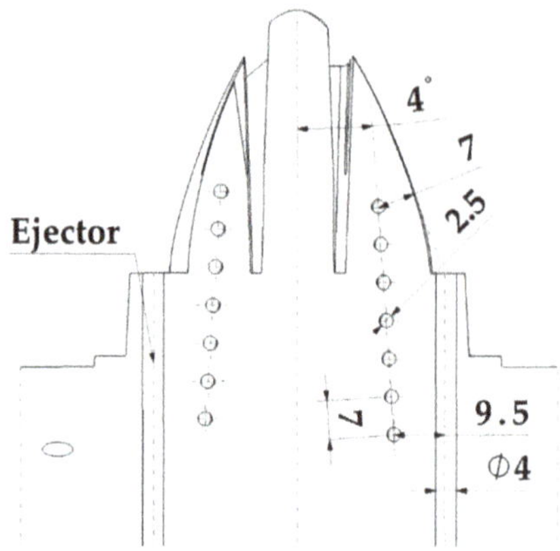

Figure 7. Hybrid Fastcool-conformal cooling system for core cooling.

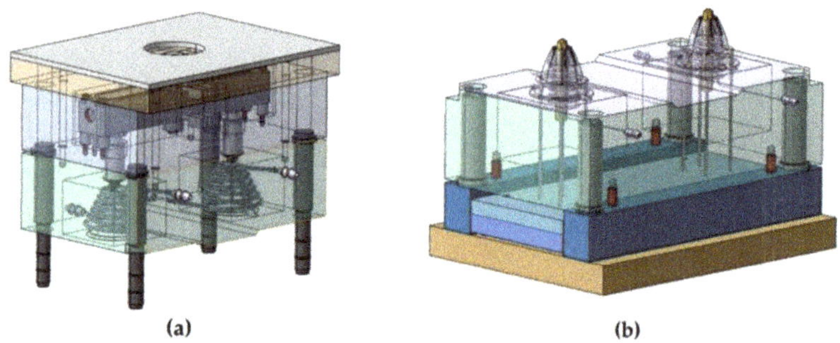

Figure 8. Final design of the mold including the hybrid cooling proposed in this paper. Mold of the plastic part injection area (**a**) and ejection area (**b**).

2.3. Materials

The piece under study was designed to be manufactured with recycled PP 108MF10 thermoplastic material from the SABIC company. This plastic material is obtained from a chemical recycling process. Therefore, the mechanical, thermal, and chemical properties of the original plastic are maintained without compromising the sustainability of the injection process. The magnitudes of the main properties of the material PP 108MF10 are indicated in Table 3. Tables 4 and 5 show the features of the metal material of Fastcool and the injection mold metal material. The material properties shown in these tables are provided by their manufacturers. Furthermore, the presented cooling system design methodology is universal; therefore, it can be applied to any thermoplastic material used in the manufacture of plastic parts through the injection mold manufacturing process.

Table 3. Magnitude of the main properties of the material PP108MF10.

Nomenclature	Units	Description	PP 108MF10
ρ_p	Kg/m^3	Density	905
C_p	J/Kg·°C	Specific heat	2704
δ_p	W/m·°C	Thermal conductivity coefficient	0.1998
T_{melt}	°C	Melt temperature (normal)	230
T_{mold}	°C	Mold temperature (normal)	40
T_{eject}	°C	Ejection temperature	100

Table 4. Mechanical, physical, and thermal properties of Fastcool-50 at 44 HRC.

Description	Units	Fastcool-50
Density	g/cm^3	7.81
Yield strength 0,2%	MPa	1070
Mechanical resistance	MPa	1400
Elongation	%	17
Specific heat capacity	J/g·K	0.47
Thermal diffusivity	mm^2/s	13.5
Thermal conductivity	(W/m·K)	50

Table 5. Mechanical, physical, and thermal properties of Steel alloy 1.2709.

Description	Units	Steel Alloy 1.2709
Density	g/cm^3	8000
Specific heat capacity	J/g·K	450
Thermal conductivity coefficient	(W/m·K)	20

3. Implementation and Results

The geometric designs of the different cooling channels proposed in this manuscript were developed and analyzed numerically in the CAD design software Catia (V5-6R2020 version, Dassault Systèmes, Vélizy-Villacoublay, Francia) [47] and numerical analysis Moldex3D (R17 version, CoreTech System Co., Ltd, Zhubei, Taiwán) [48] with an MSI notebook (Micro-Star International, Co., Ldt, Taipei, Taiwán) with an Intel (R) Core (TM) i-77700HQ CPU @ 2.80 GHz (Intel Corporation, Santa Clara, CA, US). The geometry of the cooling channels, especially the design of the conformal cooling channels, was parameterized and adapted to the geometric features of the case study and the technical and technological requirements imposed by the 3D additive manufacturing process by laser sintering (SLS). This parameterization, together with the modeling of the layout of the conformal cooling channels, was automated by generating an application in the programming environment of the CAD software Catia (V5-6R2020 version, Dassault Systèmes, Vélizy-Villacoublay, Francia) [47].

Modeling of Numerical Simulations for the Thermal Analysis of the Green Conformal Cooling System

In this section, we proceed to detail the pre-processing and modeling process of the different analyses and numerical simulations, from which the thermal performance of the designs of the conformal cooling channels and the rheological parameters associated with the process are evaluated for the manufacturing of the plastic part under study. Through this set of numerical analyses, the thermal and dynamic behavior of the coolant flow along the cooling channels is analyzed, as well as the thermal exchange produced between them, the plastic part, the Fastcool insert (see Figure 9), and the injection mold during the cooling phase of the plastic part. In this way, it can be validated if the thermal performance of the results obtained meets the technological requirements demanded by the manufacturing process using plastic injection molds. The set of numerical, thermal, and rheological analyses were carried out using the commercial simulation software CAE Moldex3D (R17 version, CoreTech System Co., Ltd, Zhubei, Taiwán) [48]. First, to describe the preprocessing modeling of this set of numerical analyses, five main 3D computation domains must be established (see Figure 9), along with the selection of the material associated with each of these. Likewise, these computational domains are established: plastic part (PP108MF10), feeding system (PP108MF10), cooling system (Water), injection mold (Steel alloy 1.2709), and Fastcool insert (Fastcool-50 at 44 HRC). Table 3, Table 4, and Table 5 show the main physical, rheological, and thermal properties of the materials used for each computational domain.

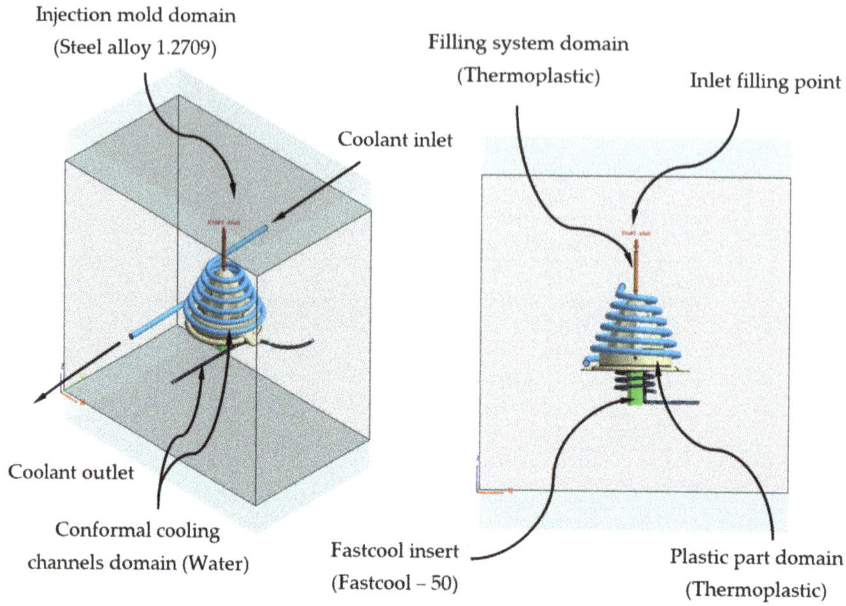

Figure 9. Domains and boundary conditions definition for the numerical simulations, conformal cooling solution with Fastcool insert.

Moreover, together with the definition of the 3D computational domains that makes up the modeling of the numerical analyses, a series of main premises that complete the pre-processing phase are determined:

- Since the complete cooling process of the plastic part is analyzed over time, the type of numerical analysis used is "Cooling transient".
- The total cooling time established for each numerical analysis is 90 s, with a time step between each time step of 10 s. For each defined time step, the numerical analysis software stores the solution obtained. Therefore, in each numerical simulation carried out, the time until reaching the ejection temperature of the plastic part and the evolution of the temperature map throughout the cooling phase can be precisely determined.
- The analysis of the behavior and evolution of the physical, dynamic and thermal properties of the coolant flow along the channels of the cooling system was modeled according to the "Run 3D cooling channels" configuration.
- The methodology used to configure the solver, in the resolution of each numerical analysis carried out, is of the type maximum variation of mold temperature, whose magnitudes established for the parameter temperature difference and maximum cycle number are 1 and 10 °C, respectively.
- The turbulence model used for the development of the numerical analyses is established using the roughness parameter. This parameter defines the interface surface between the coolant flow and the surface of the cooling channels. The magnitude defined for this technological parameter is equal to 0.02 mm

Next, we proceed to the discretization of the main geometry of the cooling channels, plastic part, and injection mold (see Figures 9 and 10) in finite volumes. The commercial software Moldex 3D R17 [48] has a Moldex Designer meshing module, from which the main parameters of the mesh are configured and defined for each numerical analysis. Table 6 shows the magnitude of the parameters defined for the meshing process, as well as its configuration.

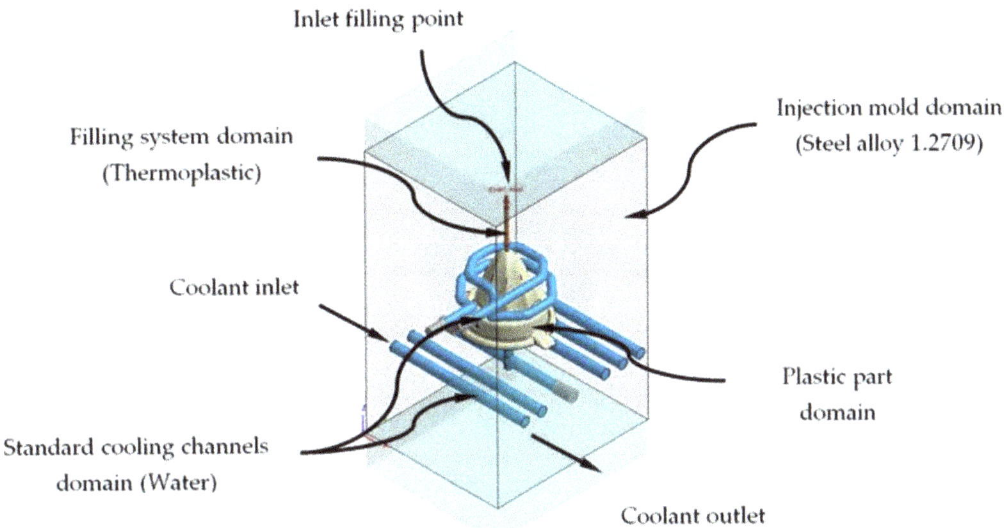

Figure 10. Domains and boundary conditions definition for the numerical simulations, traditional cooling solution.

Table 6. Mesh statistics for the standard and conformal cooling meshes.

Description	Units	Value
Part mesh node count	-	107,872
Part mesh element count	-	236,119
Part mesh volume	cm^3	33.89
Runner mesh node count	-	14,701
Runner mesh element count	-	13,440
Runner mesh volume	cm^3	0.5
Plastic part precision (ε)—Mesh sizing	mm	1.5
Element type	-	Tetrahedral (10 nodes)
Element type—Boundary layers	-	Prism (15 nodes)
Offset ratio—Boundary layers	-	0.1

Figures 11 and 12 show in detail the typology of elements used to discretize, in finite volumes, the 3D computation domains presented in Figures 9 and 10. The said elements are of the second order tetrahedron type (SOLID 186); they present 10 nodes of control of which, four are located at the vertices of the tetrahedron, and six are located at the midpoints of the edges. Being a second-order element, each node has 3 degrees of freedom, with translation in the main coordinate axes. The use of this type of element allows the resulting temperature field to be modeled with greater precision.

Furthermore, in order to improve the precision of the numerical simulations, five layers of second-order prismatic elements (SOLID 186) of the "Boundary Layer" type are defined for the interface surfaces, along with the injection mold geometry "Mesh" (see Figures 11 and 12). These elements have 15 nodes; six are located at the vertices of the tetrahedron, and nine are located at the midpoints of the edges. Being a second-order element, each node has 3 degrees of freedom, with translation in the main coordinate axes. The magnitude of these elements is set by the offset ratio parameter, which is defined as a percentage of the average size of the mesh element. Table 6 shows the mesh statistics for the standard and green conformal cooling systems.

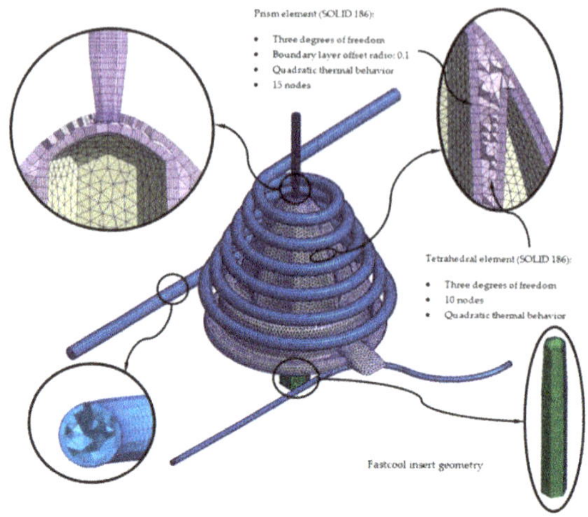

Figure 11. Mesh details for the conformal cooling solution.

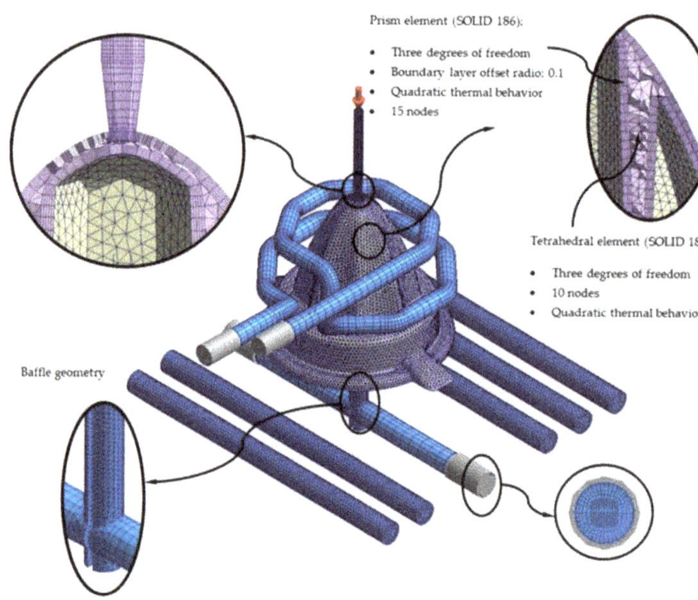

Figure 12. Mesh details for the traditional cooling solution.

For each numerical analysis performed, a set of boundary conditions related to the technological parameters that determine the molten plastic front and the coolant flow at the beginning of the plastic injection manufacturing process were established. For the cooling channels, an inlet and outlet surface of the coolant flow is established, as well as the magnitude of the technological parameters of inlet temperature and pressure of the coolant flow (see Table 7). On the other hand, for the cooling system the input surface of the molten plastic front is established, as well as the magnitude of the technological parameters of temperature, pressure and flow of the molten plastic front (see Table 7). It should be noted that, for the dimensioning of the coolant flow inlet pressure variable to the refrigeration channels, a quantity was established that ensures and allows the coolant flow front to develop in a turbulent regime. In other words, the Reynolds number of the coolant flow along the cooling channels is greater than 1.5×10^4. This condition was kept constant for each of the numerical simulations carried out, maintaining in each of them a Reynolds number equal to 5.0×10^4. Table 7 shows technological variables defined for the set-up of the filling and cooling stage for numerical simulations, the magnitude of the technological parameters presented are those recommended by the manufacturer of the plastic material.

Table 7. Technological variables defined for the setup of the filling and cooling stage for numerical simulations.

Description	Units	Study Cases—PP 108MF10 (PP)
Filling time	s	1.19
Packing time	s	5.91
Cooling time	s	90
Melt temperature	°C	230
Mold temperature	°C	40
Ejection temperature	°C	100
Coolant temperature	°C	40
Maximum injection pressure	MPa	140
Packing pressure profile	MPa	85 (0.0–3.5 s) 40 (3.5–4.7 s) 10 (4.7–5.9 s)

It should be noted that, to perform the validation of the numerical simulations carried out, the technological parameters defined as an input during the preprocessing phase (see Table 7) were determined, validated, and contrasted by means of experimental tests. These tests were performed by the manufacturer and supplier of the thermoplastic material using the standard manufacturing technique of plastic injection on industrial molds.

After describing the modeling process of each of the numerical simulations carried out for the present case study, we proceeded to present the results obtained. From the evaluation, analysis, and validation of the case study, the geometry of the conformal cooling channel that optimizes the cooling phase of the plastic part and improves the thermal performance of the cooling system can be established. Likewise, the improvement in the efficiency and thermal performance obtained by using a Fastcool insert to cool the core area of the plastic part which presents cooling difficulties due to its high temperatures is verified. It should be noted that each of the configurations of conformal cooling systems analyzed and proposed was numerically simulated including the Fastcool type for the core area of the plastic part under study.

In the first place, Table 8 and Figures 13–15 show the results obtained for the parameter time until reaching the ejection temperature of the plastic part, for each of the cooling system configurations proposed.

Table 8. Time to reach ejection temperature for each proposed cooling system configurations.

Cooling System Configurations	Time to Each Ejection Temperature [s]	Time Reduction [s]	Performance Improvement [%]
Traditional	78.149	-	-
Circular	57.426	20.723	26.5
Water drop	56.826	21.323	27.3
Elliptical	56.752	21.397	27.4
Fluted 0.6 mm	56.721	21.428	27.4
Fluted 1 mm	56.703	21.446	27.4

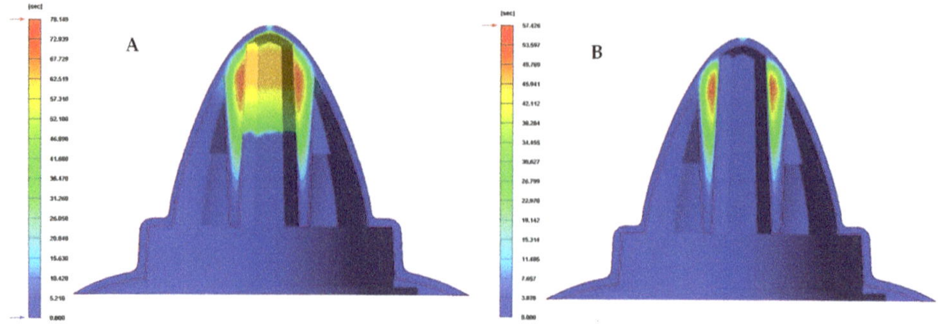

Figure 13. Time to reach ejection temperature (s). (**A**) Traditional cooling. (**B**) Circular conformal cooling channels and Fastcool insert.

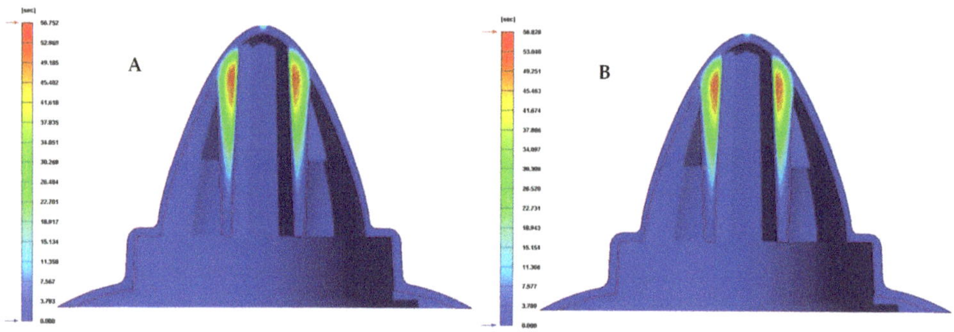

Figure 14. Time to reach ejection temperature (s). (**A**) Elliptical conformal cooling channels and Fastcool insert. (**B**) Water drop conformal cooling channels and Fastcool insert.

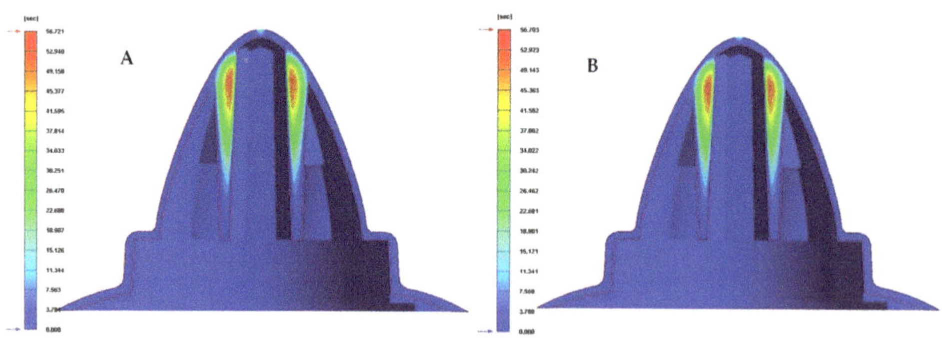

Figure 15. Time to reach ejection temperature (s). (**A**) Fluted 0.6 mm conformal cooling channels and Fastcool insert. (**B**) Fluted 1 mm conformal cooling channels and Fastcool insert.

As can be seen in Table 8 and Figures 13–15, the configurations of the conformal cooling systems that incorporate the Fastcool insert improve the time until reaching the ejection temperature in the plastic part. In particular, the conformal cooling channel solution with a fluted section of 1 mm is the one that minimizes this parameter because the time to reach the ejection temperature is reduced by 21.446 s in comparison to the traditional solution defined by straight drilled channels. This represents an improvement in the manufacturing cycle time of the plastic part under study of 27.4%. In addition, the thermal performance of the conformal cooling channel with a fluted section of 1 mm improves on the rest of the proposed conformal cooling solutions. This result is due to the geometry of the flutes along the channels surface generating vortices and turbulence in the coolant flow, which help and favor it to develop and constantly maintain a turbulent regime.

Table 9 and Figures 16–18 show the magnitude of the heat flow that is exchanged between the computational domain of the cooling channels and the rest of the domains defined for each of the numerical simulations performed.

Table 9. Heat flow for each proposed cooling system configurations.

Cooling System Configurations	Cavity Cooling [J/s·cm^2]	Core Cooling [J/s·cm^2]	Fastcool Insert [J/s·cm^2]	Total [J/s·cm^2]	Total Improvement [J/s·cm^2]	Performance Improvement [%]
Traditional	0.004	0.059	-	0.063	-	-
Circular	0.046	0.078	2.284	2.408	2.471	3722
Raindrop	0.195	0.085	2.333	2.613	2.550	4048
Elliptical	0.268	0.017	2.372	2.657	2.594	4118
Fluted 0.6 mm	0.206	0.103	2.359	2.668	2.605	4135
Fluted 1 mm	0.238	0.118	2.438	2.794	2.731	4335

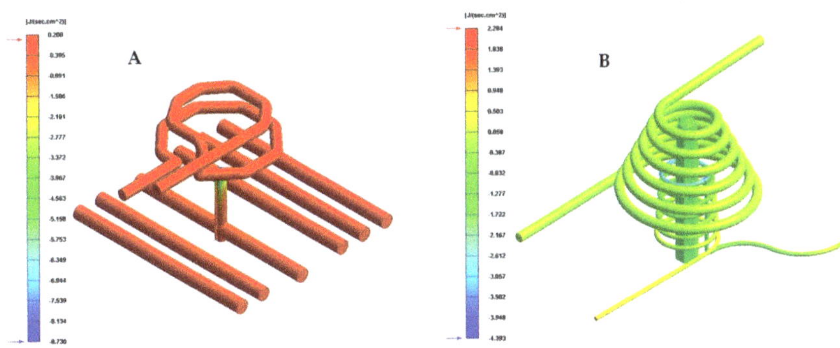

Figure 16. Heat flow (J/s·cm^2). (**A**) Traditional cooling. (**B**) Circular conformal cooling channels and Fastcool insert.

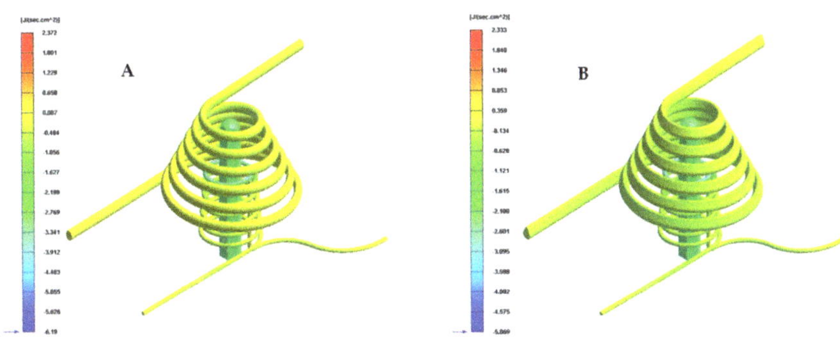

Figure 17. Heat flow (J/s·cm^2). (**A**) Elliptical conformal cooling channels and Fastcool insert. (**B**) Water drop conformal cooling channels and Fastcool insert.

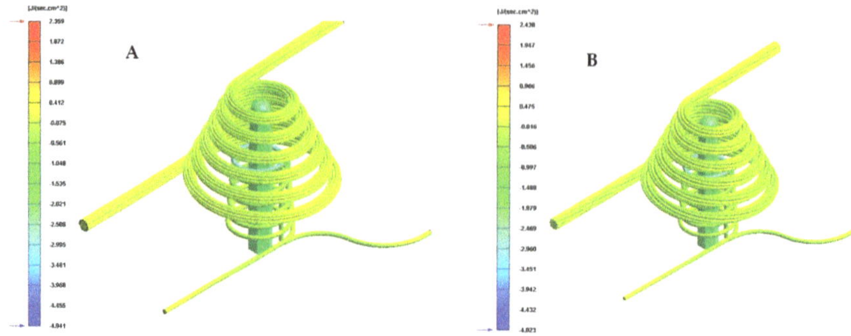

Figure 18. Heat flow (J/s·cm^2). (**A**) Fluted 0.6 mm conformal cooling channels and Fastcool insert. (**B**) Fluted 1 mm conformal cooling channels and Fastcool insert.

As can be seen in Table 9 and Figures 16–18, the design of some conformal-type cooling channels together with the implementation of a Fastcool insert, used for cooling the plastic part, represents a relevant improvement in the heat exchange that takes place between the coolant flow and the plastic part. On the one hand, for the cavity plate of the injection mold, the cooling channel geometry that exchanges the greatest amount of heat flow with the plastic part is the 1 mm fluted channel. The definition of flutes along the surface of the channel, with a dimension of 1 mm, favors the development of the coolant flow in a turbulent regime and, therefore, the thermal exchange of said fluid. On the other hand, for the core plate of the injection mold, the use of a Fastcool insert allows for the optimal evacuation and transfer of heat flow in the core of the plastic part. However, the Fastcool insert is not capable of maintaining, by itself, a constant heat exchange throughout the manufacturing cycle of the plastic part. For this reason, a conformal type cooling channel layout was defined, which allows establishing a thermal exchange between the Fastcool insert and the coolant flow. In this way, it is determined that the design of the 1 mm fluted type cooling channels together with the use of a Fastcool insert cooled by a conformal type channel improves by 4335% and 2.731 [J/s·cm^2] the thermal exchange between the cooling elements and the plastic part, with respect to the configuration of the traditional cooling system.

Finally, to complete the analysis of the results obtained in the different numerical simulations carried out, the temperature map after the cooling phase for each proposed cooling system is presented in Figures 19–21. Table 10 shows the maximum temperature difference along the surface of the plastic part under study.

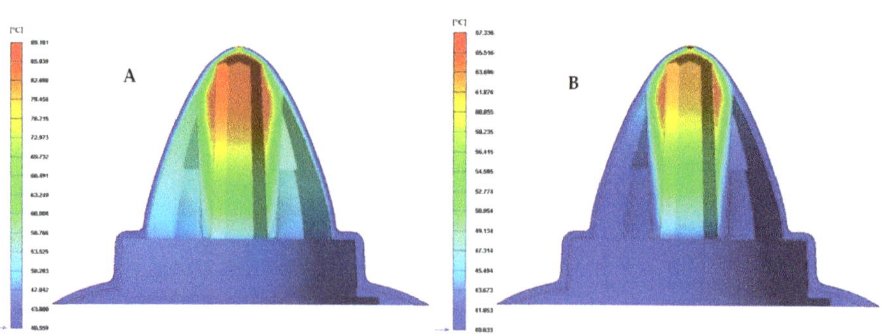

Figure 19. Cooling temperature (°C). (**A**) Traditional cooling. (**B**) Circular conformal cooling channels and Fastcool insert.

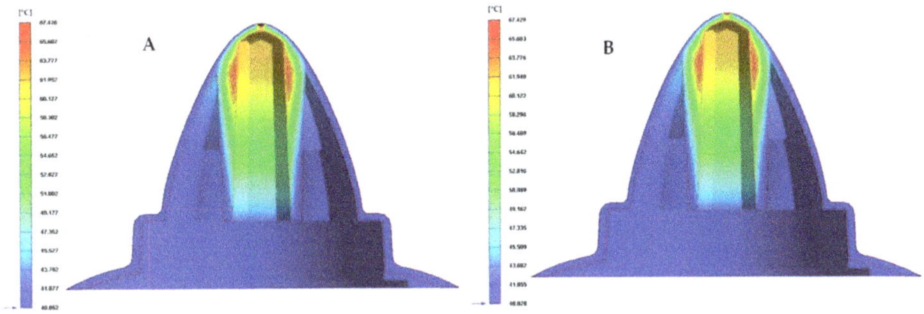

Figure 20. Cooling temperature (°C). (**A**) Elliptical conformal cooling channels and Fastcool insert. (**B**) Water drop conformal cooling channels and Fastcool insert.

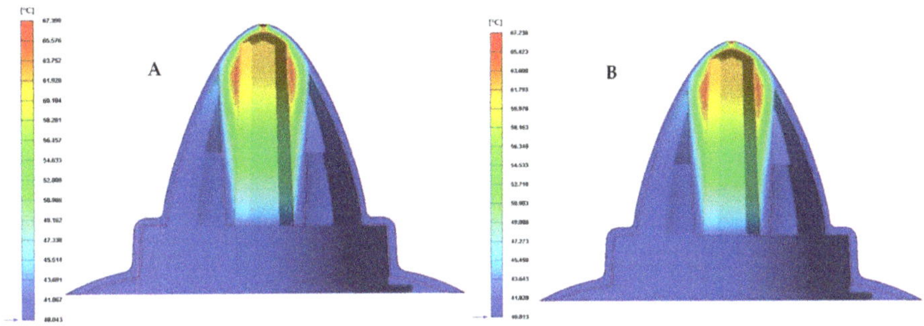

Figure 21. Cooling temperature (°C). (**A**) Fluted 0.6 mm conformal cooling channels and Fastcool insert. (**B**) Fluted 1 mm conformal cooling channels and Fastcool insert.

Table 10. Mold temperature difference [°C] for each proposed cooling system configurations.

Cooling System Configurations	Mold Temperature Difference [°C]	Total Improvement [°C]	Performance Improvement [%]
Traditional	44.728	-	-
Circular	22.635	22.093	49.4
Water drop	22.432	22.296	49.8
Elliptical	22.450	22.278	49.8
Fluted 0.6 mm	21.644	23.084	51.6
Fluted 1 mm	21.619	23.109	51.7

From the results obtained for the temperature maps, it was established that the design of a conformal type cooling system, accompanied by a Fastcool insert, improves the uniformity and gradient of the temperature map along the geometry of the plastic part object of study. Specifically, this improvement ranges from 49.4% to 51.7% and 22.093 to 23.109 °C, depending on the section of the defined conformal-type cooling channels. Being the 1 mm fluted type cooling channel, the one with the greatest uniformity of the temperature map and the lowest maximum temperature generated on the plastic piece was studied in this manuscript.

It should be noted that, the performance improvement parameter (see Table 8, Table 9, and Table 10) is computed as the percentage that represents the magnitude of the variables time reduction [s] (see Table 8), total improvement [J/s·cm^2] (see Table 9), and total improvement [°C] (see Table 10), obtained for each cooling system design proposed in this manuscript, on the magnitude of the numerical solution obtained for the traditional cooling system for each of parameters analyzed: Time to reach ejection temperature [s] (see Table 8), Heat flow [J/s·cm^2] (see Table 9), and Mold temperature difference [°C] (see Table 10).

4. Conclusions

The paper presents a hybrid cooling model based on the use of newly designed fluted conformal cooling channels in combination with inserts manufactured with Fastcool material. The hybrid cooling design was applied to an industrial part with complex geometry, high rates of thickness, and deep internal concavities. The geometry of the industrial part, besides the ejection system requirements of the mold, makes it impossible to cool it adequately using traditional or conformal standard methods.

The addition of helical flutes in the conformal cooling channel surface generates a high number of vortexes and turbulences in the coolant flow, helping and fostering the thermal exchange between the flow and plastic part. The use of a Fastcool insert allows for the optimal evacuation and transfer of the heat flow in the slender core of the plastic part. An additional conformal cooling channel layout was required, not for cooling the plastic part, but for cooling the Fastcool insert improving the thermal exchange between the Fastcool insert and the coolant flow. In this way, it is possible to maintain a constant heat exchange throughout the manufacturing cycle of the plastic part.

A transient numerical analysis carried out validates the improvements of the hybrid design, presenting reductions in cycle time by 27.442% and 21.446 s for the complex plastic part analyzed in comparison with the results obtained from traditional cooling systems. The design of the 1 mm fluted conformal cooling channels and the use of the Fastcool insert cooled by a conformal cooling channel improves, by 4334.9%, the thermal exchange between the cooling elements and the plastic part in comparison with traditional cooling systems.

From the results of the plastic part temperature map, it was established that the conformal cooling system accompanied by a Fastcool insert improves the uniformity and gradient of the temperature map in ranges from 49.394% to 51.666% and 22.093 to 23.109 °C in comparison to the traditional cooling solution. The design of a 1 mm fluted conformal cooling channel allows the greatest temperature map uniformity and the lowest maximum temperature on the plastic part studied in this manuscript.

The hybrid cooling design is shown as an alternative to traditional and standard conformal cooling systems for complex geometrical parts. In this way, it is possible to achieve a green mold with high energy savings and quality parts following the current requirements of sustainability in the plastic industry. Although there are other authors who have carried out experiments so far in the study of the application of conformal cooling circuits to improve cycle time, these authors have focused on simple geometry pieces and the use of standard conformal cooling circuits. These conformal designs are not suitable for the parts of complex geometry as presented by the authors in their research.

The results obtained by the research are in line with the sustainability criteria for green molds, centered on reducing the cycle time and improving the quality of the molded parts using recycled plastic materials, meeting the requirements established by the industrial customers.

Author Contributions: Investigation, A.T.-A., J.M.M.-C., J.D.D.C.-G., and C.M.-D.; project administration, C.M.-D.; writing—original draft, A.T.-A., J.M.M.-C., and C.M.-D.; writing—review and editing J.M.M.-C. and C.M.-D.; funding acquisition, C.M.-D. All authors have read and agreed to the published version of the manuscript.

Funding: This research work was supported by the University of Jaen through the Plan de Apoyo a la Investigación 2021-2022-ACCION1a POAI 2021-2022: TIC-159.

Institutional Review Board Statement: Not applicable.

Informed Consent Statement: Not applicable.

Data Availability Statement: All data included in this study are available upon request by contact with the corresponding author.

Acknowledgments: Authors acknowledge the support of CORETECH System Co.

Conflicts of Interest: The authors declare no conflict of interest.

References

1. Doñate, C.M.; Paramio, M.R. New methodology for demoldability analysis based on volume discretization algorithms. *Comput.-Aided Des.* **2013**, *45*, 229–240. [CrossRef]
2. Martin-Doñate, C.; Shaikheleid, S.; Torres-Alba, A.; Mercado-Colmenero, J.M. A New Smart Web Platform for Plastic Injection Molds in Industry 4.0 Environments. In *Advances on Mechanics, Design Engineering and Manufacturing III*; Springer: New York, NY, USA, 2021; pp. 309–315.
3. Mercado-Colmenero, J.M.; Muriana, J.A.M.; Rubio-Paramio, M.A.; Martín-Doñate, C. An automated manufacturing analysis of plastic parts using faceted surfaces. In *Advances on Mechanics, Design Engineering and Manufacturing*; Springer: Berlin, Germany, 2016; pp. 119–128.
4. Mercado-Colmenero, J.M.; Martin-Doñate, C.; Moramarco, V.; Attolico, M.A.; Renna, G.; Rodriguez-Santiago, M.; Casavola, C. Mechanical characterization of the plastic material GF-PA6 manufactured using FDM technology for a compression uniaxial stress field via an experimental and numerical analysis. *Polymers* **2020**, *12*, 246. [CrossRef] [PubMed]
5. Mercado-Colmenero, J.M.; La Rubia, M.D.; Mata-Garcia, E.; Rodriguez-Santiago, M.; Martin-Doñate, C. Experimental and numerical analysis for the mechanical characterization of petg polymers manufactured with fdm technology under pure uniaxial compression stress states for architectural applications. *Polymers* **2020**, *12*, 2202. [CrossRef]
6. Kuo, C.C.; Chen, W.H. Improving cooling performance of injection molding tool with conformal cooling channel by adding hybrid fillers. *Polymers* **2021**, *13*, 1224. [CrossRef]
7. Jahan, S.A.; Wu, T.; Zhang, Y.; Zhang, J.; Tovar, A.; Elmounayri, H. Thermo-mechanical design optimization of conformal cooling channels using design of experiments approach. *Procedia Manuf.* **2017**, *10*, 898–911. [CrossRef]
8. Zink, B.; Szabó, F.; Hatos, I.; Suplicz, A.; Kovács, N.K.; Hargitai, H.; Tábi, T.; Kovács, J.G. Enhanced injection molding simulation of advanced injection molds. *Polymers* **2017**, *9*, 77. [CrossRef]
9. Tang, Y.; Gao, Z.; Zhao, Y.F. Design of Conformal Porous Structures for the Cooling System of an Injection Mold Fabricated by Additive Manufacturing Process. *J. Mech. Des.* **2019**, *141*, 101702. [CrossRef]
10. Meekers, I.; Paul, R.; Arif, R. Analysis of Process Parameters affecting Energy Consumption in Plastic Injection Moulding. *Procedia CIRP* **2018**, *69*, 342–347. [CrossRef]
11. Park, H.-S.; Dang, X.P. Development of a smart plastic injection mold with conformal cooling channels. *Procedia Manuf.* **2017**. [CrossRef]
12. Torres-Alba, A.; Diaz-Perete, D.; Mercado-Colmenero, J.M. Conformal Cooling Systems Design and Dimensioning for Injection Molds. In *Advances in Design Engineering*; Springer Nature: Logroño, Spain, 2019; pp. 166–174.
13. Torres-Alba, A.; Mercado-Colmenero, J.M.; Diaz-Perete, D.; Martin-Doñate, C. A new conformal cooling design procedure for injection molding based on temperature clusters and multidimensional discrete models. *Polymers* **2020**, *12*, 154. [CrossRef]
14. Wang, Y.; Yu, K.M.; Wang, C.C.; Zhang, Y. Automatic design of conformal cooling circuits for rapid tooling. *Comput.-Aided Des.* **2011**, *43*, 1001–1010. [CrossRef]
15. Berger, G.R.; Zorn, D.; Friesenbichler, W.; Bevc, F.; Bodor, C.J. Efficient cooling of hot spots in injection molding. A biomimetic cooling channel versus a heat-conductive mold material and a heat conductive plastics. *Polym. Eng. Sci.* **2019**, *59*, E180–E188. [CrossRef]
16. Choi, J.H.; Kim, J.S.; Han, E.S.; Park, H.P.; Rhee, B.O. Study on an optimized configuration of conformal cooling channel by branching law. In *Engineering Systems Design and Analysis*; ASME: New York, NY, USA, 2014; p. V001T06A007.
17. Shinde, M.S.; Ashtankar, K.M.; Kuthe, A.M.; Dahake, S.W.; Mawale, M.B. Direct rapid manufacturing of molds with conformal cooling channels. *Rapid Proto. J.* **2018**. [CrossRef]
18. Saifullah, A.; Masood, S.H. Optimum cooling channels design and Thermal analysis of an Injection moulded plastic part mould. *Mater. Sci. Forum* **2007**. [CrossRef]
19. Wahab, M.S.; Raus, A.A.; Amir, I.; Ahmed, A.; Kamarudin, K. The thermal effect of variate cross-sectional profile on conformal cooling channels in plastic injection moulding. *Int. J. Integr. Eng.* **2018**. [CrossRef]
20. Rahim, S.Z.A.; Sharif, S.; Zain, A.M.; Nasir, S.M.; Mohd Saad, R. Improving the quality and productivity of molded parts with a new design of conformal cooling channels for the injection molding process. *Adv. Polym. Tech.* **2016**. [CrossRef]
21. Collomb, J.; Balland, P.; Francescato, P.; Gardet, Y.; Leh, D.; Saffré, P. Thermomechanical optimization and comparison of a low thermal inertia mold with rectangular heating channels and a conventional mold. *Adv. Mater. Sci. Eng.* **2019**. [CrossRef]
22. Kamat, A.M.; Pei, Y. An analytical method to predict and compensate for residual stress-induced warpage in overhanging regions of internal channels fabricated using powder bed fusion. *Addit. Manuf.* **2019**. [CrossRef]
23. Sharma, S.G.; Singraur, D.S.; Sudhakar, D.S.S. Transient analysis of an injection mould with conformal cooling channels. In *Recent Advances in Mechanical Infrastructure*; Springer: Berlin, Germany, 2020; pp. 235–244.

24. Altaf, K.; Raghavan, V.R.; Rani, A.M.A. Comparative thermal analysis of circular and profiled cooling channels for injection mold tools. *J. Appl. Sci.* **2011**, *11*, 2068–2071. [CrossRef]
25. Xi, L.; Gao, J.; Xu, L.; Zhao, Z.; Li, Y. Study on heat transfer performance of steam-cooled ribbed channel using neural networks and genetic algorithms. *Int. J. Heat Mass Transfer* **2018**, *127*, 1110–1123. [CrossRef]
26. Wang, W.; Zhang, Y.; Li, Y.; Han, H.; Li, B. Numerical study on fully-developed turbulent flow and heat transfer in inward corrugated tubes with double-objective optimization. *Int. J. Heat Mass Transfer* **2018**, *120*, 782–792. [CrossRef]
27. Jiang, G.; Gao, J.; Shi, X.; Li, F.; Xu, L. Flow and heat transfer characteristics of the mist/steam two-phase flow cooling the rectangular channel with column-row-ribs. *Int. J. Heat Mass Transfer* **2020**, *156*, 119737. [CrossRef]
28. Freitas, P.; Santos, C.; Carreira, P.; Mateus, A. High efficiency cooling and heating channels for injection moulding. *Appl. Mech. Mater.* **2019**. [CrossRef]
29. Sachs, E.; Wylonis, E.; Allen, S.; Cima, M.; Guo, H. Production of injection molding tooling with conformal cooling channels using the three dimensional printing process. *Polym. Eng. Sci.* **2000**, *40*, 1232–1247. [CrossRef]
30. Menges, G.; Michaeli, W.; Mohren, P. *How to Make Injection Molds*; Hanser Publishers: Cincinnati, OH, USA, 2001; pp. 1–612.
31. Zhong, Z.W.; Leong, M.H.; Liu, X.D. The wear rates and performance of three mold insert materials. *Mater. Des.* **2011**, *32*, 643–648. [CrossRef]
32. Valls, I.; Hamasaiid, A.; Padré, A. High thermal conductivity and high wear resistance tool steels for cost-effective hot stamping tools. *J. Phys. Conf. Ser.* **2017**. [CrossRef]
33. Mercado-Colmenero, J.M.; Rubio-Paramio, M.A.; Karlinger, P.; Martin-Doñate, C. A new procedure for calculating cycle time in injection molding based on plastic part geometry recognition. *Int. J. Adv. Manuf. Technol.* **2018**, *98*, 441–477. [CrossRef]
34. Mercado-Colmenero, J.M.; Rubio-Paramio, M.A.; de Juanes Marquez-Sevillano, J.; Martin-Doñate, C. A new method for the automated design of cooling systems in injection molds. *Comput.-Aided Des.* **2018**, *104*, 60–86. [CrossRef]
35. Vasco, J.; Barreiros, F.M.; Nabais, A.; Reis, N. Additive manufacturing applied to injection moulding: Technical and economic impact. *Rapid Proto. J.* **2019**. [CrossRef]
36. Torres-Alba, A.; Mercado-Colmenero, J.M.; Caballero-Garcia, J.d.D.; Martin-Doñate, C. Application of New Triple Hook-Shaped Conformal Cooling Channels for Cores and Sliders in Injection Molding to Reduce Residual Stress and Warping in Complex Plastic Optical Parts. *Polymers* **2021**, *13*, 2944. [CrossRef]
37. Park, H.S.; Dang, X.P.; Song, X.P. Improving the Cooling Efficiency for the Molding of a Complex Automotive Plastic Part by 3D Printing Technology. *Trans. Korean Soc. Automot.* **2017**, *25*, 508–515. [CrossRef]
38. Park, H.S.; Dang, X.P.; Nguyen, D.S.; Kumar, S. Design of advanced injection mold to increase cooling efficiency. *Int. J. Precis. Eng. Manuf.-Green Technol.* **2020**, *7*, 319–328. [CrossRef]
39. Ahn, D.G.; Park, S.H.; Kim, H.S. Manufacture of an injection mould with rapid and uniform cooling characteristics for the fan parts using a DMT process. *Int. J. Precis. Eng. Manuf.-Green Technol.* **2010**, *11*, 915–924. [CrossRef]
40. Cycle Time Reduction in Injection Moulding with Conformal Cooling Channels. Available online: https://researchbank.swinburne.edu.au/file/b9d88f55-8e50-4071-8265-498ccb3e3d45/1/PDF%20%28Published%20version%29.pdf (accessed on 10 August 2021).
41. Saifullah, A.B.M.; Masood, S.H.; Sbarski, I. Thermal–structural analysis of bi-metallic conformal cooling for injection moulds. *Int. J. Adv. Manuf. Technol.* **2012**, *62*, 123–133. [CrossRef]
42. Xu, X.; Sachs, E.; Allen, S. The design of conformal cooling channels in injection molding tooling. *Polym. Eng. Sci.* **2001**, *41*, 1265–1279. [CrossRef]
43. Mercado-Colmenero, J.M.; Torres-Alba, A.; Catalan-Requena, J.; Martin-Doñate, C. A New Conformal Cooling System for Plastic Collimators Based on the Use of Complex Geometries and Optimization of Temperature Profiles. *Polymers* **2021**, *13*, 2744. [CrossRef] [PubMed]
44. Mercado-Colmenero, J.M.; Martin-Doñate, C.; Rodriguez-Santiago, M.; Moral-Pulido, F.; Rubio-Paramio, M.A. A new conformal cooling lattice design procedure for injection molding applications based on expert algorithms. *Int. J. Adv. Manuf. Technol.* **2019**, *102*, 1719–1746. [CrossRef]
45. Brooks, H.; Brigden, K. Design of conformal cooling layers with self-supporting lattices for additively manufactured tooling. *Addit. Manuf.* **2016**, *11*, 16–22. [CrossRef]
46. Optimised Mould Temperature Control Procedure Using DMLS. Available online: https://www.3dimpuls.com/sites/default/files/download/eos_optimizedmouldtemperaturecontrolprocedureusing_dmls.pdf (accessed on 12 August 2021).
47. Dassault Systemes. Available online: https://www.3ds.com/es/productos-y-servicios/catia/ (accessed on 1 September 2021).
48. Moldex3d. Available online: https://www.moldex3d.com/ (accessed on 1 September 2021).

polymers

MDPI

Article

Application of New Triple Hook-Shaped Conformal Cooling Channels for Cores and Sliders in Injection Molding to Reduce Residual Stress and Warping in Complex Plastic Optical Parts

Abelardo Torres-Alba, Jorge Manuel Mercado-Colmenero, Juan de Dios Caballero-Garcia and Cristina Martin-Doñate *

Department of Engineering Graphics, Design and Projects, Campus Las Lagunillas s/n., University of Jaen, Building A3-210, 23071 Jaen, Spain; ata00001@red.ujaen.es (A.T.-A.); jmercado@ujaen.es (J.M.M.-C.); jdcg0004@red.ujaen.es (J.d.D.C.-G.)

* Correspondence: cdonate@ujaen.es; Tel.: +34-953-212-821; Fax: +34-953-212-334

Citation: Torres-Alba, A.; Mercado-Colmenero, J.M.; Caballero-Garcia, J.d.D.; Martin-Doñate, C. Application of New Triple Hook-Shaped Conformal Cooling Channels for Cores and Sliders in Injection Molding to Reduce Residual Stress and Warping in Complex Plastic Optical Parts. *Polymers* **2021**, *13*, 2944. https://doi.org/10.3390/polym13172944

Academic Editors: Salah Aldin Faroughi, Luís L. Ferrás, Alexandre M. Afonso and Célio Bruno Pinto Fernandes

Received: 20 August 2021
Accepted: 27 August 2021
Published: 31 August 2021

Abstract: The paper presents a new design of a triple hook-shaped conformal cooling channels for application in optical parts of great thickness, deep cores, and high dimensional and optical requirements. In these cases, the small dimensions of the core and the high requirements regarding warping and residual stresses prevent the use of traditional and standard conformal cooling channels. The research combines the use of a new triple hook-shaped conformal cooling system with the use of three independent conformal cooling sub-systems adapted to the complex geometric conditions of the sliders that completely surround the optical part under study. Finally, the new proposed conformal cooling design is complemented with a small insert manufactured with a new Fastcool material located in the internal area of the optical part beside the optical facets. A transient numerical analysis validates the set of improvements of the new proposed conformal cooling system presented. The results show an upgrade in thermal efficiency of 267.10% in comparison with the traditional solution. The increase in uniformity in the temperature gradient of the surface of the plastic part causes an enhancement in the field of displacement and in the map of residual stresses reducing the total maximum displacements by 36.343% and the Von—Mises maximum residual stress by 69.280% in comparison with the results obtained for the traditional cooling system. Additionally, the new design of cooling presented in this paper reduces the cycle time of the plastic part under study by 32.61%, compared to the traditional cooling geometry. This fact causes a very high economic and energy saving in line with the sustainability of a green mold. The improvement obtained in the technological parameters will make it possible to achieve the optical and functional requirements established for the correct operation of complex optical parts, where it is not possible to use traditional cooling channels or standard conformal cooling layouts.

Keywords: conformal cooling; sustainability; injection molding; industrial design; manufacturing; numerical simulation

1. Introduction

The use of precision polymer optics is becoming an increasing necessity today as products demand sophisticated light handling components to achieve desired results [1,2]. Even if the optical properties of materials such as glass are very stable, their manufacturing process is especially complex considering the time limitations and requirements demanded in the industry. Nevertheless, optical pieces manufactured in plastic are sturdy and low-cost, produced in one only step despite the geometric complexity. These reasons make plastic optical parts crucial for contemporary industrial development [3–5].

The diversity of optical products for the automotive sector caused the development of geometries requiring high-thickness ratios. In the injection molding process, the thickest area of the part presents heat accumulation because of slower cooling. This fact produces not only a slower cooling time but also an uneven shrinkage [6–8] that generates warping, affecting the optical properties of the plastic part [9].

Compared to the most stringent technical parts, which require tolerances of tenths of a millimeter, the requirements for optical components are up to 100 times higher, and not only with regard to a certain dimension, but also along its entire surface thus guaranteeing its correct operation. A designed geometry with a wide variation in the thickness ratios can originate thermally induced residual stress during the injection cooling process [10]. Residual stress results in a slight local reduction of the optical properties, and therefore, also of the transmitted light [11].

Achieving an even distribution of temperatures to eliminate residual stress using conventional cooling channels is very difficult. In this line, conformal cooling channels have greater flexibility to adapt their geometry to the complex topological requirements of the plastic optical part [12,13]. The use of conformal cooling channels improves the uniformity in the cavity surface temperature reducing in that way the residual stress thermally induced throughout the cooling process [14,15]. Additive technology enables the manufacturing of complex conformal cooling channels to best match the shape of the cavity and core in the mold. In this way, a uniform cooling in areas where heat can be difficult to trap, like hardly accessible and high thickness areas, can be achieved. Conformal cooling channels allow uniform heat dissipation for optical parts as well as high cooling efficiency [16].

Although some authors have made use of the advantages of employing conformal channels for cooling pieces of diverse geometry, only a few have studied their application for cooling optical parts with complex geometries. These works, mainly focus on the cooling of optical lenses with complex geometry [17]. Chung [17] combines the analysis of finite elements with an algorithm based on gradients and a robust genetic algorithm to obtain the optimal design of the cooling channels for an optical lens. According to his research, conformal channels can reduce the temperature differences on the mold surface, as well as the ejection time, and the warping.

The design of molds for complex optical parts requires a great number of mobile devices in order to manufacture all the topological part features. In complex cases, mobile devices are responsible for molding a large part of the geometry of the piece, limiting the space of the cavity plate [18]. This fact has a great influence on the design of the cooling layout since the design of the cavity cooling is performed entirely on the side cores. Side cores usually present complex geometry and reduced dimensions as they have to adapt to the geometry of the part. This precludes the design of a continuous cooling layout using straight or standard conformal channels that completely surround the part [19–21]. Likewise, the small dimensions of mobile devices limit the use of large diameter cooling channels dividing channels into individual zones adapted to the requirements of each mobile device. This fact greatly hinders the use of traditional channels since they require compliance with the high design and sizing requirements of the CNC manufacturing process.

The high optical requirements force the design of optical parts of great depth, which causes deep cores in the mold. These areas are highly difficult to cool, forcing the use of baffles normally far from the internal surface of the part due to compliance with the manufacturing criteria of the CNC process. The use of baffles in optical pieces with deep cores prevents the correct thermal exchange between coolant flow and plastic melting [22,23].

Conformal cooling channels provide greater design flexibility by reducing the distance to the part, presenting more functional layouts for parts with complex geometries [24–26]. However, the use of standard conformal channels also presents limitations in cases with a lack of accessibility and space for cooling. These boundary conditions force the use of conformal channels with very small diameters, which can cause insufficient heat exchange and obstructions due to interference from foreign objects.

The use of inserts manufactured with Fastcool [27] material is presented as an innovative solution to deal with the lack of space in very deep and difficult access regions. Nevertheless, the high conductivity of Fastcool inserts requires the use of specific layouts for their cooling, since the tool steel of the mold is not able to evacuate the heat produced by the insert quickly enough to generate a uniformity of temperatures, achieving an efficient cycle time. In this case, conformal cooling layouts are considered a feasible option to evacuate the heat generated by these inserts in those areas of the mold with a lack of accessibility and where the design requirements prevent the direct use of standard conformal channels.

To solve the problems proposed, the paper presents a new design of a triple hook-shaped conformal cooling channels for application in optical parts of great thickness, deep cores and high dimensional and optical requirements. In these cases, the small dimensions of the core and the high requirements regarding warping and residual stresses prevent the use of traditional and standard conformal cooling channels. The research combines the use of a new triple hook-shaped conformal cooling system with the use of three independent conformal cooling sub-systems adapted to the complex geometric conditions of the sliders that completely surround the optical part under study. Finally, the new proposed conformal cooling design is complemented with a small insert manufactured with a new Fastcool material located in the internal area of the optical part beside the optical facets.

In this way, it is possible to reduce warping and residual stresses in the manufacture of highly complex optical parts with high thickness ratios, meeting the demanding requirements of the automotive industry. In parallel, a reduction in the production cycle time and energy consumption of the mold is achieved, producing a sustainable mold in line with the demands of the current environment. Additionally, the uniformity in the temperatures of the surface of the piece is improved by eliminating hot spots and differential shrinkage. The paper exceeds the state of the art, being capable of cooling optical parts with deep cores, high thickness and small details, being very useful in an area as important and widespread worldwide as is the molding manufacturing of automotive optical parts.

2. Materials and Methods

2.1. Analysis and Geometrical Features for the Plastic Part Manufactured through the Injection Molding Process

In this item, the topological and technological features related to the plastic part under study are depicted. Specialized insights about the selection of the plastic material for the injection molding manufacturing process, boundary conditions, and topological features are likewise determined.

The geometry presented in the paper is a plastic optical piece, whose main function is to guide and control the luminous flux from various LED points, distributing it evenly over the illumination plane. The piece (see Figure 1), presents a complex geometric design with great influence on its manufacturing process. The piece has small dimensions with a bounding box of (67 × 52 × 127) mm and a thickness of 6 mm. The upper part of the part is characterized by including a set of 7 lugs or pins whose main function is to serve as an injection point for the part and as a means of connection to the LED optical devices and the optical control PCB. The set of pins is attached to three reinforcing ribs distributed equidistantly, forming an angle of 120° with each other. The function of the ribs is to reinforce the base of the pins against possible stresses to which the part will be subjected throughout its useful life, as well as, to reduce the effect of the plastic material shear at the injection gate. Figure 1 shows a picture of the optical piece under study as well as the topological details that characterize its geometry.

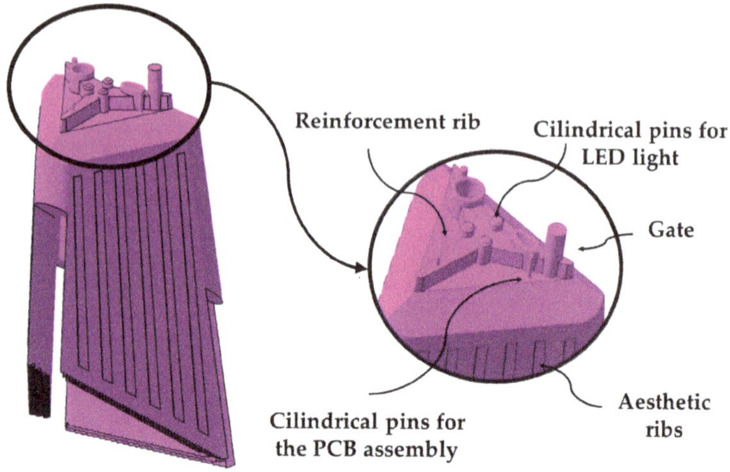

Figure 1. Optimal part under study and topological details of its geometry.

The base of the optical piece presents an irregular geometry caused by the different heights at which the three lateral surfaces that characterize the piece are found. Likewise, the lower surface of the piece is characterized by including a set of optical facets in charge of the distribution and channeling of the LED light. The lateral surfaces include a set of 23 faceted grooves of thickness 0.3 mm of an aesthetic character whose function is to avoid that the surfaces present a completely smooth appearance. The manufacture of the lateral grooves requires the use of three sliders in the injection mold since the geometries are not demoldables. The sliders of the mold completely surround the contour of the optical piece (see Figure 2).

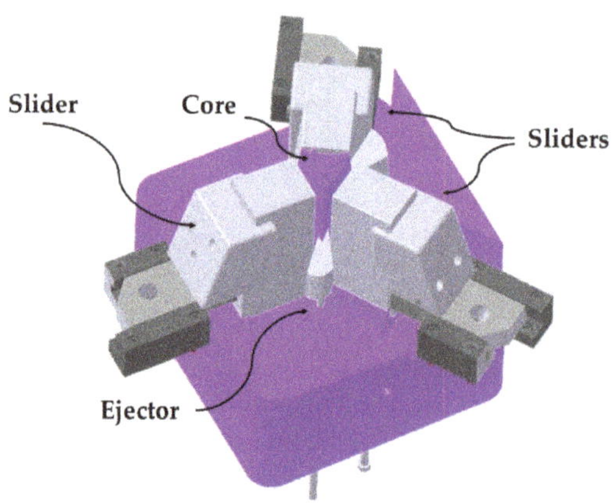

Figure 2. Side sliders.

The optical piece is characterized by the inclusion of a deep concave geometry area inside, which makes that the core plate presents a great depth that is difficult to cool. Additionally, the upper internal area of the part, close to the injection point, includes a set of faceted optical grooves. These facets require precise molding which forces to design a cooling in the core plate that avoids possible warping that could invalidate the optical function of the piece. Figure 3 presents a picture of the internal area of the optical piece under study as well as the details of the optical facets included in the upper internal surface.

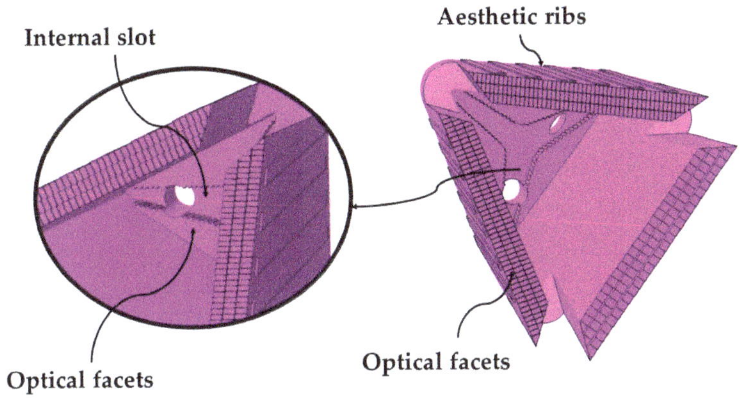

Figure 3. Geometric details of the internal area of the optical piece.

The dimensions of the core of the mold in contact with the internal area of the piece prevent the use of ejector pins to extract it from the mold. For this reason, the design of the mold requires the use of three external ejector inserts located in three lateral recesses of the optical piece analyzed. The different heights at which the lateral surfaces of the piece are located influences the parting line of the mold to be located in different planes.

The geometry of the plastic part greatly influences the design of the injection mold and fundamentally the design of its cooling system. Figure 4 shows the diagram of the cooling of the optical part with the current cooling layout of the mold, using traditional methods for its manufacture. The mold makes use of a cooling based on the design of 8 mm diameter straight cooling channels located in each of the sliders. It also uses baffle-type elements to cool the core of the mold. The small dimensions of the core and the side cores, as well as the fulfillment of the traditional manufacturing requirements that guarantee the structural safety of the mold, make the design of the cooling layout extremely difficult. Likewise, the great depth and small dimensions of the core, require the use of a single baffle with a diameter of 12 mm and a separation of 20 mm to cool the internal area of the piece. The cooling of the mold is complemented by two straight channels of 8 mm diameter in the upper part of the mold cavity in charge of cooling the material injection gate.

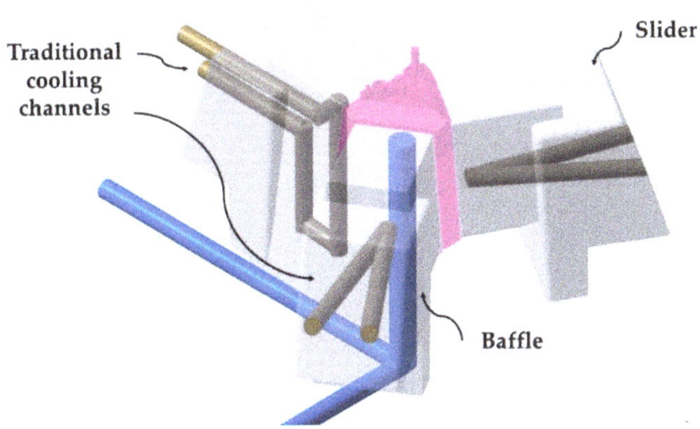

Figure 4. Current layout of the traditional cooling of the core and sliders of the mold.

The cooling of deep cores in pieces of great thickness is highly complex since it requires cooling designs capable of evacuating the heat accumulated in specific areas of the core, and of avoiding differential shrinkage and subsequent warping in the molded plastic part.

The optical piece analyzed in the paper has a slotted and faceted interior zone very close to the injection point (see Figure 3). This area, which is 6.5 mm thick and difficult to access, locally accumulates a large amount of heat, that must be removed by cooling. The use of standard conformal cooling systems is not valid for cooling small deep cores and thick molded parts. Additionally, the reduced dimensions of the core would require the use of small diameter standard conformal channels with little cooling path, preventing the correct thermal exchange between the coolant flow and the molten plastic.

To solve this problem, the paper presents a new design of triple hook-shaped conformal cooling for application in pieces of great thickness and deep cores. In these cases, the small dimensions of the core prevent the use of standard conformal cooling systems. Likewise, the research combines the use of the new triple hook-shaped conformal cooling system with the use of three independent conformal cooling channels included in the three small side cores. In this way, it is possible to adequately cool the sliders of the mold using three conformal cooling sub-systems adapted to the particular geometric conditions of each slider. Finally, the new proposed conformal cooling design is combined with a small Fastcool element located in the internal area of the piece where the optical facets are located.

The new triple hook-shaped conformal cooling design presented in this paper aims to cool the core of the mold in contact with the three side surfaces of the part, as well as to cool the Fastcool insert. The new conformal triple hook-shaped conformal design is made up of a central channel with a domed end from which three channels depart equally spaced at an angle α in the horizontal plane and an angle β of vertical slope. The triple hook design meets the design criteria in additive manufacturing [27] eliminating the use of supports in its manufacture. Equation (1) indicates the sizing criteria that the diameters of the triple hook- shaped conformal channels must meet, where ϕ_c is the diameter of the central channel and ϕ_i the diameter of the lateral channels. The central channel ϕ_c must not exceed the value of 10 mm to avoid material collapse when manufacturing the channel upper area. Figure 5 shows a picture of the new triple hook-shaped conformal channel design for cores presented in the paper.

$$\phi_c = \sum_{i=1}^{3} \phi_i \mid \phi_c \leq 10 \text{ mm} \ \ \phi_i = \phi_{i+1} \tag{1}$$

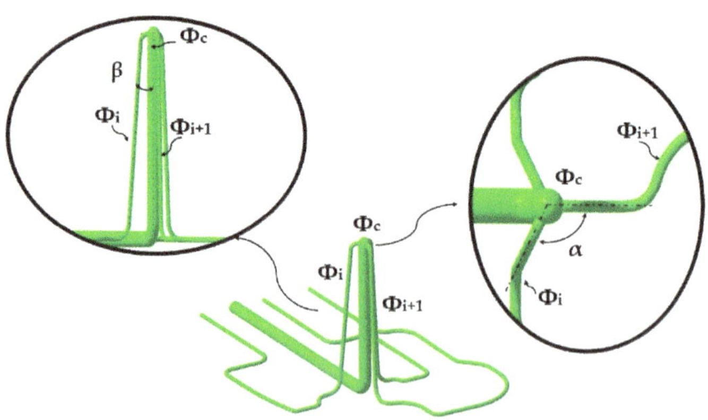

Figure 5. Conformal cooling triple hook-shaped channel.

The upper part of the core close to the injection point of the part presents a hot spot that causes a great thermal imbalance in the cooling of the part.

To eliminate this thermal imbalance, the new triple hook conformal cooling design is complemented by the use of a small flat-shaped insert of a new Fastcool material [28] located on the upper inner surface of the core. In this way, it is possible to establish a complete cooling design adapted to the geometrical and functional requirements of the part, capable of extracting the heat accumulated in the gate area, eliminating possible warping in the upper optical zone. Figure 6 indicates the location of the Fastcool insert used to cool the area of the optical facets close to the injection point.

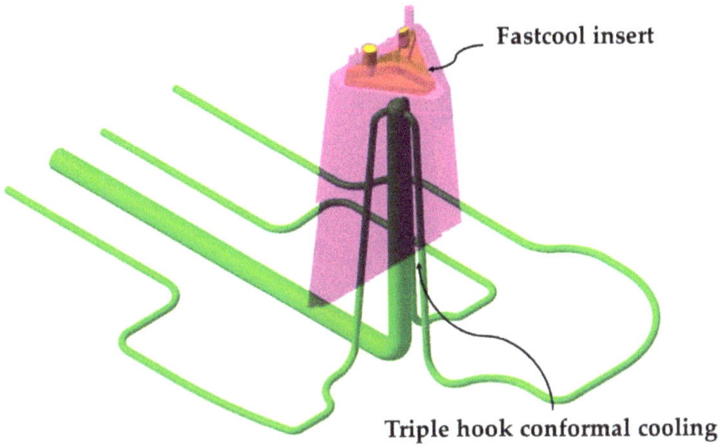

Figure 6. Fastcool insert.

The area of the three slides in charge of molding the lateral surfaces of the part is cooled using three independent conformal cooling layouts adapted to the geometry of each slide. In this way, it is possible to cool complex parts whose mold cavities are formed mostly by small sliders and in which the standard conformal spiral or zigzag layouts that surround the part are impossible to use. Figure 7 presents a picture of the conformal cooling of the core and sliders of the mold presented in the paper, formed by the use of the new conformal cooling triple hook shaped combined with a small Fastcool flat insert for cooling the mold core.

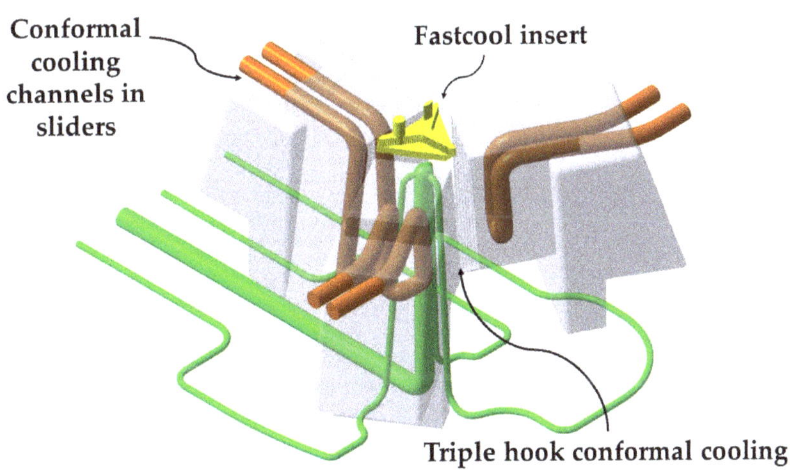

Figure 7. Conformal cooling of the core and sliders of the mold.

The mold cavity is cooled using three independent conformal cooling channels adapted to the geometric requirements of the three side sliders of the mold. Finally, the cooling of the mold cavity is complemented by two conformal cooling channels of 8 mm diameter in the upper part of the cavity plate in charge of cooling the material injection gate (see Figure 8). Table 1 indicates the dimensions of the main different elements used in the conformal cooling of the optical part.

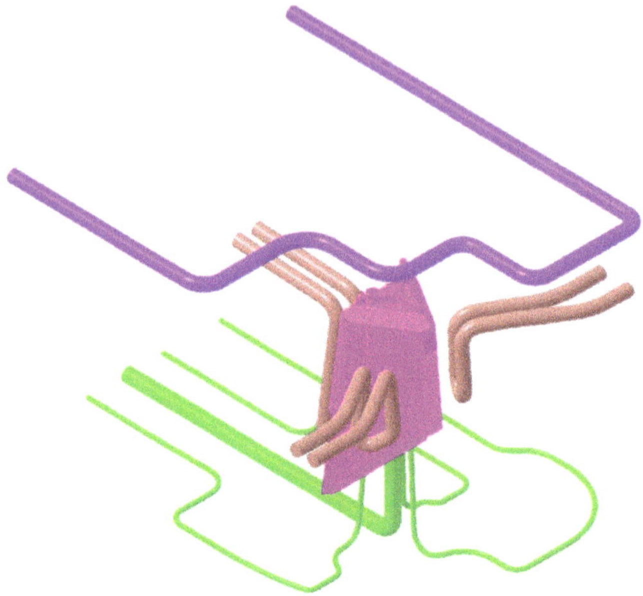

Figure 8. Complete cooling of the cavity, core and sliders of the injection mold por the presented optical part.

Table 1. Geometric parameters used in the design of conformal cooling system.

Nomenclature	Units	Description	Triple Hook	Conformal Cooling in Sliders
ϕ_c	mm	Center channel diameter	9	-
ϕ_i	mm	Side channel diameter	3	-
α		Horizontal separation angle between channels	120	-
β		Vertical slope of the channels	4.5	-
ϕ_s	mm	Diameter conformal cooling channels in sliders	-	8
s	mm	Distance channel center—sliders surface	-	16

2.2. Plastic Part Material

The optical part under study is manufactured with PC Lexan 121R plastic material from the company Sabic [29] obtained by chemical recycling. Therefore, the mechanical, thermal, and chemical properties of the original plastic are maintained without compromising the sustainability of the injection process. This thermoplastic material is a Polycarbonate that allows designers and manufacturers the facility for design freedom, aesthetics enhancements and cost reductions. Furthermore, due to its physical properties and specifications, it can be applied and used in optical plastic parts. The magnitudes of the main physic, mechanical and thermal properties of the material PC Lexan 121R are indicated in Table 2.

Table 2. Magnitude of the main properties of the material PC Lexan 121R.

Nomenclature	Units	Description	PC Lexan 121R
ρ_p	g/cm^3	Density	1.2
C_p	J/kg·°C	Specific heat	1250
δ_p	W/m·°C	Thermal conductivity coefficient	0.2
MFI	g/10 min	Melt flow index	17.5
T_{mold}	°C	Mold Temperature (normal)	40
T_{eject}	°C	Ejection temperature	100
T_{freeze}	°C	Freeze temperature	164
E_p	MPa	Elastic Modulus	2340
υ_p	-	Poisson's ratio	0.4
CLTE	1/°C	Coefficient of linear thermal expansion	6.84×10^{-5}
UOI	-	Un-oriented refractive index	1.56
FSC	$cm^2/dyne$	Flow-induced stress-optical coefficient	1.95×10^{-10}
TSC	$cm^2/dyne$	Thermally-induced stress-optical coefficient	4.50×10^{-12}

3. Implementation and Results

In this manuscript, two proposals for the design of the cooling system for the plastic part under study are presented and compared. On the one hand, the current traditional cooling system with perforated straight channels whose manufacture is carried out using traditional machining processes and tools. And, on the other hand, a new optimized cooling system that combines conformal triple hook-shaped cooling channels with a fastcool-type metal insert. So, the manufacture of this proposal is based mainly on the 3D additive manufacturing process using laser sintering (SLS). Additive manufacturing technologies are line with current sustainability requirements. The SLM additive manufacturing process allows the manufacture of conformal channels adapted to the free shape of the geometric surface of the plastic part [30,31]. The geometric CAD design of both configurations is performed using the Catia V5—6R2020 3D CAD geometric modeling software [32]. Likewise, to evaluate and analyze the thermal and technological parameters that define the cooling phase of the plastic part, numerical simulations of a thermal type are modeled using the numerical and commercial software Moldex3D R17-CoreTech System Taiwan, [33]. In this way, the results of the thermal and technological parameters obtained from both proposed cooling system configurations can be compared, establishing the one that optimizes and improves, on the one hand, the cooling phase of the plastic part, as well as the thermal efficiency. Both the 3D CAD modeling process and the numerical analysis of the different cooling system configurations proposed in this manuscript have been carried out using an MSI notebook with an Intel (R) Core- Intel corporation EEUU(TM) i-77700HQ CPU @ 2.80 GHz.

Thermal Modeling of Numerical Simulations

The definition of rheological and thermal simulations using CAE numerical software allows the analysis of the cooling phase of a plastic part and how the main elements of the injection mold and the most representative technological parameters of said phase influence and interact. Likewise, the results of the parameters obtained from the numerical simulations allow establishing whether the design of the main elements that make up the cooling system, meets the minimum industrial technical requirements that are established to validate the manufacture of the plastic part. In this section, the preprocessing configuration used for each of the different numerical simulations carried out is detailed. At the beginning of this preprocessing phase, the discretization of the different geometric elements to be analyzed numerically must be defined. That is, the three-dimensional meshes for the geometric elements of the injection mold must be defined, as well as the geometric parameters that define them. The commercial software Moldex 3D R17 [33] has a Moldex Designer meshing module, in which the geometric parameters of the meshes created can be configured and established. Table 3 shows the magnitude of the geometric parameters used during the meshing process, as well as their configuration. Said geometric parameters have been adjusted to the smallest and most relevant geometric detail or precision of the plastic part under study. However, the selection of the type of element used is important when carrying out the meshing process. In this way, three-dimensional elements of the second-order tetrahedron have been selected, called SOLID 186. These elements have 10 main nodes, located at the vertex of the tetrahedron, and 4 secondary nodes, located at the midpoint of the edges of the tetrahedron (see Figure 9). In addition, each of said nodes has 3 degrees of freedom in the main coordinate axes, notably improving the precision of the temperature field parameters and displacements in the solution obtained. Besides that, and in order to improve the precision of the numerical simulation, a series of three-dimensional elements of the prism type have been defined along the contact surfaces between the different elements that make up the injection mold. Elements are placed on the surface of the cooling channels or between the surface of the plastic part and the cavity and core surface of the injection mold cavity. These elements have 6 main nodes located at the vertices of the prism, and 9 secondary nodes located at the midpoint of the edges of the prism (see Figure 9). Furthermore, each of said nodes has 3 degrees of freedom in the main coordinate axes. The selection of this type of element is established by the "Boundary Layer Mesh" operation, which establishes a series of layers from the interface surfaces previously mentioned (see Figure 10). The average length of these elements is configured from an offset ratio or percentage of the size of the average element of the mesh. In this case, and as Table 3 shows, the offset ratio selected for the generation of the meshes is 0.1 and the number of Boundary Layers is equal to 5. The use of this meshing operation allows modeling with greater precision the roughness generated between the surface of the cooling channels and the coolant flow, as well as the layer of solidified plastic material that is generated when the molten plastic front comes into contact with the surfaces of the injection mold cavity.

Table 3. Mesh statistics for the meshes analyzed in the present manuscript.

Description	Units	Standard	Conformal and Fastcool
Part mesh node count	-	295,339	287,416
Part mesh element count	-	779,946	731,735
Part mesh volume	cm^3	77.63	70.70
Runner mesh node count	-	2892	2892
Runner mesh element count	-	2464	2464
Runner mesh volume	cm^3	0.05	0.05
Plastic part precisión (ε)—Mesh sizing	mm	1.00	1.00
Element type	-	Tetrahedral (10 nodes)	Tetrahedral (10 nodes)
Element type—Boundary layers	-	Prism (15 nodes)	Prism (15 nodes)
Offset ratioBoundary layers	-	0.1	0.1

Next, we proceed to define the material assigned to each one of the elements of the injection mold and the plastic part, as well as the physical, thermal, and rheological properties of each one of them. As shown in Figure 11, the plastic part is manufactured from the thermoplastic material PC Lexan 121 R [28]. The main feed channel, from which the filling of the mold cavity begins, has associated, like the plastic part, the PC Lexan 121 R thermoplastic material. For the cooling channels, both for the traditional configuration and for the conformal configuration, the material water is defined as the coolant flow. For the geometry defined as injection mold, the assigned metal material is a P20 steel alloy, and finally, for the Fastcool insert, used in the conformal cooling system solution, the metallic material used is a Fastcool-50 steel alloy. Table 4 shows the magnitude of the physical, thermal and rheological properties defined in the numerical simulations for each material used. As can be seen, the use of a Fastcool insert, whose metallic material is Fastcool-50 [27], considerably improves the thermal properties of the metallic material of the mold. In this way, the area of the plastic part that is in contact with the Fastcool insert will present greater heat exchange and, therefore, will improve thermal efficiency throughout the cooling phase. Therefore, as shown in Figure 11, said Fastcool insert is located in the inner central core of the plastic part. Well, in this region a large amount of residual heat accumulates and presents greater difficulty to be uniformly re-cooled, with respect to the rest of the geometric regions of the plastic part.

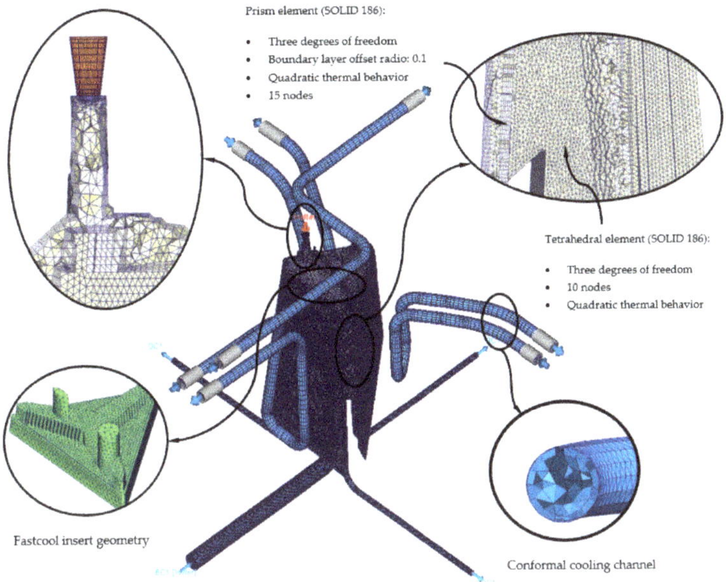

Figure 9. Mesh details for the new conformal cooling triple hook—shaped and Fastcool insert configuration.

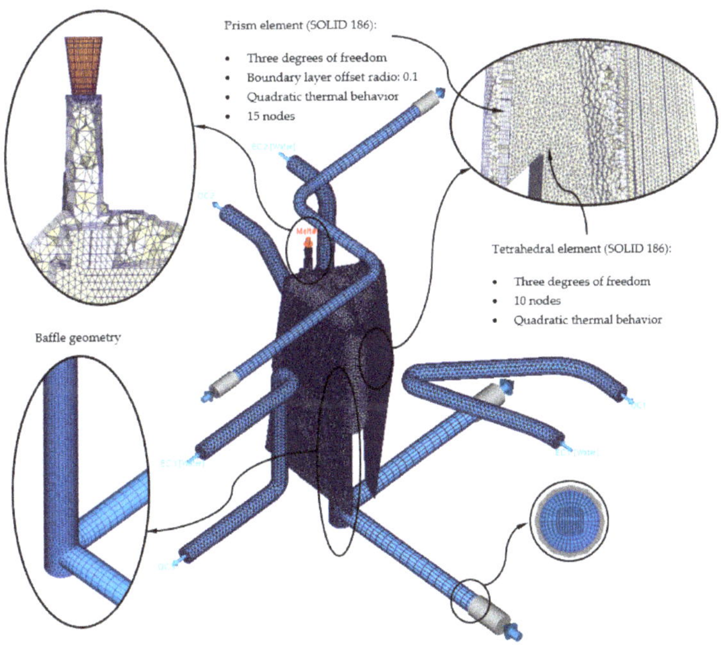

Figure 10. Mesh details for the traditional cooling configuration.

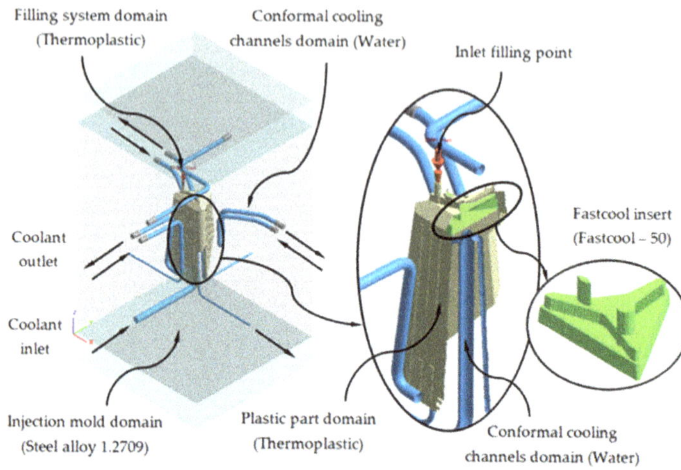

Figure 11. Materials and boundary conditions defined for the new conformal cooling triple hook—shaped and Fastcool insert configuration.

Table 4. Magnitude of the physical properties of the defined materials.

Description	Units	Water Pure	PC Lexan 121R	Steel Alloy P20	Fastcool 50
Density	kg/m^3	988	1200	7750	7810
Specific heat	J/kg·°C	4180	1250	460	470
Thermal conductivity coefficient	W/m·°C	0.643	0.200	29	50

Likewise, the definition of the thermoplastic material in the simulation software must be accompanied by numerical models that allow the modeling of both the behavior of the viscosity of the material and the behavior of its PVT curve. In addition, the manufacturer of the material [29] recommends the magnitude of a series of temperatures for each of the phases of the manufacturing cycle of the plastic part. Table 5 shows the parameters recommended by the manufacturer and the viscosity and PVT curve models of the thermoplastic material defined in the numerical simulation software.

Table 5. Magnitude of the physical properties of the defined materials.

Description	Units	Lexan 121R
Material type	-	Polycarbonate
Viscosity model	-	Modified Cross Model
PVT model	-	Modified Tait Model
Mechanical model	-	Isotropic pure polymer
Viscoelastic model	-	White-Metzner
Melt temperature	°C	280.0–310.0
Mold temperature	°C	70.0–95.0
Ejection temperature	°C	147
Freeze temperature	°C	164

As shown in Figures 11 and 12, each numerical analysis carried out has associated a set of boundary conditions, which establish the technological parameters of pressure and initial temperature for the input of the molten plastic front to the injection mold cavity and the flow of the coolant flow to the cooling channels. For the input of the molten plastic front to the injection mold cavity, the upper surface of the main feed channel (see Figure 11) is established as a boundary condition, an injection temperature of 295 °C, and a maximum injection pressure of 160 MPa. For the cooling channels, firstly, both the inlet and outlet surfaces of the coolant flow are defined (see Figure 11); secondly, an initial temperature of the coolant flow of 80 °C is determined, and finally, a pressure magnitude that allows the front of the coolant flow to develop in turbulent regime. That is, the Reynolds number of the coolant flow along the cooling channels is greater than 1.5×10^4. Likewise, and according to the recommended parameters offered by the manufacturer of the thermoplastic material, the initial temperature of the injection mold is set at 80 °C.

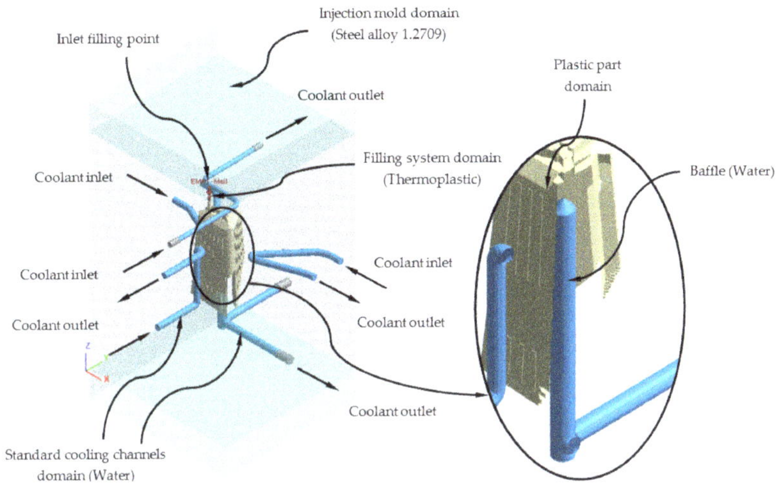

Figure 12. Materials and boundary conditions defined for the traditional cooling configuration.

Table 6 shows the magnitude of the technological parameters used in the modeling of the filling and cooling phase of the numerical simulations carried out for the present case study.

Table 6. Magnitude of the physical properties of the defined materials.

Description	Units	Study Cases—Lexan 121 R (PC)
Filling time	s	2.21
Packing time	s	15.00
Cooling time	s	90
Melt temperatue	°C	295
Mold temperature	°C	80
Coolant temperature	°C	80
Maximum injection pressure	MPa	160
Maximum packing pressure	MPa	160
Packing pressure	MPa	128

To complete the definition of the preprocessing phase of the numerical simulations carried out, the following configurations relative to the solver used to solve the numerical models of the simulations carried out are defined:

- The analysis of the cooling phase of the plastic part is carried out in a transitory regime or "Cooling transient". Given the defined cooling time (see Table 6), the solver analyzes the process and the evolution of the cooling of the plastic part over time. This type of analysis allows obtaining and saving solutions of the field of temperatures and displacements for different time intervals. The time interval defined in each numerical simulation carried out is 10 s.

- The modeling of the coolant flow along the cooling channels is done with the "Run 3D cooling channels" operation. This operation makes it possible to define a roughness magnitude on the surfaces of the cooling channels and improves the analysis of turbulence on their surface. The magnitude of the defined roughness is equal to 0.02 mm.
- The type of solver used is the maximum variation of mold temperature and its convergence criteria are temperature difference equal to 1 °C and maximum cycle number equal to 10 cycles.

After completing the definition of the pre-processing phase of the different numerical analyzes performed, the set of thermal and rheological results obtained is presented. From their analysis and evaluation, it is determined that the configuration of the triple hook-shaped conformal cooling channels for cores together with the use of conformal cooling channels adapted to the sliders and the Fastcool insert optimizes the cooling phase and improves the efficiency and thermal performance of the injection mold for the plastic part object of study.

Firstly, Table 7 and Figure 13 show the results obtained for the parameter time to reach the ejection temperature of the plastic part for each of the cooling system configurations proposed in this manuscript.

Table 7. Time to reach ejection temperature for the analyzed cooling systems.

Cooling System Typology	Time to Reach Ejection Temperature [s]	Time Reduction [s]	Improvement [%]
Traditional	41.550	-	-
Conformal and fastcool	28.187	13.363	32.161

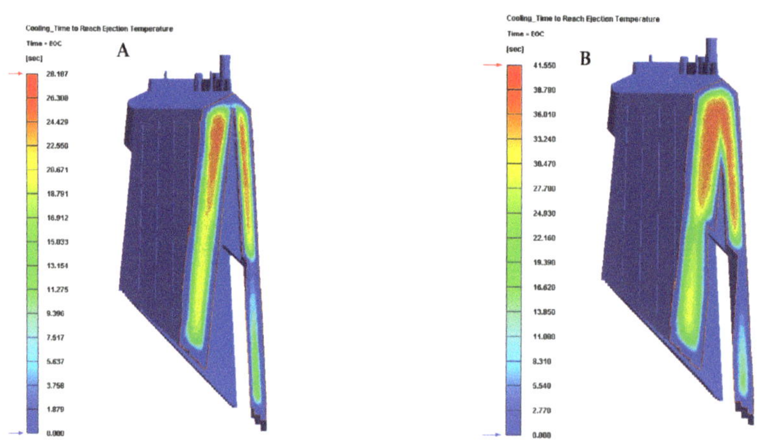

Figure 13. Time to reach ejection temperature (s). (**A**) Conformal cooling and fastcool insert. (**B**) Traditional cooling.

In this way, the results presented in Table 7 and Figure 13 show that the new configuration of the conformal cooling system presented in this paper, together with the use of a Fastcool insert, improves the time until reaching the ejection temperature of the plastic part, with respect to the configuration of the current traditional cooling system. Therefore, it can be determined that the use of this new type of conformal cooling channels, accompanied by an insert with high thermal performance, reduces the cycle time of the plastic part under study by 13.363 s or by 32.161%, compared to the classical geometry and configuration of perforated straight cooling channels.

Table 8 and Figure 14 show the results obtained for the heat flow parameter that is exchanged between the cooling mechanisms and the plastic part, for each of the cooling system configurations proposed in this manuscript.

Table 8. Heat flux (J/s·cm^2) for the analyzed cooling systems.

Cooling System Typology	Cavity Cooling	Core Cooling	Fastcool Insert	Total	Improvement [%]
Traditional	0.361	0.411	-	0.772	-
New conformal and Fastcool	0.501	0.903	1.430	2.834	267.10

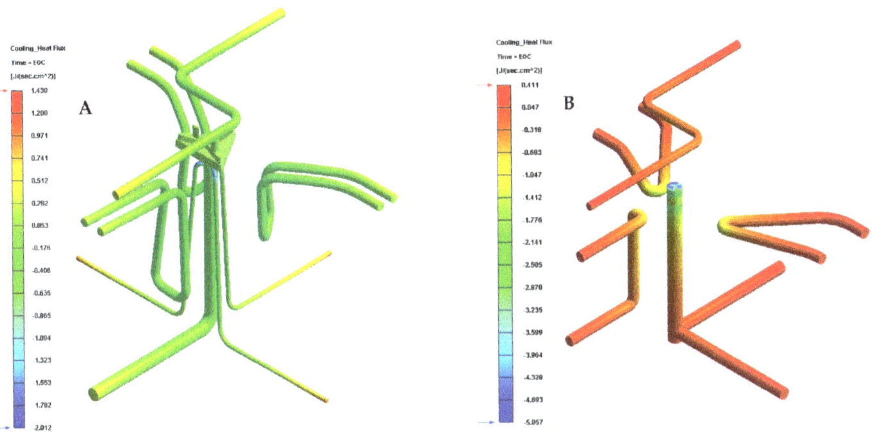

Figure 14. Heat flux (J/s·cm^2). (**A**) New conformal cooling system and Fastcool insert. (**B**) Traditional cooling.

Likewise, the results presented in Table 8 and Figure 14 show that the configuration of the new presented conformal cooling system, together with the use of a Fastcool insert, increases and optimizes the heat flow exchanged between the plastic part and the different cooling mechanisms defined for cooling the plastic part. In particular, the increase in heat exchange produced, compared to the standard cooling configuration, is 2.062 J/s·cm^2, which translates into an improvement in thermal efficiency of 267.10%. Table 9 and Figure 15 show the results obtained for the temperature gradient along the surface of the plastic piece under study in this manuscript.

Table 9. Cooling mold temperature difference for the analyzed cooling systems.

Cooling System Typology	Cooling Mold Temperature Difference (°C)	Temperature Reduction (°C)	Improvement (%)
Traditional	10.177	-	-
New conformal and Fastcool	1.557	8.620	84.701

The results presented in Table 9 and Figure 15 show that the configuration of the new conformal cooling system, together with the use of a Fastcool insert, reduces the temperature gradient along the surface of the plastic part concerning the traditional cooling system configuration. In particular, this reduction represents an improvement of 84.701% in the uniformity of the temperature map throughout the plastic part under study. In addition, the new conformal cooling channel solution, together with the Fastcool insert, meets the industrial validation requirements of the manufacture of the plastic part since the magnitude of the temperature gradient on the surface of the plastic part is less than 10 °C. However, for the traditional cooling system, this condition is very close to being fulfilled.

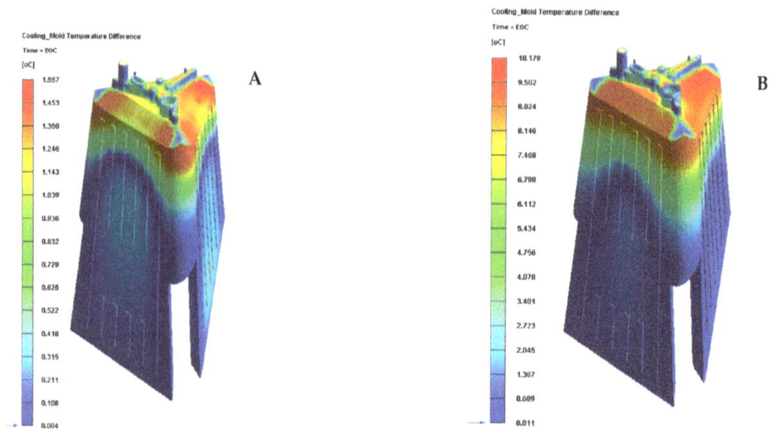

Figure 15. Cooling mold temperature difference (°C). (**A**) New conformal cooling system and Fastcool insert. (**B**) Traditional cooling.

Finally, given that the plastic part under study is an optical lighting part, it is important to analyze and check the field of displacements and the map of residual stresses resulting after the manufacturing process. Tables 10 and 11 and Figures 16 and 17 show the displacement field and the residual stress map of Von—Mises along with the geometry of the plastic part under study after the cooling phase.

Table 10. Warpage total displacement for the analyzed cooling systems.

Cooling System Typology	Warpage Total Displacement (mm)	Displacement Reduction (mm)	Improvement (%)
Traditional	0.875	-	-
New conformal and fastcool	0.557	0.318	36.343

Table 11. Warpage Von—Mises stress for the analyzed cooling systems.

Cooling System Typology	Warpage Von—Mises Stress (MPa)	Stress Reduction (MPa)	Improvement (%)
Traditional	5.319	-	-
Conformal and fastcool	1.634	3.685	69.280

As can be seen, for the new conformal cooling system with a Fastcool insert, the increase in uniformity in the temperature gradient of the surface of the plastic part causes a decrease and improvement in the field of displacement and the map of residual stresses on the plastic part. In particular, the total maximum displacements are reduced by 0.318 mm and the Von—Mises maximum residual stress by 3.685 MPa in comparison to the results obtained for the traditional cooling system. Likewise, the improvement obtained in these technological parameters makes it possible to achieve the optical and functional requirements established for the correct operation and validation of the plastic part under study in this manuscript.

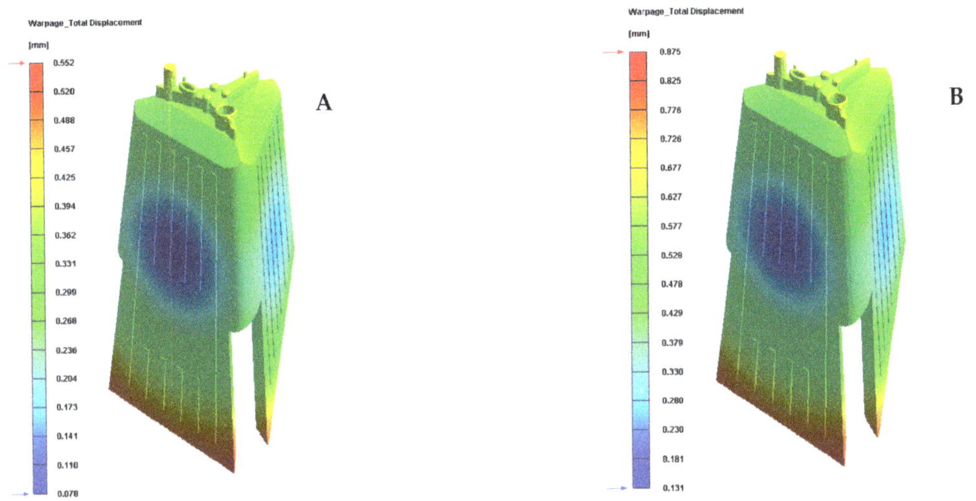

Figure 16. Warpage total displacement (mm). (**A**) New conformal cooling and Fastcool insert. (**B**) Traditional cooling.

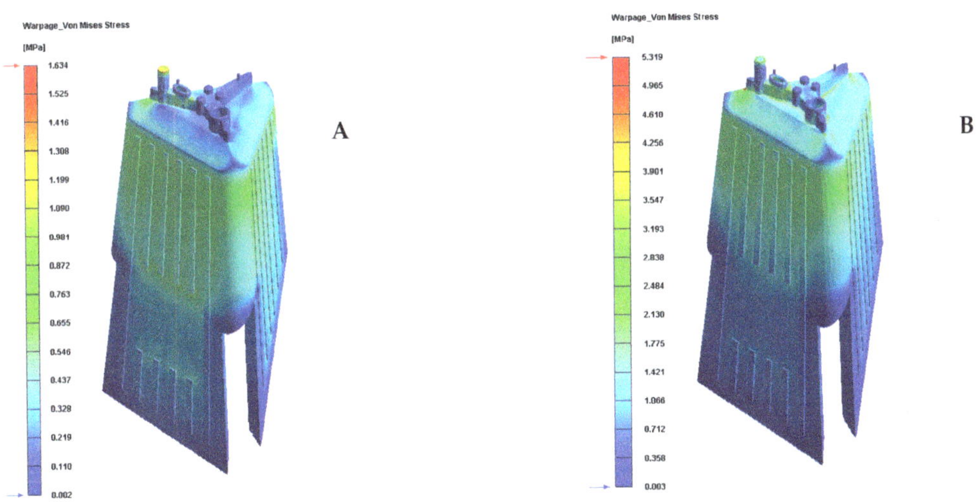

Figure 17. Warpage Von—Mises stress (MPa). (**A**) New conformal cooling system and Fastcool insert. (**B**) Traditional cooling.

4. Conclusions

The paper presents a new design of triple hook-shaped conformal cooling channels for application in optical parts of great thickness, deep cores and high dimensional and optical requirements. In these cases, the small dimensions of the core and the high requirements regarding warping and residual stresses prevent the use of traditional and standard conformal cooling channels. The research combines the use of a new triple hook-shaped conformal cooling system with the use of three independent conformal cooling sub-systems adapted to the complex geometric conditions of the sliders that surround completely the optical part under study. Finally, the new proposed conformal cooling design is completed with a small insert manufactured with a new Fastcool material and located in the internal area of the optical part where the optical facets are located. A transient numerical analysis validates the improvements of the new proposed conformal cooling system presented. The results show that the configuration of the new presented conformal cooling system, together with the use of a Fastcool insert, increases and optimizes the heat flow exchanged between the plastic part and the different cooling mechanisms defined for cooling the plastic optical part. In particular, the upgrade in heat exchange produced, compared to the traditional cooling configuration, is 2.062 $J/s{\cdot}cm^2$, which translates into an improvement in thermal efficiency of 267.10%. The enhancement in uniformity in the temperature gradient of the surface of the plastic part causes a decrease and improvement in the field of displacement and the map of residual stresses on the plastic part. In particular, the total maximum displacements are reduced by 0.318 mm or by 36.343% and the Von—Mises maximum residual stress by 3.685 MPa or by 69.280% in comparison to the results obtained for the traditional cooling system. Additionally, the new design of cooling presented in this paper reduces the cycle time of the plastic part under study by 13.363 s or by 32.161%, compared to the classical geometry and configuration of perforated straight cooling channels, which causes a very high economic and energy saving in line with the sustainability of the mold. The amellioration obtained in these technological parameters will make it possible to achieve the optical and functional requirements established for the correct operation and validation of complex optical parts, where it is not possible to use traditional cooling channels or standard conformal cooling layouts.

Author Contributions: Investigation, A.T.-A., J.M.M.-C., J.d.D.C.-G. and C.M.-D.; project administration, C.M.-D.; writing—original draft, A.T.-A., J.M.M.-C. and C.M.-D.; writing—review & editing J.M.M.-C. and C.M.-D.; funding acquisition, C.M.-D. All authors have read and agreed to the published version of the manuscript.

Funding: This research work was supported by the University of Jaen through the Plan de Apoyo a la Investigación 2021–2022-ACCION1a POAI 2021–2022: TIC-159.

Institutional Review Board Statement: Not applicable.

Informed Consent Statement: Not applicable.

Data Availability Statement: All data included in this study are available upon request by contact with the corresponding author.

Acknowledgments: Authors acknowledge the support of CORETECH System Co.

Conflicts of Interest: The authors declare no conflict of interest.

References

1. Available online: https://www.photonics.com/Articles/Plastic_Optics_Specifying_Injection-Molded/a25487 (accessed on 30 August 2021).
2. Gutierrez, G.; Hilbig, D.; Fleischmann, F.; Henning, T. Component-level test of molded freeform optics for LED beam shaping using experimental ray tracing. In *Optical Measurement Systems for Industrial Inspection X*; International Society for Optics and Photonics: San Diego, CA, USA, 2017; Volume 10329, p. 1032930.
3. Nian, S.C.; Chen, P.W.; Huang, M.S. Multilayer injection molding of high-profile convex lens. *J. Appl. Polym. Sci.* **2020**, *137*, 48600. [CrossRef]
4. Spina, R.; Walach, P.; Schild, J.; Hopmann, C. Analysis of lens manufacturing with injection molding. *Int. J. Precis. Eng. Manuf.* **2012**, *13*, 2087–2095. [CrossRef]
5. Shieh, J.Y.; Wang, L.K.; Ke, S.Y. A feasible injection molding technique for the manufacturing of large diameter aspheric plastic lenses. *Opt. Rev.* **2010**, *17*, 399–403. [CrossRef]
6. Mercado-Colmenero, J.M.; Muriana, J.A.M.; Rubio-Paramio, M.A.; Martín-Doñate, C. An automated manufacturing analysis of plastic parts using faceted surfaces. In *Advances on Mechanics, Design Engineering and Manufacturing*; Springer: Cham, Switzerland, 2017; pp. 119–128.
7. Mercado-Colmenero, J.M.; Rubio-Paramio, M.A.; Karlinger, P.; Martin-Doñate, C. A new procedure for calculating cycle time in injection molding based on plastic part geometry recognition. *Int. J. Adv. Manuf. Technol.* **2018**, *98*, 441–477. [CrossRef]
8. Kuo, C.C.; Chen, W.H.; Lin, Y.X.; Gao, Q.; Gian, S.J.; Xiao, C.X. Effects of different fillers on the silicone rubber mold with conformal cooling channels. *Int. J. Adv. Manuf. Technol.* **2020**, *108*, 1509–1525. [CrossRef]
9. Dick, L. High Precision Freeform Polymer Optics: Optical freeform surfaces–increased accuracy by 3D error compensation. *Opt. Photonik* **2012**, *7*, 33–37. [CrossRef]
10. Macías, C.; Meza, O.; Pérez, E. Relaxation of residual stresses in plastic cover lenses with applications in the injection molding process. *Eng. Fail. Anal.* **2015**, *57*, 490–498. [CrossRef]

11. Jahan, S.; El-Mounayri, H. A thermomechanical analysis of conformal cooling channels in 3D printed plastic injection molds. *Appl. Sci.* **2018**, *8*, 2567. [CrossRef]
12. Marques, S.; Souza, A.F.D.; Miranda, J.; Yadroitsau, I. Design of conformal cooling for plastic injection moulding by heat transfer simulation. *Polímeros* **2015**, *25*, 564–574. [CrossRef]
13. Martin-Doñate, C.; Shaikheleid, S.; Torres-Alba, A.; Mercado-Colmenero, J.M. A New Smart Web Platform for Plastic Injection Molds in Industry 4.0 Environments. In *Advances on Mechanics, Design Engineering and Manufacturing III*; Roucoules, L., Paredes, M., Eynard, B., Morer Camo, P., Rizzi, C., Eds.; Springer: Cham, Switzerland, 2021.
14. Wang, Y.; Yu, K.-M.; Wang, C.C.L.; Zhang, Y. Automatic design of conformal cooling channels for rapid tooling. *Comput. Aided Des.* **2011**, *43*, 1001–1010. [CrossRef]
15. Park, H.S.; Dang, X.P.; Nguyen, D.S.; Kumar, S. Design of advanced injection mold to increase cooling efficiency. *Int. J. Precis. Eng. Manuf.-Green Technol.* **2020**, *7*, 319–328. [CrossRef]
16. Park, H.S.; Dang, X.P. Development of a smart plastic injection mold with conformal cooling channels. *Procedia Manuf.* **2017**, *10*, 48–59. [CrossRef]
17. Chung, C.Y. Integrated optimum layout of conformal cooling channels and optimal injection molding process parameters for optical lenses. *Appl. Sci.* **2019**, *9*, 4341. [CrossRef]
18. Kuo, C.C.; Jiang, Z.F.; Yang, M.X.; You, B.J.; Zhong, W.C. Effects of cooling channel layout on the cooling performance of rapid injection mold. *Int. J. Adv. Manuf. Technol.* **2021**, *114*, 2697–2710. [CrossRef]
19. Rahim, S.Z.A.; Sharif, S.; Zain, A.M.; Nasir, S.M.; Mohd Saad, R. Improving the quality and productivity of molded parts with a new design of conformal cooling channels for the injection molding process. *Adv. Polym. Technol.* **2016**, *35*, 21524. [CrossRef]
20. Crema, L.; Sorgato, M.; Lucchetta, G. Thermal optimization of deterministic porous mold inserts for rapid heat cycle molding. *Int. J. Heat Mass Transf.* **2017**, *109*, 462–469. [CrossRef]
21. Torres-Alba, A.; Mercado-Colmenero, J.M.; Diaz-Perete, D.; Martin-Doñate, C. A new conformal cooling design procedure for injection molding based on temperature clusters and multidimensional discrete models. *Polymers* **2020**, *12*, 154. [CrossRef]
22. Berger, G.R.; Zorn, D.; Friesenbichler, W.; Bevc, F.; Bodor, C.J. Efficient cooling of hot spots in injection molding. A biomimetic cooling channel versus a heat-conductive mold material and a heat conductive plastics. *Polym. Eng. Sci.* **2019**, *59*, E180–E188. [CrossRef]
23. Zink, B.; Szabó, F.; Hatos, I.; Suplicz, A.; Kovács, N.K.; Hargitai, H.; Kovács, J.G. Enhanced injection molding simulation of advanced injection molds. *Polymers* **2017**, *9*, 77. [CrossRef]
24. Mercado-Colmenero, J.M.; Torres-Alba, A.; Catalan-Requena, J.; Martin-Doñate, C. A New Conformal Cooling System for Plastic Collimators Based on the Use of Complex Geometries and Optimization of Temperature Profiles. *Polymers* **2021**, *13*, 2744. [CrossRef]
25. Brooks, H.; Brigden, K. Design of conformal cooling layers with self-supporting lattices for additively manufactured tooling. *Addit. Manuf.* **2016**, *11*, 16–22. [CrossRef]
26. Tang, Y.; Gao, Z.; Zhao, Y.F. Design of conformal porous structures for the cooling system of an injection mold fabricated by Additive Manufacturing Process. *J. Mech. Des.* **2019**, *141*, 101702. [CrossRef]
27. Mayer, S. *Optimised Mould Temperature Control Procedure Using DMLS*; EOS Whitepaper; EOS GmbH—Germany Ltd.: Munich, Germany, 2005; pp. 1–10.
28. Valls, I.; Hamasaiid, A.; Padré, A. High thermal conductivity and high wear resistance tool steels for cost-effective hot stamping tools. *J. Phys. Conf. Ser.* **2017**, *896*, 012046. [CrossRef]
29. Available online: https://www.sabic.com/en/products/polymers/polycarbonate-pc/lexan-resin (accessed on 30 August 2021).
30. Mercado-Colmenero, J.M.; Martin-Doñate, C.; Moramarco, V.; Attolico, M.A.; Renna, G.; Rodriguez-Santiago, M.; Casavola, C. Mechanical characterization of the plastic material GF-PA6 manufactured using FDM technology for a compression uniaxial stress field via an experimental and numerical analysis. *Polymers* **2020**, *12*, 246. [CrossRef]
31. Mercado-Colmenero, J.M.; La Rubia, M.D.; Mata-Garcia, E.; Rodriguez-Santiago, M.; Martin-Doñate, C. Experimental and numerical analysis for the mechanical characterization of petg polymers manufactured with fdm technology under pure uniaxial compression stress states for architectural applications. *Polymers* **2020**, *12*, 2202. [CrossRef]
32. Available online: https://www.3ds.com/es/productos-y-servicios/catia/ (accessed on 30 August 2021).
33. Available online: https://www.moldex3d.com/ (accessed on 30 August 2021).

polymers

MDPI

Review

Analysis and Advances in Additive Manufacturing as a New Technology to Make Polymer Injection Molds for World-Class Production Systems

Adrian Benitez Lozano [1,2,*], Santiago Henao Álvarez [1], Carlos Vargas Isaza [1] and Wilfredo Montealegre-Rubio [2]

[1] Grupo de Investigación Calidad Metrología y Producción, Instituto Tecnológico Metropolitano, Medellín 050034, Colombia; santiagohenao@itm.edu.co (S.H.Á.); carlosvargas@itm.edu.co (C.V.I.)
[2] Grupo de Investigación Diseño y Optimización Aplicada, Universidad Nacional de Colombia, Medellín 050034, Colombia; wmontealegrer@unal.edu.co
* Correspondence: adrianbenitez@itm.edu.co; Tel.: +574-3006561492

Abstract: The currently growing demand for metallic and polymeric products has undoubtedly changed the rules of manufacturing, enabling customers to more functionally define their products based on their needs. Nowadays, a new technique for rapid tooling, Additive Manufacturing (AM), can create customized products with more complex geometries and short life cycles (flexibility) in order to keep up with the new variables imposed by the manufacturing environment. In the last two decades, the migration from subtractive manufacturing to AM has materialized such products with reduced costs and cycle times. AM has been recently promoted to develop polymer molds for product manufacturing. This paper reviews the main findings in the literature concerning polymer molds created by AM compared to conventional (metal) molds obtained by subtractive manufacturing. Information about specific topics is scarce or nonexistent, for example, about the characterization of the most commonly injected materials and molds used in this type of technology, their mechanical properties (part and mold), designs for all types of geometries, and costs. These aspects are addressed in this literature review, highlighting the advantages of this alternative manufacturing process, which is considered a desirable technology worldwide.

Keywords: mold additive manufacturing; polymer molds; subtractive manufacturing; mold characterization; rapid tooling; injection molding

Citation: Lozano, A.B.; Álvarez, S.H.; Isaza, C.V.; Montealegre-Rubio, W. Analysis and Advances in Additive Manufacturing as a New Technology to Make Polymer Injection Molds for World-Class Production Systems. *Polymers* **2022**, *14*, 1646. https://doi.org/10.3390/polym14091646

Academic Editors: Célio Bruno Pinto Fernandes, Salah Aldin Faroughi, Luís L. Ferrás and Alexandre M. Afonso

Received: 7 December 2021
Accepted: 12 February 2022
Published: 19 April 2022

1. Introduction

Over time, manufacturing industries have experienced more dynamic markets and growing competitiveness. Although Mold Additive Manufacturing (MAM) is found in high-impact scientific literature [1–7]. As a result, they need to be resilient in the face of quick changes in a market characterized by products with shorter lifecycles and great diversity in their manufacturing [8]. These changes have led to independence from conventional processes and a migration to mass production. More flexible marketing has resulted in lower-volume production with greater profitability because personalized products meet final customer requirements more precisely [9]. In addition, mass personalization has enabled a quick production of low-cost goods and services to satisfy customer needs [10], which requires flexibility and capacity to effectively respond to the demand. Additive Manufacturing (AM), a technology in line with the new requirements of global marketing, can be used for rapid tooling in order to develop high-quality products. As a result, in recent years, research into polymer injection mold design and rapid tooling by AM technologies has become more important because these innovative alternative technologies can help polymer industries achieve their objectives [11].

Previous studies [1,7,12–17] have demonstrated the potential of AM of polymer molds in the injection molding process and their impact, not only on time and cost reduction, but also on physical, mechanical, thermal, morphological, and other properties of the molded parts compared to those obtained with metal molds produced by conventional methods. Kampker et al. [1] studied the economic potential of different AM techniques with several materials to produce polymer tools, which were compared to their steel counterparts. With Selective Laser Sintering (SLS) and PA 3200 GF as mold material, they found a cost reduction of 84.2% compared to steel tools. Another study demonstrated the cost-benefit of integrating AM, using Digital Light Processing (DLP), into the conventional manufacturing process of injection molding to create mold inserts. It resulted in a cost reduction between 80% and 90% depending on the geometry of the mold insert developed for each product. In addition, a break-even point was established in [12] to determine how profitable AM is for Rapid Tooling (RT). In that case, the break-even points were 3400 and 500 for units with small and large geometries, respectively. Besides the economic aspect, another relevant field in AM is the study of the process and the characterization of the molded part and the mold obtained by different RT processes using AM. In the late 1990s, stereolithography (SLA), the first additive manufacturing technique, set a precedent in the production of injection molding tools. Authors such as Sadegh et al. [18] saw the viability of this type of mechanism to manufacture prototypes and small production series. Others delved into issues such as the capacity of the materials, the characterization of their mechanical properties, the post-treatment to increase the deflection temperature under load, and the efficiency of the manufactured tools in terms of molded parts [13,14].

In more recent studies, new AM techniques have been investigated. For instance, Triebs et al. [7] used two methodologies, i.e., PolyJet and SLS, with mold inserts created employing digital ABS and PA 3200 GF, respectively. They observed a mechanical difference in the molded parts, which was apparently due to the poor thermal conductivity and increased roughness of the polymer molds compared to their aluminum counterparts. Additionally, they discussed the crystallinity of the molded part made of polypropylene (PP) and how nucleating agents favored the crystallization rate. Another study analyzed the thermal, mechanical, and thermo-mechanical properties of epoxy-based PolyJet molds to produce small series of PLA parts [15]. Other authors have examined issues related to failure over the lifespan of the molds and established diagnoses based on their findings [16,17]. Polymer research has analyzed the thermal, mechanical, and rheological characteristics of these materials. For example, a study [19] investigated the effects of process parameters on the strength and fatigue behavior of 3D printed PLA-graphene. Its experimental results indicate that fatigue lifetime clearly depends on process parameters, as well as loading amplitude and frequency. In Fused Filament Fabrication (FFF), heat transfer plays a particular role and determines the temperature history of the merging filaments; in turn, the in-process monitoring of the temperature profile guarantees the optimization and thus the improvement of interlayer adhesion [20]. This is very important to ensure the best quality of the piece.

This article presents a comprehensive literature review of the main findings in recent research into AM (as an alternative to obtain molds for injection molding processes), a comparative analysis between AM and subtractive technologies, and research topics that should be further addressed. Section 2 below introduces the subject, the chronology of conventional manufacturing and additive manufacturing for injection molds, the state of the art, and the research approach of this paper. Subsequently, Section 3 describes the methodology of this systematic literature review and a bibliometric analysis. Section 4 details the latest techniques and guidelines applied to mold design. Section 5 deals with the characterization and performance (mechanical properties) of the materials used in AM. Section 6 discusses cost evaluation. Finally, Section 7 draws the conclusions.

2. Chronology of Conventional Manufacturing vs. Additive Mold Manufacturing

As shown in Figure 1, subtractive manufacturing dates back to 1871, with the development of the drill press with tools to make holes, nuts, tube flaring, and countersinks, which are essential for conventional cooling channels and fasteners in the mold industry.

Later, between 1940 and 1943, the first machining operations supported by Computer Numerical Control (CNC) were developed. Subsequently, in the 1960s, this technology was extended to conventional milling, a fundamental process in the conventional mold industry for metal and polymer materials. Between 1965 and 1980, advanced machining processes were developed, e.g., Electro Discharge Machining (EDM) and LASER (1980). Such processes were very useful for detailing and finishing, generally, mold cavities and vents to release the pressures generated when the molten material is compressed. Since the 1980s, there has been a "boom" in additive mold manufacturing and its variants, which are described in Figure 1 (bottom). Different additive manufacturing techniques have paved the way for the production of polymer molds, and, although they are very different in principle and execution, they have achieved significant results for this type of applications. Figure 1 shows the chronology of additive manufacturing techniques used to produce polymer molds that are commonly found in the literature. The years mark the period of commercialization of each technique [17–22].

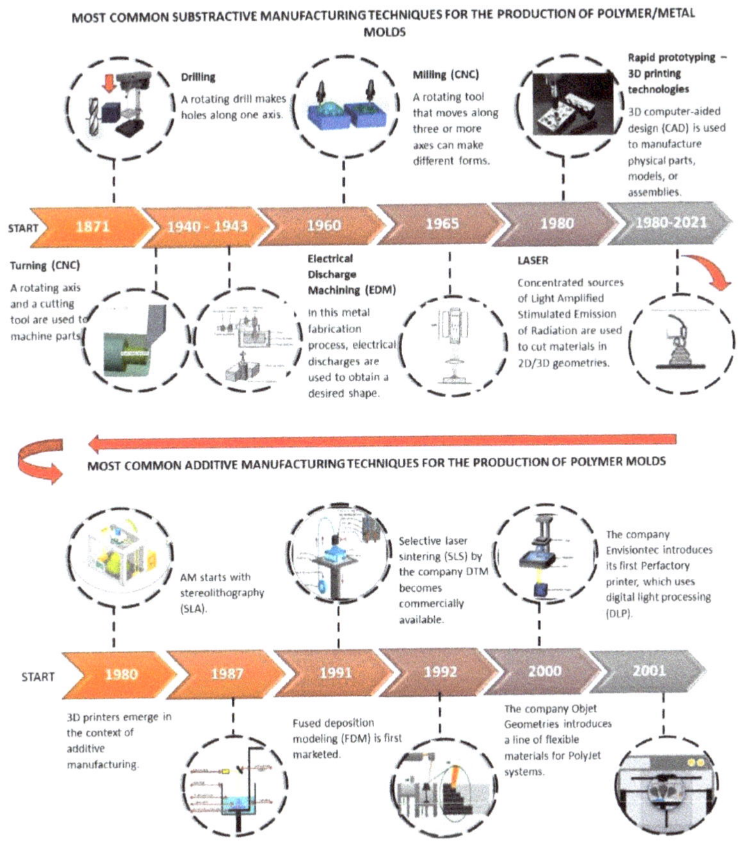

Figure 1. Chronology of the most common subtractive and additive manufacturing techniques for polymer/metal molds.

In 1980, the term 3D printing was introduced by Hideo Kodama, who invented the single-beam laser method that opened the door to the development of new 3D printing equipment and patents. In 1987, Charles W. Hull invented the first 3D printing equipment, called SLA-1, which used a technique known as stereolithography. In this technique, a photopolymer contained in a vat undergoes solidification produced by a laser, which is aimed at the cross section of the piece and gradually descends on the z-plane depending on the specified height. In 1991, the company Stratasys commercialized the first technique to extrude materials in the form of a filament; it was called Fused Deposition Modeling (FDM). In this technique, the material is melted using hot runners, which extrude the material layer by layer. In 1992, a new AM technique known as Selective Laser Sintering (SLS) entered the market. In it, the material, in powder form, is selectively sintered by a high-power CO_2 laser beam onto the cross-section of the model. The first commercially available 3D printing system, called PolyJet, was launched by the company Objet Geometries in 2000. This system uses a jetting head to inject a UV-sensitive liquid resin that solidifies on a platform until the desired object is obtained.

In 2001, Digital Light Processing (DLP) technology, developed by Texas Instruments in the field of projectors, was introduced by the company Envisiontec at the EuroMold (a trade fair for moldmaking). In DLP, multiple micromirrors reflect a light source onto the printing material contained in a vat, which is then solidified layer by layer until the part is obtained.

Thus far, many studies have investigated the performance of these techniques for injection processes because they offer alternatives to meet the new needs of the market. More specifically, the behavior and performance of PolyJet 3D printing for RT applications have been some of the most widely studied. This technique produces high-performance tools in terms of thermal and mechanical properties thanks to its multi-material technology and high resolution, which ensure a good surface finish [18,23].

3. Methodology

The most important concepts in the field of AM were used here to conduct an exhaustive search and collect information. The initial keywords were "Additive Manufacturing", "Rapid Tooling", "Injection molding", "cost", "Failure", "Polymer Mold", and other terms that fall within the scope of this review. The Scopus and ScienceDirect databases were used for this purpose because they compile a considerable amount of world-class information in different research fields. This process was complemented with a more general search on the topic using the Scopus database, which was selected because of its comprehensiveness in terms of information, abstracts, and citations. Similar terms were refined using Science Direct Topics to obtain an adequate string of keywords. A bibliometric analysis and networks were used to examine and understand trends in this field in terms of authors and countries (Figures 2 and 3).

A search string with the keywords above was used in the Scopus reference database, including Boolean operators to narrow down or filter the results as described by Burnham 2006 [23]. Once the strings shown in Table 1 were obtained, filters were used to exclude terms such as "3D printer" or "manufacture". The search was limited to documents published between 2013 and 2021, and "Rapid tooling" was taken as the key term because it is articulated with the other concepts in this review. After conducting the advanced search with each string, the list of references in the fourth column in Table 1 was compiled. These are the documents reviewed in this paper.

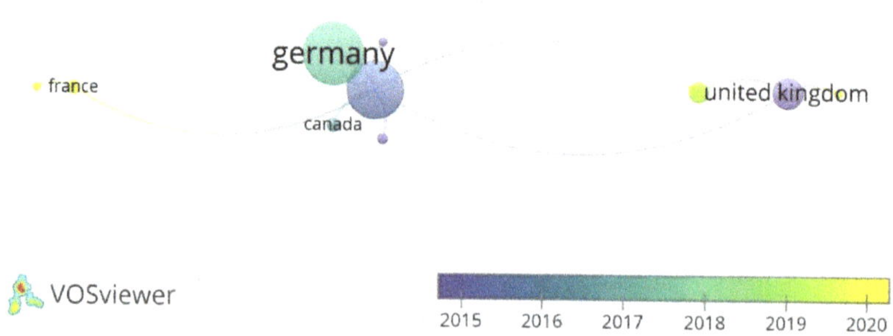

Figure 2. Network of countries based on co-authorship.

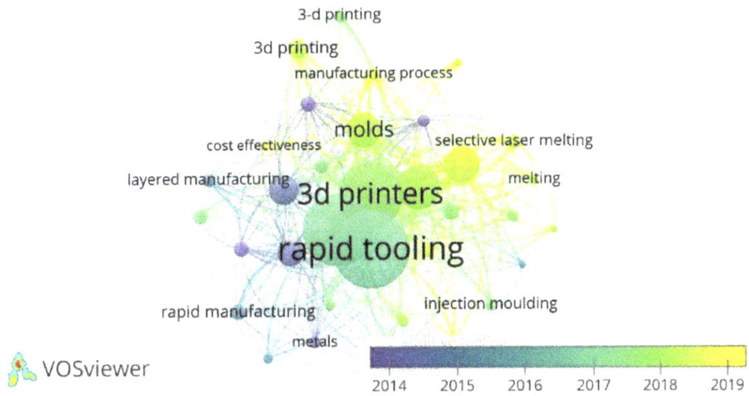

Figure 3. Network of keywords retrieved from the literature search.

Subsequently, a general search string was used in one of the reference databases (Scopus), and the results were exported to carry out a bibliometric analysis implementing VOSviewer software (version 1.6.16). The latter was employed to create networks of scientific publications, scientific journals, researchers, research organizations, countries, keywords, and terms [24,25] in order to understand current trends in the field analyzed in this review.

Table 1. Search results obtained with each string.

TITLE-ABS-KEY	Number of Retrieved Documents and Related References		
	Without Filter	Filter	References
"Additive manufacturing" AND "Rapid tooling" OR "Polymer Mould"	81	26	[1–22,26–29]
"Rapid tooling" AND "Additive manufacturing "AND "rapid manufacturing" OR "Cost model" OR "Cost Advantage" OR "Cost analysis" OR "production economics" OR "3D printing" OR "cost estimation models" OR "Injection moulding"	83	11	[1,30–39]
"Additive manufacturing" AND "Rapid tooling" AND "Injection molding" AND "Failure"	46	31	[3–31,40–42]
"Additive manufacturing" AND "Rapid tooling" AND "Injection molding" AND "Design"	15	12	[1,21,23–25,28,29,40,43–46]
Filter: Review of the abstract and relationship with the search string			

Bibliometric Analysis

As mentioned above, VOSviewer software (version 1.6.16) [13,14] was used to carry out a bibliometric analysis. This software was employed to construct networks, analyze the metadata, and establish relationships between the results of the following search string: TITTLE-ABS KEY "Additive manufacturing" AND "Rapid tooling" OR "Polymer Mold". In this case, the keyword "Rapid Tooling" was limited to publications between 2013 and 2021. The bibliographic database was exported from Scopus to create, visualize, and explore three networks of great interest for this review.

Figure 2 shows the first network, which connects countries based on co-authorship. VOSviewer was configured so that the minimum number of documents per country was 3, which resulted in a network of 11 out of the 25 countries in the bibliographic references. India, Malaysia, New Zealand, Romania, and Spain were filtered out because the total strength of their links was not significant for this review, and they did not contribute relevant information to this analysis. Figure 2 is a network of keywords represented by labeled circles, where the more weight the item has, the bigger the label and the circle. The country with the highest weight is Germany because it presents the most abundant scientific production concerning Rapid Tooling (21 documents). Countries such as India and Spain have contributed a considerable amount of scientific production (10 and 5 articles, respectively); however, in this bibliometric analysis, they are not especially relevant because they do not have a strong relationship of co-authorship with other countries. The United States has the highest level of co-authorship among the 5 countries in Figure 1, and the strongest co-authorship relationship is that between Italy and the United Kingdom, where Additive Manufacturing and Rapid Tooling have been recently explored in depth.

The links established by co-occurrences of keywords were also analyzed. The minimum number of co-occurrences of keywords was set to 5. Among the 1413 keywords in the bibliographic references, 35 were above this threshold. In this case, no keywords were eliminated to construct the network. In Figure 3, the most prominent elements in the network are the keywords "Rapid Tooling" (102 occurrences) and "Additive Manufacturing" (93 occurrences), as expected. Likewise, term "Rapid Tooling" presents a strong connection with all the keywords retrieved from the literature search.

4. Mold Design

Some of the main issues in mold injection processes are efficient material processing and obtaining products at reasonable prices that reflect a strong economy of scale [30]. Mold design and the simulation of this process are essential aspects in the product life cycle [1], quality assessment, viability, and productivity of parts manufactured by injection.

Generally, mold design is one of the most important aspects in the product life cycle because it determines the quality, viability, and productivity of parts. Mold design is necessary because parts should meet specific requirements, and, for that purpose, it is fundamental to know some characteristics of the piece to be manufactured, such as its geometry, weight, material, and volume [31]. Several mold design practices based on scientific findings represent benchmarks or references for recent research in this area. Currently, molds are designed with efficient cooling systems, air vents, and cooling channels that shorten the cycle time of injection molding processes, as shown in Figure 4. Many authors have adopted methodologies based on genetic algorithms to achieve efficiency in cooling systems that release the air trapped in injection molds, thus improving the quality, heat transfer, channel geometry, and formability of the injected product [32–35].

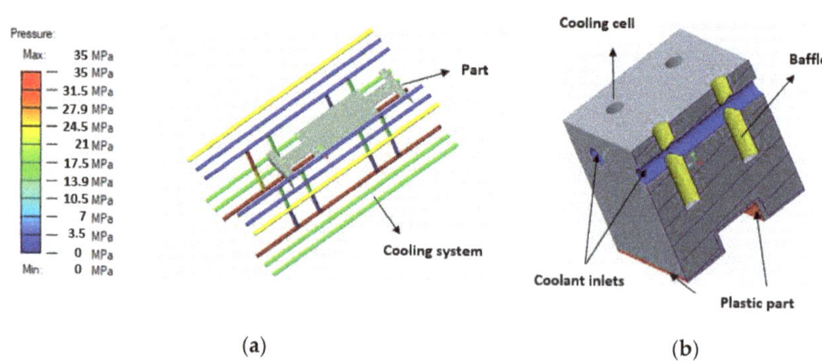

Figure 4. (**a**) Pressure field and temperatures in a refrigeration system. (**b**) Cooling system by means of baffles.

In addition, rapid prototyping technologies have been applied to manufacture molds with different types of low-pressure cooling channels for materials such as wax. Recent research in this area has focused on reducing cooling times [34]. Figure 5 compares the cooling performance of four injection molds with different cooling channels. Series conformal cooling channels (Figure 5d) are highly recommended in [34,43] to reduce the cooling time during the process because their cooling efficiency is approximately 90%.

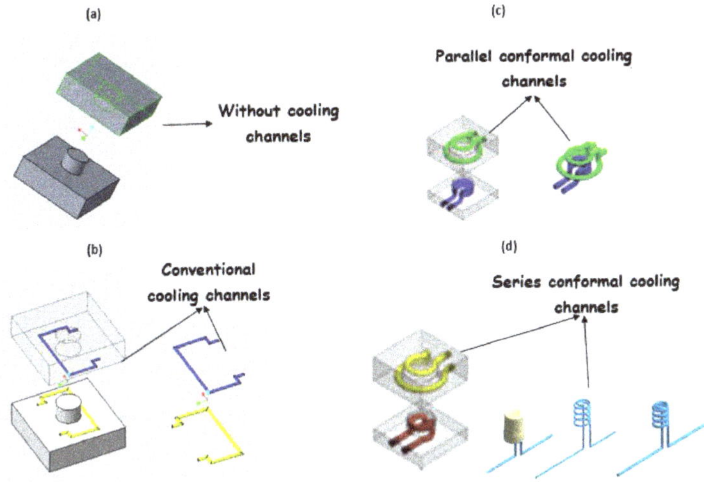

Figure 5. Solid models of cavity insert and cores (**a**) without cooling channels, (**b**) with conventional cooling channels, (**c**) with parallel conformal cooling channels, and (**d**) with series conformal cooling channels.

Mold design should observe the guidelines and best practices of traditional injection molds. These design concepts can be applied to PolyJet molds, but alterations are required to compensate for the mechanical, thermal, and dimensional characteristics of plastic molds [40,43,44]. Conformal cooling channels show great potential for substituting conventional straight-drilled cooling channels because they can provide more uniform and efficient cooling effects, and thus improve the production quality and efficiency significantly [45]. Table 2 presents a technical guide to design mold cavities.

Table 2. Technical guide to design mold cavities based on recent studies in the field [40,43–46].

Mold Cavities		
Draft	Use angles of approximately 3–5 degrees for the vertical wall. This will reduce mold damage, and the formed parts are less likely to resist ejection from the mold.	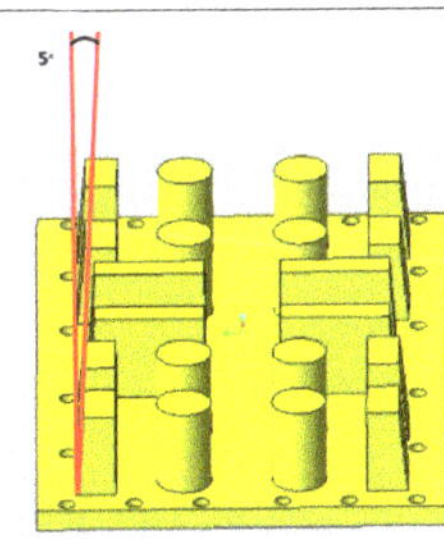
Parting surfaces	Make sure that parting surfaces have minimal flash. For this purpose, try to efficiently adjust the clamping force to compress the plastic material. Check injection parameters such as injection rate, temperature, and pressure.	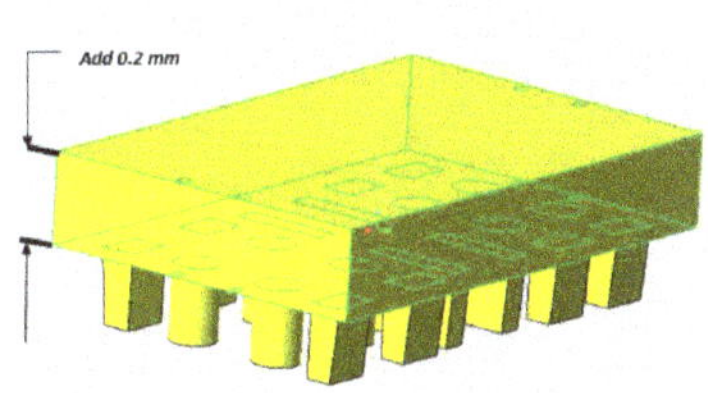

Table 2. *Cont.*

	Mold Cavities	
Core pins	Use core pins with an aspect ratio of 3:1 (height: width). Core pins could deflect due to the pressure in the filling process. A 3D printed insert can be designed to improve mold longevity.	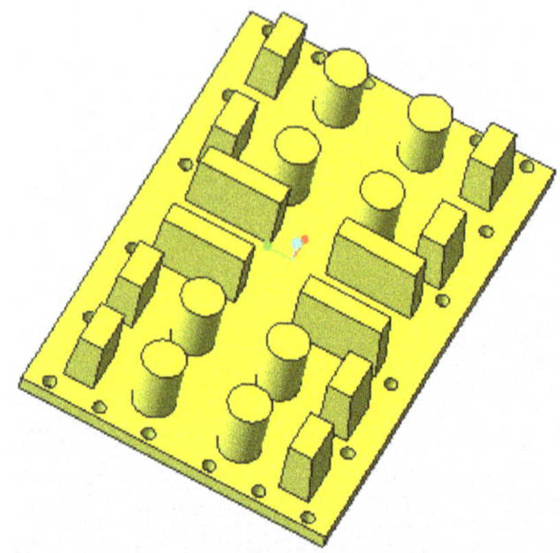
Holes	Use holes with diameters larger than 0.8 mm. Smaller holes could be machined prior to mold assembly.	
Shrinkage compensation	It is important to know the contraction or expansion of the printed material, generally in percentages. Based on these data, scale the core and the cavity to compensate for the shrinkage of the resin that occurs with conventional injection molding.	
	Mold Components	
Gates	Enlarge the gates depending on the viscosity of the plastic material used for the part and the mold's flow characteristics. Use or design gates three times larger than those used in metal molds. Make edge gate thickness equal to the wall thickness of the part at the point of injection. These measures will improve material flow and decrease pressure within the tool.	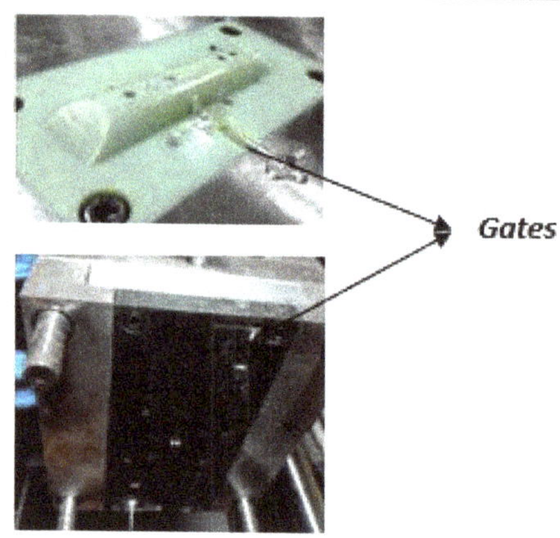
Runners	Hot runner systems are not recommended. If they are used, they do not require adjustment.	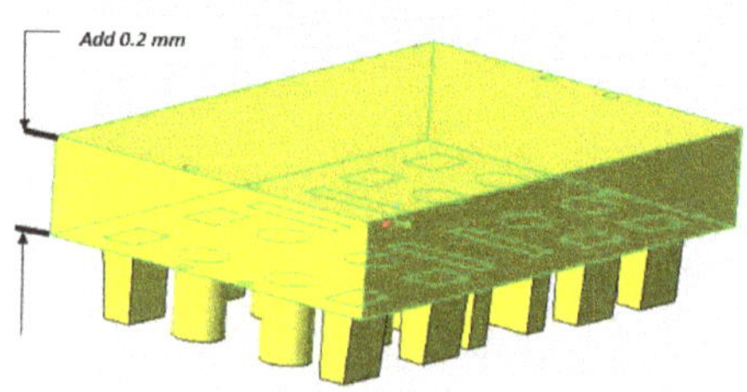

Table 2. *Cont.*

Mold Components		
Sprue	Use a sprue bushing with a minimum draft angle of 3 degrees. If a bushing is used, undersize the hole by 0.2–0.3 mm before printing and ream to size during mold assembly. Avoid physical or direct contact between the molding machine's nozzle and the mold insert.	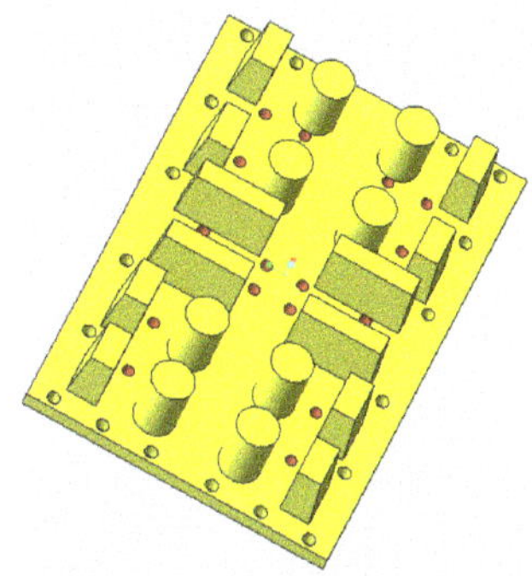
Ejection system	If an ejection system is used, add ejector holes as needed. As with the sprue bushing, undersize the holes by 0.2–0.3 mm (0.008–0.012 in) and ream to size during mold assembly. It is recommended to make sure the holes for the ejector pins will not be too close to the edges. It will weaken the mold especially after reaming.	
Cooling system	Cooling systems will not significantly affect molding cycle times or part quality thanks to the thermal characteristics of PolyJet molds. However, a cooling system can improve tool life; on average, a 20% improvement can be expected. The improvement increases as the depth of the cavity and height of the core decreases since the cooling effects reach more of the surface area of the molding cavity. In recent studies [43], the formable diameter of self-supporting channels has been significantly increased (≥20 mm). A serpentine cooling geometry [44] is able to improve process performance by imposing a cooling curve characterized by a higher slope with respect to traditional channels.	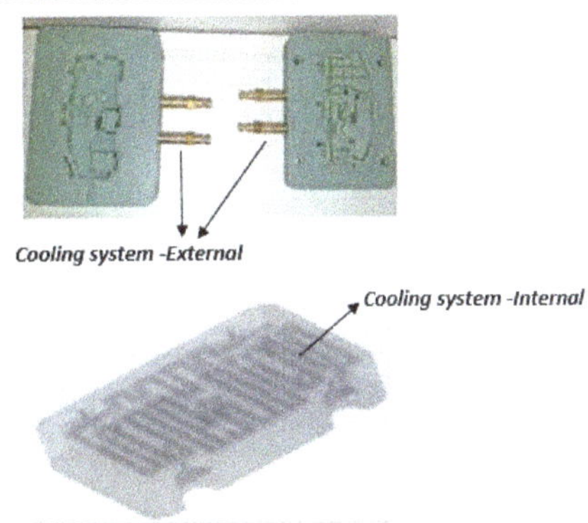 Serpentine cooling channel

5. Performance and Properties of Mold Materials and Injected Polymers

Molds or inserts used in injection molding processes can be produced by additive manufacturing, which is referred to here as Rapid Tooling for Injection Molding (abbreviated as RTIM in this paper) [47,48]. Currently, RTIM using polymeric materials is being explored thanks to the development of additive technologies for polymers, greater access to these additive technologies, and their lower costs compared to metal additive technologies [49–51]. RTIM has thus produced a new market niche in injection molding by enabling low-volume production.

Polymer RTIM poses several challenges regarding its performance and effects on the properties of the injected parts made of polymers. The performance of polymer RTIM (intended for low-volume production) has been compared to that of traditional metal molds in terms of useful life, mechanical and thermal behavior, and other characteristics; nevertheless, their performance is completely different.

Most studies into polymer RTIM have focused on the performance of the mold and the properties of the injected parts, two elements that will be discussed below.

5.1. Failures in Polymer RTIM

Failures in polymer RTIM can occur due to several factors derived from the material of the polymer mold (i.e., glass transition temperature [5,52], heat deflection temperature [1,53], thermal expansion coefficient [6,54], and its mechanical properties [3,41,55,56]); the high shrinkage of the injected polymer [41,52] or the use of fiber-loaded materials [42]; the conditions of the injection process at high injection temperatures [3,57]; the heating and cooling cycle of the process [47,54]; extreme conditions of high shear stress, shear strength, and pressures during injection; strong part ejection forces [3,6,41,58]. Mold geometry can also contribute to failures in very specific sections, such as injection points and thin mold cores or pins that are weakened when subjected to high pressures or contractions of the injected material [3,52,58]. Some of these factors may be more critical than others, or they can produce a combined effect. The Ishikawa diagram (cause-effect diagram) in Figure 6 connects details and relates different sources of crack generation and propagation in polymer RTIM that lead to subsequent failures.

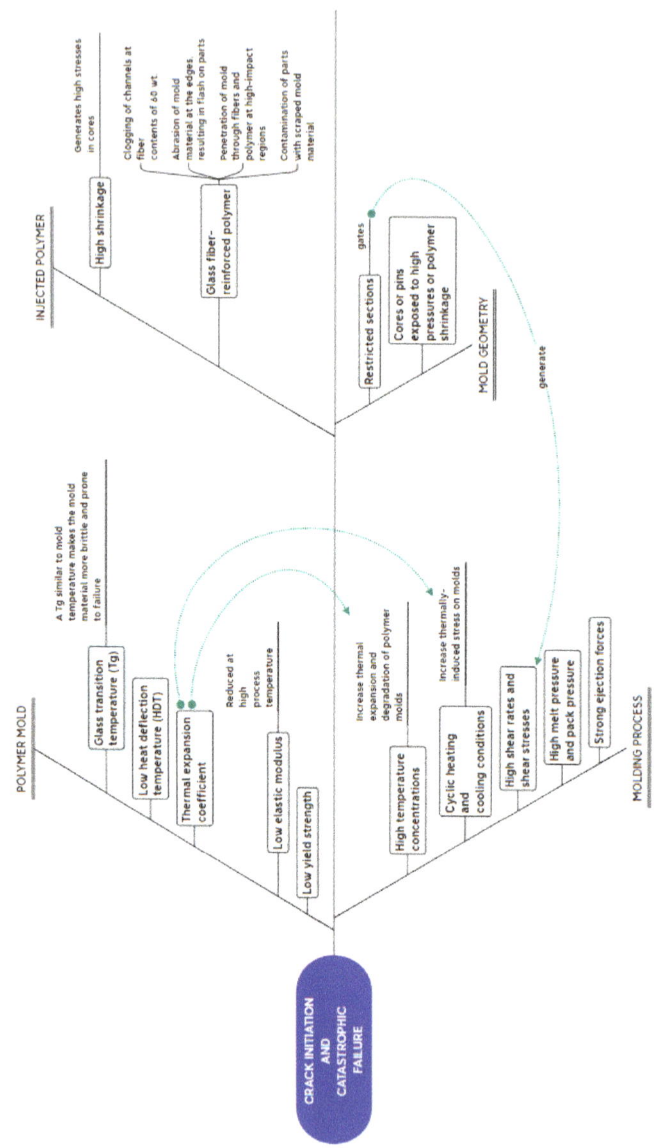

Figure 6. Ishikawa diagram of different sources of crack initiation and catastrophic failure in polymer RTIM.

5.2. Characterization of Properties of Polymer RTIM and Injected Materials

In order to extend the lifetime of polymer RTIM products, it is important to find a balance between mechanical properties, thermal properties, and injection molding process conditions for a given mold. Additionally, the injected material is affected by the characteristics and properties of the mold, as well as the conditions of the injection process. Table 3 summarizes studies that have evaluated different polymer RTIM processes and their respective injected materials. This table also includes a characterization of the properties of the mold material, the injected part, and the method adopted to evaluate the injection process (i.e., predicted by computer simulations or monitored experimentally by sensors and/or data collection equipment).

Table 3. Summary of studies that have evaluated different polymer RTIM processes.

Mold Material	Additive Technique or Machining	Method Used to Evaluate Mold Performance	Injected Polymer	Results (Mold/Part)	Ref.
Aremco 805 epoxy	FDM	Accuracy of injected parts	Polyamide	The dimensional error of the injection-molded part is less than 3%, and the warpage is almost 1 mm across the length of the component.	[59]
Acrylate-based photopolymer	Digital Light Processing (DLP)	Mold failure (# of shots vs. failure)	Liquid silicone rubber	Small-sized parts for drug-releasing (micro)implants were manufactured using micro injection molding. Molds manufactured by DLP did not suffer significant wear when they were used for a low number of microinjection molding cycles (n ~ 8).	[60]
Digital ABS, aluminum	PolyJet	Characterization of molded parts	Isotactic polypropylene	Injected parts showed slower cooling rates in Digital ABS inserts. Parts molded using aluminum tooling did not show a crystal structure. Additionally, parts molded on the digital ABS inserts exhibited higher shrinkage than those molded using aluminum tooling. The change in morphology and the presence of voids significantly affected the tensile behavior of the parts molded in Digital ABS, which broke with little cold drawing and exhibited higher tensile module a higher yield strength.	[61]
PEI (Polyetherimide)	FDM	Thermal performance (specific heat capacity, i.e., Cp, and glass transition temperature); mechanical performance (Young's modulus, loss factor, and compression tests); structural simulation	Polypropylene, POM	The PEI resulted thermally stable but not suitable for injection molding production of polypropylene parts due to prolonged cooling times and the elastic deformations of the inserts. Regarding the POM parts, the polymer insert did not present relevant damage. However, some problems occurred: difficult de-molding of the POM parts and water permeation through the inserts when the cooling system is active.	[62]
Photopolymer Rigur (RGD450)	PolyJet	Mold failure (# of shots vs. failure) and accuracy of injected parts	Polypropylene	Between 94 to 122 parts (with some geometrical parameters such as undercuts) were injected.	[63]
ABS and nylon (coated with copper)	FDM		Stainless steel powder combined with polypropylene as binder (metal injection molding, MIM)	The heat dissipation of the polymer mold was low compared to that of the metal mold. Therefore, the MIM part needs a longer cooling time inside the mold before ejection.	[64]
Epoxy Biresin, aluminum powder, SL resin, short steel fibers, and tool steel	Mold inserts manufactured by vacuum epoxy casting and stereolithography	Experimental data (pressure, temperature, ejection forces); computer simulation of injection molding (pressure, mold temperature); structural simulation	Polypropylene homopolymer	Molds made by stereolithography are viable if mold temperatures are controlled at 15 °C above the glass transition of the mold material. Otherwise, they exhibit premature failure, and their useful life is not enough for injection molds. The authors estimated the pressure associated with the high shrinkage of the injected polymer on the pins in the mold, which should be taken it into account to avoid failures in these elements.	[52]

Table 3. *Cont.*

Mold Material	Additive Technique or Machining	Method Used to Evaluate Mold Performance	Injected Polymer	Results (Mold/Part)	Ref.
Epoxy-acrylate	PolyJet	Experimental data (cavity pressure, strain–time diagram)	Polypropylene homopolymer	The authors implemented on-line monitoring of the cavity pressure during the injection process, and they determined its effect on the deformation of the polymeric inserts by finding a direct relationship between these two variables.	[6]
Different liquid photoresins	SLA 3D printing	Mechanical performance (flexural modulus); thermal performance (heating rate, maximum temperature, heat deflection temperature); cavity dimensions; experimental data (pressure, temperature)	LDPE	It was found that flexural modulus and elongation (two mechanical properties) are more relevant than deflection temperature under load to evaluate the performance of polymer molds made by additive SLA. The latter is useful to produce mold designs that require several changes in shape and dimensions.	[4]
Methacrylic photopolymer	Vat photopolymerisation	Accelerated thermal ageing (weight loss of insert); mold failure (# of shots vs. failure); and mold surface features (average roughness)	-	The application of accelerated thermal aging to polymer mold inserts is a test to evaluate and predict their behavior when they are subjected to thermal loads that determine cyclic stresses. It was found that the stresses induced by the thermal loads of the injection molding process can be reduced by increasing cooling time; however, this produces longer cycle times, thus reducing productivity.	[54]
Ceramic-filled epoxy composite, steel, and aluminum	SLA (polymeric composite)	CAE software (mold temperature); mold failure (# of shots vs. failure); mechanical performance (Young's modulus, tensile strength, elongation); fiber characterization of polymeric composite	Polypropylene	The reinforced polymer mold withstood the injection of more than 100 parts before failure. The injected parts made of long-fiber reinforced polypropylene showed good mechanical properties and good dimensional accuracy.	[42]
Digital ABS	-	Mold failure (# of shots vs. failure); thermal properties (heat capacity, heat deflection temperature, thermal expansion); CAE software (shear rate, shear stress, mold temperature)	ABS	Failure in the polymer mold was produced by a concentration of high temperatures, especially in the areas of the injection point and the mold cavity because higher shear rates and shear stresses were generated during mold filling. In polymeric molds, the solidified layer in the injected polymer is smaller than in typical steel molds, which indicates a more even cooling of the injected polymer through the flow path.	[3]
Digital ABS, polyamide (PA) 3200 GF, and aluminum	PolyJet, selective laser sintering (SLS), and milling	Surface roughness	Copolymer polypropylene	The parts injected using RTIM showed a lower percentage of elongation compared to those injected in aluminum molds. This is explained by the lower thermal conductivity and higher roughness of the cavities in polymer RTIM.	[7]
Epoxy-based resins	PolyJet	Experimental data (mold temperature distribution, mold temperature vs. time) and mechanical performance (storage modulus, loss factor)	Polypropylene and polylactic acid (PLA)	A comparative evaluation of three types of polymer mold inserts (i.e., without cooling channels, with conventional cooling channels, and with conformal cooling channels) determined that there was no difference in cooling efficiency between the insert without channels and that with conventional channels. The mold insert manufactured with conformal channels reduced the thermal load cycle by up to 70%, with good mold temperature control with respect to the glass transition temperature of the mold material.	[5]

Table 3. *Cont.*

Mold Material	Additive Technique or Machining	Method Used to Evaluate Mold Performance	Injected Polymer	Results (Mold/Part)	Ref.
Ceramic photopolymer composite	Vat photopolymerization	Dimensional accuracy (dimensions over 10,000 shots) of mold and part	-	The diameters of the cylindrical elements in the mold were much smaller than the nominal diameters, which was due to the curing process of the photopolymer at the corners and edges after the printing process. The right angles of the corners did not undergo very significant changes in a range from 500 to 1000 injections.	[49]
Digital ABS	PolyJet	Mold failure (# of shots vs. failure); CAE software (injection pressure, injection speed, shot volume, confidence of fill, ejection time, cooling time)	Polycarbonate	Reducing mold temperature and increasing melt temperature were the most important changes to delay failure in polymer rapid tooling inserts.	[57]
Form 2 high temp. resin	Stereolithography, PolyJet	Mold failure (# of shots vs. failure)	Polystyrene	Using a hybrid mold (Master Unit Dye + AM inserts), it was possible to produce up to eighty components using both SLA and PolyJet printed molds.	[65]
Digital ABS, RGD450, Accura Bluestone, Accura SL5530, Accura Xtreme, High Temp, Tough, PerForm, CE221, PA 3200 GF, and steel	PolyJet, stereolithography, CLIP, selective laser sintering, and milling	Mold failure (# of shots vs. failure)	Polypropylene, PA6, and PA6+GF30%	Two polymer reference materials, i.e., PerFrom and PA 3200 GF, offer a great technological advantage to make injection molds because all the polymers under evaluation could be injected without experiencing complete failure of the insert. They can even be used to inject high melting polymer materials.	[1]
Though resin (THO), High Temperature (HT) resins, polyamide 12 filled with 50% of aluminum (PA50Al), and photopolymerization resin (ABS-like)	Stereolithography, laser sintering, and resin photo-polymerization (3D-PolyJet)	Mechanical performance (tensile strength, Charpy impact, flexural strength) and mold failure (# of shots vs. failure)	Elastomeric polyethylene, polypropylene, and ABS	Three mold materials obtained the highest elasticity and flexural modulus: HT, PA50Al, and ABS resin. They are the most appropriate materials to manufacture polypropylene injected prototypes. However, in the injection tests, the prototype mold made of ABS was only able to resist 12 injections before it began to crack.	[66]
Acrylic-based photopolymer	PolyJet	Accuracy of injected parts; thermal properties (specific heat)	High density polyethylene	The injected parts showed large shrinkages. The mold printing material indicates that the glass transition temperature is located at 55 °C.	[67]
VisiJet FTX Green	Stereolithography	Mold failure	Polypropylene	The failure of the insert mold was due to flexural stresses exerted by melt flow on the face of features perpendicular to flow front. Longer cycle times increase the ejection forces that may damage the tool.	[68]
420 stainless steel, bronze alloy, and ABS-like photopolymer	Milling, DMLS, and PolyJet	Thermal properties (specific heat); surface features (average roughness); mold failure (# of shots vs. failure); CAE software (mold temperature, deformation, stress); experimental data (mold temperature); cavity dimensions (average dimensions); mechanical performance (tensile strength)	Polypropylene	Regarding inserts of PolyJet molds, the coefficient of thermal expansion and compressibility of the polymeric insert material should be taken into account to calculate the nominal measurement of the injected part. Additionally, during ejection, ejection force, demolding angle, and cavity surface roughness should be reduced to facilitate ejections with minimal part-to-cavity interference and avoid polymer mold failure. Mold inserts manufactured by DMLS performed similarly to inserts machined from metal, with no failure up to 500 injection cycles.	[41]
Digital ABS, SAE 1045 steel, and Zamak 8	PolyJet and milling	Surface features (average roughness); mold failure (# of shots vs. failure); cavity dimensions (average dimensions)	Polypropylene	The polymer injected in the ABS mold showed a slight increase in tensile strength and elastic modulus, and its impact resistance was increased by more than 30% compared to the parts injected in steel and Zamak. The crystallinity results of the injected polypropylene were not consistent with the cooling rate offered by the ABS mold because said polypropylene showed a lower degree of crystallinity than the parts injected in steel and Zamak.	[69]

Table 3. *Cont.*

Mold Material	Additive Technique or Machining	Method Used to Evaluate Mold Performance	Injected Polymer	Results (Mold/Part)	Ref.
Formlabs White Resin, PolyJet Objet RGD515, and PEEK	Stereolithography (SLA), PolyJet, and Fused Deposition Modelling	Accuracy of injected parts	Polylactic acid (PLA)	The molds manufactured by stereolithography and PolyJet produced better finishes on the injected parts, while the mold made of PEEK by molten filament manufacturing presented delamination. In the SLA and PolyJet molds, the accuracy of the injected parts exhibited an average variation of less than 5%.	[70]
Digital ABS, aluminum, and Very High Molecular Weight Polyethylene	PolyJet, milling	Finished mold roughness; finished mold profiles	Cyclic olefin copolymer; polypropylene	The surface finish of 3D printed molds can be improved by applying coatings on the mold surface to inject optical components.	[71]
Photopolymer R11, steel, and aluminum	Stereolithography and milling	Thermal properties (heat capacity); mechanical performance (storage modulus, dimensional change); cavity dimensions (average dimensions); mold failure (# of shots vs. failure)	Polystyrene	The cooling time of polymer rapid tooling inserts is longer than that applied to aluminum and steel inserts due to their higher heat capacity compared to metal inserts. The dimensional changes of the polystyrene moldings (concerning part design and polymeric insert) were in the range 18–4 per cent.	[72]

6. Previous Studies of the Cost Model

In recent decades, in the context of the new industrial revolution, the technological potential of AM has increased and favored the development of different technological enablers such as cloud computing, cyber-manufacturing, and augmented reality [73]. This presents an ideal scenario for the creation of intelligent companies with a high degree of efficiency in their processes. However, during this technological advancement, the field of AM has been slow in establishing accurate cost models that can support corporate decision-making. Current literature describes different cost models classified by approach, AM technique, or the field of application where they are evaluated [74–77]. As this study is focused on AM in Injection Molding (IM), the following subsections highlight the main findings and results of cost models that have been used in this area.

6.1. Cost Models for AM as a Disruptive Technology in the IM Process

According to the literature, AM has been established as a disruptive technology that seeks to replace traditional manufacturing (TM) [36] because, compared to many conventional approaches, AM offers design freedom to manufacture complex and integrated parts. Using AM, tools or other processes are not necessary to create functional parts; hence, AM reduces the time needed to introduce a product into a market and, consequently, its total costs [36]. Many studies have compared and evaluated the break-even points of cost and production times of AM and IM for certain lot sizes. For example, Hopkinson and Dickens [37] were some of the first authors who identified the main sources of costs of Rapid Manufacturing (RM) related to AM. They showed that AM can compete against IM costs in situations of relatively high volumes. In the analysis and cost model they proposed, machinery, labor, and material costs represent the most critical variables [37]. Other authors later expanded on the work of Hopkinson and Dickens because they did not take into account a series of considerations in their model. For example, subsequent studies have investigated the construction and orientation of the manufactured parts (where packaging and distribution also play an important role), recycled material, and direct and indirect costs. Additionally, other papers have analyzed the production of copies of the same part and simultaneous production of different parts by SLS [38,39].

Another study [78] aimed to reduce costs and the final redesign of a part of a component previously produced by IM. In said study, it was demonstrated that RM can have economic potential for medium-sized production lots, and a break-even point was found at a production volume of around 87,000 units. However, a critical point in the AM model was the high acquisition cost of the SLS systems, which can be a decisive factor to migrate from one technology to the other [78]. Moreover, as a result of the growing popularization of low-cost 3D printing, additive techniques have had an exponential evolution. In 2015, $4.2 billion dollars were spent on AM in the US [79], which demonstrates the expansion of these alternative manufacturing techniques.

Achillas et al. [80] furthered the work of Hopkinson and Dickens because they not only evaluated the costs associated with SLA, SLS, and FDM techniques, but also included in their model the emerging PolyJet technique. The latter was used as an RT manufacturing method whose process was complementary rather than disruptive to the IM process. In their study, the key parameters were determined by lead time and total production cost. They also included variables such as time and pre-processing and post-processing cost. In their case, using RT to make soft molds by means of PolyJet was a very cost-effective method to manufacture new products in the range from 100 to 1000 units, while SLS was the most cost-effective AM technology in terms of time and cost.

In [79], the authors calculated the break-even points of AM and TM as a function of part mass, density, and lot size. Additionally, they took into account the cost of the material, equipment purchase, initial capital cost, time constraints, waste, overhead costs, etc. in order to construct a complete and realistic model. Based on this, they carried out a sensitivity analysis that showed that, in AM, material cost and part density were the variables most susceptible to variation; in turn, in IM, material cost, and mold cost per part presented the greatest changes. A lot size of approximately 200 units was the break-even point when deciding between AM and IM [79].

6.2. Cost Model as a Complement to AM in IM

Recently, from the perspective of cost estimation, some studies have investigated how AM can create added value when it is used as a complement to the IM process. Nevertheless, there is still a gap in the literature concerning the economic aspect of this object of study; hence, the following paragraphs will detail some articles that have encouraged the combination of these two technologies.

In [12], the authors sought to create synergy between the AM technique called Digital Light Processing (DLP) and conventional manufacturing processes. In their study, the main costs were pre-processing, construction, material, post-processing, and overhead. Regarding IM, the variables that most contributed to the cost were mold, material, and production. They concluded that tool cost was decreased by 80% when AM was used instead of CNC to create molds. The reduction in tooling cost could be approximately € 3489 (US$ 3995) for the largest geometry and € 996 (US$ 1140) for the smallest geometry in their study. Later [74], the same authors continued to study the cost estimation model employing the same AM technique, finding break-even points of up to 110,000 pieces when RT was used in IM. In addition, they reported longer processing times in AM (increasing the processing cost by 4%) because a longer cooling time was needed for polymer molds [74].

Kampker et al. [1] made a technological and economic comparison of 10 AM materials in the context of RT in order to provide guidelines to select materials for this type of applications. They found that the mold material with the greatest potential was PA3200 GF using the SLS technique, which reduced costs by 84.2% compared to tool steel. A year later, Ayvaz et al. [81] created an extended model to estimate costs and lead times of AM tools in IM. In their model, tool life was a key variable. They concluded that, using AM, tooling costs, by an estimated 20% to 66%; lead time, by up to 50%.

Figure 7 shows the investments needed if AM complements IM for soft tooling and if AM replaces IM to manufacture functional parts. As AM has become more widespread in the last decade, companies such as MakerBot Inc. and Ultimaker Inc. have made parts at lower costs [82]. Nowadays, 3D printers are more affordable due to a decrease in the cost of computer processors and the expiration of patents that protected existing systems [79]. In Figure 7, the investment needed to produce units by AM is very low when production volumes are low; however, as the number of units increases, the investment is drastically affected. This is due to the longer processing and post-processing times required to improve the final properties of parts made by AM. In addition, because the raw material of AM is usually up to 10 times more expensive than that of IM [79], the break-even point of AM is found at low-volume production [74]. Nevertheless, IM represents a very high initial investment, partly due to the cost of the tooling. Although the production cost of a few parts by IM is relatively low, other alternatives could reduce it. For instance, using soft tooling by AM for IM would require a medium-sized investment and would be suitable for low- and medium-volume production, as shown in Figure 7.

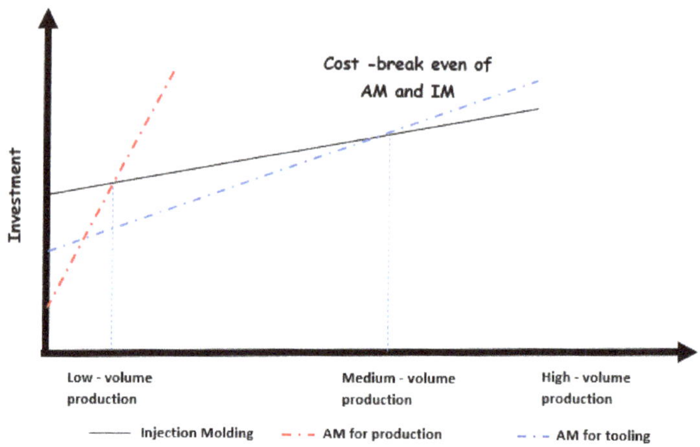

Figure 7. Cost curves of IM, AM for production, and AM for tooling.

Table 4 summarizes studies that have estimated the costs of AM as a disruptive and complementary technology to IM.

Table 4. Summary of studies into cost models of AM as a disruptive and complementary to IM.

Related Studies	Cost Approach		AM Techniques	Year
	Disruptive	Synergy between AM and IM		
[37]	X		FDM, SLS, SLA	2003
[38,39]	X		SLS	2006–2007
[78]	X		SLS	2012
[12]		X	DLP	2017
[80]	X	X	FDM, POLYJET, SLA, SLS	2017
[79]	X		FDM	2017
[74]		X	DLP	2019
[81]		X	POLYJET	2020

7. Conclusions

Mold design, an especially relevant aspect of the mold life cycle, represents around 80% of the total production cost. Therefore, designers should analyze and study technical guides (such as that in Section 4) in detail to properly design complete structures (e.g., ejection systems or ejectors, guides, cavities, runners, and gates).

Conformal cooling channels represent an innovative technique in mold design because they achieve shorter cycle times than conventional and parametric cooling channels. This technique should be further explored because it is closely related to AM.

Emerging technologies such as AM meet the new requirements imposed by the market (e.g., personalization and reduction of the product life cycle) because the products are made faster and complex parts can be freely designed. However, the costs associated with this type of technology increase exponentially because the materials, the acquisition of the machine, and processing time can affect the profitability of the process chain. Therefore, recent research has focused on the creation of cost models where AM is complemented by TM in the context of IM for low-, medium-, and high-volume production, which generates new business models and improves the efficiency of the processes.

The thermal and mechanical performance of polymer molds made by additive manufacturing for polymer injection is completely different from that of traditional metal molds. Hence, polymer molds should be previously evaluated to estimate their durability and changes caused by the injection process conditions. It is also important to quantify the properties of the polymer mold material and how they can affect the quality characteristics of the injected part, such as dimensional accuracy, shrinkage, and defects. This evaluation and economic manufacturing criteria can be used to justify the use of polymer injection molds made by additive manufacturing, which are generally well suited for medium- and low-volume production.

Author Contributions: Conceptualization, A.B.L. and S.H.Á.; methodology, A.B.L. and S.H.Á.; software, A.B.L.; validation, W.M.-R., A.B.L. and C.V.I.; formal analysis, A.B.L., S.H.Á., C.V.I.; investigation, A.B.L., S.H.Á., C.V.I., W.M.-R.; resources, Instituto Tecnológico Metropolitano; data curation, A.B.L.; writing—original draft preparation, A.B.L., S.H.Á., C.V.I.; writing—review and editing, A.B.L., S.H.Á., C.V.I.; visualization, A.B.L.; supervision, A.B.L., W.M.-R.; project administration, A.B.L.; funding acquisition, Instituto Tecnológico Metropolitano. All authors have read and agreed to the published version of the manuscript.

Funding: This work has been developed under the agreement of the collaboration agreement between Instituto Tecnológico Metropolitano—ITM and Universidad Nacional de Colombia in the framework of the research project: "Design, simulation and fabrication of plastic injection molds for production optimization" with project code P20-236.

Institutional Review Board Statement: Not applicable.

Informed Consent Statement: Not applicable. This study does not involve animals and humans and there are not exist etic conflicts.

Conflicts of Interest: The authors declare no conflict of interest.

References

1. Kampker, A.; Triebs, J.; Alves, B.; Kawollek, S.; Ayvaz, P. Potential analysis of additive manufacturing technologies for fabrication of polymer tools for injection moulding—A comparative study. In Proceedings of the 2018 IEEE International Conference on Advanced Manufacturing (ICAM), Yunlin, Taiwan, 16–18 November 2018; pp. 49–52. [CrossRef]
2. Wu, T.; Jahan, S.A.; Zhang, Y.; Zhang, J.; Elmounayri, H.; Tovar, A. Design Optimization of Plastic Injection Tooling for Additive Manufacturing. *Procedia Manuf.* **2017**, *10*, 923–934. [CrossRef]
3. Bogaerts, L.; Faes, M.; Bergen, J.; Cloots, J.; Vasiliauskaite, E.; Vogeler, F.; Moens, D. Influence of thermo-mechanical loads on the lifetime of plastic inserts for injection moulds produced via additive manufacturing. *Procedia CIRP* **2021**, *96*, 109–114. [CrossRef]
4. Walsh, E.; ter Horst, J.H.; Markl, D. Development of 3D printed rapid tooling for micro-injection moulding. *Chem. Eng. Sci.* **2021**, *235*, 116498. [CrossRef]
5. Zink, B.; Kovács, N.K.; Kovács, J.G. Thermal analysis based method development for novel rapid tooling applications. *Int. Commun. Heat Mass Transf.* **2019**, *108*, 104297. [CrossRef]
6. Krizsma, S.; Kovács, N.; Kovács, J.; Suplicz, A. In-situ monitoring of deformation in rapid prototyped injection molds. *Addit. Manuf.* **2021**, *42*, 102001. [CrossRef]
7. Kampker, A.; Triebs, J.; Kawollek, S.; Ayvaz, P.; Beyer, T. Direct polymer additive tooling—Effect of additive manufactured polymer tools on part material properties for injection moulding. *Rapid Prototyp. J.* **2019**, *25*, 1575–1584. [CrossRef]
8. Kampker, A.; Burggräf, P.; Wesch-Potente, C.; Petersohn, G.; Krunke, M. Life cycle oriented evaluation of flexibility in investment decisions for automated assembly systems. *CIRP J. Manuf. Sci. Technol.* **2013**, *6*, 274–280. [CrossRef]
9. Hu, S.J. Evolving Paradigms of Manufacturing: From Mass Production to Mass Customization and Personalization. *Procedia CIRP* **2013**, *7*, 3–8. [CrossRef]
10. Heizer, J.; Render, B. *Principios de Administración de Operaciones*; Pearson education: Ciudad de México, México, 2009.
11. Farotti, E.; Natalini, M. Injection molding. Influence of process parameters on mechanical properties of polypropylene polymer. A first study. *Procedia Struct. Integr.* **2018**, *8*, 256–264. [CrossRef]

12. Charalambis, A.; Kerbache, L.; Tosello, G.; Pedersen, D.B.; Mischkot, M.; Hansen, H.N. Dimensional Accuracy and Surface Finish in Additive Manufacturing, Economic Trade-Offs of Additive Manufacturing Integration in Injection Moulding Process Chain. Available online: https://www.euspen.eu/knowledge-base/AM17101.pdf (accessed on 1 December 2021).
13. Rahmati, S.; Dickens, P. Stereolithography for injection mould tooling. *Rapid Prototyp. J.* **1997**, *3*, 53–60. [CrossRef]
14. Jayanthi, S.; Bokuf, B.; Mcconnell, R.; Speer, R.J.; Fussell, P.S. Stereolithographic Injection Molds for Direct Tooling. In Proceedings of the International Solid Freeform Fabrication Symposium; 1997; pp. 275–286. Available online: https://repositories.lib.utexas.edu/handle/2152/70595 (accessed on 1 January 2022).
15. Tábi, T.; Kovács, N.K.; Sajó, I.; Czigany, T.; Hajba, S.; Kovács, J.G. Comparison of thermal, mechanical and thermomechanical properties of poly(lactic acid) injection-molded into epoxy-based Rapid Prototyped (PolyJet) and conventional steel mold. *J. Therm. Anal. Calorim.* **2016**, *123*, 349–361. [CrossRef]
16. Hofstätter, T.; Mischkot, M.; Pedersen, D.B.; Tosello, G.; Hansen, H.N. Evolution of surface texture and cracks during injection molding of fiber-reinforced, additively-manufactured, injection molding inserts. In Proceedings of the ASPE Summer Topical Meeting 2016: Dimensional Accuracy and Surface Finish in Additive Manufacturing, Raleigh, NC, USA, 27–30 June 2016; pp. 38–43.
17. Bagalkot, A.; Pons, D.; Symons, D.; Clucas, D. Categorization of Failures in Polymer Rapid Tools Used for Injection Molding. *Processes* **2019**, *7*, 17. [CrossRef]
18. Gill, S.S.; Tuli, S.; Xu, M.; Singh, I.; Singh, K.V.; Lindsay, D.; Tuli, S.; Smirnova, D.; Singh, M.; Jain, U.; et al. Transformative effects of IoT, Blockchain and Artificial Intelligence on cloud computing: Evolution, vision, trends and open challenges. *Internet Things* **2019**, *8*, 100118. [CrossRef]
19. EL Magri, A.; Vanaei, S.; Shirinbayan, M.; Vaudreuil, S.; Tcharkhtchi, A. An Investigation to Study the Effect of Process Parameters on the Strength and Fatigue Behavior of 3D-Printed PLA-Graphene. *Polymers* **2021**, *13*, 3218. [CrossRef] [PubMed]
20. Vanaei, H.R.; Shirinbayan, M.; Deligant, M.; Khelladi, S.; Tcharkhtchi, A. In-Process Monitoring of Temperature Evolution during Fused Filament Fabrication: A Journey from Numerical to Experimental Approaches. *Thermo* **2021**, *1*, 332–360. [CrossRef]
21. Udroiu, R.; Braga, I.C. Polyjet technology applications for rapid tooling. *MATEC Web Conf.* **2017**, *112*, 03011. [CrossRef]
22. Díaz, D.T. Tecnologías de Fabricación Digital Aditiva, ventajas para la construcción de modelos, prototipos y series cortas en el proceso de diseño de productos. *Iconofacto* **2016**, *12*, 118–143. [CrossRef]
23. Burnham, J.F. Scopus database: A review. *Biomed. Digit. Libr.* **2006**, *3*, 1. [CrossRef]
24. Van Eck, N.J.; Waltman, L. Software survey: VOSviewer, a computer program for bibliometric mapping. *Scientometrics* **2010**, *84*, 523–538. [CrossRef]
25. van Eck, N.J.; Waltman, L. VOSviewer Manual version 1.6.16, Univeristeit Leiden. 25 November 2020, pp. 1–52. Available online: https://www.vosviewer.com/download/f-33t2.pdf (accessed on 19 July 2021).
26. Qaud, N. Additive Manufacturing Technologies at Sulzer. In *Sulzer Technical Review*; Sulzer Management AG: Singapore, 2018; Volume 100, pp. 4–7.
27. Odremán, R.J. *Impresión 3D en la Industria: Un Acercamiento a la Tecnología y su Influencia en la Industria Petrolera*; Universidad Nacional Experimental Politécnica Antonio José de Sucre; Guayana, Venezuela, Cienc. y Tecnol.: Bolívar, Venezuela, 2014; Volume 18, p. 166.
28. Gornet, T. History of Additive Manufacturing. In *3D Printing and Its Impact on the Production of Fully Functional Components: Emerging Research and Opportunities*; IGI Global: Hershey, PA, USA, 2017; pp. 1–24. [CrossRef]
29. Jiménez, M.; Romero, L.; Domínguez, I.A.; Espinosa, M.D.M.; Domínguez, M. Additive Manufacturing Technologies: An Overview about 3D Printing Methods and Future Prospects. *Complexity* **2019**, *2019*, 9656938. [CrossRef]
30. Benitez, A.; Montealegre, W.; Vargas, C. Simulación de moldeo por inyección basado en el método de volúmenes finitos (FVM). In *Actualidad de la Ingeniería Mecánica en Iberoamérica—XIV CIBIM, XIV*; de Federación Iberoamericana de Ingeniería Mecánica, Ed.; Manuel del Jesús Martínez, división de publicaciones UIS: Cartagena, Colombia, 2019; p. 283.
31. Biron, M. Thermoplastic Processing. In *Plastics Design Library, Thermoplastics and Thermoplastic Composites*, 2nd ed.; William Andrew Publishing: New York, NY, USA, 2013; pp. 715–768. [CrossRef]
32. Mercado-Colmenero, J.M.; Rubio-Paramio, M.A.; Marquez-Sevillano, J.D.J.; Martin-Doñate, C. A new method for the automated design of cooling systems in injection molds. *Comput. Des.* **2018**, *104*, 60–86. [CrossRef]
33. Zhang, Y.; Hou, B.; Wang, Q.; Huang, Z.; Zhou, H. Automatic generation of venting system on complex surfaces of injection mold. *Int. J. Adv. Manuf. Technol.* **2018**, *98*, 1379–1389. [CrossRef]
34. Kuo, C.-C.; Xu, W.C. Effects of different cooling channels on the cooling efficiency in the wax injection molding process. *Int. J. Adv. Manuf. Technol.* **2018**, *98*, 887–895. [CrossRef]
35. Altaf, K.; Qayyum, J.A.; Rani, A.M.A.; Ahmad, F.; Megat-Yusoff, P.S.M.; Baharom, M.; Aziz, A.R.A.; Jahanzaib, M.; German, R.M. Performance Analysis of Enhanced 3D Printed Polymer Molds for Metal Injection Molding Process. *Metals* **2018**, *8*, 433. [CrossRef]
36. Hague, R.J.; Campbell, R.; Dickens, P.M. Implications on design of rapid manufacturing. *Proc. Inst. Mech. Eng. Part C J. Mech. Eng. Sci.* **2003**, *217*, 25–30. [CrossRef]
37. Hopkinson, N.; Dicknes, P. Analysis of rapid manufacturing—Using layer manufacturing processes for production. *Proc. Inst. Mech. Eng. Part C J. Mech. Eng. Sci.* **2003**, *217*, 31–39. [CrossRef]
38. Ruffo, M.; Tuck, C.; Hague, R. Cost estimation for rapid manufacturing-laser sintering production for low to medium volumes. *Proc. Inst. Mech. Eng. Part B J. Eng. Manuf.* **2006**, *220*, 1417–1427. [CrossRef]

39. Ruffo, M.; Hague, R. Cost estimation for rapid manufacturing simultaneous production of mixed components using laser sintering. *Proc. Inst. Mech. Eng. Part B J. Eng. Manuf.* **2007**, *221*, 1585–1591. [CrossRef]
40. StrataSys. Technical application guide. Polyjet for Injection Molding. *3D Print. Solut.* 2015, pp. 218–223. Available online: https://www.google.com.tw/url?sa=t&rct=j&q=&esrc=s&source=web&cd=&ved=2ahUKEwjHwKfUg5j2AhWbk1YBHemhBQoQFnoECAUQAQ&url=https%3A%2F%2Fwww.stratasys.com%2F-%2Fmedia%2Ffiles%2Ftechnical-application-guides%2Ftag_pj_injectionmolding.pdf&usg=AOvVaw2dhobNf6nuDXSYkYIut1oC (accessed on 1 January 2022).
41. Antonio, G.; Jack, M.; Stephen, A.R. Comparative study of rapid and conventional tooling for plastics injection molding. *Rapid Prototyp. J.* **2017**, *23*, 344–352.
42. Knorr, L.; Setter, R.; Rietzel, D.; Wudy, K.; Osswald, T. Comparative Analysis of the Impact of Additively Manufactured Polymer Tools on the Fiber Configuration of Injection Molded Long-Fiber-Reinforced Thermoplastics. *J. Compos. Sci.* **2020**, *4*, 136. [CrossRef]
43. Tan, C.; Wang, D.; Ma, W.; Chen, Y.; Chen, S.; Yang, Y.; Zhou, K. Design and additive manufacturing of novel conformal cooling molds. *Mater. Des.* **2020**, *196*, 109147. [CrossRef]
44. Tomasoni, D.; Colosio, S.; Giorleo, L.; Ceretti, E. Design for Additive Manufacturing: Thermoforming Mold Optimization via Conformal Cooling Channel Technology. *Procedia Manuf.* **2020**, *47*, 1117–1122. [CrossRef]
45. Feng, S.; Kamat, A.M.; Pei, Y. Design and fabrication of conformal cooling channels in molds: Review and progress updates. *Int. J. Heat Mass Transf.* **2021**, *171*, 121082. [CrossRef]
46. Godec, D.; Breški, T.; Katalenić, M. Additive Manufacturing of Polymer Moulds for Small-Batch Injection Moulding. *Teh. Glas.* **2020**, *14*, 218–223. [CrossRef]
47. Dizon, J.R.C.; Valino, A.D.; Souza, L.R.; Espera, A.; Chen, Q.; Advincula, R.C. Three-dimensional-printed molds and materials for injection molding and rapid tooling applications. *MRS Commun.* **2019**, *9*, 1267–1283. [CrossRef]
48. Martinho, P.G.; Bartolo, P.; Pouzada, A. Hybrid moulds: Effect of the moulding blocks on the morphology and dimensional properties. *Rapid Prototyp. J.* **2009**, *15*, 71–82. [CrossRef]
49. Mischkot, M.; Davoudinejad, A.; Charalambis, A.; Tosello, G.; Pedersen, D.B.; Nørgaard, H. Dimensional accuracy of Acrylonitrile Butadiene Styrene injection molded parts produced in a pilot production with an additively manufactured insert. In Proceedings of the 33rd Conference of the Polymer Processing Society, Cancun, Mexico, 10–14 December 2017; pp. 1–6.
50. Formlabs. Low-Volume Rapid Injection Molding with 3D Printed Molds. 2020. Available online: https://f.hubspotusercontent20.net/hubfs/7196278/Formlabs%20injection%20molding/Low%20volume%20rapid%20injection%20m.pdf (accessed on 15 September 2021).
51. Stratasys. Working Together: 3D Printing Expands Efficiencies in Traditional Manufacturing. 2022. Available online: https://www.stratasysdirect.com/manufacturing-services/subtractive-manufacturing/3d-printing-expands-traditional-manufacturing (accessed on 23 January 2022).
52. Martinho, P.G.; Pouzada, A.S. Alternative materials in moulding elements of hybrid moulds: Structural integrity and tribological aspects. *Int. J. Adv. Manuf. Technol.* **2021**, *113*, 351–363. [CrossRef]
53. Qayyum, J.A.; Altaf, K.; Rani, A.M.A.; Ahmad, F.; Jahanzaib, M. Performance of 3D printed polymer mold for metal injection molding process. *ARPN J. Eng. Appl. Sci.* **2017**, *12*, 6430–6434.
54. Davoudinejad, A.; Khosravani, M.R.; Pedersen, D.B.; Tosello, G. Influence of thermal ageing on the fracture and lifetime of additively manufactured mold inserts. *Eng. Fail. Anal.* **2020**, *115*, 104694. [CrossRef]
55. Kumar, R.; Kumar, M.; Chohan, J.S. Material-specific properties and applications of additive manufacturing techniques: A comprehensive review. *Bull. Mater. Sci.* **2021**, *44*, 1–19. [CrossRef]
56. Bin Hussin, R.; Bin Sharif, S.; Rahim, S.Z.B.A.; Bin Suhaimi, M.A.; Khushairi, M.T.B.M.; El-Hadj, A.A.; Bin Shuaib, N.A. The potential of metal epoxy composite (MEC) as hybrid mold inserts in rapid tooling application: A review. *Rapid Prototyp. J.* **2021**, *27*, 1069–1100. [CrossRef]
57. Bagalkot, A.; Pons, D.; Clucas, D.; Symons, D. A methodology for setting the injection moulding process parameters for polymer rapid tooling inserts. *Rapid Prototyp. J.* **2019**, *25*, 1493–1505. [CrossRef]
58. Bagalkot, A.; Pons, D.; Symons, D.; Clucas, D. Analysis of Raised Feature Failures on 3D Printed Injection Moulds. *Polymers* **2021**, *13*, 1541. [CrossRef] [PubMed]
59. Pasha, A.D.; Urbanic, J.; Hedrick, B.; Ramezani, H.; Jianu, O. Leveraging Advanced Design and Novel Rapid Manufacturing Solutions to Respond to the COVID-19 Pandemic. *Comput. Aided. Des. Appl.* **2022**, *19*, 755–778. Available online: https://scopus.bibliotecaitm.elogim.com/record/display.uri?eid=2-s2.0-85122291519&origin=resultslist&sort=plf-f&src=s&st1=%2522rapid+tooling%2522&nlo=&nlr=&nls=&sid=ff72908c072a212eb940dac0b887115b&sot=b&sdt=cl&cluster=scoexactkeywords%252C%2522Injection+ (accessed on 29 January 2022). [CrossRef]
60. Mau, R.; Jüttner, G.; Gao, Z.; Matin, F.; Labrador, D.A.; Repp, F.; John, S.; Scheper, V.; Lenarz, T.; Seitz, H. Micro injection molding of individualised implants using 3D printed molds manufactured via digital light processing. *Curr. Dir. Biomed. Eng.* **2021**, *7*, 399–402. [CrossRef]
61. Mendible, G.A.; Saleh, N.; Barry, C.; Johnston, S.P. Mechanical properties and crystallinity of polypropylene injection molded in polyjet and aluminum tooling. *Rapid Prototyp. J.* **2021**. *Epub ahead of printing*. [CrossRef]
62. Farioli, D.; Strano, M.; Vangosa, F.B.; Zaragoza, V.G.; Aicardi, A. Rapid tooling for injection molding inserts. In Proceedings of the ESAFORM 2021—24th International Conference on Material Forming, Liege, Belgium, 16–14 April 2021; Volume 4186, pp. 1–10.

63. Schuh, G.; Bergweiler, G.; Lukas, G.; Abrams, J.A. Feasibility and Process capability of polymer additive injection molds with slide technology. *Procedia CIRP* **2020**, *93*, 102–107. [CrossRef]
64. Altaf, K.; Majdi, A.; Rani, A.; Ahmad, F.; Ahmad, J. Enhanced polymer rapid tooling for metal injection moulding process. *Proc. Int. Conf. Prog. Addit. Manuf.* **2018**, *2018-May*, 656–661. [CrossRef]
65. Whlean, C.; Sheahan, C. Using Additive Manufacturing to Produce Injection Moulds Suitable for Short Series Production. *Procedia Manuf.* **2019**, *38*, 60–68. [CrossRef]
66. León-Cabezas, M.; García, A.M.; Varela-Gandía, F. Innovative advances in additive manufactured moulds for short plastic injection series. *Procedia Manuf.* **2017**, *13*, 732–737. [CrossRef]
67. Mitterlehner, T.; Beisteiner, C.; Rieger, H.; Dunzendorfer, P.; Steinbichler, G. Back Injection Molding with Additive Manufactured Mold Inserts Using All-Inkjet Printed Substrates. 2017. Available online: https://www.researchgate.net/profile/Thomas-Mitterlehner-2/publication/317013699_Back_injection_molding_with_additive_manufactured_mold_inserts_using_all-inkjet_printed_substrates/links/5aaa21450f7e9b88266fcc87/Back-injection-molding-with-additive-manufactured-mold-inserts-using-all-inkjet-printed-substrates.pdf (accessed on 1 January 2022).
68. Gheisari, R.; Bartolo, P.J.D.S.; Goddard, N.; Domingos, M.A.D.N. An experimental study to investigate the micro-stereolithography tools for micro injection molding. *Rapid Prototyp. J.* **2017**, *23*, 711–719. [CrossRef]
69. Volpato, N.; Solis, D.M.; Costa, C.A. An analysis of Digital ABS as a rapid tooling material for polymer injection moulding. *Int. J. Mater. Prod. Technol.* **2016**, *52*, 3–16. [CrossRef]
70. Dizon, J.R.C.; Valino, A.D.; Souza, L.R.; Espera, A.H.; Chen, Q.; Advincula, R.C. 3D Printed Injection Molds Using Various 3D Printing Technologies. *Mater. Sci. Forum* **2020**, *1005*, 150–156. [CrossRef]
71. Noble, J.; Walczak, K.; Dornfeld, D. Rapid Tooling Injection Molded Prototypes: A Case Study in Artificial Photosynthesis Technology. *Procedia CIRP* **2014**, *14*, 251–256. [CrossRef]
72. Griffiths, C.A.; Dimov, S.S.; Fischer, S.; Spitzbart, M.; Lacan, F. Micro-stereolithography tools for small-batch manufacture of polymer micro-parts. *Proc. Inst. Mech. Eng. Part B J. Eng. Manuf.* **2011**, *226*, 708–721. [CrossRef]
73. Zuhra, A.; Kadir, A. Additive manufacturing cost estimation models—A classification review. *Int. J. Adv. Manuf. Technol.* **2020**, *107*, 4033–4053.
74. Tosello, G.; Charalambis, A.; Kerbache, L.; Mischkot, M.; Pedersen, D.B.; Calaon, M.; Hansen, H.N. Value chain and production cost optimization by integrating additive manufacturing in injection molding process chain. *Int. J. Adv. Manuf. Technol.* **2018**, *100*, 783–795. [CrossRef]
75. Gisario, A.; Kazarian, M.; Martina, F.; Mehrpouya, M. Metal additive manufacturing in the commercial aviation industry: A review. *J. Manuf. Syst.* **2019**, *53*, 124–149. [CrossRef]
76. Xu, F.; Wong, Y.; Loh, H. Toward generic models for comparative evaluation and process selection in rapid prototyping and manufacturing. *J. Manuf. Syst.* **2001**, *19*, 283–296. [CrossRef]
77. Busachi, A.; Erkoyuncu, J.; Colegrove, P.; Drake, R.; Watts, C.; Martina, F.; Tapoglou, N.; Lockett, H. A System Approach for Modelling Additive Manufacturing in Defence Acquisition Programs. *Procedia CIRP* **2018**, *67*, 209–214. [CrossRef]
78. Atzeni, E.; Iuliano, L.; Minetola, P.; Salmi, A. Redesign and cost estimation of rapid manufactured plastic parts. *Rapid Prototyp. J.* **2010**, *16*, 308–317. [CrossRef]
79. Franchetti, M.; Kress, C. An economic analysis comparing the cost feasibility of replacing injection molding processes with emerging additive manufacturing techniques. *Int. J. Adv. Manuf. Technol.* **2016**, *88*, 2573–2579. [CrossRef]
80. Achillas, C.; Tzetzis, D.; Raimondo, M.O. Alternative production strategies based on the comparison of additive and traditional manufacturing technologies. *Int. J. Prod. Res.* **2016**, *55*, 3497–3509. [CrossRef]
81. Kampker, A.; Ayvaz, P.; Lukas, G. Direct Polymer Additive Tooling-Economic Analysis of Additive Manufacturing Technologies for Fabrication of Polymer Tools for Injection Molding. *Key Eng. Mater.* **2020**, *843*, 9–18. [CrossRef]
82. Malone, E.; Lipson, H. Fab@Home: The personal desktop fabricator kit. *Rapid Prototyp. J.* **2007**, *13*, 245–255. [CrossRef]

polymers

Article

An Effective Interface Tracking Method for Simulating the Extrudate Swell Phenomenon

Ahmad Fakhari [1], Željko Tukovic [2,*], Olga Sousa Carneiro [1] and Célio Fernandes [1]

[1] Institute for Polymers and Composites, Department of Polymer Engineering, Campus of Azurém, University of Minho, 4800-058 Guimarães, Portugal; ahmadfakhari@gmail.com (A.F.); olgasc@dep.uminho.pt (O.S.C.); cbpf@dep.uminho.pt (C.F.)
[2] Faculty of Mechanical Engineering and Naval Architecture, University of Zagreb, Ivana Lučića 5, 10000 Zagreb, Croatia
* Correspondence: zeljko.tukovic@fsb.hr

Abstract: The extrudate swell, i.e., the geometrical modifications that take place when the flowing material leaves the confined flow inside a channel and moves freely without the restrictions that are promoted by the walls, is a relevant phenomenon in several polymer processing techniques. For instance, in profile extrusion, the extrudate cross-section is subjected to a number of distortions that are motivated by the swell, which are very difficult to anticipate, especially for complex geometries. As happens in many industrial processes, numerical modelling might provide useful information to support design tasks, i.e., to allow for identifying the best strategy to compensate the changes promoted by the extrudate swell. This study reports the development of an improved interface tracking algorithm that employs the least-squares volume-to-point interpolation method for the grid movement. The formulation is enriched further with the consistent second-order time-accurate non-iterative Pressure-Implicit with Splitting of Operators (PISO) algorithm, which allows for efficiently simulating free-surface flows. The accuracy and robustness of the proposed solver is illustrated through the simulation of the steady planar and asymmetric extrudate swell flows of Newtonian fluids. The role of inertia on the extrudate swell is studied, and the results that are obtained with the newly improved solver show good agreement with reference data that are found in the scientific literature.

Keywords: extrusion; extrudate swell; interface tracking; least-squares volume-to-point interpolation; consistent PISO; finite volume method; OpenFOAM

Citation: Fakhari, A.; Tukovic, Ž.; Carneiro, O.S.; Fernandes, C. An Effective Interface Tracking Method for Simulating the Extrudate Swell Phenomenon. *Polymers* **2021**, *13*, 1305. https://doi.org/10.3390/polym13081305

Academic Editor: Serge Bourbigot

Received: 30 March 2021
Accepted: 11 April 2021
Published: 16 April 2021

1. Introduction

Free-surface flows are encountered in many polymer processing and environmental applications [1,2]. Nevertheless, the variety of the analytical solutions for free-surface flows is usually very limited, even for very simple cases [3]. On the other hand, the experimental observations of real phenomena are onerous [3], and many experimental techniques are suitable for single-phase flows and undergo many difficulties to be extended to two-phase flows [4]. For these reasons, the use of numerical simulations to provide useful information about free-surface flows would be of great advantage. Flows with a free-surface are difficult to be modeled since the free-surface is a moving boundary, whose location is merely known initially, and it has to be determined later during the simulation [5]. There are different ways of modeling free-surface flows: the Interface Tracking (IT) approach, in which the free-surface is tracked using a sharp interface, and a dynamic computational grid is applied to follow the movement of the free-surface; and, the Interface Capturing (IC) approach, in which the free-surface is not treated as a sharp interface and, generally, the computational grid is static. Among the methods following the IC approach, the Marker-And-Cell (MAC) [6] method is based on a finite difference scheme applied to an Eulerian grid to solve the Navier–Stokes equations for the fluid flow motion, and resort to Lagrangian virtual particles to impose the movement at the free-surface, which is based on the velocity interpolated from the Eulerian grid [7,8]. Although the MAC method provides accurate information on the free-surface location [8,9], the computational cost is enormous, because the number of particles needed to reconstruct the free-surface is vast [5]. Instead of considering virtual Lagrangian particles to reconstruct the free-surface, the Volume-of-Fluid (VoF) method [10] solves a transport equation to calculate the volume fraction of each fluid present in the interface cells. Although the VoF method is more efficient than the MAC method, and is more practical for complex interface shapes, it considers the interface as a layer that usually covers one to three computational cells [5]; therefore, it does not provide an exact location of the free-surface. Despite many attempts to obtain the precise local curvature of the free-surface, when using the VoF method [9,11–14], this disadvantage still remains. One of the attempts employed to compute sharp interfaces for free-surface flows was implemented by Roenby et al. [15], by using the so-called isoAdvector algorithm, which follows two-steps: first, it computes an isosurface to evaluate the distribution of fluids inside the cells (known as the surface reconstruction step); and second, it advects the face–interface intersection line to obtain the time evolution within a time step of the submerged face area (known as the advection step). The method provided very satisfactory results in terms of volume conservation, boundedness, surface sharpness, and efficiency for two-dimensional and three-dimensional problems on both structured and unstructured meshes. The Level-Set (LS) method, which was proposed by Osher and Sethian [16], is another method that considers the contour of a smooth scalar function to specify the location of the free-surface. In this method, the value of the scalar function at a computational grid cell is often calculated based on the signed distance function [11,17,18]. Although the transition of fluid properties through the interface is smooth, it experiences difficulty if the curvature of the free-surface undergoes rapid changes [8]; and, also there is a need to define a transition region with a finite thickness [5]. Furthermore, mass conservation is an issue when using the LS method [5,8].

Because the VoF and LS methods are based on an implicit identification of the free-surface through the volume fraction and distance function, respectively, they are commonly called IC methods. From the reasoning that is explained above, the lack of prediction of the exact location of the free-surface, the high computational cost, and the precise calculation of the representation of surface forces (for example, surface tension) are the general disadvantages of the IC methods. On the other hand, IT methods use an explicit discretization of the interfacial discontinuity [19,20], which applies a body-fitted (boundary-fitted) grid and the free-surface boundary is tracked using mesh movement. Because the free-surface is treated as a sharp interface, it is the most accurate approach, albeit with limitations on the deformation of the free-surface.

Although limited in their application, IT methods have an important role in the numerical analyses of fundamental multiphase flows, such as the extrudate swell (or die swell) phenomenon exhibited by viscous fluids exiting long slits or capillary dies [21–30]. Extrudate swell, which is also known as Barus effect [31], occurs when melted polymer comes out of a die, where the size of the emerged polymer becomes different from the size of the die. This even happens in Newtonian fluids due to the streamline rearrangements at the die exit [32], which is around 13% for cylindrical channels and 19% for sheets at very low Reynolds numbers, while, at high Reynolds numbers, the swell shrinks and, finally, the Newtonian liquid comes out like a thinning jet. The extrudate swell of polymers is usually in a very low range of Reynolds numbers from 10^{-4} to 10^{-2}, and the swell ratio can reach as high as 400% in specific cases, which are related to the viscoelastic character of polymers. Experiments have shown that, when the die length is short enough, the extrudate swell grows when compared to a case with the same mass flow rate, but a longer die, which is commonly attributed to the memory of entrance. Thus, the swell in short dies is a consequence of two components, the memory of entrance and also the normal stress release at the die exit. In addition to these parameters, temperature also influences extrudate swell. The thermal effects can increase the extrudate swell up to 15%. If the die temperature is lower than the melted polymer, the viscosity of the melted polymer increases at the wall and, therefore, the flow becomes limited and it undergoes a lower swell when compared to the case where the die temperature is higher than the melted polymer, where the liquid is less viscous at the wall, which lubricates the flow and results in a larger swell. Another important parameter for the polymer extrudate swell is the molecular structure of the polymer, since the first normal stress difference that appears at the die exit is affected by the molecular weight distribution. Furthermore, long chain branching enhances extrudate swell in polymers [33].

It is crucial to know the dimension of the emerging polymer from the die, since it determines the exact size of the extruded products; therefore, many attempts have been made to obtain equations for anticipating the swell ratio [31,34]. As an example, Tanner [35] introduced an equation for the swell ratio of a Maxwell-type constitutive equation of a viscoelastic fluid, and then revised it several years later [36] while considering that the first normal stress difference obeys $N_1 = k\tau^m$, where τ is the fluid shear stress and k and m are model parameters, instead of the initial assumption $N_1 = \text{constant} \times \tau^2$. A large number of investigations of the die swell phenomenon are reported in the literature, which goes from two-dimensional (2D) to three-dimensional (3D) flows, from Newtonian to viscoelastic fluids rheology or from isothermal to non-isothermal flows. Crochet and Keunings [21] presented a 2D numerical investigation on slit die swell with three finite element meshes and two different techniques, a mixed and an extended $u - v - p$ methods, and concluded that the swelling ratio depends upon the method that is used for its calculation and, for a given method, it is highly dependent upon mesh refinement. Subsequently, Mitsoulis et al. [22] used viscometric flow equations to simulate the swelling of viscoelastic fluids from long slit and capillary dies. The obtained results showed that, even when using this simple model for viscoelastic calculations, they were in good qualitative agreement with other numerical simulations, which employ a constitutive equation satisfying tensorial invariance, but accelerate the breakdown of the iterative scheme. Another work from the group of J. Vlachopoulos was presented by Karagiannis et al. [23], which studied the 3D free surface die swell of a Newtonian fluid in different geometries, specifically square, rectangular, equilateral triangular, bow-tie, and key-hole-shaped geometries. The obtained results were compared with experimental measurements and other numerical calculations with favourable agreement. The swelling ratios were found to strongly depend on the die geometry. The group of Crouchet also presented calculations for steady state 3D free surfaces of Newtonian and power-law fluids in the work of Wambersie and Crochet [24]. They combined a pseudo-transient marching technique, a decoupling algorithm, and a conjugate gradient solver to reduce the cost of the 3D calculations. The method was employed to study the circular, square, and rectangular die-swell problems, where the effects of inertia and shear-thinning were revealed. The works of Legat and Marchal [25,26] addressed the prediction of 3D free surface extrudate flows with a fully implicit finite element algorithm, in the sense that a Newton–Raphson scheme was applied to all variables and is geometrically general. The algorithm was employed to compute the extrudate swell of a rectangular die and in various complex sections containing multiple corners. The obtained results showed that the extrudate shape exhibits large deformations in the vicinity of all re-entrant corners, which would not be possible to predict in 2D simulations. Subsequently, the works of Georgiou and Boudouvis [27] and Mitsoulis et al. [28] studied the effects of inertia, surface tension, gravity, slip, and compressibility for both the 2D planar and axisymmetric extrudate-swell flows of Newtonian fluids. Recently, 3D isothermal and non-isothermal viscoelastic flow calculations with a transient finite element method for predicting extrudate swell of domains containing sharp edges were conducted by Spanjaards et al. [29,30]. The obtained results showed that the extrudate swelling is highly dependent on the rheological parameters and the constitutive model used, and that the wall temperature of the die can lead to a change in the bending direction. All of the presented works employed the finite element method to solve the problem of the extrudate swell for Newtonian and viscoelastic fluids. The present work aims to revisit the

Newtonian extrudate swell flow problem by using the finite volume method, which is the core of the open-source computational library OpenFOAM [37].

When dealing with a steady state process, which is the case of profile extrusion, IT is usually the best alternative, since it does not present the problems that are related to interface diffusion inherent to the IC methods. OpenFOAM [37] comprises a solver to simulate free-surface flows following an IT approach, which was proposed by Tuković and Jasak [38]. One of the disadvantages generally raised to the IT methods is their computational efficiency, and many attempts have been carried out in order to increase the convergence rate of the pressure–velocity calculations [39–41]. Tuković et al. [41] proposed a non-iterative Pressure-Implicit with Splitting of Operators (PISO) algorithm based on extrapolation of mass flux, nodal velocity, and pressure from two previous time steps in order to have an approximation of these quantities in the new time step, and obtained a second order temporal accuracy in the cases with static and dynamic meshes.

This work aims to assess the capability of the solver that was developed by Tuković and Jasak [38] with the non-iterative PISO algorithm proposed by Tuković et al. [41] to efficiently simulate the extrudate swell phenomenon. For this purpose, the developed solver couples the *interfaceTrackingFvMesh* and *interTrackMeshMotion* libraries that are available in OpenFOAM [37] with the consistent second-order time-accurate non-iterative PISO algorithm. A least-squares volume-to-point interpolation method for the grid movement, which enables an efficient and accurate tracking of the free-surface motion, is employed. The enhanced algorithm is used to simulate the steady-state Newtonian extrudate swell problem in both planar and axisymmetric geometries for a parametric study of the effects of inertia, and the obtained results are compared with the reference data of Mitsoulis et al. [28]. Notice that, although the results presented here are limited to Newtonian fluids, the effects which are discussed can be qualitatively applied to all fluids (e.g., viscoelastic fluids). The main aim of this work is to present an open source finite volume numerical framework that can handle the extrudate swell problem in an efficient way, which can be extended in the future to allow simulating other fluids rheology.

2. Mathematical Formulation

In this section, the governing equations and numerical method employed for simulating the two-phase fluid flow with a sharp interface are described, while using the finite-volume method and an interface tracking algorithm for the moving mesh. The numerical scheme developed in this work is enhanced with the consistent second-order time-accurate non-iterative PISO algorithm [39–41] to reduce the computational wall time of the simulations and, for the moving mesh calculation, a Laplacian scheme is used with a least-squares interpolation, which allows for robust and stable deformation of the interface.

2.1. Governing Equations

The mass and linear momentum conservation laws are the equations of motion governing the isothermal flow of incompressible Newtonian fluids inside an arbitrary volume V bounded by a closed surface S,

$$\oint_S \mathbf{n} \cdot \mathbf{u}\, dS = 0, \tag{1}$$

$$\frac{d}{dt}\int_V \mathbf{u}\, dV + \oint_S \mathbf{n} \cdot (\mathbf{u} - \mathbf{u}_s)\mathbf{u}\, dS = \oint_S \mathbf{n} \cdot (\nu \nabla \mathbf{u})\, dS - \int_V \nabla P\, dV, \tag{2}$$

where $\mathbf{n}$ is the outward pointing unit normal vector on S, $\mathbf{u}$ is the fluid velocity, $\mathbf{u}_s$ is the surface S velocity, ν is the fluid kinematic viscosity, and P is the kinematic pressure obtained by subtracting the hydrostatic kinematic pressure $P_{hydrostatic} = \mathbf{g} \cdot \mathbf{r}$ from the absolute pressure, where $\mathbf{g}$ is the gravitational acceleration and $\mathbf{r}$ is the position vector.

For an arbitrary moving volume, the relationship between the rate of change of the volume V and the velocity $\mathbf{u}_s$ is defined by the *geometrical (space) conservation law* [42],

$$\frac{d}{dt}\int_V dV - \oint_S \mathbf{n}\cdot\mathbf{u}_s\, dS = 0. \tag{3}$$

When considering that the fluid phases are immiscible, the fluid flow Equations (1) and (2) can be used for each phase individually and, at the interface, the proper boundary conditions must be used.

2.1.1. Kinematic Condition

The *kinematic condition* states that the fluid velocities on the two sides of the interface, $\mathbf{u}_{1f}$ and $\mathbf{u}_{2f}$, must be continuous (see Figure 1),

$$\mathbf{u}_{1f} = \mathbf{u}_{2f}. \tag{4}$$

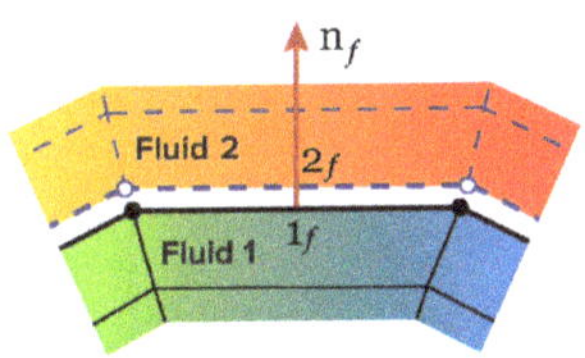

Figure 1. Representation of the interface with the mesh boundary faces.

2.1.2. Dynamic Condition

From the momentum conservation law follows the *dynamic condition*, which states that forces acting on the fluid at the interface are in equilibrium. The general form of the *dynamic condition* at the interface, which gives the fundamental relationship between the *jump* in stress across an interface and the surface tension force, is given by,

$$[\mathbf{T}_2 - \mathbf{T}_1]\cdot\mathbf{n} = \nabla_s\sigma - \sigma\kappa\mathbf{n}, \tag{5}$$

where $\mathbf{T}_1$ and $\mathbf{T}_2$ are the stress tensors that are defined in terms of the local fluid pressure and velocity fields, as $\mathbf{T}_1 = -p_1\mathbf{I} + \nu_1[\nabla\mathbf{u}_1 + (\nabla\mathbf{u}_1)^T]$ and $\mathbf{T}_2 = -p_2\mathbf{I} + \nu_2[\nabla\mathbf{u}_2 + (\nabla\mathbf{u}_2)^T]$, respectively, σ is the interfacial tension and $\nabla_s = [\mathbf{I} - \mathbf{nn}]\cdot\nabla = \nabla - \mathbf{n}\frac{\partial}{\partial n}$ is the tangential gradient operator, which appears because σ and $\mathbf{n}$ are only defined on the surface. Equation (5) is a vectorial equation, which is often written in terms of its normal and tangential components. We proceed by deriving the normal and tangential force balances appropriately at a fluid–fluid interface that is characterized by an interfacial tension σ.

From the normal force balance follows the *pressure jump* across the interface [43],

$$p_2 - p_1 = \sigma\kappa - 2(\nu_2 - \nu_1)\nabla_s\cdot\mathbf{u}, \tag{6}$$

where $\kappa = -\nabla_s\cdot\mathbf{n}$ is twice the mean curvature of the interface.

The tangential force balance yields a relation between the normal derivative of tangential velocity on the two sides of the interface [43],

$$\nu_2[\mathbf{n}\cdot(\nabla\mathbf{u}_t)_2] - \nu_1[\mathbf{n}\cdot(\nabla\mathbf{u}_t)_1] = -\nabla_s\sigma - \mathbf{n}(\nu_2 - \nu_1)\nabla_s\cdot\mathbf{u} - (\nu_2 - \nu_1)(\nabla_s\mathbf{u})\cdot\mathbf{n}, \tag{7}$$

where $\mathbf{u}_t = (I - \mathbf{nn})\cdot\mathbf{u}$ is the tangential velocity component.

2.2. Numerical Method

The numerical integration in time of the mathematical model that is described in Section 2.1 is performed using a second order accurate implicit method, and the integral form of the fluid flow equations are discretized in space using a second order accurate cell-centred unstructured finite volume method. Detailed information of the computational domain discretization, of the mathematical model, and interface tracking procedure can be found in Tuković and Jasak [38], being out of the scope of this work. Here, we devote our attention to the improvements performed in the numerical algorithm related to both efficiency and robustness of the calculations. For this purpose, the consistent non-iterative PISO algorithm [39–41] was employed to assure the pressure–velocity coupling in the calculation of the free surface flow studied in this work, and a least-squares volume-to-point interpolation method to compute the interface motion, were newly-implemented in the two-phase fluid-flow solver with a sharp interface.

2.2.1. Consistent Non-Iterative PISO Algorithm

The rate of convergence of the collocated PISO algorithm is known to be problem dependent. In this segregated algorithm, a velocity correction term is neglected, which affects the path to convergence or may either cause the divergence of the numerical simulation, as a result of an exaggerated pressure correction field. Nevertheless, a common remedy for alleviating this problem is to under relax the pressure field. However, the rate of convergence remains a problem. In this work, we modified the original two-phase flow interface tracking solver, which is based in the segregated PISO algorithm, by approximating the velocity correction at the main grid point by a weighted average of the velocity corrections at the neighboring locations, the so-called consistent PISO algorithm [39–41].

A brief summary of the collocated PISO algorithm is described hereafter, along with the modification performed to the original formulation, by using the consistent counterpart of the algorithm, to improve the convergence rate of the calculations. A detailed analysis of both algorithms can be found in Van Doormaal et al. [39], Issa [40] and Tuković et al. [41]. First, the discretized momentum equations are given by

$$\mathbf{u}_C + \mathbf{H}_C[\mathbf{u}] = -\mathbf{D}_C^{\mathbf{u}}(\nabla P)_C + \mathbf{B}_C^{\mathbf{u}}, \tag{8}$$

with

$$\mathbf{H}_C[\mathbf{u}] = \sum_{f=NB(C)} \frac{a_F^{\mathbf{u}}}{a_C^{\mathbf{u}}}\mathbf{u}_F, \tag{9}$$

which is a weighted average that consists of the contribution of the neighbor cell with centroid F, $a_F^{\mathbf{u}}$, and the contribution of the current cell with centroid C, $a_C^{\mathbf{u}}$, to the velocity of the neighbour cells $\mathbf{u}_F$. Notice that f refers to a face of the current cell, which shares it with a neighbor cell. The transient term contribution $\mathbf{D}_C^{\mathbf{u}}$ and the source term contribution $\mathbf{B}_C^{\mathbf{u}}$ are defined as the vector operators:

$$\mathbf{D}_C^{\mathbf{u}} = \frac{V_C}{a_C^{\mathbf{u}}}, \tag{10}$$

$$\mathbf{B}_C^{\mathbf{u}} = \frac{b_C^{\mathbf{u}}}{a_C^{\mathbf{u}}}, \tag{11}$$

where V_C is the volume of current cell C. Equation (8) is solved to obtain a momentum conserving velocity field $\mathbf{u}^*$. Subsequently, the mass flow rate, $\dot{m}_f^*$, at the computational element faces should be updated using the Rhie–Chow [44] interpolation,

$$\dot{m}_f^* = \mathbf{u}_f^* \cdot \mathbf{S}_f = \overline{\mathbf{u}_f^*} \cdot \mathbf{S}_f - \overline{\mathbf{D}_f^{\mathbf{u}}}\left(\nabla P_f^{(n)} - \overline{\nabla P_f^{(n)}}\right) \cdot \mathbf{S}_f, \tag{12}$$

which allows for obtaining a momentum satisfying the mass flow rate $\dot{m}^*$. Here, $\mathbf{S}_f$ denotes the normal face area vector and all of the values with an over bar are obtained by linear interpolation between the values at points C and F. Subsequently, we assemble the pressure correction equation [40],

$$\sum_{f=nb(C)} \left(\overline{\mathbf{D}_f^{\mathbf{u}}} (\nabla P')_f \cdot \mathbf{S}_f \right) = - \sum_{f=nb(C)} \dot{m}_f^* + \sum_{f=nb(C)} \left(\overline{\sum_{F=NB(C)} \frac{a_F^{\mathbf{u}}}{a_C^{\mathbf{u}}} \mathbf{u}'_F} \right) \cdot \mathbf{S}_f, \tag{13}$$

and solve it to obtain a pressure correction field P'. In the original PISO algorithm, the last term in the RHS of Equation (13) is neglected, which affects the convergence rate, because, the larger this term, the higher the error present in the approximation at each iteration is. Finally, the mass flow rate at the element faces ($\dot{m}_f^{**}$) and the pressure (P_C^*) and velocity ($\mathbf{u}_C^{**}$) at the element centroids are updated with the corrected pressure field P' by,

$$\dot{m}_f^{**} = \dot{m}_f^* + \dot{m}'_f, \; \dot{m}'_f = -\overline{\mathbf{D}_f^{\mathbf{u}}} \nabla P'_f \cdot \mathbf{S}_f, \tag{14}$$

$$\mathbf{u}_C^{**} = \mathbf{u}_C^* + \mathbf{u}'_C, \; \mathbf{u}'_C = -\overline{\mathbf{D}_C^{\mathbf{u}}} (\nabla P')_C, \tag{15}$$

$$P_C^* = P_C^{(n)} + \alpha^p P'_C \tag{16}$$

where the superscript $^{(n)}$ denotes the solution at time $t = n$ and α^p is the under relaxation factor for the pressure correction values, which increases the robustness and convergence behavior of the PISO algorithm.

Nevertheless, even when using under-relaxation factors in the PISO algorithm, the convergence rate is problem dependent. To improve the efficiency of the two-phase fluid flow calculations, we modified the original PISO algorithm by simply assuming that the velocity correction at point C is the weighted average of the corrections at the neighboring points,

$$\mathbf{u}'_C \approx \frac{\sum_{F=NB(C)} a_F^{\mathbf{u}} \mathbf{u}'_F}{\sum_{F=NB(C)} a_F^{\mathbf{u}}} \Rightarrow \sum_{F=NB(C)} a_F^{\mathbf{u}} \mathbf{u}'_F \approx \mathbf{u}'_C \sum_{F=NB(C)} a_F^{\mathbf{u}}, \tag{17}$$

which can be written as,

$$\sum_{F=NB(C)} \frac{a_F^{\mathbf{u}} \mathbf{u}'_F}{a_C^{\mathbf{u}}} \approx \mathbf{u}'_C \sum_{F=NB(C)} \frac{a_F^{\mathbf{u}}}{a_C^{\mathbf{u}}}. \tag{18}$$

Hence, the neglected term in the PISO algorithm (last term in the RHS of Equation (13)) is replaced by the approximate value that is given by Equation (18). Thus, in the consistent PISO algorithm a smaller term is discarded, which allows for obtaining more accurate velocity corrections with the momentum equations. Therefore, the convergence rate of the consistent PISO algorithm is higher than the one of the original PISO algorithm. Notice that our approach to simulate steady state free surfaces is, in fact, a time-dependent marching technique that allows separately calculating the free surface movement and the other fields at the different time steps, which reduces the number of iterations needed to converge to a steady solution [24].

A detailed analysis regarding the numerical setup for the PISO algorithm allowed for concluding that 10 outer corrector loops and three pressure–velocity correctors were needed to obtain stable and converged iterative solutions for all the cases simulated in this work. Additionally, for the PISO algorithm, the simulations only converged at maximum with a Courant number of 0.2, while using the consistent PISO algorithm the simulations were performed with a Courant number of 1.

2.2.2. Least-Squares Volume-to-Point Interpolation

The mesh deformation is calculated using the Laplace mesh motion equation with variable diffusivity [45]. The method discretizes the motion equation using the cell-centred FV method, by which vertex displacements are obtained using a reconstruction procedure, instead of the commonly employed FE discretization. Detailed information regarding this procedure can be found in Tuković and Jasak [38] and Jasak and Tuković [45].

Following the work of Tuković et al. [46], we employ the weighted least-squares method and linear fitting function to reconstruct the vertex displacements from the cell-centre displacements of the cells surrounding the vertex, which increased the robustness of the free-surface flows calculation. It is worth noting that, without the least-squares volume-to-point interpolation, the original mesh motion algorithm available in OpenFOAM [37] was not able to compute the mesh deformation that occurs in the simulations of the extrudate swell flows presented in Section 3.

Consider the interpolation stencils for the internal (i) and the boundary (b) vertices that are given in Figure 2. The former is constituted of all cells sharing the vertex, while, in the latter, the boundary faces sharing the corresponding vertex are also included into the stencil. In the vicinity of each vertex i (or boundary b), a linear interpolation function is considered:

$$\phi(\mathbf{r}) = \phi_{i0} + \mathbf{C}_i \cdot (\mathbf{r} - \mathbf{r}_{i0}), \tag{19}$$

where $\mathbf{C}_i$ is the unknown coefficient vector, and the field value ϕ_{i0} and the reference position $\mathbf{r}_{i0}$ are obtained as the weighted average of cell-centre field and positions values, respectively:

$$\phi_{i0} = \frac{\sum_{j=1}^{n} w_{ij}\phi_{ij}}{\sum_{j=1}^{n} w_{ij}}, \tag{20}$$

$$\mathbf{r}_{i0} = \frac{\sum_{j=1}^{n} w_{ij}\mathbf{r}_{ij}}{\sum_{j=1}^{n} w_{ij}}, \tag{21}$$

where ϕ_{ij} is the field value in the centre of cell j in the interpolation stencil of the vertex i, $\mathbf{r}_{ij}$ is the centre of cell j in the interpolation stencil of the vertex i, and w_{ij} is the weighting factor calculated as the inverse square distance between the position of vertex i and the centre of cell j. Finally, to obtain the unknown coefficient vector, $\mathbf{C}_i$, the weighted least-squares method is employed:

$$\mathbf{C}_i = \left[\left(\mathbf{X}^T\mathbf{W}\mathbf{X}\right)^{-1}\mathbf{X}^T\mathbf{W}\right] \cdot \Phi_i, \tag{22}$$

where $\mathbf{X}$ is the $n \times 3$ matrix whose row j is the position vector $(\mathbf{r}_{ij} - \mathbf{r}_{i0})$ of the cell j in the interpolation stencil of the vertex i, $\mathbf{W}$ is the diagonal matrix whose elements are the weighting factors for all cells in the interpolation stencil of the vertex i, and Φ_i is the vector constituted by the elements $(\phi_{ij} - \phi_{i0})$ for all cells in the interpolation stencil of the vertex i.

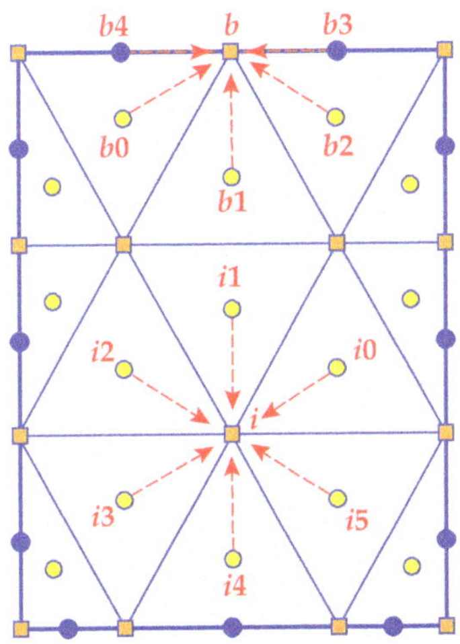

Figure 2. Finite volume mesh with an interpolation stencil for the vertex field value reconstruction. The interpolation stencil is given for the internal vertex i and boundary vertex b.

3. Results and Discussion

3.1. Planar Extrudate Swell of Newtonian Fluids

The first benchmark case study that will be discussed is the planar extrudate swell of Newtonian fluids. Figure 3 shows a schematic representation of the computational flow domain, the boundary faces, and an indicative discretization mesh for the initial time-step ($t = 0$) and at steady-state. Cartesian coordinates are employed for the description of the planar flow domain, thus $\mathbf{x} = (x, y)$. The half width of the planar channel is denoted as H, which is considered as the scaling length. The inlet plane is taken sufficiently far upstream from the exit so that the flow is fully developed with a mean velocity U. Along the axis of symmetry the standard symmetry conditions are imposed. At the solid die wall the no-slip (tangential velocity is zero) and no penetration (normal velocity is zero) conditions are imposed. At the free-surface the kinematic, Equation (4), and dynamic conditions, Equations (6) and (7), are imposed. Finally, the outflow plane is taken sufficiently far downstream from the exit, so that the flow is uniform. The die exit of the planar domain is located at $x = 5H$ from the inlet, and the outflow is located at $x = 25H$ from the die exit.

In this section, we compare the results that were obtained with the newly-improved interface tracking algorithm with those given by Tanner [1], Georgiou and Boudouvis [27] and Mitsoulis et al. [28]. First, a mesh convergence sensitivity analysis is performed. Table 1 summarizes the main characteristics of the meshes employed in this preliminary study. The rectangular domain was initially discretized with 320 cells in the streamwise (x), 37 cells in wall normal (y), and one cell in span-wise directions, with this mesh being named M1. A linear stretch was employed in streamwise and wall normal directions to have the highest resolution at downstream edge of the die (at S) with the largest to smallest cell aspect ratio being equal to 2.5. An extensive investigation was carried out on the necessary grid spacing aspect ratio, and it was concluded that this value was efficient enough to track the free-surface. Five different meshes, designated $M1$, $M2$, $M3$, $M4$, and $M5$, were found to be sufficient to obtain accurate results for this problem. Table 1 shows the number of mesh elements used in each direction of the computational domain, the total number of mesh elements, and the degree of freedom (dof) of the numerical algorithm for each level of mesh refinement. Notice that each mesh is obtained from the preceding one by multiplying the number of cells in each direction by a factor of 1.5.

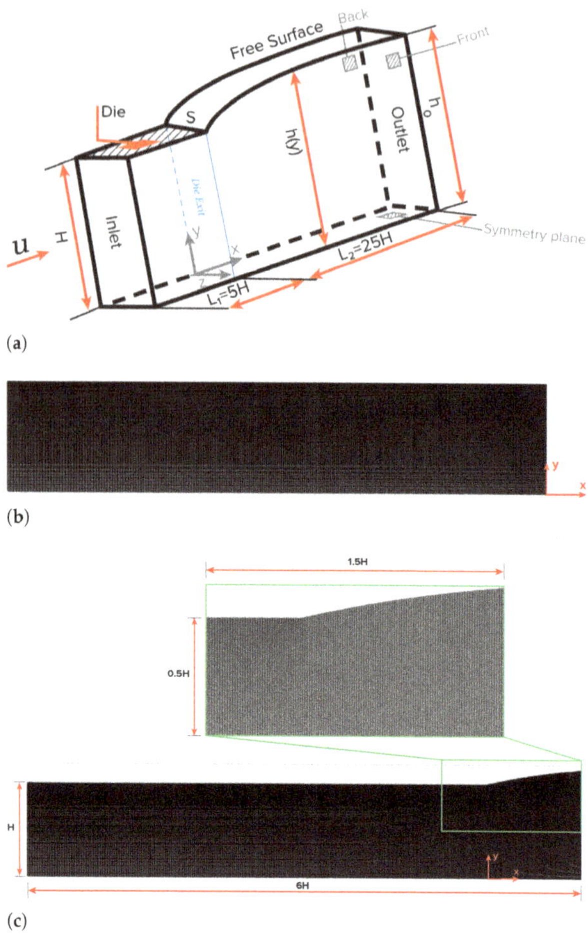

Figure 3. Schematic representation of the planar extrudate swell domain geometry and boundary faces (**a**), of an indicative discretization mesh at the initial time-step $t = 0$ (**b**), and at steady-state (**c**).

Table 1. Finite volume mesh characteristics used in the simulations and Newtonian base results for the extrudate swell ratio, χ, at $Re = 0.1$ while using the planar extrudate swell domain geometry.

Mesh	$N_x \times N_y$	**No. of Elements**	**No. of Dof**	χ	**Error (%)**
M1	320 × 37	11,840	59,200	1.214	2.3
M2	480 × 56	26,880	134,400	1.207	1.7
M3	720 × 84	60,480	302,400	1.201	1.2
M4	1080 × 126	136,080	680,400	1.198	0.9
M5	1620 × 189	306,180	1,530,900	1.195	0.7
Extrapolated	-	-	-	1.187	
Mitsoulis et al. [28]	-	-	-	1.186	
Tanner [1]	-	-	-	1.190 ± 0.002	
Georgiou and Boudouvis [27]	-	-	-	1.186	

First, we compare our mesh convergence predictions of the steady-state swell ratio, χ, with the results found in the scientific literature [1,27,28]. The swell ratio is defined as the height of the free-surface away from the die exit, where the plug flow has been established, divided by the die height, i.e., $\chi = h_0/H$. For the mesh convergence analysis, we employed the Newtonian fluid at creeping flow conditions ($Re = 0.1$). Table 1 shows the extrudate swell ratio for the different mesh resolutions. The accuracy of the developed code is estimated via the application of Richardson's extrapolation [47] to the limit, by using the three most refined levels of mesh discretization. The extrapolated value of the swell ratio is equal to $\chi = 1.187$, similar to the reference data of Mitsoulis et al. [28], which reported a value of $\chi = 1.186$ using a FEM numerical algorithm, to the estimated extrapolated value of Tanner [1] with $\chi = 1.190 \pm 0.002$, and to the converged results obtained by Georgiou and Boudouvis [27] of $\chi = 1.186$, as shown in Table 1. Additionally, notice that the relative error calculated between the extrudate swell ratio obtained using M5 (the most refined mesh) and the one extrapolated using the Richardson's technique [47] is equal to $\left(\chi_{M5} - \chi_{Extrapolated}\right)/\chi_{Extrapolated} \times 100 \approx 0.7\%$. Additionally, notice that, even when using M1 (the coarsest mesh), the relative error is only 2.3%. In terms of computational wall time, the simulation of the planar extrudate swell with M1 and M5 took approximately 0.94 and 24.3 hours, respectively, where all of the computations were performed in parallel using 80 processors on a computer with a 2.70 GHz Intel Xeon CPU E5-2680.

Subsequently, calculations for studying the robustness of the newly-improved interface tracking algorithm were pursued, by increasing the Reynolds number ($Re \leq 10$) in the planar extrudate swell domain geometry. Figure 4 depicts the extrudate swell ratio, χ, as a function of the Re number, where the blue square symbols are the results obtained by the newly-improved interface tracking code using M5, and the dashed lines and red and green square symbols are the results found in the scientific literature [1,27,28]. As can be seen, the numerical results obtained are in very good agreement with the reference data.

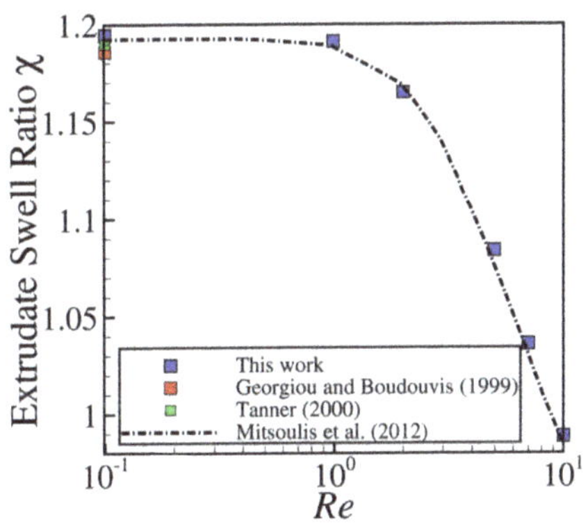

Figure 4. Steady state extrudate swell ratio χ for the simulations using the planar extrudate swell domain geometry of Newtonian fluids at $Re = \{0.1, 1, 2, 5, 7, 10\}$. Dashed lines, and red and green square symbols represent the results obtained by Mitsoulis et al. [28], Georgiou and Boudouvis [27] and Tanner [1], respectively, and the blue square symbols represent the results obtained by the newly-improved interface tracking code.

Figure 5 shows the transient evolution of χ for the planar extrudate swell domain geometry at $Re = \{0.1, 1, 2, 5, 7, 10\}$. The numerical results obtained for $Re < 7$ show an undershoot in the values of χ, before reaching the steady-state value. For $Re \geq 7$, after the undershoot in the values of χ, an overshoot is present, and, ultimately, the steady-state value is reached, being approximately 0.99. This increase of inertia allowed for verifying the robustness of the improved interface-tracking algorithm, namely the Laplacian solver with least-squares volume-to-point interpolation, to handle abrupt changes in the mesh motion.

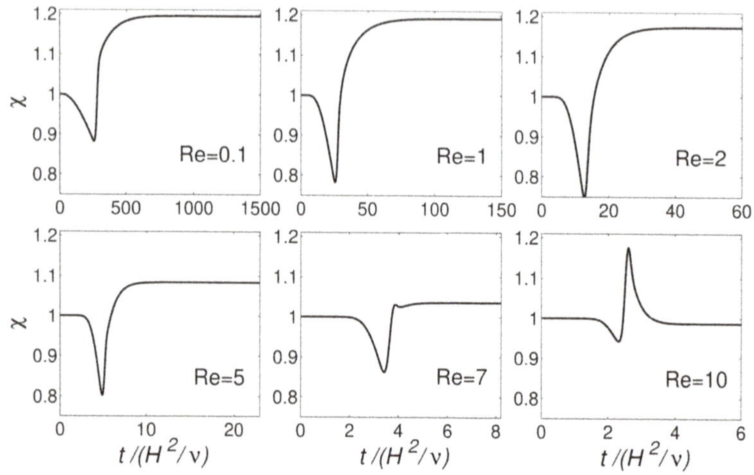

Figure 5. Transient evolution of extrudate swell ratio (χ) with dimensionless time ($t/(H^2/\nu)$ for the simulation of the planar extrudate swell domain geometry of Newtonian fluids at $Re = \{0.1, 1, 2, 5, 7, 10\}$.

The steady-state results for the primary field variables, magnitude of the velocity vector $|\mathbf{u}|$, and pressure field p, in the form of contours, are shown in Figures 6 and 7, respectively, for the planar extrudate swell domain geometry, and at different Re numbers. From a detailed inspection of Figures 6 and 7, we can see that inertia substantially reduces the free-surface height, as already shown in Figure 4. This fact occurs, because inertia pushes the material to the center of the domain, generating a peak of negative pressure in the upper point near the die exit. This result was expected to occur due to the Poiseuille flow that developed in the upstream region of the domain geometry [48]. Additionally, for $Re = 10$, the contour of the magnitude of the velocity vector changes its behavior, when compared to the lower Re number cases. It can be seen that the maximum values of the velocity vector magnitude are extended after the die exit for the higher Re case. Finally, regarding the pressure contours, we can see that the increase of inertia does not change abruptly the pressure contours, and, at $Re \geq 7$, we see an extension of the minimum pressure values from the top corner of the die exit to the center of the channel, which seems to cause the reduction in the free-surface height.

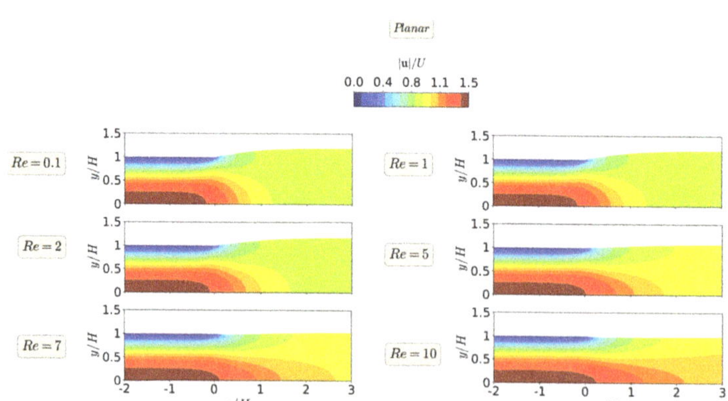

Figure 6. Steady state velocity vector magnitude contours for the planar extrudate swell flow of Newtonian fluids, at $Re = 0.1$ (**top**), 1, 2, 5, 7, and $Re = 10$ (**bottom**).

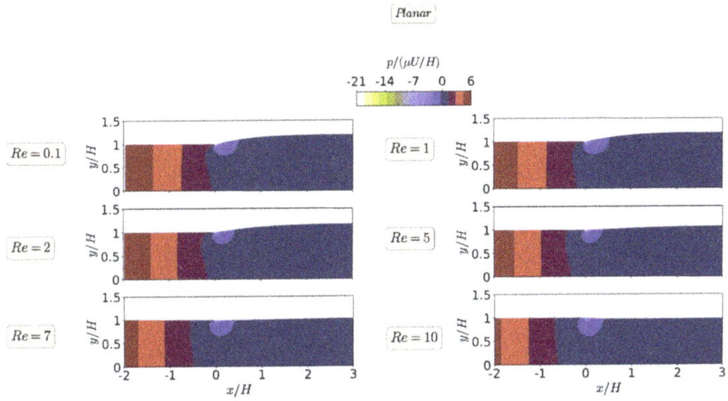

Figure 7. Steady state pressure contours for the planar extrudate swell flow of Newtonian fluids, at $Re = 0.1$ (**top**), 1, 2, 5, 7, and $Re = 10$ (**bottom**).

Finally, the efficiency of the newly-improved interface-tracking solver was assessed by comparing the required CPU wall time per time-step when using the original PISO algorithm or the enhanced consistent-PISO algorithm. Table 2 shows a comparison of the dimensionless time-step $\Delta t/(H^2/\nu)$ employed in the simulations of the planar extrudate swell and the elapsed time per time-step that is required by both PISO and consistent PISO algorithms for $Re = \{0.1, 1, 10\}$, while using mesh M5. For $Re = 0.1$, the consistent-PISO algorithm allows using a time-step 21.4 times higher than the one that is used by the PISO algorithm, and the CPU wall time per time-step that elapsed using the consistent-PISO algorithm is approximately half of the one taken by the PISO algorithm. For $Re = 1$ and $Re = 10$, the scenario is also favorable to the consistent-PISO algorithm to the detriment of the PISO algorithm, which allows for concluding that the enhanced correction procedure for the $p - U$ fields is advantageous for efficiently simulating the extrudate swell problem. Table 2 also shows the total calculation time in hours when using M5 and the consistent PISO algorithm. As can be seen, all of the calculations finished in less than 1.5 days.

Table 2. Comparison of the dimensionless time-step $\Delta t/(H^2/\nu)$ employed in the simulations and the CPU wall time (s) per time-step required by the PISO and consistent PISO algorithms for all *Re*, using mesh M5, for the planar extrudate swell flow of Newtonian fluids.

Re	$\Delta t/(H^2/\nu)$			**CPU Wall Time (s) Per Time-Step**			
	C-PISO	PISO	C-PISO/PISO	C-PISO	PISO	C-PISO/PISO	Total calculation time [h]
0.1	0.0685	0.0032	21.4	8	15	0.53	24.3
1	0.0069	0.0014	4.9	8	11	0.72	32.2
10	0.0007	0.0001	7.0	7	11	0.63	13.9

3.2. Axisymmetric Extrudate Swell of Newtonian Fluids

The second benchmark case study that is discussed in this work is the axisymmetric extrudate swell of Newtonian fluids. Figure 8 shows a schematic representation of the computational flow domain, the boundary faces, and an indicative discretization mesh for the initial time-step ($t = 0$) and at steady-state. Polar coordinates are employed for the description of the axisymmetric flow domain, thus $\mathbf{x} = (r, z)$. The half width of the axisymmetric channel is denoted as R, which is considered to be the scaling length. The boundary conditions imposed in the boundary faces are similar to the ones presented for the planar extrudate swell case study, with the exception that, in the axisymmetric domain, the two lateral boundary sides are considered to be `wedge` patches (i.e., the cylinder is specified as a wedge of small angle, e.g., 5° and one cell thick running along the plane of symmetry, straddling one of the coordinate planes), while, in the planar case study, they were considered as `empty` patches (i.e., this condition applies on each patch whose plane is normal to the third dimension for which no solution is required); and, instead of `symmetryPlane` at the bottom of the planar case, the `axis` of symmetry is considered as `empty` patch. The die exit of the axisymmetric domain is located at $z = 5R$ from the inlet, and the outflow is located at $z = 25R$ from the die exit.

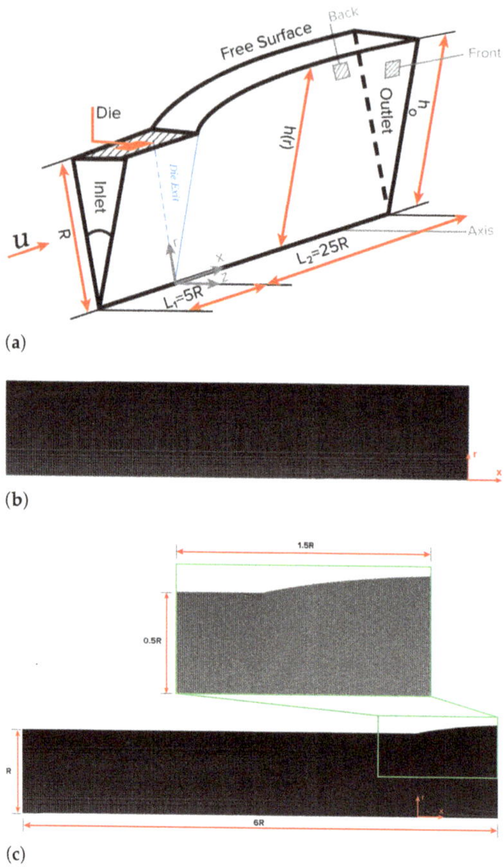

Figure 8. Schematic representation of the axisymmetric extrudate swell domain geometry and boundary faces (**a**), of an indicative discretization mesh at the initial time-step $t = 0$ (**b**), and at steady-state (**c**).

Figure 9 illustrates the extrudate swell ratio, χ, for different *Re* with mesh resolution *M*5 (see Table 1) using the axisymmetric domain geometry. Similarly to the planar domain geometry, generally, the extrudate swell ratio that is obtained by the newly-improved interface tracking solver using the axisymmetric domain is in good agreement with the reference data [1,27,28].

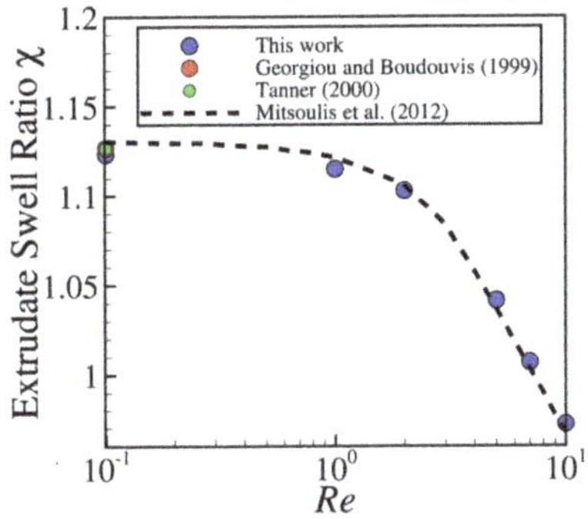

Figure 9. Steady state extrudate swell ratio χ for the simulations using the axisymmetric extrudate swell domain geometry of Newtonian fluids at $Re = \{0.1, 1, 2, 5, 7, 10\}$. Dashed lines, and red and green circle symbols represent the results that were obtained by Mitsoulis et al. [28], Georgiou and Boudouvis [27] and Tanner [1], respectively, and the blue circle symbols represent the results obtained by the newly-improved interface tracking code.

Figure 10 shows the transient evolution of χ for the axisymmetric extrudate swell domain geometry at $Re = \{0.1, 1, 10\}$. Again, the numerical results obtained for $Re < 7$ show an undershoot in the values of χ, before reaching the steady-state value. When comparing with the planar case (see Figure 5), the magnitude of the undershoot is smaller for the present case, which can be attributed to the round boundary surface of the axisymmetric domain geometry. For $Re \geq 7$, an almost imperceptible undershoot is present when using the axisymmetric domain geometry (contrarily to the planar domain geometry), followed by an overshoot, and ultimately the steady-state χ value is reached, being approximately 0.97 (2% less than the value that is obtained for the planar domain geometry).

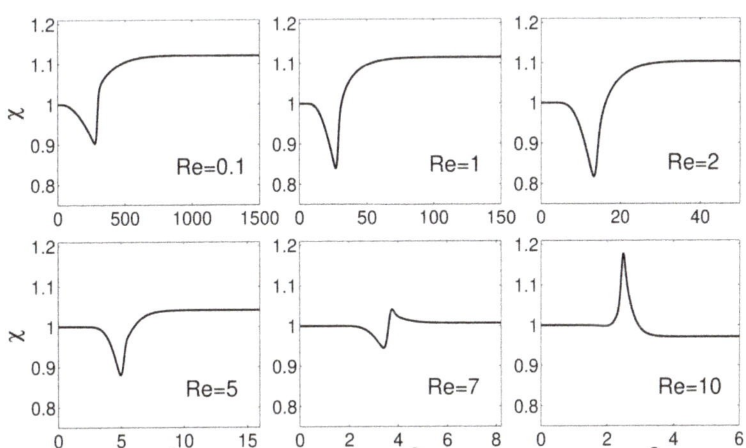

Figure 10. Transient evolution of extrudate swell ratio (χ) with dimensionless time ($t/(R^2/\nu)$) for the simulation of the axisymmetric extrudate swell domain geometry of Newtonian fluids at $Re = \{0.1, 1, 2, 5, 7, 10\}$.

The steady-state results for the primary field variables, magnitude of the velocity vector $|\mathbf{u}|$, and pressure field p, in the form of contours are shown in Figures 11 and 12, respectively, for the axisymmetric extrudate swell domain geometry and at different Re numbers. From a detailed inspection of Figures 11 and 12, we can also see here that inertia substantially reduces the free-surface height, as already shown in Figure 9. Additionally, when compared to the lower *Re* number cases, we can see that, only for $Re = 10$, the contour of the magnitude of the velocity vector changes its behavior, where the maximum values are extended until the die exit for the latter case. Again, the increase of inertia, does not change abruptly the pressure contours, and only at $Re = 10$, we see an extension of the minimum pressure contours from the top corner of the die exit to the center of the channels, which seems to cause the reduction in the free-surface height (but with a lesser effect in the axisymmetric domain geometry).

When comparing the contour results from both planar and axisymmetric extrudate swell domain geometries, we can state that the maximum velocity vector magnitude is obtained for the axisymmetric domain, being two times the magnitude of the inflow average velocity, in contrast with the planar case, where it is only 1.5 times higher than the magnitude of the inflow average velocity. This result was expected to occur due to the Poiseuille flow, which is developed in the upstream channel of both domains [48]. Finally, regarding the pressure contours, we can see that, for the axisymmetric domain, the computed maximum and minimum pressure values are symmetric, as expected, due to domain symmetry, in contrast with the planar domain case, where the minimum pressure value that is obtained is 3.5 times lesser than the maximum one. In both planar and axisymmetric cases, as the *Re* increases the extrudate swell ratio and pressure decreases, which physically states that the inertia forces stretch out the material and prevent the swelling, with an immediate effect on the reduction of pressure losses [28].

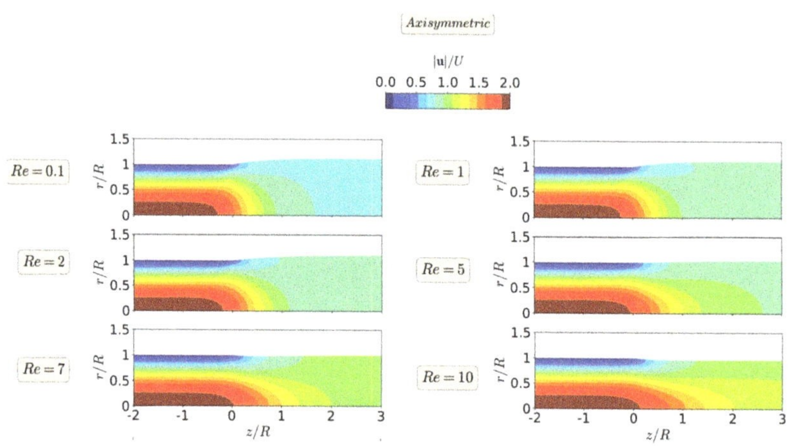

Figure 11. Steady state velocity magnitude contour for axisymmetric swell flow of Newtonian fluids, at $Re = 0.1$ (**top**), 1, 2, 5, 7 and $Re = 10$ (**bottom**).

Finally, Table 3 shows a comparison of the dimensionless time-step $\Delta t/(H^2/\nu)$ employed in the simulations of the axisymmetric extrudate swell and the CPU wall time (s) per time-step required by both PISO and consistent PISO algorithms for $Re = \{0.1, 1, 10\}$ using mesh M5. For $Re = 0.1$, the consistent-PISO algorithm allows for using a time-step 21.0 times higher than the one used by the PISO algorithm, and the CPU wall time per time-step elapsed using the consistent-PISO algorithm is 0.58 times the one taken by the PISO algorithm. Again, for $Re = 1$ and $Re = 10$, the scenario is also favorable to the consistent-PISO algorithm in detriment of PISO algorithm. Additionally, notice that, when comparing with the values that are shown in Table 2, the ratios of CPU wall time per time-step (C-PISO/PISO) are lower for the planar extrudate swell, which mean that the CPU wall time gains using the C-PISO algorithm seems to be higher for non-smooth geometries. Table 2 also shows the total calculation time in hours when using M5 and the consistent PISO algorithm. As for the planar extrudate swell, all of the calculations finished in less than 1.5 days.

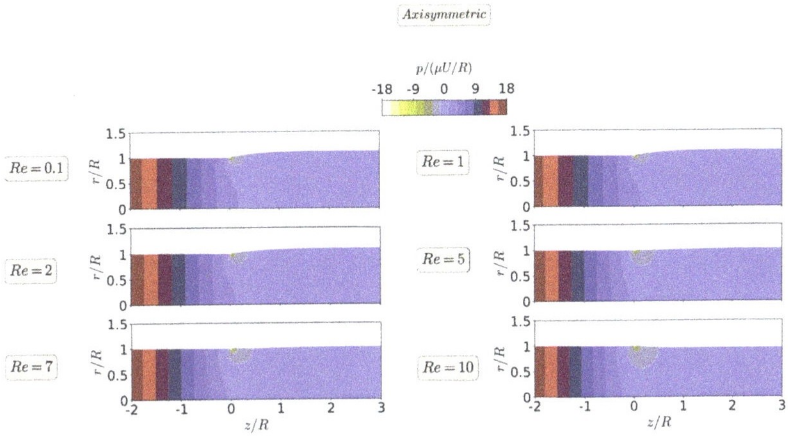

Figure 12. Steady state pressure contour for axisymmetric extrudate swell flow of Newtonian fluids, at $Re = 0.1$ (**top**), 1, 2, 5, 7, and $Re = 10$ (**bottom**).

Table 3. A comparison of the dimensionless time-step $\Delta t/(R^2/\nu)$ employed in the simulations and the CPU wall time (s) per time-step required by the PISO and consistent PISO algorithms for all *Re*, using mesh M5, for the axisymmetric extrudate swell of Newtonian fluids.

Re	$\Delta t/(R^2/\nu)$			CPU Wall Time (s) Per Time-Step			
	C-PISO	PISO	C-PISO/PISO	C-PISO	PISO	C-PISO/PISO	Total calculation time [h]
0.1	0.0652	0.0031	21.0	7	12	0.58	22.4
1	0.0064	0.0013	4.9	7	9	0.78	30.4
10	0.0006	0.0001	6	5	9	0.56	11.6

4. Conclusions

A numerical formulation for efficient moving mesh interface tracking simulations of free-surface flows was presented and implemented using the finite-volume method. The implementation was performed in the open-source OpenFOAM framework [37], where the interface is tracked in a semi-implicit manner inside the consistent second-order time-accurate non-iterative Pressure-Implicit with Splitting of Operators (consistent PISO) algorithm for the numerical solution of incompressible fluid flows. Additionally, the moving mesh was adjusted to the time varying shape of the interface, using a Laplacian scheme with a least-squares volume-to-point interpolation, which allowed for robust and stable deformations of the interface.

The improved algorithm was assessed in terms of the accuracy and efficiency for the fluid flow simulations in planar and axisymmetric extrudate swell of Newtonian fluids. A mesh sensitivity analysis allowed for obtaining a grid refinement level from which the calculated extrudate swell ratio differs from the extrapolated value of 0.7%. Subsequently, the robustness of the numerical algorithm was pursued, increasing the Reynolds number from $Re = 0.1$ to $Re = 10$. The extrudate swell ratio that was obtained for both domains compared well with the results found in the scientific literature for that range of *Re*. Additionally, the contours for the magnitude of the velocity vector and pressure fields are also shown, and a detailed study of the contours reveals that the obtained results are physically meaningful. Finally, the efficiency of the improved numerical solver was evaluated by comparing the CPU wall time (s) per time-step for both the PISO and consistent PISO algorithms. The results obtained allowed for concluding that the consistent-PISO is, at maximum, 47% and 42% faster than the PISO algorithm for the planar and axisymmetric extrudate swell domain geometries, respectively.

In summary, the results presented here show that the newly-improved interface tracking code, developed using an open-source framework, can accurately and efficiently predict the Newtonian extrudate swell. The code that was implemented here is being currently extended to handle viscoelastic fluid flow calculations and non-isothermal processes. For the viscoelastic fluid flow calculations, we will use the quasi-linear Oldroyd-B and exponential PTT rheological models. The former will be used due to the numerical instabilities that are caused by the infinite polymeric stresses generated at singular points, which will verify the robustness of the numerical implementation; and the latter will be used because it is more suitable for approximating the behavior of polymer melts, where the extensibility parameter introduces elongational and shear thinning in the fluid model. Additionally, the solvent and polymeric viscosities and the relaxation time of the viscoelastic fluid will be considered as temperature dependent, by employing the Williams–Landel–Ferry relation.

Author Contributions: Conceptualization, A.F. and C.F.; methodology, A.F., C.F. and Ž.T.; software, C.F. and Ž.T.; validation, A.F., C.F. and Ž.T.; formal analysis, C.F. and Ž.T.; investigation, A.F. and C.F.; resources, C.F. and Ž.T.; writing—original draft preparation, A.F. and C.F.; writing—review and editing, A.F., O.S.C. and C.F.; visualization, A.F., Ž.T., O.S.C. and C.F.; supervision, Ž.T., O.S.C. and C.F.; project administration, O.S.C.; funding acquisition, O.S.C. All authors have read and agreed to the published version of the manuscript.

Funding: This research was funded by FEDER through the COMPETE 2020 Programme and National Funds through FCT—Portuguese Foundation for Science and Technology under the projects UIDB/05256/2020 and UIDP/05256/2020, TSSiPRO—Technologies for Sustainable and Smart Innovative Products (NORTE-01-0145-FEDER-000015) and FAMEST—Footwear, Advanced Materials, Equipment's and Software Technologies (POCI-01-0247-FEDER-024529).

Institutional Review Board Statement: Not applicable.

Informed Consent Statement: Not applicable.

Data Availability Statement: Not applicable.

Acknowledgments: The authors acknowledge the support of the computational clusters Search-ON2 (NORTE-07-0162-FEDER-000086) the HPC infrastructure of UMinho under NSRF through ERDF; and FCT I.P. through the Advanced Computing Project CPCA/A00/6057/2020 using the Minho Advanced Computing Center (MACC).

Conflicts of Interest: The authors declare no conflict of interest.

References

1. Tanner, R.I. *Engineering Rheology*, 2nd ed.; Oxford University Press: Oxford, UK, 2000.
2. Tadmor, Z.; Gogos, C.G. *Principles of Polymer Processing, SPE Monograph Series*; Wiley: New York, NY, USA, 1979.
3. Mirjalili, S.; Jain, S.S.; Dodd, M. Interface-capturing methods for two-phase flows: An overview and recent developments. *Cent. Turbul. Res. Annu. Res. Briefs* **2017**, *2017*, 117–135.
4. Ferziger, J.H.; Peric, M. *Computational Methods for Multiphase Flow*; Cambridge University Press: Cambridge, UK, 2009.
5. Ferziger, J.H.; Peric, M. *Computational Methods for Fluid Dynamics*, 3rd ed.; Springer: Berlin, Germany, 2002.
6. Harlaw, F.H.; Welch, J.E. Numerical calculation of time-dependent viscous incompressible flow of fluid with free-surface. *Phys. Fluids* **1965**, *8*, 2182. [CrossRef]
7. McKee, S.; Tomé, M.F.; Cuminato, J.A.; Castelo, A.; Ferreira, V.G. Recent Advances in the Marker and Cell Method. *Arch. Comput. Meth. Eng.* **2004**, *11*, 107–142. [CrossRef]
8. McKee, S.; Tomé, M.F.; Ferreira, V.G.; Cuminato, J.A.; Castelo, A.; Sousa, F.S.; Mangiavacchi, N. The MAC method. *Comput. Fluids* **2008**, *37*, 907–930. [CrossRef]
9. Sussman, M.; Smereka, P. Axisymmetric free boundary problems. *J. Fluid Mech.* **1997**, *341*, 269–294. [CrossRef]
10. Hirt, C.W.; Nichols, B.D. Volume of fluid (VOF) method for the dynamics of free boundaries. *J. Comput. Phys.* **1981**, *39*, 201–225. [CrossRef]
11. Sussman, M.; Fatemi, E. An efficient interface-preserving level set redistancing algorithm and its application to interfacial incompressible fluid flow. *J. Sci. Comput.* **1999**, *20*, 1165–1191. [CrossRef]
12. Sussman, M.; Puckett, E.G. A coupled level set and volume-Of-fluid method for Computing 3D and axisymmetric incompressible Two-Phase Flows. *J. Comput. Phys.* **2000**, *162*, 301–337. [CrossRef]
13. Gueyffier, D.; Li, J.; Scardovelli, R.; Zaleski, S. Volume-of-fluid interface tracking with smoothed surface stress methods for three-dimensional flows. *J. Comput. Phys.* **1999**, *152*, 423–456. [CrossRef]

14. Queutey, P.; Visonneau, M. An interface capturing method for free-surface hydrodynamic flows. *Comput. Fluids* **2007**, *36*, 1481–1510. [CrossRef]
15. Roenby, J.; Bredmose, H.; Jasak, A computational method for sharp interface advection. *R. Soc. Open Sci.* **1988**, *3*, 160405. [CrossRef]
16. Osher, S.; Sethian, A. Fronts propagating with curvature dependent speed: algorithms based on Hamilton-Jacobi formulations. *J. Comput. Phys.* **1988**, *79*, 12–49. [CrossRef]
17. Losasso, F.; Fedkiw, R.; Osher, S. Spatially adaptive techniques for level set methods and incompressible flow. *Comput. Fluids* **2006**, *35*, 995–1010. [CrossRef]
18. Sethian, J.A. *Level Set Methods and Fast Marching Methods Evolving Interfaces in Computational Geometry, Fluid Mechanics, Computer Vision, and Materials Science*; Cambridge University Press: Cambridge, UK, 1996.
19. Tryggvason, G.; Unverdi, S.O. Computations of three-dimensional Rayleigh-Taylor instability. *Phys. Fluids* **1990**, *2*, 656–659. [CrossRef]
20. Unverdi, S.O.; Tryggvason, G. A front-tracking method for viscous incompressible, multi-fluids flows. *J. Comput. Phys.* **1992**, *100*, 25–37. [CrossRef]
21. Crochet, M.J.; Keunings, R. On numerical die swell calculation. *J. Non-Newton. Fluid Mech.* **1982**, *10*, 85–94. [CrossRef]
22. Mitsoulis, E.; Vlachopoulos, J.; Mirza, F.A. Simulation of extrudate swell from long slit and capillary dies. *Polym. Process. Eng.* **1984**, *2*, 153–177.
23. Karagiannis, A.; Hrymak, A.N.; Vlachopoulos, J. Three-dimensional extrudate swell of creeping Newtonian jets. *AIChE J.* **1988**, *34*, 2088–2094. [CrossRef]
24. Wambersie, O.; Crochet, M.J. Transient finite element method for calculating steady state three-dimensional free surfaces. *Int. J. Numer. Methods Fluids* **1992**, *14*, 343–360. [CrossRef]
25. Legat, V.; Marchal, J-M. Prediction of three-dimensional general shape extrudates by an implicit iterative scheme. *Int. J. Numer. Methods Fluids* **1992**, *14*, 609–625. [CrossRef]
26. Legat, V.; Marchal, J-M. Die design: An implicit formulation for the inverse problem. *Int. J. Numer. Methods Fluids* **1993**, *16*, 29–42. [CrossRef]
27. Georgiou, G.C.; Boudouvis, A.G. Converged solutions of the Newtonian extrudate-swell problem. *Int. J. Numer. Methods Fluids* **1999**, *29*, 363–371. [CrossRef]
28. Mitsoulis, E.; Georgiou, G.C.; Kountouriotis, Z. A study of various factors affecting Newtonian extrudate swell. *Comput. Fluids* **2012**, *57*, 195–207. [CrossRef]
29. Spanjaards, M.M.A.; Hulsen, M.A.; Anderson, P.D. Transient 3D finite element method for predicting extrudate swell of domain containing sharp edges. *J. Non-Newton. Fluid Mech.* **2019**, *270*, 79–95. [CrossRef]
30. Spanjaards, M.M.A.; Hulsen, M.A.; Anderson, P.D. Computational analysis of the extrudate shape of three-dimensional viscoelastic, non-isothermal extrusion flows. *J. Non-Newton. Fluid Mech.* **2020**, *282*, 104310. [CrossRef]
31. Vlachopoulos, J. Extrudate Swell in Polymer Rheology. *Rev. Def. Beh. Mat.* **1981**, *3*, 219–248.
32. Vlachopoulos, J.; Polychronopoulos, N. Basic Concepts in Polymer Melt Rheology and Their Importance in Processing. In *Applied Polymer Rheology: Polymeric Fluids with Industrial Applications*; Kontopoulou M., Ed.; John Wiley & Sons: Hoboken, NJ, USA, 2012.
33. Hamielec, L.A.; Vlachopoulos J. Influence of Long Chain Branching on Extrudate Swell of Low-Density Polyethylenes. *J. Appl. Polym. Sci.* **1983**, *28*, 2389–2392. [CrossRef]
34. Vlachopoulos, J.; Horie M.; Lidorikis S. An Evaluation of Expressions Predicting Die Swell. *J. Rheol.* **1972**, *16*, 669. [CrossRef]
35. Tanner, R.I. A Theory of Die-Swell. *J. Polym. Sci. Part A-2 Polym. Phys.* **1970**, *8*, 2067. [CrossRef]
36. Tanner, R.I. A Theory of Die-Swell Revisited. *J. Non-Newt. Fluid Mech.* **2005**, *129*, 85–87. [CrossRef]
37. OpenCFD. *OpenFOAM—The Open Source CFD Toolbox—User's Guide*; OpenCFD Ltd.: London, UK, 2007.
38. Tuković, Ž.; Jasak, H. A moving mesh finite volume interface tracking method for surface tension dominated interfacial fluid flow. *Comput. Fluids* **2012**, *55*, 70–84. [CrossRef]
39. Van Doormaal, J.P.; Raithby, G.D. Enhancement of the SIMPLE method for predicting incompressible fluid flows. *Numer. Heat Transf.* **1984**, *7*, 147–163. [CrossRef]
40. Issa, R.I. Solution of the implicitly discretised fluid flow equations by operator-splitting. *J. Comput. Phys.* **1986**, *62*, 40–65. [CrossRef]
41. Tuković, Ž.; Perić, M.; Jasak, H. Consistent second-order time-accurate non-iterative PISO-algorithm. *Comput. Fluids* **2018**, *166*, 78–85. [CrossRef]
42. Demirdžić, I.; Perić, M. Space conservation law in finite volume calculations of fluid flow. *Int. J. Numer. Methods Fluids* **1988**, *8*, 1037–1050. [CrossRef]
43. Tuković Z.; Jasak, H. Simulation of free-rising bubble with soluble surfactant using moving mesh finite volume/area method. In Proceedings of the 6th International Conference on CFD in Oil & Gas, Metallurgical and Process Industries, Trondheim, Norway, 10–12 June 2008.
44. Rhie, C.M.; Chow, W.L. Numerical study of turbulent flow past an isolated airfol with trailing edge separation. *AIAA J.* **1983**, *21*, 1525–1532. [CrossRef]
45. Jasak, H.; Tuković, Ž. Automatic mesh motion for the unstructured finite volume method. *Trans. FAMENA* **2006**, *30*, 1–20.

46. Tuković, Ž.; Karač, A.; Cardiff, P.; Jasak, H. OpenFOAM Finite Volume Solver for Fluid-Solid Interaction. *Trans. FAMENA* **2018**, *42*, 1–31. [CrossRef]
47. Celik, I.; Ghia, U.; Freitas, C.J.; Coleman, H.; Raad, P.E. Procedure for Estimation and Reporting of Uncertainty Due to Discretization in CFD Applications. *J. Fluids Eng.* **2008**, *130*, 1–4. [CrossRef]
48. Bird, R.B.; Armstrong, R.C.; Hassager, O. *Dynamics of Polymeric Liquids*; Wiley-Interscience 2nd Ed.; Wiley: New York, NY, USA, 1987; Volume 1.

polymers

MDPI

Article

Thermo-Mechanical Behavior of Poly(ether ether ketone): Experiments and Modeling

A. D. Drozdov * and J. deClaville Christiansen

Department of Materials and Production, Aalborg University, Fibigerstraede 16, 9220 Aalborg, Denmark; jc@mp.aau.dk
* Correspondence: aleksey@m-tech.aau.dk

Abstract: Observations are reported on poly(ether ether ketone) (PEEK) in uniaxial tensile tests, relaxation tests and creep tests with various stresses in a wide interval of temperatures ranging from room temperature to 180 °C. Constitutive equations are developed for the thermo–mechanical behavior of PEEK under uniaxial deformation. Adjustable parameters in the governing equations are found by matching the experimental data. Good agreement is demonstrated between the observations and results of numerical simulation. It is shown that the activation energies for the elastoplastic, viscoelastic and viscoelastoplastic responses adopt similar values at temperatures above the glass transition point.

Keywords: poly(ether ether ketone); thermo–mechanical response; constitutive modeling

Citation: Drozdov, A.D.; deClaville Christiansen, J. Thermo-Mechanical Behavior of Poly(ether ether ketone): Experiments and Modeling. *Polymers* **2021**, *13*, 1779. https://doi.org/10.3390/polym13111779

Academic Editor: Célio Bruno Pinto Fernandes

Received: 27 April 2021
Accepted: 26 May 2021
Published: 28 May 2021

1. Introduction

Poly(ether ether ketone) (PEEK) is a semicrystalline thermoplastic homopolymer with a linear molecular structure and relatively stiff backbone chains. This polymer belongs to the family of poly(ether ketones) whose ethers functional groups are linked together through aromatic groups. PEEK is a high performance polymer that displays a unique combination of toughness, stiffness, strong abrasion resistance and tribological performance, low moisture absorption, thermo-oxidative stability, chemical and solvent resistance, biocompatibility, flame retardancy, and retention of physical properties at elevated (up to 200 °C) temperatures [1]. Due to the excellent balance of mechanical and physical properties, this polymer and composites with PEEK matrices are widely used in aerospace [2,3] and automobile industries [4], energy technologies [5,6] and biomedicine [7,8].

Due to the importance of mechanical properties for application of poly(ether ether ketone) as a load-bearing material, a number of studies have dealt with the experimental investigation of the thermo–mechanical response of PEEK. Results of DMA analysis in the temperature-sweep mode are presented in [9–14]. Observations in tests with various strain rates are reported in [10,15–19] for tensile deformation, in [10,20–22] for compressive deformation, and in [23,24] for biaxial loading. Experimental data in cyclic and fatigue tests are presented in [10,14,25–29]. Observations in the Izod impact tests are given in [13,14,30], whereas those in the Taylor impact tests are provided in [31–33]. Observations in nanoindentation tests are discussed in [34,35]. Experimental data in relaxation tests are reported in [11,36,37], and those in creep tests are presented in [11,38–40].

As PEEK is a semicrystalline polymer, its time- and rate-dependent behavior can be described by conventional models in viscoelasticity and viscoplasticity of semicrystalline polymers, see [41–43], to mention a few. Constitutive equations accounting for the peculiarities in the thermo–mechanical response of PEEK induced by stiffness of its backbone chains were developed in [18,19,22,24,29,44–46].

Previous studies on the thermo–mechanical behavior of PEEK focused on its viscoelastic and viscoplastic responses below the glass transition temperature $T_c \approx 150$ °C [24,29,44–46]. Above this temperature, only observations in tensile tests with constant strain rates were reported and analyzed in [18,19,22]. The mechanical behavior of PEEK and its micro- and nanocomposites at temperatures exceeding T_c have recently attracted substantial attention due to applications of sulfonated PEEK-based polymers as membrane materials in polymer electrolyte membrane fuel cells and direct methanol fuel cells with the interval of working temperatures up to 180 °C [47–49]. The aim of this study is to perform a thorough investigation of the mechanical behavior of PEEK both below and above its glass transition temperature. In particular, we concentrate on the activation energies for the viscoelastic and viscoplastic processes in the sub-T_g and post-T_g intervals. These characteristics allow correlations to be established between mobility of chains and the structure of polymer networks, on the one hand, and physical properties of PEEK membranes in high-temperature electrochemical cells (ionic conductivity, methanol permeability, dielectric permittivity, thermal stability, chemical resistance), on the other [50].

The objective of this paper is three-fold: (i) to analyze experimentally the thermo-mechanical response of PEEK in uniaxial tensile tests with a constant strain rate, relaxation tests with a constant strain, and creep tests with various stresses in a wide interval of temperatures from room temperature up to 180 °C, (ii) to develop constitutive equations for the thermo–elastoplastic, thermo–viscoelastic and thermo–viscoelastoplastic responses of PEEK and to find material constants in these relations by matching the observations, and (iii) to compare activation energies for the elastoplastic behavior (tensile tests with small strains), viscoelastic response (short-term relaxation tests and creep tests in the linear regime of deformation) and viscoelastoplastic behavior (creep tests with relatively large stresses above the glass transition temperature).

Unlike previous studies on modeling the time- and rate-dependent behavior of PEEK subjected to arbitrary 3D deformations with finite strains [19,24,29,44,46], we confine ourselves to the analysis of its thermo-mechanical response under uniaxial deformation with small strains. This allows the number of adjustable coefficients in the constitutive equations to be reduced noticeably (compared with conventional models). As a result, the effect of temperature on the material parameters can be determined with high accuracy. The latter is of primary importance for (i) the design of PEEK implants produced by additive manufacturing, 3D printing and fused deposition modeling (FDM) technology, and (ii) prediction of their microstructure, tribological performance and mechanical properties [51–53].

2. Materials and Methods

Poly(ether ether ketone) KETRON 1000 PEEK FKM NATUR (density 1.31 g/cm^3, tensile modulus 4.34 GPa, ultimate tensile strength 110 MPa) was supplied as extruded sheets by Vink Plast ApS (Denmark). Dumbbell specimens for tensile tests (ASTM standard D638) with length in the active zone 50 mm, width 5.1 mm, and thickness 4.5 mm were machined from the sheet. To exclude the effect of stresses developed under preparation, tests were conducted a week after machining samples.

Differential scanning calorimetry (DSC) measurements were carried out by means of STA 449/Netzsch apparatus at the heating and cooling rate of 20 K/min. Specimens with mass of about 10 mg were tested in alumina pans covered by lid under argon atmosphere. The experimental program involves: heating from the initial temperature $T_i = 104$ up to the final temperature $T_f = 400$ °C, followed by cooling to the initial temperature T_i, and re-heating up to the final temperature T_f. Experimental data are depicted in Figure 1 which shows that the DSC scans for the first and second heating coincide. Figure 1 reveals that the glass transition temperature T_g equals 151 °C, the crystallization temperature T_c is 293 °C, and the melting temperature T_m equals 339 °C. These values are in good agreement with observations in DSC tests on PEEK 90G ($T_g = 155$, $T_c = 317$, $T_m = 345$ °C) [14] and PEEK 450G ($T_g = 158$, $T_m = 341$ °C) [10], as well as with the data in DMA test on PEEK 15G ($T_g = 144$ °C) [11].

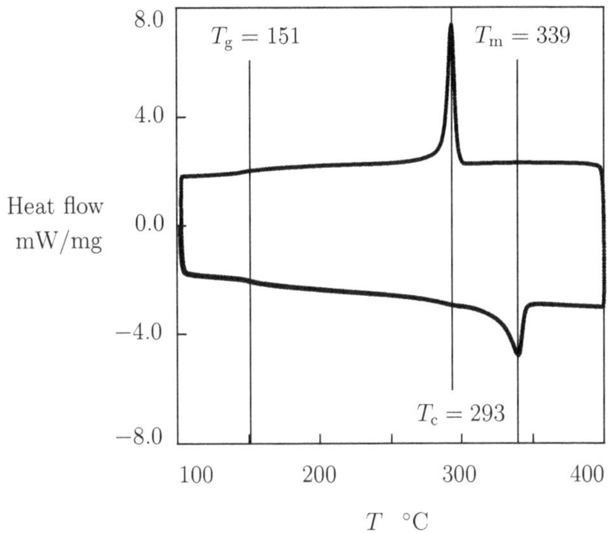

Figure 1. DSC thermogram of PEEK. Solid line: experimental data in DSC test with the heating and cooling rates of 20 K/min.

Mechanical tests were performed by means of a universal testing machine Instron–5568 equipped with a thermal chamber and an electro–mechanical sensor (Instron Static 2630113) for control of longitudinal strain in the active zone of samples. Tensile force was measured by a 50 kN load cell. The engineering stress σ was calculated as the ratio of axial force to the cross-sectional area of specimens in the undeformed state.

The experimental program included three series of tests at temperatures T ranging from room temperature to 180 °C. Each test was conducted on a virgin specimen. For each deformation program, tests were repeated three times of different samples to assess repeatability of measurements. The accuracy of measurements is estimated in Supplementary Material (Figures S1–S4) where experimental data are depicted (with their standard deviations) in selected uniaxial tensile tests, relaxation tests, and creep tests together with results of numerical analysis.

The first series involved uniaxial tensile tests with a cross-head speed of 20 mm/min (which corresponded to the strain rate $\dot{\epsilon} = 3.1 \times 10^{-3}$ s^{-1}) up to breakage of specimens. The experimental stress–strain diagrams at temperatures $T = 20, 80, 120, 130, 140, 150, 160,$ 170 and 180 °C are depicted in Figure 2, where tensile stress σ is plotted versus engineering strain ϵ. We confine ourselves to the interval $0 \le \epsilon \le 0.06$ for necking of specimens occur under stretching in the post-yield region at temperatures below the glass transition point T_g, whereas we focus on the analysis of homogeneous uniaxial deformation with small strains.

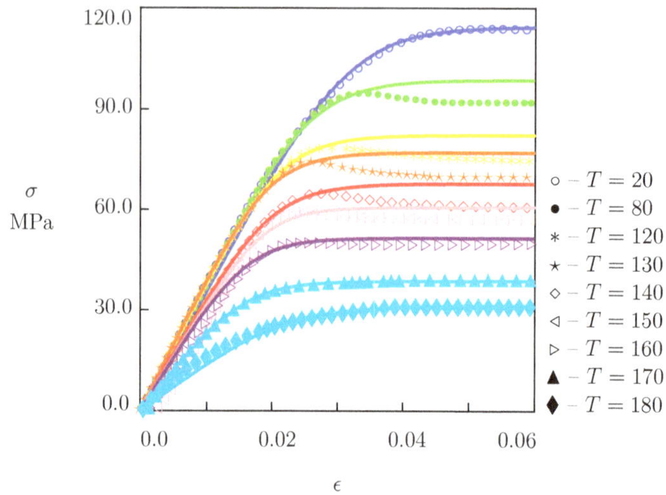

Figure 2. Stress σ versus strain ϵ. Symbols: experimental data in tensile tests with strain rate $\dot{\epsilon} = 3.1 \times 10^{-3}$ s^{-1} at temperatures T °C. Solid lines: results of simulation.

For each set of data, the maximum stress $\sigma_{\max}$ on the stress–strain curve was measured and associated with the yield stress. The effect of temperature on $\sigma_{\max}$ is illustrated in Figure 3, where $\sigma_{\max}$ is plotted versus T. With reference to [54], the data are approximated by the linear equation

$$\sigma_{\max} = \sigma_{\max 0} - \sigma_{\max 1} T, \tag{1}$$

with the coefficients calculated by the least-squares technique. Figure 3 shows good agreement between the observations and their approximation by Equation (1) with different coefficients below and above the glass transition temperature T_g.

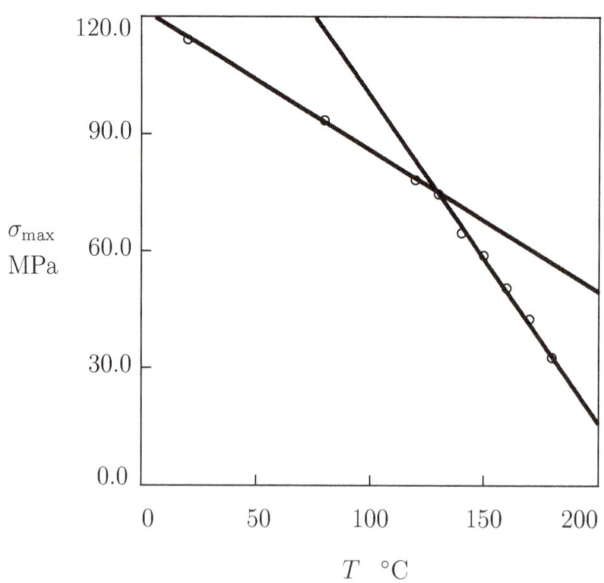

Figure 3. Tensile strength $\sigma_{\max}$ versus temperature T. Circles: experimental data in tensile tests with strain rate $\dot{\epsilon} = 3.1 \times 10^{-3}$ s^{-1}. Solid lines: approximation of the data by Equation (1).

Our findings are in accord with observations on PEEK 450G reported in [10,21], which revealed changes in slope of the dependence $\sigma_{max}(T)$ in the interval of temperatures between 135 and 140 °C.

The other series of experiments involved tensile relaxation tests with a fixed strain $\epsilon_0 = 0.01$. In each test, a specimen was stretched with a cross-head speed of 20 mm/min up to the strain ϵ_0. Afterwards, the strain was preserved constant, and the tensile stress σ was monitored as a function of time t. Following the protocol ASTM E-328, a duration of 20 min was chosen for the short-term relaxation tests. Experiments were carried out at temperatures $T = 20$, 80, 120, 130, 140, 150, 160, 170 and 180 °C. Selected relaxation curves are reported in Figure 4, where tensile stress σ is plotted versus relaxation time $t_{rel} = t - t_0$ (t_0 stands for the time needed to stretch samples up to the strain ϵ_0). Following common practice, observations are presented by means of the semi-logarithmic plots with $\log = \log_{10}$.

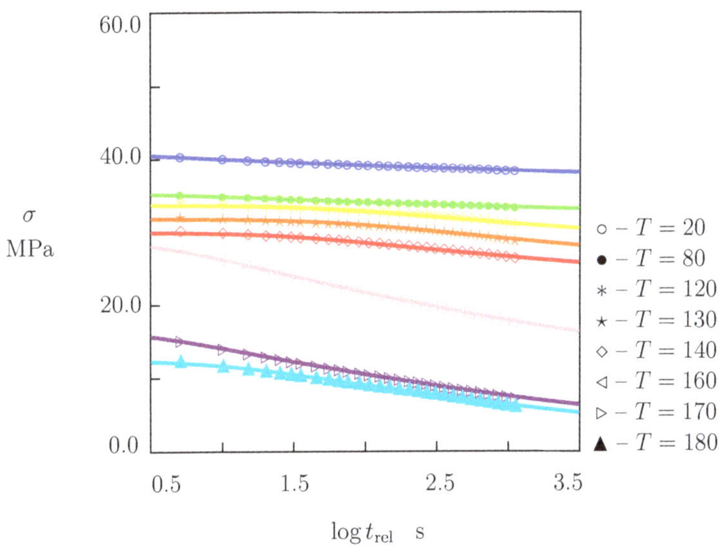

Figure 4. Stress σ versus relaxation time t_{rel}. Symbols: experimental data in tensile relaxation tests with strain $\epsilon_0 = 0.01$ at temperatures T °C. Solid lines: results of simulation.

In the last series of experiments, tensile creep tests were performed with various tensile stresses σ_0 at various temperatures T. In each test, a specimen was stretched with a cross-head speed of 20 mm/min up to the required stress σ_0. Afterwards, the stress was preserved constant, and an increase in strain ϵ was monitored a function of time t. Following the protocol ASTM D-2990, a duration of 20 min was chosen for the short-term creep tests.

Two types of creep tests were conducted. Experimental data in these tests are reported in Figures 5 and 6, where tensile strain ϵ is plotted versus creep time $t_{cr} = t - t_0$ (t_0 stands for the time needed to reach the stress σ_0 under stretching).

In experiments of the first type, tensile stresses σ_0 were chosen to be relatively low. Observations in these tests are used to validate our model in linear viscoelasticity. Creep curves in selected tests (with $\sigma_0 = 70$ MPa at $T = 20$ °C, $\sigma_0 = 40$ MPa at $T = 120$ °C and $\sigma_0 = 30$ MPa at $T = 150$ °C) are depicted in Figure 5, and those with $\sigma_0 = 10$ MPa at $T = 160$, 170 and 180 °C are presented in Figure 6.

Tests of the other type were performed with relatively large tensile stresses σ_0 to evaluate the viscoplastic flow under creep conditions at temperatures above T_g. Experimental data in these tests are reported in Figure 6, where creep diagrams are depicted at temperatures $T = 160$ °C (with $\sigma_0 = 20, 25, 30$ and 35 MPa), $T = 170$ °C (with $\sigma_0 = 20$ MPa) and $T = 180$ °C (with $\sigma_0 = 30$ MPa).

The following conclusions are drawn from Figures 2, 4–6: (i) under tension, stress σ decreases monotonically with temperature T, (ii) relaxation of stresses is negligible at temperatures below 130 °C and becomes noticeable at temperatures above T_g, (iii) creep flow below the glass transition temperature is weak (an increase in ϵ in short-term creep tests does not exceed 0.5%), while this flow becomes pronounced in tests with relatively large stresses above T_g (tensile strain grows by several times).

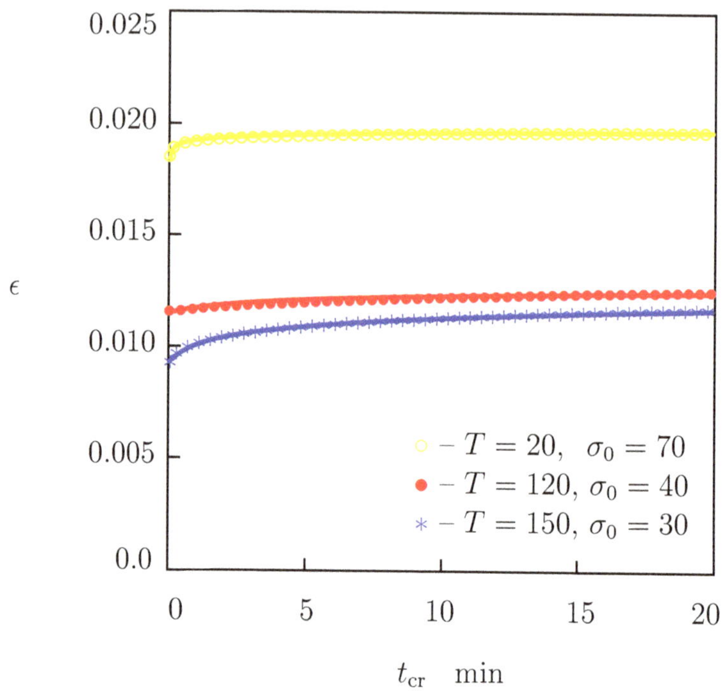

Figure 5. Strain ϵ versus creep time t_{cr}. Symbols: experimental data in creep tests with various stresses σ_0 MPa at temperatures T °C. Solid lines: results of simulation.

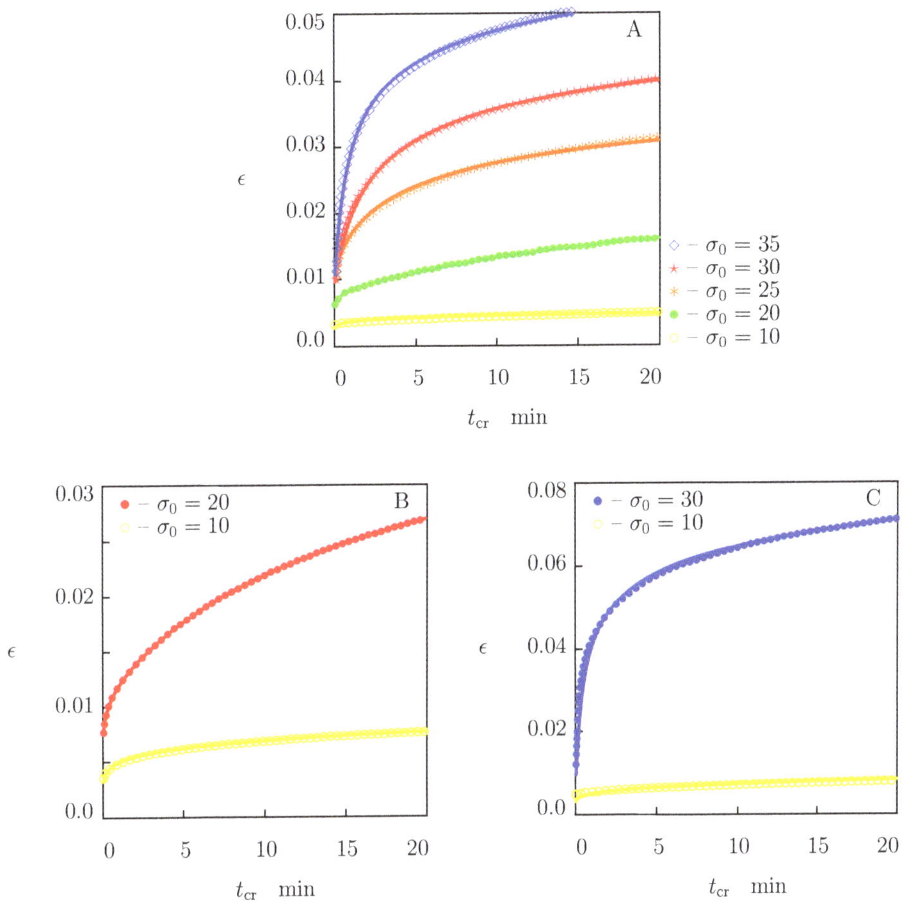

Figure 6. Strain ϵ versus creep time t_{cr}. Circles: experimental data in creep tests with various stresses σ_0 MPa at temperatures $T = 160$ °C (**A**), $T = 170$ °C (**B**) and $T = 180$ °C (**C**). Solid lines: results of simulation.

3. Results and Discussion

We now develop simple constitutive equations for the thermo–mechanical response of PEEK and determine adjustable parameters in these relations by matching the experimental data in Figures 2, 4–6.

3.1. Thermo–Elastoplasticity

Under tension with a constant strain rate $\dot{\epsilon}$, ductile failure (necking) of specimens is observed in the post-yield region. As the failure process is unstable, we study the mechanical behavior of PEEK under tensile deformation up to the points of maximum on the stress–strain diagrams in Figure 2. At all temperatures T, tensile stress reaches its ultimate value at strains ϵ below 0.05, which corresponds to the duration of stretching of about 16 s. Figure 4 shows that relaxation of stresses during this period does not exceed 7%. This implies that the viscoelastic effects can be disregarded in the analysis of the stress–strain diagrams, and the response of PEEK can be described within the theory of elastoplasticity.

According to this concept, the total strain ϵ under uniaxial deformation equals the sum of the elastic, $\epsilon_{\rm e}$, and plastic, $\epsilon_{\rm p}$, components:

$$\epsilon = \epsilon_{\rm e} + \epsilon_{\rm p}. \tag{2}$$

The stress σ is connected with the elastic strain $\epsilon_{\rm e}$ by the linear equation

$$\sigma = E\epsilon_{\rm e}, \tag{3}$$

where the Young's modulus E is treated as a function of temperature T. The plastic strain $\epsilon_{\rm p}$ is connected with the stress σ by the flow rule

$$\dot{\epsilon}_{\rm p} = A\sinh(B\sigma)\sigma^2, \tag{4}$$

where A is a function of temperature T, B is a temperature-independent material parameter, and the superscript dot stands for the derivative with respect to time t. The multiplier σ^2 in Equation (4) provides the simplest version of the Bailey–Norton law (its presence means that the rate of plastic deformation is governed by the stored mechanical energy [41]). The multiplier $\sinh(B\sigma)$ is introduced to avoid the use of a yield surface (plastic deformation is presumed to occur at any stress, but its rate is negligible at small stresses due to the properties of the hyperbolic sinus). Another explanation for this term is based on the Eyring theory of thermally activated processes, according to which B is proportional to the activation volume for cooperative motion of polymer chains [55].

Equations (2)–(4) provide constitutive relations in thermo–elastoplasticity of PEEK under uniaxial deformation. These equations involve two functions of temperature, $E(T)$ and $A(T)$, and a constant B to be found by approximation of experimental data in Figure 2.

We begin with matching observations in tensile tests at room temperature, calculate E by fitting the data at $0 \le \epsilon \le 0.01$, and determine A and B from the best-fit condition for the entire stress–strain diagram. Then, the coefficient $B = 0.03$ MPa^{-1} is fixed, and the stress–strain curves at elevated temperatures are matched by means of two parameters, E and A, only. Each set of observations is approximated separately. Figure 2 shows an acceptable agreement between the data and results of simulation.

To evaluate the activation volume $V_{\rm a}$ associated with the coefficient B and to compare its values with results of other studies, the theory of thermally activated processes is applied. According to this concept,

$$B = \frac{V_{\rm a}\sqrt{3}}{k_{\rm B}T_0}, \tag{5}$$

where $k_{\rm B}$ is the Boltzmann constant, $T_0 = 293$ K stands for room temperature, and the coefficient $\sqrt{3}$ appears due to transformation of tensile stress into the equivalent shear stress. For B found by fitting observations in Figure 2, it follows from Equation (5) that $V_{\rm a} = 7.0 \times 10^{-2}$ nm^3. This value is substantially (by two to three orders of magnitude) lower than $V_{\rm a} = 1$ nm^3 [16], $V_{\rm a} = 1$ to 7 nm^3 [21], $V_{\rm a} = 3.4$ nm^3 [35], and $V_{\rm a} = 12.6$ nm^3 [30]. This difference can be attributed to the fact that the activation volumes were calculated in the above works as measures of sensitivity of the yield stress to changes in the strain rate [55]. This explains also large (by an order of magnitude) deviations between the reported values of $V_{\rm a}$.

The situation changes drastically if we associate $V_{\rm a}$ with volumes of holes measured by means of the positron annihilation lifetime spectroscopy (PALS) and diffusivity of gases (these two methods lead to similar results [56]). PALS measurements of free volume imply that $V_{\rm a} = 7.15 \times 10^{-2}$ nm^3 for PEEK specimens [57], and this parameter varies between 7.2×10^{-2} to 8.4×10^{-2} nm^3 depending on degree of crystallinity [58]. Both estimates are in good accord with the value of $V_{\rm a}$ obtained in our analysis of observations.

The effects of temperature T on the elastic modulus E and the rate of plastic flow A are illustrated in Figures 7 and 8. By analogy with Equation (1), the data in Figure 7 are approximated by the linear function

$$E = E_0 - E_1 T \tag{6}$$

with the coefficients accepting different values below and above the glass transition temperature T_g. Figure 7A shows that the modulus E remains practically constant below the glass transition temperature and decreases strongly above T_g. This behavior resembles that observed for the shear storage modulus [10] and tensile storage modulus [9,11] in DMA tests on PEEK.

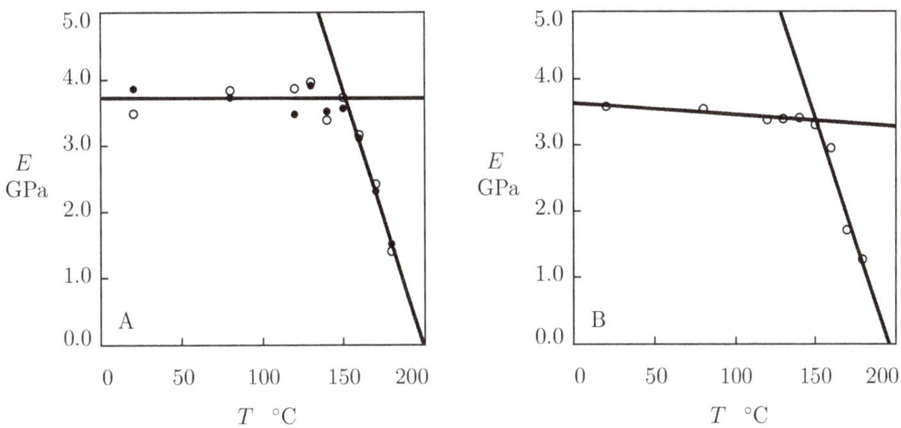

Figure 7. The Young's modulus E versus temperature T. Symbols: (**A**)—treatment of observations in tensile tests (○) and creep tests (•). (**B**)—treatment of observations in relaxation tests. Solid lines: approximation of the data by Equation (6).

The coefficient A is plotted versus reciprocal absolute temperature in Figure 8. The data are approximated by the Arrhenius dependence

$$A = A_0 \exp\Big(-\frac{E_a}{RT}\Big), \tag{7}$$

where A_0 is a pre-factor, R is the universal gas constant, and E_a stands for the activation energy. Figure 8 shows that the observations are correctly described by Equation (7) when different activation energies are used below and above the glass transition temperature T_g.

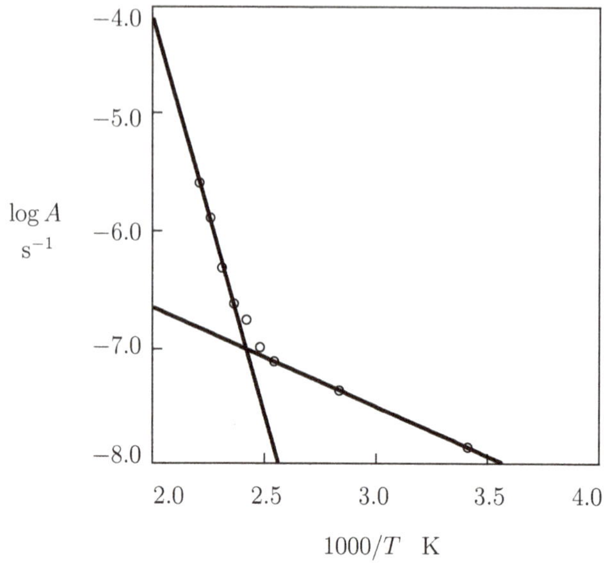

Figure 8. Coefficient A versus temperature T. Circles: treatment of observations in tensile tests. Solid lines: approximation of the data by Equation (7) with $E_a = 16.7$ kJ/mol (low temperatures) and $E_a = 133.3$ kJ/mol (high temperatures).

3.2. Thermo–Viscoelasticity

The experimental data in tensile relaxation tests with a small strain $\epsilon_0 = 0.01$ at various temperatures T are described by the constitutive equations in linear viscoelasticity of semicrystalline polymers [59]. A polymer is thought of as a network with two types of chains: permanent (whose ends are bridged by covalent cross-links) and temporary (able to separate from their junctions and to merge with the network at random times being driven by thermal fluctuations). The heterogeneous network is composed of meso-domains with various activation energies for rearrangement. The rate Γ for separation of temporary chains from their junctions in meso-domains with a dimensionless activation energy u (normalized by $k_B T_0$) is governed by the Eyring equation

$$\Gamma = \gamma \exp(-u), \tag{8}$$

where γ is a pre-factor. A quasi-Gaussian expression is adopted for the distribution function $f(u)$ of meso-domains with various activation energies u,

$$f(u) = f_0 \exp\Big(-\frac{u^2}{2\Sigma}\Big) \qquad (u \geq 0). \tag{9}$$

The dimensionless parameter Σ characterizes inhomogeneity of an ensemble of meso-domains. The coefficient f_0 is determined by the normalization condition

$$\int_0^\infty f(u)\mathrm{d}u = 1. \tag{10}$$

Under uniaxial tension with an arbitrary deformation program $\epsilon(t)$, tensile stress $\sigma(t)$ obeys the constitutive equation

$$\sigma(t) = E\Big[\epsilon(t) - \kappa \int_0^\infty \Gamma(u) f(u)\mathrm{d}u \int_0^t \exp\Big(-\Gamma(u)(t-\tau)\Big)\epsilon(\tau)\mathrm{d}\tau\Big], \tag{11}$$

where E stands for the Young's modulus, and κ denotes the ratio of the number of temporary chains to the total number of chains per unit volume in the initial state. Unlike conventional (the Maxwell–Wiechert type) models for the linear viscoelastic response of polymers (involving a large number of material constants), Equation (11) is entirely determined by four parameters, E, κ, γ and Σ. With reference to [60], we suppose that Σ is independent of temperature, κ increases linearly with temperature and reaches its ultimate value $\kappa = 1$ at relatively high temperatures $T > T_g$, and γ obeys the Arrhenius law. Equation (11) implies that, in tensile relaxation tests with a fixed strain ϵ_0, the stress σ decreases with relaxation time t_{rel} following the pattern

$$\sigma(t_{rel}) = \sigma_0 \Big\{ 1 - \kappa \int_0^\infty f(u) \Big[1 - \exp\Big(-\gamma \exp(-u) t_{rel} \Big) \Big] du \Big\}, \tag{12}$$

where

$$\sigma_0 = E\epsilon_0 \tag{13}$$

stands for the stress at the beginning of the relaxation process.

Each set of data in Figure 4 is matched separately by means with Equation (12) with three parameters σ_0, κ and γ (since Σ is independent of temperature, we find its value $\Sigma = 7.0$ by fitting observations at $T = 160$ °C and use it at all temperatures T under consideration). The best-fit coefficient γ is determined by means of the nonlinear regression algorithm, while σ_0 and κ are determined by the least-squares technique. Given σ_0, the modulus E is calculated from Equation (13) and plotted versus temperature T in Figure 7B. This figure demonstrates that the data are described adequately by Equation (6) with different coefficients E_0 and E_1 below and above the glass transition temperature T_g.

The coefficients E_0 and E_1 found by approximation of experimental data in Figure 7A,B above T_g coincide practically: the difference is less than 7% for E_0 and 5% for E_1. The discrepancies between the coefficients E_0 and E_1 calculated by matching observations in tensile and relaxation tests below T_g are higher (but do not exceed 14%). They may be explained by local variations in thicknesses of specimens machined from an extruded sheet.

The parameter κ is plotted versus temperature in Figure 9A. This figure demonstrates that κ increases monotonically with T (the growth of intensity of thermal fluctuations results in transformation of some permanent chains into transient chains). The influence of temperature T on κ is described by the linear equation

$$\kappa = \kappa_0 + \kappa_1 T, \tag{14}$$

where κ_0 and κ_1 are calculated by the least-squares method. Figure 9A shows that Equation (14) describes correctly the function $\kappa(T)$ when different coefficients κ_0, κ_1 are used below and above the glass transition temperature T_g.

The effect of temperature T on the rate of relaxation γ is illustrated in Figure 9B. The data (in the region of temperatures $T \geq 120$ °C) are approximated by the Arrhenius equation

$$\gamma = \gamma_0 \exp\Big(-\frac{E_a}{RT} \Big), \tag{15}$$

where γ_0 stands for a pre-factor, and E_a denotes the activation energy. Comparison of Figures 8 and 9B shows that the activation energies found by matching observations in tensile tests and relaxation tests in the high temperature region adopt similar values (deviations between $E_a = 133.3$ kJ/mol in Figure 8 and $E_a = 113.4$ kJ/mol in Figure 9B do not exceed 15 %).

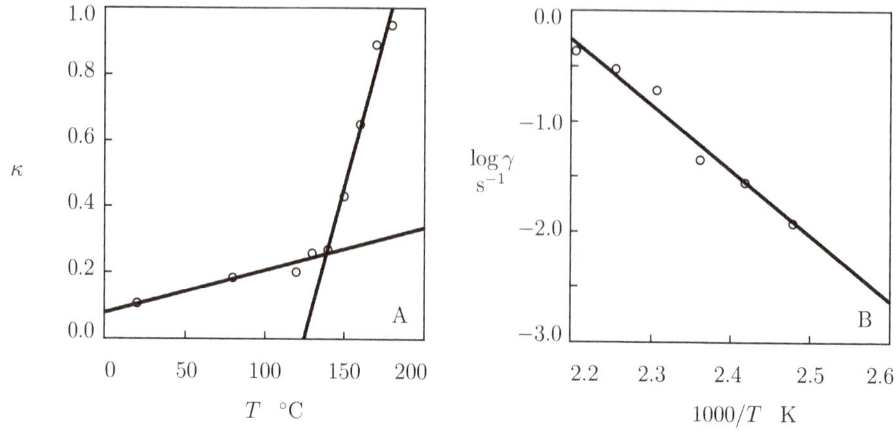

Figure 9. Parameters κ and γ versus temperature T. Circles: treatment of observations in relaxation tests. Solid lines: (**A**)—approximation of the data by Equation (14). (**B**)—approximation of the data by Equation (15) with $E_a = 113.4$ kJ/mol.

It is worth noting that these activation energies differ pronouncedly from the activation energies for α-relaxation reported in previous studies on the time-dependent response of PEEK ($E_a = 377$ [38], $E_a = 494$ [37] $E_a = 582$ [61], $E_a = 810$ to 1000 [9], $E_a = 1094$ [39], and $E_a = 1100$ kJ/mol [62]). To explain this difference, experimental data in relaxation tests are treated by the conventional method: all relaxation curves at elevated temperatures T are shifted horizontally (along the time-axis) to construct a master-curve at room temperature T_0. Figure 10A confirms that a smooth master-curve is formed by means of this technique. The shift factor a is plotted versus temperature T in Figure 10B. The data are approximated by the Arrhenius equation

$$\log a = \log a_0 + \frac{E_a}{RT}, \qquad (16)$$

where a_0 and E_a are calculated by the least-squares technique.

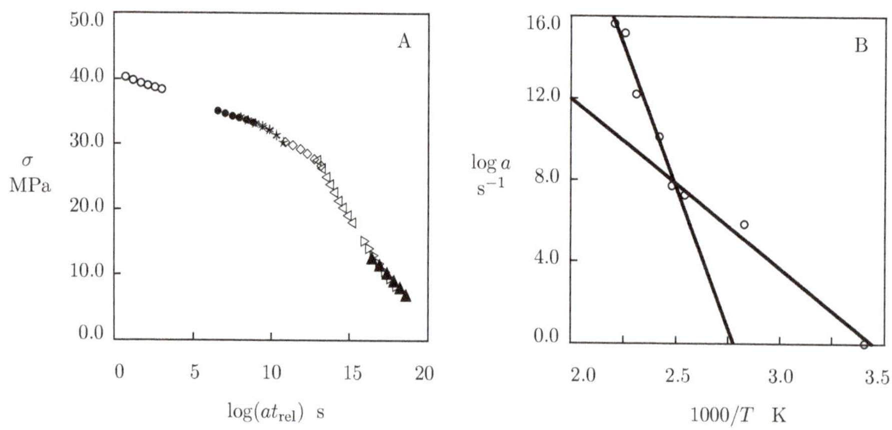

Figure 10. (**A**)—Stress σ versus relaxation time t_{rel}. Symbols: master-curve at room temperature constructed by shift of experimental data in relaxation tests. (**B**)—Shift factor a versus temperature T. Circles: treatment of observations in relaxation tests. Solid lines: approximation of the data by Equation (16) with $E_a = 159.4$ kJ/mol (low temperatures) and $E_a = 531.3$ kJ/mol (high temperatures).

Figure 10B shows good agreement between the data and predictions of Equation (16) with different coefficients below and above the glass transition temperature T_g. The activation energy in the low-temperature region $E_a = 159.4$ kJ/mol is in accord with the activation energy $E_a = 177$ kJ/mol for the β-relaxation in PEEK [9], whereas the activation energy in the high-temperature region $E_a = 531.3$ kJ/mol is close to the activation energies $E_a = 494$ kJ/mol [37] and $E_a = 582$ kJ/mol [61] for the α-relaxation. These results demonstrate that the method based on the time-temperature superposition principle (shifts of observations in relaxation and dynamic mechanical tests) overly estimates the activation energies since this approach disregards evolution of E and κ with temperature T.

To confirm validity of our model for the linear viscoelastic behavior of PEEK, we apply Equation (11) to describe the time-dependent response of specimens in tensile creep tests and compare results of simulation with experimental data depicted in Figures 5 and 6. Resolving Equation (11) with respect to ϵ, we find that the growth of tensile strain with time in a creep test with a constant stress σ_0 is governed by the equation

$$\epsilon(t_{\rm cr}) = \epsilon_0 + \kappa \int_0^\infty f(u) s(t_{\rm cr}, u) {\rm d}u. \tag{17}$$

Here $t_{\rm cr} = t - t_0$, where t_0 is the moment when tensile stress reaches the required value σ_0,

$$\epsilon_0 = \frac{\sigma_0}{E} \tag{18}$$

is the strain at the beginning of the creep process, and the function $s(t_{\rm cr}, u)$ obeys the differential equation

$$\frac{\partial s}{\partial t_{\rm cr}}(t_{\rm cr}, u) = \gamma \exp(-u) \Big[\epsilon(t_{\rm cr}) - s(t_{\rm cr}, u)\Big], \qquad s(0, u) = 0. \tag{19}$$

For each temperature T and stress σ_0, Equations (17)–(19) are integrated numerically by the Runge–Kutta method with the material constants γ, κ and Σ determined by matching observations in relaxation tests (Figure 9). To ensure good agreement between the observations in Figures 5 and 6 and results of simulation, the coefficient E is treated as an adjustable parameter. Its best-fit values at various temperatures T are reported in Figure 7A. This figure shows that the Young's moduli found by matching experimental data in tensile tests and creep tests coincide practically. Comparison of experimental data with results of simulation (Figures 5 and 6) confirms the ability of the model (17)–(19) to predict the response of PEEK in short-term creep tests with small strains when material parameters are determined in relaxation tests.

3.3. *Thermo–Viscoelastoplasticity*

The time-dependent response of PEEK in creep tests with relatively high stresses σ_0 (beyond the interval of linear viscoelasticity) is described within the concept of viscoelastoplasticity [63]. In accord with Equation (2), the total strain ϵ is split into the sum of the viscoelastic strain, $\epsilon_{\rm ve}$, and plastic strain, $\epsilon_{\rm p}$,

$$\epsilon = \epsilon_{\rm ve} + \epsilon_{\rm p}. \tag{20}$$

The viscoelastic strain $\epsilon_{\rm ve}$ is connected with tensile stress σ by Equation (11),

$$\sigma(t) = E\left[\epsilon_{\rm ve}(t) - \kappa \int_0^\infty \Gamma(u) f(u) {\rm d}u \int_0^t \exp\Big(-\Gamma(u)(t-\tau)\Big) \epsilon_{\rm ve}(\tau) {\rm d}\tau\right]. \tag{21}$$

The evolution of the plastic strain $\epsilon_{\rm p}$ with time t is governed by an analog of Equation (4),

$$\dot{\epsilon}_{\rm p} = \bar{A} \sinh(\epsilon)(\sigma - \sigma_{\rm b})^m, \qquad \epsilon_{\rm p}(0) = 0, \tag{22}$$

where $\bar{A}$ stands for a pre-factor, the term $\sinh(\epsilon)$ is introduced to avoid the growth of plastic strain at small strains (far below the yield point), and m is an exponent in the Bailey–Norton law. The back-stress σ_b accounting for strain hardening under plastic flow [64] reads

$$\sigma_b = E_b \epsilon_p, \tag{23}$$

where E_b is an analog of the elastic modulus.

Equations (20)–(23) together with Equations (8) and (9) provide a constitutive model in viscoelastoplasticity of semicrystalline polymers under uniaxial deformation. These relations involve four parameters, E, κ, γ and Σ, that characterize the linear viscoelastic response, and three extra coefficients, $\bar{A}$, E_b and m. To reduce the number of parameters, we set $m = 7$ (a typical value of the Bailey–Norton exponent) in matching experimental creep curves in Figure 6. The other two quantities, $\bar{A}$ and E_b, are considered as functions of temperature T only.

Under tensile creep with a fixed stress σ_0, Equation (21) takes a form similar to Equation (17),

$$\epsilon_{ve}(t_{cr}) = \epsilon_0 + \kappa \int_0^\infty f(u) s(t_{cr}, u) \mathrm{d}u, \tag{24}$$

while Equation (22) yields

$$\frac{\mathrm{d}\epsilon_p}{\mathrm{d}t_{cr}} = \bar{A} \sinh(\epsilon)(\sigma_0 - \sigma_b)^m, \qquad \epsilon_p(0) = 0. \tag{25}$$

The following procedure is applied to fit observations in Figure 6 at each temperature under consideration. First, the creep diagram with the highest stress σ_0 is approximated with the help of two parameters, $\bar{A}$ and E_b (the quantities E, κ, γ and Σ are found by matching experimental data in the corresponding relaxation test). Afterwards, the quantities $\bar{A}$, E_b, m, κ, γ, Σ are fixed, and each remaining creep curve is fitted with the only coefficient E. We treat E as an adjustable parameter to account for deviations in thicknesses of specimens machined from an extruded sheet. Discrepancies between the best-fit values of E determined by matching creep curves with various σ_0 have the same order of magnitude as those between the data in tensile and creep tests in Figure 7 (they do not exceed 12 % at all temperatures).

The effects of temperature on coefficients $\bar{A}$ and E_b are illustrated in Figure 11.

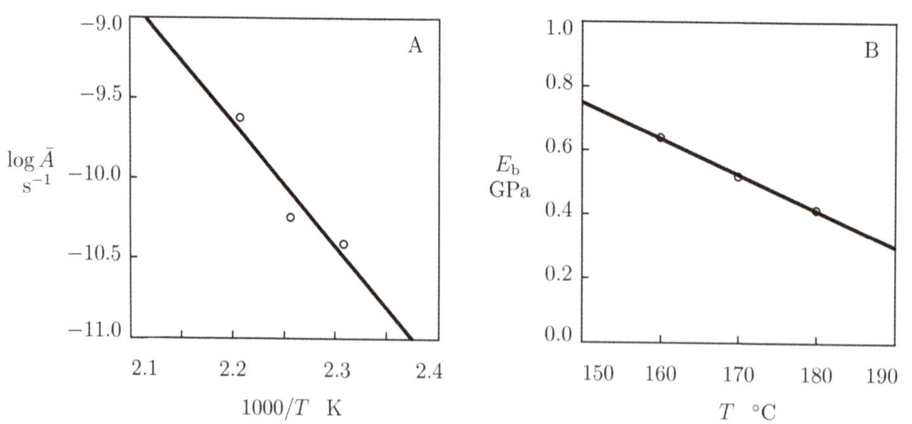

Figure 11. Parameters $\bar{A}$ and E_b versus temperature T. Circles: treatment of observations in creep tests. Solid lines: (**A**)—approximation of the data by Equation (26) with $E_a = 147.6$ kJ/mol. (**B**)—approximation of the data by Equation (27).

The growth of the kinetic parameter $\bar{A}$ with T is described by the Arrhenius dependence,

$$\bar{A} = \bar{A}_0 \exp\Big(-\frac{E_a}{RT}\Big). \tag{26}$$

Figure 11A shows an acceptable agreement between the data and their predictions by Equation (26) with the activation energy $E_a = 147.6$ kJ/mol. The latter value is in agreement with the activation energy $E_a = 133.3$ kJ/mol determined by matching observations in Figure 8 (the difference is less than 10 %).

The decay in E_b with temperature is described by the equation analogous to Equation (6),

$$E_b = E_{b0} - E_{b1} T, \tag{27}$$

where E_{b0} and E_{b1} are found by the least-squares method. Figure 11B demonstrates good agreement between the data and results of simulation. A similarity should be stressed between the effect of temperature on the Young's modulus E and the modulus E_b that characterize the back-stress. Above the glass transition temperature T_g, the dimensionless ratios E_1/E_0 adopt the values 5.09×10^{-3} (Figure 7A) and 5.01×10^{-3} (Figure 7B), whereas the ratio E_{b1}/E_{b0} equals 4.62×10^{-3} (Figure 11B), which confirms that changes in E and E_b with temperature are governed by the same physical mechanism.

4. Conclusions

Observations are reported on poly(ether ether ketone) KETRON 1000 PEEK in DSC tests (with a constant rate of 20 K/min under heating and cooling), as well as in uniaxial tensile tests with constant strain rate $\dot{\epsilon} = 3.1 \times 10^{-3}$ s^{-1}, relaxation tests with a constant strain $\epsilon_0 = 0.01$, and creep tests with various stresses σ_0 (ranging from 10 to 70 MPa) at temperatures ranging from room temperature to 180 °C.

Constitutive equations are developed for the thermo–elastoplastic, thermo–viscoelastic and thermo–viscoelastoplastic responses of PEEK under uniaxial deformation with small strains. An advantage of these relations is that they involve relatively small numbers of adjustable parameters (three for the elastoplastic behavior, four for the linear viscoelastic response, and an extra three for the viscoelastoplastic flow). Good agreement is demonstrated between the experimental data and results of simulation.

Analysis of the effect of temperature on the rate of elastoplastic deformation A, the relaxation rate γ, and the rate of viscoplastic flow $\bar{A}$ reveals that the growth of these quantities with T above the glass transition temperature T_g obeys the Arrhenius law with similar activation energies E_a (in the range between 113 and 148 kJ/mol).

The study of the elastoplastic and viscoelastoplastic responses of PEEK shows that (i) the activation volume for plastic deformation coincides with the free volume found in PALS and diffusion tests and (ii) the elastic moduli E (stress) and E_b (back-stress) decrease similarly with temperature.

Supplementary Materials: The following are available online at https://www.mdpi.com/article/10.3390/polym13111779/s1, Figure S1: Stress σ versus strain ϵ. Symbols: experimental data in tensile tests at temperatures T = 20 (○) and 170 (•) °C. Bars stand for the standard deviations. Solid lines: results of simulation, Figure S2: Stress σ_0 versus relaxation time t_{rel}. Symbols: experimental data in tensile relaxation tests with strain ϵ_0 = 0:01 at temperatures T = 20 (○) and 170 (•) °C. Bars stand for the standard deviations. Solid lines: results of simulation, Figure S3: Strain σ versus creep time t_{cr}. Circles: experimental data in creep test with stress σ_0 = 70:0 MPa at temperature T = 20 °C. Bars stand for the standard deviations. Solid line: results of simulation, Figure S4: Strain ϵ versus creep time tcr. Symbols: experimental data in creep tests with stresses σ_0 = 10 (○) and 20 (•) MPa at temperature T = 170 °C. Bars stand for the standard deviations. Solid lines: results of simulation.

Author Contributions: A.D.D.: Conceptualization, methodology, software, formal analysis, data curation, writing—original draft preparation, writing—review and editing. J.d.C.: Conceptualization, methodology, investigation, writing—original draft preparation, writing—review and editing. All authors have read and agreed to the published version of the manuscript.

Funding: This research was supported Innovationsfonden (Innovation Fund Denmark).

Institutional Review Board Statement: not applicable

Informed Consent Statement: not applicable

Data Availability Statement: The data that support the findings of this study are available from the corresponding author upon reasonable request.

Acknowledgments: Financial support by Innovationsfonden (Innovation Fund Denmark, project 9091-00010B) is gratefully acknowledged.

Conflicts of Interest: The authors declare no conflict of interest.

References

1. Diez-Pascual, A.M.; Naffakh, M.; Marco, C.; Ellis, G.; Gomez-Fatou, M.A. High-performance nanocomposites based on polyetherketones. *Prog. Mater. Sci.* **2012**, *57*, 1106–1190. [CrossRef]
2. Flanagan, M.; Grogan, D.M.; Goggins, J.; Appel, S.; Doyle, K.; Leen, S.B.; O'Bradaigh, C.M. Permeability of carbon fibre PEEK composites for cryogenic storage tanks of future space launchers. *Composites A* **2017**, *101*, 173–184. [CrossRef]
3. Kalra, S.; Munjal, B.S.; Singh, V.R.; Mahajan, M.; Bhattacharya, B. Investigations on the suitability of PEEK material under space environment conditions and its application in a parabolic space antenna. *Adv. Space Res.* **2019**, *63*, 4039–4045. [CrossRef]
4. Lin, L.; Schlarb, A.K. Effect of the varied load conditions on the tribological performance and the thermal characteristics of PEEK-based hybrid composites. *Tribol. Int.* **2016**, *101*, 218–225. [CrossRef]
5. Kim, D.J.; Lee, B.-N.; Nam, S.Y. Characterization of highly sulfonated PEEK based membrane for the fuel cell application. *Int. J. Hydrogen Energy* **2017**, *42*, 23768–23775. [CrossRef]
6. Lei, L.; Zhu, X.; Xu, J.; Qian, H.; Zou, Z.; Yang, H. Highly stable ionic-covalent cross-linked sulfonated poly(ether ether ketone) for direct methanol fuel cells. *J. Power Sources* **2017**, *350*, 41–48. [CrossRef]
7. Panayotov, I.V.; Orti, V.; Cuisinier, F.; Yachouh, J. Polyetheretherketone (PEEK) for medical applications. *J. Mater. Sci. Mater. Med.* **2016**, *27*, 2–11. [CrossRef] [PubMed]
8. Verma, S.; Sharma, N.; Kango, S.; Sharma, S. Developments of PEEK (polyetheretherketone) as a biomedical material: A focused review. *Eur. Polym. J.* **2021**, *147*, 110295. [CrossRef]
9. Krishnaswamy, R.K.; Kalika, D.S. Dynamic mechanical relaxation properties of poly(ether ether ketone). *Polymer* **1994**, *35*, 1157–1165. [CrossRef]
10. Rae, P.J.; Brown, E.N.; Orler, E.B. The mechanical properties of poly(ether-ether-ketone) (PEEK) with emphasis on the large compressive strain response. *Polymer* **2007**, *48*, 598–615. [CrossRef]
11. Guo, Y.; Bradshaw, R.D. Isothermal physical aging characterization of polyether-ether-ketone (PEEK) and polyphenylene sulfide (PPS) films by creep and stress relaxation. *Mech. Time-Depend. Mater.* **2007**, *11*, 61–89. [CrossRef]
12. Yuan, M.; Galloway, J.A.; Hoffman, R.J.; Bhatt, S. Influence of molecular weight on rheological, thermal, and mechanical properties of PEEK. *Polym. Eng. Sci.* **2011**, *51*, 94–102. [CrossRef]
13. Garcia-Leiner, M.; Reitman, M.T.F.; El-Hibri, M.J.; Roeder, R.K. Structure-property relationships in commercial polyetheretherketone resins. *Polym. Eng. Sci.* **2017**, *57*, 955–964. [CrossRef]
14. Abbasnezhad, N.; Khavandi, A.; Fitoussi, J.; Arabi, H.; Shirinbayan, M.; Tcharkhtchi, A. Influence of loading conditions on the overall mechanical behavior of polyether-ether-ketone (PEEK). *Int. J. Fatigue* **2018**, *109*, 83–92. [CrossRef]
15. Ha, S.Y.; Ang, Q.; Chang, F.-K. Modeling the viscoplastic behavior of fiber-reinforced thermoplastic matrix composites at elevated temperatures. *J. Compos. Mater.* **1991**, *25*, 334–374. [CrossRef]
16. Hamdan, S.; Swallowe, G.M. The strain-rate and temperature dependence of the mechanical properties of polyetherketone and polyetheretherketone. *J. Mater. Sci.* **1996**, *31*, 1415–1423. [CrossRef]
17. Chen, F.; Ou, H.; Gatea, S.; Long, H. Hot tensile fracture characteristics and constitutive modelling of polyether-ether-ketone (PEEK). *Polym. Test.* **2017**, *63*, 168–179. [CrossRef]
18. Zheng, B.; Wang, H.; Huang, Z.; Zhang, Y.; Zhou, H.; Li, D. Experimental investigation and constitutive modeling of the deformation behavior of poly-ether-ether-ketone at elevated temperatures. *Polym. Test.* **2017**, *63*, 349–359. [CrossRef]
19. Barba, D.; Arias, A.; Garcia-Gonzalez, D. Temperature and strain rate dependences on hardening and softening behaviours in semi-crystalline polymers: Application to PEEK. *Int. J. Solids Struct.* **2020**, *182–183*, 205–217. [CrossRef]
20. El-Qoubaa, Z.; Othman, R. Characterization and modelling of the strain rate sensitivity of polyetheretherketones compressive yield stress. *Mater. Des.* **2015**, *66*, 336–345. [CrossRef]
21. El-Qoubaa, Z.; Othman, R. Strain rate sensitivity of polyetheretherketones compressive yield stress at low and high temperatures. *Mech. Mater.* **2016**, *95*, 15–27. [CrossRef]
22. Chen, F.; Ou, H.; Lu, B.; Long, H. A constitutive model of polyether-ether-ketone (PEEK). *J. Mech. Behav. Biomed. Mater.* **2016**, *53*, 427–433. [CrossRef]
23. Turner, J.A.; Menary, G.H.; Martin, P.J. Biaxial characterization of poly(ether-ether-ketone) for thermoforming: A comparison between bulge and in-plane biaxial testing. *Polym. Eng. Sci.* **2019**, *59*, 1853–1865. [CrossRef]

24. Turner, J.A.; Menary, G.H.; Martin, P.J.; Yan, S. Modelling the temperature dependent biaxial response of poly(ether-ether-ketone) above and below the glass transition for thermoforming applications. *Polymers* **2019**, *11*, 1042. [CrossRef] [PubMed]
25. Sobieraj, M.C.; Murphy, J.E.; Brinkman, J.G.; Kurtz, S.M.; Rimnac, C.M. Notched fatigue behavior of PEEK. *Biomaterials* **2010**, *31*, 9156–9162. [CrossRef] [PubMed]
26. Berer, M.; Tscharnuter, D.; Pinter, G. Dynamic mechanical response of polyetheretherketone (PEEK) exposed to cyclic loads in the high stress tensile regime. *Int. J. Fatigue* **2015**, *80*, 397–405. [CrossRef]
27. Shrestha, R.; Simsiriwong, J.; Shamsaei, N.; Moser, R.D. Cyclic deformation and fatigue behavior of polyether ether ketone (PEEK). *Int. J. Fatigue* **2016**, *82*, 411–427. [CrossRef]
28. Shrestha, R.; Simsiriwong, J.; Shamsaei, N. Mean strain effects on cyclic deformation and fatigue behavior of polyether ether ketone (PEEK). *Polym. Test.* **2016**, *55*, 69–77. [CrossRef]
29. Li, W.; Gazonas, G.; Brown, E.N.; Rae, P.J.; Negahban, M. Thermomechanical model for monotonic and cyclic loading of PEEK. *Mech. Mater.* **2019**, *129*, 113–138. [CrossRef]
30. Kemmish, D.J.; Hay, J.N. The effect of physical ageing on the properties of amorphous PEEK. *Polymer* **1985**, *26*, 905–912. [CrossRef]
31. Millett, J.C.F.; Bournel, N.K.; Stevens, G.S. Taylor impact of polyether ether ketone. *Int. J. Impact Eng.* **2006**, *32*, 1086–1094. [CrossRef]
32. Garcia-Gonzalez, D.; Rusinek, A.; Jankowiak, T.; Arias, A. Mechanical impact behavior of polyether-ether-ketone (PEEK). *Compos. Struct.* **2015**, *124*, 88–99. [CrossRef]
33. Bourne, N.K.; Garcea, S.C.; Eastwood, D.S.; Parry, S.; Rau, C.; Withers, P.J.; McDonald, S.A.; Brown, E.N. On compression and damage evolution in two thermoplastics. *Proc. R. Soc. A* **2017**, *473*, 20160495. [CrossRef] [PubMed]
34. Yang, Y. Sensitivity of nanoindentation strain rate in poly(ester-ester-ketone) using atomic force microscopy. *Polym. Test.* **2016**, *53*, 85–88. [CrossRef]
35. Voyiadjis, G.Z.; Samadi-Dooki, A.; Malekmotiei, L. Nanoindentation of high performance semicrystalline polymers: A case study on PEEK. *Polym. Test.* **2017**, *61*, 57–64. [CrossRef]
36. Kim, W.; Sun, C.T. Modeling relaxation of a polymeric composite during loading and unloading. *J. Compos. Mater.* **2002**, *36*, 745–755. [CrossRef]
37. Jiang, Z.; Liu, P.; Sue, H.-J.; Bremner, T. Effect of annealing on the viscoelastic behavior of poly(ether-ether-ketone). *Polymer* **2019**, *160*, 231–237. [CrossRef]
38. Ogale, A.A.; McCullough, R.L. Influence of microstructure on elastic and viscoelastic properties of polyether ether ketone. *Compos. Sci. Technol.* **1987**, *30*, 185–201. [CrossRef]
39. Xiao, X.R.; Hiel, C.C.; Cardon, A.H. Characterization and modeling of nonlinear viscoelastic response of PEEK resin and PEEK composites. *Compos. Eng.* **1994**, *4*, 681–702. [CrossRef]
40. Katouzian, M.; Bruller, O.S.; Horoschenkoff, A. On the effect of temperature on the creep behavior of neat and carbon fiber reinforced PEEK and epoxy resin. *J. Compos. Mater.* **1995**, *29*, 372–387. [CrossRef]
41. Drozdov, A.D.; Gupta, R.K. Constitutive equations in finite viscoplasticity of semicrystalline polymers. *Int. J. Solids Struct.* **2003**, *40*, 6217–6243. [CrossRef]
42. Drozdov, A.D.; Klitkou, R.; Christiansen, J.D. Cyclic viscoplasticity of semicrystalline polymers with finite deformations. *Mech. Mater.* **2013**, *56*, 53–64. [CrossRef]
43. Yu, C.; Kang, G.; Chen, K.; Lu, F. A thermo-mechanically coupled nonlinear viscoelastic-viscoplastic cyclic constitutive model for polymeric materials. *Mech. Mater.* **2017**, *105*, 1–15. [CrossRef]
44. Garcia-Gonzalez, D.; Zaera, R.; Arias, A. A hyperelastic–thermoviscoplastic constitutive model for semi-crystalline polymers: application to PEEK under dynamic loading conditions. *Int. J. Plast.* **2016**, *88*, 27–52. [CrossRef]
45. Chang, B.; Wang, X.; Long, Z.; Li, Z.; Gu, J.; Ruan, S.; Shen, C. Constitutive modeling for the accurate characterization of the tension behavior of PEEK under small strain. *Polym. Test.* **2018**, *69*, 514–521. [CrossRef]
46. Lei, M.; Hamel, C.M.; Chen, K.; Zhao, Z.; Lu, H.; Yu, K.; Qi, H.J. Thermomechanical behaviors of polyether ether ketone (PEEK) with stretch-induced anisotropy. *J. Mech. Phys. Solids* **2021**, *148*, 104271. [CrossRef]
47. Iulianelli, A.; Basile, A. Sulfonated PEEK based polymers in PEMFC and DMFC applications: A review. *Int. J. Hydrogen Energy* **2012**, *37*, 15241–15255. [CrossRef]
48. Yang, J.; Jiang, H.; Wang, J.; Xu, Y.; Pan, C.; Li, Q.; He, R. Dual cross-linked polymer electrolyte membranes based on poly(aryl ether ketone) and poly(styrene-vinylimidazole-divinylbenzene) for high temperature proton exchange membrane fuel cells. *J. Power Sources* **2020**, *480*, 228859. [CrossRef]
49. Liu, D.; Dong, B.; Zhang, H.; Xie, Y.; Pang, J.; Jiang, Z. High methanol resistant polyelectrolyte membrane based on semi-crystalline poly(ether ketone) with densely sulfonated side chain for direct methanol fuel cell. *J. Power Sources* **2021**, *482*, 228982. [CrossRef]
50. Wu, G.; Lin, S.-J.; Hsu, I.-C.; Su, J.-Y.; Chen, D.W. Study of high performance sulfonated polyether ether ketone composite electrolyte membranes. *Polymers* **2019**, *11*, 1177. [CrossRef] [PubMed]
51. Haleem, A.; Javaid, M. Polyether ether ketone (PEEK) and its 3D printed implants applications in medical field: An overview. *Clin. Epidemiol. Glob. Health* **2019**, *7*, 571–577. [CrossRef]
52. Singh, S.; Prakash, C.; Ramakrishna, S. 3D printing of polyether-ether-ketone for biomedical applications. *Eur. Polym. J.* **2019**, *114*, 234–248. [CrossRef]

53. Song, X.; Shi, D.; Song, P.; Han, X.; Wei, Q.; Huang, C. Fused deposition modeling of poly(ether ether ketone) scaffolds. *High Temp. Mater. Process.* **2021**, *40*, 1–11. [CrossRef]
54. Bauwens-Crowet, C.; Bauwens, J.C.; Homes, G. The temperature dependence of yield of polycarbonate in uniaxial compression and tensile tests. *J. Mater. Sci.* **1972**, *7*, 176–183. [CrossRef]
55. Richeton, J.; Ahzi, S.; Vecchio, K.S.; Jiang, F.C.; Adharapurapu, R.R. Influence of temperature and strain rate on the mechanical behavior of three amorphous polymers: Characterization and modeling of the compressive yield stress. *Int. J. Solids Struct.* **2006**, *43*, 2318–2335. [CrossRef]
56. Kobayashi, Y.; Haraya, K.; Hattori, S.; Sasuga, T. Evaluation of polymer free volume by positron annihilation and gas diffusivity measurements. *Polymer* **1994**, *35*, 925–928. [CrossRef]
57. Jean, Y.C.; Nakanishi, H.; Hao, L.Y.; Sandreczki, T.C. Anisotropy of free-volume-hole dimensions in polymers probed by positron-annihilation spectroscopy. *Phys. Rev. B* **1990**, *42*, 9705–9708. [CrossRef]
58. Nakanishi, H.; Jean, Y.C. Dynamics of excess free volume in semicrystalline PEEK studied by positron annihilation. *Macromolecules* **1991**, *24*, 6618–6621. [CrossRef]
59. Drozdov, A.D. Effect of temperature on the viscoelastic and viscoplastic behavior of polypropylene. *Mech. Time-Depend. Mater.* **2010**, *14*, 411–434. [CrossRef]
60. Drozdov, A.D.; Christiansen, J.D. Modeling non-isothermal viscoelastic response of glassy polymers. *Acta Mech.* **2018**, *229*, 1137–1156. [CrossRef]
61. Sauer, B.B.; Avakian, P.; Starkweather, H.W.; Hsiao, B.S. Thermally stimulated current and dielectric studies of poly(aryl ether ketone ketone). *Macromolecules* **1990**, *23*, 5119–5126. [CrossRef]
62. Goodwin, A.A.; Hay, J.N. Dielectric and dynamic mechanical relaxation studies on poly(aryl ether ketone)s. *J. Polym. Sci. B Polym. Phys.* **1998**, *36*, 851–859. [CrossRef]
63. Drozdov, A.D.; Kalamkarov, A.L. A constitutive model for nonlinear viscoelastic behavior of polymers. *Polym. Eng. Sci.* **1996**, *36*, 1907–1919. [CrossRef]
64. Henann, D.L.; Anand, L. A large deformation theory for rate-dependent elastic-plastic materials with combined isotropic and kinematic hardening. *Int. J. Plast.* **2009**, *25*, 1833–1878. [CrossRef]

polymers

MDPI

Article

Effect of Bacterial Cellulose Plasma Treatment on the Biological Activity of Ag Nanoparticles Deposited Using Magnetron Deposition

Alexander Vasil'kov [1,*], Alexander Budnikov [1], Tatiana Gromovykh [2], Marina Pigaleva [3], Vera Sadykova [4], Natalia Arkharova [5] and Alexander Naumkin [1]

1 A. N. Nesmeyanov Institute of Organoelement Compounds, Russian Academy of Sciences, 28 Vavilov St., Moscow 119991, Russia
2 Department of Biotechnology, Moscow Polytechnic University, Bolshaya Semyonovskaya Str., 38, Moscow 107023, Russia
3 Faculty of Physics, Lomonosov Moscow State University, Leninskie Gory 1–2, Moscow 119991, Russia
4 G. F. Gause Institute of New Antibiotics, 11 Bol'shaya Pirogovskaya St., Moscow 119021, Russia
5 FSRC "Crystallography and Photonics" RAS, 59 Leninsky Prospekt, Moscow 119333, Russia
* Correspondence: alexandervasilkov@yandex.ru

Abstract: New functional medical materials with antibacterial activity based on biocompatible bacterial cellulose (BC) and Ag nanoparticles (Ag NPs) were obtained. Bacterial cellulose films were prepared by stationary liquid-phase cultivation of the *Gluconacetobacter hansenii* strain GH-1/2008 in Hestrin–Schramm medium with glucose as a carbon source. To functionalize the surface and immobilize Ag NPs deposited by magnetron sputtering, BC films were treated with low-pressure oxygen–nitrogen plasma. The composition and structure of the nanomaterials were studied using transmission (TEM) and scanning (SEM) electron microscopy and X-ray photoelectron spectroscopy (XPS). Using electron microscopy, it was shown that on the surface of the fibrils that make up the network of bacterial cellulose, Ag particles are stabilized in the form of aggregates 5–35 nm in size. The XPS C 1s spectra show that after the deposition of Ag NPs, the relative intensities of the C-OH and O-C-O bonds are significantly reduced. This may indicate the destruction of BC oxypyran rings and the oxidation of alcohol groups. In the Ag $3d_{5/2}$ spectrum, two states at 368.4 and 369.7 eV with relative intensities of 0.86 and 0.14 are distinguished, which are assigned to Ag^0 state and Ag acetate, respectively. Nanocomposites based on plasma-treated BC and Ag nanoparticles deposited by magnetron sputtering (BCP-Ag) exhibited antimicrobial activity against *Aspergillus niger*, *S. aureus* and *Bacillus subtilis*.

Keywords: bacterial cellulose; plasma treatment; magnetron sputtering; silver nanoparticles; antimicrobial activity; X-ray photoelectron spectroscopy

Citation: Vasil'kov, A.; Budnikov, A.; Gromovykh, T.; Pigaleva, M.; Sadykova, V.; Arkharova, N.; Naumkin, A. Effect of Bacterial Cellulose Plasma Treatment on the Biological Activity of Ag Nanoparticles Deposited Using Magnetron Deposition. *Polymers* **2022**, *14*, 3907. https://doi.org/10.3390/polym14183907

Academic Editor: Célio Bruno Pinto Fernandes

Received: 5 September 2022
Accepted: 12 September 2022
Published: 19 September 2022

1. Introduction

The use of green technologies for the production of medical materials can significantly reduce or eliminate their negative impact on the environment. Simultaneously, there is a steady trend towards the widespread use of renewable resources (cellulose, chitosan, etc.) for medical purposes. Dressings based on biodegradable bacterial cellulose are a good alternative to synthetic polymeric materials [1]. Indeed, sugar and molasses, the production of which is effectively established in the food industry, are used as raw materials for BC production [2,3]. Technologies for the production of BC fibers, tubes and films have been developed, which makes it possible to obtain nonwoven cover bandages for the treatment of wounds and burns [4,5].

Burned patients are especially at risk of bacterial infections due to local and systemic immune dysfunctions. Fungal infection risk factors (prolonged intensive-care-unit stay, mechanical ventilation, broad spectrum antibiotherapy, central venous lines, systemic antibiotics) are frequent in severe burn care and predispose the patient to potentially serious fungal wound infections, whether associated with bacterial infection or not [6]. Although filamentous fungal infections (FFIs) caused by *Aspergillus niger* are not frequent, this fungal infection can cause severe wound infections in patients with extensive burns. Moreover, *Aspergillus* seems to be related with the higher mortality [7,8].

However, BC itself does not have an antibacterial effect and cannot prevent a possible secondary bacterial infection of the skin. To impart antibacterial properties to BC, the material is impregnated with antibiotics [6] and/or with Ag, Cu, ZnO, or TiO_2 nanoparticles [9–14]. The latter are considered more preferable because pathogenic organisms are rapidly developing resistance to antibiotics, which presents a huge problem for modern medicine [15].

Ag NPs/bacterial cellulose composites represent an active area of research for medical applications as evidenced by some recent reviews [16–19]. An important role in their preparation is played by the development of new environmentally friendly technologies. Some of them are based on physical approaches without introducing any chemical reagents. Plasma treatment, magnetron sputtering, hydrothermal synthesis, photoirradiation, etc., are widely used [20–26].

When $AgNO_3$ is used as a source of silver, it is supposed to be reduced by bacterial cellulose. However, as a rule, there are no data on the degree of Ag reduction and elemental composition, which can be used to estimate the content of unreacted $AgNO_3$, while pure silver is deposited during magnetron sputtering.

The modification of the polymer surface by low-temperature low-pressure cold plasma treatment using inert or active gases promotes the formation of functional groups on the surface that can stabilize metal nanoparticles. Plasma treatment changes the surface morphology, increasing the roughness of the material used, and also improves its biocompatibility with human tissues [27,28]. A significant number of works have been devoted to the effects of processing polymers for medical and food purposes with various types of plasma [29–33]. The mechanisms of modification of inert -CH_2 groups for various polymers by their oxidation with active plasma particles and vacuum UV radiation were investigated [27]. However, the use of preliminary plasma treatment of the BC surface with mixtures of oxidizing gases to stabilize metal nanoparticles deposited by magnetron sputtering has been little studied [34]. Oxygen-nitrogen plasma treatment allows the surface of polymers to be modified to form active oxygen-containing groups through direct oxidation with active gases, or rearrangement of charged oxygen-containing groups on the surface into oxidized forms [35]. The interaction of nanoparticles of biologically active metals with such groups leads to their immobilization and effective chemical stabilization, which contributes to the manifestation of bactericidal properties.

It is known that with an increase in the etching time, the surface roughness of fibrous polymers increases, numerous "sharp peaks" of considerable height appear and the surface area increases, which promotes the adsorption of nanoparticles. For etching polymers of different nature, the most effective and affordable method is the use of plasma of atmospheric composition in various instrumentations [36,37].

To activate cellulose against pathogenic microorganisms, nanoparticles of various metals and oxides were applied to it both by traditional chemical methods and by magnetron sputtering. The surface was modified by magnetron co-deposition of Ag and SiO_2 nanoparticles without preliminary plasma treatment of cellulose [38]. The resulting material showed activity against *S. aureus*, *E. coli*, and *C. albicans* strains. Magnetron deposition allows one to control the process parameters and set the size and composition of nanoparticles. Compared to traditional methods using silver salts, plasma processing allows materials to be obtained free of toxic precursor residues [39].

In this work, new hybrid materials with antibacterial activity against pathogens have been prepared by plasma treatment of bacterial cellulose film (BCP) and subsequent deposition of Ag NPs by magnetron sputtering (BCP-Ag). The materials were studied by XPS, a surface-sensitive tool, which provides elemental and chemical compositions.

2. Materials and Methods

2.1. Production and Characterization of Bacterial Cellulose

To prepare bacterial cellulose, the *Gluconacetobacter hansenii* GH-1/2008 strain from the collection of VKPM B-10547 (Gause Institute of New Antibiotics, Moscow, Russia) was used. The strain is non-toxic and non-pathogenic to humans [40]. The used medium was composed of (g/L): sucrose (20.0), peptone (5.0), yeast extract (5.0), Na_2HPO_4 (2.7) and citric acid monohydrate (1.15). The inoculum was prepared by growing *G. hansenii* on this medium with the use of a rotary shaker ThermoStable IS-20 DAIHAN Scientific Co., Seoul, Korea at 30 °C for 3 days. Upon completion of the cultivation, the prepared bacterial cellulose films were separated from the culture broth via filtration, repeatedly washed with distilled water to remove the medium components, treated with 1.0 M NaOH solution at 80 °C for 2 h to remove cells and other impurities immobilized on the films, and finally washed with distilled water until a neutral pH of wash liquid was reached. The detailed preparation of the BC films was described elsewhere [41].

The degree of BC polymerization was determined by the intrinsic viscosity measurement of its solutions in cadoxen according to ASTM D1795-96 and ASTM D4243-99 standards [42]. The average value of polymerization degree was 900.

2.2. Modification of BC Surface by Plasma Treatment

For plasma treatment of the films, a modified high-voltage converter of a VUP-5 vacuum station (SELMI Ltd., Sumy, Ukraine) was used. It was connected to a gas-discharge glass chamber placed under a vacuum cap with a flat stainless steel electrode at the base and a flat aluminum electrode fixed in the upper part of the chamber. The scheme of the installation is shown in Figure 1.

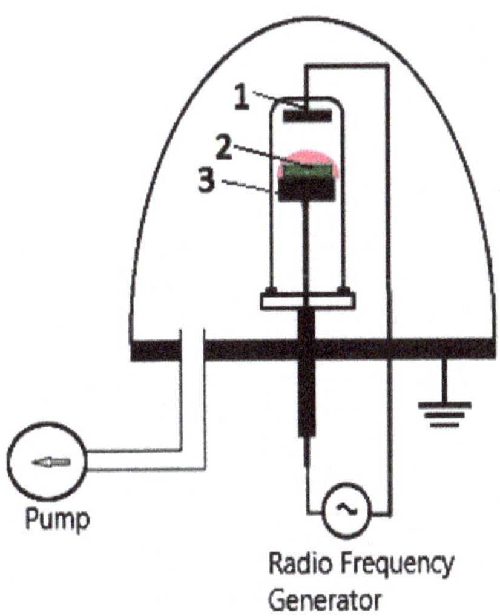

Figure 1. The installation for plasma treatment of the BC films: 1—an aluminum electrode; 2—a film placed on a grounded electrode; 3—a base of the grounded electrode.

The high-voltage discharge frequency was 15 ± 5 kHz, the source voltage was 3 kV, and the generator power was 100 W. The working zone was evacuated to a working pressure of 10 Pa, the composition of the active gas corresponded to atmospheric air. The samples were exposed to oxidizing plasma for 1, 3 and 5 min.

2.3. Metallization of BC Surface by Magnetron Sputtering

To modify BC with silver, a laboratory setup for DC magnetron sputtering built into the VUP-5 was used. A target made of compact silver (99.99%) 10 cm in diameter was mounted on the cathode, and BC samples were placed on the anode with the working side facing the target. The distance between the target and the sample was 40 mm. Argon (99.99%) was used as a source of bombardment gas ions. Before deposition, the target was degassed in vacuum for 5 min, until a base pressure of 5.0×10^{-4} Pa was reached. The metal was deposited at a pressure of 0.2 Pa, a voltage of 700 V, and a power of 400 W. The coating time was 60 s.

2.4. Scanning Electron Microscopy

The surface morphology of the BC films was studied by low-voltage scanning electron microscopy (LVSEM) with a Scios (FEI, Waltham, MA, USA) microscope at an accelerating voltage of 1 kV. EDX studies were carried out using Oxford Instruments X-max EDX system, (Abingdon, Oxfordshire, UK).

2.5. Transmission Electron Microscopy

The BCP-Ag samples were studied using Hitachi transmission electron microscope HT7700 (Tokyo, Japan). Images were acquired at 100 kV accelerating voltage. Before measurements the samples were mounted on a 3 mm copper grid with a carbon film and fixed in a grid holder.

2.6. X-ray Photoelectron Spectroscopy

X-ray photoelectron spectra were acquired with an Axis Ultra DLD (Kratos, UK) spectrometer using monochromatized Al Kα (1486.6 eV) radiation at an operating power of 150 W of the X-ray tube. Survey and high-resolution spectra of appropriate core levels were recorded at pass energies of 160 and 40 eV and with step sizes of 1 and 0.1 eV, respectively. Sample area of 300 μm × 700 μm contributed to the spectra. The samples were mounted on a sample holder with a two-sided adhesive tape, and the spectra were collected at room temperature. The base pressure in the analytical UHV chamber of the spectrometer during measurements did not exceed 10^{-8} torr. The energy scale of the spectrometer was calibrated to provide the following values for reference samples (i.e., metal surfaces freshly cleaned by ion bombardment): Au $4f_{7/2}$–83.96 eV, Cu $2p_{3/2}$–932.62 eV, Ag $3d_{5/2}$–368.21 eV. The electrostatic charging effects were compensated by using an electron neutralizer. Sample charging was corrected by referencing to the C-OH peak identified in the C ls spectrum to which a binding energy of 286.73 eV was assigned [43]. The background of inelastic electron energy losses was subtracted by the Shirley method. The elemental composition was calculated using atomic sensitivity factors included in the software of the spectrometer corrected for the transfer function of the instrument.

2.7. Antimicrobial Activity Assay

The antimicrobial activity of the BC films was measured by the disc-diffusion method. Disks of 150 μm thickness without nanoparticles (control) and coated with Ag NPs were used. Antifungal activity was assessed using test strain of yeast *C. albicans* ATCC 2091 and opportunistic filamentous fungi *Aspergillus niger* ATCC 16404. The spectrum of antibacterial activity was studied using test cultures of gram-positive bacteria strains—*Bacillus subtilis* ATCC 6633, *S. aureus* FDA 209 ATCC 6538 and gram-negative bacteria—*Escherichia coli* ATCC 25922 from the collection of cultures of the Gause Institute of New Antibiotics (Moscow, Russia). Bacteria and fungi were incubated at 37 °C for 24 h. Standard discs with antibiotics (amoxicillin—20 μg/disc, amphotericin B—40 μg/disc, manufactured by Institut Pasteur, (Saint-Petersburg, Russia) and sterile discs with BC without antibiotics were used as negative and positive controls.

3. Results and Discussion

The development of new effective and environmentally friendly methods for the synthesis of antibacterial drugs and materials is a research priority. An increase in the number of strains of microorganisms that are resistant to most antibiotics makes it worthless to apply previously developed and intensively used antibacterial drugs.

The use of metal nanoparticles, which have high biological activity and do not cause resistance of microorganisms, to obtain medical materials is one of the actively developing areas of research. Currently, silver nanoparticles are the preferred candidates for incorporation into medical polymeric materials to impart antibacterial properties to them [44]. As a rule, methods of the chemical reduction of metal salts are used to obtain a biopolymer matrix with embedded Ag NPs [45,46].

Nanoparticles can be prepared by reducing silver salt on the surface of BC with visible radiation [47]. In all cases, the researchers postulated a good antimicrobial activity of the obtained materials; however, the cytotoxic effect was not tested. These methods have a number of limitations, which significantly complicate the use of the obtained materials for biomedical purposes. These are the presence of a significant amount of impurities of surfactants and residues of synthesis products, as well as the difficulty of controlling the completeness of metal reduction [48]. It should also be noted that in some cases, significant thermal heating is required during the recovery process. This can lead to partial degradation of the biopolymer and, accordingly, to a change in its molar mass.

One of the promising methods for the synthesis of metal nanoparticles is magnetron sputtering, which makes it possible to apply Ag NPs of a given composition under controlled conditions. However, metal nanoparticles deposited onto the BC films are known to have poor adhesion to the surface [49]. To increase the adhesion properties of the surface and the ability to retain nanoparticles, polymers are exposed to various oxidizing agents [50].

In the literature, the processing time that is traditionally used for surface plasma treatment of the various polymer films and fibers is about 1–30 min at 10–100 W. A processing time of less than 1 min does not increase the hydrophilicity of cotton, cotton-PET and PP at a given power [51]. Our experiments reveal the same results for 1 min BC treatment, while an exposure time of more than 5 min significantly reduces the mechanical properties of the films. Taking into account the results obtained, samples treated with plasma for 3 min were examined. A mixture of gases of atmospheric composition was used as an active plasma, since pure oxygen burns out the BC surface, increasing the roughness, but not the number of oxidized groups required to stabilize deposited Ag NPs [52]. BC films were modified using combination of low-pressure low-temperature atmospheric plasma treatment and Ag magnetron sputtering. To assess the surface morphology of the prepared nanomaterials, SEM was implicated.

Figure 2 shows SEM images of the initial BC (a), treated with plasma (BCP) (b) and a nanocomposite covered with agglomerates of Ag nanoparticles (BCP-Ag) with different magnifications (c,d). Compared to the original BC, the plasma-treated polymer surface underwent significant changes. Its roughness and heterogeneity increased due to the destruction of the fibers. SEM analysis of BCP-Ag films showed a uniform distribution of silver particles over the surface of the plasma-treated polymer that can be seen on the microphotograph as a white network of dots (Figure 2c,d).

Figure 3 displays the distribution maps of C, O, Ag and the energy dispersive X-ray spectrum of the BCP-Ag nanocomposite. It can be concluded that silver present in the nanocomposite and is not uniformly distributed on the surface at this length scale. Elemental analysis through EDX confirmed the presence of silver on the coated fabric.

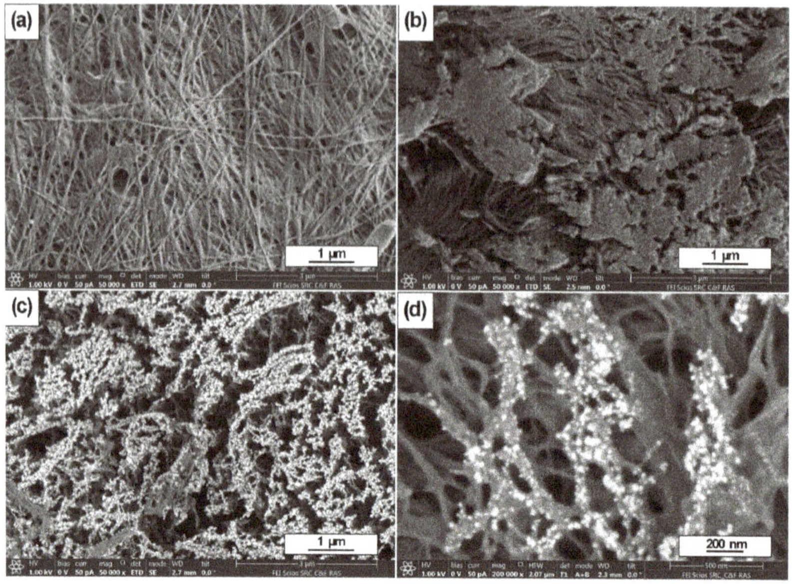

Figure 2. SEM images of the morphology of the initial bacterial cellulose film (BC) (**a**), BC after plasma treatment (BCP) (**b**) and BSE images of the BCP with agglomerates of Ag nanoparticles lying on its surface after Ag magnetron sputtering (**c**,**d**).

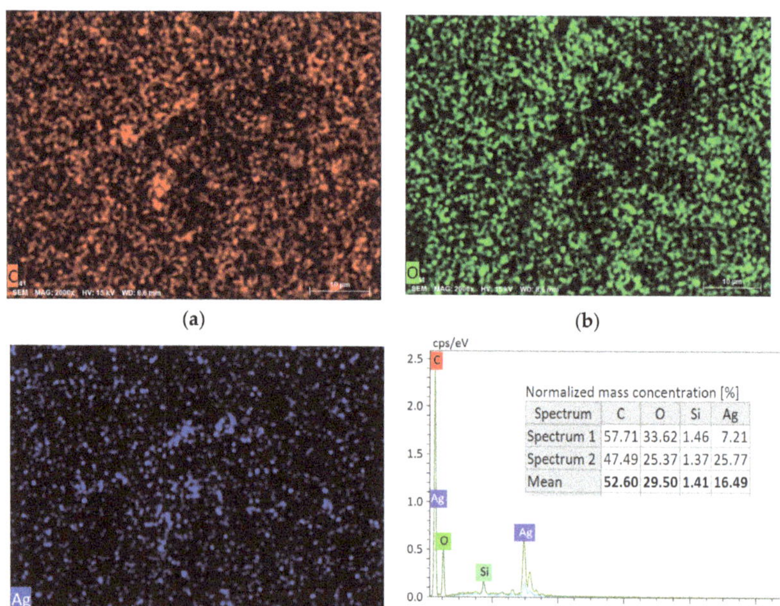

Figure 3. The elemental distribution of C (**a**), O (**b**), Ag (**c**) and energy dispersive X-ray spectrum (**d**) of the BCP-Ag nanocomposite.

It was shown that during magnetron deposition of metal nanoparticles on a plasma-modified surface, the surface morphology is intact [53]. Our results are same. Apparently, the distribution of nanoparticles over the nanocomposite surfaces is likely to be greatly influenced by the contour of inhomogeneities and oxidized regions formed during etching of BC, where they are stabilized. The SEM micrographs showed that Ag NPs deposited on the BC surface are located mainly in places with surface inhomogeneities—ridges, roughness, etc.

This may be influenced by an uncompensated charge on these surface fragments, which leads to the coalescence of metal particles formed during magnetron evaporation. It was shown [54] that Ag NPs coalesce into worm-like structures on inert (not plasma-treated) surfaces upon prolonged deposition. When the surface is treated with plasma, this effect becomes even more pronounced. The particle pattern took the form of clusters based on partially destroyed networks of BC fibers.

In order to estimate the size of the deposited Ag NPs, a set of TEM images was obtained. Figure 4 shows a TEM micrograph of BCP-Ag nanocomposite fibers with the particle size distribution. Ag nanoparticles with sizes ranging from 5 to 35 nm were recorded. Larger particles represent aggregates in the form of a "bunch of grapes", consisting of smaller particles. For statistical analysis hundreds of nanoparticles were taken into the account.

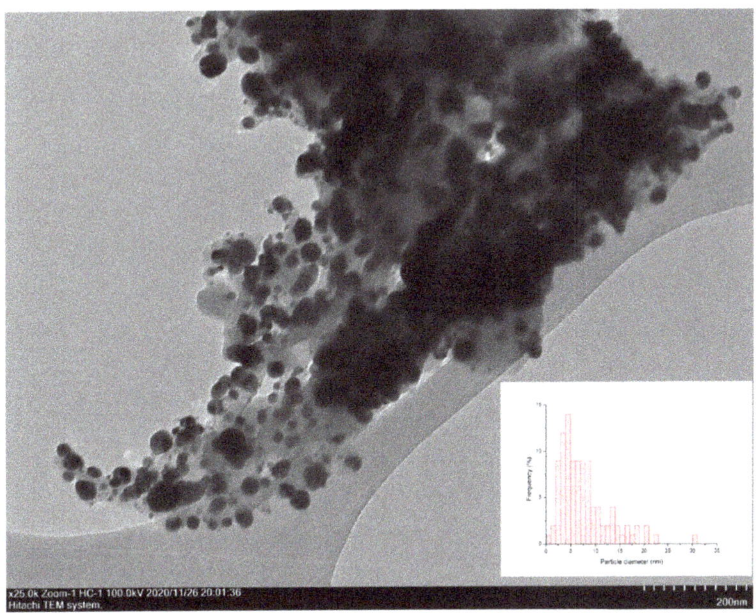

Figure 4. Bright field TEM image of the surface of BCP-Ag composite, obtained by plasma treatment of the BC and subsequent Ag nanoparticle deposition. In the right bottom corner is a histogram of the nanoparticles' size distribution.

To evaluate chemical transformations of the BC surface induced by the plasma treatment, a comparative XPS study of untreated samples BC and BC-Ag (BC with deposited Ag) and treated for 3 min BCP and BCP-Ag was performed.

Figure 5 shows the C 1s photoelectron spectra, fitted with some Gaussian profiles using the reliable chemical shifts [43]. The peaks at ~285, 286.73, 288.06 and 288.6 eV correspond to the C-C/C-H, C-OH/C-O-C, O-C-O and C(O)O groups, respectively. The relative intensities of different groups and a characterization of the photoelectron peaks are presented in Table 1. It is clearly seen that plasma treatment and Ag deposition lead to their significant transformation. After modification of the original BC, an increase in relative intensity of C-C/C-H groups was observed. There is also an increase in the signal at 288.06 eV, which may be assigned to C=O groups formed as a result of the oxidation of BC by plasma. The relative intensities of C(O)O groups are rather similar. This may indicate the destruction of BC oxypyran rings and the oxidation of alcohol groups.

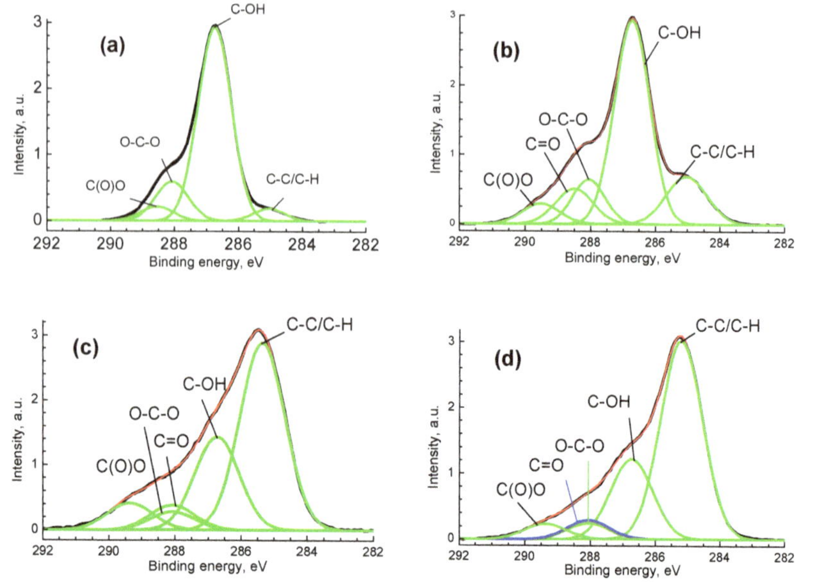

Figure 5. The C 1s spectra for pristine BC (**a**), BCP (**b**), BC-Ag (**c**) and BCP-Ag (**d**), black—experimental, red—fitting, green—individual peaks, blue—peak assigned to C=O group, which ovelapped with peak assigned to O-C-O group.

Table 1. Characteristics of the C 1s photoelectron spectra: binding energies (E_b), Gaussian widths (W), and relative intensities (I_{rel}) of photoelectron peaks belonging to different chemical groups.

Sample	Group	C-C/C-H	C-OH	O-C-O	C=O	C(O)O
BC	E_b, eV	285.03	286.73	288.08		288.56
	W, eV	0.96	1.04	1.03		1.20
	I_{rel}	0.05	0.75	0.15		0.05
BCP	E_b, eV	285.03	286.73	288.06	288.5	289.51
	W, eV	1.26	1.09	1.0	1.15	1.15
	I_{rel}	0.15	0.57	0.11	0.10	0.06
BC-Ag	E_b,eV	285.4	286.73	288.06	288.06	289.37
	W, eV	1.36	1.34	1.34	1.34	1.37
	I_{rel}	0.54	0.26	0.05	0.07	0.08
BCP-Ag	E_b, eV	285.19	286.73	288.06	288.06	289.37
	W, eV	1.30	1.30	1.30	1.30	1.30
	I_{rel}	0.60	0.24	0.05	0.06	0.05

The elemental compositions of samples BC, BCP, BC-Ag and BCP-Ag obtained from high-resolution spectra atoms without taking into account impurity atoms are shown in Table 2. It shows significant decrease in surface oxygen content and O/C ratio during the treatment.

Table 2. XPS quantification data (at. %) determined from the high-resolution spectra.

Samples	C	O	Ag	Ag/O	O/C	Ag/C
BC	42.3	57.7			1.36	
BCP	45.5	54.5			1.2	
BC-Ag	47.8	17.8	34.3	1.93	0.37	0.72
BCP-Ag	69.8	21.2	9.0	0.42	0.30	0.13

Figure 6 shows the Ag 3d photoelectron spectra and MNN Auger spectra of the composites, and Table 3 shows their characteristics. The binding energies and the widths of the photoelectron peaks differ markedly, which indicate different chemical states of Ag atoms and the presence of at least two chemical states of atoms in BCP-Ag nanocomposite.

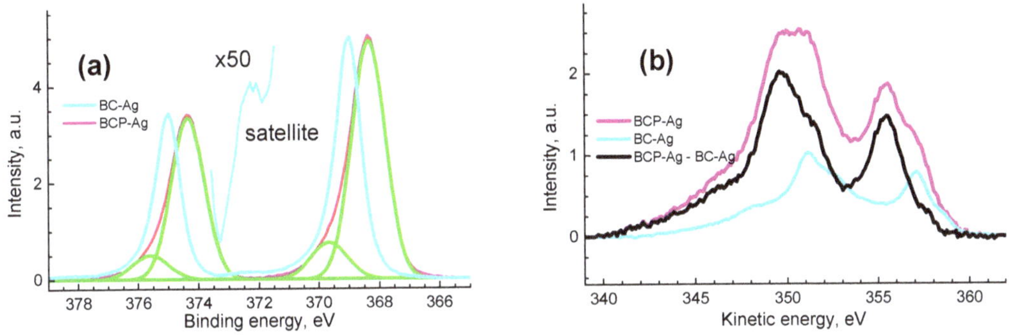

Figure 6. Ag 3d photoelectron spectra (**a**) and MNN Auger spectra (**b**) for BC-Ag and BCP-Ag nanocomposites.

Table 3. Characteristics of photoelectron and Auger spectra for Ag and the nanocomposite: binding energies (E_b), spin-orbit splitting (SOS), kinetic energies (E_k) and Auger parameters.

	E_b		SOS	E_k		AP	
Sample	**$3d_{5/2}$, eV**	**$3d_{3/2}$, eV**	**$3d_{3/2}$- $3d_{5/2}$, eV**	**$M_4N_{45}N_{45}$, eV**	**$M_5N_{45}N_{45}$, eV**	**$3d_{5/2}$+ $M_4N_{45}N_{45}$, eV**	**State**
Ag	368.327			357.855		726.182	Ag^0 [55]
BC-Ag	369.0	375.0	6.0	357.1		726.1	Ag^0
BCP-Ag	368.4 369.7	374.4 375.6	6.0 5.9	355.5 357.1	349.7 351.2	723.8 726.8	$Ag^{\delta+}$ Ag^0

The presence of a plasmon loss peak at 372.3 eV in the spectrum of the BC-Ag sample (Figure 6a) indicates the Ag^0 state [56], which is also confirmed by the value of 726.1 eV of the modified Auger parameter. The difference between the measured and literature values of the binding energies of photoelectron peaks, kinetic energies and Auger parameters should be attributed to both the size effect caused by the small size of Ag nanoparticles and the manifestation of local differential charging near Ag nanoparticles and the inhomogeneous distribution of Ag over the depth of the samples.

The use of the MNN Auger spectrum of the BC-Ag sample as a reference made it possible to discriminate two states in the Auger spectrum of the BCP-Ag sample, one of which is Ag^0, and the second at kinetic energies of 349.7 and 351.2 eV should be assigned to $Ag^{\delta+}$ (Figure 6). To represent the spectrum as two states, the spectra were normalized by intensity in the region of high kinetic energies to achieve their best match, and the spectrum of the BC-Ag sample was subtracted from the Auger spectrum of the BCP-Ag sample, as shown in Figure 6. The area ratio of the Auger spectra BCP-Ag/BC-Ag is ~2.

Corresponding correlations are also observed in the kinetic energies and in the shape of Auger peaks. In the Ag $3d_{5/2}$ spectrum, two states at 368.4 and 369.7 eV with relative intensities of 0.86 and 0.14 are distinguished, which are assigned to Ag^0 state and Ag acetate, respectively. The difference observed in the data obtained from the photoelectron and Auger spectra indicates a nonuniform distribution of the Ag^0 state over the depth of the sample. This follows from the fact that the inelastic mean free path of photoelectrons is greater than that of Auger electrons in view of its dependence on kinetic energy. That is, in the case under consideration, Ag MNN Auger electrons are more surface-sensitive than Ag 3d photoelectrons. It is not strange that outer layers contain silver presumably in oxide form.

Figure 7 shows the O 1s spectra of the studied samples with selected chemical states of oxygen atoms, and Table 4 shows their characteristics.

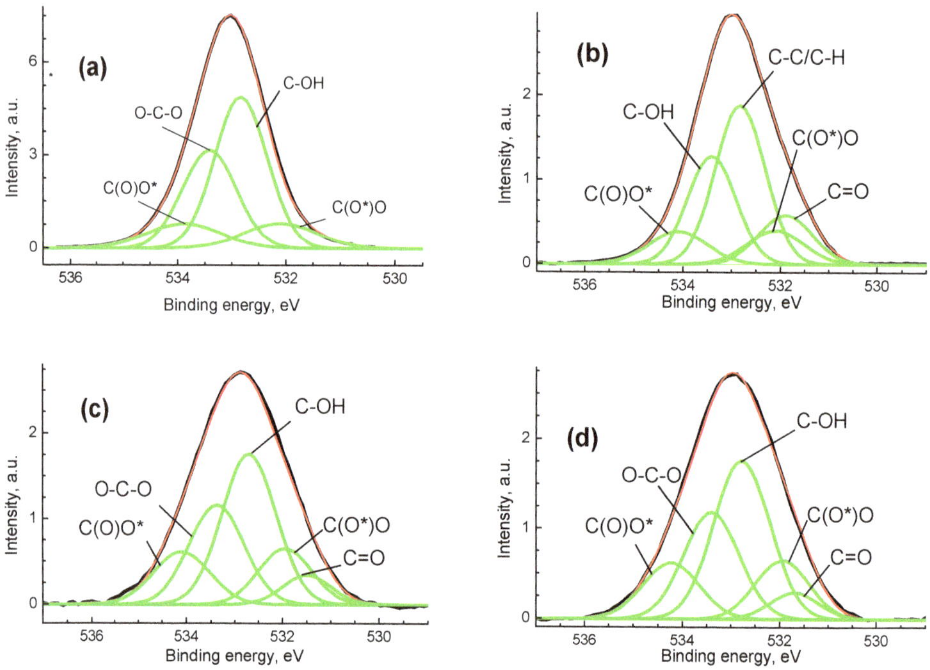

Figure 7. The O 1s photoelectron spectra for the BC (**a**), BCP (**b**), BC-Ag (**c**) and BCP-Ag (**d**), black—experimental, red—fitting, green—individual peaks.

The O 1s spectrum of pristine cellulose is characterized with four peaks at 532.1, 532.8, 533.4 and 534.0 eV which are assigned to C(O*)O, C-OH, O-C-O and C(O)O* groups. The peaks at 532.8 and 533.4 with an intensity ratio of 3:2 are characteristic of cellulose. The peaks related to carboxylate groups have equal intensities. After the plasma treatment and deposition of Ag, a new peak at 531.7 eV related to C=O bond was recorded. In the O 1s spectra of the BC-Ag and BCP-Ag samples, an increase in the relative concentrations of C(O)O groups is observed in comparison with that of BC and BCP samples. As a result of plasma treatment of the original cellulose, a carbonyl group appears, while in the case of BC-Ag and BCP-Ag, the samples' spectra are similar. In other words, the silver present protects the cellulose from potential damage. Therefore, we can conclude that it is mainly located in the upper layers and does not penetrate deep into the material. This is also confirmed by the quantification data (Table 4), from which it follows that the deposition of silver leads to a sharp decrease in the oxygen content.

A significant decrease in the oxygen content in metallized samples and an increase in the carbon content can be explained by the screening effect of nanoparticle aggregates concentrated around oxygen-containing groups. In this case, the depths of secondary electrons escape for oxygen and carbon are different, for the O 1s photoelectrons it is smaller and can be screened to a certain extent by Ag. In addition, part of the oxygen can be bound by metal nanoparticles.

Table 4. Characteristics of the O 1s photoelectron spectra: binding energies (E_b), Gaussian widths (W) and relative intensities (I_{rel}) of photoelectron peaks belonging to different chemical groups.

Sample	Group	C=O	C(O*)O	C-OH	O-C-O	C(O)O*
BC	E_b, eV		532.1	532.8	533.4	534.0
	W, eV		1.00	0.91	0.91	1.15
	I_{rel}		0.1	0.48	0.32	0.1
BCP	E_b, eV	531.9	532.1	532.8	533.4	534.1
	W, eV	1.17	1.1	1.02	1.0	1.2
	I_{rel}	0.13	0.1	0.4	0.27	0.1
BC-Ag	E_b, eV	531.5	532.0	532.7	533.4	534.1
	W, eV	1.0	1.1	1.13	1.14	1.17
	I_{rel}	0.07	0.14	0.39	0.26	0.14
BCP-Ag	E_b, eV	531.7	532.0	532.8	533.4	534.2
	W, eV	1.00	1.10	1.15	1.14	1.15
	I_{rel}	0.06	0.14	0.40	0.26	0.14

The moderate antifungal activity of the obtained samples was investigated in relation to the test strain of opportunistic filamentous fungi *Aspergillus niger* ATCC 16404. After incubation, the inhibition zones for *Aspergillus niger* were found to be 8 ± 0.1 for BCP-Ag and 10 ± 0.4 mm for AmpB, respectively. Plasma treatment of the BC surface affects the antifungal activity of magnetron-deposited silver against the *A. niger* ATCC 16404 strain. Wang et al. [57] showed that this species easily degrades cellulose-containing materials in a humid environment, which causes certain difficulties when it was used for wound healing or product packaging.

The antibacterial activity was studied using test cultures of Gram-positive bacteria strains—*Bacillus subtilis* ATCC 6633, *S. aureus* FDA 209 ATCC 6538—and Gram-negative bacteria—*Escherichia coli* ATCC 25922. 25922. Table 5 presents data on the antimicrobial activity of materials based on BC. The pristine BC and that treated with plasma showed no biological activity.

Table 5. Antimicrobial activity of the bacterial cellulose samples with Ag nanoparticle and plasma treatment.

Sample	Inhibition Zone, mm.				
	Bacillus subtilis ATCC 6633	*S. aureus* FDA 209 ATCC 6538	*Escherichia coli* ATCC 25922.	*Aspergillus niger* ATCC 16404	*C. albicans* ATCC 2091
BCP	0	0	0	0	0
BC-Ag	16 ± 0.4	0	0	0	0
BCP-Ag	18 ± 0.3	9 ± 0.1	0	8 ± 0.1	0
Amoxicillin 20 μg.	42 ± 0.7	27 ± 0.7	29 ± 0.2	not tested	not tested
Amphotericin B 40 μg.	not tested	not tested	not tested	10 ± 0.4	12 ± 0.6

The BCP-Ag material exhibits moderate antibacterial activity against Gram-positive bacteria *B. subtilis* ATCC 6633 and weak activity towards *S. aureus* FDA 209 ATCC 6538 (Figure 8). The antimicrobial activities for BCP-Ag and BC-Ag towards *Bacillus subtilis* were equal, whereas BC-Ag was not active on *S. aureus*. However, no activity against Gram-negative bacteria was found for BC-based composites.

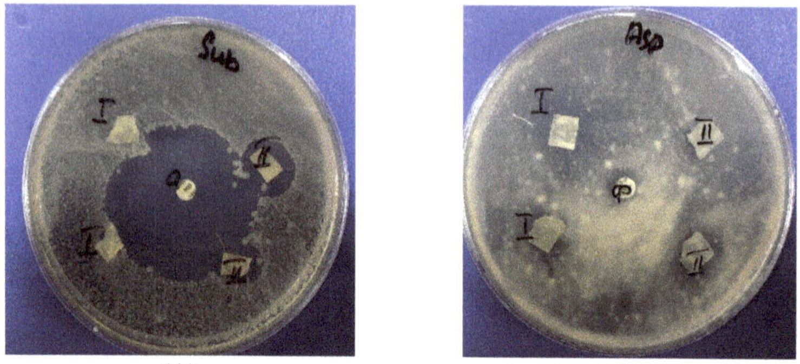

(A) **(B)**

Figure 8. Antimicrobial activity of bacterial cellulose films with Ag and plasma treatment detected by disc-diffusion method: I—BC-Ag; II—BCP-Ag; (**A**) *B. subtilis* ATCC 6633; (**B**) *A. niger* ATCC 16404.

The zone of inhibition of the *A. niger* ATCC 16404 for the BCP-Ag composite is comparable to the amphotericin B-polyene macrocyclic antibiotic. Thus, the combination of BC plasma treatment and magnetron sputtering of metal can be used to impart new functional properties to the material.

4. Conclusions

By treating cellulose with low-frequency plasma and the simultaneous deposition of silver nanoparticles, a new material with antimicrobial properties was obtained. XPS analysis of the material surface made it possible to isolate various chemical groups and characterize the charge states of silver atoms using the Auger parameter. It was established that the combination of methods of plasma processing materials based on bacterial cellulose, synthesized by the producer *G. hansenii* GH-1/2008, with magnetron sputtering of Ag NPs, makes it possible to obtain medical and other products resistant to destruction by the mold fungi *Aspergillus niger* and possessing antibacterial activity against the Gram-positive bacteria *Bacillus subtilis* and *Staphylococcus aureus*.

XPS revealed that BC plasma treatment leads to the formation of new oxygen-containing functional groups on the polymer surface, which effectively stabilize Ag nanoparticles with a diameter of 5–35 nm. Ag nanoparticles exist on the surface of BC pre-treated by plasma as metal Ag and Ag acetate.

The films with plasma treatment and Ag nanoparticles demonstrated antimicrobial action against the Gram-positive bacteria *Bacillus subtilis* and *Staphylococcus aureus* and filamentous fungi. These findings might provide insights into the possible therapeutic application of BCP-Ag films as an agent to treat bacterial and fungal wound infections.

The proposed complex method for obtaining biologically active materials is an environmentally friendly technology that does not introduce contamination into the material, which is confirmed by XPS analysis at all stages of its production.

Author Contributions: Conceptualization, supervision, A.V.; methodology, A.B., T.G. and V.S.; investigation, A.N., A.B., N.A. and A.V.; resources, T.G. and V.S.; writing—original draft preparation, A.B. and M.P.; writing—review and editing, A.V. and A.N. All authors have read and agreed to the published version of the manuscript.

Funding: The APC was funded by MDPI. This work was supported by the Ministry of Science and Higher Education of the Russian Federation (Contract/agreement No. 075-00697-22-00). SEM investigations were supported by the Ministry of Science and Higher Education of the Russian Federation using the equipment of Shared Research Center FSRC "Crystallography and Photonics" of RAS.

Institutional Review Board Statement: Not applicable.

Data Availability Statement: Not applicable.

Conflicts of Interest: The authors declare no conflict of interest.

References

1. Moniri, M.; Boroumand, M.A.; Azizi, S.; Abdul, R.R.; Bin Ariff, A.; Zuhainis, S.W.; Navaderi, M.; Mohamad, R. Production and status of bacterial cellulose in biomedical engineering. *Nanomaterials* **2017**, *7*, 257. [CrossRef] [PubMed]
2. Han, Y.-H.; Mao, H.-L.; Wang, S.-S.; Deng, J.-C.; Chen, D.-L.; Li, M. Ecofriendly green biosynthesis of bacterial cellulose by Komagataeibacter xylinus B2-1 using the shell extract of Sapindus mukorossi Gaertn. as culture medium. *Cellulose* **2019**, *27*, 1255–1272. [CrossRef]
3. Tsouko, E.; Maina, S.; Ladakis, D.; Kookos, I.K.; Koutinas, A. Integrated biorefinery development for the extraction of value-added components and bacterial cellulose production from orange peel waste streams. *Renew. Energy* **2020**, *160*, 944–954. [CrossRef]
4. Liu, W.; Du, H.; Zhang, M.; Liu, K.; Liu, H.; Xie, H.; Zhang, X.; Si, C. Bacterial Cellulose-Based Composite Scaffolds for Biomedical Applications: A Review. *ACS Sustain. Chem. Eng.* **2020**, *8*, 7536–7562. [CrossRef]
5. Tseomashko, N.E.; Rai, M.; Vasil'kov, A.Y. New hybrid materials for wound cover dressings. In *Bio-Polymer-Based Nano Films*; Rai, M., Dos Santos, C.A., Eds.; Elsevier Inc.: Cambridge, MA, USA, 2021; pp. 203–245. [CrossRef]
6. Pruskowski, K.A.; Mitchell, T.A.; Kiley, J.L.; Wellington, T.; Britton, G.W.; Cancio, L.C. Diagnosis and Management of Invasive Fungal Wound Infections in Burn Patients. *Eur. Burn J.* **2021**, *2*, 168–183. [CrossRef]
7. Aries, P.; Hoffmann, C.; Schaal, J.-V.; Leclerc, T.; Donat, N.; Cirodde, A.; Masson, Y.; Renner, J.; Soler, C. *Aspergillus tamarii*: An uncommon burn wound infection. *J. Clin. Pathol.* **2018**, *71*, 379–380. [CrossRef]
8. Struck, M.F.; Gille, J. Fungal infection in burns: Acomprehensive reviwer. *Ann. Burn. Fire Disasters* **2013**, *26*, 147–153.
9. Zhang, Z.-Y.; Sun, Y.; Zheng, Y.-D.; He, W.; Yang, Y.-Y.; Xie, Y.-J.; Feng, Z.-X.; Qiao, K. A biocompatible bacterial cellulose/tannic acid composite with antibacterial and anti-biofilm activities for biomedical applications. *Mater. Sci. Eng. C.* **2020**, *106*, 110249. [CrossRef]
10. Gromovykh, T.I.; Vasil'kov AYu Sadykova, V.S.; Feldman, N.B.; Demchenko, A.G.; Lyundup, A.V.; Butenko, I.E.; Lutsenko, S.V. Creation of composites of bacterial cellulose and silver nanoparticles: Evaluation of antimicrobial activity and cytotoxicity. *Int. J. Nanotechnol.* **2019**, *16*, 408–420. [CrossRef]
11. Volova, T.; Shumilova, A.; Shidlovsky, I.; Nikolaeva, E.; Sukovatiy, A.; Vasilieva, A.; Shishatskaya, E. Antibacterial properties of films of cellulose composites with silver nanoparticles and antibiotics. *Polym. Test.* **2018**, *65*, 54–68. [CrossRef]
12. He, W.; Huang, X.; Zheng, Y.; Sun, Y.; Xie, Y.; Wang, Y.; Yue, L. In situ synthesis of bacterial cellulose/copper nanoparticles composite membranes with long term antibacterial property. *J. Biomater. Sci. Polym. Ed.* **2018**, *29*, 2137–2153. [CrossRef]
13. Mocanu, A.; Isopencu, G.; Busuioc, C.; Popa, O.-M.; Dietrich, P.; Socaciu-Siebert, L. Bacterial cellulose films with ZnO nanoparticles and propolis extracts: Synergistic antimicrobial effect. *Sci. Rep.* **2019**, *9*, 17687. [CrossRef]
14. Malmira, S.; Karbalaeia, A.; Pourmadadip, M.; Hamedid, I.; Yazdianc, F.; Navaeef, M. Antibacterial properties of a bacterial cellulose CQD-TiO_2 nanocomposite. *Carbohydr. Polym.* **2020**, *234*, 115835. [CrossRef]
15. Peterson, E.; Kaur, P. Antibiotic Resistance Mechanisms in Bacteria: Relationships Between Resistance Determinants of Antibiotic Producers, Environmental Bacteria, and Clinical Pathogens. *Front. Microbiol.* **2018**, *9*, 2928. [CrossRef]
16. Benetti, G.; Cavaliere, E.; Banfi, F.; Gavioli, L. Antimicrobial nanostructured coatings: A gas phase deposition and magnetron sputtering perspective. *Materials* **2020**, *13*, 784. [CrossRef]
17. Azizi-Lalabadi, M.; Garavand, F.; Jafari, S.M. Incorporation of silver nanoparticles into active antimicrobial nanocomposites: Release behavior, analyzing techniques, applications and safety issues. *Adv. Colloid Interface Sci.* **2021**, *293*, 102440. [CrossRef]
18. Khan, S.; Ul-Islam, M.; Ullah, M.W.; Zhu, Y.; Narayanan, K.B.; Han, S.S.; Park, J.K. Fabrication strategies and biomedical applications of three-dimensional bacterial cellulose-based scaffolds: A review. *Int. J. Biol. Macromol.* **2022**, *209*, 9–30. [CrossRef]
19. Popa, L.; Ghica, M.V.; Tudoroiu, E.E.; Ionescu, D.G.; Dinu-Pîrvu, C.E. Bacterial Cellulose—A Remarkable Polymer as a Source for Biomaterials Tailoring. *Materials* **2022**, *15*, 1054. [CrossRef]
20. Radetić, M.; Marković, D. A review on the role of plasma technology in the nano-finishing of textile materials with metal and metal oxide nanoparticles. *Plasma Process. Polym.* **2022**, *19*, 2100197. [CrossRef]
21. Dallas, P.; Sharma, V.K.; Zboril, R. Silver polymeric nanocomposites as advanced antimicrobial agents: Classification, synthetic paths, applications, and perspectives. *Adv. Colloid Interface Sci.* **2011**, *166*, 119–135. [CrossRef]
22. Mejía, M.I.; Restrepo, G.; Marín, J.M.; Sanjines, R.; Pulgarín, C.; Mielczarski, E.; Mielczarski, J.; Kiwi, J. Magnetron-sputtered Ag surfaces. New evidence for the nature of the Ag ions intervening in bacterial inactivation. *ACS Appl. Mater. Interfaces* **2010**, *2*, 230–235. [CrossRef] [PubMed]
23. Katouah, H.; El-Metwaly, N.M. Plasma treatment toward electrically conductive and superhydrophobic cotton fibers by in situ preparation of polypyrrole and silver nanoparticles. *React. Funct. Polym.* **2021**, *159*, 104810. [CrossRef]
24. Atta, A.M.; Abomelka, H.M. Multifunctional finishing of cotton fibers using silver nanoparticles via microwave-assisted reduction of silver alkylcarbamate. *Mater. Chem. Phys.* **2021**, *260*, 124137. [CrossRef]
25. Yang, G.; Xie, J.; Deng, Y.; Bian, Y.; Hong, F. Hydrothermal synthesis of bacterial cellulose/AgNPs composite: A "green" route for antibacterial application. *Carbohydr. Polym.* **2012**, *87*, 2482–2487. [CrossRef]

26. Zahoor, M.; Nazir, N.; Iftikhar, M.; Naz, S.; Zekker, I.; Burlakovs, J. Faheem Uddin Abdul Waheed Kamran AKallistova NPimenov Ali Khan, F. A review on silver nanoparticles: Classification, various methods of synthesis, and their potential roles in biomedical applications and water treatment. *Water* **2021**, *13*, 2216. [CrossRef]
27. Shalia, T.; Sudarsan, N. Tailoring the Adhesion of Polymers using Plasma for Biomedical Applications: A Critical Review. *Rev. Adhes. Adhes.* **2015**, *3*, 53–97.
28. Pertile, R.; Andrade, F.; Alves, C.; Gama, M. Surface modification of bacterial cellulose by nitrogen-containing plasma for improved interaction with cells. *Carbohydr. Polym.* **2010**, *82*, 692–698. [CrossRef]
29. Bhanthumnavin, W.; Wanichapichart, P.; Taweepreeda, W.; Sirijarukula, S.; Paosawatyanyong, B. Surface modification of bacterial cellulose membrane by oxygen plasma treatment. *Surf. Coat. Technol.* **2016**, *306*, 272–278. [CrossRef]
30. Wei, P.; Lou, H.; Xu, X.; Xu, W.; Yang, H.; Zhang, W.; Zhang, Y. Preparation of PP non-woven fabric with good heavy metal adsorption performance via plasma modification and graft polymerization. *Appl. Surf. Sci.* **2021**, *539*, 148195. [CrossRef]
31. Learn, G.; Lai, E.; von Recum, H. Nonthermal Plasma Treatment Improves Uniformity and Adherence of Cyclodextrin-Based Coatings on Hydrophobic Polymer Substrates. *Coatings* **2020**, *10*, 1056. [CrossRef]
32. Gama, L.; Duque, T.; Özcan, M.; Philippi, A.; Mezzomo, L.; Gonçalves, T. Adhesion to high-performance polymers applied in dentistry: A systematic review. *Dent. Mater.* **2020**, *36*, e93–e108. [CrossRef]
33. Haji, A.; Naebe, M. Cleaner dyeing of textiles using plasma treatment and natural dyes: A review. *J. Clean. Prod.* **2020**, *265*, 121866. [CrossRef]
34. Lv, P.; Xu, W.; Li, D.; Feng, Q.; Yao, Y.; Pang, Z.; Lucia, L.; Wei, Q. Metal-based bacterial cellulose of sandwich nanomaterials for anti-oxidation electromagnetic interference shielding. *Mater. Des.* **2016**, *112*, 374–382. [CrossRef]
35. Vosmanska, V.; Kolarova, K.; Rimpelova, S.; Svorcik, V. Surface modification of oxidized cellulose haemostat by argon plasma treatment. *Cellulose* **2014**, *21*, 2445–2456. [CrossRef]
36. Ribeiro, A.I.; Senturk, D.; Silva, K.S.; Modic, M.; Cvelbar, U.; Dinescu, G.; Mitu, B.; Nikiforov, A.; Leys, C.; Kuchakova, I.; et al. Efficient silver nanoparticles deposition method on DBD plasma-treated polyamide 6,6 for antimicrobial textiles. *IOP Conf. Ser. Mater. Sci. Eng.* **2018**, *460*, 012007. [CrossRef]
37. Tomsic, B.; Vasiljević, J.; Simoncic, B.; Radoičić, M.; Radetić, M. The influence of corona treatment and impregnation with colloidal TiO2 nanoparticles on biodegradability of cotton fabric. *Cellulose* **2017**, *24*, 4533–4545. [CrossRef]
38. Irfan, M.; Perero, S.; Miola, M.; Maina, G.; Ferri, A.; Ferraris, M.; Balagna, C. Antimicrobial functionalization of cotton fabric with silver nanoclusters/silica composite coating via RF co-sputtering technique. *Cellulose* **2017**, *24*, 2331–2345. [CrossRef]
39. Olenin, A.; Lisichkin, G. Metal nanoparticles in condensed media: Preparation and the bulk and surface structural dynamics. *Russ. Chem. Rev.* **2011**, *80*, 605–630. [CrossRef]
40. Gromovykh, T.; Phan, M.; Danil'chuk, T. Gluconacetobacter Hansenii GH1/2008 Bacterial Strain—Bacterial Cellulose Producer. RF Patent 2464307, 20 October 2012.
41. Gromovykh, T.; Lutcenko, S.; Feldman, N.; Kashirin, V.; Sadykova, V.; Dmitrenok, A.; Kashirin, V. Bacterial cellulose synthesized by *Gluconacetobacter hansenii* for medical applications. *Appl. Biochem. Microbiol.* **2017**, *53*, 60–67. [CrossRef]
42. Makarov, I.; Shambilova, G.; Vinogradov, M.; Zatonskih, P.; Gromovykh, T.; Lutsenko, S.; Arkharova, N.; Kulichikhin, V. Films of Bacterial Cellulose Prepared from Solutions in N-Methylmorpholine-N-Oxide: Structure and Properties. *Processes* **2020**, *8*, 171. [CrossRef]
43. Beamson, G.; Briggs, D. *High Resolution XPS of Organic Polymers: The Scienta ESCA 300 Database*; Wiley: Chichester, UK, 1992; 280p. [CrossRef]
44. Vasil'kov, A.Y.; Abd-Elsalam, K.A.; Olenin, A.Y. Biogenic Silver Nanoparticles: New Trends and Applications. In *Nanobiotechnology for Plant Protection. Green Synthesis of Silver Nanomaterials*; Abd-Elsalam, K.A., Ed.; Elsevier Inc.: Cambridge, MA, USA, 2021; pp. 241–281. [CrossRef]
45. Chen, Q.-Y.; Xiao, S.-L.; Shi, S.; Cai, L.-P. A One-Pot Synthesis and Characterization of Antibacterial Silver Nanoparticle–Cellulose Film. *Polymers* **2020**, *12*, 440. [CrossRef] [PubMed]
46. Pugazhendhi, A.; Prabakar, D.; Jacob, J.; Karuppusamy, I.; Saratale, R. Synthesis and characterization of silver nanoparticles using Gelidium amansii and its antimicrobial property against various pathogenic bacteria. *Microb. Pathog.* **2018**, *114*, 41–45. [CrossRef] [PubMed]
47. Pal, S.; Nisi, R.; Stoppa, M.; Licciulli, A. Silver-Functionalized Bacterial Cellulose as Antibacterial Membrane for Wound-Healing Applications. *ACS Omega* **2017**, *2*, 3632–3639. [CrossRef] [PubMed]
48. Hu, B.; Yin, N.; Yang, R.; Liang, S.; Liang, S.; Faiola, F. Silver nanoparticles (AgNPs) and AgNO3 perturb the specification of human hepatocyte-like cells and cardiomyocytes. *Sci. Total Environ.* **2020**, *725*, 138433. [CrossRef]
49. Purniawan, A.; Hermastuti, R.; Purwaningsih, H.; Atmono, T.M. Effect of deposition time of sputtering Ag-Cu thin film on mechanical and antimicrobial properties. *AIP Conf. Proc.* **2018**, *1945*, 020008. [CrossRef]
50. Elayaraja, S.; Liu, G.; Zagorsek, K.; Mabrok, M.; Ji, M.; Ye, Z.; Zhu, S.; Rodkhum, C. TEMPO-oxidized biodegradable bacterial cellulose (BBC) membrane coated with biologically-synthesized silver nanoparticles (AgNPs) as a potential antimicrobial agent in aquaculture (In vitro). *Aquaculture* **2021**, *530*, 735–746. [CrossRef]
51. Aksit, A.; Camlibel, N.; Zeren, E.; Kutlu, B. Development of antibacterial fabrics by treatment with Ag-doped TiO2 nanoparticles. *J. Text. Inst.* **2017**, *108*, 2046–2056. [CrossRef]

52. Leal, S.; Cristelo, C.; Silvestre, S.; Fortunato, E.; Sousa, A.; Alves, A.; Correia, D.M.; Lanceros-Mendez, S.; Gama, M. Hydrophobic modification of bacterial cellulose using oxygen plasma treatment and chemical vapor deposition. *Cellulose* **2020**, *27*, 10733–10746. [CrossRef]
53. Meng, L.; Wei, Q.; Li, Y.; Xu, W. Effects of plasma pre-treatment on surface properties of fabric sputtered with copper. *Int. J. Cloth. Sci. Technol.* **2014**, *26*, 96–104. [CrossRef]
54. Asanithi, P.; Chaiyakun, S.; Limsuwan, P. Growth of Silver Nanoparticles by DC Magnetron Sputtering. *J. Nanomater.* **2012**, *2012*, 963609. [CrossRef]
55. Seah, M.P. AES: Energy calibration of electron spectrometers. IV. A reevaluation of the reference energies. *J. Electron Spectrosc. Relat. Phenom.* **1998**, *97*, 235–241. Available online: https://goo.gl/ekaCwB (accessed on 1 September 2022). [CrossRef]
56. Moulder, J.F.; Stickle, W.F.; Sobol, P.E.; Bomben, K.D. *Handbook of X-ray Photoelectron Spectroscopy*; Perkin Elmer Corporation: Eden Prairie, MN, USA, 1995; 260p.
57. Wang, B.-T.; Hu, S.; Yu, X.-Y.; Jin, L.; Zhu, Y.-J.; Jin, F.-J. Studies of Cellulose and Starch Utilization and the Regulatory Mechanisms of Related Enzymes in Fungi. *Polymers* **2020**, *12*, 530. [CrossRef]

polymers

MDPI

Article

A Multi-Rheology Design Method of Sheeting Polymer Extrusion Dies Based on Flow Network and the Winter-Fritz Design Equation

Amin Razeghiyadaki [1], Dongming Wei [1,*], Asma Perveen [2] and Dichuan Zhang [3]

1 Department of Mathematics, School of Sciences & Humanities, Nazarbayev University, Kabanbay Batyr 53, Nur-Sultan 0100000, Kazakhstan; amin.razeghiyadaki@nu.edu.kz
2 Department of Mechanical & Aerospace Engineering, School of Engineering & Digital Sciences, Nazarbayev University, Kabanbay Batyr 53, Nur-Sultan 0100000, Kazakhstan; asma.perveen@nu.edu.kz
3 Department of Civil & Environmental Engineering, School of Engineering & Digital Sciences, Nazarbayev University, Kabanbay Batyr 53, Nur-Sultan 0100000, Kazakhstan; dichuan.zhang@nu.edu.kz
* Correspondence: dongming.wei@nu.edu.kz

Abstract: In the polymer sheet processing industry, the primary objective when designing a coat-hanger die is to achieve a uniform velocity distribution at the exit of the extrusion die outlet. This velocity distribution depends on the internal flow channels of the die, rheological parameters and extrusion process conditions. As a result, coat-hanger dies are often designed for each polymer based on its individual rheological data and other conditions. A multi-rheology method based on a flow network model and the Winter–Fritz equation is proposed and implemented for the calculation, design and optimization of flat sheeting polymer extrusion dies. This method provides a fast and accurate algorithm to obtain die design geometries with constant wall-shear rates and optimal outlet velocity distributions. The geometric design when complemented and validated with fluid flow simulations could be applied for multi-rheological fluid models such as the power-law, Carreau–Yasuda and Cross. This method is applied to sheet dies with both circular- and rectangular-shaped manifolds for several rheological fluids. The designed geometrical parameters are obtained, and the associated fluid simulations are performed to demonstrate its favorable applicability without being limited to only the power-law rheology. The two such designed dies exhibit 32.9 and 21.5 percent improvement in flow uniformity compared to the previous methods for dies with circular and rectangular manifolds, respectively.

Keywords: polymer processing; sheet die design; manufacturing process design; coat-hanger die; modeling; rheology; constant shear-rate die; non-Newtonian fluids

Citation: Razeghiyadaki, A.; Wei, D.; Perveen, A.; Zhang, D. A Multi-Rheology Design Method of Sheeting Polymer Extrusion Dies Based on Flow Network and the Winter-Fritz Design Equation. *Polymers* **2021**, *13*, 1924. https://doi.org/10.3390/polym13121924

Academic Editors: Célio Bruno Pinto Fernandes, Salah Aldin Faroughi, Luís L. Ferrás and Alexandre M. Afonso

Received: 20 April 2021
Accepted: 29 May 2021
Published: 10 June 2021

1. Introduction

One of the most important processes in the polymer processing industry is the extrusion of sheets. Coat-hanger dies are widely used in the polymer processing industry for the production of sheets and films. The desired design provides a uniform polymer sheet with constant thickness at the exit of the die. Uniform thickness only can be achieved when the polymer melt exits the die uniformly. Therefore, the design of a die can be accomplished by evaluating the design parameters that affect the velocity distribution at the exit of the die and by optimizing these parameters to obtain the desired uniform outlet velocity distribution. The latter depends on the rheological properties of the polymer melt entering the die, the process conditions such as flow rate and the internal geometry of the flow distribution in the manifold, as well as in the slit [1]. Therefore, finding the optimized internal design for sheeting extrusion die is of great importance.

In the literature, different designs of coat-hanger dies have been suggested and proposed by using analytical, numerical and experimental techniques with various degrees of uniform outputs. Reid et al. [2] presented an analytical method for the design of tapered coat-hanger dies for power-law fluids. A study by Winter and Fritz [3] proposed an analytical design method for circular- and rectangular-shaped manifolds for power-law fluids. Awe et al. [4] modified the Winter–Fritz method and applied it to rectangular-shaped manifolds with shortened die pre-land depth for the extrusion of degradable materials. Igali et al. [5] derived design equations of Casson fluids based on the Winter–Fritz model.

Other approaches for the design of extrusion dies include the integration of a computational fluid dynamics (CFD) method, such as finite volume or the finite element method with an optimization algorithm for the evaluation of the desired design parameters. Han and Wang [6] integrated a CFD simulation with a genetic algorithm for the optimization of velocity distribution and residence time of a tapered coat-hanger die. Lie et al. [7] proposed a modified finite element method, namely finite piece method, for the simulation of a tapered coat-hanger die. Their method proved to save up to 80% of the CPU time. Smith [8,9] presented a design methodology based on the finite element and the adjoint-variable optimization method for power-law, Carreau–Yasuda, Cross, Elis and Bingham fluids. An optimal geometry was obtained from their proposed methodology, although their results showed an insignificant dependency of optimal geometry on the fluid model. Lebaal [10–12] studied the capability of different optimization algorithms such as response surface method, Kriging interpolation, and the sequential quadratic method for the optimization of geometry and processing parameters. Razeghiyadaki et al. [13] developed a design method based on the response surface method and B-spline method for the optimization of die geometries, specifically for die profiles. However, the majority of previous studies have been focused on power-law fluids.

Another possible approach is devising a semi-analytical method, such as the flow network method, for the fast and accurate design of an extrusion die. A flow network analysis, also known as a hydraulic-electric circuit analogy, uses the conventional concept of electric circuit theory for an analysis of fluid flow problems [14]. The literature demonstrates [14,15] the application of this method to design coat-hanger dies with general rheology. Michaeli et al. [16] combined the finite element method with the network theory to find the optimum velocity distribution. They demonstrated this network as being optimal for the design of rectangular- and circular-shaped manifold extrusion dies for Carreau–Yasuda and power-law fluids. Yilmaz and Kadikkopru's model was successful in identifying only geometrical parameters in manifold (i.e., height of the manifold).

Analytical methods tend to oversimplify the extrusion die problem, and these are limited to simple rheological models such as the power-law [3] or Elis model [17]. However, the power-law model is not applicable in a situation with very low shear rates [18,19]. More sophisticated rheology models such as Carreau–Yasuda represents the apparent viscosity at low shear rates more accurately. Nevertheless, analytical pressure-drop/flow-rate relation is not available for these rheology models. Since previous studies such the one of Liu and Liu [17] require an analytical pressure-drop/flow-rate relation, these methods have limited applicability. Therefore, there is always a need for a single design method that satisfies multiple generalized rheology fluids. Secondly, CFD-based design methods require numerous simulations [16]; as a result, in term of computational time, these methods are computationally expensive. In addition, due to the high cost of die block fabrication, a trial-and-error approach for the design of extrusion dies is very costly. On the other hand, numerical and analytical methods are either computationally expensive or lack accuracy due to unrealistic assumptions. Hence, a fast and accurate model for finding the optimized internal geometry of extrusion dies while being flexible in term of rheology is of great importance. In this study, a computationally fast semi-analytical method is established by constructing a flow network model.

Based on the circuit method introduced in [14] for fluid networks and the work shown in [15] for the application of such methods to coat-hanger die design, a multi-rheology design method is proposed and implemented for sheeting die design for general rheology fluids. This method has been demonstrated to be computationally efficient and effective through the design cases of dies with rectangular and circular manifolds. The obtained results are validated and compared against the existing results in the literature. Yet, it is shown to have an extended capability for finding both geometrical parameters of the manifold and the slit region for a uniform outlet velocity distribution. In addition, it can be adapted to any rheological model with few modifications. The method utilizes a modified discretized form of the Winter–Fritz equation on a flow network, which is considered a novel extension and contribution to the circuit network method for die design. Moreover, this method enforces additional uniform velocity distribution constraints, resulting in a significant improvement in the velocity distribution without conducting full-fledged CFD analyses on the manifold.

2. Methodology

As shown in Figure 1, a coat-hanger die consists of two parts: the slit and the manifold. The design of a die can be reduced to the evaluation of the curve for the manifold and the variance in the cross-sectional area along the manifold length. The manifold cross-section area determines the flow resistance at various locations along the curve of the manifold. In addition, the manifold cross-section area specifies how much of the polymer gets to be distributed into the slit at each point along the curve. In the case of a manifold with a circular cross-section, radius (R) is considered the geometrical parameter, while in the case of a rectangular manifold, the height of the manifold (H) is considered the geometrical parameter of the manifold, and each determines the flow distribution. Moreover, the die profile, which is also known as the preland area, determines flow resistance in the slit. The goal of the die design is to find an optimized flow distribution manifold curve that yields constant flow rates at the die exit through the slit region. To demonstrate the proposed flow network method, circular and rectangular cross-sections of manifolds are selected as examples. With slight modifications, the cross-sectional shapes can be modified between these two or any other desired geometry, which are not considered here.

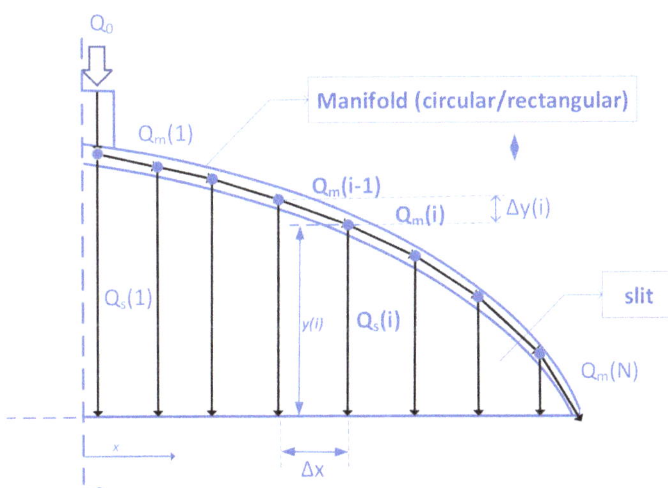

Figure 1. Schematic of a sheeting extrusion die and the corresponding flow network.

Due to the symmetry in geometry, only half of the die is modeled. The die is divided into a finite number (N) of segments along the *x*-axis. The proposed method can be utilized for a simple prediction of the flow distribution at the outlet or the design of the extrusion die by withholding certain constraints. In both cases, following assumptions are made for the proposed method:

1. Isothermal incompressible flow

2. Streamlined flow
3. Steady and fully-developed flows in both slit and manifold
4. Uniform pressure at the exit of the slit
5. Unidirectional flow in both manifold and slit network segments (ignoring traverse flow in the manifold)

In a previous study, it was observed that, due to the laminar nature of the flow in an extrusion die, the flow field is streamlined, and each fluid particle's path does not cross any other particle's path [13]. In addition, the flow field can be discretized as a summation of a finite number of separate flow lines, as shown in Figure 1. Polymer melt with flow rate of Q_0 enters at the die. At each node of the manifold, a portion of polymer flow goes into the slit, while the rest is sent forth to the next manifold segment. At each segment, flow rates inside the slit and the manifold are denoted by $Q_s(i)$ and $Q_m(i)$ (i = 1 to N), respectively. The vertical distance between the manifold and the die exit is denoted by $y(i)$. The vertical and horizontal distances between two adjacent manifold by $Q_s(i)$ segments are Δy and Δx, respectively.

2.1. Calculation of the Flow Distribution

When the geometry of the die is known, the model can predict the flow distribution at the die exit. The conservation of mass for each node on the manifold (see Figure 1) gives the following:

$$Q_m(i-1) = Q_m(i) + Q_s(i) \qquad 1 \leq i \leq N \tag{1}$$

with the condition:

$$Q_m(0) = Q_0 \tag{2}$$

It is known that the Hagen–Poiseuille equation relates pressure drop and flow rate in a pipe or a duct [$Q = Q(\Delta P)$]. For illustration, Hagen–Poiseuille equations of a power-law fluid are given as follows:

$$Q_s(i) = \frac{2(2n+1)}{n\Delta x h_s^2}\left(\frac{\Delta P_s(i)}{2k\,y(i)}\right)^{1/n} \tag{3}$$

$$Q_m(i) = \frac{n\pi R(i)^3}{3n+1}\left(\frac{\Delta P_m(i) R(i)}{2k\Delta\zeta(i)}\right)^{1/n} \tag{4}$$

$$Q_m(i) = \frac{2(2n+1)}{nWH(i)^2}\left(\frac{\Delta P_m(i)}{2k\Delta\zeta(i)}\right)^{1/n} \tag{5}$$

where Equation (3) applies to the slit, Equation (4) applies to the circular-shaped manifold, and Equation (5) applies to the rectangular-shaped manifold.

By substituting Equation (3) and either Equation (4) or Equation (5) into Equation (1), a set of equations with the unknowns of pressure at the manifold segments is produced. Since the pressures at the exit of the slit are assumed to be equal to zero (atmospheric), pressures at the end of the slit are removed in Equation (3) and therefore, the only remaining unknowns are the pressure at manifold nodes. In summary, solving Equation (1) with the appropriate Hagen–Poiseuille equations, the pressure at the manifold nodes can be calculated. With calculated pressure at manifold nodes, the flow distribution at the exit can be calculated from Equation (3). The system of equations is numerically solved by the secant method. Figure 2a shows an algorithm for this method of calculation for flow distribution. Convergence criterion is set as:

$$max\left|\frac{P_{new}(i) - P_{old}(i)}{P_{old}(i)}\right| < \varepsilon \quad 1 \leq i \leq N \tag{6}$$

where convergence is achieved when ε is to 10^{-4}.

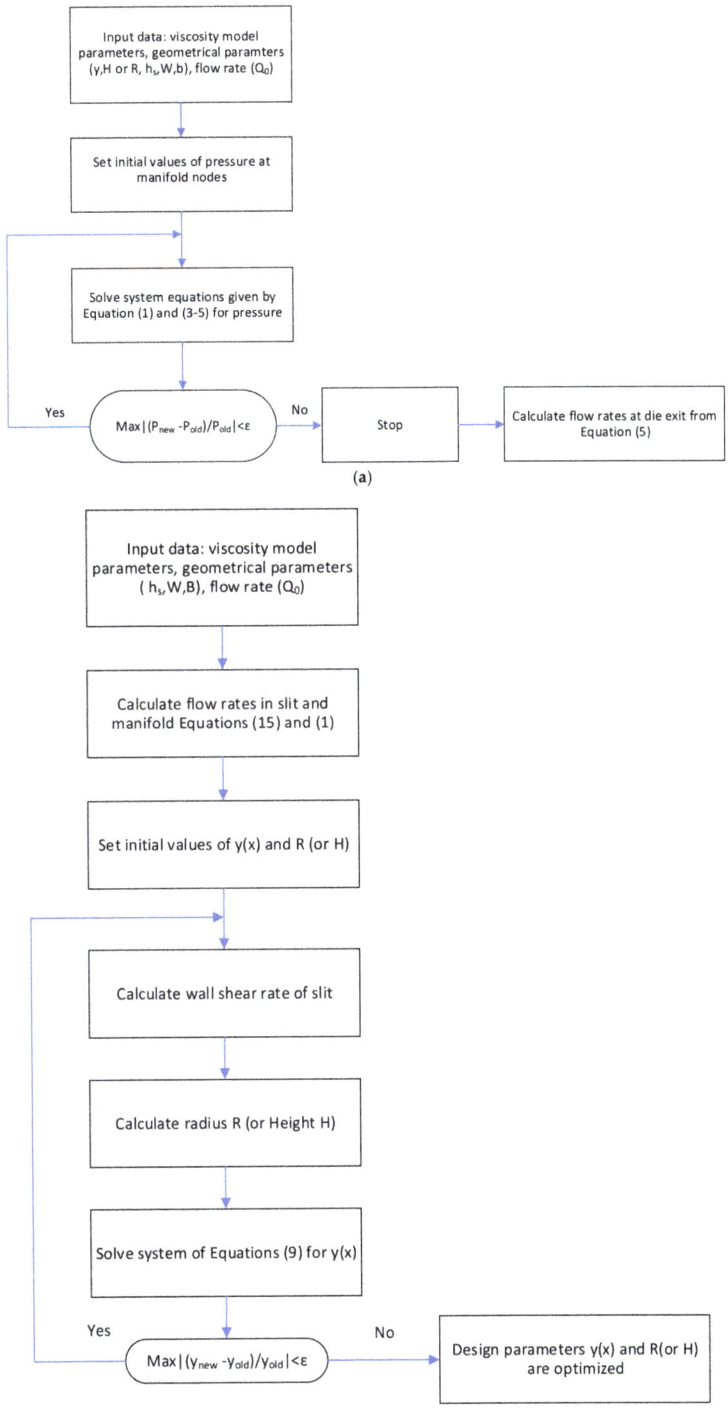

Figure 2. Flow charts for (**a**) flow distribution calculation and (**b**) proposed extrusion die design method.

2.2. Design of the Manifold Curve

A design procedure is required to find geometrical parameters for given process conditions and polymer melt rheology. Geometrical parameters defining an extrusion dies are radius $R(x)$ (circular manifold) or height $H(x)$ (rectangular manifold) and $y(x)$, defined as the distance between the manifold and the die outlet at each x along the width of the die. For each segment shown in Figure 1, two equations are required. Besides the assumptions made for Equation (2), the following constraints are selected to determine the curve for the manifold:

1. Constant shear rates between the slit and the manifold and
2. Uniform velocity distribution at the exit of the slit.

The uniform shear-rate assumption gives the following:

$$\gamma_m(R \text{ or } H) = \gamma_s(y) \quad (7)$$

The uniform shear-rate dies satisfying the above constraints are also known as the Winter–Fritz dies [3], which satisfy the following differential equation:

$$\frac{dy}{dx} = -\left[\left(\frac{\left(\frac{dP}{dy}\right)_s}{\left(\frac{dP}{d\zeta}\right)_m}\right)^2 - 1\right]^{-1/2} \quad (8)$$

for the determination of the curve of the manifold as it is a solution of $y(x)$.

In Equation (8), ζ is the distance between two adjacent nodes along the manifold ($d\zeta^2 = dx^2 + dy^2$). Subscripts m and s refer to the manifold and the slit, respectively. In a discretized finite difference form, Equation (8) can be written as follows:

$$(y_{i+1} - y_i) = -(x_{i+1} - x_i)\left[\left(\frac{\left(\frac{dP}{dy}\right)_s(i)}{\left(\frac{dP}{d\zeta}\right)_m(i)}\right)^2 - 1\right]^{-1/2} \quad (9)$$

The ratio of pressure gradients $\frac{\left(\frac{dP}{dy}\right)_s(i)}{\left(\frac{dP}{d\zeta}\right)_m(i)}$ in Equation (9) depends on both radius R (or height H) and $y(x)$. Therefore, with known relations for wall shear rates and pressure gradients in the slit and the manifold, geometrical parameters R (or H) and $y(x)$ can be calculated iteratively by solving Equations (7) and (9). For illustration, the wall shear rates and pressure gradients of a power-law fluid with a circular manifold are shown below (see [1] for similar formulas):

$$\gamma_w(i) = \frac{2(2n+1)Q_s(i)}{n\Delta x h_s^2}, 1 \le i \le N \text{ slit} \quad (10)$$

$$\gamma_w(i) = \frac{(3n+1)Q_m(i)}{n\pi R(i)^3} 1 \le i \le N \quad \text{manifold} \quad (11)$$

$$\left(\frac{dP}{d\zeta}\right)_m(i) = \frac{\Delta P_m(i)}{\Delta\zeta(i)} = 2k\left(\frac{(3n+1)Q_m(i)}{n\pi R(i)^{\frac{3n+1}{n}}}\right)^n \quad (12)$$

$$\left(\frac{dP}{dy}\right)_s(i) = \frac{\Delta P_s(i)}{\Delta y(i)} = 2k\left(\frac{2(2n+1)Q_s(i)}{n\Delta x\, h_s^2}\right)^n \quad (13)$$

The substitution of Equations (10) and (11) in Equation (7) gives the following equation of the radius:

$$R(i) = \left(\frac{3n+1}{2(2n+1)}\frac{Q_m(i)}{Q_s(i)}\Delta x h_s^2\right)^{1/3} \quad (14)$$

The second constraint of the model states a uniform velocity distribution at the exit of the die. With this constraint, flow rates at the slit are as follows:

$$Q_s(i) = \frac{Q_0}{N+1} \quad 1 \le i \le N \tag{15}$$

With known flow rates of the slit, flow rates in the manifold can be obtained by the conservation of mass given in Equation (1).

Similarly, two systems of Equations (14) and (9) with the pressure gradient Equations (12) and (13) and flow rates from Equations (1) and (15) are solved by an iterative numerical scheme, such as the secant method. The convergence criterion is defined as follows:

$$max\left|\frac{y_{new}(i) - y_{old}(i)}{y_{old}(i)}\right| < \varepsilon = 10^{-4} \quad 1 \le i \le N \tag{16}$$

Figure 2a shows the algorithm used for calculations.

2.3. Calculation of Wall Shear Rates and Pressure Gradients

The solution of the design Equation (8) requires information on shear rates at the walls of the manifold and the slit, and the pressure gradients respectively. This section provides a method for the evaluation of these values using a general non-Newtonian rheology equation:

$$\mu = \mu(\gamma) \tag{17}$$

Most Poiseuille and shear-rate relations given in the literature have limited capacity to only one (generalized) non-Newtonian fluid, such as the power-law fluid [1,20,21].

Sochi, based on the Weissenberg-Rabinowitsch-Mooney-Schofield (WRMS) method, gives a relation for the wall shear rate of a fluid flowing between two parallel plates [18]:

$$Q = \frac{2\Delta x h_s^2 I_{pp}}{\tau_w^2} \tag{18}$$

where

$$I_{pp} = \int_0^{\tau_w} \gamma\tau d\tau = \int_0^{\gamma_w} \left[\gamma^3 \mu \frac{d\mu}{d\gamma} + \mu^2\gamma^2\right] d\gamma \tag{19}$$

$$\tau_w = \mu(\gamma_w)\gamma_w = \frac{h_s \Delta P}{y(x)} \tag{20}$$

In these equations, h_s, and Δx are the thickness and width, respectively. From Equation (18), the wall shear rates can be calculated for given pressure gradient ΔP. Meanwhile, for a given flow rate there is no explicit relation for the calculation of the pressure drop ΔP. Therefore, an iterative numerical scheme is required. The same WRMS method has been adapted for the flow in a circular duct:

$$Q = \frac{\pi R^3 I_c}{\tau_w^3} \tag{21}$$

$$I_c = \int_0^{\tau_w} \gamma\tau^2 d\tau = \int_0^{\gamma_w} \left[\gamma^4 \mu^3 \frac{d\mu}{d\gamma} + \mu^3\gamma^3\right] d\gamma \tag{22}$$

$$\tau_w = \mu(\gamma_w)\gamma_w = \frac{R\Delta P}{2\Delta\zeta} \tag{23}$$

where $\Delta\zeta$ and R are the length and radius of the manifold segment. Equations (18)–(20) give the wall shear rate in a slit segment, while Equations (21)–(23) give the wall shear rate in a circular manifold. The design curve $y(x)$ is obtained by solving these equations and then substituting pressure gradients in Equation (9), respectively, for the polymer melt rheology models. It is important to notice that Equations (19)–(23) are valid for the general rheology model in the form of Equation (17). Therefore, by adapting them in the proposed method, the restriction and limitation of the traditional constant shear-rate dies applicable to the power-law fluids can be relaxed.

3. Results and Discussion

3.1. Validation

For validation of the proposed method, the fluid distribution at the exit of a die obtained from it is compared with the experimental data of Meng et al. [22] and with the results of a CFD analysis. Meng provides the flow distribution for a tapered extrusion die with power-law consistency factor (k) and index (n) of 0.799 Pa s^n and 0.696, respectively. Details of the process conditions and geometrical parameters can be found in Meng [22]. The flow distribution is calculated (Figure 2a) for the given geometry and process conditions of Meng's tapered die. Figure 3 depicts the comparison of the velocity at the outlet for experimental data [19] and the CFD simulation performed on Ansys Fluent 19.1. As it can be seen, our method shows good agreement with the experimental data and the CFD simulation results. The main difference between the obtained results and the CFD could be due to the three-dimensional effects. In each segment, part of the flow enters the slit. Consequently, the flow field in the manifold is not completely unidirectional (assumption e), and the flow field in the manifold has a transverse component. In addition, the carboxymethyl cellulose (CMC) water solution used in Meng's study has lower power-law consistency factor (k) than common polymer melts. As a result, the fluid flow of the CMC solution may enter the transitional flow regime, which in turn defies the stratified flow pattern in assumption b.

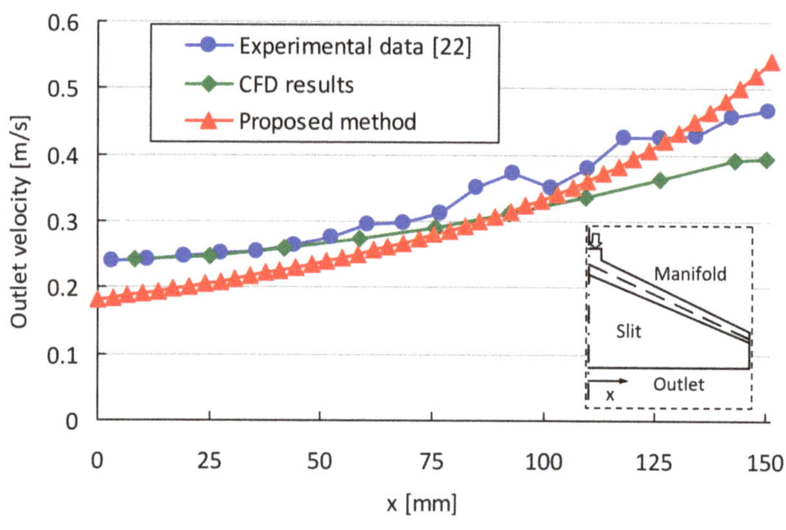

Figure 3. Validation of the model and its comparison with CFD and experimental data [22].

3.2. Sensitivity Analysis

In the sensitivity analysis, the optimized design of a power-law fluid with input design parameters given in Table 1 and a power-law index of 0.38 is performed by the design methodology (algorithm in Figure 2b). Subsequently, for the obtained design profiles of $y(x)$ and radius (R), the flow distribution is calculated (algorithm in Figure 2a) for fluids with different power-law index values. As discussed earlier, an extrusion die with a uniform velocity at the outlet is desired. Therefore, velocity variance is defined as follows:

$$\varphi = \frac{\sum_0^N (v_i - v_{ave})^2}{N} \tag{24}$$

where φ, v_i and v_{ave} are velocity variance, velocity at slit segment i and the average velocity of all segments. Figure 4 depicts the velocity variance for fluids with different values of the power-law index (n) running through a die whose geometries were optimized ($y(x)$ and manifold radius R) for a power-law index (n) of 0.38. In Figure 4, the flow rate at entry Q_0, slit height h_s and half of die width b are 5×10^{-5} m^3/s, 1.5 mm and 360 mm, respectively. Since a uniform velocity is assumed in the design method (algorithm in Figure 2b), the velocity variance calculated from the proposed method (algorithm in Figure 2a) for the fluid with a power-law index (n) of 0.38 is nearly zero, since a uniform flow distribution is assumed (Equation (15)). However, this die geometry gives higher velocity variance for other fluids with different power-law indices. This observation shows that an optimized extrusion die design is dependent on the fluid rheological parameters.

Table 1. Rheology models and corresponding parameters [20].

Rheology Model	Equation	Parameters
Power-law	$\eta = k\dot{\gamma}^{n-1}$	$k = 8.125 \times 10^3$ Pa s^n $n = 0.38$
Carreau–Yasuda	$\eta = \mu_\infty + (\mu_0 - \mu_\infty)(1 + \lambda^2\gamma^2)^{\frac{n-1}{2}}$	$\mu_0 = 1326$ Pa s^n $\mu_\infty = 0$ $\lambda = 0.12$ s $n = 0.35$
Cross	$\mu = \mu_\infty + \frac{\mu_0 - \mu_\infty}{1+\lambda^m\gamma^m}$	$\mu_0 = 564.4$ Pa s^n $\mu_\infty = 0$ $\lambda = 0.017$ s $m = 0.749$

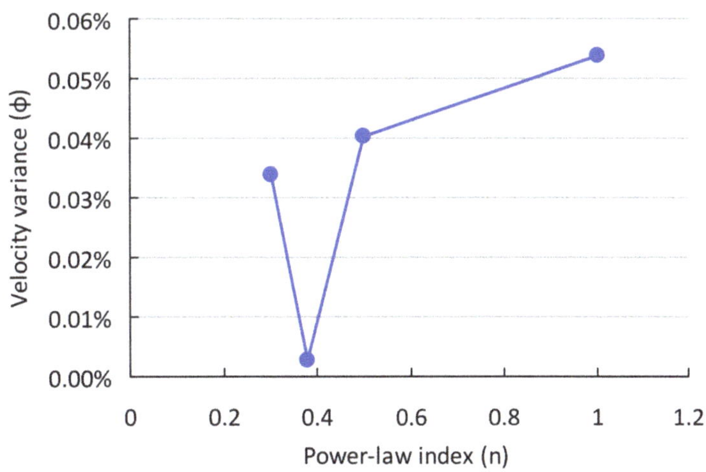

Figure 4. Sensitivity analysis of the model for a power-law fluid (Q_0 of 5×10^{-5} m^3/s, h_s of 1.5 mm and b of 360 mm).

3.3. Design of a Power-Law Fluid with Circular Manifold

Three rheological models are considered in this study as examples. Based on the constant shear-rate assumption, Winter and Fritz [3] provided a design model for a power-law fluid that flows in an extrusion die. Their model gives the following equations for the radius $R(x)$ of the manifold and its profile curve y:

$$R(x) = h_s\left[\frac{(b-x)(1+3n)}{\pi h(1+2n)}\right]^{\frac{1}{3}} \quad (25)$$

$$y(x) = \frac{3Bb}{2}\left[\frac{\sqrt{1+g(x)}}{g(x)} - \frac{1}{2}\ln\frac{\sqrt{1+g(x)}-1}{\sqrt{1+g(x)}+1}\right] \tag{26}$$

where

$$g(x) = \left(\left(R/h_s\right)^2 - 1\right)^{-1} \tag{27}$$

$$B = \frac{\pi h}{b}\frac{1+2n}{1+3n} \tag{28}$$

Figure 5 shows two designs from Winter and Fritz and the proposed method for a power-law rheological model. The consistency factor and power-law index of the rheology model is given in Table 1. Input design parameters are also shown in Table 1. $y(x)$ is a distance between the manifold and the die outlet, and R denotes the manifold radius. As is shown, the proposed method gives higher values for both $y(x)$ and the manifold radius (R). Larger $y(x)$ and manifold radius correspond to deeper die cavity and a larger cross-sectional area of the manifold, respectively. In addition to the shared assumption of a constant wall shear rate in both the Winter–Fritz and the proposed design methods, a new assumption of a uniform flow rate at the exit exerts an additional constraint on the proposed method. This new assumption provides a different profile for $y(x)$ and the manifold radius (R) compared to the Winter–Fritz method.

Table 2. Input design parameters of the model.

Parameter	Value
Flow rate in the entry of the die, Q_0	5×10^{-5} m^3/s
Land height, h_s	1.5 mm
Total die width at exit, b	360 mm

The input of the design model is given in Table 2. After some preliminary calculations, the calculated results were found to be invariant for 50 or more number of segments (N).

In order to validate the performance of two available designs, a CFD package, Ansys Fluent 19.1, was used to show the flow distribution inside the die. Due to symmetry, only a quarter of the die is considered as the simulation domain. A python code was developed to build a mathematically accurate geometrical representation of die designs.

The CFD package of Ansys Fluent 19.1 solves Navier–Stokes equations via the finite volume method. Continuity and momentum equations for a fluid with density of ρ without body force are as follows:

$$\nabla\cdot\left(\rho\vec{v}\right) = 0 \tag{29}$$

$$\nabla\cdot\left(\rho\vec{v}\vec{v}\right) = -\nabla p + \nabla\cdot\overline{\tau} \tag{30}$$

where v is velocity, p is static pressure and τ is stress tensor.

The stress tensor is given by:

$$\overline{\tau} = \mu\left[\left(\nabla\cdot\vec{v} + \nabla\cdot\vec{v}^{T}\right) - \frac{2}{3}\nabla\cdot\vec{v}I\right] \tag{31}$$

where I is the unit identity tensor and μ is viscosity.

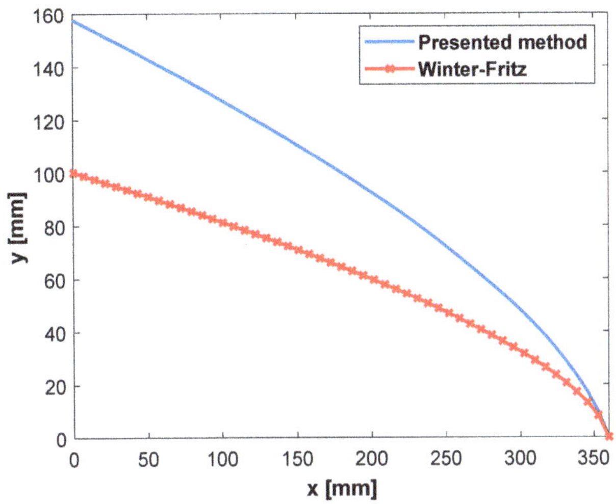

(a)

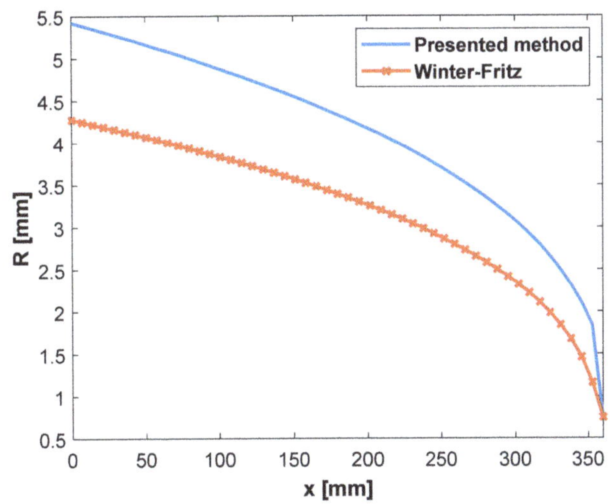

(b)

Figure 5. (**a**) Change of design parameter y with respect to x (**b**) Change of manifold radius R with respect to x for power-law polypropylene (parameters in Table 2) with a circular manifold.

Figures 6 and 7 show the velocity distribution at the center plane for the Winter–Fritz and the proposed design methods based on the geometric parameters shown in Figure 5, respectively. As shown in Figure 5b, the proposed method gives a larger manifold radius compared to the Winter–Fritz method. As Figure 7 shows, due to the larger cross-sectional area, the proposed method gives more uniform velocity distribution in the manifold compared to the Winter–Fritz method (Figure 6). In addition, $y(x)$ of the proposed method (shown in Figure 5a) results in a smoother velocity distribution at the edge of the die.

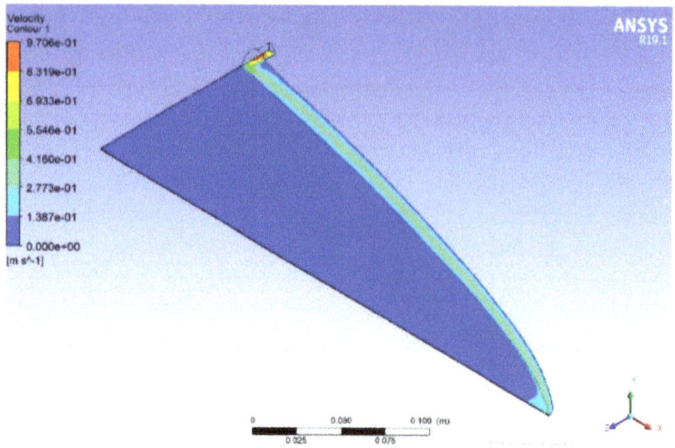

Figure 6. Velocity distribution of Winter–Fritz coat-hanger die design with a circular manifold (flow rate $Q_0 = 5 \times 10^{-5}$ m^3/s, power-law index n = 0.38, die width b = 360 mm and land height h_s = 1.5 mm).

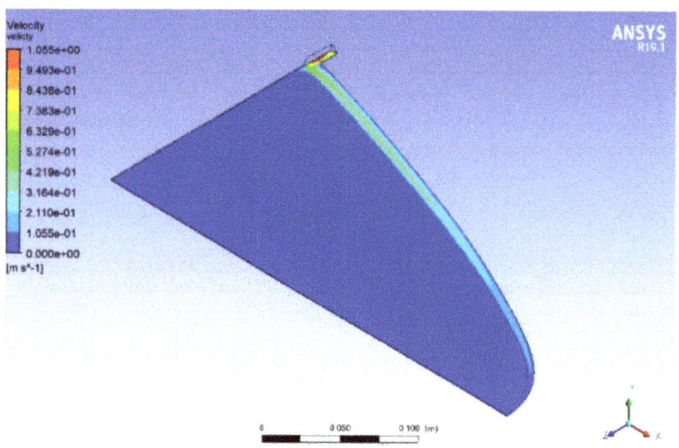

Figure 7. Velocity distribution of an optimized coat-hanger die using the proposed model for a power-law fluid with a circular manifold (power-law index n = 0.38, flow rate $Q_0 = 5 \times 10^{-5}$ m^3/s, die width b = 360 mm and land height h_s = 1.5 mm).

For a quantitative comparison, velocity variance that is defined by Equation (24) is adapted. Velocity variances (φ) for the Winter–Fritz and the proposed methods are 0.00116 and 0.00078, respectively. The velocity distribution shows a 32.9% [(0.00116–0.00078)/0.00116] improvement in the proposed method.

3.4. Design of a Carreau-Yasuda Fluid

For other rheological models, two relations between flow rate and pressure drop for both the manifold and the slit are required. In the literature, different relations are available for this purpose [23–26]. Sochi, based on the Weissenberg-Rabinowitsch-Mooney-Schofield (WRMS) method, gives a relation for pressure-flow rate and shear rate. The Carreau–Yasuda model is defined as follows:

$$\mu = \mu_\infty + (\mu_0 - \mu_\infty)\left(1 + \lambda^2\gamma^2\right)^{\frac{n-1}{2}} = \mu_\infty + \delta\left(1 + \lambda^2\gamma^2\right)^{\frac{n'}{2}} \tag{32}$$

For a Carreau–Yasuda fluid flowing in a circular duct:

$$Q = \frac{\pi R^3 I_c}{\tau_w^3} \tag{33}$$

where I_c and τ_w are given by: [26]

$$\begin{aligned} I_c &= \int_0^{\tau_w} \gamma\tau^2 d\tau = \int_0^{\gamma_w}\left[\gamma^4\mu^3\frac{d\mu}{d\gamma} + \mu^3\gamma^3\right]d\gamma \\ &= \frac{\delta^3\left[3\lambda^4\left(3n'^2+5n'+2\right)\gamma_w^4 - 3n'\lambda^2\gamma_w^2+2\right](1+\lambda^2\gamma_w^2)^{3n'/2}}{3\lambda^4(9n'^2+18n'+8)} \\ &+ \frac{\mu_i\delta^2\left[\lambda^4\left(2n'^2+5n'+3\right)\gamma_w^4 - n'\lambda^2\gamma_w^2+1\right](1+\lambda^2\gamma_w^2)^{n'}}{2\lambda^4(n'+1)(n'+2)} \\ &+ \frac{\mu_i^2\delta\left[\lambda^4\left(n'^2+5n'+6\right)\gamma_w^4 - n'\lambda^2\gamma_w^2+2\right](1+\lambda^2\gamma_w^2)^{n'/2}}{\lambda^4(n'+2)(n'+4)} + \frac{\mu_i^3\gamma_w^4}{4} - \left(\frac{2\delta^3}{3\lambda^4(9n'^2+18n'+8)}\right. \\ &\left.+ \frac{\mu_i\delta^2}{2\lambda^4(n'+1)(n'+2)} + \frac{2\mu_i^2\delta}{\lambda^4(n'+2)(n'+4)}\right) \end{aligned} \tag{34}$$

where only the shear rate at the wall γ_w is unknown. By definition, shear stress and shear rate are related by $\tau = \mu(\gamma)\gamma$. Momentum balance at wall gives the following:

$$\tau_w = \mu(\gamma_w)\gamma_w = \frac{R\Delta P}{2L} \tag{35}$$

Due to the non-linearity of Equation (35), a numerical method is required for the shear rate at the wall (γ_w). In a similar way as Sochi provided, the following relation for the flow of a Carreau-Yasuda fluid in a slit with a width of W and thickness of H is given:

$$Q = \frac{2WH^2 I_{pp}}{\tau_w^2} \tag{36}$$

$$\begin{aligned} I_{pp} &= \int_0^{\tau_w}\gamma\tau d\tau = \int_0^{\gamma_w}\left[\gamma^3\mu\frac{d\mu}{d\gamma} + \mu^2\gamma^2\right]d\gamma \\ &= \frac{\mu_i^2\gamma_w^3}{3} + 2\mu_i\delta\left(\frac{\sqrt{\pi}\Gamma\left(-\frac{n'}{2}-\frac{3}{2}\right)}{4\lambda^3\Gamma\left(-\frac{n'}{2}\right)}\right) + 2\mu_i\delta\gamma_w^{n'}\left(\frac{(\lambda^2)^{\frac{n'}{2}}\gamma_w^3}{n'+3} + \frac{(\lambda^2)^{\frac{n'}{2}-1}n'\gamma_w}{2n'+2} + \frac{(\lambda^2)^{\frac{n'}{2}}(n'-2)n'}{8\lambda^4(n'-1)\gamma_w}\right. \\ &\left.+ \frac{(\lambda^2)^{\frac{n'}{2}}(n'-4)(n'-2)n'}{48\lambda^6(n'-3)\gamma_w^{n'}}\right) \\ &+ \mu_i\delta n'\left(\frac{\sqrt{\pi}\left(\Gamma\left(-\frac{n'}{2}-\frac{3}{2}\right)\Gamma\left(1-\frac{n'}{2}\right) - 2\Gamma\left(-\frac{n'}{2}-\frac{1}{2}\right)\Gamma\left(1-\frac{n'}{2}\right) + 2\Gamma\left(\frac{1}{2}-\frac{n'}{2}\right)\Gamma\left(-\frac{n'}{2}\right)\right)}{4\lambda^3\Gamma\left(1-\frac{n'}{2}\right)\Gamma\left(-\frac{n'}{2}\right)}\right) \\ &+ \mu_i\delta n'\gamma_w^{n'}\left(\frac{\lambda^{n'}\gamma_w^3}{n'+3} + \frac{\lambda^{n'-2}(n'-2)\gamma_w}{2(n'+1)} + \frac{\lambda^{n'}(n'-4)(n'-2)}{8\lambda^4(n'-1)\gamma_w}\right. \\ &\left.+ \frac{\lambda^{n'}(n'-6)(n'-4)(n'-2)}{48\lambda^6(n'-3)\gamma_w^3} + \frac{\lambda^{n'}(n'-8)(n'-6)(n'-4)(n'-2)}{348\lambda^8(n'-5)\gamma_w^5}\right) \\ &+ \delta^2\left(\frac{\sqrt{\pi}\Gamma\left(-n'-\frac{3}{2}\right)}{4\lambda^3\Gamma(-n')}\right) + \delta^2\gamma_w^{2n'}\left(\frac{(\lambda^2)^{n'}\gamma_w^3}{2n'+3} + \frac{(\lambda^2)^{n'-1}n'\gamma_w}{2n'+1} + \frac{(\lambda^2)^{n'}(n'-1)n'}{2\lambda^4(2n'-1)\gamma_w}\right. \\ &\left.+ \frac{(\lambda^2)^{n'}(n'-2)(n'-1)n'}{6\lambda^6(n'-3)\gamma_w^3}\right) \\ &+ \delta^2 n'\left(\frac{\sqrt{\pi}\left(\Gamma\left(-n'-\frac{3}{2}\right)\Gamma(1-n') - 2\Gamma\left(-n'-\frac{1}{2}\right)\Gamma(1-n') + 2\Gamma\left(\frac{1}{2}-n'\right)\Gamma(-n')\right)}{4\lambda^3\Gamma(1-n')\Gamma(-n')}\right) \\ &+ \delta^2 n'\gamma_w^{2n'}\left(\frac{(\lambda^2)^{n'}\gamma_w^3}{2n'+3} + \frac{(\lambda^2)^{n'-1}(n'-1)\gamma_w}{2n'+1} + \frac{(\lambda^2)^{n'}(n'-2)(n'-1)}{2\lambda^4(2n'-1)\gamma_w}\right. \\ &\left.+ \frac{(\lambda^2)^{n'}(n'-3)(n'-2)(n'-1)}{6\lambda^6(2n'-3)\gamma_w^3} + \frac{(\lambda^2)^{n'}(n'-4)(n'-3)(n'-2)(n'-1)}{24\lambda^8(22n'-5)\gamma_w^5}\right) \end{aligned} \tag{37}$$

Similarly, shear rate at wall is given by:

$$\left[\mu_i + \delta(1+\lambda^2\gamma_w^2)^{\frac{n'}{2}}\right]\gamma_w = \frac{H\Delta P}{L} \tag{38}$$

With relations provided in this section, a design for the extrusion die of a Carreau–Yasuda fluid can be given by Equation (9). Figure 8 depicts the radius and $y(x)$ of Carreau fluid in a circular manifold for λ = 2.5 s, n = 0.75, μ_i = 0.009 Pa s and μ_0 = 0.17 Pa s.

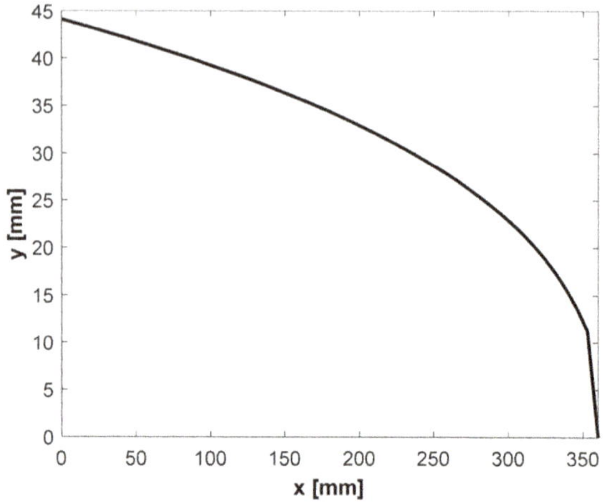

(a)

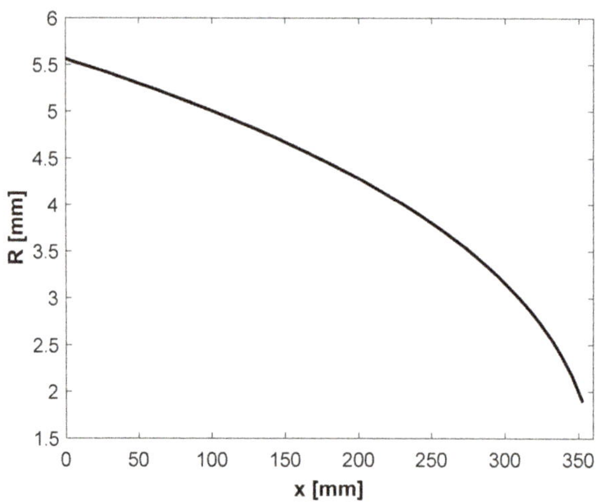

(b)

Figure 8. (**a**) Change of design parameter $y(x)$ with respect to x and (**b**) change of the manifold radius R with respect to x for Carreau–Yasuda polypropylene (parameters in Table 2) using the proposed model with a circular manifold.

3.5. Design of a Cross Fluid

A similar methodology can be applied for a Cross fluid. A Cross fluid is defined as follows:

$$\mu = \mu_\infty + \frac{\mu_0 - \mu_\infty}{1 + \lambda^m \gamma^m} \quad (39)$$

For the cross fluid, the Weissenberg-Rabinowitsch-Mooney-Schofield (WRMS) method described by Sochi gives no analytical solution. Therefore, a numerical integration was used for Equations (19) and (22). Figure 9 depicts the radius and $y(x)$ of a Cross fluid in a circular manifold for $m = 0.5$, $\mu_0 = 0.15$ Pa s, $\mu_i = 0.009$ Pa s and $\lambda = 7.9$ s. In both cases, the flow rate at entry (Q_0) and die width (b) are 5×10^{-5} m^3/s and 360 mm, respectively.

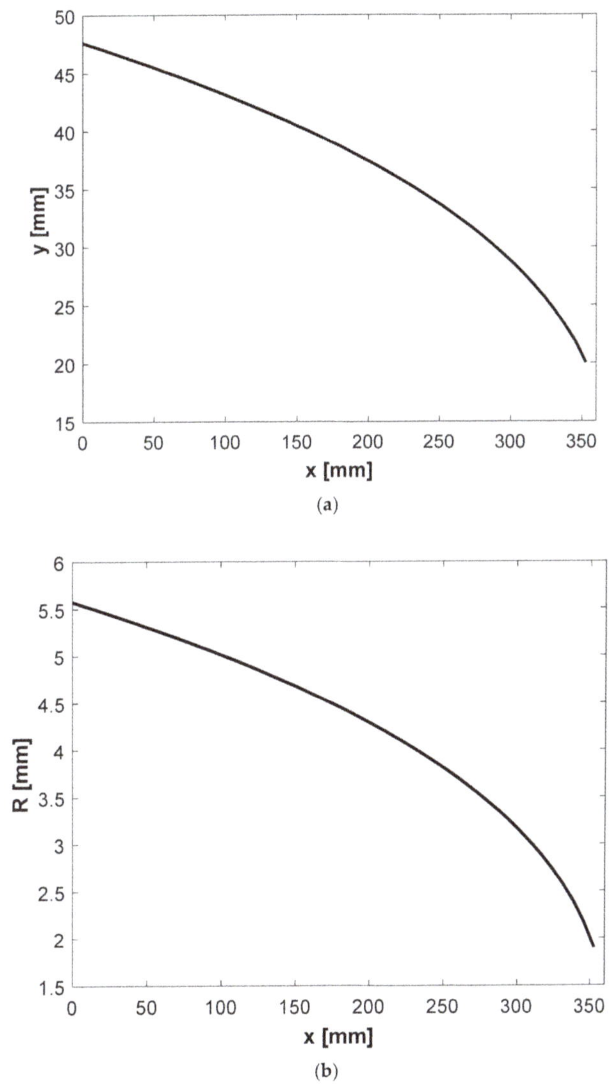

Figure 9. (**a**) Change of design parameter $y(x)$ with respect to x and (**b**) change of manifold radius R with respect to x for a Cross polypropylene (parameters in Table 1) using the proposed model with a circular manifold.

In the last three subsections, the proposed method was used for designing the die for power-law, Carreau–Yasuda and Cross fluids. Polypropylene (PP) as an extrusion fluid is used for this study. Rheological models and the corresponding parameters of polypropylene are adapted from Rudolph [19] and are shown in Table 1. Figure 10 shows a comparison of apparent viscosity based on these models. An effort to design for a PP polymer using these three rheological models is performed. The parameters of these models for PP are given in Table 2 and shown in Figure 3. Both designs based on the flow equations of the Carreau–Yasuda and Cross rheological model converge to the same design as the power-law fluid, as shown in Figure 5a,b. Winter and Fritz [3] concluded that their design only depends on the power-law index and is independent of the power-law consistency factor. For a given operational condition, wall shear rates are approximately 800–900 s^{-1}. The slope of apparent viscosity versus shear rate in the logarithmic scale is the representative power-law index factor of the rheological model. As it is clearly shown in Figure 10, all rheological models of power-law, Carreau–Yasuda and Cross have a same representative power-law index, in the range of 800–900 s^{-1}. As a result, it can be concluded that the proposed method is independent of all rheological parameters except the power-law index shown as n or m for power-law, Carreau–Yasuda and Cross models.

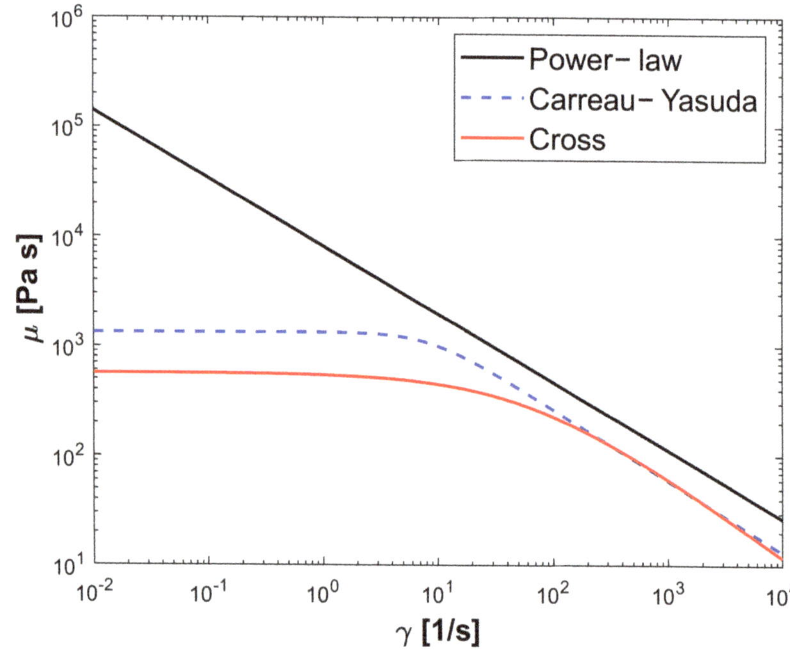

Figure 10. Shear rate versus the apparent viscosity of polypropylene for power-law, Cross and Carreau–Yasuda rheology models [19].

3.6. *Desgin of a Power-Law Fluid with Rectangular Manifold*

Two cases for a rectangular manifold are considered: constant manifold width (W) and constant manifold width to die width (W/b). The geometry for constant width proposed by Winter and Fritz is given by Equations (40) and (41):

$$H(x) = h\sqrt{\frac{b-x}{Wf_p(x)}} \tag{40}$$

$$y(x) = 2W\sqrt{\frac{b-x}{W} - 1} \tag{41}$$

In the case of a constant $a = W/b$ ratio, the following equations are given by Winter and Fritz:

$$H(x) = h[(b-x)/(af_p h)]^{\frac{1}{3}} \tag{42}$$

$$W(x) = aH(x) = [a^2h^2(b-x)/f_p]^{\frac{1}{3}} \tag{43}$$

$$y(x) = \frac{3Bb}{2}\left[\frac{\sqrt{1+g(x)}}{g(x)} - \frac{1}{2}\ln\frac{\sqrt{1+g(x)}-1}{\sqrt{1+g(x)}+1}\right] \quad (44)$$

where $g(x)$ and B are given by:

$$g(x) = \left[(H(x)/h)^2 - 1\right]^{-1} \quad (45)$$

$$B = \frac{ahf_p}{b} \quad (46)$$

Figure 11 shows the comparison of manifold height and die land for the case of constant W. Figure 12 shows the comparison of the velocity at the exit obtained from CFD for both designs. A comparison of velocity variance defined in Equation (24) gives values of 0.005 and 0.0039 for the Winter–Fritz and the proposed methods, respectively. This result shows a 21.5 percent improvement in the velocity variance. By comparing velocity variances, it is clear that the die designed with a circular manifold is approximately three-fold more efficient in terms of velocity uniformity than the die designed with a rectangular manifold.

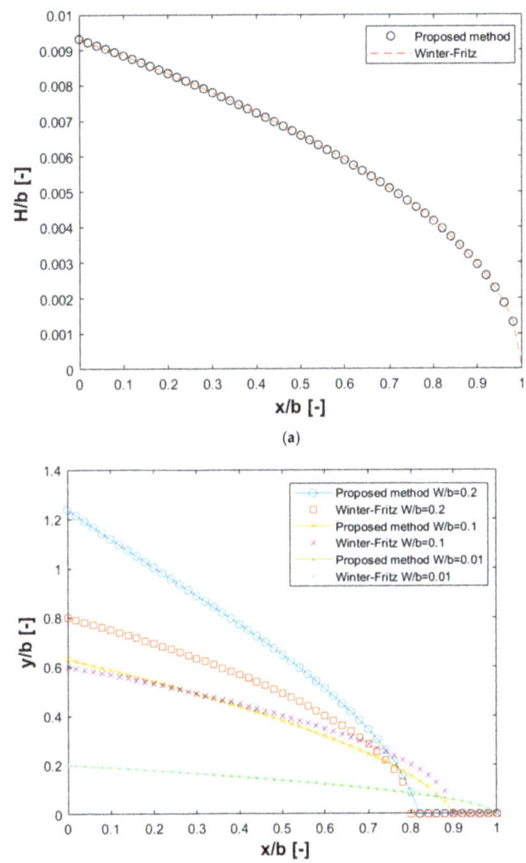

Figure 11. (**a**) Change of normalized design parameter H with respect to normalized x and (**b**) change of normalized design parameter y with respect to normalized x for power-law polypropylene (parameters in Table 1) having a rectangular manifold with constant W/b ratio, (W/b = 0.01, 0.1 and 0.2, N = 50).

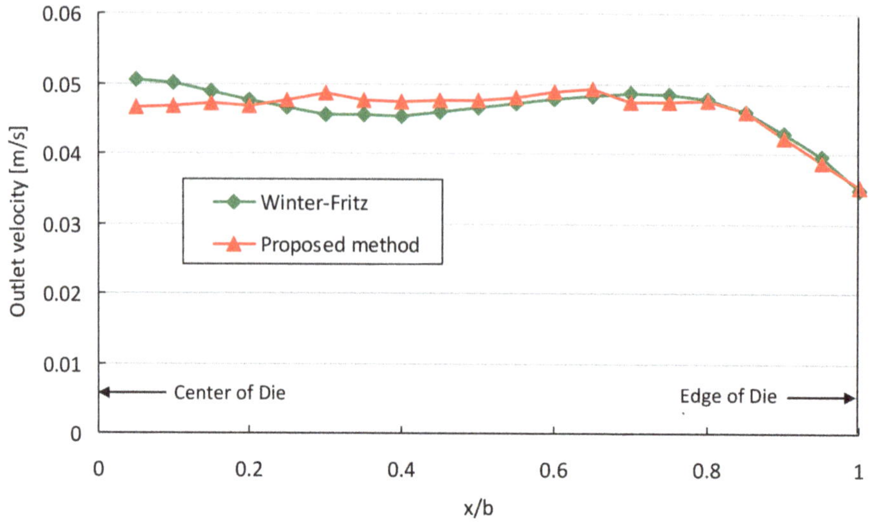

Figure 12. Comparison of velocity distribution obtained from CFD at the slit exit of die for a power-law fluid with $W/b = 0.19$.

4. Conclusions

In this study, a general rheology network method for the design of sheeting extrusion dies in the polymer processing industry is proposed based on the Winter–Fritz constant wall shear-rate conditions and conservation laws of fluid flows in the network. Design parameters based on this method for dies with both circular- and rectangular-shaped manifolds are presented and validated as examples of demonstrations of the method. Computational fluid dynamics (CFD) was adapted as a means for comparison of the presented design models in this study with the purely analytical model of Winter and Fritz. The CFD results show 32.9% and 21.5% improvement in the velocity distribution for dies with circular and rectangular manifolds, respectively, over similar dies designed by other methods. In addition, a method for calculation of wall-flow rates/pressure drops is presented based on the work of Sochi [26]. Since this method is inherently flexible, it can be easily extended to other rheological models. It is demonstrated that the proposed method can be broaden and applied to other rheological models such as Carreau–Yasuda and Cross models, which are more accurate at low shear rates. Thus, the proposed method utilizes constant shear-rate constraint of the Winter–Fritz for power-law and virtually for any other rheology model. Furthermore, it can be extended to take into account temperature effects, such as viscous dissipation, as well as the elastic deformation of the die slit, with some modifications. The proposed method is a fast and computationally efficient method for the first design iteration of sheeting extrusion dies with constant shear rates, as CPU time for every design calculation takes less than a few seconds on a PC (Intel Core i7 2.6 GHz).

The uniform velocity distribution leads to less material loss, which is both economically and environmentally desirable. The proposed semi-analytical design method, which was developed for the fast and accurate pre-design of coat hanger dies with applications in the film and sheeting industries, can be further extended to blow molding, wire-coating and cable sheathing, and the textile and food industries.

Author Contributions: Conceptualization, A.P. and D.W.; writing—original draft preparation, data curation, formal analysis, A.R.; Supervision, A.P., D.W. and D.Z. All authors have read and agreed to the published version of the manuscript.

Funding: This research was funded by Nazarbayev University under the Faculty Development Competitive Research Grant Program No 021220FD4651, "Design & Development of Multiphysics Algorithm for Polymer Sheet Processing Die Design".

Institutional Review Board Statement: Not applicable.

Informed Consent Statement: Not applicable.

Data Availability Statement: The data presented in this study are available on request from the corresponding author.

Acknowledgments: The authors would like to express their deepest gratitude to Otilia Nuta for checking the English of this paper.

Conflicts of Interest: The authors declare no conflict of interest.

Nomenclature

a	Ratio of manifold width to half die thickness (W/b)
b	Half die thickness [m]
B	Parameter in Winter–Fritz equation
f_p	Shape factor in Winter–Fritz model
g	Parameter in Winter–Fritz equation
H	Manifold height [m]
h_s	Slit height [m]
I	Identity tensor
I_c	Definite integral definition for circular duct in Equation (22) [$Pa^3\ s^{-1}$]
I_{pp}	Definite integral definition for parallel plates Equation (19) [$Pa^2\ s^{-1}$]
k	Power-law consistency factor
m	Cross model parameter
n	Power-law index
N	Number of manifold and slit segments
Q_0	Flow rate in entry [m^3/s]
Q_m	Flow rate in the manifold [m^3/s]
Q_s	Flow rate in the slit [m^3/s]
R	Radius of manifold [m]
v	Velocity [m/s]
φ	Velocity variance
W	Width of channel [m]
x	Coordinate along die [m]
y	Distance between manifold and die outlet [m]
ΔP_m	Incremental pressure drops in manifold [Pa]
ΔP_s	Incremental pressure drops in slit [Pa]
μ	Apparent viscosity [Pa s]
μ_0	Rheological parameter in Cross and Carreau–Yasuda models [Pa s]
μ_∞	Rheological parameter in Cross and Carreau–Yasuda models [Pa s]
γ_m	Shear rate in manifold [1/s]
γ_s	Shear rate in slit [1/s]
γ_w	Shear rate at wall [1/s]
ζ	Coordinate in direction of manifold [m]
λ	Rheological parameter in Cross and Carreau–Yasuda models [s]
τ	Shear stress [Pa]
τ_w	Shear stress at wall [Pa]
φ	Velocity variance [m^2/s^2]
dP_m	Differential pressure drops in manifold [Pa]
dP_s	Differential pressure drops in slit [Pa]

References

1. Hopmann, C.; Michaeli, W. *Extrusion Dies for Plastics and Rubber: Design and Engineering Computations*; Carl Hanser Verlag GmbH Co KG: Munich, Germany, 2016.
2. Reid, J.; Campanella, O.; Corvalan, C.; Okos, M. The influence of power-law rheology on flow distribution in coathanger manifolds. *Polym. Eng. Sci.* **2003**, *43*, 693–703. [CrossRef]
3. Winter, H.; Fritz, H. Design of dies for the extrusion of sheets and annular parisons: The distribution problem. *Polym. Eng. Sci.* **1986**, *26*, 543–553. [CrossRef]
4. Awe, T.J.; Eligindi, M.; Langer, R. Internal Design of Uniform Shear Rate Dies. *Morehead Electron. J. Appl. Math.* **2005**.
5. Igali, D.; Perveen, A.; Zhang, D.; Wei, D. Shear Rate Coat-Hanger Die Using Casson Viscosity Model. *Processes* **2020**, *8*, 1524. [CrossRef]
6. Han, W.; Wang, X. Multi-objective optimization of the coat-hanger die for melt-blowing process. *Fibers Polym.* **2012**, *13*, 626–631. [CrossRef]
7. Li, Y.; Zheng, J.; Xia, C.; Zhou, W. A finite piece method for simulating polymer melt flows in extrusion sheet dies. *J. Appl. Polym. Sci.* **2012**, *123*, 3189–3198. [CrossRef]
8. Smith, D.E.; Wang, Q. Optimization-based design of polymer sheeting dies using generalized Newtonian fluid models. *Polym. Eng. Sci.* **2005**, *45*, 953–965. [CrossRef]
9. Smith, D.E. Design sensitivity analysis and optimization for polymer sheet extrusion and mold filling processes. *Int. J. Numer. Methods Eng.* **2003**, *57*, 1381–1411. [CrossRef]
10. Nadhir, L.; Fabrice, S.; Stephan, P. Design and optimization of three-dimensional extrusion dies, using constraint optimization algorithm. *Finite Elem. Anal. Des.* **2009**, *45*, 333–340.
11. Lebaal, N.; Puissant, S.; Schmidt, F. Application of a response surface method to the optimal design of the wall temperature profiles in extrusion die. *Int. J. Mater. Form.* **2010**, *3*, 47–58. [CrossRef]
12. Lebaal, N.; Puissant, S.; Schmidt, F. Rheological parameters identification using in situ experimental data of a flat die extrusion. *J. Mater. Process. Technol.* **2005**, *164*, 1524–1529. [CrossRef]
13. Razeghiyadaki, A.; Zhang, D.; Wei, D.; Perveen, A. Optimization of Polymer Extrusion Die Based on Response Surface Method. *Processes* **2020**, *8*, 1043. [CrossRef]
14. Oh, K.W.; Lee, K.; Ahn, B.; Furlani, E.P. Design of pressure-driven microfluidic networks using electric circuit analogy. *Lab A Chip* **2012**, *12*, 515–545. [CrossRef]
15. Yilmaz, O.; Kirkkopru, K. Inverse Design and Flow Distribution Analysis of Carreau Type Fluid Flow through Coat-Hanger Die. *Fibers Polym.* **2020**, *21*, 204–215. [CrossRef]
16. Michaeli, W.; Kaul, S.; Wolff, T. Computer-aided optimization of extrusion dies. *J. Polym. Eng.* **2001**, *21*, 225–238. [CrossRef]
17. Liu, T.J.; Liu, L.D.; Tsou, J.D. A unified lubrication approach for the design of a coat-hanger die. *Polym. Eng. Sci.* **1994**, *34*, 541–550. [CrossRef]
18. Dealy, J.M.; Wissbrun, K.F. *Melt Rheology and Its Role in Plastics Processing: Theory and Applications*; Springer Science & Business Media: Berlin/Heidelberg, Germany, 2012.
19. Rudolph, N.; Osswald, T.A. *Polymer Rheology: Fundamentals and Applications*; Carl Hanser Verlag GmbH Co KG: Berlin, Germany, 2014.
20. Chhabra, R.P.; Richardson, J.F. *Non-Newtonian Flow and Applied Rheology: Engineering Applications*; Butterworth-Heinemann: Oxford, UK, 2011.
21. Amangeldi, M.; Wei, D.; Perveen, A.; Zhang, D. Numerical Modeling of Thermal Flows in Entrance Channels for Polymer Extrusion: A Parametric Study. *Processes* **2020**, *8*, 1256. [CrossRef]
22. Meng, K.; Wang, X.; Chen, Q. Fluid flow in coat-hanger die of melt blowing process: Comparison of numerical simulations and experimental measurements. *Text. Res. J.* **2011**, *81*, 1686–1693. [CrossRef]
23. Kozicki, W.; Chou, C.; Tiu, C. Non-Newtonian flow in ducts of arbitrary cross-sectional shape. *Chem. Eng. Sci.* **1966**, *21*, 665–679. [CrossRef]
24. Xiao, B.; Huang, Q.; Chen, H.; Chen, X.; Long, G. A fractal model for capillary flow through a single tortuous capillary with roughened surfaces in fibrous porous media. *Fractals* **2021**, *29*, 2150017. [CrossRef]
25. Xiao, B.; Liu, Y.; Chen, H.; Chen, X.; Long, G. A Novel Fractal Solution for Laminar Flow Resistance in Roughened Cylindrical Microchannels. *Fractals* **2020**, *28*, 2050097–2050968. [CrossRef]
26. Sochi, T. Analytical solutions for the flow of Carreau and Cross fluids in circular pipes and thin slits. *Rheol. Acta* **2015**, *54*, 745–756. [CrossRef]

polymers

MDPI

Article

Influence of Processing Glass-Fiber Filled Plastics on Different Twin-Screw Extruders and Varying Screw Designs on Fiber Length and Particle Distribution

Annette Rüppel [1], Susanne Wolff [1], Jan Philipp Oldemeier [2,*], Volker Schöppner [2] and Hans-Peter Heim [1]

[1] Institut für Werkstofftechnik, Kunststofftechnik, University of Kassel, 34125 Kassel, Germany; annette.rueppel@uni-kassel.de (A.R.); susanne.wolff@uni-kassel.de (S.W.); heim@uni-kassel.de (H.-P.H.)
[2] Kunststofftechnik Paderborn, Paderborn University, 33098 Paderborn, Germany; volker.schoeppner@ktp.upb.de
* Correspondence: jan.oldemeier@ktp.uni-paderborn.de; Tel.: +49-(0)5251-60-5824

Abstract: Due to their valuable properties (low weight, and good thermal and mechanical properties), glass fiber reinforced thermoplastics are becoming increasingly important. Fiber-reinforced thermoplastics are mainly manufactured by injection molding and extrusion, whereby the extrusion compounding process is primarily used to produce fiber-filled granulates. Reproducible production of high-quality components requires a granulate in which the fiber length is even and high. However, the extrusion process leads to the fact that fiber breakages can occur during processing. To enable a significant quality enhancement, experimentally validated modeling is required. In this study, short glass fiber reinforced thermoplastics (polypropylene) were produced on two different twin-screw extruders. Therefore, the machine-specific process behavior is of major interest regarding its influence. First, the fiber length change after processing was determined by experimental investigations and then simulated with the SIGMA simulation software. By comparing the simulation and experimental tests, important insights could be gained and the effects on fiber lengths could be determined in advance. The resulting fiber lengths and distributions were different, not only for different screw configurations (SC), but also for the same screw configurations on different twin-screw extruders. This may have been due to manufacturer-specific tolerances.

Keywords: extrusion; polypropylene; glass fiber; fiber reinforced; simulation; fiber shortening; compound; SIGMA; dynamic image analysis

Citation: Rüppel, A.; Wolff, S.; Oldemeier, J.P.; Schöppner, V.; Heim, H.-P. Influence of Processing Glass-Fiber Filled Plastics on Different Twin-Screw Extruders and Varying Screw Designs on Fiber Length and Particle Distribution. *Polymers* **2022**, *14*, 3113. https://doi.org/10.3390/polym14153113

Academic Editors: Célio Bruno Pinto Fernandes, Salah Aldin Faroughi, Luís L. Ferrás, Alexandre M. Afonso and Lilia Sabantina

Received: 30 May 2022
Accepted: 28 July 2022
Published: 30 July 2022

1. Introduction

Due to their valuable properties, fiber-reinforced thermoplastics have a wide range of applications. Possible areas of application can be found in the fields of the automotive industry, aeronautic industry, and the household sector. In particular, polypropylene composites have been established as an alternative for a wide range of applications, due to their positive properties. In addition to their lightweight potential, the materials offer good mechanical and thermal properties, as well as a low weight and good mechanical strength, at low production costs [1–4]. As reinforcing materials, glass fibers are mainly used. This can significantly increase the mechanical properties (notched impact strength, tensile strength). Furthermore, the addition of glass fibers enhances the heat deflection temperature. In addition, the use of fibers is becoming increasingly popular, as the materials produced with them have a low density. Moreover, environmental awareness is becoming increasingly important, which is why the automotive industry is using fiber-reinforced materials in automotive interiors [5–8].

Fiber-reinforced thermoplastics are mainly obtained by injection molding and extrusion processes. However, an extrusion compounding process is primarily used to produce fiber-filled granulates, while the injection molding process is used to manufacture the products. In addition to the fiber orientation, process-related damage to the fibers (shortening of the fibers) can also have a decisive influence on the subsequent component properties. In particular, this can lead to undesirable changes in mechanical properties. [9–12].

In order to produce components with the best possible properties, a flawless granulate from the extrusion process is required for the subsequent injection molding process. One aspect to consider, as described previously, is that the original fiber length can be significantly reduced in some process steps during extrusion compounding. This can lead to a reduction in fiber reinforcement. Due to different extrusion lines and screw configurations, the process parameters may vary, leading to different resulting fiber lengths and compound properties. As a result, the subsequent component properties can be very different from each other.

The influence of extrusion parameters on fiber length reduction in a twin-screw extruder for fiber-reinforced plastics was previously fundamentally investigated. The change from a laboratory extruder to an industrial extruder led to differences in the fiber length distributions. With the industrial extruder, longer fibers and a more widely distributed fiber length could be identified in the compound. Under the conditions tested, the laboratory extruder was more fiber-damaging than the industrial extruder [13]. Gamon et al. investigated the influence of different parameters (rotational speed, feed rate, shear rate) in the extrusion process of natural fiber-reinforced PLA, concerning the fiber morphology (fiber breaks, fiber length change). Here, an increase in screw speed with a simultaneous increase in feed rate could contribute to maintaining the fiber length. In addition, it was shown that fiber breaks occur more frequently with longer fibers than with short fibers [14]. Durmaz et al. examined the properties of carbon fiber-reinforced bio-based polyamide in the extrusion and injection molding process. The fiber content ranged from 20 to 40%. It was found that the average fiber length decreased significantly with increasing fiber content [15]. In their study, Wang et al. investigated fiber orientation, as well as fiber breaks, after processing glass fiber-reinforced polypropylene. They used either short or long glass fibers. The results showed that, due to the shear in the processing step, long glass fiber-reinforced polypropylene leads to more fiber breakages than short glass fiber-reinforced polypropylene. [16].

In order to comprehensively determine the correlation of process parameters during compounding experimentally, a high level of testing, personnel and material expenditure is required [17,18]. To date, fiber length change in composites has primarily been determined by microscopic observation and X-ray microtomography of the samples [19–23]. To enable significant quality improvements, experimentally validated modeling is required that reaches far beyond the current state of technology and considers the details of the used machines and materials. This would also significantly reduce the experimental effort. The higher the prediction quality of the fiber length distribution of the software used, the more accurate the prediction that can be made about the resulting mechanical behavior for the subsequent application areas.

In this study, short glass fiber reinforced thermoplastics (polypropylene) were produced on two different twin-screw extruders at the University of Paderborn (Kunststofftechnik Paderborn, Paderborn, Germany) and the University of Kassel (Institut für Werkstofftechnik, Kunststofftechnik, Kassel, Germany). First, a comparison of the real and simulated fiber length changes was necessary, based on which, the simulation models and their parameters were adjusted. Subsequently, the results could be adapted to real structures. The influences of the compounding parameters, in particular the different machine and screw configurations, on the real and simulated fiber length change are presented, whereby the real fiber length change was determined by means of dynamic image analysis (Section 2.3). Since the processing was carried out on two different machines, the machine-specific process behavior was of major interest regarding its influence. With the simulation software SIGMA (Simulation of co-rotating twin-screw machines), fiber length, in general, can be calculated. However, the software utilizes idealized machines. As a result, important knowledge could be obtained by comparing the simulation and experimental studies, and the effects of process and material changes on fiber lengths could be determined in advance. This allows a specific adjustment of the machine settings, whereby the components to be manufactured subsequently can be produced without any defects.

2. Materials and Methods

2.1. Used Materials

For the manufacturing of the test specimens, polypropylene (SABIC® PP 520P, SABIC Polymers, Genk, Belgium) provided by SABIC Polymers (Genk, Belgium) with a glass fiber (e-glass fiber FGCS 3540, Schwarzwälder Textil-Werke, Schenkenzell, Germany) provided by Schwarzwälder Textil-Werke (Schenkenzell, Germany) compounds with fiber content of 20, 30, and 40 wt.% content were manufactured. No additional additives were added to the polymer and fibers in the compounding process. According to the manufacturer, SABIC® PP 520P is a polypropylene (PP) homopolymer grade and can be used for the production of injection-molded or extruded products or can be used as a component of other industrial products. It provides an excellent combination of transparency, stiffness, and high heat resistance. The material has a density of 0.905 to 0.930 g/cm^3. The melting temperature (Tm) of the material is 160 to 170 °C (peak at 164 °C), and the melt volume rate (MVR) is 10.5 g/10 min (230 °C/2.16 kg) according to ISO 1133.

The short cut glass fibers FGCS 3540 show a diameter of approx. d_f = 10 µm and a mean glass fiber length of approx. l_f = 3 mm. The glass fiber has a density of 2.53 to 2.55 g/cm^3. The moisture content in delivery form is a maximum of 0.3%. The fiber sizing content is approx. 1.1%, according to DIN ISO 1887 (sizing content determination by loss on ignition (625 °C)). The moisture content of the materials was checked before processing using a Sartorius Moisture Analyzer MA 100 (MA 100, Satorius, Goettingen, Germany).

2.2. Processing

Compounding of the materials was carried out on a Leistritz ZSE 18 (ZSE 18, LEISTRITZ Extrusionstechnik, Nürnberg, Germany) and a Coperion ZSK 25 WLE (ZSK 25 WLE, Coperion, Stuttgart, Germany). Both are co-rotating twin-screw extruders. The Leistritz ZSE 18 was used with two screw configurations. This extruder had a screw diameter of 18 mm, with an L/D ratio of 40. The material was compounded without a nozzle, to avoid further shortening of the fibers due to different nozzle configurations. Therefore, no downstream equipment needed to be considered in the simulation. The screw configuration (SC1) included kneading discs and mixing elements, to mix and distribute the fibers after they had passed through the feeding zone. The screw configuration (SC2) simply consisted of the conveying elements after the fiber feeding zone. The screw speed was set to 200 rpm, and the material throughput was set to 3 kg/h.

A Coperion ZSK 25 was used with one screw configuration. This extruder had a screw diameter of 25 mm with an L/D ratio of 41. The used screw configuration (SC3) simply consisted of the conveying elements after the fiber feeding zone. The screw speed was set to 200 rpm, and the material throughput was set to 9 kg/h. The different screw designs are shown in Figure 1. The kneading discs are shown in blue and the mixing elements in green. The temperatures of the heating zones are listed in Table 1.

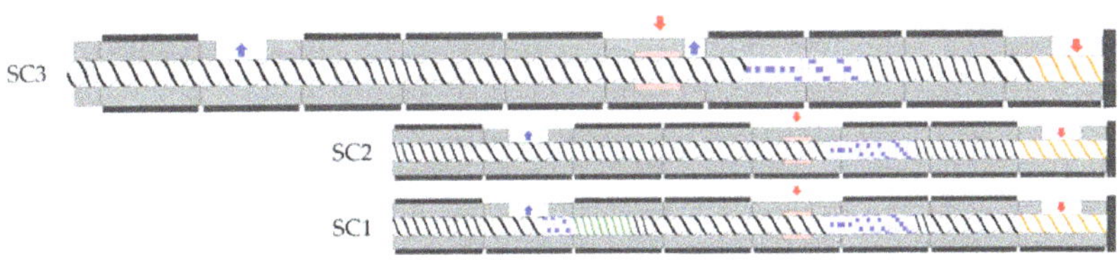

Figure 1. Used screw configurations in the different twin-screw extruders.

Table 1. Temperatures used in the compounding process in the different twin-screw extruders.

Process Temperatures (°C)	Leistritz ZSE 18	Coperion ZSK 25
Feeding Section	50	25
Zone 1	160	160
Zone 2	180	180
Zone 3	180	180
Zone 4	180	180
Zone 5	180	180
Zone 6	180	180
Zone 7	180	180
Zone 8		180
Zone 9		180

2.3. *Characterization*

2.3.1. Dynamic Image Analysis

Dynamic image analysis was used to investigate the influence of fiber length during compounding on various compounders. For the fiber length measurement, the produced pellets were ashed in an oven (600 °C, 6 h). This removed the matrix material polypropylene from the glass fibers for subsequent image analysis. The fiber length distribution was measured using a Sympatec QICPIC/R06 (QICPIC/R06, Sympatec, Clausthal-Zellerfeld, Germany) dynamic image analysis with a wet disperser (MIXCEL unit, Sympatec, Clausthal-Zellerfeld, Germany). The images were acquired at a frame rate of 175 Hz with an M7 magnification. The ISO measurement range varied from 88 μm to 2888 μm. The results represent the change in fiber length after compounding.

2.3.2. Material Characterization for Simulation

The rheological material data analysis required for the simulation, such as the viscosity, thermal conductivity, and p-v-T behavior, were determined using a high-pressure capillary rheometer (RG25, Göttfert, Buchen, Germany). The melting point, enthalpy, and specific heat capacity were determined by differential scanning calorimetry (DSC Q2000, TA Instruments, New Castle, DE 19720, USA). The determined data for the used polymer and the glass fiber were entered into the database PAM (Paderborn Material Database)(PAM 3.0.0, Kunststofftechnik Paderborn, Paderborn, Germany) and exported into the simulation software Sigma (SIGMA 13.0.1, Kunststofftechnik Paderborn, Paderborn, Germany).

2.3.3. Simulation of the Compounding Process

The SIGMA simulation software was used to simulate the compounding process. The simulation software simulates compounding and processing operations on co-rotating twin-screw extruders. In addition to the material data, the barrels and screw configurations of the used extruders (see above) were created in SIGMA. Using models for polymer-filler compounds, the compounding processes were simulated for the matrix–fiber designs and various screw configurations of the different extruders used.

2.4. *Extrusion Simulation of Fiber Length*

The SIGMA simulation software was developed for the design of co-rotating twin-screw extruders. For the further development of SIGMA, Kunststofftechnik Paderborn cooperates with the leading machine and material manufacturers and continuously optimizes the simulation software. For the design of a twin-screw extruder, the user has at his disposal, for example, not only the calculated melting, pressure, and filling degree curves, but also the fiber length degradation of fillers, such as glass fibers or carbon fibers, and the dwell time. These curves can be used to optimize the basic process. For this publication, the calculation of the expected fiber length was of relevance. For this purpose, the real processes with the machines and material data used were entered into SIGMA and subsequently simulated.

3. Results and Discussion

3.1. Influence of Screw Configuration on the Fiber Length

Figure 2 shows the influence on the fiber length during the production of polypropylene + 20 wt.% glass fibers (PP GF 20) with different screw configurations (SC1, SC2, SC3) in the extrusion process. The distribution density is given as a relative frequency. The results show that the production of the materials with the SC1 and SC3 screw configurations led to a significant shortening of the glass fibers. Considering the relative fiber distribution, these two screw configurations showed the shortest fibers after the extrusion process, with a length of 10–30 μm. Analogous to this, the results of SC2 showed similar tendencies. In the range from 400 to 600 μm, a further local maximum in the fiber length was detectable for the SC1 screw configuration. The local maximum for screw configuration SC2 was around 800 μm, and there was no further local maximum for screw configuration SC3. Nevertheless, the total fiber shortening was slightly lower in comparison with the other two screw configurations.

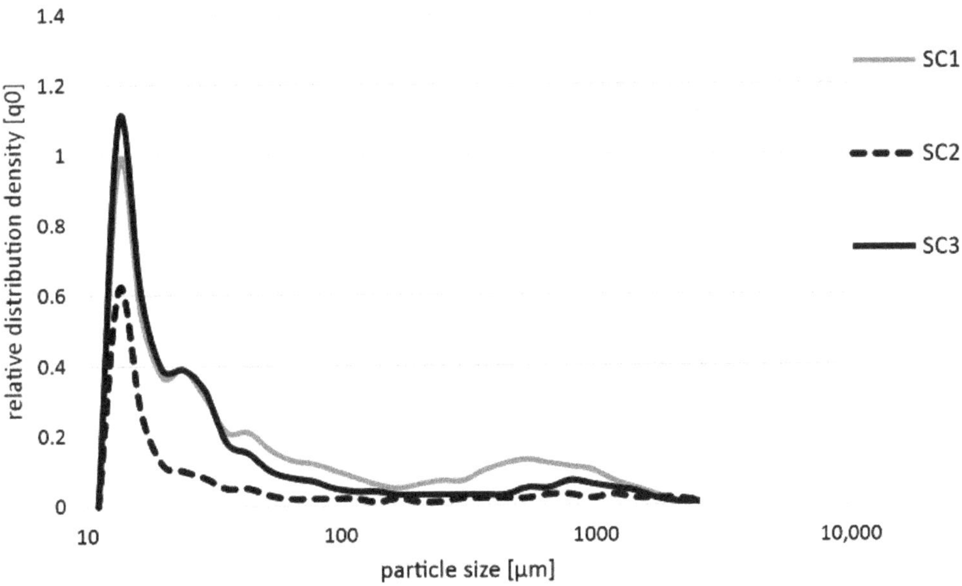

Figure 2. Influence on fiber length during the production of PP GF 20 with different screw configurations (SC1, SC2, SC3) in the extrusion process.

In a comparison of the results shown in Figure 3 (PP + GF 20), the results of polypropylene with 30 wt.% glass fibers also showed that the highest shortening of the fibers occurred in the production with the screw configurations SC1 and SC3. In addition, it is also shown here that there were no significant differences between these two screw configurations. Moreover, it can be seen here that the fiber shortening with the SC2 screw configuration was slightly lower, and significantly fewer short fibers resulted after processing compared to the SC1 and SC3 screws. Overall, the behavior and fiber distribution were very similar to the values of PP + GF 20 (cf. Figure 2).

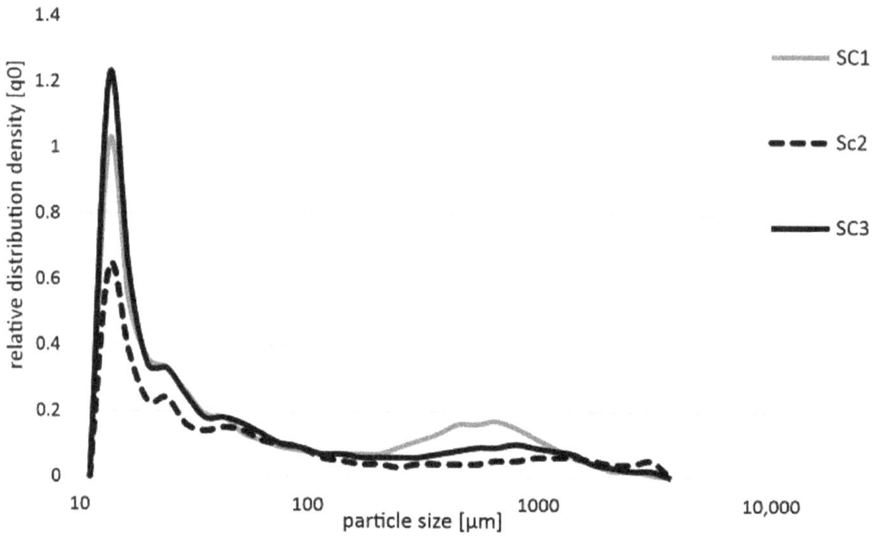

Figure 3. Influence on fiber length during the production of PP GF 30 with different screw configurations (SC1, SC2, SC3) in the extrusion process.

Figure 4 shows the influence on the fiber length during the production of polypropylene + 40 wt.% glass fibers (PP GF 40) with different screw configurations (SC1, SC2, SC3). The larger amount of glass fibers resulted in a further shortening of the fibers, so that overall shorter fibers were measured for the screw configurations SC1 and SC3. Again, significantly less fiber shortening resulted when using screw SC2 compared to both screws SC1 and SC3.

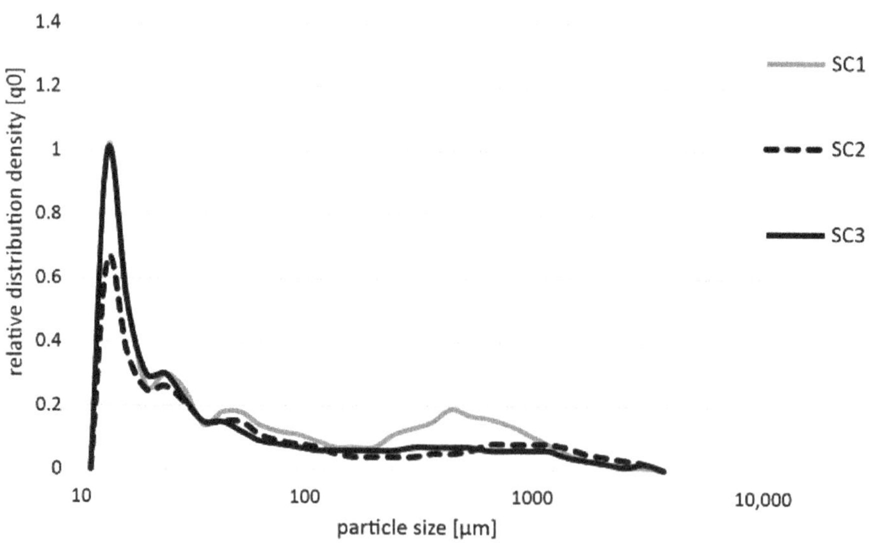

Figure 4. Influence on fiber length during the production of PP GF 40 with different screw configurations (SC1, SC2, SC3) in the extrusion process.

3.2. Simulated Fiber Length

The simulated fiber length is shown graphically in Figure 5. The simulated fiber length decreased with increasing glass fiber content. With a glass fiber content of 20 wt.%, the minimum was 516 μm for screw configuration SC1 and the maximum was 551 μm for screw configuration SC3. If the glass fiber content was increased to 30 wt.%, a maximum fiber length of 490 μm was achieved for screw configuration SC3. The minimum fiber length was 470 μm with screw configuration SC1. With a glass fiber content of 40 wt.%, only a maximum fiber length of 439 μm was achieved with screw configuration SC3. The minimum fiber length was about 426 μm and was calculated for both screw configurations SC1 and SC2. On average, the fiber lengths of screw configuration SC3 were 3.73% longer than the fiber lengths of screw configuration SC2. With screw configuration SC1, the resulting fiber lengths were on average 0.71% shorter than the fiber lengths of screw configuration SC2. The differences between the simulated fiber lengths decreased with increasing fiber content. At a fiber content of 40 wt.%, the simulated fiber lengths of screw configurations SC1 and SC2 no longer differed.

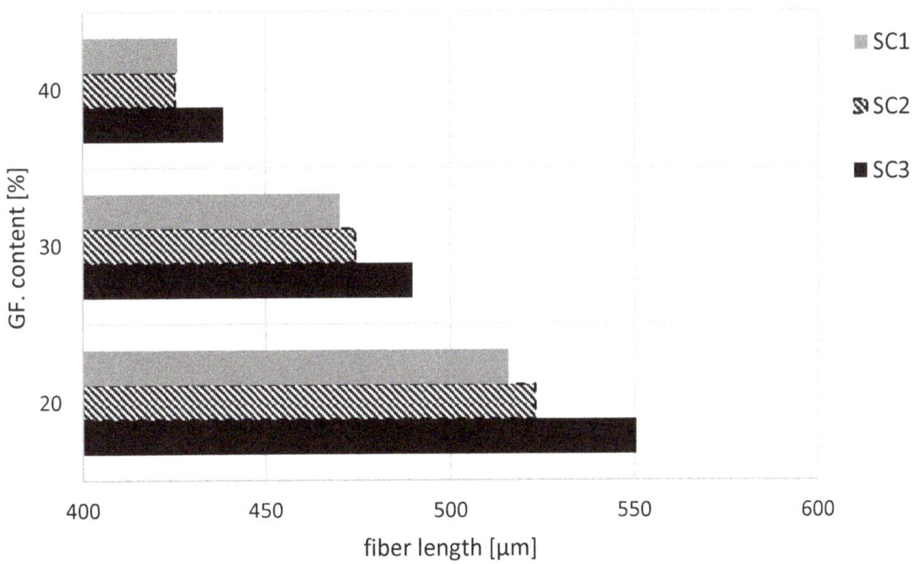

Figure 5. Results of the simulation of the fiber lengths for the glass fiber contents of 20, 30, and 40 wt.% for the three different screw configurations.

In comparison to Figure 5, Figure 6 shows the simulated fiber length at a 20 wt.% fiber content along the axial extruder position. The fiber length decreased rapidly shortly after being added to the extruder. Already, after approximately 15 mm, the fiber length had decreased from 3 mm to 2 mm. After approximately 82.5 mm, the fiber length was already 1 mm for screw configurations SC1 and SC2. With screw configuration SC3, the fiber length dropped below 1 mm only after approx. 107.5 mm. The fiber length decreased slightly at the end of the compounding process.

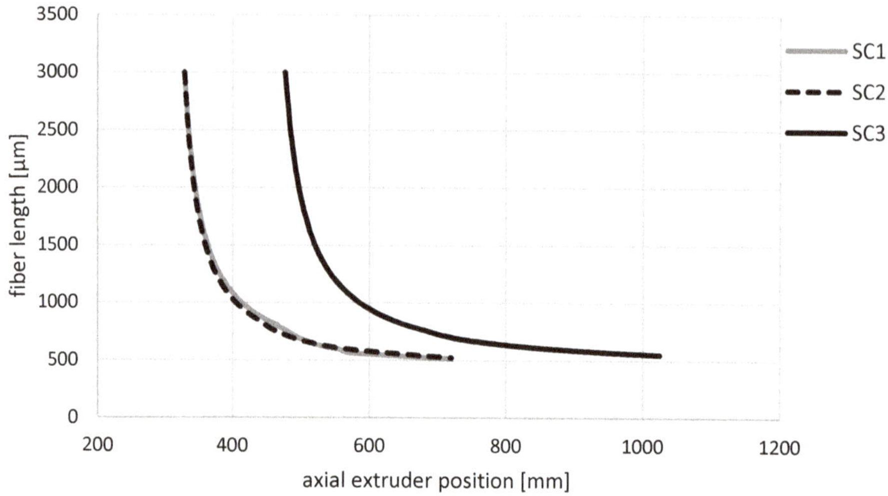

Figure 6. Simulated fiber length curves for the glass fiber content of 20 wt.% for the three different screw configurations along the axial extruder position.

The effect of varying the fiber content with otherwise constant process parameters on the curve of the simulated fiber length is shown in Figure 7. As in Figure 6, the curve is very steep at the beginning and becomes flatter towards the end of the screw. The fiber content influences the slope of the fiber length curve. With a larger fiber content, the fiber length curve is steeper at the beginning and is below the values of the curves with a smaller fiber content.

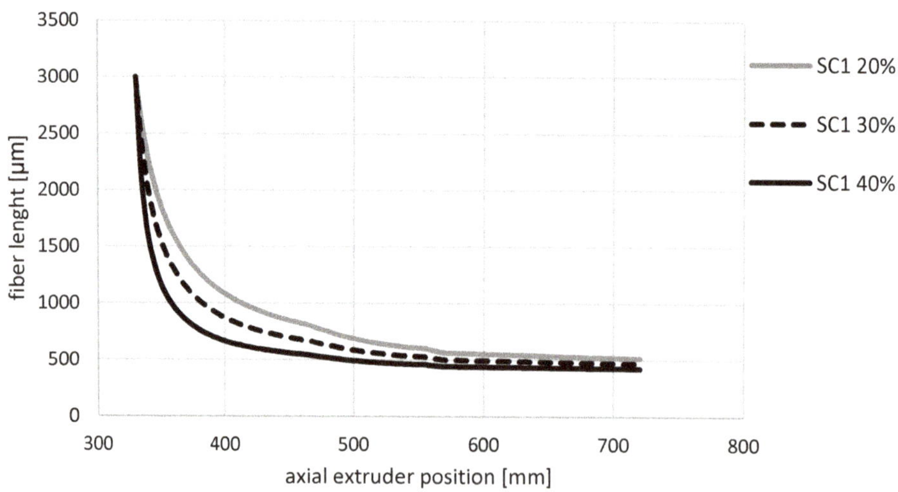

Figure 7. Simulated fiber length curves for the glass fiber content 20, 30, and 40 wt.% for the screw configuration SC1 along the axial extruder position.

Figure 8 shows the different fiber length distributions for the screw configuration SC1 with different fiber contents. The local maximum in fiber length distribution changed from approx. 550 µm to approx. 450 µm with a higher fiber content.

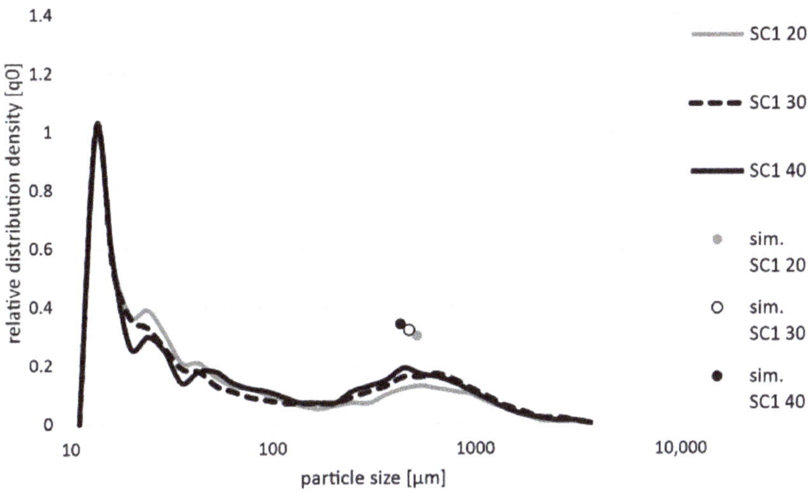

Figure 8. Illustration of the fiber length distributions as a function of fiber content for screw configuration SC1.

The results suggest that the measured particle sizes do not correspond to reality. The large proportion of particle sizes shorter than 50 µm may contain not only glass fibers but also remnants of ashing particles and dust particles. If the particle sizes in the iso-measurement range between 100 and 2888 µm are examined more closely, it is noticeable that both the SC2 and SC3 screw configurations have a significantly smaller proportion of fiber lengths over the entire range than the SC1 screw configuration. This effect is independent of the fiber content. Nevertheless, the resulting local maxima of screw configuration SC1 are ahead of the local maxima of screw configurations SC2 and SC3. These results correspond to the simulation results of SIGMA. The simulation results do not exactly match the experimental results for smaller fiber fractions. However, as the fiber content increases, the difference between the simulation result and the experimental result decreases. The difference between the screw configurations SC2 and SC3 is not as clear in the measurement as it is in the simulation. The resulting fiber length also decreased with increasing fiber content. This effect was more pronounced for the three screw configurations when the glass fiber content was increased from 30 to 40 wt.% then it was with the increase from 20 to 30 wt.%.

The resulting fiber lengths and distributions in this study were slightly above the fiber lengths and distributions documented in the literature. Ghanbari et al. [24] investigated the fiber lengths and distributions at 75 and 400 1/min, using a similar screw design. The measured fiber lengths ranged from about 260 µm at 75 1/min to about 50 µm at 400 1/min. Whereas, in Bumm et al. [25], the resulting fiber lengths were much closer to the values from this study. Nevertheless, the fiber lengths were slightly above the expected range specified in the literature and those calculated in the SIGMA simulation software. The decrease in glass fiber length due to the kneading blocks after addition of the glass fibers can be clearly seen in the resulting fiber lengths. As mentioned in [25], fiber length decreases during compounding with kneading blocks. The SC1 screw configuration consistently produced lower resulting fiber lengths than the other two screw configurations. Furthermore, the resulting fiber length decreased with increasing glass fiber content. This behavior was previously investigated by Durmaz and Aytac [15] and was consistent with the experimental results in this study. The simulated curve of the fiber lengths along the axial extruder position is similar to the curves presented in the literature. The fiber lengths decreased rapidly directly after the addition of the fibers. In the further progression, the fiber length continued to decrease, but no longer as rapidly as at the beginning [13,25].

Thus, the resulting fiber lengths differed for the same process and boundary conditions, not only due to a different screw configuration, but also for identical screw configurations on different twin-screw extruders. These results were evident in both the simulation and experimental results.

4. Conclusions

In this article, short glass fiber reinforced thermoplastics (polypropylene) were produced on two different twin-screw extruders, with three different screw configurations (SC1, SC2, SC3) at the University of Paderborn (Kunststofftechnik Paderborn, Paderborn, Germany) and the University of Kassel (Institut für Werkstofftechnik, Kunststofftechnik, Kassel, Germany). Fiber shortening, which can result during processing, was simulated using the SIGMA simulation software. In addition, real experiments were carried out, in order to determine the fiber length changes due to the compounding process.

The different particle size distributions showed that the resulting particle size distribution depended, not only on the screw configuration, but also on the machine size. The screw configuration with kneading blocks (SC1) led to the smallest fiber lengths. Despite an identical screw configuration of SC2 and SC3, the results of the fiber lengths of both screw configurations were different from one to the other. Both in simulation and in practice, the resulting fiber lengths differed. Nevertheless, the resulting fiber lengths of SC2 and SC3 were both greater than those of screw configuration SC1. With the different particle distributions of the SC2 and SC3 screw configurations, the question arises as to whether the different course of the particle distribution is also dependent on other machine parameters, in addition to the change in machine size. For example, manufacturer-specific gap dimensions between the barrel wall and the screw could have a recognizable influence on the resulting particle size. In order to be able to exclude these machine-related influences or, if necessary, to specify them in more detail, further investigations on several twin-screw extruders with different manufacturer-specific gap dimensions are necessary.

Author Contributions: Investigation, S.W. and J.P.O.; resources, H.-P.H.; data curation, S.W., A.R. and J.P.O.; writing—original draft preparation, S.W., A.R. and J.P.O.; writing—review and editing, S.W., A.R. and J.P.O.; visualization, S.W. and J.P.O.; supervision, H.-P.H. and V.S.; project administration, H.-P.H. and V.S.; All authors have read and agreed to the published version of the manuscript.

Funding: This research was funded by the Deutsche Forschungsgemeinschaft (DFG, German Research Foundation) grant number HE 3020/19-1.

Institutional Review Board Statement: Not applicable.

Informed Consent Statement: Not applicable.

Data Availability Statement: Not applicable.

Conflicts of Interest: The authors declare no conflict of interest.

References

1. Delli, E.; Giliopoulos, D.; Bikiaris, D.; Chrissafis, K. Fibre length and loading impact on the properties of glass fibre reinforced polypropylene random composites. *Compos. Struct.* **2021**, *263*, 113678. [CrossRef]
2. Nuruzzaman, D.M.; Kusaseh, N.M.; Ismail, N.M.; Iqbal, A.A.; Rahman, M.; Azhari, A.; Hamedon, Z.; ShinYi, C. Influence of glass fiber content on tensile properties of polyamide-polypropylene based polymer blend composites. *Mater. Today Proc.* **2020**, *29*, 133–137. [CrossRef]
3. Fang, J.; Zhang, L.; Li, C. The combined effect of impregnated rollers configuration and glass fibers surface modification on the properties of continuous glass fibers reinforced polypropylene prepreg composites. *Compos. Sci. Technol.* **2020**, *197*, 108259. [CrossRef]
4. Hariprasad, K.; Ravichandran, K.; Jayaseelan, V.; Muthuramalingam, T. Acoustic and mechanical characterisation of polypropylene composites reinforced by natural fibres for automotive applications. *J. Mater. Res. Technol.* **2020**, *9*, 14029–14035. [CrossRef]
5. Wie, L.; Donald, A.; Freitag, C.; Morell, J. Effects of wood fiber esterification on properties, weatherability and biodurability of wood plastic composites. *Polym. Degrad. Stab.* **2013**, *98*, 1348–1361.
6. Jamririchova, Z.; Akova, E. Mechanical Testing of natural fiber composites for automotive industry. *Univ. Rev.* **2013**, *7*, 20–25.
7. Langer, B.; Bierögel, C.; Grellmann, W. Eigenschaften von Polypropylenen gezielt bewerten. *Kunststoffe* **2008**, *5*, 87–93.

8. Hellerich, W.; Harsch, G.; Baur, E. *Werkstoff-Führer Kunststoffe: Eigenschaften-Prüfungen-Kennwerte*; Hanser: München, Germany, 2010.
9. Lizama-Camara, Y.A.; Pinna, C.; Lu, Z.; Blagdon, M. Effect of the injection moulding fibre orientation distribution on the fatigue life of short glass fibre reinforced plastics for automotive applications. *Procedia CIRP* **2019**, *85*, 255–260. [CrossRef]
10. Bondy, M.; Pinter, P.; Altenhof, W. Experimental characterization and modelling of the elastic properties of direct compounded compression molded carbon fibre/polyamide 6 long fibre thermoplastic. *Mater. Des.* **2017**, *122*, 184–196. [CrossRef]
11. Sui, G.; Wong, S.-C.; Yue, C.-Y. Effect of extrusion compounding on the mechanical properties of rubber-toughened polymers containing short glass fibers. *J. Mater. Processing Technol.* **2001**, *113*, 167–171. [CrossRef]
12. Yan, X.; Cao, S. Structure and interfacial shear strength of polypropylene-glass fiber/carbon fiber hybrid composites fabricated by direct fiber feeding injection molding. *Compos. Struct.* **2018**, *185*, 362–372. [CrossRef]
13. Ville, J.; Inceoglu, F.; Ghamri, N.; Pradel, J.L.; Durin, A.; Valette, R.; Vergnes, B. Influence of Extrusion Conditions on Fiber Breakage along the Screw Profile during TwinScrew Compounding of Glass Fiber-reinforced PA. *Int. Polym. Processing* **2013**, *28*, 49–57. [CrossRef]
14. Gamon, G.; Evon, P.; Riga, L. Twin-screw extrusion impact on natural fibre morphology and material properties in poly(lactic acid) based biocomposites. *Ind. Crops Prod.* **2013**, *46*, 173–185. [CrossRef]
15. Durmaz, B.; Aytac, A. Investigation of the mechanical, thermal, morphological and rheological properties of bio-based polyamide11/poly(lactic acid) blend reinforced with short carbon fiber. *Mater. Today Commun.* **2022**, *30*, 103030. [CrossRef]
16. Wang, J.; Geng, C.; Luo, F.; Liu, Y.; Wang, K.; Fu, Q.; He, B. Shear induced fiber orientation, fiber breakage and matrix molecular orientation in long glass fiber reinforced polypropylene composites. *Mater. Sci. Eng. A* **2011**, *528*, 3169–3176. [CrossRef]
17. Durin, A.; Micheli, P.; Ville, J.; Inceoglu, F.; Valette, R.; Vergnes, B. A matricial approach of fibre breakage in twin-screw extrusion of glass fibres reinforced thermoplastics. *Compos. Part A Appl. Sci. Manuf.* **2013**, *48*, 47–56. [CrossRef]
18. Zurawik, R.; Vilke, J.; Zarges, J.-C.; Heim, H.-P. Comparison of real and Simulated Fiber Orientations in Injection Molded Short Glass Fiber Reinforced Polyamide by X-ray Microtomography. *Polymers* **2022**, *14*, 29. [CrossRef]
19. Ketata, N.; Seantier, B.; Guermazi, N.; Grohens, Y. Processing and properties of flax fibers reinforced PLA/PBS biocomposites. *Mater. Today Proc.* **2022**, *53*, 228–236. [CrossRef]
20. Hietala, M.; Oksman, K. Pelletized cellulose fibres used in twin-screw extrusion for biocomposite manufacturing: Fibre breakage and dispersion. *Compos. Part A Appl. Sci. Manuf.* **2018**, *109*, 538–545. [CrossRef]
21. Oksman, K.; Mathew, A.; Långström, R.; Nyström, B.; Joseph, K. The influence of fibre microstructure on fibre breakage and mechanical properties of natural fibre reinforced polypropylene. *Compos. Sci. Technol.* **2009**, *69*, 1847–1853. [CrossRef]
22. Castellani, R.; Giuseppe, E.; Beagrand, J.; Dobosz, S.; Berzin, F.; Vergnes, B.; Budtova, T. Lignocellulosic fiber breakage in a molten polymer. Part 1. Qualitative analysis using rheo-optical observations. *Compos. Part A Appl. Sci. Manuf.* **2016**, *91*, 229–237. [CrossRef]
23. Kahl, C.; Feldmann, M.; Sälzer, P.; Heim, H.-P. Advanced short fiber composites with hybrid reinforcement and selective fiber-matrix-adhesion based on polypropylene—Characterization of mechanical properties and fiber orientation using high-resolution X-ray tomography. *Compos. Part A* **2018**, *111*, 54–61. [CrossRef]
24. Ghanbari, A.; Jalili, N.S.; Haddadi, S.A.; Arjmand, M.; Nofar, M. Mechanical properties of extruded glass fiber reinforced thermoplastic polyolefin composites. *Polym. Compos.* **2020**, *41*, 3748–3757. [CrossRef]
25. Bumm, S.H.; White, J.L.; Isayev, A.I. Glass Fiber Breakup in Corotating Twin Screw Extruder: Simulation and Experiment. *Polym. Compos.* **2012**, *33*, 2147–2158. [CrossRef]

polymers

MDPI

Article

A Hierarchical Grid Solver for Simulation of Flows of Complex Fluids

Antonio Castelo [1,*,†], Alexandre M. Afonso [2,†] and Wesley De Souza Bezerra [3,†]

1 Departamento de Matemática Aplicada e Estatística, Instituto de Ciências Matemáticas e de Computação, Universidade de São Paulo, Cx.P. 668, São Carlos 13560-970, SP, Brazil
2 Centro de Estudos de Fenómenos de Transporte, Departamento de Engenharia Mecânica, Faculdade de Engenharia da Universidade do Porto, 4200-465 Porto, Portugal; aafonso@fe.up.pt
3 Faculdade de Ciências Exatas e Tecnologia, Universidade Federal da Grande Dourados, Cx.P. 364, Dourados 79804-970, MS, Brazil; bezerra.ifsc@gmail.com
* Correspondence: castelo@icmc.usp.br
† These authors contributed equally to this work.

Abstract: Tree-based grids bring the advantage of using fast Cartesian discretizations, such as finite differences, and the flexibility and accuracy of local mesh refinement. The main challenge is how to adapt the discretization stencil near the interfaces between grid elements of different sizes, which is usually solved by local high-order geometrical interpolations. Most methods usually avoid this by limiting the mesh configuration (usually to graded quadtree/octree grids), reducing the number of cases to be treated locally. In this work, we employ a moving least squares meshless interpolation technique, allowing for more complex mesh configurations, still keeping the overall order of accuracy. This technique was implemented in the HiG-Flow code to simulate Newtonian, generalized Newtonian and viscoelastic fluids flows. Numerical tests and application to viscoelastic fluid flow simulations were performed to illustrate the flexibility and robustness of this new approach.

Keywords: finite difference methods; meshless interpolation; numerical solution; polymer flows; viscoelastic flows

Citation: Castelo, A.; Afonso, A.M.; De Souza Bezerra, W. A Hierarchical Grid Solver for Simulation of Flows of Complex Fluids. *Polymers* **2021**, *13*, 3168. https://doi.org/10.3390/polym13183168

Academic Editors: Roland G. Winkler and Carlos A. García-González

Received: 7 July 2021
Accepted: 10 September 2021
Published: 18 September 2021

1. Introduction

Many researchers are constantly working on improving numerical solution techniques for partial differential equations that govern the flow of Newtonian and non-Newtonian fluids. One of the major problems faced is the part that generates the geometry of the problem to be simulated.

Cartesian hierarchical grids, or tree-based grids are the most common choices for discretizing the spatial domain. This choice allows the implementation of the finite difference method, while avoiding working with more complicated stencils, which occurs for example in curved meshes. Thus, it becomes easier to process the flow properties when a refinement of the mesh is desired in a determined region of the domain, since in a Cartesian grid, the flows are calculated in facets parallel to the Cartesian axes, favoring the implementation of the numerical method [1]. In the literature, quadtree and octree are 2D and 3D meshes, respectively, generated to perform these problems. One can say that a hierarchical grid is a generalization of quadtree and octree. In this sense, the choice of hierarchical grids is convenient to address the problem of flows in complex geometries [2–6].

Regarding the mesh refinement, one of the difficulties is to calculate the flow property value on these interfaces. High-order interpolations are commonly used. Several improvements of the interpolation techniques have been studied [7–11], in order to optimize the number of cells used in calculations, since this influences the computational time and storage over simulations.

In this way, the HiG-Fow system makes interpolations using the method of moving least squares, adapting the stencil according to the interface between the fine and coarse grids. Sousa et al. developed this methodology and compared it with non-graded methods by using the new system to simulate Newtonian flows [12].

Our interest is to use the HiG-Flow for the simulation of non-Newtonian flows. In this way, a code module for simulations of non-Newtonian flows was implemented, taking into account considerations shown in Section 3.

Depending on the temperature or mixture in liquid solvents, polymeric materials behave similar to viscoelastic fluids [13]. In this work we show that a new computer system is able to perform numerical simulations of viscoelastic fluid flows in two and three dimensions in channels with complex geometries. In one of the most common applications, polymers are used to construct electronic devices. Thus, the study of viscoelastic fluids is important due to applications in science and technology and the use of numerical simulators can be useful for support in important decisions in the engineering design of any device. In general, the behavior of viscoelastic fluid can be described using an appropriate constitutive model. So, in addition to using HiG-Flow in Newtonian flows, the system has implemented a module to solve the constitutive equations through the kernel-conformation technique. [14]. Different constitutive models are implemented, among them the Oldroyd-B [15] and Phan-Thien–Tanner (PTT) [16], which we used as reference in this work. In the last 20 years several works involving the solution of these constitutive models have been published. In 1999, Dou and Phan-Thien [17] used the finite volume method to solve the flow in a channel of an Oldroyd-B fluid past a circular cylinder. Then, Alves et al. [18] showed the effect of a high-resolution scheme MINMOD [19] on an upper-convected Maxwell fluid solution, improving accuracy and increasing the convergence rate of the finite volume method and then they proposed a new high resolution scheme [20]. Later the article was published [21] with benchmark solutions for the flow of Oldroyd-B and PTT fluids in planar contractions. In the year 2005 Chinyoka et al. [22] studied the deformation of a circular drop of an Oldroyd-B fluid by applying the volume-of-fluid method for two-dimensional interfaces. Later, Tomé et al. [23] applied the finite difference method to simulate free surface flow of PTT fluid in three dimensional geometry. Then, Mompean et al. [24] investigated fluid flows using the Upper-Convected Maxwell (UCM) constitutive equation and an explicit algebraic model to develop an approximation that could be applied to the extrudate-swell problem. In 2012, Tom é et al. [25] applied the log-conformation technique to study three-dimensional viscoelastic flows for jet buckling analysis and later Oishi et al. [26] and Paulo et al. [27] continued studies in this same way.

In 2019, Tomé et al. [28] presented a solution method for the Giesekus model flow and proposed a new analytical solution for this problem. In 2019, Bezerra et al. [29] used HiG-Flow to perform the solution of electro-osmotic flow of a viscoelastic fluid, where they proposed an approximation for the vortices simulation in a nozzle. Shojaei et al. [30] investigated a generalized finite difference method using the weighted moving least squares procedure, in the same way of our proposed numerical solution. Corresponding with one of the proposals of this work, [31] used stabilization techniques in 2D and 3D viscoelastic fluid flows. In 2020, Guan et al. proposed a improved finite difference method and they checked its convergence. Recently, [32] presented a implementation and computational verification of KBKZ integral constitutive equations in hierarchical grids. More recently, ref. [33] performed a generalized finite differences method for flows in a dam.

The finite difference method was used in the discretization of equations. The HiG-Flow system was also implemented taking into account advances in the MAC-Marker and Cell method [34], allowing the implementation of several solution methods for the different terms of the equation of motion as well as the constitutive model solution. Convective terms in equations can be solved by high-accuracy methods. Moreover, we can say that the main novelty for the simulation of viscoelastic fluids is the kernel-conformation technique. The technique is already known, however, the differential is the manner it was implemented in, which allows the user to choose a numerical stabilizer easily—one just needs to enter the desired mathematical function, the derivative of this function and its inverse function. More details can be found in the Governing Equations section. Here, numerical stabilizers were used for Oldroyd-B flow solution in a 2D cavity and for a PTT fluid in a complex 3D geometry.

In Section 2, we show the finite differences method of the approximation used. Then, the governing equations and the constitutive models are presented in Section 3, as well as the explanation of the kernel-conformation technique. In Section 4, we present the validation tests for a PTT fluid flow in a pipe and to an Oldroyd-B fluid flow in a 2D-lid-driven cavity. Finally, we performed simulations of a PTT flow in a complex 3D geometry and the results are shown in Section 5.

2. Finite Difference Approximation in Tree-Based Grids

In the HiG-Flow code, the equations are solved using finite difference approach in hierarchical meshes. Figure 1a shows a representative type of mesh and the dependency structure (tree data structure) is presented in Figure 1b. In this approach, cells can be partitioned into different geometrical shapes. Such generalization leads to the difficult task of finding an accurate approximation to the different differential and integral operators.

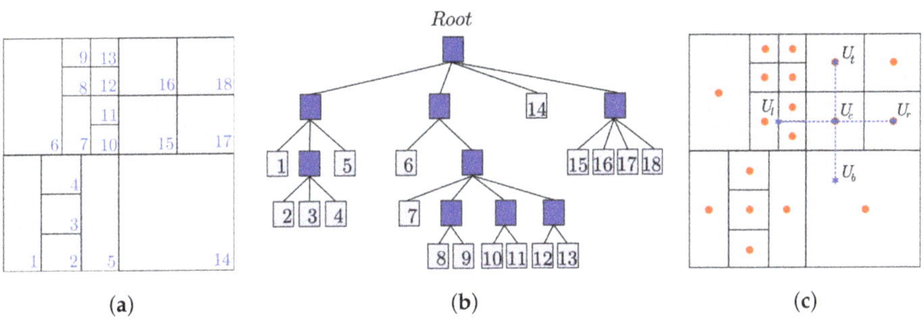

Figure 1. (**a**) Example of hierarchical grid. (**b**) Tree data structure. (**c**) Finite difference method.

Looking at Figure 1c, a second-order approximation to $\frac{\partial^2 U_c}{\partial y^2}$ can be given by (we assume the y axis is in the direction bottom $\rightarrow$ top):

$$\frac{\partial^2 U_c}{\partial y^2} \approx \frac{1}{\delta y}(U_t - 2U_c + U_b). \tag{1}$$

Note that U_b is not known and must be obtained by interpolation (the same applies to U_r) using the following formula:

$$U_b = \sum_{k=1}^{\mathcal{V}_b} w_k^b \, U_k, \tag{2}$$

where $\mathcal{V}_b$ is the number of neighbor cells, which depends on the imposed accuracy of the method.

The weights $w_k^b = w_k^b(\mathbf{x})$ are obtained through the moving least squares (MLS) method. In a set of n smooth interpolating functions that are linearly independent $\mathbf{\Phi}_i : \mathbb{R}^d \rightarrow \mathbb{R}$ $(d = 2, 3)$, we want to obtain the interpolated value u such that $U_b = U(\mathbf{x}) = \sum_{k=1}^{n} c_i \mathbf{\Phi}_i(\mathbf{x}) = \mathbf{c}^t \mathbf{\Phi}$.

Given m points $\mathbf{x}_1, \mathbf{x}_2, \ldots, \mathbf{x}_m \in \mathbb{R}^d$ with $m > n$ and m values $u_1 = u(\mathbf{x}_1), u_2 = u(\mathbf{x}_2), \ldots,$ $u_m = u(\mathbf{x}_m)$, to interpolate u in $\mathbf{x}$ using MLS consists in minimizing the error $E(\mathbf{c})$

$$E(\mathbf{c}) = \|U - u\|_2^2 = \sum_{i=1}^{m} (U(\mathbf{x}_i) - u_i)^2 \frac{1}{\|\mathbf{x} - \mathbf{x}_i\|_2}. \tag{3}$$

Or,

$$E(\mathbf{c}) = \|WP\mathbf{c} - W\mathbf{u}\|_2^2, \tag{4}$$

where $W = W(\mathbf{x}) = \left\{\delta_{ij}\sqrt{\frac{1}{\|\mathbf{x}-\mathbf{x}_i\|_2}}\right\} \in \mathbb{R}^{m\times m}$, $P = \{\mathbf{\Phi}_i(\mathbf{x}_j)\} \in \mathbb{R}^{m\times n}$ and $\mathbf{u} = (u_1, u_2, \ldots, u_m)$.

The solution is given by

$$\mathbf{c}(\mathbf{x}) = (WP)^{\dagger} W\mathbf{u} \tag{5}$$

where $(\cdot)^{\dagger}$ is the Moore–Penrose pseudo-inverse.

Decomposing (WP) into QR we have that

$$WP = Q\begin{bmatrix} R \\ 0 \end{bmatrix} = \begin{bmatrix} Q_{\parallel} & Q_{\perp} \end{bmatrix}\begin{bmatrix} R \\ 0 \end{bmatrix}, \tag{6}$$

where $Q \in \mathbb{R}^{m\times m}$, $Q_{\parallel} \in \mathbb{R}^{m\times n}$, $Q_{\perp} \in \mathbb{R}^{m\times m-n}$, and $R \in \mathbb{R}^{n\times n}$. This decomposition is then used to finally compute

$$\mathbf{c}(\mathbf{x}) = R^{-1} Q_{\parallel}^{t} W\mathbf{u}\,. \tag{7}$$

Then

$$U(\mathbf{x}) = \mathbf{c}^t\mathbf{\Phi} = \mathbf{u}^t \underbrace{W Q_{\parallel} R^{-t}\mathbf{\Phi}}_{\mathbf{w}} = \sum_{k=1}^{m} w_k u_k\,, \tag{8}$$

that is $\mathbf{w} = \mathbf{w}(\mathbf{x}) = WQ_{\parallel}R^{-t}\mathbf{\Phi}$.

The procedure to calculate $\mathbf{w}(\mathbf{x})$ must be performed for each approximation $U(\mathbf{x})$. This is performed only once since we are using a static mesh.

3. Governing Equations

The flow is assumed to be isothermal, laminar and the fluids incompressible. The governing equations are those expressing conservation of mass

$$\nabla \cdot \mathbf{u} = 0, \tag{9}$$

and conservation of momentum

$$\frac{\partial \mathbf{u}}{\partial t} + \mathbf{u}\cdot\nabla\mathbf{u} = -\nabla p + \frac{1}{Re}\nabla^2\mathbf{u} + \nabla\cdot\mathbf{S} + \frac{1}{Fr^2}\mathbf{g} + \mathbf{F}, \tag{10}$$

$$\mathbf{T} = \frac{2(1-\beta)}{Re}\mathbf{D} + \mathbf{S}, \tag{11}$$

where $\mathbf{u}$ is the velocity field, t is time, p is the pressure, Re is the Reynolds number, Fr is the Froude number, $\mathbf{g}$ is the gravity force and $\mathbf{F}$ is the surface tension force and source force. The symbol $\mathbf{D} = \frac{1}{2}\left(\nabla\mathbf{u} + (\nabla\mathbf{u})^T\right)$ is the rate of deformation tensor, $\mathbf{T}$ is the elastic stress. The amount of Newtonian solvent is controlled by the dimensionless solvent viscosity coefficient, $\beta = \frac{\mu_S}{\mu_0}$, where $\mu_0 = \mu_S + \mu_P$ denotes the total shear viscosity. Several polymeric constitutive equations are implemented in the current version of the solver: the upper-convected Maxwell model, the Oldroyd-B model, the linear form of the Phan-Thien/Tanner (LPTT) model [35] and the Giesekus model [36]. For an isothermal flow, these five rheological equations of state can be written in a compact form as:

$$\frac{\partial \mathbf{T}}{\partial t} + (\mathbf{u}\cdot\nabla)\mathbf{T} - \left[(\nabla\mathbf{u})^T\cdot\mathbf{T} + \mathbf{T}\cdot\nabla\mathbf{u}\right] = \frac{1}{De}\mathbf{M}(\mathbf{T}). \tag{12}$$

where $\mathbf{M}(\mathbf{T})$ is defined by the viscoelastic model

$$\mathbf{M}(\mathbf{T}) = \begin{cases} \frac{2(1-\beta)}{Re}\mathbf{D} - \mathbf{T} & \text{Oldroyd-B,} \\ \frac{2(1-\beta)}{Re}\mathbf{D} - \mathbf{T} - \frac{\alpha\, Re\, De}{1-\beta}\mathbf{T}\cdot\mathbf{T} & \text{Giesekus,} \\ \frac{2(1-\beta)}{Re}\mathbf{D} - \left(1 + \frac{\epsilon\, Re\, De}{1-\beta}\mathrm{tr}(\mathbf{T})\right)\mathbf{T} - \xi\, De(\mathbf{T}\cdot\mathbf{D} + \mathbf{D}\cdot\mathbf{T}) & \text{LPTT,} \end{cases} \tag{13}$$

where De is the Deborah number. The stress coefficient function of the LPTT model depends on the trace of $\mathbf{T}$, $\text{tr}(\mathbf{T})$ and introduces the dimensionless parameter ϵ, which is closely related to the steady-state elongational viscosity in extensional flows. The slip parameter, ξ, takes into account the non-affine motion between the polymer molecules and the continuum. The polymer strands embedded in the medium may slip with respect to the deformation of the macroscopic medium, thus each strand may transmit only a fraction of its tension to the surrounding continuum. When $\xi = 0$ there is no slip and the motion becomes affine. Parameter ξ is responsible for a non-zero second normal-stress difference in shear, leading to secondary flows in ducts having non-circular cross-sections, which is superimposed on the streamwise flow. In the non-linear term of the Giesekus model, α represents a dimensionless "mobility factor".

An alternative form to describe viscoelastic models is by using the conformation tensor, $\mathbf{A}$. This tensor is Symmetric and Positive Definite (SPD), which is an important mathematical property for the construction of matrix transformations and/or decompositions. In general, the equation for $\mathbf{A}$ can be written as

$$\frac{\partial \mathbf{A}}{\partial t} + (\mathbf{u} \cdot \nabla)\mathbf{A} - \left[\mathbf{A}\nabla\mathbf{u} + \nabla\mathbf{u}^{\mathrm{T}}\mathbf{A}\right] = \frac{1}{De}\mathcal{M}(\mathbf{A}), \tag{14}$$

where $\mathcal{M}(\mathbf{A})$ is a function that depends on the specific constitutive model. The relation between stress tensor $\mathbf{T}$ and $\mathbf{A}$ is given by

$$\mathbf{T} = \frac{1-\beta}{Re\, De}(\mathbf{A} - \mathbf{I}), \tag{15}$$

that can rewritten as a relation between the tensor $\mathbf{S}$ and $\mathbf{A}$ given by

$$\mathbf{S} = \frac{1-\beta}{Re\, De}(\mathbf{A} - \mathbf{I} - 2De\, \mathbf{D}). \tag{16}$$

A problem that challenges many researchers in computational rheology is solving Equation (12) or Equation (14) for high values of the Deborah number, $De = \lambda / t_c$, where t_c is a characteristic time of the flow. This problem occurs because all numerical methods are unstable for certain critical values of De. In order to overcome such failure, Fattal and Kupferman [37] proposed a reformulation of the differential constitutive equations into a equation for the matrix logarithm of the conformation tensor. Extending the ideas proposed by [37,38], ref. [14] presented a generic kernel-conformation tensor transformation that allows us to apply different kernel functions to the matrix transformation.

The reformulation of the tensor conformation was possible by the decomposition of the velocity gradient proposal by [37,38]

$$\nabla\mathbf{u}^{\mathrm{T}} = \mathbf{\Omega} + \mathbf{B} + \mathbf{N}\mathbf{A}^{-1}, \tag{17}$$

where $\mathbf{\Omega}$ and $\mathbf{N}$ are anti-symmetric tensors, $\mathbf{B}$ is symmetric and commutes with $\mathbf{A}$. Thus, the constitutive equation based on the conformation tensor can be rewritten using the decomposition (17) as

$$\frac{\partial \mathbf{A}}{\partial t} + (\mathbf{u} \cdot \nabla)\mathbf{A} - (\mathbf{\Omega}\mathbf{A} - \mathbf{A}\mathbf{\Omega}) - 2\mathbf{B}\mathbf{A} = \frac{1}{De}\mathcal{M}(\mathbf{A}), \tag{18}$$

where $\mathcal{M}(\mathbf{A})$ is defined according to the viscoelastic model,

$$\mathcal{M}(\mathbf{A}) = \begin{cases} \mathbf{I} - \mathbf{A} & \text{Oldroyd-B,} \\ \mathbf{I} - \mathbf{A} - \alpha(\mathbf{A} - \mathbf{I}) \cdot (\mathbf{A} - \mathbf{I}) & \text{Giesekus,} \\ \left(1 + \frac{\epsilon\, Re\, De}{1-\beta}\text{tr}(\mathbf{S})\right)(\mathbf{I} - \mathbf{A}) - 2\xi\, De(\mathbf{B} - \mathbf{B}\mathbf{A}) & \text{PTT.} \end{cases} \tag{19}$$

Fattal and Kupferman showed that the matrix logarithm of the conformation tensor is a linear transformation of $\mathbf{A}$ and derived a constitutive equation from the Equation (18) in the function of the matrix logarithm. Afonso et al. proposed a generic *kernel-conformation* tensor transformation for a large class of differential constitutive models, in which the evolution equation for $\mathbb{k}(\mathbf{A})$, can be expressed in its tensorial formulations as

$$\frac{\mathrm{D}\mathbb{k}(\mathbf{A})}{\mathrm{D}t} = \mathbf{\Omega}\mathbb{k}(\mathbf{A}) - \mathbb{k}(\mathbf{A})\mathbf{\Omega} + 2\mathbb{B} + \frac{1}{De}\mathbb{M} \tag{20}$$

where $\mathbb{B}$ and $\mathbb{M}$ are symmetric tensors constructed by the orthogonalization of the diagonal tensors $\mathbf{D}_{\mathbb{B}}$ and $\mathbf{D}_{\mathbb{M}}$, respectively. These tensors can be constructed as

$$\begin{aligned} \mathbb{B} &= \mathbf{O}\mathbf{D}_{\mathbb{B}}\mathbf{O}^T = \mathbf{O}\tilde{\mathbf{B}}\mathbf{\Lambda}\mathbf{J}\mathbf{O}^T \\ \mathbb{M} &= \mathbf{O}\mathbf{D}_{\mathbb{M}}\mathbf{O}^T = \mathbf{O}\mathcal{M}(\mathbf{\Lambda})\mathbf{J}\mathbf{O}^T. \end{aligned} \tag{21}$$

In Equations (21), $\mathbf{J}$ is the gradient matrix, a diagonal matrix of the form,

$$\mathbf{J} = \mathrm{diag}\left(\frac{\partial\mathbb{k}(\lambda_1)}{\partial\lambda_1}; \frac{\partial\mathbb{k}(\lambda_2)}{\partial\lambda_2}; \frac{\partial\mathbb{k}(\lambda_3)}{\partial\lambda_3}\right). \tag{22}$$

4. Verification Tests

In this section, we address two test problems in terms of checking the HiG-Flow code for simulations of viscoelastic flows. One of the problems is the flow of a Phan-Thien–Tanner model fluid in a circular cross-section channel. The other test problem concerns the constitutive model of Oldroyd-B. The geometry used for this test was a driven cavity in two dimensions.

4.1. Phan-Thien–Tanner Model Fluid Flow in a Pipe

We consider a flow into a circular cylinder of radius R, where there exists only the axial velocity component u, which depends on the radial coordinate r. In addition, we consider that the fluid obeys the PTT fluid model [16] and the flow occurs in the x direction, the same as the cylinder axis. Here, we consider the known solutions available of the literature to the flow properties, namely velocity u, shear stress T_{rx} and normal stress T_{xx}. More detailed treatment about the analytical solution to this problem in a steady state, as well as the results verified here, can be found in [39–41]. Essentially, to obtain the viscoelastic component T_{rx}, it is necessary to solve a cubic equation $T_{rx}^3 + 3AT_{rx} - 2B = 0$, whose its solution is given by

$$T_{rx} = \left[B + \left(A^3 + B^2\right)^{1/2}\right]^{1/3} + \left[B - \left(A^3 + B^2\right)^{1/2}\right]^{1/3}, \tag{23}$$

where A and B depends on the set of know parameters of flow:

$$A = \frac{\eta_p^2}{6\varepsilon\lambda^2\beta}, \tag{24}$$

$$B = -\frac{\eta_p^3 u_N}{\varepsilon\lambda^2 R^2\beta} r. \tag{25}$$

In Equations (24) and (25), η_p is the polymer viscosity, ε is the PTT parameter, λ is the relaxation time and R is the cylinder radius. The amount of solvent contribution is given by $\beta = \eta_s/\eta_0$, the reference velocity is u_N and r is the radial coordinate. After obtaining T_{rx}, one can calculate the normal stress T_{xx} and also integrate the equation of motion to determine the velocity field. The corresponding expressions are given by:

$$T_{xx} = \frac{2\lambda}{\eta_p} T_{rx}^2, \tag{26}$$

$$u(r) = \frac{2u_N}{\beta}\left[1 - \left(\frac{r}{R}\right)^2\right] + f(A, B), \tag{27}$$

where $f(A, B)$ is a function that depends on the parameters A and B given in (24) and (25), respectively. These simulation parameters can be adjusted when the polymer viscosity is fixed, then by varying β it is possible to control the amount of the solvent contribution. In addition, just as β, ε and De are input viscoelastic parameters, λ is adjusted by the Deborah number. To verify that the results are in agreement with [41], we set $\varepsilon = 0.25$ and $De = 6.3$, which corresponds to the reference $De_N = 1.0$. Non-slip boundary conditions were used for the velocity in the cylinder wall. At the channel inlet, we imposed a parabolic velocity profile and at the outlet, the homogeneous Neummann boundary condition, that is, spatial variations in velocity are not allowed at the outflow. For pressure, a zero gradient was imposed on the wall and at the channel inlet while the outflow was fixed at a constant value. The initial conditions for the bulk domain is zero velocity.

Figures 2–4 show the velocity field, shear stress and normal stress, respectively, as a function of the amount of solvent, which is controlled by β parameter. When $\beta \approx 1$, the polymer concentration is approximately zero and the fluid has Newtonian behavior. On the other hand, if $\beta \approx 0$, the behavior of the PTT fluid resembles that of the Oldroyd-B model. The curves represented by down triangles corresponds to $\beta = 0.9$, up triangles to $\beta = 0.5$, circles to $\beta = 0.2$ and squares to $\beta = 0.01$. All these results are obtained by HiG-Flow simulation. They are in perfect agreement with the analytical curves represented by solid lines in Figures 2–4, which corresponds to the solutions given by Equations (23), (27) and (26) for u, T_{rx} and T_{xx}, respectively.

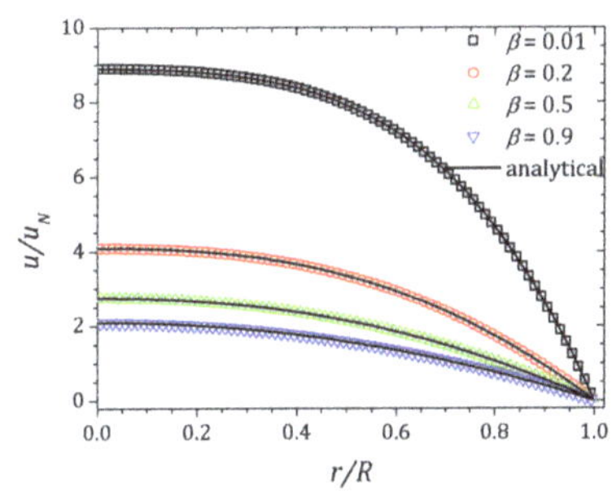

Figure 2. Velocity field for a PTT flow in a pipe.

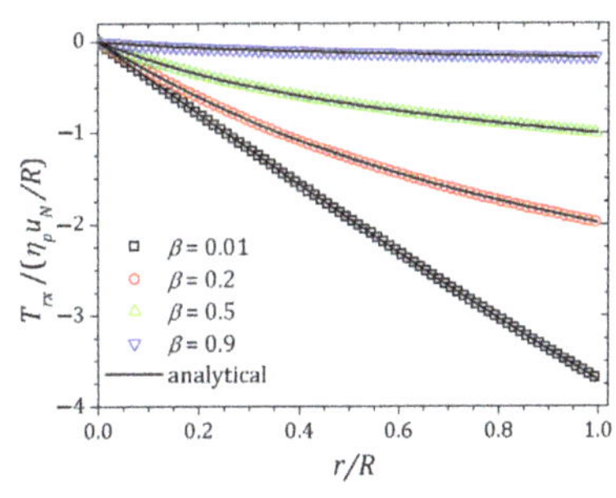

Figure 3. Shear stress T_{rx} for a PTT flow in a pipe.

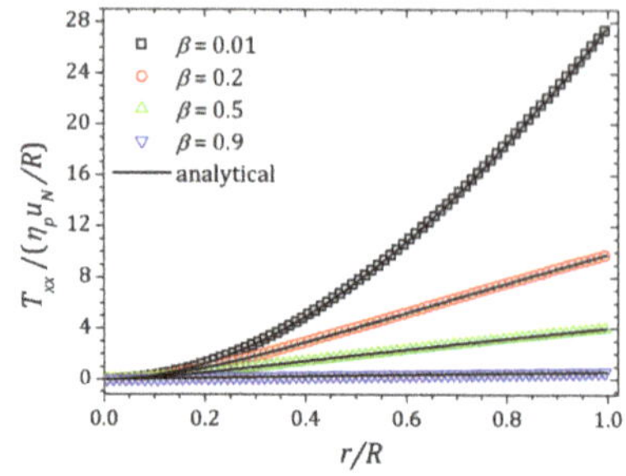

Figure 4. Normal stress for a PTT flow in a pipe.

4.2. 2D-Driven Cavity with Oldroyd-B Flow

Flows in rectangular cavities have been studied since 1967 when the article [42] was published. In the 21st century, several studies of this type for viscoelastic fluids have been published [38,43–47]. The problem studied here has no analytical solution; however, there are results obtained by the cited authors that can be used for comparison. The data used for comparison in this study were provided by Palhares Junior et al. [47].

Figure 5 shows the schematic of lid-driven cavity. A parabolic profile velocity is imposed on the top. The aspect ratio is defined as $\Lambda \equiv H/L$. Some concerned works are listed in Table 1.

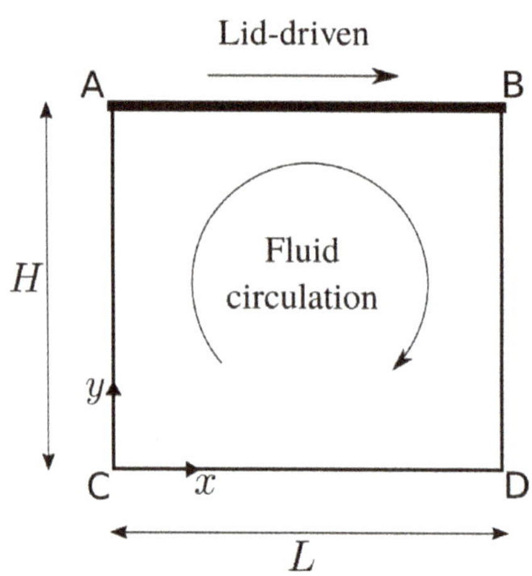

Figure 5. Illustration of lid-driven cavity. The parabolic velocity profile is imposed on the top. The aspect ratio is defined as $\Lambda \equiv H/L$.

The use of stabilization methods within the HiG-Flow system can be considered simple from the coder point of view because the code has been implemented in such a way that one can make a choice directly in the main simulation file. In this sense, the kernel conformation tensor is used to perform this operation, as previously described in the Section 3. Generically, the user simply writes the kernel $\mathbb{k}$, the kernel derivative $d\mathbb{k}/d\mathbf{x}$ for the Jacobian transformation calculation and the kernel inverse $\mathbb{k}^{-1}$ correspondents. For the square root stabilizer used in these simulations, one can write Equations (28)–(30), respectively, as follows:

$$\mathbb{k}_{ij} = \sqrt{S_{ij}}, \tag{28}$$

$$\frac{d\mathbb{k}_{ij}}{dx} = \frac{1}{2\sqrt{S_{ij}}}, \tag{29}$$

$$\mathbb{k}_{ij}^{-1} = S_{ij}^{2}, \tag{30}$$

Table 1. Previous and current numerical studies concerned with lid-driven cavity flow of constant viscosity viscoelastic fluids.

Reference	Aspect Ratios	Constitutive Equation	*De*	Regularization	Notes
Grillet et al. [48]	0.5, 1.0, 3.0	FENE-CR, $L^2 = 25, 100, 400$	≤ 0.24	Leakage at corners A and B	FE
Fattal and Kupferman [38]	1.0	Oldroyd-B, $\beta = 0.5$	1.0, 2.0, 3.0, 5.0	$u(x) = 16Ux^2(1-x)^2$	FD, Log conformation technique
Pan et al. [43]	1.0	Oldroyd-B, $\beta = 0.5$	0.5, 1.0	$u(x) = 16Ux^2(1-x)^2$	FE, Log conformation technique
Yapici et al. [44]	1.0	Oldroyd-B, $\beta = 0.3$	≤ 1.0	No	FV, First-order upwind
Habla et al. [46]	1.0	Oldroyd-B, $\beta = 0.5$	0 to 2	$u(x,z) = 128[1 + \tanh 8(t - 1/2)]$ $x^2(1-x)^2z^2(1-z)^2$	FV, 3D, Log conformation technique, CUBISTA
Comminal et al. [49]	1.0	Oldroyd-B, $\beta = 0.5$	0.25 to 10	$u(x) = 16Ux^2(1-x)^2$	FD/FV, Log-conformation, stream function
Martins et al. [50]	1.0	Oldroyd-B, $\beta = 0.5$	0.5, 1.0, 2.0	$u(x) = 16Ux^2(1-x)^2$	FD, Kernel-conformation technique
Dalal et al. [51]	1.0	Oldroyd-B, $\beta = 0.5$	1.0	$u(x) = 16Ux^2(1-x)^2$	FD, Symmetric square root
Palhares Junior et al. [47]	1.0	Oldroyd-B, $\beta = 0.5$	1.0, 2.0	$u(x,t) = 8[1 + \tanh(8t-4)]x^2$ $(1-x^2)^2$	FD, Symmetric square root
Current work	1.0	Oldroyd-B, $\beta = 0.5$	1.0, 2.0	$u(x,t) = 8[1 + \tanh(8t-4)]x^2$ $(1-x^2)^2$	FD, Kernel-conformation technique

For the simulations, the Reynolds number was fixed as $Re = 0.01$. The proportion of solvent in Oldroyd-B fluid was also fixed as $\beta = 0.5$. Simulations were performed for two different Deborah numbers, $De = 1.0$ and $De = 2.0$. On the top lid-driven section of the cavity, we imposed a parabolic velocity profile given by

$$u(x,t) = 8[1 + \tanh(8t-4)]x^2\left(1 - x^2\right)^2. \tag{31}$$

The other cavity walls are stationary and the non-slip condition is imposed over all of them. A regular mesh of 256x256 cells was used. The velocity component u and the normal stress T_{xx} were plotted along the vertical line $x = 0.5$ while the velocity component v was obtained on the horizontal line $y = 0.75$. The (x, y) coordinates are scaled by the cavity side size $L = 1$ unit of length. The results are shown in Figures 6–8. The curves are represented by squares and circles corresponding to the HiG-Flow and Palhares Junior et al. results, respectively. The graphs indicate that the results are in good agreement.

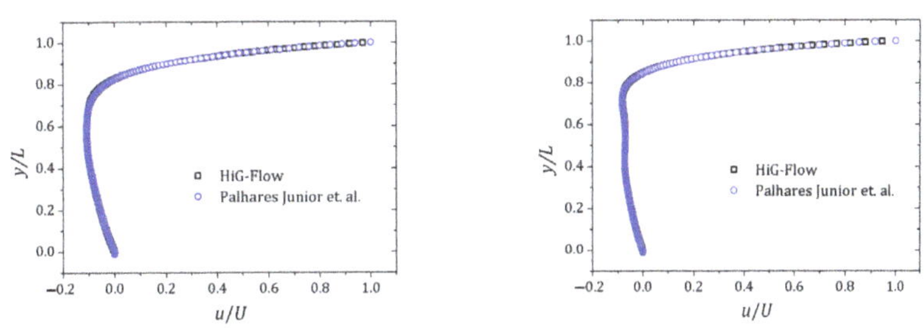

(a) (b)

Figure 6. Velocity field u obtained along the vertical line $x = 0.5$: (**a**) $De = 1.0$; (**b**) $De = 2.0$.

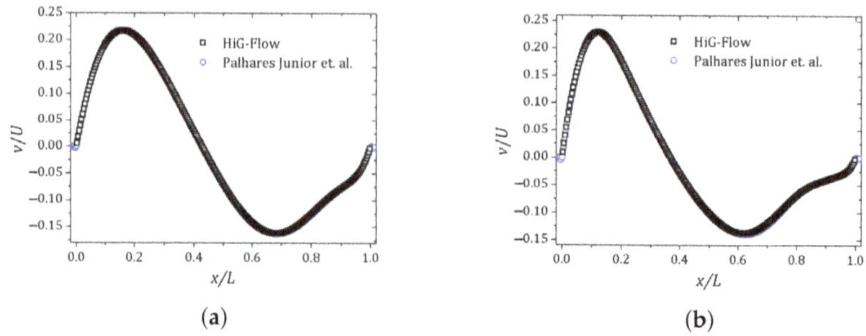

Figure 7. Velocity field v obtained along the horizontal line $y = 0.75$: (**a**) $De = 1.0$; (**b**) $De = 2.0$.

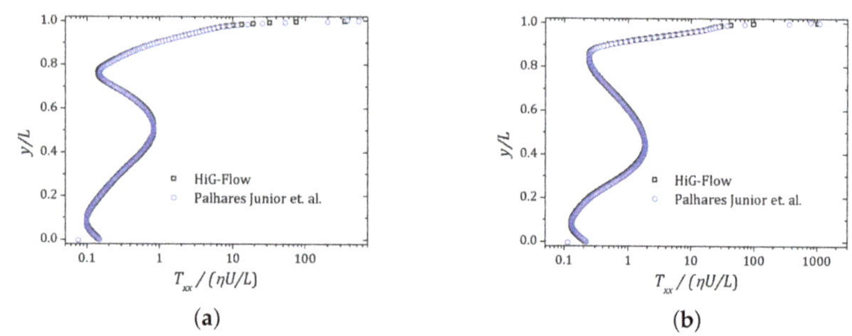

Figure 8. Normal stress T_{xx} obtained along the vertical line $x = 0.5$: (**a**) $De = 1.0$; (**b**) $De = 2.0$.

5. Simulation in Complex 3D Array of Channels

In this section, we present the results obtained with the HiG-Flow system for flows in complex domains. We simulated an incompressible viscoelastic fluid flow in a complex array of microchannels, introducing some level of geometric complexity in the three-dimensional flow domain.

The geometry, as well as boundary conditions, can be seen in Figure 9. The total width, length and height are set to be $W = 0.8$ mm, $L = 2.4$ mm and $H = 0.4$ mm, respectively. The inlet is a channel of 0.1 mm $\times$ 0.1 mm, where polymer at temperature is injected with a constant velocity of $U_{\text{in}} = 0.1$ mm/s. Scaling this geometry by $\ell = 0.1$ mm, and using $\nu \approx 10^{-4}$ m^2/s as the kinematic viscosity of polymer at room temperature, we end up with a Reynolds number of $Re = \ell\, U_{\text{in}}/\nu = 1.0$. In this test, we used the PTT model with $\beta = 0.5$, $\varepsilon = 0.25$ and $\xi = 0.0$ for several values of $De = [0 - 500]$ as viscoelastic parameters.

Streamlines for the flow of a Newtonian fluid can be observed in Figure 10. The result is qualitative, but demonstrates the robustness and applicability of this newly developed methodology. Several simulations using viscoelastic fluids for $De = [0 - 500]$ were performed on the 3D complex geometry. We analyzed the complex fluid flow by observing the profiles of the polymeric stresses along the probe line near the 3D channel exits, as shown in Figure 11. The probe is aligned on the y direction at half channel height (along the z direction), orthogonal to the main flow direction near the channel exits.

The increasing values of elasticity, reflected on the value of Deborah number represented in Figure 12, affects the six components of the non-dimensional extra stress tensor along the probe line, with higher impact for the normal components, as the T_{zz} profiles. Nevertheless, given that no geometrical singularity is presented along the probe line, the maximum value for all extra stress components is not significant and slightly affected by the increase in elasticity.

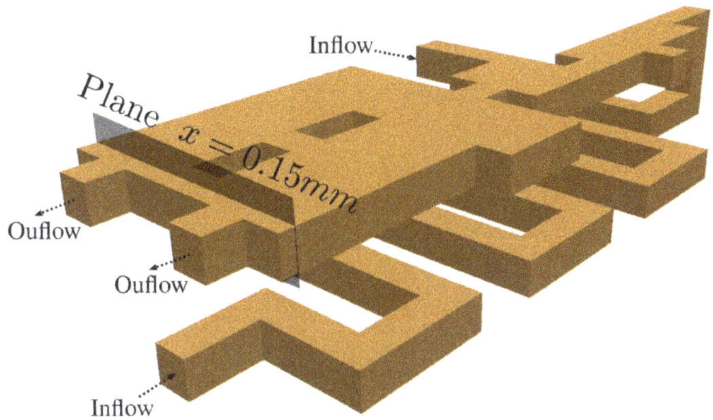

Figure 9. Geometry for the complex 3D array of channels.

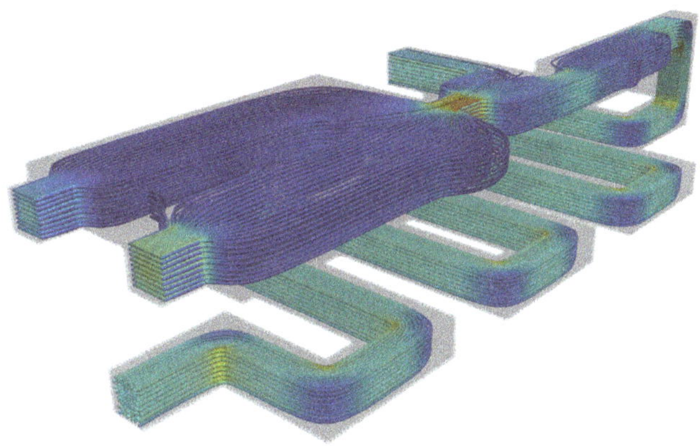

Figure 10. Streamlines for the complex 3D array of channels. The color scale varies from smallest (blue) to largest (red) velocity magnitude.

We used a computer with a 3.1 GHz Intel Core i7 Quad-Core Processor and 16 GB 2133 MHz LPDDR3 memory. The HiG-Flow software was used with four cores for all the calculation, and each simulation took 14 h of processing.

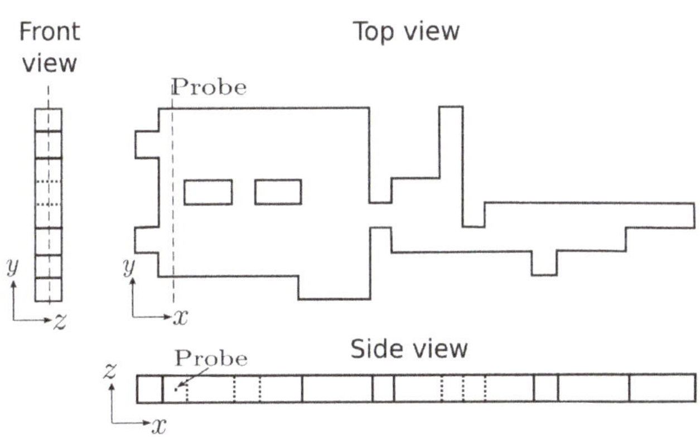

Figure 11. Probe views.

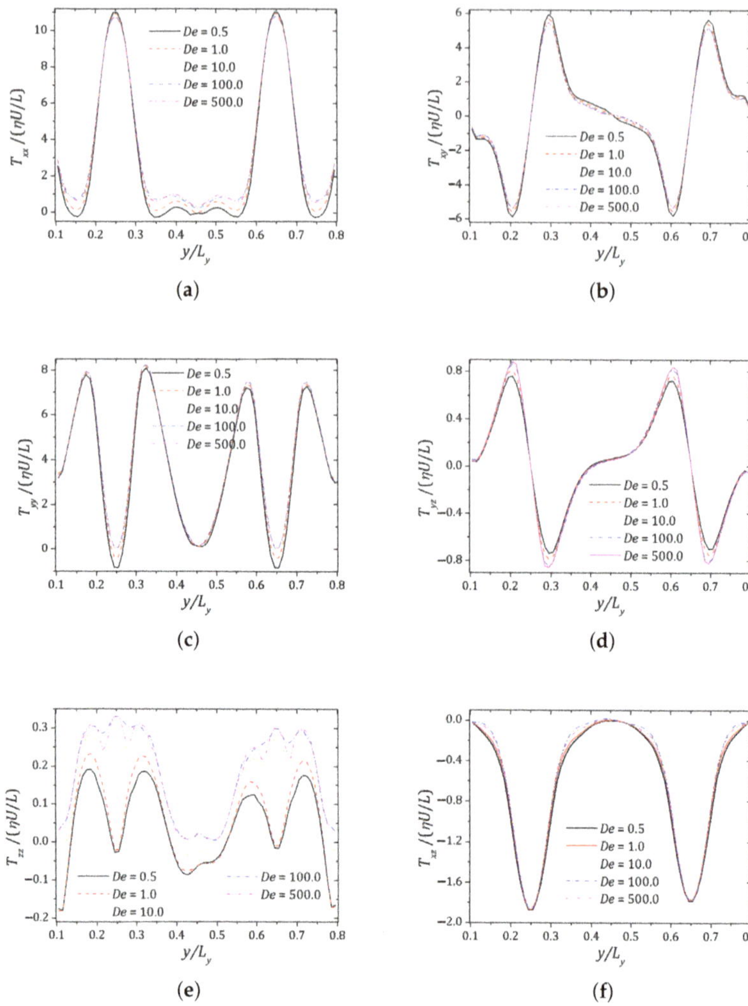

Figure 12. Tensor components: (**a**) T_{xx}; (**b**) T_{xy}; (**c**) T_{yy}; (**d**) T_{yz}; (**e**) T_{zz}; (**f**) T_{xz}.

6. Conclusions

Tree-based grids bring the advantage of using fast Cartesian discretizations, such as finite differences, and the flexibility and accuracy of local mesh refinement. Most methods usually avoid this by limiting the mesh configuration (usually to graded quadtree/octree grids), reducing the number of cases to be treated locally. In this work, we employ a moving least squares meshless interpolation technique, allowing for more complex mesh configurations, while still keeping the overall order of accuracy. This technique was implemented in the HiG-Flow code to simulate Newtonian, generalized Newtonian and viscoelastic fluids flows. The code verification and testing was performed using numerical stabilizers for the Oldroyd-B flow solution in a 2D cavity and for a PTT fluid in a complex 3D geometry.

Author Contributions: Funding acquisition, A.C.; Methodology, A.C., A.M.A. and W.D.S.B.; Project administration, A.C.; Software, A.C., A.M.A. and W.D.S.B.; Supervision, A.C.; Validation, A.C., A.M.A. and W.D.S.B.; Writing—Original draft, A.C., A.M.A. and W.D.S.B. All authors have read and agreed to the published version of the manuscript.

Funding: This research was funded by Fundação para a Ciência e a Tecnologia I.P. (FCT), CEFT (Centro de Estudos de Fenómenos de Transporte) and by FEDER funds through COMPETE2020. Grants: PTDC/EMS-ENE/3362/2014, POCI-01-0145-FEDER-016665, UIDB/00013/2020 and UIDP/00013/2020. All authors are grateful for the financial support from the Brazilian Petroleum Agency (ANP)/Petrobras, grant 0050.0075367.12.9, and from the São Paulo Research Foundation (FAPESP), grants 2013/07375-0, 2017/21105-6 and 2020/02990-1. A. Castelo is also grateful for the financial support from the National Council for Scientific and Technological Development (CNPq), grant 307483/2017-7. Research was carried out using the computational resources of the Center for Mathematical Sciences Applied to Industry (CeMEAI), funded by FAPESP grant 2013/07375-0.

Acknowledgments: Castelo, A. acknowledges the support of the ICMC-Instituto de Ciências Matemáticas e de Computação, Departamento de Matemática e Estatística, USP, São Carlos, SP. Afonso, A.M. acknowledges the support of the Faculdade de Engenharia da Universidade do Porto and Bezerra. W. S. acknowledges the support of the UFGD-Universidade Federal da Grande Dourados, Instituto de Ciências Exatas e Tecnologia.

Conflicts of Interest: The authors declare no conflict of interest.

References

1. Berger, M.J.; Oliger, J. Adaptive mesh refinement for hyperbolic partial differential equations. *J. Comput. Phys.* **1984**, *53*, 484–512. [CrossRef]
2. Popinet, S. Gerris: A tree-based adaptive solver for the incompressible Euler equations in complex geometries. *J. Comput. Phys.* **2003**, *190*, 572–600. [CrossRef]
3. Losasso, F.; Gibou, F.; Fedkiw, R. Simulating Water and Smoke with an Octree Data Structure. *ACM Trans. Graph.* **2004**, *23*, 457–462. [CrossRef]
4. Olshanskii, M.A.; Terekhov, K.M.; Vassilevski, Y.V. An octree-based solver for the incompressible Navier Stokes equations with enhanced stability and low dissipation. *Comput. Fluids* **2013**, *84*, 231–246. http://dx.doi.org/10.1016/j.compfluid.2013.04.027. [CrossRef]
5. Guittet, A.; Theillard, M.; Gibou, F. A stable projection method for the incompressible Navier-Stokes equations on arbitrary geometries and adaptive Quad/Octrees. *J. Comput. Phys.* **2015**, *292*, 215–238. [CrossRef]
6. Batty, C. A cell-centred finite volume method for the Poisson problem on non-graded quadtrees with second order accurate gradients. *J. Comput. Phys.* **2017**, *331*, 49–72. [CrossRef]
7. Ding, H.; Shu, C.; Yeo, K.S.; Xu, D. Simulation of incompressible viscous flows past a circular cylinder by hybrid FD scheme and meshless least square-based finite difference method. *Comput. Methods Appl. Mech. Eng.* **2004**, *193*, 727–744. [CrossRef]
8. Chew, C.S.; Yeo, K.S.; Shu, C. A generalized finite-difference (GFD) ALE scheme for incompressible flows around moving solid bodies on hybrid meshfree-Cartesian grids. *J. Comput. Phys.* **2006**, *218*, 510–548. [CrossRef]
9. Min, C.; Gibou, F.; Ceniceros, H.D. A supra-convergent finite difference scheme for the variable coefficient Poisson equation on non-graded grids. *J. Comput. Phys.* **2006**, *218*, 123–140. [CrossRef]
10. Ding, H.; Shu, C.; Cai, Q.D. Applications of stencil-adaptive finite difference method to incompressible viscous flows with curved boundary. *Comput. Fluids* **2007**, *36*, 786–793. [CrossRef]
11. Wang, X.Y.; Yeo, K.S.; Chew, C.S.; Khoo, B.C. A SVD-GFD scheme for computing 3D incompressible viscous fluid flows. *Comput. Fluids* **2008**, *37*, 733–746. [CrossRef]
12. Sousa, F.; Lages, C.; Ansoni, J.; Castelo, A.; Simao, A. A finite difference method with meshless interpolation for incompressible flows in non-graded tree-based grids. *J. Comput. Phys.* **2019**, *396*, 848–866. doi: 10.1016/j.jcp.2019.07.011. [CrossRef]
13. Bird, R.B.; Armstrong, R.C.; Hassager, O. *Dynamics of Polymeric Liquids*, 2nd ed.; Fluid Mechanics; John Wiley and Sons Inc.: New York, NY, USA, 1987; Volume 1, p. 784. ISBN 0-471-80245-X.
14. Afonso, A.M.; Pinho, F.T.; Alves, M.A. The kernel-conformation constitutive laws. *J. Non-Newton. Fluid Mech.* **2012**, *167–168*, 30–37. [CrossRef]
15. Oldroyd, J.G. On the formulation of rheological equations of state. *Proc. R. Soc. Lond. Ser. A Math. Phys. Sci.* **1950**, *200*, 523–541.
16. Thien, N.P.; Tanner, R.I. A new constitutive equation derived from network theory. *J. Non-Newton. Fluid Mech.* **1977**, *2*, 353–365. [CrossRef]
17. Dou, H.S.; Phan-Thien, N. The flow of an Oldroyd-B fluid past a cylinder in a channel: adaptive viscosity vorticity (DAVSS-ω) formulation. *J. Non-Newton. Fluid Mech.* **1999**, *87*, 47–73. [CrossRef]
18. Alves, M.; Pinho, F.; Oliveira, P. Effect of a high-resolution differencing scheme on finite-volume predictions of viscoelastic flows. *J. Non-Newton. Fluid Mech.* **2000**, *93*, 287–314. [CrossRef]

19. Harten, A. High resolution schemes for hyperbolic conservation laws. *J. Comput. Phys.* **1983**, *49*, 357–393. [CrossRef]
20. Alves, M.; Oliveira, P.; Pinho, F. A convergent and universally bounded interpolation scheme for the treatment of advection. *Int. J. Numer. Methods Fluids* **2003**, *41*, 47–75. [CrossRef]
21. Alves, M.A.; Oliveira, P.J.; Pinho, F.T. Benchmark solutions for the flow of Oldroyd-B and PTT fluids in planar contractions. *J. Non-Newton. Fluid Mech.* **2003**, *110*, 45–75. [CrossRef]
22. Chinyoka, T.; Renardy, Y.; Renardy, M.; Khismatullin, D. Two-dimensional study of drop deformation under simple shear for Oldroyd-B liquids. *J. Non-Newton. Fluid Mech.* **2005**, *130*, 45–56. [CrossRef]
23. Tomé, M.F.; Paulo, G.S.d.; Pinho, F.; Alves, M. Numerical solution of the PTT constitutive equation for unsteady three-dimensional free surface flows. *J. Non-Newton. Fluid Mech.* **2010**, *165*, 247–262. [CrossRef]
24. Mompean, G.; Thais, L.; Tomé, M.F.; Castelo, A. Numerical prediction of three-dimensional time-dependent viscoelastic extrudate swell using differential and algebraic models. *Comput. Fluids* **2011**, *44*, 68–78. [CrossRef]
25. Tomé, M.F.; Castelo, A.; Afonso, A.M.; Alves, M.A.; Pinho, F.T.d. Application of the log-conformation tensor to three-dimensional time-dependent free surface flows. *J. Non-Newton. Fluid Mech.* **2012**, *175*, 44–54. [CrossRef]
26. Oishi, C.M.; Martins, F.P.; Tomé, M.F.; Alves, M.A. Numerical simulation of drop impact and jet buckling problems using the eXtended Pom–Pom model. *J. Non-Newton. Fluid Mech.* **2012**, *169*, 91–103. [CrossRef]
27. Paulo, G.S.d.; Oishi, C.M.; Tomé, M.F.; Alves, M.A.; Pinho, F. Numerical solution of the FENE-CR model in complex flows. *J. Non-Newton. Fluid Mech.* **2014**, *204*, 50–61. [CrossRef]
28. Tomé, M.F.; Araujo, M.T.d.; Evans, J.D.; McKee, S. Numerical solution of the Giesekus model for incompressible free surface flows without solvent viscosity. *J. Non-Newton. Fluid Mech.* **2019**, *263*, 104–119. [CrossRef]
29. Bezerra, W.D.S.; Castelo, A.; Afonso, A.M. Numerical Study of Electro-Osmotic Fluid Flow and Vortex Formation. *Micromachines* **2019**, *10*, 796. [CrossRef]
30. Shojaei, A.; Galvanetto, U.; Rabczuk, T.; Jenabi, A.; Zaccariotto, M. A generalized finite difference method based on the Peridynamic differential operator for the solution of problems in bounded and unbounded domains. *Comput. Methods Appl. Mech. Eng.* **2019**, *343*, 100–126. [CrossRef]
31. Varchanis, S.; Syrakos, A.; Dimakopoulos, Y.; Tsamopoulos, J. A new finite element formulation for viscoelastic flows: Circumventing simultaneously the LBB condition and the high-Weissenberg number problem. *J. Non-Newton. Fluid Mech.* **2019**, *267*, 78–97. [CrossRef]
32. Bertoco, J.; de Araújo, M.S.; Leiva, R.T.; Sánchez, H.A.; Castelo, A. Numerical Simulation of KBKZ Integral Constitutive Equations in Hierarchical Grids. *Appl. Sci.* **2021**, *11*, 4875. [CrossRef]
33. Chávez-Negrete, C.; Santana-Quinteros, D.; Domínguez-Mota, F. A Solution of Richards' Equation by Generalized Finite Differences for Stationary Flow in a Dam. *Mathematics* **2021**, *9*, 1604. [CrossRef]
34. McKee, S.; Tomé, M.F.; Ferreira, V.G.; Cuminato, J.A.; Castelo, A.; Sousa, F.; Mangiavacchi, N. The MAC method. *Comput. Fluids* **2008**, *37*, 907–930. [CrossRef]
35. Phan-Thien, N.; Tanner, R. A non-linear network viscoelastic model. *J. Rheol.* **1978**, *22*, 259–283. [CrossRef]
36. Giesekus, H. A simple constitutive equation for polymer fluids based on the concept of defor mation-dependent tensional mobility. *J. Non-Newton. Fluid Mech.* **1982**, *11*, 69–109. [CrossRef]
37. Fattal, R.; Kupferman, R. Constitutive laws for the matrix-logarithm the conformation tensor. *J. Non-Newton. Fluid Mech.* **2004**, *123*, 281–285. [CrossRef]
38. Fattal, R.; Kupferman, R.Time-dependent simulation of viscoelastic flows at high Weissenberg number using the log-conformation representation. *J. Non-Newton. Fluid Mech.* **2005**, *126*, 23–37. [CrossRef]
39. Oliveira, P.J.; Pinho, F.T. Analytical solution for fully developed channel and pipe flow of Phan-Thien–Tanner fluids. *J. Fluid Mech.* **1999**, *387*, 271–280. [CrossRef]
40. Alves, M.A.; Pinho, F.T.; Oliveira, P.J. Study of steady pipe and channel flows of a single-mode Phan-Thien–Tanner fluid. *J. Non-Newton. Fluid Mech.* **2001**, *101*, 55–76. [CrossRef]
41. Cruz, D.; Pinho, F.T.d.; Oliveira, P. Analytical solutions for fully developed laminar flow of some viscoelastic liquids with a Newtonian solvent contribution. *J. Non-Newton. Fluid Mech.* **2005**, *132*, 28–35. [CrossRef]
42. Pan, F.; Acrivos, A. Steady flows in rectangular cavities. *J. Fluid Mech.* **1967**, *28*, 643–655. [CrossRef]
43. Pan, T.W.; Hao, J.; Glowinski, R. On the simulation of a time-dependent cavity flow of an Oldroyd-B fluid. *Int. J. Numer. Methods Fluids* **2009**, *60*, 791–808. [CrossRef]
44. Yapici, K.; Karasozen, B.; Uludag, Y. Finite volume simulation of viscoelastic laminar flow in a lid-driven cavity. *J. Non-Newton. Fluid Mech.* **2009**, *164*, 51–65. [CrossRef]
45. Poole, R.; Afonso, A.; Pinho, F.; Oliveira, P.; Alves, M. Scaling of purely-elastic instabilities in viscoelastic lid-driven cavity flow. In Proceedings of the XVIth International Workshop for Numerical Methods in Non-Newtonian Flows, Northhampton, MA, USA, 13–16 June 2010.
46. Habla, F.; Tan, M.W.; Haßlberger, J.; Hinrichsen, O. Numerical simulation of the viscoelastic flow in a three-dimensional lid-driven cavity using the log-conformation reformulation in OpenFOAM®. *J. Non-Newton. Fluid Mech.* **2014**, *212*, 47–62. [CrossRef]
47. Junior, I.L.P.; Oishi, C.M.; Afonso, A.M.; Alves, M.A.; Pinho, F.T. Numerical study of the square-root conformation tensor formulation for confined and free-surface viscoelastic fluid flows. *Adv. Model. Simul. Eng. Sci.* **2016**, *3*, 2. [CrossRef]

48. Grillet, A.M.; Yang, B.; Khomami, B.; Shaqfeh, E.S. Modeling of viscoelastic lid driven cavity flow using finite element simulations. *J. Non-Newton. Fluid Mech.* **1999**, *88*, 99–131. [CrossRef]
49. Comminal, R.; Spangenberg, J.; Hattel, J.H. Robust simulations of viscoelastic flows at high Weissenberg numbers with the streamfunction/log-conformation formulation. *J. Non-Newton. Fluid Mech.* **2015**, *223*, 37–61. [CrossRef]
50. Martins, F.; Oishi, C.M.; Afonso, A.M.; Alves, M.A. A numerical study of the kernel-conformation transformation for transient viscoelastic fluid flows. *J. Comput. Phys.* **2015**, *302*, 653–673. [CrossRef]
51. Dalal, S.; Tomar, G.; Dutta, P. Numerical study of driven flows of shear thinning viscoelastic fluids in rectangular cavities. *J. Non-Newton. Fluid Mech.* **2016**, *229*, 59–78. [CrossRef]

polymers

MDPI

Article

Structural and Energetic Affinity of Annocatacin B with ND1 Subunit of the Human Mitochondrial Respiratory Complex I as a Potential Inhibitor: An In Silico Comparison Study with the Known Inhibitor Rotenone

Camilo Febres-Molina [1], Jorge A. Aguilar-Pineda [1,2], Pamela L. Gamero-Begazo [1], Haruna L. Barazorda-Ccahuana [1], Diego E. Valencia [1], Karin J. Vera-López [2,3], Gonzalo Davila-Del-Carpio [3,4] and Badhin Gómez [1,3,*]

1 Centro de Investigación en Ingeniería Molecular—CIIM, Universidad Católica de Santa María, Urb. San José s/n—Umacollo, Arequipa 04013, CP, Peru; cfebres@ucsm.edu.pe (C.F.-M.); jaguilar@ucsm.edu.pe (J.A.A.-P.); pgamero@ucsm.edu.pe (P.L.G.-B.); hbarazorda@ucsm.edu.pe (H.L.B.-C.); dvalenciac@ucsm.edu.pe (D.E.V.)
2 Laboratory of Genomics and Neurovascular Diseases, Universidad Católica de Santa María, Arequipa 04013, CP, Peru; kvera@ucsm.edu.pe
3 Facultad de Ciencias Farmacéuticas, Bioquímicas y Biotecnológicas, Universidad Católica de Santa María, Urb. San José s/n—Umacollo, Arequipa 04013, CP, Peru; gdavilad@ucsm.edu.pe
4 Vicerrectorado de Investigación, Universidad Católica de Santa María, Arequipa 04013, CP, Peru
* Correspondence: bgomez@ucsm.edu.pe; Tel.: +51-982895967

Citation: Febres-Molina, C.; Aguilar-Pineda, J.A.; Gamero-Begazo, P.L.; Barazorda-Ccahuana, H.L.; Valencia, D.E.; Vera-Lopez, K.J.; Dávila-Del-Carpio, G.; Gómez, B. Structural and Energetic Affinity of Annocatacin B with ND1 Subunit of the Human Mitochondrial Respiratory Complex I as a Potential Inhibitor: An In Silico Comparison Study with the Known Inhibitor Rotenone. *Polymers* **2021**, *13*, 1840. https://doi.org/10.3390/polym13111840

Academic Editors: Salah Aldin Faroughi, Luís L. Ferrás, Alexandre M. Afonso and Célio Bruno Pinto Fernandes

Received: 29 April 2021
Accepted: 28 May 2021
Published: 2 June 2021

Abstract: ND1 subunit possesses the majority of the inhibitor binding domain of the human mitochondrial respiratory complex I. This is an attractive target for the search for new inhibitors that seek mitochondrial dysfunction. It is known, from in vitro experiments, that some metabolites from *Annona muricata* called acetogenins have important biological activities, such as anticancer, antiparasitic, and insecticide. Previous studies propose an inhibitory activity of bovine mitochondrial respiratory complex I by bis-tetrahydrofurans acetogenins such as annocatacin B, however, there are few studies on its inhibitory effect on human mitochondrial respiratory complex I. In this work, we evaluate the in silico molecular and energetic affinity of the annocatacin B molecule with the human ND1 subunit in order to elucidate its potential capacity to be a good inhibitor of this subunit. For this purpose, quantum mechanical optimizations, molecular dynamics simulations and the molecular mechanics/Poisson–Boltzmann surface area (MM/PBSA) analysis were performed. As a control to compare our outcomes, the molecule rotenone, which is a known mitochondrial respiratory complex I inhibitor, was chosen. Our results show that annocatacin B has a greater affinity for the ND1 structure, its size and folding were probably the main characteristics that contributed to stabilize the molecular complex. Furthermore, the MM/PBSA calculations showed a 35% stronger binding free energy compared to the rotenone complex. Detailed analysis of the binding free energy shows that the aliphatic chains of annocatacin B play a key role in molecular coupling by distributing favorable interactions throughout the major part of the ND1 structure. These results are consistent with experimental studies that mention that acetogenins may be good inhibitors of the mitochondrial respiratory complex I.

Keywords: annocatacin B; ND1 subunit; mitochondrial respiratory complex I; MRC-I; molecular dynamics simulations; MD; Hirshfeld charges; MM/PBSA

1. Introduction

It has been almost 100 years since Warburg presented the first connection between the mitochondria and tumors appearance [1]. The mitochondria fulfill an energetic role in cells, specifically in cancer cells; this role is essential for developing tumors through glycolysis [2,3]. On that basis, several mechanisms associated with tumor generation, such as loss of enzymatic function, mitochondrial genome mutation, reprogramming of mitochondrial metabolism, have been studied [4,5]. Although controversial [6], some hypotheses and studies show that, to a greater or lesser extent, neoplastic cells have many phenotypes related to their energy production, from high aerobic glycolysis, through a partially active oxidative phosphorylation, to a highly productive one [7,8].

For instance, the mitochondrial respiratory complex I (MRC-I) is directly involved in the appearance of colorectal cancer [9], prostate cancer [10], endometrial cancer [11], breast cancer [12], and melanoma [13]. Thus, this complex protein has become a therapeutic target to develop anticancer drugs. Besides, the MRC-I catalyze the formation of reactive oxygen species (ROS).

MRC-I, also named ubiquinone oxidoreductase , has a molecular mass of approximately 1 MDa; its structural conformation is composed of fourteen central subunits. ND1 subunit is one of those and has most of the inhibitor binding domain in the ubiquinone oxidoreductase. To date, one the main known inhibitors of the MRC-I is the rotenone molecule [14]. Rotenone is an isoflavone compound and has been found in many *Fabaceae* plants. Furthermore, it was used as a pesticide and piscicide [15] due to its high toxicity [16,17]. Its effect on cancer cell lines has been evaluated in vitro, showing the inhibition of proliferation and induction of apoptosis [18,19]. Nevertheless, its toxicity in cells complicates its use as an anticancer drug, mainly because it is highly neurotoxic due to its lipophilic nature and the fact that it does not need an extra metabolism to be active or transporter to enter neurons [14,20,21]. Consequently, the challenge is to find new inhibitors that could be less toxic than rotenone.

Murai et al. analyzed rotenone and a synthetic acetogenin as an inhibitor of the bovine heart MRC-I [22]. They revealed that acetogenins are involved in the binding domain of several inhibitors as rotenone does. In fact, acetogenins with two adjacent tetrahydrofurans (THF) rings were reported to show higher antitumor activity and toxicity than those that had only one THF [23], and have been found in the family of *Annonaceae*, i.e., soursop (*Annonamuricata*) [24].

In traditional medicine, soursop also has important uses, including anticonvulsant, antiarthritic, antiparasitic, hepatoprotective, etc. Many of these beneficial attributes have been ascribed to acetogenins [24,25]. One of the most studied properties in soursop is its potential anticarcinogenic effect due, in a way, to its powerful cytotoxic features [25,26]. It has been possible to isolate more than 100 acetogenins from different parts of the *Annonaceae* plants [24,25,27]. The effect of acetogenins as inhibitors of the MRC-I has been suggested and demonstrated for more than 20 years [28].

Acetogenins have showed important behaviors when evaluating their potential cytotoxic activity against cancer cells; some of these molecules already have proven anti-cancer properties, such as bullatacin, motrilin, assimin, trilobacin, annonacin, gianttronenin, and squamocin. However, we still do not have enough information about most of the acetogenins [29]. The main characteristic of acetogenins' molecular structure is their linear 32 to 34 carbon chains containing oxygen-containing functional groups. Annocatacin B is an acetogenin with two adjacent THF rings and has been identified in the leaves of soursop; it has also been reported that it possesses toxicity against human hepatoma cells [24,30]. Currently, there is not much information about annocatacin B; so, it has a great potential for new research. In that sense, the objective of this work was to determine the plausible inhibitory role of annocatacin B with the ND1 subunit compared with rotenone as a control, considering all this as a challenge in the search for new inhibitors of MRC-I. To accomplish this, we applied computational techniques as quantum mechanical (QM) calculations, molecular dynamics (MD) simulations, and molecular mechanics/Poisson–Boltzmann surface area (MM/PBSA) calculations.

2. Computational Details

2.1. Structural Preparation

We analyzed two molecules as ligands to the ND1 complex, rotenone (PubChem ID 6758) and annocatacin B (PubChem ID 10483312) (Figure 1a). The structures of both molecules were built using the GaussView v.6 software package [31], and optimized by DFT calculations using Gaussian 16 software package [32] (Figure 1b). The optimization process were performed using the CAM-B3LYP exchange-correlation functional [33], and the TZVP basis set [34]. The vibrational frequencies were calculated to ensure that the geometries were those of the minimum energy. In order to investigate the electrostatic effect of the ligands on the ND1 complex, atomic charges were calculated using the Hirshfeld population analysis [35–37] with implicit solvent effect (*SCRF = (SMD, Solvent = Water)*), and molecular electrostatic potential (ESP) surfaces were used to visualize the polar and non-polar regions of these ligands. To obtain the MD parameters and topologies of the ligands, we used the TPPMKOP server [38], which uses the parameters of the OPLS-AA force field to generates them [39,40]. These topologies were reparametrized using the optimized structures and atomic charges obtained in previous quantum calculations.

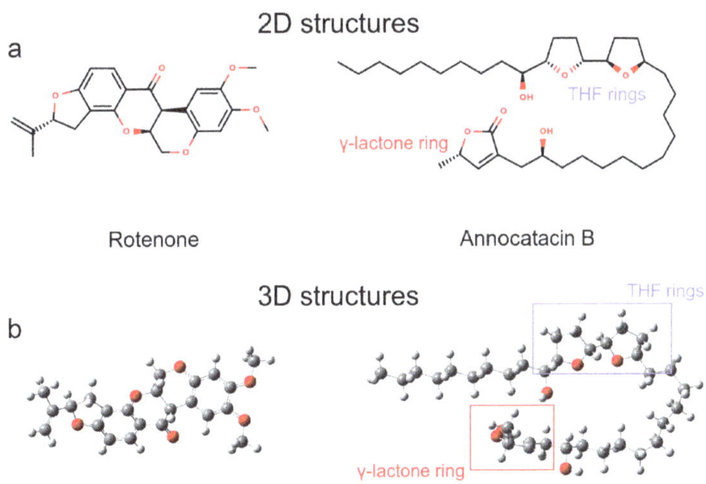

Figure 1. Ligand molecules used in this work. (**a**) 2D representations. (**b**) 3D representations obtained after QM optimization.

On the other hand, the phospholipid bilayer membrane was built with 512 dipalmitoyl-phosphatidylcholine (DPPC) molecules. A 128-DPPC bilayer with 64 lipid molecules in each layer was replicated four times (twice in both the x and y directions), to obtain the membrane model. The InflateGRO methodology was used for the embedding of ND1 protein in the lipid membrane [41].

The three-dimensional crystallographic structure of the human MRC-I was considered for this study and obtained from the Protein Data Bank (PDB) by the PDB ID: 5XTD [42]. Crystallographic water molecules were removed in Chimera UCSF 1.11.2 [43]. From this MRC-I, the structure of the ND1 subunit was extracted, since it largely possesses the quinone-binding domain between residues Y127 and K262 (according to the ND1 subunit nomenclature).

2.2. MD Simulations

Molecular dynamics (MD) simulations were carried out in Gromacs 2019 [44] with the OPLS-AA force field. Firstly, we performed an energy minimization of the whole protein in the vacuum with the steepest descent algorithm with a maximum of 50,000 steps.

Then, the DPPC parameters for the lipid bilayer were obtained from the work of Peter Tieleman et al. [45]. The new system (protein + DPPC membrane) was located in the center of a cubic box with a 1.0 nm distance between the system-surface and the box edge on z axis. SPC water-model molecules and ions were added to neutralize the systems. Next, we proceeded with another energy minimization with a maximum of 50,000 steps. The equilibrium MD simulation was realized with position restraint in two ensembles. The first was the canonical ensemble (NVT) at 323.15 K with a trajectory of 50 ps using a V-rescale thermostat. The second was the isobaric-isothermal ensemble (NPT) at 309.65 K, with semi-isotropic pressure coupling, the compressibility of 4.5×10^{-5}, and a reference pressure of 1.0 bar for along the 50 ps of the trajectory using the Nosé–Hoover thermostat and the Parrinello–Rahman barostat. The production of MD without position restrain was calculated in the isobaric-isothermal ensemble at 309.65 K and semi-isotropic pressure coupling (same equilibrium condition of NPT ensemble) for 500 ns of trajectory. Periodic boundary conditions (PBC) in all directions, particle mesh Ewald (PME) algorithm for long-range electrostatics with cubic interpolation with a cut-off of 0.9 nm, and linear constraint solver (LINCS) with all bonds constrained were applied for all MD simulations.

2.3. Molecular Docking Calculations

First, the coupling was made between the ND1 subunit and rotenone, and then, between ND1 and annocatacin B. To accomplish this, we used PATCHDOCK server [46,47], a molecular docking algorithm based on shape complementarity principles, and we selected the top score solution for each of the two systems, because these top score structures were in agreement with the experimental data [48]. 4.0Å clustering RMSD and default mode parameters were used. Later, we took these top score solution complexes and introduced them into the lipid bilayer/water systems. Subsequently, we carried out the MD simulations of the systems: ND1—rotenone and ND1—annocatacin B, following the aforementioned steps.

2.4. MM/PBSA Calculations

To evaluate the binding affinities of ND1-ligand interactions, we performed the molecular mechanics Poisson–Boltzmann surface area (MM/PBSA) calculations [49]. This was made using the g_mmpbsa program [50], which calculates components of binding energy using the MM/PBSA method except the entropic term using a energy decomposition scheme. Despite g_mmpbsa not including the calculation of entropic terms and therefore not being able to calculate the absolute binding free energies (BFE), as Kumari et al. stated [50], it does calculate the relative BFE. So, we used this tool to compare different ligands that bind to the same receptor protein. Calculations of free energies and energy contributions by residue were carried out in order to localize the main residue interactions and to assess the effect of each residue on the ND1—ligand complexes. The last 200 ns of the MD trajectories were analyzed at a 1 ns time interval to estimate the binding free energy (ΔG_{bind}), which was calculated using the following equation:

$$\Delta G_{bind} = G_{complex} - (G_{ND1} + G_{lig}) = \Delta E_{MM} + \Delta G_{sol} - T\Delta S \quad (1)$$

where $G_{complex}$ is the total free energy of the ND1-ligand complexes; G_{ND1} and G_{lig}, are the free energies of isolated ND1 structure and rotenone or annocatacin B in solvent. ΔE_{MM}, represents the molecular mechanics energy contributions; ΔG_{sol} is the free energy solvation required to transfer a solute from vacuum into the solvent. The TΔS term refers to the entropic contribution and was not included in this calculation due to the computational costs [50–52]. Therefore, individual E_{MM}, and G_{sol} terms were calculated as follows:

$$E_{MM} = E_{bonded} + E_{vdW} + E_{elec} \quad (2)$$

$$G_{sol} = G_p + G_{np} = G_p + \gamma A \quad (3)$$

In Equation (2), the bonded interactions are represented by the E_{bonded} term, and in the single-trajectory approach, ΔE_{bonded} is taken as zero [49]. The non-bonded interactions are represented by the E_{vdW} and E_{elec} terms. In Equation (3), the solvation free energy of (G_{sol}), is the sum of the polar (G_p) and non polar (G_{np}) contributions. The G_p term is calculated by solving the Poisson–Boltzmann equation, while for the G_{np} term, we used the SASA nonpolar model, where γ (0.0226778 kJ/mol A2) is a coefficient related to the surface tension of the solvent, and A is SASA value. In order to ensure the convergence of our MM/PBSA results, we have considered only the last stable 200 ns (20 frames) of the MD trajectories and were assessed using the FEL analyses from each complex. The frames were selected at a regular interval of 1 ns for better structure–function correlation. In addition, we used the bootstrap analysis to calculate the average binding energy included in the g_mmpbsa tools. All calculations were obtained at 309.65 K, and default parameters were used to calculate molecular mechanics potential energy and solvation free energy [50]. Finally, the binding free energy by residue was obtained using:

$$\Delta G_{bind}^{res} = \Delta E_{MM}^{res} + G_p^{res} + G_{np}^{res} \tag{4}$$

2.5. Structure and Data Analysis

Statistical results, root mean squared deviation (RMSD), root mean squared fluctuation (RMSF), radii of gyration (RG), solvent accessible surface area (SASA), hydrogen bonds (HB), binding free energies (BFE), matches, structures, trajectories, B-factor maps, were obtained using Gromacs modules. An analysis of structure properties was performed using the MD trajectories of the last 200 ns of each simulations, then visualized using Visual Molecular Dynamics (VMD) software [53] and UCSF Chimera v.1.14 [43]. The graphs were plotted using XMGrace software [54]. Moreover, 2D representations of electrostatic and hydrophobic interactios were built using LigPlot program [55]. The ESP surfaces within the molecular mechanics framework were calculated in APBS (Adaptive Poisson Boltzmann Surface) software v.1.4.1, [56] and the pqr entry was created in the PDB2PQR server [57]. Free Energy Landscape (FEL) maps were used to visualize the energy associated with the protein conformation of the different models during the MD simulations. These maps are usually represented by two variables related to atomic position and one energetic variable, typically Gibbs free energy. In this work, we considered two substructures of ND1 protein for the FEL map analysis, Site A (Y127 to F198) and Site B (D199 to K262). These two regions were adopted from the work of Kakutani et al. [48]. The FEL maps were plotted using *gmx sham* module, while the RMSD and RG were considered as the atomic position variables with respect to its average structure and figures were constructed using Wolfram Mathematica 12.1 [58].

3. Results and Discussion

The human MRC-I belongs to a highly organized supercomplex, named respirasome. The complexes I, III, and IV arise in a more stable fashion at that supercomplex and have the special task of channeling electrons effectively through the electron transport chain [59]. Nevertheless, Guo et al. proposed an even larger system called megacomplex that includes complex II at the previous respirasome [42]. They suggested that a quinone/quinol (oxidized/reduced forms of the same molecule) pool maximizes the oxide-reduction reactions. Recent studies suggest that there are around 100 Å between complex I and complex III when actively translocating electrons, proposing with this that there is no need for a mediating protein to help the electron channeling through these complexes [42,60].

MRC-I is the first in the mega-complex that encounters the quinone site to start the oxide-reduction process. This complex is composed of several subunits, and mainly the ND1 subunit is the one that possesses the majority of the quinone binding domain and, to a lesser extent, the ND3, PSST, and 49 kDa subunits. Fiedorczuk et al. studied the open and close positions of the above-mentioned complex I to be active and inactive, respectively [61]. The ND1 subunit has a predominantly structural conformation of alpha-helices that provides the hydrophobic environment expected of a membrane protein and owns the quinone binding domain which is in its core (Table 1).

Table 1. Secondary structure summary.

System	Strand	Alpha Helix	3–10 Helix	Other	Total Res.
ND1	0 (0.00%)	154 (48.40%)	9 (2.80%)	155 (48.70%)	318
ND1—Annocatacin B	0 (0.00%)	129 (39%)	28 (8.80%)	166 (52.20%)	318
ND1—Rotenone	0 (0.00%)	148 (46.50%)	11 (3.50%)	159 (50.00%)	318

3.1. Structural Analysis

3.1.1. Rotenone and Annocatacin B

Before performing the MD simulations, we carried out QM calculations to obtain the optimized structures and analyze the electrostatic properties of the ligand molecules. Figure 1a shows the 2D representation of the ligands, where we can visualize that annocatacin B is larger than rotenone. The optimized structure of annocatacin B shows a closed isoform between the THF rings and the γ-lactone ring (Figure 1b). This result is in agreement with that observed by Nakanishi et al., who reported that the hydrophobic alkyl tail of the acetogenins, in general, looks to serve as a spacer to accommodate the polar hydroxylated bis-THF motif to the polar-membrane part, and its apolar counterpart, the γ-lactone ring, into the core of the lipid bilayer [62].

Both ligand molecules have an electrophilic character and one of the major goals of this study aimed to assess the electrostatic effect of the ligands on the ND1 structure. Figure 2 shows the quantum and classical ESP surfaces of annocatacin B and rotenone molecules obtained from Hirshfeld population analysis. We can observe that the annocatacin B structure has a high electron density region over the γ-lactone ring and it decreases at the THF rings (Figure 2a). On the other hand, as can be seen from the ESP surface of the rotenone molecule, the high electron density sites are close to the carbonyl group, and the oxygen atoms, as expected (Figure 2b). With these charges, and using the OPLS/AA parameters, we built the annocatacin B and rotenone force fields for the MD simulations. Hirshfeld's atomic charges calculation and their use in molecular mechanics (MM) force fields has been employed in many liquid solvents studies [63–67]. The main advantages of these atomic charges are not to overestimate the electrostatic properties and accelerate the MD calculations.

Additionally, the drug-like properties of annocatacin B and rotenone have the following values: six hydrogen bond acceptors in both of them; hydrogen bond donors of 2 and 0; molecular weight of 578.875 g/mol and 394.423 g/mol; the number of rotational bonds of 23 and 3; partition coefficient LogP of 8.1069 and 3.7033, and a surface area of 250.531 $Å^2$ and 168.525 $Å^2$, respectively. These results confirmed that both molecules are very hydrophobic, annocatacin B being more lipophilic than rotenone, due mostly to its alkyl chain.

The pharmacokinetic properties of absorption, distribution, metabolism, excretion, and toxicity (ADMET) are in Table 2. The absorption is similar in both compounds; however, rotenone is not a P-glycoprotein substrate giving a slim advantage to the other molecule. The distribution property is slightly higher for rotenone, which implies that its distribution in the human body (tissues) is a bit greater than annocatacin B. Regarding metabolism, both could be substrates of the CYP3A4 protein, but only rotenone could act as an inhibitor. The excretion and toxicity of these molecules are similar in both cases. In general terms, this description shows that both rotenone and annocatacin B have very similar properties. The pharmacokinetic and toxicological properties of these compounds were analyzed through the pkCSM server [68].

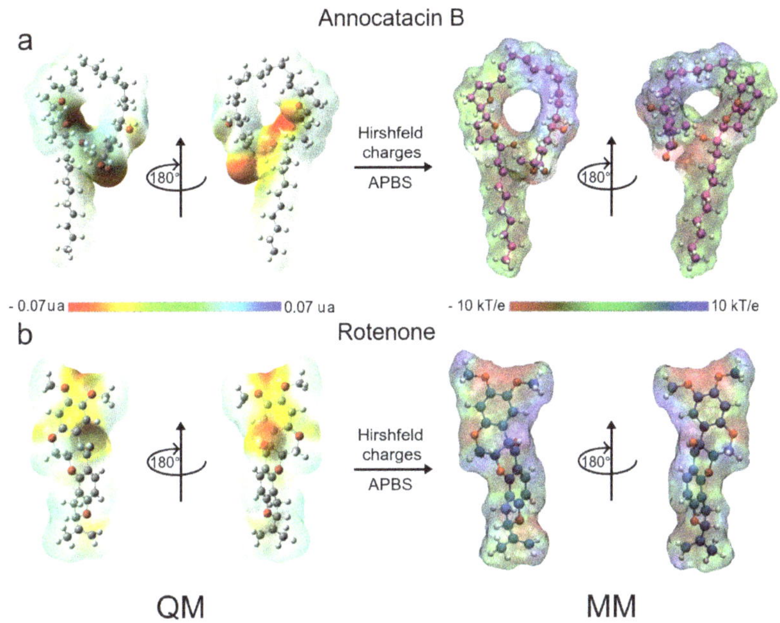

Figure 2. Chemical structure of Annocatacin B and Rotenone molecules. Calculated molecular ESP surfaces of (**a**) annocatacin B, and (**b**) rotenone. In the left panel, ESP surfaces obtained at the DFT level using the CAM-B3LYP/TZVP method. In the right panel, ESP surfaces are obtained with APBS methodology and the Hirshfeld's atomic charges. On all surfaces, the different colors indicate their molecular electrostatic properties; red for the most nucleophilic zones; dark blue for the most electrophilic zones, and green for the neutral zones.

Table 2. ADMET prediction of annocatacin B and rotenone by pkCSM server.

ADMET			
Property	**Model Name**	**Predicted Value**	
		Annocatacin B	**Rotenone**
Absorption	Water solubility [a]	−5.85	−5.05
Absorption	Caco2 permeability [b]	0.40	1.31
Absorption	Intestinal absorption [c]	86.98	99.63
Absorption	Skin Permeability [d]	−2.70	−2.75
Absorption	P-glycoprotein substrate	Yes	No
Absorption	P-glycoprotein I inhibitor	Yes	Yes
Absorption	P-glycoprotein II inhibitor	Yes	Yes
Distribution	VDss (human) [e]	−0.29	−0.04
Distribution	Fraction unbound (human) [f]	0.05	0
Distribution	BBB permeability [g]	−0.95	−0.87
Distribution	CNS permeability [h]	−2.90	−2.82
Metabolism	CYP2D6 substrate	No	No
Metabolism	CYP3A4 substrate	Yes	Yes
Metabolism	CYP1A2 inhibitior	No	Yes
Metabolism	CYP2C19 inhibitior	No	Yes
Metabolism	CYP2C9 inhibitior	No	Yes
Metabolism	CYP2D6 inhibitior	No	No

Table 2. *Cont.*

ADMET			
Property	**Model Name**	**Predicted Value**	
		Annocatacin B	**Rotenone**
Metabolism	CYP3A4 inhibitior	No	Yes
Excretion	Total Clearance [i]	1.601	0.195
Excretion	Renal OCT2 substrate	No	No
Toxicity	AMES toxicity	No	No
Toxicity	Max. tolerated dose (human) [j]	−0.64	0.16
Toxicity	hERG I inhibitor	No	No
Toxicity	hERG II inhibitor	No	No
Toxicity	Oral Rat Acute Toxicity (LD50) [k]	3.03	2.87
Toxicity	Oral Rat Chronic Toxicity (LOAEL) [l]	0.79	1.43
Toxicity	Hepatotoxicity	No	No
Toxicity	Skin Sensitisation	No	No
Toxicity	T.Pyriformis toxicity [m]	0.31	0.35
Toxicity	Minnow toxicity [n]	−1.89	−0.33

[a] In log mol/L; [b] In log Papp in 10^{-6} cm/s; [c] In % Absorbed; [d] In log Kp; [e] In log L/kg; [f] In Fu; [g] In log BB; [h] In log PS; [i] In log mL/min/kg; [j] In log mg/kg/day; [k] In mol/kg; [l] In log mg/kg_bw/day; [m] In log ug/L; [n] In log mM.

3.1.2. ND1—Ligand Complexes and Stability Descriptors

As we said earlier, ND1 subunit is located in the transmembrane region of human MRC-I [42]. In order to understand the ligand effect on its structure, we carried out MD simulations of a full-length ND1 subunit and its ND1-ligand complexes. To obtain the molecular systems, we isolated the ND1 protein of the MRC-I and this was embedded inside a phospholipid bilayer (Figure 3a). As said before, according to Kakutani et al., the ND1 subunit has two regions in the active site, Y127 to F198 (site A) and D199 to K262 (site B), which are involved in the quinone binding domain (Figure 3b) [48]. The authors suggest that natural acetogenins prefer to accommodate more likely in site A and synthetic molecules in site B.

Before studying the structural and energy changes of ND1 protein, it was necessary to assess the stability of the molecular complexes during MD simulations. For this purpose, we calculated and plotted the root mean square deviation (RMSD, for additional information, see figure S1) of the ND1 subunit for all complexes, with respect to its equilibrated structure. With the best molecular docking results (for additional information, see Figure S2 and Table S1), we carried out 500 ns of MD simulations, and we observed that due to the movement restrictions of the lipid bilayer on the ND1 atoms, there is no significant difference between the protein containing the ligands and the one that does not have them. Specifically, the average RMSD of the last 300 ns of the ND1 without ligands was 0.40 ± 0.04 nm, and the average RMSD of the last 200 ns of ND1 with ligands was 0.48 ± 0.02 nm and 0.44 ± 0.02 nm for the systems ND1-rotenone and ND1-annocatacin B, respectively. At a glance, we can notice that the last 200 ns in the three systems is specially stabilized, that is, within the range of the 0.2 nm (2 Å) of deviation permitted. However, when we analyzed the final MD structures, we observed a structural impact of the ligands in the active site (Figure 4a). In both ND1-ligand complexes, the ND1 subunit shows an open conformation to allow ligand stability (Figure 4b,c). In the case of the rotenone complex, the addition of this ligand caused a structural instability observed in the RMSD of the active site, 0.41 ± 0.11 nm against 0.30 ± 0.06 nm of the annocatacin B complex (Table 3 and Figure 5a).

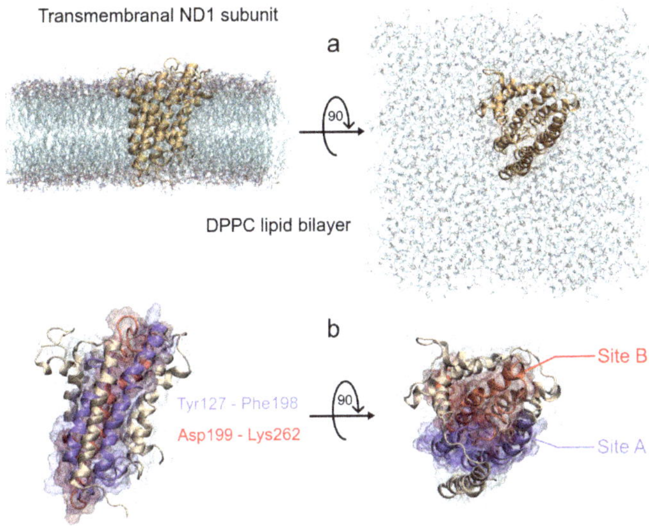

Figure 3. Schematic representation of the ND1—DPPC membrane complex, (**a**) initial distribution of the simulated model used in the MD simulation, the lengths of the simulation box are lx = 11.66 nm, ly = 11.71 nm and lz = 13.02 nm. (**b**) Front and top views of the active site of ND1 protein, in purple color, the active site A, and in red color, the active site B.

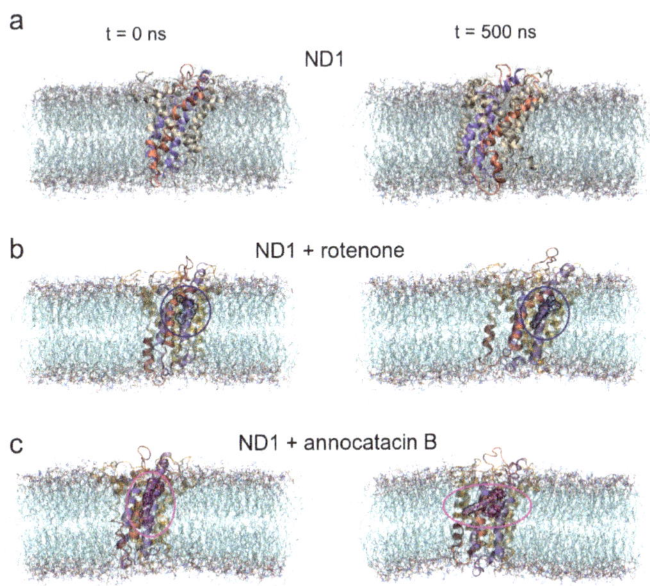

Figure 4. Overall structural organization of ND1 protein and its ligands in a membrane-embedded condition. The left panel shows the molecular complexes at initial conditions (0 ns). The right panel shows the complexes at 500 ns. (**a**) ND1 + DPPC membrane, (**b**) ND1 + DPCC + rotenone, and (**c**) ND1 + DPPC + annocatacin B.

Table 3. Stability Descriptors of the ND1 complexes.

System	Region	RMSD [a]	RMSF [a]	RG [a]	HB		
					Intra	Inter/Solv	Inter/Mem
ND1	whole prot	0.40 ± 0.02	0.19 ± 0.09	2.12 ± 0.01	209 ± 9 (200)	338 ± 15 (336)	33 ± 5 (39)
	Active site	0.30 ± 0.05	0.13 ± 0.04	1.91 ± 0.01	89 ± 6 (77)	166 ± 10 (165)	6 ± 2 (6)
	y-axis			1.74 ± 0.03			
ND1 + rotenone	whole prot	0.48 ± 0.02	0.20 ± 0.10	2.13 ± 0.01	203 ± 8 (202)	356 ± 13 (356)	31 ± 5 (38)
	Active site	0.41 ± 0.11	0.14 ± 0.07	1.90 ± 0.02	84 ± 6 (88)	176 ± 10 (174)	8 ± 3 (10)
	y-axis			1.70 ± 0.04			
ND1 + annocatacin B	whole prot	0.44 ± 0.01	0.20 ± 0.09	2.13 ± 0.01	213 ± 11 (207)	341 ± 14 (330)	26 ± 5 (23)
	Active site	0.30 ± 0.03	0.13 ± 0.05	1.90 ± 0.01	89 ± 6 (81)	167 ± 10 (161)	5 ± 3 (2)
	y-axis			1.77 ± 0.02			

[a] In nanometers. For HB calculations, we considered determining those formed between ND1 subunit itself (intra); the ND1 subunit and solvent molecules (inter/solv); and the ND1 subunit and lipid bilayer membrane (inter/mem). Values between parenthesis were calculated on global minimum energy structures obtained in the FEL analysis. All values were obtained from the last 300 ns of the MD simulations.

In the case of the radii of gyration (RG), close values were obtained for the ND1-ligand complexes (~2.12 nm), due to the stability provided by the lipid membrane (Table 3). Similarly, calculations performed in the active site showed few variations among these zones in the three structures (~1.90 nm, Figure 5b). However, calculations around the *y*-axis, showed that the most opened structure was the ND1—annocatacin B complex (1.77 ± 0.02 nm), being the ND1—rotenone complex the most compacted structure (1.70 ± 0.04 nm, Figure 5c).

Using both results, the active site RMSDs and *y*-axis RG, we performed a free energy landscape (FEL) analysis to obtain the minimal global energy conformations of the ND1 complexes. The FEL maps showed the impact of ligands on the ND1 structure stabilization. In Figure 5d, we can observe that there is only a single conformation cluster (dashed circles in the 2D maps), which indicates the great stability of the ND1 subunit in the lipid membrane. In the case of the ligand complexes, there are four conformation clusters that indicate the destabilization caused by the ligand molecules. However, the location of these clusters was more close in the ND1—annocatacin B complex (Figure 5e,f). According to the results above, the 3D maps showed a large top area in the ND1—rotenone complex and a less top area in the annocatacin B complex, which suggests a more profound stabilization effect by annocatacin B on the ND1 subunit.

3.1.3. Hydrogen Bond Analysis

To elucidate this apparent contradiction between the results obtained from the RMSD and RG analyses, we performed a hydrogen bonds (HB) analysis. Using the *hbond* tool of Gromacs for the MD simulations, and the Hydrogen bonds plugin of VMD for the global minimum energy structures, we obtained the HB interactions based on a cutoff distance of 0.35 nm and a cutoff angle of 30°. Initially, we determined the HB formation of the ND1 subunit, both intra and intermolecular (Figure 6a, Table 3). The results showed a greater intramolecular HB formation in the ND1—annocatacin B complex (~216) and its active site (~89), but a decrease in the intermolecular interactions, mainly with the lipid bilayer (~26 and ~5 for the active site). On the other hand, the ND1-rotenone complex shows a maximum number of interactions with the solvent molecules (~356) and the most formation of HB with the DPPC molecules (~31 and 8 to active site). The same trend was presented in the case of the minimal energy structures (parenthesis results). These results suggest that the annocatacin B increases the intramolecular stability of the ND1 subunit contrary to the rotenone molecule, which increases the intermolecular interactions mainly with the solvent molecules that are the main cause of protein instability.

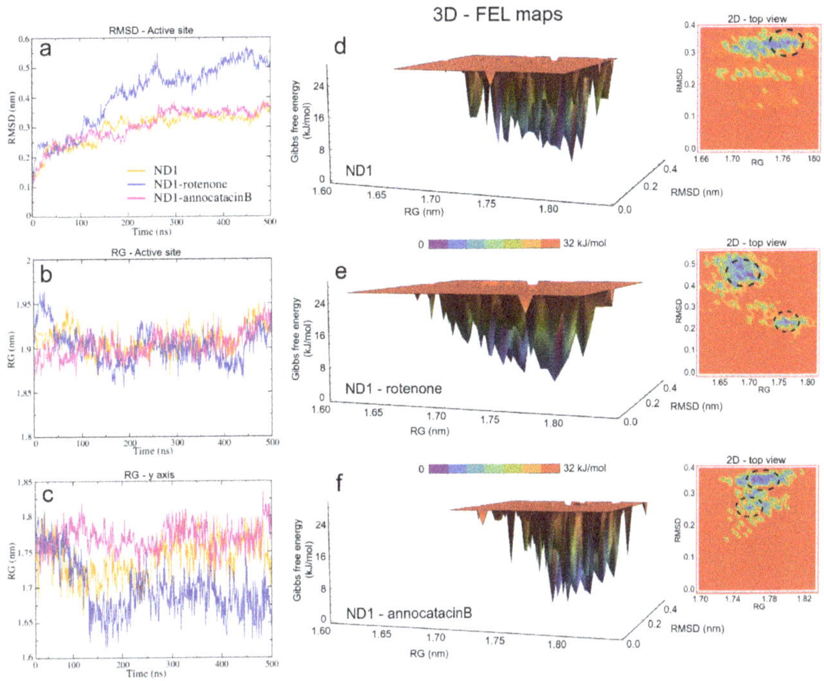

Figure 5. Structural analysis of the ND1 subunit active site. (**a**) RMSD; (**b**) RG; (**c**) RG on of *y*-axis of the ND1 protein. The FEL maps (**d**–**f**) were built using structural coordinates from RMSD results and RG on the *y*-axis. The dash circles in the 2D plots indicate the global minimum energy structures showed in purple color in the 3D plots.

To clarify these stability behaviors, we carried out HB calculations between the ligand molecules and the system components. In Figure 6b, we can see the HB formations of these ligands and all atoms in the molecular complexes. From a statistical perspective at the last 200 ns of MD simulations, there are more HB formations in the annocatacin B complex (~0.80) than the rotenone complex (~0.42). Furthermore, the analysis of the ND1—ligand interactions (Figure 6c), showed almost exclusively ligand interactions by annocatacin B on the ND1 subunit (~0.15) versus rotenone interactions (~0.01). The results confirm that the annocatacin B stabilizes, in part, the ND1 structure by polar interactions with its nearby residues. In order to identify the active site residues involved in the stabilization interactions, we calculated the HB occupancies in the MD simulations, and the Figure 6d shows the results obtained. In the case of the ND1—rotenone complex, we can observe the greater occupancy value (9.95%) due to the F223. However, only four residues were involved in the polar interactions (L222, F223, A226, and T229). On the other hand, the ND1—annocatacin B complex showed a major number of polar interactions (14), being W185, F223, M233, and L237 residues that had the highest number of occupancy values. Despite the hydrophobic character of the ligand molecules, our hydrogen bonds analysis showed the importance of polar interactions in the ND1 stabilization.

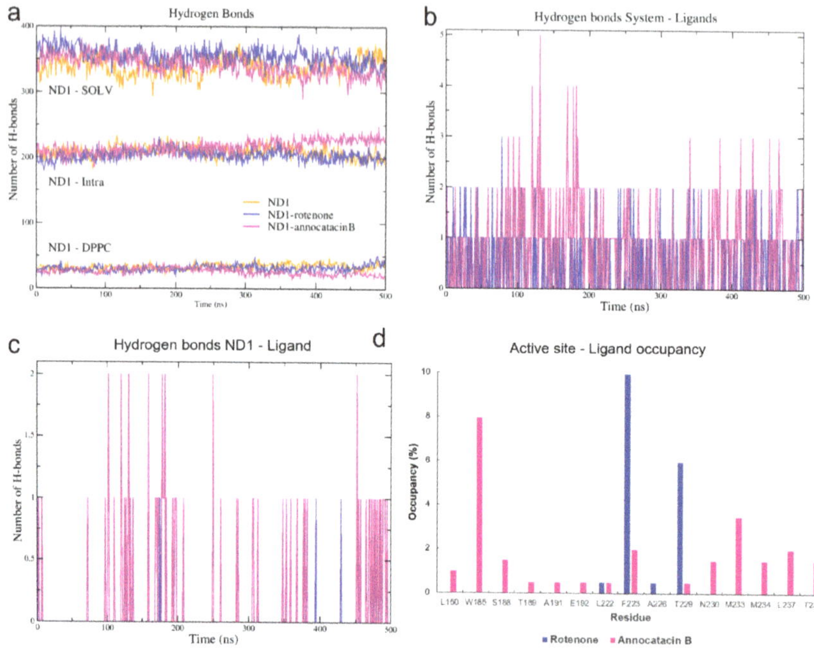

Figure 6. H-bond analysis of the ND1 complexes during 500 ns MD simulations. (**a**) ND1-intramolecular and ND1-intermolecular interactions (solvent and lipid-bilayer). (**b**) H-bond formation between the molecular systems and the ligand molecules (**c**) H-bond formation taking into account just the ND1 subunit and the ligand molecules. (**d**) H-bond occupancy of the active site residues interacting with the ligand molecules.

3.1.4. RMSF and B-Factor Analysis

To evaluate local flexibilities of the ND1 subunit and describe the deviations of residues from the average position due to the ligand effects, we performed the root mean square fluctuation (RMSF) analysis. The main fluctuations of the ND1 protein were observed at the unembedded-loop regions, as expected (Table 3). In particular, high RMSF values were located between L33-G36 residues (ND1—annocatacin B complex, ~0.84 nm), and A249-S251 residues (ND1—rotenone complex, ~0.86 nm).

Despite the high stability of the active site, a fluctuation analysis was performed to understand the ligand effect in these region. For this purpose, in addition to RMSF calculations, we analyzed the B-factor, also called thermal factor or Debye–Walle factor [69] and we mapped the values on the active site surfaces.

The stable regions in the MD trajectories were used and the nearest neighbor residues B-factor values are shown in Table 4. The RMSF values of the three systems exhibit similar fluctuation values (see Table 3), showing a high stable behavior. However, when we analyze the fluctuation in site B of the active site, we observed more instability in the rotenone complex, which is reflected in its dispersion value (0.17 ± 0.08 nm) as compared with the ND1 and ND1—annocatacin B values (0.14 ± 0.04 nm and 0.13 ± 0.04 nm, respectively). The highest fluctuations were located between A201-F211, and D248-E253 residues (Figure 7a). In embedded active site regions, the ND1 subunit presents high stability, denoted by the green color of the B-factor surface (Figure 7b). However, this stability is altered by the presence of the ligand molecules, making these regions more flexible. Figure 7c shows the rotenone effects on the neighbor residue fluctuations. The presence of white and red zones on the B-factor surface denotes a flexibility increase, especially, the F223 residue shows a high fluctuation value (144.4 Å). Figure 7d shows a zoom of rotenone and its influence zone on the ND1 protein, calculated at a minor distance of 0.5 nm. The interactions with 23 residues are observed in Table 4. On the other hand, the annocatacin B effects on the B-factor surface shows an increase in the number of residues with high fluctuation, being the L79 (179.7 Å) and M225 (231.7 Å) residues that had the highest fluctuation values (Figure 7e). Hence, the total number of residues interacting with the annocatacin B ligand were 37 (Figure 7f). The results suggest that the annocatacin B stabilizes the ND1 structure by size effect and by interaction with different domains out of the active site.

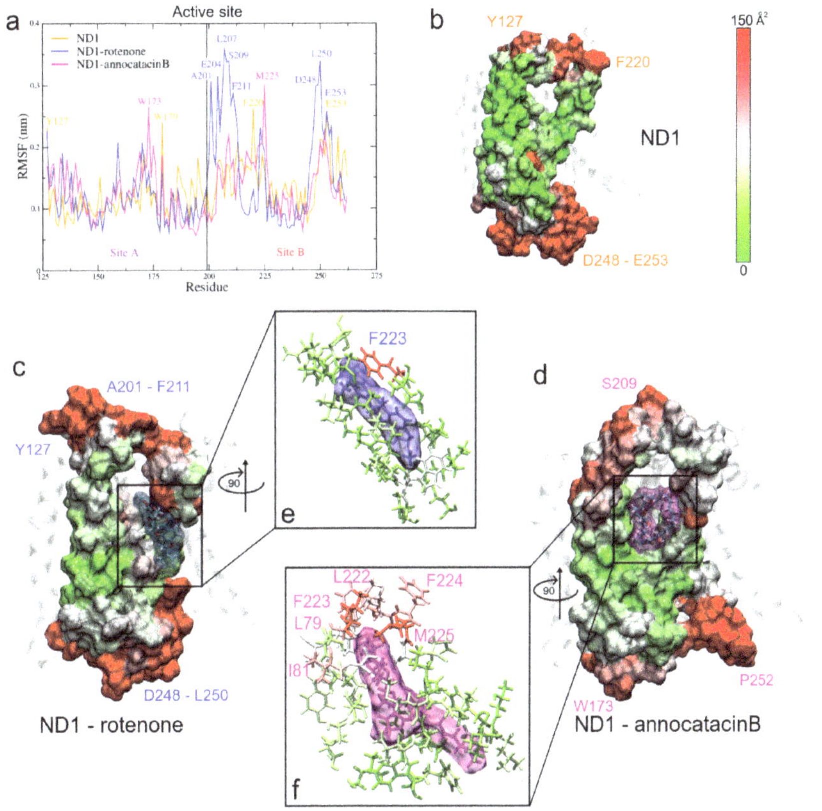

Figure 7. Fluctuation analysis of the ND1 residues. (**a**) RMSF plot of the active site residues obtained during the last 200 ns of the MD trajectories. (**b–d**) B-factor plotted on the molecular surface of the active site. The red color indicates high B factor values, whereas the green, low values. Ligands are shown in translucid surface, rotenone in blue color, and annocatacin B in magenta color. (**e**,**f**) Zoom of the ligand interactions.

Finally, we included the solvent accessible surface area (SASA) value, that is an important descriptor of the ligand effects over the structure, in which ND1—rotenone has a higher value (179.29 ± 1.47 nm^2) than ND1—annocatacin B (177.67 ± 1.53 nm^2) and ND1 without ligand (174.84 ± 1.37 nm^2).

Table 4. Contact analysis.

System	Active Site				Other Sites		
	Site A		Site B				
ND1 + rot + mem					T73(21)	A74(19)	L77(37)
			L22(23)	F223(144)	A78(15)	L79(45)	I81(20)
			A226(14)	T229(17)	A82(18)	L83(37)	L85(19)
			N230(36)	I232(10)	W86(68)	L89(30)	M91(23)
			M233(20)		S115(23)	I116(38)	W118(41)
					S119(37)		
ND1 + ann + mem			L222(94)	F223(103)	A78(102)	L79(179)	I81(101)
	E143(39)	L146(27)	F224(106)	M225(232)	A82(76)	L85(43)	S109(48)
	W185(14)	F186(41)	A226(70)	E227(48)	A112(40)	V113(53)	Y114(52)
	S188(13)	T189(14)	T229(27)	N230(25)	S115(39)	I116(70)	L117(74)
	A191(14)	E192(15)	I231(37)	M233(49)	L266(28)	T267(19)	L269(46)
			M234(36)	L237(50)	F270(26)	I273(49)	

Residues close to ligand at distance minor to 0.5 nm obtained at the last 200 ns of the MD trajectories. In parenthesis, B-factor values.

3.1.5. MM Electrostatic Potential Surfaces

As mentioned above, ND1—ligand interactions mainly have a hydrophobic character and that is demonstrated by their drug-like properties. However, our results show an electrostatic contribution to structure stabilization. Thus, using the molecular mechanics adaptive Poisson–Boltzmann solver (APBS) approximation [56], we calculated the ESP surfaces of ND1 subunit and their ligand complexes. For this purpose, we used the minimum energy structures and the Hirshfeld's atomic charges of the ligand molecules obtained in FEL analysis and QM calculations, respectively.

The electrostatic map of the ND1 structure shows a well-defined charged core surrounded by hydrophobic alpha-helices substructures. The core is formed mainly by the active site residues that confer a high electrophilic character to this region (T73-L117 and L266-I273, Figure 8a). We have observed that the binding domain comprises residues out of the active site and the electrostatic properties of these residues are affected by the ligand interactions. Figure 8b shows the drastic variations in the polar properties of the binding domain due to rotenone, increasing the positively charged regions. In addition, the binding domain seems to close, which would explain the more compactness observed in the radii of gyration analysis in this complex.

On the other hand, the electrostatic changes observed by the annocatacin B presence in the binding domain were less dramatic, yet, conserving the electrophilic character in most of its structure (Figure 8c). The main polar variations were located on the A78, S115, I116, L222, F223, N230, M233, and M234 residues, which increased their nucleophilic character. These electrostatic variations suggest that the structural instability observed in the ND1—rotenone complex can be due to structural changes in the active site.

3.2. *Binding Free Energy*

To analyze the energy properties of rotenone and annocatacin B when forming the ND1—ligand complexes, we carried out MM/PBSA calculations based on the last 200 ns of the MD trajectories. In addition, an energy decomposition analysis per residue was performed to highlight the main residues that contribute to the stability of the complexes. As shown in Table 5, the binding free energy (BFE) of the two complexes was energetically favorable, however,the interaction energy of the ND1—annocatacin B complex (-333.18 ± 2.14 kJ/mol) was lower than that of the ND1—rotenone complex (-218.15 ± 1.78 kJ/mol), indicating that the complexation reaction is more spontaneous, which is according to that reported by Murai et al., where they say that the inhibition potency of natural acetogenins is stronger than that of common synthetic inhibitors [22]. Due to the hydrophobic character of the ligand interactions, the main contributions to $\Delta G_{binding}$ energy were the van der Waals (vdW) and nonpolar solvation terms. In both of them, the binding energy was more favorable to the annocatacin B interactions with the ND1 subunit (~39%). Furthermore, the electrostatic energy term confirms the polar contribution to the stability of the ND1—ligand complexes as seen in the HB analysis, being higher in the annocatacin B complex. These results suggest that annocatacin B has a better stabilization effect on the whole ND1 structure.

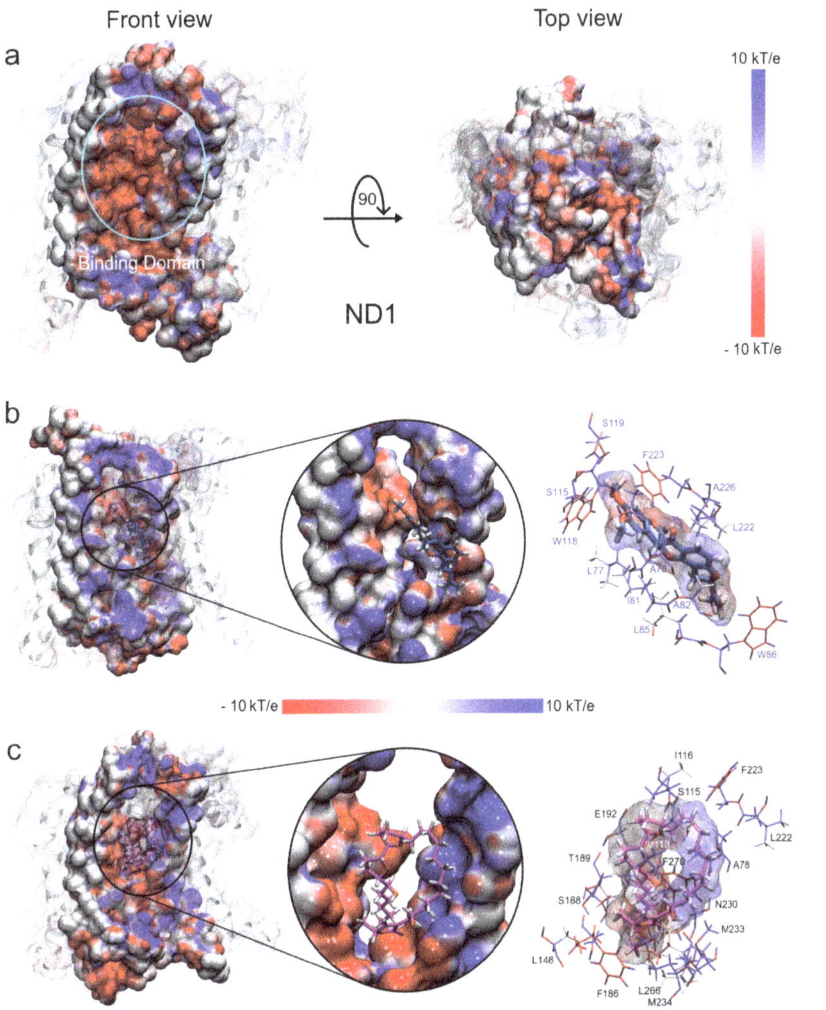

Figure 8. Electrostatic potential surfaces of molecular systems obtained with APBS. (**a**) ND1 protein structure. (**b**) ND1—rotenone complex. (**c**) ND1—annocatacin B complex. The red color indicates negatively charged regions and blue, positively charged. White color denotes hydrophobic regions.

Table 5. Average MM/PBSA free energies of ND1 complexes.

System	ΔE^{a}_{VW}	ΔE^{b}_{Elec}	ΔE^{c}_{PS}	ΔE^{d}_{SASA}	BFE^{e}
ND1-Annocatacin B	−358.76 ± 1.26	−23.04 ± 0.69	85.00 ± 0.92	−36.38 ± 0.10	−333.18 ± 2.14
ND1-Rotenone	−219.81 ± 0.89	−21.14 ± 0.39	45.01 ± 0.34	−22.21 ± 0.06	−218.15 ± 1.78

[a] Van der Waals energy.; [b] Electrostatic Energy,[c] Polar Solvation Energy; [d] SASA Energy; [e] Binding Free Energy. All values are in kJ·mol^{-1}.

As mentioned above, the ligand interactions involve, besides the active site, residues in other regions of the ND1 subunit allowing its structural stability. The large size of the annocatacin B molecule allows a greater number of energetically favorable contacts with these residues compared to those with which the rotenone molecule contacts (Figure 9a). The energy per residue decomposition shows the different contributions to the binding strength in the ND1—ligand complexes (Figure 9b). The highest binding free energy contribution was presented at active site B in the ND1—rotenone complex with the A226 residue (−19.53 ± 0.24 kJ/mol). This complex showed two regions that favored the interactions with rotenone, namely r_1 (L79-L85) and r_2 (A221-M234), being r_2 a zone that involves HB interactions, which would explain its high contributions to the BFE. On the other hand, in the ND1—annocatacin B complex, the greatest BFE contribution was with Val113 (−14.84 ± 0.54 kJ/mol), residue located outside the active site and denoted as a_1. In addition, three other favorable regions were obtained in this complex, namely a_2 (M184-T193), a_3 (A226-L237), and a_4 (L266-L271), which suggest a better molecular coupling of annocatacin B into the ND1 protein. Table 6 shows the residues that contributed the most to the BFE for both complexes.

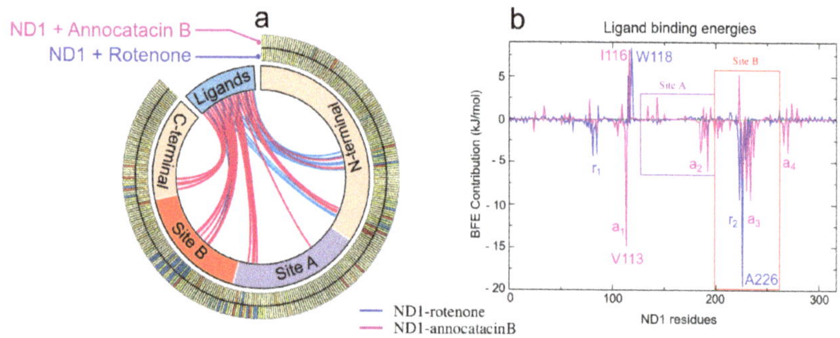

Figure 9. MM/PBSA analysis. (**a**) Circos plot representation of the ND1–ligand structures,where the interactions of the ND1 subunit residues to the ligand molecules are shown as links inthe middle of the plot. The blue lines indicate interactions with rotenone and the magenta lines, with annocatacin B. The outer part of this plot shows the heat map of the BFE per residue, where blue color represents favorable BFE, red color unfavorable BFE, and yellow color indicates neutral energies. (**b**) Energy per-residue contributions plot.

The positive energies in the BFE calculations are associated with unfavorable energy interactions between the protein-ligand complexes, Table 7 shows the residues with the highest positive values. In the rotenone complex, these residues were located previous to the active site, namely S115, W118, and S119, being the W118 residue with the highest energy value (8.39 ± 0.16 kJ/mol). On the other hand, in the annocatacin B complex, we observed a greater amount of residues with positive values, being A116 (8.14 ± 0.15 kJ/mol), F223 (5.35 ± 0.19 kJ/mol), and E143 (2.65 ± 0.18 kJ/mol), the residues with the most significant values. These results seem to indicate that the size of annocatacin B could also have a slight destabilizing effect on the ND1 structure, however, this effect is counteracted by the favorable contributions that stabilize it.

Table 6. Top 10 residues that does contribute to the binding free energy.

ND1—Rotetone	ΔE_{MM}	ΔE_{PS}	ΔE_{AS}	BFE
A226	−8.18 ± 0.12	0.92 ± 0.06	−12.27 ± 0.19	−19.53 ± 0.24
F223	−14.37 ± 0.20	3.94 ± 0.08	1.06 ± 0.63	−9.41 ± 0.60
I81	−9.21 ± 0.11	0.66 ± 0.02	4.19 ± 0.14	−4.36 ± 0.19
T229	−4.32 ± 0.11	5.20 ± 0.13	−5.06 ± 0.32	−4.17 ± 0.35
L85	−5.97 ± 0.11	0.72 ± 0.07	1.10 ± 0.13	−4.15 ± 0.16
N230	−1.35 ± 0.06	0.72 ± 0.06	−2.86 ± 0.14	−3.49 ± 0.16
L222	−9.03 ± 0.10	2.46 ± 0.07	3.79 ± 0.18	−2.77 ± 0.22
M225	−3.74 ± 0.13	1.19 ± 0.07	0.24 ± 0.12	−2.31 ± 0.13
M233	−2.77 ± 0.07	2.23 ± 0.07	−1.51 ± 0.11	−2.05 ± 0.13
E192	−1.61 ± 0.08	−0.17 ± 0.15	−0.22 ± 0.01	−2.00 ± 0.16
ND1—Annocatacin B				
V113	−12.39 ± 0.22	2.74 ± 0.12	−5.20 ± 0.34	−14.84 ± 0.54
M234	−10.30 ± 0.16	3.67 ± 0.08	−2.80 ± 0.16	−9.43 ± 0.24
N230	−10.59 ± 0.19	6.10 ± 0.13	−3.82 ± 0.17	−8.30 ± 0.25
A226	−4.94 ± 0.22	0.60 ± 0.08	−2.59 ± 0.46	−6.92 ± 0.61
E192	−8.48 ± 0.20	1.29 ± 0.35	1.11 ± 0.22	−6.09 ± 0.33
I231	−4.26 ± 0.10	−0.19 ± 0.02	−0.74 ± 0.09	−5.19 ± 0.13
E227	−0.98 ± 0.26	−3.77 ± 0.29	−0.35 ± 0.10	−5.09 ± 0.30
M233	9.20 ± 0.16	4.13 ± 0.08	0.61 ± 0.09	−4.46 ± 0.18
L237	−6.74 ± 0.20	1.18 ± 0.03	1.15 ± 0.07	−4.40 ± 0.19
F270	−7.34 ± 0.16	1.39 ± 0.06	2.06 ± 0.10	−3.90 ± 0.15

All values are in kJ·mol^{-1}

Continuing with the BFE analysis, we used the minimum energy structures obtained in the FEL analysis and the 2D ligand-protein interaction diagrams, in order to visualize the moiety interactions of the ligand molecules. For this task, we plotted the BFE values on the ND1 surfaces and analyzed the energy interactions on the ligand structures. In the rotenone complex, we observed that the main residue interactions were located at the dimethoxychromene moiety of rotenone. The greatest favorable interactions were A226 and F223 residues (dark blue color in Figure 10a), being this last one the residue that more interactions showed with rotenone (up to 7 direct interactions with ND1 residues, Figure 10b). Despite the numerous interactions (5), the energy of A82 residue was just -1.04 ± 0.23 kJ/mol. Another important contributions to the BFE were I81, L85, T229, and N230 residues that interacted with the pyranol and the dihydrofuran moieties of the rotenone molecule, respectively. The rotenone's atom that involved the main energy contributions was the oxygen of the hydroxyl group located at the pyranoloid ring. Residues with the lowest contribution to the BFE were located on the methoxy groups, being S115, W118, and S119 residues with high positive values. The non-polar solvation contribution was the predominant energy term in these residues.

In the case of the annocatacin B complex, the aliphatic chains of this ligand involve the majority of the interactions with its amino acid environment. This includes the interactions with V113 and I116, that were the main contribution and no-contribution residues to the binding energy, respectively (Figure 10c,d). The bis-THF rings only showed one interaction with S188 residue, being an unfavorable one to the total BFE, nevertheless, the energy value was low (1.55 ± 0.54 kJ/mol). The γ-lactone ring involved five interactions, three of them important contributors, namely N230 and M234 (-8.30 ± 0.25 and -9.43 ± 0.24 kJ/mol, respectively). The other two interactions with E143 and S188 residues represented unfavorable energies. It is interesting to note that although the F223 residue showed high fluctuations in both molecular complexes, different energy behavior was observed at the interactions with the ligand molecules. At the rotenone complex, the interaction contributed favorably to the BFE with a significant value (-9.41 ± 0.60 kJ/mol), while at the annocatacin B complex, this contribution was unfavorable (5.35 ± 0.30 kJ/mol).

Table 7. Top 10 residues that does not contribute to the binding free energy.

ND1—Rotetone	ΔE_{MM}	ΔE_{PS}	ΔE_{AS}	*BFE*
W118	−7.31 ± 0.12	2.65 ± 0.07	13.05 ± 0.30	8.39 ± 0.30
S119	−1.11 ± 0.04	1.12 ± 0.07	6.11 ± 0.23	6.12 ± 0.22
S115	−4.37 ± 0.11	1.67 ± 0.06	7.78 ± 0.20	5.09 ± 0.20
W86	−4.52 ± 0.10	0.93 ± 0.05	5.21 ± 0.25	1.61 ± 0.25
K262	1.14 ± 0.01	−0.13 ± 0.01	0.22 ± 0.00	1.23 ± 0.01
R195	1.61 ± 0.03	−0.58 ± 0.04	0.00 ± 0.00	1.03 ± 0.05
E214	1.53 ± 0.03	−0.75 ± 0.02	0.07 ± 0.00	0.85 ± 0.02
Y215	−0.47 ± 0.02	0.41 ± 0.03	0.91 ± 0.12	0.84 ± 0.13
A78	−9.30 ± 0.11	2.64 ± 0.06	7.37 ± 0.23	0.71 ± 0.29
E59	1.05 ± 0.02	−0.30 ± 0.01	−0.17 ± 0.00	0.58 ± 0.01
ND1—Annocatacin B				
I116	−3.78 ± 0.09	0.51 ± 0.03	11.40 ± 0.24	8.14 ± 0.23
F223	−4.83 ± 0.13	2.67 ± 0.15	7.50 ± 0.25	5.35 ± 0.30
E143	−7.08 ± 0.23	8.98 ± 0.56	0.70 ± 0.10	2.65 ± 0.52
A78	−3.20 ± 0.08	2.23 ± 0.09	3.42 ± 0.15	2.44 ± 0.16
R274	−1.97 ± 0.13	4.04 ± 0.10	−0.08 ± 0.01	1.98 ± 0.13
S109	−5.27 ± 0.17	5.55 ± 0.15	1.55 ± 0.10	1.82 ± 0.21
R134	1.12 ± 0.02	0.65 ± 0.03	−0.16 ± 0.00	1.61 ± 0.04
R195	0.83 ± 0.04	0.74 ± 0.03	0.01 ± 0.00	1.59 ± 0.04
R34	−0.24 ± 0.06	1.88 ± 0.06	−0.08 ± 0.00	1.55 ± 0.08
S188	−6.55 ± 0.12	6.38 ± 0.14	1.72 ± 0.09	1.55 ± 0.19

All values are in kJ·mol^{-1}.

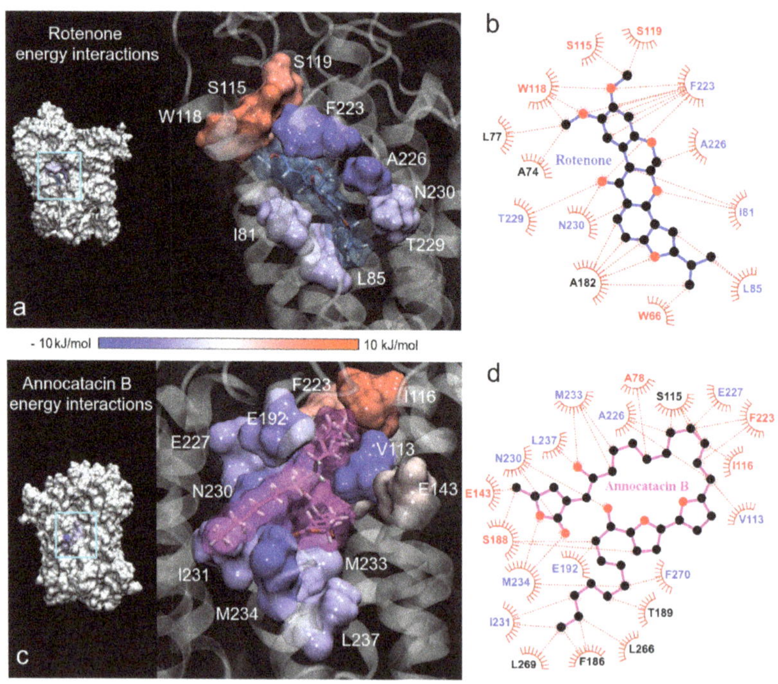

Figure 10. Binding free energy (BFE) plotted on the ND1 surfaces. (**a**,**c**) 3D representation of the main residue contributions to BFE in the ND1 complexes. Blue color indicates favorable energies and red color unfavorable energies. (**b**,**d**) 2D representation of all interactions obtained in contact analysis in the minimum energy structures. The color of residue labels is the same as energy values.

4. Conclusions

QM optimizations, MD simulations, and MM/PBSA analyses were performed to evaluate the molecular and energetic complementarity of the annocatacin B molecule in the ND1 subunit of the human mitochondrial respiratory complex I (MRC-I). We compared these results with those obtained through the analysis of the ND1—rotenone complex, owing to the fact that the rotenone molecule is a powerful inhibitor of this MRC-I. ND1 subunit is a transmembrane protein, thus we used a DPPC lipid bilayer as a membrane model for the entire simulations of the ND1 complexes in this work.

The overall analysis revealed a stabilizing effect of the annocatacin B molecule over the ND1 protein structure. Mainly, this could be due to the size and its capability of folding of annocatacin B, which ultimately allowed it to have more interactions with the ND1 nearby residues. We observe an increase in the formation of hydrogen bonds in the ND1—annocatacin B complex through the whole MD trajectory, which suggests a considerable electrostatic contribution to the stability of the complex. Thus, with respect to the ESP surfaces at the active site, while annocatacin B retained the electrophilic pattern of the native active site of ND1, rotenone largely changed it to a nucleophilic one.

An analysis of MM/PBSA showed that hydrophobic interactions were the main energetic component of the relative binding free energy (BFE), hence the important role of the aliphatic annocatacin B chains in their affinity for the ND1 subunit. Several favorable interactions were observed on these chains in the ND1—annocatacin B complex, including residues outside the active site (V113 and L266-L271), which allowed a 35% better energetic coupling than those observed in the ND1—rotenone complex. Despite the high structural fluctuations of the F223 residue in both complexes, a significant energy interaction was observed, favorable to the BFE (-9.41 ± 0.60 kJ/mol) for the rotenone complex, and unfavorable (5.35 ± 0.30 kJ/mol) for the annocatacin B complex. The reason for this behavior could be that the residue F223 prefers electrostatic to hydrophobic interactions.

Our results suggest that the natural annocatacin B molecule could display better inhibitory capabilities than the rotenone molecule, an issue to be taken into account for future research.

Supplementary Materials: The following are available online at https://www.mdpi.com/article/10.3390/polym13111840/s1, Figure S1: Stability indicators used in this work. Plots were obtained in full MD simulations. (a) Root mean square deviation (RMSD). (b) The total radius of gyration (RG). (c) Root mean square fluctuation of Cα atoms. The A and B sites comprise the active site of the ND1 subunit. (d) Solvent accessible area (SAS) of the ND1 protein in the molecular complexes. Figure S2: Top 10 structure complexes of the PatchDock docking results. Above figures correspond to the ND1-Annocatacin B complexes. Below figures correspond to the ND1-Rotenone complexes. Table S1: System details in the MD simulations.

Author Contributions: Conceptualization, B.G. and C.F.-M.; Methodology, J.A.A.-P. and P.L.G.-B.; Software, C.F.-M.; Validation, H.L.B.-C. and D.E.V. and K.J.V.-L.; Formal Analysis, B.G. and C.F.-M.; Investigation, P.L.G.-B. and G.D.-D.-C.; Resources, C.F.-M. and H.L.B.-C.; Data Curation, J.A.A.-P.; Writing—Original Draft Preparation, C.F.-M.; Writing—Review and Editing, B.G.; Visualization, K.J.V.-L. and G.D.-D.-C.; Supervision, B.G. and J.A.A.-P. and H.L.B.-C.; Project Administration, B.G.; Funding Acquisition, B.G. All authors have read and agreed to the published version of the manuscript.

Funding: The author thanks the financial support of the UCSM grant under Project 20170116.

Institutional Review Board Statement: Not applicable.

Informed Consent Statement: Not applicable.

Acknowledgments: For the development of this research, we have the support of workstations by the internal Project 23824-2016VRINV-UCSM.

Conflicts of Interest: The authors declare no conflict of interest.

References

1. Warburg, O.H.; Dickens, F. *Metabolism of Tumours*; Constable & Co. Ltd.: London, UK, 1930.
2. Ganapathy-Kanniappan, S.; Geschwind, J.F.H. Tumor glycolysis as a target for cancer therapy: Progress and prospects. *Mol. Cancer* **2013**, *12*, 152. [CrossRef]
3. Vander Heiden, M.G.; Cantley, L.C.; Thompson, C.B. Understanding the Warburg effect: the metabolic requirements of cell proliferation. *Science* **2009**, *324*, 1029–1033. [CrossRef]
4. Wallace, D.C. Mitochondria and cancer. *Nat. Rev. Cancer* **2012**, *12*, 685. [CrossRef]
5. Koppenol, W.H.; Bounds, P.L.; Dang, C.V. Otto Warburg's contributions to current concepts of cancer metabolism. *Nat. Rev. Cancer* **2011**, *11*, 325. [CrossRef]
6. Jose, C.; Bellance, N.; Rossignol, R. Choosing between glycolysis and oxidative phosphorylation: A tumor's dilemma? *Biochim. Biophys. Acta (BBA) Bioenergy* **2011**, *1807*, 552–561. [CrossRef] [PubMed]
7. Smolková, K.; Plecitá-Hlavatá, L.; Bellance, N.; Benard, G.; Rossignol, R.; Ježek, P. Waves of gene regulation suppress and then restore oxidative phosphorylation in cancer cells. *Int. J. Biochem. Cell Biol.* **2011**, *43*, 950–968. [CrossRef]
8. Moreno-Sánchez, R.; Rodríguez-Enríquez, S.; Marín-Hernández, A.; Saavedra, E. Energy Substrate Modulates in tumor cells. *FEBS J.* **2007**, *274*, 1393–1418. [CrossRef] [PubMed]
9. Burlaka, A.; Ganusevich, I.; Vovk, A.; Burlaka, A.; Gafurov, M.; Lukin, S. Colorectal Cancer and Mitochondrial Dysfunctions of the Adjunct Adipose Tissues: A Case Study. *BioMed Res. Int.* **2018**, *2018*, 1–7. [CrossRef] [PubMed]
10. Philley, J.V.; Kannan, A.; Qin, W.; Sauter, E.R.; Ikebe, M.; Hertweck, K.L.; Troyer, D.A.; Semmes, O.J.; Dasgupta, S. Complex-I alteration and enhanced mitochondrial fusion are associated with prostate cancer progression. *J. Cell. Physiol.* **2016**, *231*, 1364–1374. [CrossRef]
11. Cormio, A.; Musicco, C.; Gasparre, G.; Cormio, G.; Pesce, V.; Sardanelli, A.M.; Gadaleta, M.N. Increase in proteins involved in mitochondrial fission, mitophagy, proteolysis and antioxidant response in type I endometrial cancer as an adaptive response to respiratory complex I deficiency. *Biochem. Biophys. Res. Commun.* **2017**, *491*, 85–90. [CrossRef] [PubMed]
12. Santidrian, A.F.; Matsuno-Yagi, A.; Ritland, M.; Seo, B.B.; LeBoeuf, S.E.; Gay, L.J.; Yagi, T.; Felding-Habermann, B. Mitochondrial complex I activity and NAD+/NADH balance regulate breast cancer progression. *J. Clin. Investig.* **2013**, *123*, 1068–1081. [CrossRef]
13. Basit, F.; Van Oppen, L.M.; Schöckel, L.; Bossenbroek, H.M.; Van Emst-de Vries, S.E.; Hermeling, J.C.; Grefte, S.; Kopitz, C.; Heroult, M.; Willems, P.H. Mitochondrial complex I inhibition triggers a mitophagy-dependent ROS increase leading to necroptosis and ferroptosis in melanoma cells. *Cell Death Dis.* **2017**, *8*, e2716. [CrossRef]
14. Pamies, D.; Block, K.; Lau, P.; Gribaldo, L.; Pardo, C.A.; Barreras, P.; Smirnova, L.; Wiersma, D.; Zhao, L.; Harris, G. Rotenone exerts developmental neurotoxicity in a human brain spheroid model. *Toxicol. Appl. Pharmacol.* **2018**, *354*, 101–114. [CrossRef] [PubMed]
15. Chadderton, L.; Kelleher, S.; Brow, A.; Shaw, T.; Studholme, B.; Barrier, R. Testing the efficacy of rotenone as a piscicide for New Zealand pest fish species. In *Managing Invasive Freshwater Fish in New Zealand, Proceedings of the Workshop Hosted by Department of Conservation, Hamilton, New Zeland, 10–12 May 2001*; Department of Conservation: Wellington, New Zealand, 2001; pp. 10–12.
16. Biñas Jr, Enrique E. A Review on Tubli Plant used as Organic Pesticide: Input toward Sustainable Agriculture *Int. J. Res. Appl. Sci. Biotechnol.* **2021**, *8*, 107–115.
17. Betarbet, R.; Sherer, T.B.; MacKenzie, G.; Garcia-Osuna, M.; Panov, A.V.; Greenamyre, J.T. Chronic systemic pesticide exposure reproduces features of Parkinson's disease. *Nat. Neurosci.* **2000**, *3*, 1301. [CrossRef] [PubMed]
18. Li, N.; Ragheb, K.; Lawler, G.; Sturgis, J.; Rajwa, B.; Melendez, J.A.; Robinson, J.P. Mitochondrial complex I inhibitor rotenone induces apoptosis through enhancing mitochondrial reactive oxygen species production. *J. Biol. Chem.* **2003**, *278*, 8516–8525. [CrossRef] [PubMed]
19. Cunningham, M.L.; Soliman, M.S.; Badr, M.Z.; Matthews, H. Rotenone, an anticarcinogen, inhibits cellular proliferation but not peroxisome proliferation in mouse liver. *Cancer Lett.* **1995**, *95*, 93–97. [CrossRef]
20. Bhurtel, S.; Katila, N.; Srivastav, S.; Neupane, S.; Choi, D.Y. Mechanistic comparison between MPTP and rotenone neurotoxicity in mice. *Neurotoxicology* **2019**, *71*, 113–121. [CrossRef]
21. Nopparat, C.; Porter, J.E.; Ebadi, M.; Govitrapong, P. 1-Methyl-4-phenylpyridinium-induced cell death via autophagy through a Bcl-2/Beclin 1 complex-dependent pathway. *Neurochem. Res.* **2014**, *39*, 225–232. [CrossRef]
22. Murai, M.; Ishihara, A.; Nishioka, T.; Yagi, T.; Miyoshi, H. The ND1 subunit constructs the inhibitor binding domain in bovine heart mitochondrial complex I. *Biochemistry* **2007**, *46*, 6409–6416. [CrossRef]
23. Chen, Y.; Chen, J.W.; Zhai, J.H.; Wang, Y.; Wang, S.L.; Li, X. Antitumor activity and toxicity relationship of annonaceous acetogenins. *Food Chem. Toxicol.* **2013**, *58*, 394–400. [CrossRef]
24. Patel, M.S.; Patel, J.K. A review on a miracle fruits of Annona muricata. *J. Pharmacogn. Phytochem.* **2016**, *5*, 137.
25. Moghadamtousi, S.; Fadaeinasab, M.; Nikzad, S.; Mohan, G.; Ali, H.; Kadir, H. Annona muricata (Annonaceae): A review of its traditional uses, isolated acetogenins and biological activities. *Int. J. Mol. Sci.* **2015**, *16*, 15625–15658. [CrossRef]
26. Prasad, S.K.; Varsha, V.; Devananda, D. Anti-cancer properties of Annona muricata (L.): A Review. *Med. Plants Int. J. Phytomed. Relat. Ind.* **2019**, *11*, 123–134. [CrossRef]

27. Gavamukulya, Y.; Wamunyokoli, F.; El-Shemy, H.A. Annona muricata: Is the natural therapy to most disease conditions including cancer growing in our backyard? A systematic review of its research history and future prospects. *Asian Pac. J. Trop. Med.* **2017**, *10*, 835–848. [CrossRef]
28. Zafra-Polo, M.C.; González, M.C.; Estornell, E.; Sahpaz, S.; Cortes, D. Acetogenins from Annonaceae, inhibitors of mitochondrial complex I. *Phytochemistry* **1996**, *42*, 253–271. [CrossRef]
29. Mangal, M.; Imran Khan, M.; Mohan Agarwal, S. Acetogenins as potential anticancer agents. *Anti-Cancer Agents Med. Chem. Formerly Curr. Med. Chem. Anti-Cancer Agents* **2016**, *16*, 138–159. [CrossRef]
30. Chang, F.R.; Liaw, C.C.; Lin, C.Y.; Chou, C.J.; Chiu, H.F.; Wu, Y.C. New adjacent bis-tetrahydrofuran annonaceous acetogenins from Annona muricata. *Planta Med.* **2003**, *69*, 241–246. [CrossRef] [PubMed]
31. Dennington, R.; Keith, T.A.; Millam, J.M. *GaussView 6.0. 16*; Semichem Inc.: Shawnee Mission, KS, USA, 2016.
32. Frisch, M.J.; Trucks, G.W.; Schlegel, H.B.; Scuseria, G.E.; Robb, M.A.; Cheeseman, J.R.; Scalmani, G.; Barone, V.; Petersson, G.A.; Nakatsuji, H.; et al. *Gaussian 16 Revision C.01*; Gaussian Inc.: Wallingford, CT, USA, 2016.
33. Yanai, T.; Tew, D.P.; Handy, N.C. A new hybrid exchange-correlation functional using the Coulomb-attenuating method (CAM-B3LYP). *Chem. Phys. Lett.* **2004**, *393*, 51–57. doi:10.1016/j.cplett.2004.06.011. [CrossRef]
34. Schäfer, A.; Huber, C.; Ahlrichs, R. Fully optimized contracted Gaussian basis sets of triple zeta valence quality for atoms Li to Kr. *J. Chem. Phys.* **1994**, *100*, 5829–5835. doi:10.1063/1.467146. [CrossRef]
35. Hirshfeld, F.L. Bonded-atom fragments for describing molecular charge densities. *Theor. Chim. Acta* **1977**, *44*, 129–138. [CrossRef]
36. Ritchie, J.P. Electron density distribution analysis for nitromethane, nitromethide, and nitramide. *J. Am. Chem. Soc.* **1985**, *107*, 1829–1837. [CrossRef]
37. Ritchie, J.P.; Bachrach, S.M. Some methods and applications of electron density distribution analysis. *J. Comput. Chem.* **1987**, *8*, 499–509. [CrossRef]
38. Mikhailovich, K.P. ERG Research Group. 2008. Available online: http://erg.biophys.msu.ru/tpp/ (accessed on 28 April 2021).
39. Jorgensen, W.L.; Maxwell, D.S.; Tirado-Rives, J. Development and testing of the OPLS all-atom force field on conformational energetics and properties of organic liquids. *J. Am. Chem. Soc.* **1996**, *118*, 11225–11236. [CrossRef]
40. Kaminski, G.A.; Friesner, R.A.; Tirado-Rives, J.; Jorgensen, W.L. Evaluation and reparametrization of the OPLS-AA force field for proteins via comparison with accurate quantum chemical calculations on peptides. *J. Phys. Chem. B* **2001**, *105*, 6474–6487. [CrossRef]
41. Kandt, C.; Ash, W.L.; Tieleman, D.P. Setting up and running molecular dynamics simulations of membrane proteins. *Methods* **2007**, *41*, 475–488. [CrossRef]
42. Guo, R.; Zong, S.; Wu, M.; Gu, J.; Yang, M. Architecture of human mitochondrial respiratory megacomplex I2III2IV2. *Cell* **2017**, *170*, 1247–1257. [CrossRef]
43. Pettersen, E.F.; Goddard, T.D.; Huang, C.C.; Couch, G.S.; Greenblatt, D.M.; Meng, E.C.; Ferrin, T.E. UCSF Chimera—A visualization system for exploratory research and analysis. *J. Comput. Chem.* **2004**, *25*, 1605–1612. [CrossRef]
44. Van Der Spoel, D.; Lindahl, E.; Hess, B.; Groenhof, G.; Mark, A.E.; Berendsen, H.J. GROMACS: Fast, flexible, and free. *J. Comput. Chem.* **2005**, *26*, 1701–1718. [CrossRef] [PubMed]
45. Tieleman, D.P.; Berendsen, H. Molecular dynamics simulations of a fully hydrated dipalmitoylphosphatidylcholine bilayer with different macroscopic boundary conditions and parameters. *J. Chem. Phys.* **1996**, *105*, 4871–4880. [CrossRef]
46. Duhovny, D.; Nussinov, R.; Wolfson, H.J. Efficient unbound docking of rigid molecules. In Proceedings of the International Workshop on Algorithms in Bioinformatics, Rome, Italy, 17–21 September 2002; Springer: Berlin/Heidelberg, Germany, 2002; pp. 185–200.
47. Schneidman-Duhovny, D.; Inbar, Y.; Nussinov, R.; Wolfson, H.J. PatchDock and SymmDock: Servers for rigid and symmetric docking. *Nucleic Acids Res.* **2005**, *33*, W363–W367. [CrossRef]
48. Kakutani, N.; Murai, M.; Sakiyama, N.; Miyoshi, H. Exploring the Binding Site of Δlac-Acetogenin in Bovine Heart Mitochondrial NADH- Ubiquinone Oxidoreductase. *Biochemistry* **2010**, *49*, 4794–4803. [CrossRef]
49. Homeyer, N.; Gohlke, H. Free energy calculations by the molecular mechanics Poisson- Boltzmann surface area method. *Mol. Inform.* **2012**, *31*, 114–122. [CrossRef]
50. Kumari, R.; Kumar, R.; Consortium, O.S.D.D.; Lynn, A. g_mmpbsa A GROMACS tool for high-throughput MM-PBSA calculations. *J. Chem. Inf. Model.* **2014**, *54*, 1951–1962. [CrossRef]
51. Brown, S.P.; Muchmore, S.W. Large-scale application of high-throughput molecular mechanics with Poisson- Boltzmann surface area for routine physics-based scoring of protein- ligand complexes. *J. Med. Chem.* **2009**, *52*, 3159–3165. [CrossRef]
52. Rastelli, G.; Rio, A.D.; Degliesposti, G.; Sgobba, M. Fast and accurate predictions of binding free energies using MM-PBSA and MM-GBSA. *J. Comput. Chem.* **2010**, *31*, 797–810. [CrossRef]
53. Humphrey, W.; Dalke, A.; Schulten, K. VMD: Visual molecular dynamics. *J. Mol. Graph.* **1996**, *14*, 33–38. [CrossRef]
54. Turner, P.; Grace, X. *Version 5.1. 19. Center for Coastal and Land-Margin Research*; Oregon Graduate Institute of Science and Technology: Beaverton, OR, USA, 2005.
55. Wallace, A.C.; Laskowski, R.A.; Thornton, J.M. LIGPLOT: A program to generate schematic diagrams of protein-ligand interactions. *Protein Eng. Des. Sel.* **1995**, *8*, 127–134. [CrossRef]
56. Baker, N.A.; Sept, D.; Joseph, S.; Holst, M.J.; McCammon, J.A. Electrostatics of nanosystems: application to microtubules and the ribosome. *Proc. Natl. Acad. Sci. USA* **2001**, *98*, 10037–10041. [CrossRef]

57. Dolinsky, T.J.; Nielsen, J.E.; McCammon, J.A.; Baker, N.A. PDB2PQR: an automated pipeline for the setup of Poisson–Boltzmann electrostatics calculations. *Nucleic Acids Res.* **2004**, *32*, W665–W667. [CrossRef]
58. WR Inc. *Mathematica, Version 12.1*; WR Inc.: Champaign, IL, USA, 2020.
59. Lobo-Jarne, T.; Ugalde, C. Respiratory chain supercomplexes: Structures, function and biogenesis. In *Seminars in Cell & Developmental Biology*; Elsevier: Amsterdam, The Netherlands, **2018**; Volume 76, pp. 179–190.
60. Fedor, J.G.; Hirst, J. Mitochondrial supercomplexes do not enhance catalysis by quinone channeling. *Cell Metab.* **2018**, *28*, 525–531. [CrossRef]
61. Fiedorczuk, K.; Letts, J.A.; Degliesposti, G.; Kaszuba, K.; Skehel, M.; Sazanov, L.A. Atomic structure of the entire mammalian mitochondrial complex I. *Nature* **2016**, *538*, 406. [CrossRef]
62. Nakanishi, S.; Abe, M.; Yamamoto, S.; Murai, M.; Miyoshi, H. Bis-THF motif of acetogenin binds to the third matrix-side loop of ND1 subunit in mitochondrial NADH-ubiquinone oxidoreductase. *Biochim. Biophys. Acta (BBA) Bioenergy* **2011**, *1807*, 1170–1176. [CrossRef]
63. Pérez de la Luz, A.; Aguilar-Pineda, J.A.; Méndez-Bermúdez, J.G.; Alejandre, J. Force field parametrization from the Hirshfeld molecular electronic density. *J. Chem. Theory Comput.* **2018**, *14*, 5949–5958. [CrossRef]
64. Riquelme, M.; Lara, A.; Mobley, D.L.; Verstraelen, T.; Matamala, A.R.; Vohringer-Martinez, E. Hydration free energies in the FreeSolv database calculated with polarized iterative Hirshfeld charges. *J. Chem. Inf. Model.* **2018**, *58*, 1779–1797. [CrossRef]
65. Ishizuka, R.; Matubayasi, N. Effective charges of ionic liquid determined self-consistently through combination of molecular dynamics simulation and density-functional theory. *J. Comput. Chem.* **2017**, *38*, 2559–2569. [CrossRef]
66. Gastegger, M.; Marquetand, P. Molecular dynamics with neural network potentials. In *Machine Learning Meets Quantum Physics*; Springer: Berlin/Heidelberg, Germany, **2020**; pp. 233–252.
67. Llanos, A.G.; Vöhringer-Martinez, E. pKa Calculations of Asp26 in Thioredoxin with Alchemical Free Energy Simulations and Hirshfeld-I Atomic Charges. *ChemRxiv* **2018**. [CrossRef]
68. Pires, D.E.; Blundell, T.L.; Ascher, D.B. pkCSM: predicting small-molecule pharmacokinetic and toxicity properties using graph-based signatures. *J. Med. Chem.* **2015**, *58*, 4066–4072. [CrossRef] [PubMed]
69. Pang, Y.P. Use of multiple picosecond high-mass molecular dynamics simulations to predict crystallographic B-factors of folded globular proteins. *Heliyon* **2016**, *2*, e00161. [CrossRef]

polymers

MDPI

Article

A Fully Implicit Log-Conformation Tensor Coupled Algorithm for the Solution of Incompressible Non-Isothermal Viscoelastic Flows

Célio Fernandes

LASI—Associate Laboratory of Intelligent Systems, IPC—Institute for Polymers and Composites, Polymer Engineering Department, School of Engineering, Campus de Azurém, University of Minho, 4800-058 Guimarães, Portugal; cbpf@dep.uminho.pt

Abstract: In this work, a block-coupled algorithm is presented, which can compute laminar, incompressible, non-isothermal, viscoelastic flow problems based on the log-conformation tensor approach. The inter-equation coupling of the incompressible Cauchy linear momentum and mass conservation equations is obtained in a procedure based on the Rhie–Chow interpolation. The divergence of the log-conformation tensor term in the linear momentum equations is implicitly discretized in this work. In addition, the velocity field is considered implicitly in the log-conformation tensor constitutive equations by expanding the advection, rotation and the rate of deformation terms with a Taylor series expansion truncated at the second-order error term. Finally, the advection and diffusion terms in the energy equation are also implicitly discretized. The mass, linear momentum, log-conformation tensor constitutive model and energy-discretized linear equations are joined into a block-matrix following a monolithic framework. Validation of the newly developed algorithm is performed for the non-isothermal viscoelastic matrix-based Oldroyd-B fluid flow in the axisymmetric 4:1 planar sudden contraction benchmark problem.

Keywords: fully implicit coupled solver; viscoelastic flow; log-conformation tensor approach; non-isothermal effects; finite volume method; OpenFOAM

Citation: Fernandes, C. A Fully Implicit Log-Conformation Tensor Coupled Algorithm for the Solution of Incompressible Non-Isothermal Viscoelastic Flows. *Polymers* **2022**, *14*, 4099. https://doi.org/10.3390/polym14194099

Academic Editors: Patrick Ilg and Maria Bercea

Received: 5 September 2022
Accepted: 26 September 2022
Published: 30 September 2022

1. Introduction

The polymers' processing techniques are predominantly non-isothermal, such as injection molding [1–3], heat exchange problems [4,5], or in plastication, including heating and cooling sequences [6,7]. The thermal conductivity and heat transfer are usually low in this processes; however, due to the heating or cooling of the machine's operations, large temperature gradients arise in the fluid [4,8]. In addition, the viscoelastic behavior of polymers acts on the temperature field as well as on the fluid deformation [4,8]. Therefore, flow properties are strongly dependent on both rheology and temperature; and, thus, it is essential to understand and make predictions regarding non-isothermal viscoelastic fluid flows.

The temperature dependence of linear viscoelastic properties (such as the relaxation time λ) can be included in constitutive equations using the time-temperature superposition principle [9]. In this way, the material properties can be defined through a function of the temperature, the so-called shift factor [10]. Two empirical correlations of the shift factor are widely employed: the William–Landel-Ferry (WLF) [11] and Arrhenius [12] models. Thus, the temperature is considered an independent variable in the constitutive equations employed to compute the components of the polymeric stress tensor (see the work of Peters and Baaijens [13] for a detailed discussion on this topic). In addition, when solving non-isothermal viscoelastic flows, the internal energy of the fluid is not only a function of the temperature [13]. The conversion mechanisms of internal energy need to be taken into account for non-isothermal viscoelastic flows; specifically, the thermal energy is partly dissipated and partly stored in the fluid. Therefore, the energy equation should predict which part of the mechanical power is dissipated and which part is accumulated as elastic energy [4,8]. For that purpose, an additional term is needed in the energy equation [14]. Peters and Baaijens [13] developed an internal energy equation for multiple rate-type fluids based on a constant weighting factor that characterizes the ratio of entropy to energy elasticity [15]. Several numerical studies have also used this concept [16–18] and we will employ this in the current work.

Numerical simulations can describe these complex flow mechanisms and help to gain a better understanding of, and improvements in, the processes where they occur. For that purpose, Computational Fluid Dynamics (CFD) are used to guide the theoretical researchers and practitioner engineers, through the use of both open-source [3–5,7,19] and proprietary software [20]. In the last decade, a significant effort has been made in research on the non-isothermal flows of viscoelastic fluids. A survey of the scientific literature finds different works that describe the non-isothermal viscoelastic fluid flows based on iterative numerical algorithms. For example, Shahbani-Zahiri et al. [21] studied the recirculation and thermal regions of viscoelastic flow in the symmetric planar problem for different expansion angles. Kunisch and Marduel [22] employed the finite element (FE) method to study the optimal control of non-isothermal viscoelastic fluids to minimize vortices and control the heat flux. Spanjaards et al. [23] performed a 3D transient non-isothermal simulation to predict the extrudate shape of viscoelastic fluids emerging from an asymmetric keyhole-shaped die. However, the current state-of-the-art codes depend on iterative algorithms, such as the Semi-Implicit Method for Pressure Linked Equations (SIMPLE) procedure [24], which are known to delay the convergence of the problem of interest when compared to monolithic or coupled algorithms [25–30]. The iterative algorithms, also known as segregated algorithms, are characterized to provide a separate solution of the linear momentum, mass, viscoelastic polymer stress tensor and energy conservation equations, which are then iterated until convergence. Recently, with the increase in computational resources and due to scalability problems in the segregated algorithms, the monolithic approach has been used, with the advantage of decreasing the computational wall time of the simulation, particularly for finer meshes, as shown by Fernandes et al. [28]. Thus, a methodology based on the monolithic approach for the simulation of non-isothermal viscoelastic flows would be of major importance.

In addition, the benchmark problems of both planar and axisymmetric contraction flows are also extensively studied to evaluate the stability of newly developed numerical algorithms [31–33]. These benchmark problems are especially important because, near the contraction, complex flow profiles are generated, and thus large stress gradients are developed, which can cause numerical difficulties, leading to the overall failure of the algorithms. Bearing this in mind, we will revisit the axisymmetric sudden contraction benchmark flow to validate the newly developed, fully implicit, coupled solver for non-isothermal viscoelastic fluid flows.

The rheological data available in the literature for the validation of non-isothermal viscoelastic fluids are scarse due to the complex fluid behaviour, which generally requires several modes to capture the full range of operating conditions. In this work, we will employ the highly elastic, polyisobutylene-based polymer solution (PIB-Boger fluid), which is typically described by the quasi-linear Oldroyd-B viscoelastic fluid model. Another important issue to consider when solving viscoelastic fluid flows is the elastic effects, specifically, the flow at high Weissenberg numbers [34]. It is well known that numerical simulations tend to become unstable at increased Weissenberg numbers, the so-called High Weissenberg Number Problem (HWNP). The seminal work of Fattal and Kupferman [35] proposed a reformulation of the viscoelastic stress-tensor-based formulation to solve the HWNP, where the logarithm of the conformation tensor is used as the main variable in the constitutive transport equation. Different methods have been also used to solve the HWNP, such as the square root of the conformation tensor [36]. A detailed discussion of this topic can be found in Afonso et al. [37]. In this work, we will employ the log-conformation tensor approach to handle the HWNP.

In this manuscript, a new numerical code is developed in the context of the Finite Volume Method (FVM), following a monolithic framework to compute the non-isothermal flow of viscoelastic fluids. To the author's knowledge, the other CFD codes, which provide a fully implicit block-coupled solution for a discretized, log-conformation, viscoelastic, fluid-flow system, were developed by Knechtges [38] using the Finite Element Method (FEM) and Spahn [39] using FVM; however, these studies did not consider the non-isothermal effects. In this work, the solution to the enlarged system of equations, composed of continuity, linear momentum, log-conformation tensor constitutive equation and energy, is obtained using a Bi-Conjugate Gradient Stabilized solver. The validation of the fully implicit, block-coupled, non-isothermal, viscoelastic, log-conformation tensor-based solver is performed for the Oldroyd-B fluid flow in the axisymmetric 4:1 planar sudden-contraction benchmark problem. For assessment purposes, the results obtained with the newly-developed code are compared with numerical results found in the scientific literature. We study flows at a high Weissenberg number and we investigate the limits of pure energy elasticity and pure entropy elasticity. Lastly, we also analyzed the effect of the jump in wall temperature near the contraction for positive and negative increments.

The remaining sections of the manuscript are organized as follows. In Section 2, the governing equations for the stress tensor and log-conformation tensor-based formulations of non-isothermal viscoelastic flows are presented. Subsequently, in Section 3, the numerical procedure of the block-coupled algorithm will be described in detail, including the finite-volume discretization process for all the implicit terms considered in the governing equations. In Section 4, the validation of the newly-developed numerical algorithm is performed, and in Section 5 the main conclusions of the work are addressed.

2. Governing Equations

In this section, the equations that involve non-isothermal viscoelastic fluid flow problems are presented for both stress-tensor- and log-conformation tensor-based formulations.

2.1. Stress-Tensor-Based Formulation

The governing equations for laminar, incompressible, non-isothermal viscoelastic flows are the conservation of mass and linear momentum, together with a constitutive equation modeling the polymer stress behavior and the energy equation to account for thermodynamical effects.

The conservation of mass and linear momentum equations read as follows:

$$\frac{\partial u_i}{\partial x_i} = 0, \tag{1}$$

$$\rho\left(\frac{\partial u_i}{\partial t} + u_j\frac{\partial u_i}{\partial x_j}\right) + \frac{\partial p}{\partial x_i} - \frac{\partial \tau_{ij}}{\partial x_j} = 0, \tag{2}$$

where Einstein's summation convention applies, u_i are the velocity components along the Cartesian co-ordinates x_i, ρ is the fluid density, t is the time, p is the pressure and τ_{ij} are the components of the total extra-stress tensor ($i,j = 1,2$ for 2D flows), which is split into Newtonian (solvent), $(\tau_N)_{ij}$, and elastic (polymeric), $(\tau_E)_{ij}$, contributions, such that $\tau_{ij} = (\tau_N)_{ij} + (\tau_E)_{ij}$.

The calculation of the stress terms is completed using the following relations:

$$(\tau_N)_{ij} = 2\eta_N(T)\mathrm{D}_{ij} = \eta_N(T)\left(\frac{\partial u_i}{\partial x_j} + \frac{\partial u_j}{\partial x_i}\right), \tag{3}$$

$$\frac{1}{\eta_E(T)}(\mathrm{I}_{ij}+\mathrm{h}((\tau_E)_{ij}))(\tau_E)_{ij}+\frac{\lambda(T)}{\eta_E(T)}\left(\frac{\partial(\tau_E)_{ij}}{\partial t}+u_k\frac{\partial(\tau_E)_{ij}}{\partial x_k}\right)-\left(\frac{\partial u_i}{\partial x_j}+\frac{\partial u_j}{\partial x_i}\right)-$$
$$\frac{\lambda(T)}{\eta_E(T)}\left((\tau_E)_{ik}\frac{\partial u_j}{\partial x_k}+(\tau_E)_{jk}\frac{\partial u_i}{\partial x_k}\right)=0, \tag{4}$$

where $\eta_N(T)$ and $\eta_E(T)$ are the temperature-dependent solvent and polymeric viscosities, respectively, D_{ij} is the strain rate tensor, which describes the rate of stretching and shearing, $\lambda(T)$ is the temperature-dependent polymer relaxation time, I_{ij} is the identity tensor and $\mathrm{h}((\tau_E)_{ij})$ is a tensor that can be given by different expressions, related to the constitutive equation chosen to model the viscoelastic fluid. For the Oldroyd-B model [40], $\mathrm{h}((\tau_E)_{ij}) = 0$. For the Giesekus model [41] $\mathrm{h}((\tau_E)_{ij}) = \frac{\kappa\lambda}{\eta_E}(\tau_E)_{ij}$, where κ is a positive constant, the so-called mobility factor, which is related to the elongational behavior of the fluids. For the Phan–Thien–Tanner (PTT) model [42,43], the tensor h is of the form $\mathrm{h}((\tau_E)_{ij}) = \frac{\epsilon\lambda}{\eta_E}\mathrm{tr}((\tau_E)_{ij})\mathrm{I}_{ij}$, where $\mathrm{tr}((\tau_E)_{ij}) = (\tau_E)_{ii}$ is the trace of the polymeric stress tensor and ϵ is a material parameter called the extensibility factor, related to the fluid behavior in extensional flow. In addition, the Giesekus and PTT models present one more non-linearity, which is given by the product $\mathrm{h}((\tau_E)_{ij})(\tau_E)_{ij}$. This term is responsible for the shear-thinning, the non-zero second normal stress coefficient and the stress overshoot in transient shear flows of viscoelastic fluids. In this work, we will provide a preliminary assessment of the merits of the fully implicit, block-coupled, non-isothermal, viscoelastic, log-conformation tensor-based algorithm for calculations using the Oldroyd-B fluid model, which is commonly used to validate newly-developed viscoelastic codes due to the stress singular behavior near sharp corners or at stagnation points. For these models, a characteristic (solvent) viscosity ratio can be defined by $\beta = \eta_N/(\eta_N+\eta_E) = \eta_N/\eta_0$, known as the retardation ratio, where η_0 is the total viscosity in the limit of vanishing shear rate.

Following the work of Peters and Baaijens [13] the energy balance equation for the case of viscoelastic flows is as follows:

$$\rho C_p\left(\frac{\partial T}{\partial t}+u_i\frac{\partial T}{\partial x_i}\right)-k\frac{\partial^2 T}{\partial x_i^2}=(\tau_N)_{ij}\mathrm{D}_{ji}+\alpha(\tau_E)_{ij}\mathrm{D}_{ji}+(1-\alpha)\frac{(\tau_E)_{ii}}{2\bar{\lambda}(T)}, \tag{5}$$

where k is the thermal conductivity of the fluid, without dependence on temperature T and polymer orientation, C_p is the specific heat capacity, also without temperature and polymer orientation dependence [44], and α is the energy partitioning coefficient. When $\alpha = 1$, all mechanical energy is dissipated as heat (pure entropy elasticity), and if $\alpha = 0$, all mechanical energy is stored as elastic energy (pure elastic material) [13,18]. Habla et al. [18] concluded that the effect of the parameter α is negligible because, with a fully developed shear flow, only stress work occurs, and the internal energy storage is absent. In addition, $\bar{\lambda}(T) = \lambda(T)\left(1+\frac{\lambda(T)\,\epsilon\,(\tau_E)_{ii}}{\eta_E(T)}\right)^{-1}$ for the PTT model and $\bar{\lambda}(T) = \lambda(T)\left(1+\frac{\lambda(T)\,\kappa\,(\tau_E)_{ij}}{\eta_E(T)}\right)^{-1}$ for the Giesekus model. For the Oldroyd-B model calculations considered in the validation section of this work (Section 4), $\bar{\lambda}(T) = \lambda(T)$. The temperature dependencies of the relaxation time, $\lambda(T)$, solvent and polymeric viscosities, $\eta_N(T)$ and $\eta_E(T)$, respectively, are given by

$$\lambda(T) = a_T\lambda(T_0), \tag{6}$$

$$\eta_N(T) = a_T\eta_N(T_0), \tag{7}$$

$$\eta_E(T) = a_T\eta_E(T_0), \tag{8}$$

where T_0 is a reference temperature and a_T is a shift factor obeying the Williams–Landel–Ferry (WLF) relation:

$$a_T = \exp\left(\frac{-C_1(T-T_0)}{C_2+T-T_0}\right), \quad (9)$$

in which C_1 and C_2 are the WLF parameters and T_0 is the reference temperature. Frequently used sets of WLF parameters (C_1, C_2) are (5, 150) for temperatures relatively far from the glass transition temperature T_g, enabling the thermorheological coupling, and (15, 50) for temperatures near T_g [18].

2.2. Log-Conformation Tensor-Based Formulation

In this section, we write the viscoelastic stress tensor-based formulation in terms of the log-conformation tensor variable, which was proposed by Fattal and Kupferman [35] to address the HWNP. For that purpose, the polymeric stress tensor, $(\tau_E)_{ij}$, is related to the conformation tensor, σ_{ij}, by the following equation

$$(\tau_E)_{ij} = \frac{\eta_E(T)}{\lambda(T)}(\sigma_{ij} - \mathrm{I}_{ij}). \quad (10)$$

Subsequently, the conformation tensor σ_{ij} is replaced by its matrix logarithmic $\Psi_{ij} = \log(\sigma_{ij})$, and Equations (2), (4) and (5) are substituted by

$$\rho\left(\frac{\partial u_i}{\partial t} + u_j\frac{\partial u_i}{\partial x_j}\right) + \frac{\partial p}{\partial x_i} - 2\eta_N(T)\frac{\partial \mathrm{D}_{ij}}{\partial x_j} - \frac{\eta_E(T)}{\lambda(T)}\frac{\partial\left(\mathrm{e}^{\Psi_{ij}} - \mathrm{I}_{ij}\right)}{\partial x_j} = 0, \quad (11)$$

$$\frac{1}{\lambda(T)}\left[\left(\mathrm{I}_{ij} + \mathrm{h}(\mathrm{e}^{\Psi_{ij}})\right)(\mathrm{e}^{\Psi_{ij}} - \mathrm{I}_{ij})\right] + \frac{\partial \mathrm{e}^{\Psi_{ij}}}{\partial t} + u_k\frac{\partial \mathrm{e}^{\Psi_{ij}}}{\partial x_k} - \mathrm{e}^{\Psi_{ik}}\frac{\partial u_j}{\partial x_k} - \mathrm{e}^{\Psi_{jk}}\frac{\partial u_i}{\partial x_k} = 0, \quad (12)$$

$$\rho C_p\left(\frac{\partial T}{\partial t} + u_i\frac{\partial T}{\partial x_i}\right) - k\frac{\partial^2 T}{\partial x_i^2} = (\tau_N)_{ij}\mathrm{D}_{ji} + \frac{\eta_E(T)}{\lambda(T)}\left[\alpha(\mathrm{e}^{\Psi_{ij}} - \mathrm{I}_{ij})\mathrm{D}_{ji} + (1-\alpha)\frac{(\mathrm{e}^{\Psi} - \mathrm{I})_{ii}}{2\bar{\lambda}(T)}\right], \quad (13)$$

with $\bar{\lambda}(T) = \lambda(T)\left(1 + \epsilon\,(\mathrm{e}^{\Psi} - \mathrm{I})_{ii}\right)^{-1}$ for the PTT model, $\bar{\lambda}(T) = \lambda(T)\left(1 + \kappa\,(\mathrm{e}^{\Psi} - \mathrm{I})_{ij}\right)^{-1}$ for the Giesekus model and $\bar{\lambda}(T) = \lambda(T)$ for the Oldroyd-B model. Here, $\mathrm{e}^{\Psi_{ij}}$ is the matrix exponential function, defined as $\mathrm{e}^{\Psi_{ij}} = \sum_{m=1}^{d} \mathrm{e}^{\xi_m}\mathbf{P}_m$, with d as the dimension of the physical space ($d = 2$ for the calculations performed in this work), ξ_m the eigenvalues of Ψ_{ij} and $\mathbf{P}_m$ is the projection matrix onto the corresponding eigenspace. Therefore, if $\hat{\mathbf{e}}_m$ is the eigenvector corresponding to ξ_m, then $\mathbf{P}_m = \hat{\mathbf{e}}_m \otimes \hat{\mathbf{e}}_m$ [38]. In addition, note that $\lambda(T_0)$ and $\eta_P(T_0)$ are known values for the reference temperature T_0, and, therefore, the quotient $\frac{\eta_E(T)}{\lambda(T)}$ in Equations (11) and (13) is a constant, considering that λ and η_E scale in the same way [45].

3. Numerical Method

In this section, we will describe a finite volume numerical method to set up a block-coupled solver procedure to simultaneously solve the continuity (Equation (1)), linear momentum (Equation (11)), log-conformation tensor (Equation (12)) and energy (Equation (13)) equations.

Within the framework of the block-coupled solver developed in this work, the advection, pressure gradient, diffusion and log-conformation tensor terms in the conservation of linear momentum equations are implicitly discretized (see Section 3.1). Subsequently, the velocity field term in the conservation of mass equation is also treated in an implicit manner (see Section 3.2). In addition, and as an extension to our previous work [28], where we have discretized implicitly the advection term in the stress constitutive equation, here the rotation and the rate of deformation terms are implicitly discretized (see Section 3.3). Lastly, the advection and diffusion terms in the energy equation are also implicitly discretized (see Section 3.4). The rate of change term in all the equations is implicitly discretized using the backward implicit Euler scheme.

3.1. Discretization of the Equations for Conservation of Linear Momentum

In the framework of the FVM, the discretization process starts by integrating the conservation of linear momentum equations (Equation (11)) over a general control volume (also called *representative volume* or *cell*) V_P, where the subscript P refers to values of the variables at cell with centroid P, as shown in Figure 1, to yield

$$\rho\left(\int_{V_P}\frac{\partial u_i}{\partial t}\,\mathrm{d}V_P+\int_{V_P}u_j\frac{\partial u_i}{\partial x_j}\,\mathrm{d}V_P\right)+\int_{V_P}\frac{\partial p}{\partial x_i}\,\mathrm{d}V_P-\int_{V_P}(\eta_N(T)+\eta^\star)\frac{\partial^2 u_i}{\partial x_j^2}\,\mathrm{d}V_P- \\ -\int_{V_P}\frac{\eta_E(T)}{\lambda(T)}\frac{\partial(\mathrm{e}^{\Psi}-\mathrm{I})_{ij}}{\partial x_j}\,\mathrm{d}V_P=-\int_{V_P}\frac{\partial}{\partial x_j}\left(\eta^\star\frac{\partial u_i}{\partial x_j}\right)\mathrm{d}V_P, \tag{14}$$

where the additional terms involving $\eta^\star$ are related to the improved both-side diffusion technique [46], which can solve the checkerboard pattern due to numerical instabilities caused by a velocity–stress decoupling. Note that we also used the identity $\frac{\partial}{\partial x_j}\left(\frac{\partial u_j}{\partial x_i}\right)=0$.

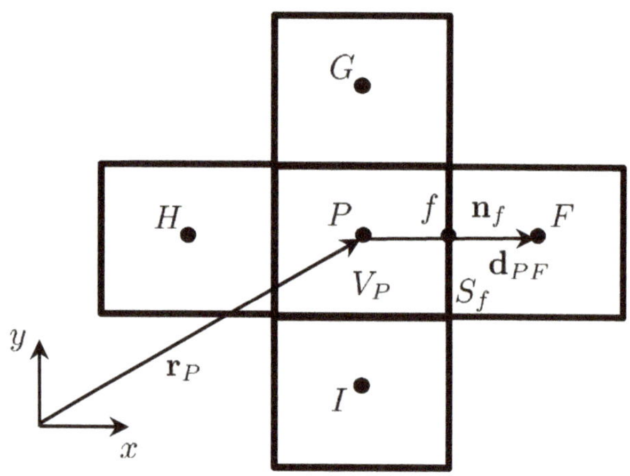

Figure 1. Schematic representation of the control volume V_P with centroid P (owner), with distance vector to the origin $\mathbf{r}_P$, and neighboring control volumes with centroids F, G, H and I. The face shared by the control volumes with centroids P and F is represented by f, with area S_f and face unit normal vector $\mathbf{n}_f$. The distance vector between P and F is represented by $\mathbf{d}_{PF}$.

The following step of the discretization process is to apply the Gauss divergence theorem to transform the volume integrals of the advection, pressure and diffusion terms in Equation (14) into surface integrals as follows:

$$\rho\left(\int_{V_P}\frac{\partial u_i}{\partial t}\,\mathrm{d}V_P+\oint_S n_j(u_j u_i)\,\mathrm{d}S\right)+\oint_S n_i p\,\mathrm{d}S-\oint_S n_j\left[(\eta_N(T)+\eta^\star)\frac{\partial u_i}{\partial x_j}\right]\mathrm{d}S- \\ -\int_{V_P}\frac{\eta_E(T)}{\lambda(T)}\frac{\partial(\mathrm{e}^{\Psi}-\mathrm{I})_{ij}}{\partial x_j}\,\mathrm{d}V_P=-\oint_S n_j\eta^\star\frac{\partial u_i}{\partial x_j}\,\mathrm{d}S, \tag{15}$$

where S is the boundary of control volume V_P and n_i are the components of the outward pointing unit vector normal to S.

Subsequently, a second-order integration scheme is employed to approximate the surface integrals and the following linear momentum semi-discretized equations are obtained:

$$
\begin{aligned}
&V_P\rho_P\frac{(u_i)_P-(u_i)_P^0}{\Delta t}+\sum_{f=nb(P)}(s_j\rho u_j u_i)_f+\sum_{f=nb(P)}(s_i p)_f-\\
&-\sum_{f=nb(P)}\left[s_j(\eta_N(T)+\eta^\star)\frac{\partial u_i}{\partial x_j}\right]_f-\int_{V_P}\frac{\eta_E(T)}{\lambda(T)}\frac{\partial(\mathrm{e}^\Psi-\mathrm{I})_{ij}}{\partial x_j}\,\mathrm{d}V_P\\
&=-\sum_{f=nb(P)}\left(s_j\eta^\star\frac{\partial u_i}{\partial x_j}\right)_f,
\end{aligned}
\tag{16}
$$

where Δt is the time-step, the superscript 0 represents the previous time step value, nb refers to values at the faces f, obtained by interpolation between P and its neighbors, and s_i are the components of the area normal vector to face f.

Finally, the linear momentum semi-discretized equations are transformed into an algebraic linear system of equations by expressing the variation in the dependent variable and its derivatives in terms of the control volume P and its neighbors' values at the respective centroids, such as

$$
\begin{cases}
a_P^{uu}u_P+a_P^{uv}v_P+a_P^{up}p_P+a_P^{u\Psi_{xx}}(\Psi_{xx})_P+a_P^{u\Psi_{xy}}(\Psi_{xy})_P+\\
+\sum_{F=NB(P)}a_F^{uu}u_F+\sum_{F=NB(P)}a_F^{uv}v_F+\sum_{F=NB(P)}a_F^{up}p_F+\\
+\sum_{F=NB(P)}a_F^{u\Psi_{xx}}(\Psi_{xx})_F+\sum_{F=NB(P)}a_F^{u\Psi_{xy}}(\Psi_{xy})_F=b_P^u,\\
a_P^{vu}u_P+a_P^{vv}v_P+a_P^{vp}p_P+a_P^{v\Psi_{xy}}(\Psi_{xy})_P+a_P^{v\Psi_{yy}}(\Psi_{yy})_P+\\
+\sum_{F=NB(P)}a_F^{vu}u_F+\sum_{F=NB(P)}a_F^{vv}v_F+\sum_{F=NB(P)}a_F^{vp}p_F+\\
\sum_{F=NB(P)}a_F^{v\Psi_{xy}}(\Psi_{xy})_F+\sum_{F=NB(P)}a_F^{v\Psi_{yy}}(\Psi_{yy})_F=b_P^v,
\end{cases}
\tag{17}
$$

where $a_P^{u_i\phi}$ and $a_F^{u_i\phi}$ are the owner and neighbor coefficients in the discretized linear momentum equation representing the velocity component u_i and the variable ϕ interactions, respectively; $b_P^{u_i}$ is the source term, where $NB(P)$ refers to the neighbors of the control-volume with centroid P.

For the sake of conciseness, the contributions of the rate of change, advection, pressure gradient, implicit diffusion and explicit diffusion terms shown in Equation (16) can be found in Fernandes et al. [28]. Regarding the contribution of the log-conformation tensor term, $\frac{\partial(\mathrm{e}^\Psi-\mathrm{I})_{ij}}{\partial x_j}$, to the linear momentum equations, we will employ an implicit discretization by considering the following Taylor approximation for the functional $(\mathrm{e}^\Psi-\mathrm{I})_{ij}$ [28,39]

$$
\begin{aligned}
(\mathrm{e}^\Psi-\mathrm{I})_{ij}&\approx(\mathrm{e}^\Psi-\mathrm{I})_{ij}^0+\left.\frac{\partial(\mathrm{e}^\Psi-\mathrm{I})_{ij}}{\partial\Psi_{kl}}\right|_{\Psi_{kl}^0}\left(\Psi_{kl}-\Psi_{kl}^0\right)\\
&=(\mathrm{e}^\Psi-\mathrm{I})_{ij}^0+\sum_{i,j}^{d}\mathrm{e}^{\lambda_m/2}\mathrm{e}^{\lambda_n/2}\frac{\sinh((\lambda_m-\lambda_n)/2)}{(\lambda_m-\lambda_n)/2}\sum_{k,l}p_{ik}^m p_{lj}^n\left(\Psi_{kl}-\Psi_{kl}^0\right),
\end{aligned}
\tag{18}
$$

where the derivative of the functional $(e^{\Psi} - I)_{ij}$ in order to the log-conformation tensor variable is substituted by the expression found in Knechtges [38] for the finite element method, and p_{ik}^{m} and p_{lj}^{n} are the coefficients of the projector matrix belonging to the m-th and n-th eigenvalues (λ_m, λ_n) of Ψ_{ij}^{0}. Therefore, the divergence of the functional $(e^{\Psi} - I)_{ij}$ can be written as:

$$\begin{aligned} &\int_{V_P} \left(\frac{\eta_E(T)}{\lambda(T)} \frac{\partial (e^{\Psi} - I)_{ij}}{\partial x_j} \right) dV_P \\ \approx &\frac{\eta_E(T)}{\lambda(T)} \int_{V_P} \left(\frac{\partial (e^{\Psi} - I)_{ij}}{\partial x_j} \right)^0 dV_P + \frac{\eta_E(T)}{\lambda(T)} \int_{V_P} \partial \left[\left(\frac{\partial (e^{\Psi} - I)_{ij}}{\partial \Psi_{kl}} \right)^0 \Psi_{kl} \right] / \partial x_j \, dV_P \\ &- \frac{\eta_E(T)}{\lambda(T)} \int_{V_P} \partial \left[\left(\frac{\partial (e^{\Psi} - I)_{ij}}{\partial \Psi_{kl}} \right)^0 \Psi_{kl}^0 \right] / \partial x_j \, dV_P, \end{aligned} \tag{19}$$

Subsequently, applying the Gauss divergence theorem we can transform the volume integrals with derivatives into surface integrals

$$\int_{V_P} \partial \left[\left(\frac{\partial (e^{\Psi} - I)_{ij}}{\partial \Psi_{kl}} \right)^0 \Psi_{kl} \right] / \partial x_j \, dV_P = \oint_S n_j \left(\frac{\partial (e^{\Psi} - I)_{ij}}{\partial \Psi_{kl}} \right)^0 \Psi_{kl} \, dS, \tag{20a}$$

$$\int_{V_P} \partial \left[\left(\frac{\partial (e^{\Psi} - I)_{ij}}{\partial \Psi_{kl}} \right)^0 \Psi_{kl}^0 \right] / \partial x_j \, dV_P = \oint_S n_j \left(\frac{\partial (e^{\Psi} - I)_{ij}}{\partial \Psi_{kl}} \right)^0 \Psi_{kl}^0 \, dS, \tag{20b}$$

and obtain the discretized form for the divergence of the functional $(e^{\Psi} - I)_{ij}$ (i.e., the log-conformation tensor term) in the linear momentum equations as follows:

$$\begin{aligned} &\int_{V_P} \left(\frac{\eta_E(T)}{\lambda(T)} \frac{\partial (e^{\Psi} - I)_{ij}}{\partial x_j} \right) dV_P \\ \approx &\frac{\eta_E(T)}{\lambda(T)} \left(\frac{\partial (e^{\Psi} - I)_{ij}}{\partial x_j} \right)^0 V_P + \frac{\eta_E(T)}{\lambda(T)} \sum_{f=nb(P)} \left(s_j \left(\frac{\partial (e^{\Psi} - I)_{ij}}{\partial \Psi_{kl}} \right)^0 \Psi_{kl} \right)_f \\ &- \frac{\eta_E(T)}{\lambda(T)} \sum_{f=nb(P)} \left(s_j \left(\frac{\partial (e^{\Psi} - I)_{ij}}{\partial \Psi_{kl}} \right)^0 \Psi_{kl}^0 \right)_f . \end{aligned} \tag{21}$$

Lastly, the contributions of the divergence of the log-conformation (DLC) tensor term for the linear momentum equations are given by

$$a_{F,DLC}^{u_j \psi_{kl}} = -\frac{\eta_E(T)}{\lambda(T)} \left[s_j \left(\frac{\partial (e^{\Psi} - I)_{ij}}{\partial \Psi_{kl}} \right)^0 \right]_f (1 - w_f), \tag{22a}$$

$$a_{P,DLC}^{u_j \psi_{kl}} = -\frac{\eta_E(T)}{\lambda(T)} \sum_{f=nb(P)} \left[s_j \left(\frac{\partial (e^{\Psi} - I)_{ij}}{\partial \Psi_{kl}} \right)^0 \right]_f w_f, \tag{22b}$$

$$b_{P,DLC}^{u_j} = -\frac{\eta_E(T)}{\lambda(T)} \left[\left(\frac{\partial (e^{\Psi} - I)_{ij}}{\partial x_j} \right)^0 V_P - \sum_{f=nb(P)} \left(s_j \left(\frac{\partial (e^{\Psi} - I)_{ij}}{\partial \Psi_{kl}} \right)^0 \Psi_{kl}^0 \right)_f \right], \tag{22c}$$

where w_f are the interpolation weights.

It should be noted that the total contribution for the owner and neighbor coefficients related to the linear momentum equations' interactions is given by the sum of the rate of change, advection, pressure gradient, implicit diffusion and divergence of log-conformation tensor terms. In addition, the total contribution for the explicit coefficient related to the linear momentum equations is given by the sum of the rate of change, explicit diffusion and divergence of log-conformation tensor terms.

3.2. Discretization of the Equation for Conservation of Mass

Following the same steps as presented in Section 3.1, we begin the discretization of the continuity equation, Equation (1), with the integration over the control volume V_P as follows:

$$\int_{V_P} \frac{\partial u_i}{\partial x_i}\, \mathrm{d}V_P = 0. \tag{23}$$

Subsequently, by employing the divergence theorem, the volume integral is transformed into a surface integral as follows:

$$\oint_S n_i u_i \,\mathrm{d}S = 0. \tag{24}$$

Then, by applying a second-order integration scheme to approximate the surface integral, we can write the semi-discretized form of the continuity equation as:

$$\sum_{f=nb(P)} (s_i u_i)_f = 0. \tag{25}$$

In a collocated framework, the velocity at the face is obtained by reconstructing a pseudo-momentum equation at the face from the linear momentum equations of the straddling cells P and F, known as the Rhie–Chow interpolation [47]. For the sake of conciseness, the derivation of the discretized form for the equation of conservation of mass will not be given in detail, but it can be found in our previous work [28], which reads as

$$\sum_{f=nb(P)} \left[s_i \left(-\frac{\partial p}{\partial x_i} \overline{D}_i \right) \right]_f + \sum_{f=nb(P)} (s_i \overline{u}_i)_f = \sum_{f=nb(P)} \left[s_i \left(-\overline{\frac{\partial p}{\partial x_i}} \overline{D}_i \right) \right]_f. \tag{26}$$

Finally, we can write the mass conservation algebraic equation as:

$$a_P^{pp} p_P + a_P^{pu} u_P + a_P^{pv} v_P + \sum_{F=NB(P)} a_F^{pp} p_F + \sum_{F=NB(P)} a_F^{pu} u_F + \sum_{F=NB(P)} a_F^{pv} v_F = b_P^p, \tag{27}$$

where $a_P^{p\phi}$ and $a_F^{p\phi}$ are the owner and neighbor coefficients in the discretized mass conservation equation representing the pressure field and the variable ϕ interactions, respectively; b_P^p is the source term.

The implicit pressure gradient term is discretized (see page 86 of [48]) as follows:

$$\left[s_i \left(-\frac{\partial p}{\partial x_i} \overline{D}_i \right) \right]_f = -\frac{\left[s_i (s_i \overline{D}_i) \right]_f}{(d_{PF})_i (s_i)_f} (p_F - p_P), \tag{28}$$

where the coefficients of the implicit pressure gradient term for the mass conservation equation are given by

$$a_F^{pp} = -\frac{\left[s_i(s_i\overline{D}_i)\right]_f}{(d_{PF})_i(s_i)_f}, \tag{29a}$$

$$a_P^{pp} = -\sum_{F=NB(P)} a_F^{pp}. \tag{29b}$$

The implicit coefficients for the second term in Equation (26) (corresponding to the velocity contribution) are given by

$$a_F^{pu} = s_f^x(1-w_f), \qquad a_F^{pv} = s_f^y(1-w_f), \tag{30a}$$

$$a_P^{pu} = \sum_{f=nb(P)} s_f^x w_f, \qquad a_P^{pv} = \sum_{f=nb(P)} s_f^y w_f. \tag{30b}$$

Lastly, the coefficients of the explicit pressure gradient term contribution for the mass conservation equation are given by

$$b_P^p = \sum_{f=nb(P)} \left[s_i \left(-\overline{\frac{\partial p}{\partial x_i}} \overline{D}_i \right) \right]_f. \tag{31}$$

3.3. Discretization of the Log-Conformation Tensor Constitutive Equations

The constitutive equations, Equation (4), can be written, without loss of generality, for a Giesekus fluid model by (see Theorem A.42 in [38])

$$\begin{aligned} &\frac{\partial \Psi_{ij}}{\partial t} + u_k \frac{\partial \Psi_{ij}}{\partial x_k} + [\Psi_{ij}, \Omega_{ij}] - 2f(\mathrm{ad}(\Psi_{ij}))\mathrm{D}_{ij} \\ &= -\frac{1}{\lambda(T)}\left(\mathrm{I}_{ij} + \kappa\left(\mathrm{e}^{\Psi_{ij}} - \mathrm{I}_{ij}\right)\right)\left(\mathrm{e}^{\Psi_{ij}} - \mathrm{I}_{ij}\right)\mathrm{e}^{-\Psi_{ij}}, \end{aligned} \tag{32}$$

where $\Omega_{ij} = \left(\frac{\partial u_i}{\partial x_j} - \frac{\partial u_j}{\partial x_i}\right)/2$ is the vorticity tensor and $[\Psi_{ij}, \Omega_{ij}] = \Psi_{ij}\Omega_{ij} - \Omega_{ij}\Psi_{ij}$ is the commutator term. Following Knechtges [38], $f(\mathrm{ad}(\Psi_{ij}))$ is defined by the application of the function $f(x) = \frac{x/2}{\tanh(x/2)}$ to the adjoint operator $\mathrm{ad}(\Psi_{ij})$, such as

$$f(\mathrm{ad}(\Psi_{ij}))y = \sum_{m,n=1}^{d} f(\lambda_m - \lambda_n)\mathbf{P}_m y \mathbf{P}_n, \tag{33}$$

where y is a symmetric matrix satisfying $\mathrm{ad}(\Psi_{ij})y = [\Psi_{ij}, y]$, and y is continuously differentiable (C^1). Notice that, Equation (32) simplifies to the constitutive equation for an Oldroyd-B fluid model when $\kappa = 0$.

The discretization of the log-conformation tensor constitutive equations, Equation (32), starts with the integration over the control volume V_P to yield

$$\begin{aligned} &\int_{V_P} \frac{\partial \Psi_{ij}}{\partial t}\,\mathrm{d}V_P + \int_{V_P} u_k \frac{\partial \Psi_{ij}}{\partial x_k}\,\mathrm{d}V_P + \int_{V_P} [\Psi_{ij}, \Omega_{ij}]\,\mathrm{d}V_P - \int_{V_P} 2f(\mathrm{ad}(\Psi_{ij}))\mathrm{D}_{ij}\,\mathrm{d}V_P = \\ &\quad -\int_{V_P} \frac{1}{\lambda(T)}\left(\mathrm{I}_{ij} + \kappa\left(\mathrm{e}^{\Psi_{ij}} - \mathrm{I}_{ij}\right)\right)\left(\mathrm{e}^{\Psi_{ij}} - \mathrm{I}_{ij}\right)\mathrm{e}^{-\Psi_{ij}}\,\mathrm{d}V_P. \end{aligned} \tag{34}$$

This leads to the algebraic equation with the following form:

$$\begin{cases} a_P^{\Psi_{xx}\Psi_{xx}}(\Psi_{xx})_P + a_P^{\Psi_{xx}u}u_P + a_P^{\Psi_{xx}v}v_P + \sum\limits_{F=NB(P)} a_F^{\Psi_{xx}\Psi_{xx}}(\Psi_{xx})_F + \\ + \sum\limits_{F=NB(P)} a_F^{\Psi_{xx}u}u_F + \sum\limits_{F=NB(P)} a_F^{\Psi_{xx}v}v_F = b_P^{\Psi_{xx}}, \\ \\ a_P^{\Psi_{xy}\Psi_{xy}}(\Psi_{xy})_P + a_P^{\Psi_{xy}u}u_P + a_P^{\Psi_{xy}v}v_P + \sum\limits_{F=NB(P)} a_F^{\Psi_{xy}\Psi_{xy}}(\Psi_{xy})_F + \\ + \sum\limits_{F=NB(P)} a_F^{\Psi_{xy}u}u_F + \sum\limits_{F=NB(P)} a_F^{\Psi_{xy}v}v_F = b_P^{\Psi_{xy}}, \\ \\ a_P^{\Psi_{yy}\Psi_{yy}}(\Psi_{yy})_P + a_P^{\Psi_{yy}u}u_P + a_P^{\Psi_{yy}v}v_P + \sum\limits_{F=NB(P)} a_F^{\Psi_{yy}\Psi_{yy}}(\Psi_{yy})_F + \\ + \sum\limits_{F=NB(P)} a_F^{\Psi_{yy}u}u_F + \sum\limits_{F=NB(P)} a_F^{\Psi_{yy}v}v_F = b_P^{\Psi_{yy}}, \end{cases} \tag{35}$$

where $a_P^{\Psi_{ij}\phi}$ and $a_F^{\Psi_{ij}\phi}$ are the owner and neighbor coefficients in the discretized log-conformation tensor constitutive equations representing the tensor component Ψ_{ij} and the variable ϕ interactions, respectively, and $b_P^{\Psi_{ij}}$ is the source term.

Again, for the sake of conciseness, the discretization of the rate of change $\frac{\partial \Psi_{ij}}{\partial t}$ and advective $u_k \frac{\partial (\Psi_{ij})}{\partial x_k}$ terms in Equation (32) are not detailed here because similar discretizations were performed for the polymeric stress-tensor $(\tau_E)_{ij}$ (see previous work [28]).

Regarding the commutator term $[\Psi_{ij}, \Omega_{ij}]$, we can write the following expanded form:

$$[\Psi_{ij}, \Omega_{ij}] = \frac{1}{2}\left(\Psi_{ik}\frac{\partial u_k}{\partial x_j} - \Psi_{ik}\frac{\partial u_j}{\partial x_k} - \frac{\partial u_i}{\partial x_k}\Psi_{kj} + \frac{\partial u_k}{\partial x_i}\Psi_{kj} \right), \tag{36}$$

and, subsequently, we can use Taylor approximations such as

$$\begin{aligned} & -\frac{1}{2}\left(\Psi_{ik}\frac{\partial u_j}{\partial x_k} + \frac{\partial u_i}{\partial x_k}\Psi_{kj} \right) \\ & \approx +\frac{1}{2}\left[\Psi_{ik}^0 \left(\frac{\partial u_j}{\partial x_k} \right)^0 + \left(\frac{\partial u_i}{\partial x_k} \right)^0 \Psi_{kj}^0 \right] - \frac{1}{2}\left[\Psi_{ik} \left(\frac{\partial u_j}{\partial x_k} \right)^0 + \left(\frac{\partial u_i}{\partial x_k} \right)^0 \Psi_{kj} \right] \\ & -\frac{1}{2}\left[\Psi_{ik}^0 \frac{\partial u_j}{\partial x_k} + \frac{\partial u_i}{\partial x_k}\Psi_{kj}^0 \right], \end{aligned} \tag{37}$$

and

$$\begin{aligned} & +\frac{1}{2}\left(\Psi_{ik}\frac{\partial u_k}{\partial x_j} + \frac{\partial u_k}{\partial x_i}\Psi_{kj} \right) \\ & \approx -\frac{1}{2}\left[\Psi_{ik}^0 \left(\frac{\partial u_k}{\partial x_j} \right)^0 + \left(\frac{\partial u_k}{\partial x_i} \right)^0 \Psi_{kj}^0 \right] + \frac{1}{2}\left[\Psi_{ik} \left(\frac{\partial u_k}{\partial x_j} \right)^0 + \left(\frac{\partial u_k}{\partial x_i} \right)^0 \Psi_{kj} \right] \\ & +\frac{1}{2}\left[\Psi_{ik}^0 \frac{\partial u_k}{\partial x_j} + \frac{\partial u_k}{\partial x_i}\Psi_{kj}^0 \right]. \end{aligned} \tag{38}$$

Starting with the terms in Equation (37), the negative commutator terms, the first contribution is explicit, being given by

$$(b_P^{\Psi_{ij}})_{neg} = -\frac{1}{2}V_P\left[\Psi_{ik}^0\left(\frac{\partial u_j}{\partial x_k}\right)^0 + \left(\frac{\partial u_i}{\partial x_k}\right)^0\Psi_{kj}^0\right]. \tag{39}$$

The second contribution is implicit in Ψ_{ij} and explicit in $\frac{\partial u_i}{\partial x_j}$, being given by

$$(a_P^{\Psi_{ij}\Psi_{ik}})_{neg,1} = -\frac{1}{2}V_P\left(\frac{\partial u_j}{\partial x_k}\right)^0, \tag{40a}$$

$$(a_P^{\Psi_{ij}\Psi_{kj}})_{neg,2} = -\frac{1}{2}V_P\left(\frac{\partial u_i}{\partial x_k}\right)^0. \tag{40b}$$

The third contribution is implicit in $\frac{\partial u_i}{\partial x_j}$ and explicit in Ψ_{ij}; therefore, we need an implicit discretization of the gradient operator for velocity, which requires the integration by parts of this term, such as:

$$\begin{aligned}&\int_{V_P}\left(\Psi_{ik}^0\frac{\partial u_j}{\partial x_k} + \frac{\partial u_i}{\partial x_k}\Psi_{kj}^0\right)\mathrm{d}V_P\\ &= \oint_S n_k\left(\Psi_{ik}^0 u_j + u_i\Psi_{kj}^0\right)\mathrm{d}S - \int_{V_P}\left[\left(\frac{\partial \Psi_{ik}}{\partial x_k}\right)^0 u_j + u_i\left(\frac{\partial \Psi_{kj}}{\partial x_k}\right)^0\right]\mathrm{d}V_P.\end{aligned} \tag{41}$$

Applying the Gauss divergence theorem, the discretized form of the terms on the right-hand side of Equation (41) is

$$\sum_{f=nb(P)}\left[s_k\left(\Psi_{ik}^0 u_j + u_i\Psi_{kj}^0\right)\right]_f - V_P\left[\left(\frac{\partial \Psi_{ik}}{\partial x_k}\right)^0 u_j + u_i\left(\frac{\partial \Psi_{kj}}{\partial x_k}\right)^0\right]. \tag{42}$$

Using linear weighted interpolation, we can write the contributions of the third term as:

$$(a_F^{\Psi_{ij}u_j})_{neg,1} = -\frac{1}{2}(s_k\Psi_{ik}^0)_f(1-w_f), \tag{43a}$$

$$(a_P^{\Psi_{ij}u_j})_{neg,1} = -\frac{1}{2}\left[\sum_{f=nb(P)}(s_k\Psi_{ik}^0)_f w_f - V_P\left(\frac{\partial \Psi_{ik}}{\partial x_k}\right)^0\right], \tag{43b}$$

$$(a_F^{\Psi_{ij}u_i})_{neg,2} = -\frac{1}{2}(\Psi_{kb}^0\mathbf{s}_k)_f(1-w_f), \tag{43c}$$

$$(a_P^{\Psi_{ij}u_i})_{neg,2} = -\frac{1}{2}\left[\sum_{f=nb(P)}(s_k\Psi_{kj}^0)_f w_f - V_P\left(\frac{\partial \Psi_{kj}}{\partial x_k}\right)^0\right]. \tag{43d}$$

Following the same reasoning as given above, the terms in Equation (38), the positive commutator terms, generate the following coefficients:

$$(b_P^{\Psi_{ij}})_{pos} = \frac{1}{2}V_P\left[\Psi_{ik}^0\left(\frac{\partial u_k}{\partial x_j}\right)^0 + \left(\frac{\partial u_k}{\partial x_i}\right)^0\Psi_{kj}^0\right], \tag{44}$$

$$(a_P^{\Psi_{ij}\Psi_{ik}})_{pos,1} = \frac{1}{2} V_P \left(\frac{\partial u_k}{\partial x_j}\right)^0, \tag{45a}$$

$$(a_P^{\Psi_{ij}\Psi_{kj}})_{pos,2} = \frac{1}{2} V_P \left(\frac{\partial u_k}{\partial x_i}\right)^0, \tag{45b}$$

$$(a_F^{\Psi_{ij}u_k})_{pos,1} = \frac{1}{2}(s_j \Psi_{ik}^0)_f (1 - w_f), \tag{46a}$$

$$(a_P^{\Psi_{ij}u_k})_{pos,1} = \frac{1}{2}\left[\sum_{f=nb(P)} (s_j \Psi_{ik}^0)_f w_f - V_P \left(\frac{\partial \Psi_{ik}}{\partial x_j}\right)^0\right], \tag{46b}$$

$$(a_F^{\Psi_{ij}u_k})_{pos,2} = \frac{1}{2}(s_i \Psi_{kj}^0)_f (1 - w_f), \tag{46c}$$

$$(a_P^{\Psi_{ij}u_k})_{pos,2} = \frac{1}{2}\left[\sum_{f=nb(P)} (s_i \Psi_{kj}^0)_f w_f - V_P \left(\frac{\partial \Psi_{kj}}{\partial x_i}\right)^0\right]. \tag{46d}$$

Lastly, we implicitly discretize the adjoint operator $f(\mathrm{ad}(\Psi))_{ijkl}\mathrm{D}_{kl}$ by considering the following Taylor approximation [39]

$$\begin{aligned} & f(\mathrm{ad}(\Psi))_{ijkl}\mathrm{D}_{kl} \\ \approx & f(\mathrm{ad}(\Psi^0))_{ijkl}\mathrm{D}_{kl}^0 + \sum_{kl} \frac{\partial\left(f(\mathrm{ad}(\Psi^0))_{iqrj}\mathrm{D}_{qr}^0\right)}{\partial \Psi_{kl}} (\Psi_{kl} - \Psi_{kl}^0) \\ & + \sum_{kl} \frac{\partial\left(f(\mathrm{ad}(\Psi^0))_{iqrj}\mathrm{D}_{qr}^0\right)}{\partial \mathrm{D}_{kl}} (\mathrm{D}_{kl} - \mathrm{D}_{kl}^0), \end{aligned} \tag{47}$$

where $\frac{\partial\left(f(\mathrm{ad}(\Psi^0))_{iqrj}\mathrm{D}_{qr}^0\right)}{\partial \mathrm{D}_{kl}}$ is the derivative of the adjoint operator with respect to D_{kl}, given by [38]

$$\begin{aligned} & \frac{\partial\left(f(\mathrm{ad}(\Psi^0))_{iqrj}\mathrm{D}_{qr}\right)}{\partial \mathrm{D}_{kl}} \\ = & f(\mathrm{ad}(\Psi^0))_{ijkl} \\ = & \sum_{m,n=1}^{d} f(\lambda_m - \lambda_n) p_{ik}^m p_{lj}^n, \end{aligned} \tag{48}$$

and $\frac{\partial\left(f(\mathrm{ad}(\Psi^0))_{iqrj}\mathrm{D}_{qr}^0\right)}{\partial \Psi_{kl}}$ is the derivative of the adjoint operator with respect to Ψ_{kl}, given by [49]

$$\begin{aligned} & \frac{\partial\left(f(\mathrm{ad}(\Psi^0))_{iqrj}\mathrm{D}_{qr}^0\right)}{\partial \Psi_{kl}} \\ = & \sum_{m,n,o=1}^{d} \frac{f(\lambda_m - \lambda_o) - f(\lambda_n - \lambda_o)}{\lambda_m - \lambda_n} \left(p_{ik}^m p_{lq}^n \mathrm{D}_{qr} p_{rj}^o + p_{iq}^o \mathrm{D}_{qr} p_{rk}^n p_{lj}^m\right). \end{aligned} \tag{49}$$

Note that, if λ_m is equal to λ_n, then the denominator of Equation (49) needs to be replaced by $f'(\lambda_m - \lambda_n)$ [38].

Subsequently, we integrate the adjoint term over a control volume V_P as follows:

$$\int_{V_P} 2f(\mathrm{ad}(\Psi))_{ijkl}\mathrm{D}_{kl}\,\mathrm{d}V_P$$
$$\approx \int_{V_P} 2f(\mathrm{ad}(\Psi^0))_{ijkl}\mathrm{D}^0_{kl}\,\mathrm{d}V_P + \int_{V_P} 2\left[\frac{\partial\left(f(\mathrm{ad}(\Psi))_{iqrj}\mathrm{D}_{qr}\right)}{\partial\Psi_{kl}}\right]^0 \Psi_{kl}\,\mathrm{d}V_P$$
$$- \int_{V_P} 2\left[\frac{\partial\left(f(\mathrm{ad}(\Psi))_{iqrj}\mathrm{D}_{qr}\right)}{\partial\Psi_{kl}}\right]^0 \Psi^0_{kl}\,\mathrm{d}V_P + \int_{V_P} 2\left[\frac{\partial\left(f(\mathrm{ad}(\Psi))_{iqrj}\mathrm{D}_{qr}\right)}{\partial\mathrm{D}_{kl}}\right]^0 \mathrm{D}_{kl}\,\mathrm{d}V_P \tag{50}$$
$$- \int_{V_P} 2\left[\frac{\partial\left(f(\mathrm{ad}(\Psi))_{iqrj}\mathrm{D}_{qr}\right)}{\partial\mathrm{D}_{kl}}\right]^0 \mathrm{D}^0_{kl}\,\mathrm{d}V_P,$$

and substituting $\mathrm{D}_{kl} = \frac{1}{2}\left(\frac{\partial u_k}{\partial x_l} + \frac{\partial u_l}{\partial x_k}\right)$ in Equation (50), we obtain

$$\int_{V_P} 2f(\mathrm{ad}(\Psi))_{ijkl}\mathrm{D}_{kl}\,\mathrm{d}V_P$$
$$\approx \int_{V_P} 2f(\mathrm{ad}(\Psi^0))_{ijkl}\mathrm{D}^0_{kl}\,\mathrm{d}V_P + \int_{V_P} 2\left[\frac{\partial\left(f(\mathrm{ad}(\Psi))_{iqrj}\mathrm{D}_{qr}\right)}{\partial\Psi_{kl}}\right]^0 \Psi_{kl}\,\mathrm{d}V_P$$
$$- \int_{V_P} 2\left[\frac{\partial\left(f(\mathrm{ad}(\Psi))_{iqrj}\mathrm{D}_{qr}\right)}{\partial\Psi_{kl}}\right]^0 \Psi^0_{kl}\,\mathrm{d}V_P \tag{51}$$
$$+ \int_{V_P} \left[\frac{\partial\left(f(\mathrm{ad}(\Psi))_{iqrj}\mathrm{D}_{qr}\right)}{\partial\mathrm{D}_{kl}}\right]^0 \left(\frac{\partial u_k}{\partial x_l} + \frac{\partial u_l}{\partial x_k}\right)\mathrm{d}V_P$$
$$- \int_{V_P} 2\left[\frac{\partial\left(f(\mathrm{ad}(\Psi))_{iqrj}\mathrm{D}_{qr}\right)}{\partial\mathrm{D}_{kl}}\right]^0 \mathrm{D}^0_{kl}\,\mathrm{d}V_P.$$

As the fourth term on the right hand side of Equation (51) contains implicit velocity gradients, we employ integration by parts to linearize them, obtaining

$$\int_{V_P} 2f(\mathrm{ad}(\Psi))_{ijkl}\mathrm{D}_{kl}\,\mathrm{d}V_P$$
$$\approx \int_{V_P} 2f(\mathrm{ad}(\Psi^0))_{ijkl}\mathrm{D}^0_{kl}\,\mathrm{d}V_P + \int_{V_P} 2\left[\frac{\partial\left(f(\mathrm{ad}(\Psi))_{iqrj}\mathrm{D}_{qr}\right)}{\partial\Psi_{kl}}\right]^0 \Psi_{kl}\,\mathrm{d}V_P$$
$$- \int_{V_P} 2\left[\frac{\partial\left(f(\mathrm{ad}(\Psi))_{iqrj}\mathrm{D}_{qr}\right)}{\partial\Psi_{kl}}\right]^0 \Psi^0_{kl}\,\mathrm{d}V_P$$
$$+ \oint_{S} \left[\frac{\partial\left(f(\mathrm{ad}(\Psi))_{iqrj}\mathrm{D}_{qr}\right)}{\partial\mathrm{D}_{kl}}\right]^0 (u_k n_l + u_l n_k)\,\mathrm{d}S \tag{52}$$
$$- \int_{V_P} \left[\frac{\partial\left[\frac{\partial\left(f(\mathrm{ad}(\Psi))_{iqrj}\mathrm{D}_{qr}\right)}{\partial\mathrm{D}_{kl}}\right]^0}{\partial x_l} u_k + \frac{\partial\left[\frac{\partial\left(f(\mathrm{ad}(\Psi))_{iqrj}\mathrm{D}_{qr}\right)}{\partial\mathrm{D}_{kl}}\right]^0}{\partial x_k} u_l\right]\mathrm{d}V_P$$
$$- \int_{V_P} 2\left[\frac{\partial\left(f(\mathrm{ad}(\Psi))_{iqrj}\mathrm{D}_{qr}\right)}{\partial\mathrm{D}_{kl}}\right]^0 \mathrm{D}^0_{kl}\,\mathrm{d}V_P.$$

Thus, the discretized form of Equation (52) can then be written as:

$$
\begin{aligned}
&\int_{V_P} 2f(\mathrm{ad}(\Psi))_{ijkl}\mathrm{D}_{kl}\,\mathrm{d}V_P \\
&\approx 2f(\mathrm{ad}(\Psi^0))_{ijkl}\mathrm{D}^0_{kl}V_P + 2\left[\frac{\partial\big(f(\mathrm{ad}(\Psi))_{iqrj}\mathrm{D}_{qr}\big)}{\partial\Psi_{kl}}\right]^0\Psi_{kl}V_P \\
&-2\left[\frac{\partial\big(f(\mathrm{ad}(\Psi))_{iqrj}\mathrm{D}_{qr}\big)}{\partial\Psi_{kl}}\right]^0\Psi^0_{kl}V_P + \sum_{f=nb(P)}\left[\left[\frac{\partial\big(f(\mathrm{ad}(\Psi))_{iqrj}\mathrm{D}_{qr}\big)}{\partial\mathrm{D}_{kl}}\right]^0(u_k s_l + u_l s_k)\right]_f \\
&-\left[\frac{\partial\left[\frac{\partial(f(\mathrm{ad}(\Psi))_{iqrj}\mathrm{D}_{qr})}{\partial\mathrm{D}_{kl}}\right]^0}{\partial x_l}u_k + \frac{\partial\left[\frac{\partial(f(\mathrm{ad}(\Psi))_{iqrj}\mathrm{D}_{qr})}{\partial\mathrm{D}_{kl}}\right]^0}{\partial x_k}u_l\right]V_P \\
&-2\left[\frac{\partial\big(f(\mathrm{ad}(\Psi))_{iqrj}\mathrm{D}_{qr}\big)}{\partial\mathrm{D}_{kl}}\right]^0\mathrm{D}^0_{kl}V_P.
\end{aligned}
\tag{53}
$$

The contributions of the adjoint term for the log-conformation tensor constitutive equations, using linear weighted interpolation, read as

$$
(a_P^{\Psi_{ij}\Psi_{kl}})_{adj} = 2\left[\frac{\partial\big(f(\mathrm{ad}(\Psi))_{iqrj}\mathrm{D}_{qr}\big)}{\partial\Psi_{kl}}\right]^0 V_P, \tag{54a}
$$

$$
(a_F^{\Psi_{ij}u_k})_{adj} = \left[\left[\frac{\partial\big(f(\mathrm{ad}(\Psi))_{iqrj}\mathrm{D}_{qr}\big)}{\partial\mathrm{D}_{kl}}\right]^0 s_l\right]_f (1-w_f), \tag{54b}
$$

$$
(a_P^{\Psi_{ij}u_k})_{adj} = \sum_{f=nb(P)}\left[\left[\frac{\partial\big(f(\mathrm{ad}(\Psi))_{iqrj}\mathrm{D}_{qr}\big)}{\partial\mathrm{D}_{kl}}\right]^0 s_l\right]_f w_f - \frac{\partial\left[\frac{\partial(f(\mathrm{ad}(\Psi))_{iqrj}\mathrm{D}_{qr})}{\partial\mathrm{D}_{kl}}\right]^0}{\partial x_l}V_P, \tag{54c}
$$

$$
(a_F^{\Psi_{ij}u_l})_{adj} = \left[\left[\frac{\partial\big(f(\mathrm{ad}(\Psi))_{iqrj}\mathrm{D}_{qr}\big)}{\partial\mathrm{D}_{kl}}\right]^0 s_k\right]_f (1-w_f), \tag{54d}
$$

$$
(a_P^{\Psi_{ij}u_l})_{adj} = \sum_{f=nb(P)}\left[\left[\frac{\partial\big(f(\mathrm{ad}(\Psi))_{iqrj}\mathrm{D}_{qr}\big)}{\partial\mathrm{D}_{kl}}\right]^0 s_k\right]_f w_f - \frac{\partial\left[\frac{\partial(f(\mathrm{ad}(\Psi))_{iqrj}\mathrm{D}_{qr})}{\partial\mathrm{D}_{kl}}\right]^0}{\partial x_k}V_P, \tag{54e}
$$

$$
\begin{aligned}
(b_P^{\Psi_{ij}})_{adj} = {}& 2f(\mathrm{ad}(\Psi^0))_{ijkl}D^0_{kl}V_P - 2\left[\frac{\partial\big(f(\mathrm{ad}(\Psi))_{iqrj}\mathrm{D}_{qr}\big)}{\partial\Psi_{kl}}\right]^0\Psi^0_{kl}V_P \\
&-2\left[\frac{\partial\big(f(\mathrm{ad}(\Psi))_{iqrj}\mathrm{D}_{qr}\big)}{\partial\mathrm{D}_{kl}}\right]^0\mathrm{D}^0_{kl}V_P.
\end{aligned}
\tag{54f}
$$

Again, it should be noted that the total contribution of the owner and neighbor coefficients related to the log-conformation tensor components' interactions is given by the sum of the log-conformation tensor, rate of change, advection, commutator and adjoint terms. In addition, the total contribution for the explicit coefficient related to the log-conformation tensor constitutive equations is given by the sum of the rate of change, commutator and adjoint terms.

3.4. Discretization of the Equation for Conservation of Energy

The discretization starts by integrating the equation for the conservation of energy (Equation (13)) over a general control volume V_P, to yield

$$\begin{aligned}\rho C_p\left(\int_{V_P}\frac{\partial T}{\partial t}\,\mathrm{d}V_P+\int_{V_P}u_i\frac{\partial T}{\partial x_i}\,\mathrm{d}V_P\right)-\int_{V_P}k\frac{\partial^2 T}{\partial x_i^2}\,\mathrm{d}V_P\\=\int_{V_P}\left[(\tau_N)_{ij}\mathrm{D}_{ji}+\frac{\eta_E(T)}{\lambda(T)}\left(\alpha(\mathrm{e}^{\Psi}-\mathrm{I})_{ij}\mathrm{D}_{ji}+(1-\alpha)\frac{(\mathrm{e}^{\Psi}-\mathrm{I})_{ii}}{2\bar{\lambda}(T)}\right)\right]\mathrm{d}V_P.\end{aligned}\tag{55}$$

Using the Gauss divergence theorem, the volume integrals of the advection and diffusion terms in Equation (55) are transformed into surface integrals as:

$$\begin{aligned}\rho C_p\left(\int_{V_P}\frac{\partial T}{\partial t}\,\mathrm{d}V_P+\oint_S n_i u_i T\,\mathrm{d}S\right)-\oint_S n_i\left(k\frac{\partial T}{\partial x_i}\right)\mathrm{d}S\\=\int_{V_P}\left[(\tau_N)_{ij}\mathrm{D}_{ji}+\frac{\eta_E(T)}{\lambda(T)}\left(\alpha(\mathrm{e}^{\Psi}-\mathrm{I})_{ij}\mathrm{D}_{ji}+(1-\alpha)\frac{(\mathrm{e}^{\Psi}-\mathrm{I})_{ii}}{2\bar{\lambda}(T)}\right)\right]\mathrm{d}V_P.\end{aligned}\tag{56}$$

The semi-discretized equation for the conservation of energy is obtained by evaluating the surface integrals using a second-order integration scheme and approximating the rate of change term with a backward implicit Euler scheme, such as

$$\begin{aligned}V_P\rho_P(C_p)_P\frac{T_P-T_P^0}{\Delta t}+\sum_{f=nb(P)}(s_i\rho C_p u_i T)_f-\sum_{f=nb(P)}\left(s_i k\frac{\partial T}{\partial x_i}\right)_f\\=V_P\left[(\tau_N)_{ij}\mathrm{D}_{ji}+\frac{\eta_E(T)}{\lambda(T)}\left(\alpha(\mathrm{e}^{\Psi}-\mathrm{I})_{ij}\mathrm{D}_{ji}+(1-\alpha)\frac{(\mathrm{e}^{\Psi}-\mathrm{I})_{ii}}{2\bar{\lambda}(T)}\right)\right].\end{aligned}\tag{57}$$

This leads to the algebraic equation for the energy balance with the following form:

$$a_P^{TT}T_P+a_P^{Tu}u_P+a_P^{Tv}v_P+\sum_{F=NB(P)}a_F^{TT}T_F+\sum_{F=NB(P)}a_F^{Tu}u_F+\sum_{F=NB(P)}a_F^{Tv}v_F=b_P^T,\tag{58}$$

where $a_P^{T\phi}$ and $a_F^{T\phi}$ are the owner and neighbor coefficients in the discretized conservation of energy equation, representing the temperature T and the variable ϕ interactions, respectively, and b_P^T is the source term.

The rate of change ($rchg$) term in Equation (57) (first term), contributes to both the diagonal of the system of equations and to the explicit term as:

$$a_{P,rchg}^{TT}=\frac{V_P\,\rho_P\,(C_p)_P}{\Delta t},\tag{59a}$$

$$b_{P,rchg}^{T}=\frac{V_P\,\rho_P\,(C_p)_P T_P^0}{\Delta t}.\tag{59b}$$

Then, and in the framework of the FVM, the advection term in Equation (57) (second term) is linearized by computing the mass flow rate at control-volume face f ($\dot{m}_f=(s_i\rho u_i)_f$) using the previous iteration values. Here, we used the UDS differentiating scheme to approximate the advection term. However, many high-order schemes could be used, such as the MINMOD or CUBISTA schemes. For the sake of readability, the discretization procedure will be presented for the UDS scheme, but it is important to stress that the methodology is independent of the adopted discretization scheme. In this way, the coefficients of the advection (adv) term contribution for the conservation of energy equation are given by:

$$a_{F,adv}^{Tu}=a_{F,adv}^{Tv}=(C_p)_f\max(\dot{m}_f,\,0),\tag{60a}$$

$$a_{P,adv}^{Tu}=-\sum_{F=NB(P)}a_{F,adv}^{Tu},\qquad a_{P,adv}^{Tv}=-\sum_{F=NB(P)}a_{F,adv}^{Tv},\tag{60b}$$

where the superscript $T\phi$ means the influence of the ϕ variable in the T energy equation. The term $\max(\dot{m}_f, 0)$ represents the maximum of $\dot{m}_f$ and 0, where the mass flux is positive if it goes from owner to neighbor cells, i.e., leaves the control-volume V_P.

The implicit diffusion ($idiff$) contribution, third term of Equation (57), is discretized, taking a linear profile (see page 86 of [48]) as

$$\left(s_i k \frac{\partial T}{\partial x_i}\right)_f = k_f \frac{(s_i s_i)_f}{(d_{PF})_i (s_i)_f}(T_F - T_P), \tag{61}$$

where $(d_{PF})_i$ is the vector joining the centroids P and F (see Figure 1). The coefficients of the implicit diffusion term for the conservation of energy equation are given by

$$a^{TT}_{F,idiff} = -k_f \frac{(s_i s_i)_f}{(d_{PF})_i (s_i)_f}, \tag{62a}$$

$$a^{TT}_{P,idiff} = -\sum_{F=NB(P)} a^{TT}_{F,idiff}. \tag{62b}$$

Finally, the coefficients of the explicit term contribution (right-hand side of Equation (57)) for the conservation of energy equation are given by

$$b^T_{P,source} = V_P\left[(\tau_N)_{ij} \mathrm{D}_{ji} + \frac{\eta_E(T)}{\lambda(T)}\left(\alpha(\mathrm{e}^{\Psi} - \mathrm{I})_{ij}\mathrm{D}_{ji} + (1-\alpha)\frac{(\mathrm{e}^{\Psi} - \mathrm{I})_{ii}}{2\tilde{\lambda}(T)}\right)\right]. \tag{63}$$

3.5. Block-Coupled Algorithm

Combining the discretized conservation of linear momentum (Equation (17)), conservation of mass (Equation (27)), log-conformation tensor (Equation (35)) and conservation of energy equations (Equation (58)), the following linear system of equations, written in matrix form, is obtained for each control volume:

$$\begin{bmatrix} a_P^{uu} & a_P^{uv} & a_P^{up} & a_P^{u\Psi_{xx}} & a_P^{u\Psi_{xy}} & a_P^{u\Psi_{yy}} & a_P^{uT} \\ a_P^{vu} & a_P^{vv} & a_P^{vp} & a_P^{v\Psi_{xx}} & a_P^{v\Psi_{xy}} & a_P^{v\Psi_{yy}} & a_P^{vT} \\ a_P^{pu} & a_P^{pv} & a_P^{pp} & a_P^{p\Psi_{xx}} & a_P^{p\Psi_{xy}} & a_P^{p\Psi_{yy}} & a_P^{pT} \\ a_P^{\Psi_{xx}u} & a_P^{\Psi_{xx}v} & a_P^{\Psi_{xx}p} & a_P^{\Psi_{xx}\Psi_{xx}} & a_P^{\Psi_{xx}\Psi_{xy}} & a_P^{\Psi_{xx}\Psi_{yy}} & a_P^{\Psi_{xx}T} \\ a_P^{\Psi_{xy}u} & a_P^{\Psi_{xy}v} & a_P^{\Psi_{xy}p} & a_P^{\Psi_{xy}\Psi_{xx}} & a_P^{\Psi_{xy}\Psi_{xy}} & a_P^{\Psi_{xy}\Psi_{yy}} & a_P^{\Psi_{xy}T} \\ a_P^{\Psi_{yy}u} & a_P^{\Psi_{yy}v} & a_P^{\Psi_{yy}p} & a_P^{\Psi_{yy}\Psi_{xx}} & a_P^{\Psi_{yy}\Psi_{xy}} & a_P^{\Psi_{yy}\Psi_{yy}} & a_P^{\Psi_{yy}T} \\ a_P^{Tu} & a_P^{Tv} & a_P^{Tp} & a_P^{T\Psi_{xx}} & a_P^{T\Psi_{xy}} & a_P^{T\Psi_{yy}} & a_P^{TT} \end{bmatrix} \begin{bmatrix} u_P \\ v_P \\ p_P \\ (\Psi_{xx})_P \\ (\Psi_{xy})_P \\ (\Psi_{yy})_P \\ T_P \end{bmatrix} +$$
$$+ \sum_{F=nb(P)} \begin{bmatrix} a_F^{uu} & a_F^{uv} & a_F^{up} & a_F^{u\Psi_{xx}} & a_F^{u\Psi_{xy}} & a_F^{u\Psi_{yy}} & a_F^{uT} \\ a_F^{vu} & a_F^{vv} & a_F^{vp} & a_F^{v\Psi_{xx}} & a_F^{v\Psi_{xy}} & a_F^{v\Psi_{yy}} & a_F^{vT} \\ a_F^{pu} & a_F^{pv} & a_F^{pp} & a_F^{p\Psi_{xx}} & a_F^{p\Psi_{xy}} & a_F^{p\Psi_{yy}} & a_F^{pT} \\ a_F^{\Psi_{xx}u} & a_F^{\Psi_{xx}v} & a_F^{\Psi_{xx}p} & a_F^{\Psi_{xx}\Psi_{xx}} & a_F^{\Psi_{xx}\Psi_{xy}} & a_F^{\Psi_{xx}\Psi_{yy}} & a_F^{\Psi_{xx}T} \\ a_F^{\Psi_{xy}u} & a_F^{\Psi_{xy}v} & a_F^{\Psi_{xy}p} & a_F^{\Psi_{xy}\Psi_{xx}} & a_F^{\Psi_{xy}\Psi_{xy}} & a_F^{\Psi_{xy}\Psi_{yy}} & a_F^{\Psi_{xy}T} \\ a_F^{\Psi_{yy}u} & a_F^{\Psi_{yy}v} & a_F^{\Psi_{yy}p} & a_F^{\Psi_{yy}\Psi_{xx}} & a_F^{\Psi_{yy}\Psi_{xy}} & a_F^{\Psi_{yy}\Psi_{yy}} & a_F^{\Psi_{yy}T} \\ a_F^{Tu} & a_F^{Tv} & a_F^{Tp} & a_F^{T\Psi_{xx}} & a_F^{T\Psi_{xy}} & a_F^{T\Psi_{yy}} & a_F^{TT} \end{bmatrix} \begin{bmatrix} u_F \\ v_F \\ p_F \\ (\Psi_{xx})_F \\ (\Psi_{xy})_F \\ (\Psi_{yy})_F \\ T_F \end{bmatrix} = \begin{bmatrix} b_P^u \\ b_P^v \\ b_P^p \\ b_P^{\Psi_{xx}} \\ b_P^{\Psi_{xy}} \\ b_P^{\Psi_{yy}} \\ b_P^T \end{bmatrix} \tag{64}$$

The linear systems (Equation (64)) obtained for each control volume of the computational domain are merged in a full system of equations, which can be written in the form $\mathbf{A\Phi} = \mathbf{b}$ where all variables $\mathbf{\Phi} = (u_i, p, \Psi_{ij}, T)$ are solved simultaneously. In this procedure, all variables in the different equations are treated implicitly, which is expected to be advantageous to the stability of the overall calculation process. The fully implicit coupled algorithm can be summarized into the following steps:

1. Initialize the fluid variables with the latest known values $(u_i^n, p^n, \Psi_{ij}^n, T^n)$.

2. Assemble the discretized equations for the conservation of linear momentum, conservation of mass, log-conformation tensor and conservation of energy (see Equations (17), (27), (35), (58)) and solve for u_i, p, Ψ_{ij} and T (Equation (64)).
3. Iterate until convergence.

For the solution of the global system of discretized algebraic equations, it is fundamental that an efficient linear solver is used to obtain the best overall convergence. In this work, the iterative solver Bi-Conjugate Gradient Stabilized (BiCGStab) [50], combined with an LU preconditioner, was used to retrieve the solution of the global system of discretized algebraic equations (see detailed discussion in Pimenta and Alves [30]). The initial residual for each iteration is evaluated based on the current values of the field, before solving the block-coupled system. After each block solver linear iteration, the residual is re-evaluated (final residual). When the maximum number of linear iterations (in this work defined as 100) or the final residual falls below the solver absolute tolerance (set as 10^{-6}), the block-coupled system current iteration stops and advances in time. All computations are performed on a computer with a 2.00-GHz 64 cores AMD EPYC 7662 CPU processor and 128 GB of RAM.

4. Results and Discussion

The validation of the newly-developed, fully implicit, block-coupled, non-isothermal, viscoelastic, log-conformation tensor-based algorithm was performed for the laminar, incompressible, non-isothermal viscoelastic flow of an Oldroyd-B fluid in an axisymmetric 4:1 sudden contraction geometry. For assessment purposes, the results computed with the newly-developed code were compared with the available data from the scientific literature [18].

4.1. Geometry, Meshes, and Initial and Boundary Conditions

An axisymmetric 4:1 sudden contraction with ratio of the radii $R_1/R_2 = 4$ was chosen as test geometry (Figure 2a), because of the availability of numerical data in the literature [18]. The upstream length was $l_1 = 80R_2$ and the downstream length was $l_2 = 50R_2$. The downstream channel height was chosen as $R_2 = 0.0020604$ m.

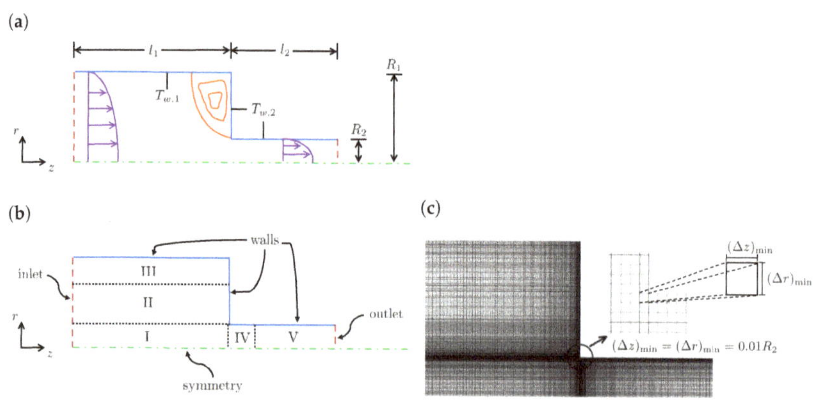

Figure 2. (**a**) Schematic of the 4:1 planar sudden contraction cross-section used to simulate the non-isothermal flow of a viscoelastic matrix fluid described by the Oldroyd-B constitutive model. The upstream length was l_1 and the downstream length was l_2. The upstream channel height was $R1$ and the downstream channel height was $R2$. The temperature at the upstream wall is $T_{w,1}$, while, for the downstream wall, the temperature was $T_{w,2}$. (**b**) The geometrical domain was divided into five blocks for mesh discretization. (**c**) A detailed view of the contraction area with the minimum normalized cell size at the corner $(\Delta z)_{\text{min}} = (\Delta r)_{\text{min}} = 0.01R_2$.

The flow had a rotational symmetry that was normal to the rz−plane and, to save computational resources and reduce the CPU times, only half of the domain was considered. The characteristics of the meshes used to discretize the geometrical domain are presented in Table 1. The meshes employed in the current work (M1 and M2) resulted from a mesh convergence analysis performed by Habla et al. [18], and had a similar refinement level to the two most refined meshes (M3 and M4) therein. The expansion or contraction geometrical factors were defined for each direction as the ratio of two consecutive cells lengths ($f_z = \Delta z_{i+1}/\Delta z_i$,with Δz_i being the length of the cell i in the z-direction). In this way, since $f_z > 1$ in Block V (see Table 1), in the z direction, the cells expanded from left to right. Figure 2c shows the details of the level of mesh refinement at the contraction region for M2. The higher refinement that occurs near the walls and in the contraction region allowed for the highest gradients of the computed flow variables in these locations to be captured.

Table 1. Characteristics of the meshes employed to simulate the non-isothermal viscoelastic flow of a viscoelastic matrix fluid described by the Oldroyd-B constitutive model in the axisymmetric 4:1 planar sudden contraction geometry.

Block	Mesh 1 (M1)			Mesh 2 (M2)		
	Nz × Nr	f_z	f_r	**Nz × Nr**	f_z	f_r
Block I	61 × 20	0.9061	0.9206	122 × 40	0.9519	0.9595
Block II	61 × 25	0.9061	1.0996	122 × 50	0.9519	1.0486
Block III	61 × 8	0.9061	0.8593	122 × 16	0.9519	0.9270
Block IV	40 × 20	1.1036	0.9206	80 × 40	1.0505	0.9595
Block V	13 × 20	1.1740	0.9206	26 × 40	1.0835	0.9595
NC	4293				17172	
$\Delta z_{\min} = \Delta r_{\min}$	$0.02R_2$			$\Delta z_{\min} = \Delta r_{\min}$	$0.01R_2$	

Nz and Nr are number of cells along z and r directions, respectively, inside each block. f_z and f_r are the expansion/contraction ratios inside each block. NC is the number of cells in the mesh. $\Delta z_{\min}$ and $\Delta r_{\min}$ are the minimum cell size in each direction.

The following boundary and initial conditions were used for all the runs that were performed:

- For velocity, no-slip at the walls, symmetry at the centerline, parabolic velocity profile at the inlet (with average velocity $\overline{U}_{z,1} = 0.00129$ m/s), and a zero-gradient condition at the outlet, i.e., assuming a fully developed flow;
- For pressure, the inlet and wall boundary conditions were set as zero-gradient and the centerline as symmetry boundary condition. At the outlet Dirichlet boundary condition was used, with a fixed value $p = 0$. Notice that, although the zero-pressure gradient specified at the inlet did not match the fully developed Poiseuille flow with the average velocity $\overline{U}_{z,1}$, this inconsistency did not affect the results, because the length of the upstream channel was sufficiently large to achieve fully developed flow conditions;
- For the log-conformation tensor components, zero values were assumed at the inlet, a symmetry boundary condition was used at the centerline, a linear extrapolation of the tensor components to the boundary was used at the walls, and a zero-gradient condition was imposed at the outlet;
- For the temperature, a Dirichlet condition was imposed at the inlet ($T_{inl} = 462$ K), a symmetry boundary condition was used at the centerline, at the upstream wall, $(z < l_1)$, $T_{w,1} = 462$ K, while, for the downstream wall, $(z \geq l_1)$ the temperature $T_{w,2}$ was chosen such as to give temperature jumps of $\Delta T = T_{w,2} - T_{w,1} = -30\,\text{K}, 0\,\text{K}, 30\,\text{K}$. A zero-gradient condition was imposed at the outlet;
- All fields were set to zero at the initial time.

4.2. Numerical Parameters

The dimensionless numbers governing the flow are the Reynolds number Re, the Weissenberg number Wi, the Peclet number Pe and the retardation ratio β, defined as:

$$Re = \frac{\rho R_2 \overline{U}_{z,2}}{\eta_0} \tag{65}$$

$$Wi = \frac{\lambda \overline{U}_{z,2}}{R_2} \tag{66}$$

$$Pe = \frac{\lambda R_2 \overline{U}_{z,2} c_P}{k} \tag{67}$$

where $\overline{U}_{z,2}$ is the mean velocity in the axial direction in the downstream channel, and λ is the relaxation time. In this case study, $Re = 3.9 \times 10^{-5}$, corresponding to creeping flow conditions. The retardation ratio was equal to $\beta = \eta_S/\eta_0 = 19/20$, thus assuming the solvent contribution to be negligibly small, which approximately recovered an Upper-Convected-Maxwell model. The Peclet number was kept constant at $Pe = 345$ by setting $c_P = 1500$ J/kg K and $k = 0.17$ W/mK. The WLF parameters were $C_1 = 4.54$ and $C_2 = 150.36$. The split coefficient varied between pure energy elasticity and entropy elasticity, $\alpha = 0$ or 1, respectively.

The use of a normalized time-step $\Delta t/(R_2/\overline{U}_{z,2})$ of 10^{-4} allowed for converged solutions to be obtained for all the runs performed. The maximum local Courant number corresponding to the normalized time-step 10^{-4} obtained for the axisymmetric 4:1 sudden contraction is 0.07.

4.3. Effects of the Energy Partitioning Parameter α

In Figure 3, the temperature profile on the line $r = 0.97R_2$ (Figure 3a) and the temperature contour plots (Figure 3b) are shown as a function of the axial position z/R_1 for $\alpha = 0$ and $\alpha = 1$ at $Wi = 5$ and $\Delta T = 0$ K. As illustrated in Figure 3a, the temperature profile computed by the newly-developed, fully implicit, block-coupled, non-isothermal, viscoelastic, log-conformation tensor-based algorithm is in agreement with the results presented by Habla et al. [18]. Due to the increase in the deformation rate near the contraction, the dissipation also increases, which leads to a temperature rise shortly before the contraction. At the contraction, the fluid moves to the wall with the imposed fixed wall temperature $T_{w,2} = 462$ K and, therefore, due to heat conduction towards the wall, a fast decrease in temperature is observed. Note that this decrease is remarkably higher for entropy elasticity ($\alpha = 1$) due to its higher temperature, which promotes a larger heat conduction rate. Subsequently, just after the re-entrant corner, due to the larger normal stresses developed here (see Figure 4), which lead to an increase of dissipation, the temperature rises again at the steepest rate. Further along the downstream channel, the temperatures also increase, but now at a smaller rate and, ultimately, the difference in the temperatures between energy elasticity and entropy elasticity cases is small, because the energy is now more released as pure energy elasticity ($\alpha = 0$) [18]. The temperature contour plots shown in Figure 3b are similar for both the energy and entropy elasticities, as expected from the marginal differences shown in Figure 3a for the temperature profile at $r = 0.97R_2$. In both cases, we see the formation of a larger temperature-rising region for $z/R_1 > 1$, which is also extended through the downstream channel radial direction.

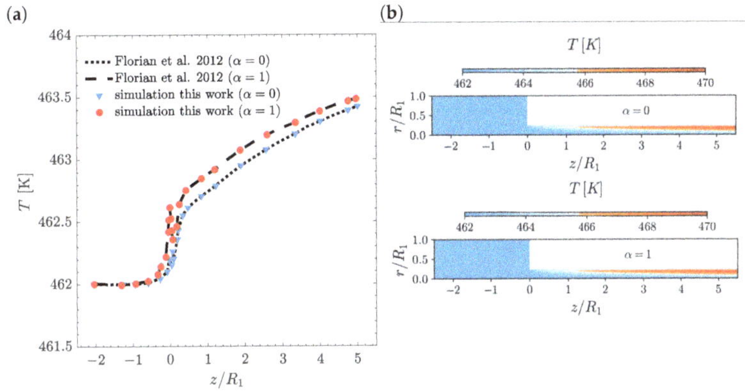

Figure 3. (**a**) Temperature profiles on the line $r = 0.97R_2$ [18] and (**b**) temperature contours as a function of the axial position z/R_1 for $\alpha = 0$ and $\alpha = 1$ at $Wi = 5$ and $\Delta T = 0$ K using M2.

In Figure 4, the axial normal stress profile on the line $r = 0.97R_2$ (Figure 4a) and the axial normal stress contour plots (Figure 4b) as a function of the axial position z/R_1 for $\alpha = 0$ and $\alpha = 1$ at $Wi = 5$ and $\Delta T = 0$ K, are shown. As illustrated in Figure 4a, the axial normal stress profile computed by the newly-developed, fully implicit, block-coupled, non-isothermal, viscoelastic, log-conformation tensor-based algorithm and the isothermal version is in agreement with the results presented by Habla et al. [18]. In the non-isothermal cases, the axial normal stress is smaller than the one obtained for the isothermal calculation. This behaviour can be attributed to the fact that increasing the temperature leads to a reduction in the viscosity value (see Equation (9)), which decreases the stress values, and also leads to a reduction in the relaxation time, resulting in a smaller local Weissenberg number and, therefore, fewer elastic effects (i.e., decrease in stresses). In addition, just after the re-entrant corner, we see an abrupt increase of the normal stresses due to the increase in the fluid deformation in this region, followed by a fast relaxation, before it starts to increase further down the downstream channel, but now at a smaller rate. The axial normal stress contour plots shown in Figure 4b are again similar for both the energy and entropy elasticities. In both cases, we see the formation of a thinner layer of normal stresses rising region at $0 < z/R_1 < 0.2$, which then increases in width, but with smaller normal stress values, at $0.2 < z/R_1 < 1$.

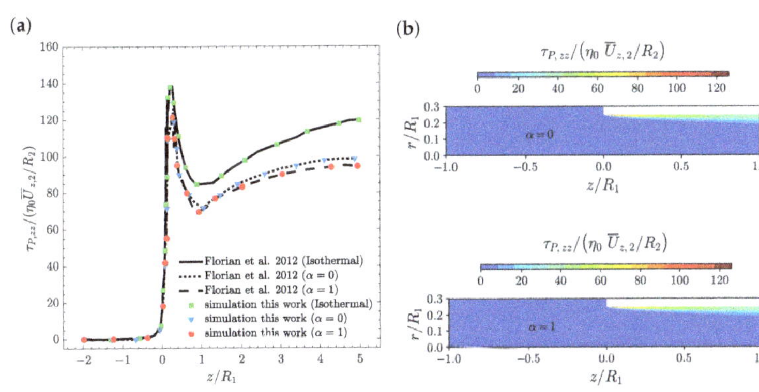

Figure 4. (**a**) Axial normal stress profiles $\tau_{P,zz}/(\eta_0\overline{U}_{z,2}/R_2)$ on the line $r = 0.97R_2$ [18] and (**b**) axial normal stress contours as a function of the axial position z/R_1 for $\alpha = 0$ and $\alpha = 1$ at $Wi = 5$ and $\Delta T = 0$ K using M2.

Additionally, Figure 5 shows the contours of the first normal stress difference and shear stress predicted on M2 at $\alpha = 0$, $Wi = 5$ and $\Delta T = 0$ K, for a zoomed region around the re-entrant corner. The maximum first normal stress difference is located on the downstream channel wall near the re-entrant corner (see Figure 5a). Moreover, at the contraction vertical wall, the first normal stress difference is negative, which is responsible for fluid re-circulation and the formation of the corner vortex. The maximum magnitude of the shear stress component is also located on the downstream channel wall near the re-entrant corner (see Figure 5b); however, in that case, the shear stress value is negative, pushing the fluid towards the symmetry line at $r/R_1 = 0$. Finally, at the contraction vertical wall, the shear stress is positive, allowing for the extension of the corner vortex size.

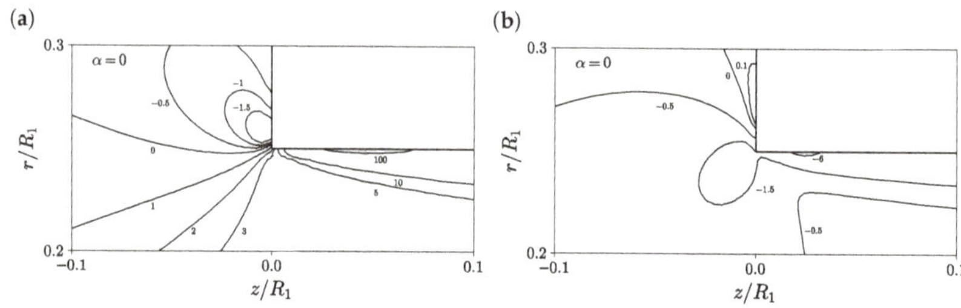

Figure 5. Contours of (**a**) first normal stress difference $(\tau_{P,zz} - \tau_{P,rr})/(\eta_0 \overline{U}_{z,2}/R_2)$ and (**b**) shear stress $\tau_{P,rz}/(\eta_0 \overline{U}_{z,2}/R_2)$ for $\alpha = 0$, $Wi = 5$ and $\Delta T = 0$ K using M2.

In Figure 6, the axial velocity profile on the line $r = 0.97R_2$ (Figure 6a) and the axial velocity contour plots (Figure 6b) as a function of the axial position z/R_1 for $\alpha = 0$ and $\alpha = 1$ at $Wi = 5$ and $\Delta T = 0$ K are shown. As illustrated in Figure 6a, the axial velocity profiles computed by the newly-developed, fully implicit, block-coupled, non-isothermal, viscoelastic, log-conformation tensor-based algorithm and with the isothermal version are in agreement with the results presented by Habla et al. [18]. In addition, we can see that the influence of temperature on the local velocity profile for both cases of entropy elasticity and energy elasticity is negligible, being similar to the axial velocity of the isothermal case. The axial velocity contour plots shown in Figure 6b are, again, similar for both the energy and entropy elasticities, showing the accommodation of the fluid near the re-entrant corner, i.e., the fluid is accelerated in the center while decelerating in the wall-near regions, and a fully developed velocity profile at the downstream channel.

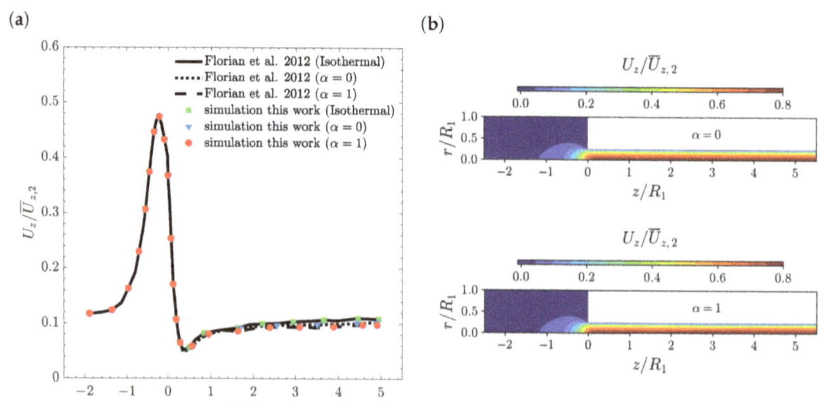

Figure 6. (**a**) Axial velocity profiles $U_z/\overline{U}_{z,2}$ on the line $r = 0.97R_2$ [18] and (**b**) axial velocity contours as a function of the axial position z/R_1 for $\alpha = 0$ and $\alpha = 1$ at $Wi = 5$ and $\Delta T = 0$ K using M2.

In Table 2, we provide a summary of the number of iterations and execution time of the segregated/iterative and coupled/monolithic approaches for the calculation of the non-isothermal, viscoelastic matrix-based Oldroyd-B fluid flow in the two-dimensional, axisymmetric 4:1 planar sudden-contraction geometry, using the two different meshes, M1 and M2, with $\alpha = 0$ at $Wi = 5$ and $\Delta T = 0$ K. The ratio of the number of iterations required by the segregated algorithm to that required by the coupled one increases from 432 to 524 for M1 and M2, respectively, with accompanying computational cost ratios of 17 and 19, respectively, which clearly shows the benefits obtained by using the fully implicit coupled approach.

Table 2. Comparison of the number of iterations and CPU time required by the segregated (S) and coupled (C) solvers for the calculation of the non-isothermal viscoelastic matrix-based Oldroyd-B fluid flow in the two-dimensional axisymmetric 4:1 planar sudden-contraction geometry, using the two different meshes M1 and M2, with $\alpha = 0$ at $Wi = 5$ and $\Delta T = 0$ K.

Mesh	Number of Iterations			Execution Time [s]		
	C	S	S/C	C	S	S/C
M1	827	357,681	432	103	1760	17
M2	1484	778,344	524	689	13,102	19

M1: 4293 CV; M2: 17172 CV.

4.4. Effects of Wall Temperature Jumps

The temperature and axial velocity profiles at the outlet of the downstream section are shown in Figure 7 for $Wi = 5$, $\alpha = 0$ and temperature jumps $\Delta T = -30$ K, $\Delta T = 0$ K and $\Delta T = +30$ K. The results for the case of energy elasticity at the outlet of the downstream section, for all temperature jumps, did not present any differences [18]. As shown in Figure 7a, the wall ($r/R_1 = 0.25$) temperature changes by more than 60 K and the centerline ($r/R_1 = 0$) temperature varied less than 5 K from $\Delta T = -30$ K to $\Delta T = +30$ K, in agreement with the results obtained by Habla et al. [18]. These temperature changes are responsible for the limited effect of external heating or cooling in the thermal control of the flow at the bulk region. In Figure 7b, the axial velocity profile at the centerline increases with the decrease in temperature jump, due to the smaller viscosity in the near-wall region.

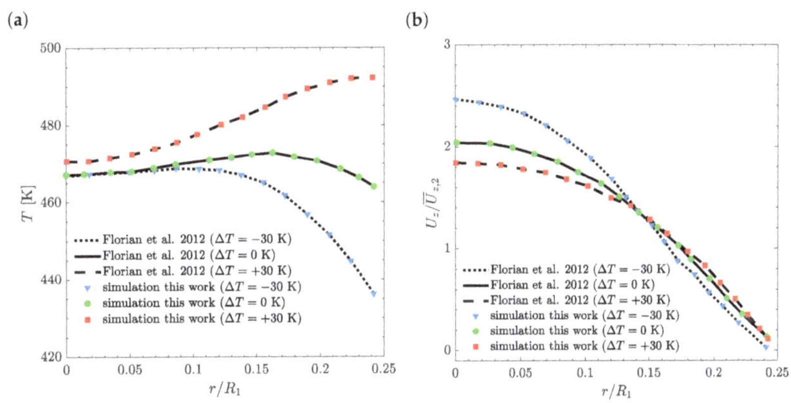

Figure 7. (**a**) Temperature T and (**b**) axial velocity $U_z/\overline{U}_{z,2}$ as a function of the radial distance r/R_1 at the outlet, with $Wi = 5$, $\alpha = 0$ and temperature jumps $\Delta T = -30$ K, $\Delta T = 0$ K and $\Delta T = +30$ K using M2 [18].

The temperature and axial velocity contours are shown in Figure 8 for $Wi = 5$, $\alpha = 0$ and temperature jumps $\Delta T = -30$ K and $\Delta T = +30$ K. For the cooling case, i.e.,

$\Delta T = -30$ K (top figures in Figure 8), the reduction in temperature on the downstream channel walls promotes an increase in fluid centerline velocity of approximately 3.1 times more than the one obtained for the case $\Delta T = 0$ K (see Figure 6). For the heating case, i.e., $\Delta T = +30$ K (bottom figures in Figure 8), the increase in temperature on the downstream channel walls promotes an increase in fluid centerline velocity of approximately 2.3 times more than the one obtained for the case $\Delta T = 0$ K (see Figure 6). Notice that the increase in centerline velocity for the heating case is approximately 75% smaller than the one obtained for the cooling case.

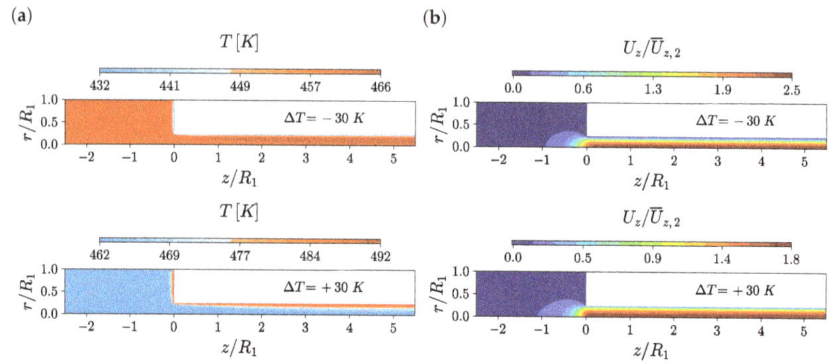

Figure 8. (**a**) Temperature T and (**b**) axial velocity $U_z/\overline{U}_{z,2}$ contours at $Wi = 5$, $\alpha = 0$ and temperature jumps $\Delta T = -30$ K and $\Delta T = +30$ K using M2.

Additionally, Figure 9 shows the contours of the first normal stress difference and shear stress predicted on M2 at $\alpha = 0$, $Wi = 5$, $\Delta T = -30$ K (top) and $\Delta T = +30$ K (bottom), for a zoomed region around the re-entrant corner. The maximum first normal stress difference was found to decrease by 35% and increase by 30% for the cases $\Delta T = -30$ K and $\Delta T = +30$ K, respectively, when compared to the case $\Delta T = 0$ K. The minimum value of the shear stress is found to both increase and decrease by 50% for the cases $\Delta T = -30$ K and $\Delta T = +30$ K, respectively, when compared to the case $\Delta T = 0$ K.

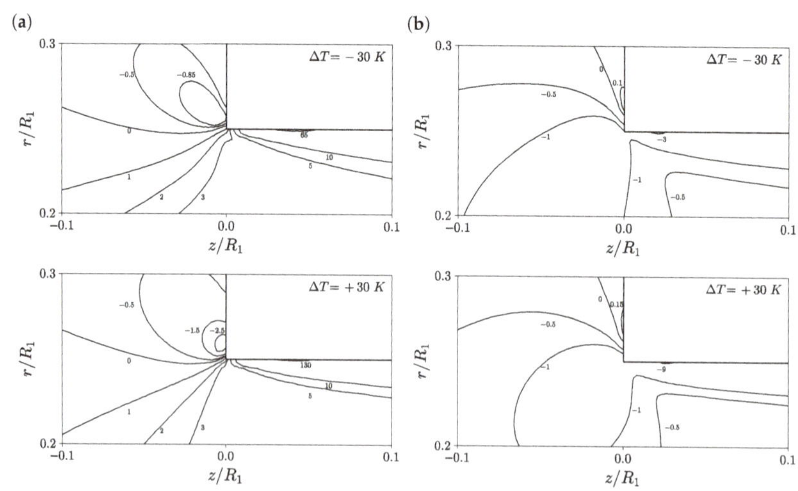

Figure 9. Contours of (**a**) first normal stress difference $(\tau_{P,zz} - \tau_{P,rr})/(\eta_0\overline{U}_{z,2}/R_2)$ and (**b**) shear stress $\tau_{P,rz}/(\eta_0\overline{U}_{z,2}/R_2)$ for $\alpha = 0$, $Wi = 5$, $\Delta T = -30$ K (top) and $\Delta T = +30$ K (bottom) using M2.

5. Conclusions

This paper presented a fully implicit coupled method (also known as a monolithic approach) for the solution of laminar, incompressible, non-isothermal, viscoelastic flows based on the log-conformation tensor framework for high Weissenberg number problems. The fully implicit coupled solver is a pressure-based method in which the pressure equation is derived through the Rhie–Chow interpolation, allowing for coupling between the pressure and velocity fields. In addition, the explicit diffusion term added by the improved both-sides-diffusion (iBSD) technique to the linear momentum equations is discretized with a special second-order derivative of the velocity field, allowing for coupling between the velocity and log-conformation tensor fields. Subsequently, the divergence of the log-conformation tensor term in the linear momentum equations is implicitly discretized, and the velocity field is considered implicitly in the log-conformation tensor constitutive equations by expanding the advection, rotation and the rate of deformation terms, all by considering a Taylor series expansion truncated at the second-order error term. Finally, the advection and diffusion terms in the energy equation are also implicitly discretized.

The validation of the newly-developed algorithm was performed for the non-isothermal viscoelastic matrix-based Oldroyd-B fluid flow in the benchmark problem of a two-dimensional axisymmetric 4:1 planar sudden-contraction geometry, and the results obtained by the fully implicit coupled algorithm were shown to be in remarkable agreement with other results found in the scientific literature (less than 5% error differences), which validated the present implementation. The results were obtained at a high Weissenberg number, and allowed to study the influence of the energy splitting factor at the limit of pure energy elasticity and pure entropy elasticity, and the effect of the wall temperature jump near the contraction for positive and negative temperature increments.

In future works, the algorithm will be further assessed by looking at 3D case studies and employing non-linear viscoelastic fluid behaviors, such as the shear-thinning phenomenon predicted by the Giesekus fluid model.

Funding: This work is funded by FEDER funds through the COMPETE 2020 Programme and National Funds through FCT (Portuguese Foundation for Science and Technology) under the projects UID-B/05256/2020 and UID-P/05256/2020.

Institutional Review Board Statement: Not applicable.

Data Availability Statement: Not applicable.

Acknowledgments: The authors would like to acknowledge the Minho University cluster under the project NORTE-07-0162-FEDER-000086 (URL: http://search6.di.uminho.pt (accessed on 1 February 2022)) for providing HPC resources that have contributed to the research results reported within this paper. The author thank M. Spahn from RWTH Aachen University for insightful comments regarding this work, specifically during the implementation of the isothermal log-conformation discretization schemes.

Conflicts of Interest: The funders had no role in the design of the study; in the collection, analyses, or interpretation of data; in the writing of the manuscript, or in the decision to publish the results.

References

1. Chang, R.Y.; Yang, W.H. Numerical simulation of mold filling in injection molding using a three-dimensional finite volume approach. *J. Numer. Methods Fluids* **2001**, *37*, 125–148. [CrossRef]
2. Zhou, J.; Turng, L.S. Three-dimensional numerical simulation of injection mold filling with a finite-volume method and parallel computing. *Adv. Polym. Technol.* **2007**, *25*, 247–258. [CrossRef]
3. Pedro, J.; Ramôa, B.; Nóbrega, J.; Fernandes, C. Verification and Validation of openInjMoldSim, an Open-Source Solver to Model the Filling Stage of Thermoplastic Injection Molding. *Fluids* **2020**, *5*, 84. [CrossRef]
4. Habla, F.; Fernandes, C.; Maier, M.; Densky, L.; Ferrás, L.; Rajkumar, A.; Carneiro, O.; Hinrichsen, O.; Nóbrega, J.M. Development and validation of a model for the temperature distribution in the extrusion calibration stage. *Appl. Therm. Eng.* **2016**, *100*, 538–552. [CrossRef]
5. Pimenta, F.; Alves, M. Conjugate heat transfer in the unbounded flow of a viscoelastic fluid past a sphere. *Int. J. Heat Fluid Flow* **2021**, *89*, 108784. [CrossRef]

6. Fernandes, C.; Pontes, A.J.; Viana, J.C.; Nóbrega, J.M.; Gaspar-Cunha, A. Modeling of Plasticating Injection Molding-Experimental Assessment. *Int. Polym. Proc.* **2014**, *29*, 558–569. [CrossRef]
7. Fernandes, C.; Fakhari, A.; Tukovic, Ž. Non-Isothermal Free-Surface Viscous Flow of Polymer Melts in Pipe Extrusion Using an Open-Source Interface Tracking Finite Volume Method. *Polymers* **2021**, *13*, 4454. [CrossRef]
8. Meburger, S.; Niethammer, M.; Bothe, D.; Schäfer, M. Numerical simulation of non-isothermal viscoelastic flows at high Weissenberg numbers using a finite volume method on general unstructured meshes. *J.-Non-Newton. Fluid Mech.* **2021**, *287*, 104451. [CrossRef]
9. Bird, R. Constitutive equations for polymeric liquids. *Annu. Rev. Fluid Mech.* **1995**, *27*, 169–193. [CrossRef]
10. Williams, M.; Landel, R.; Ferry, J. The temperature dependence of relaxation mechanisms in amorphous polymers and other glass-forming liquids. *J. Am. Chem. Soc.* **1955**, *77*, 3701–3707. [CrossRef]
11. Ferry, J. *Viscoelastic Properties of Polymers*, 3rd ed.; John Wiley & Sons: New York, NY, USA, 1980.
12. Bird, R.; Armstrong, R.; Hassager, O. *Dynamics of Polymeric Liquids, Volume 1: Fluid Mechanics*, 2nd ed.; John Wiley & Sons: New York, NY, USA, 1987.
13. Peters, G.; Baaijens, F. Modelling of non-isothermal viscoelastic flows. *J.-Non-Newton. Fluid Mech.* **1997**, *68*, 205–224. [CrossRef]
14. Hütter, M.; Luap, C.; Öttinger, H. Energy elastic effects and the concept of temperature in flowing polymeric liquids. *Rheol. Acta* **2009**, *48*, 301–316. [CrossRef]
15. Braun, H. A model for the thermorheological behavior of viscoelastic fluids. *Rheol. Acta* **1991**, *30*, 523–529. [CrossRef]
16. Wachs, A.; Clermont, J.R. Non-isothermal viscoelastic flow computations in an axisymmetric contraction at high Weissenberg numbers by a finite volume method. *J.-Non-Newton. Fluid Mech.* **2000**, *95*, 147–184. [CrossRef]
17. Wachs, A.; Clermont, J.R.; Khalifeh, A. Computations of non-isothermal viscous and viscoelastic flows in abrupt contractions using a finite volume method. *Eng. Comput.* **2002**, *19*, 874–901. [CrossRef]
18. Habla, F.; Woitalka, A.; Neuner, S.; Hinrichsen, O. Development of a methodology for numerical simulation of non-isothermal viscoelastic fluid flows with application to axisymmetric 4:1 contraction flows. *Chem. Eng. J.* **2012**, *207–208*, 772–784. [CrossRef]
19. Fernandes, C.; Faroughi, S.A.; Ribeiro, R.; Isabel, A.; McKinley, G.H. Finite volume simulations of particle-laden viscoelastic fluid flows: Application to hydraulic fracture processes. *Eng. Comput.* **2022**. [CrossRef]
20. Zheng, Z.Y.; Li, F.C.; Yang, J.C. Modeling Asymmetric Flow of Viscoelastic Fluid in Symmetric Planar Sudden Expansion Geometry Based on User-Defined Function in FLUENT CFD Package. *Adv. Mech. Eng.* **2013**, *Volume 2013*, 795937. [CrossRef]
21. Shahbani-Zahiri, A.; Shahmardan, M.; Hassanzadeh, H.; Norouzi, M. Effects of fluid inertia and elasticity and expansion angles on recirculation and thermal regions of viscoelastic flow in the symmetric planar gradual expansions. *J. Braz. Soc. Mech. Sci. Eng.* **2018**, *40*, 480–492. [CrossRef]
22. Kunisch, K.; Marduel, X. Optimal control of non-isothermal viscoelastic fluid flow. *J.-Non-Newton. Fluid Mech.* **2000**, *88*, 261–301. [CrossRef]
23. Spanjaards, M.; Hulsen, M.; Anderson, P. Computational analysis of the extrudate shape of three-dimensional viscoelastic, non-isothermal extrusion flows. *J.-Non-Newton. Fluid Mech.* **2020**, *282*, 104310. [CrossRef]
24. Patankar, S.; Spalding, D. A calculation procedure for heat, mass and momentum transfer in three-dimensional parabolic flows. *Int. Heat Mass Transf.* **1972**, *15*, 1787–1806. [CrossRef]
25. Darwish, M.; Sraj, I.; Moukalled, F. A coupled incompressible flow solver on structured grid. *Numer. Heat Transf. Part B Fundam.* **2007**, *52*, 353–371. [CrossRef]
26. Darwish, M.; Sraj, I.; Moukalled, F. A coupled finite volume solver for the solution of incompressible flows on unstructured grids. *J. Comput. Phys.* **2009**, *228*, 180–201. [CrossRef]
27. Mangani, L.; Buchmayr, M.; Darwish, M. Development of a novel fully coupled solver in OpenFOAM: Steady-state incompressible turbulent flows. *Numer. Heat Transf. Part B Fundam.* **2014**, *66*, 1–20. [CrossRef]
28. Fernandes, C.; Cević, V.V.; Uroić, T.; Simoes, R.; Carneiro, O.; Jasak, H.; Nóbrega, J. A coupled finite volume flow solver for the solution of incompressible viscoelastic flows. *J.-Non-Newton. Fluid Mech.* **2019**, *265*, 99–115. [CrossRef]
29. Ferreira, G.; Lage, P.; Silva, L.; Jasak, H. Implementation of an implicit pressure-velocity coupling for the Eulerian multi-fluid model. *Comput. Fluids* **2002**, *181*, 188–207. [CrossRef]
30. Pimenta, F.; Alves, M.A. A coupled finite-volume solver for numerical simulation of electrically-driven flows. *Comput. Fluids* **2019**, *193*, 104279. [CrossRef]
31. Kim, J.; Kim, C.; Kim, J.; Chung, C.; Ahn, K.; Lee, S. High-resolution finite element simulation of 4:1 planar contraction flow of viscoelastic fluid. *J.-Non-Newton. Fluid Mech.* **2005**, *129*, 23–37. [CrossRef]
32. Oliveira, M.; Oliveira, P.; Pinho, F.; Alves, M. Effect of contraction ratio upon viscoelastic flow in contractions: The axisymmetric case. *J.-Non-Newton. Fluid Mech.* **2007**, *147*, 92–108. [CrossRef]
33. Xue, S.C.; Phan-Thien, N.; Tanner, R. Three dimensional numerical simulations of viscoelastic flows through planar contractions. *J.-Non-Newton. Fluid Mech.* **1998**, *74*, 195–245. [CrossRef]
34. Shaw, M. *Introduction to Polymer Rheology*; John Wiley & Sons: New York, NY, USA, 2011. [CrossRef]
35. Fattal, R.; Kupferman, R. Time-dependent simulation of viscoelastic flows at high Weissenberg number using the log-conformation representation. *J.-Non-Newton. Fluid Mech.* **2005**, *126*, 23–37. [CrossRef]
36. Balci, N.; Thomases, B.; Renardy, M.; Doering, C. Symmetric factorization of the conformation tensor in viscoelastic fluid models. *J.-Non-Newton. Fluid Mech.* **2011**, *166*, 546–553. [CrossRef]

37. Afonso, A.; Pinho, F.; Alves, M. The kernel-conformation constitutive laws. *J.-Non-Newton. Fluid Mech.* **2012**, *167–168*, 30–37. [CrossRef]
38. Knechtges, P. Simulation of Viscoelastic Free-Surface Flows. Ph.D. Thesis, RWTH Aachen University, Aachen, Germany, 2018.
39. Spahn, M. Modeling High Weissenberg Number Flows in OpenFOAM: Implementation of a Novel Log-Conf Approach in the Context of Finite Volumes. Master's Thesis, RWTH Aachen University, Aachen, Germany, 2019.
40. Oldroyd, J.G. On the formulation of rheological equations of state. *Proc. R. Soc. Lond. Ser. A Math. Phys. Sci.* **1950**, *200*, 523–541.
41. Giesekus, H. A simple constitutive equation for polymer fluids based on the concept of deformation-dependent tensorial mobility. *J.-Non-Newton. Fluid Mech.* **1982**, *11*, 69–109. [CrossRef]
42. Phan-Thien, N.; Tanner, R. A new constitutive equation derived from network theory. *J.-Non-Newton. Fluid Mech.* **1977**, *2*, 353–365. [CrossRef]
43. Phan-Thien, N. A nonlinear network viscoelastic model. *J. Rheol.* **1978**, *22*, 259–283. [CrossRef]
44. Wapperom, P.; Hulsen, M. Thermodynamics of viscoelastic fluids: The temperature equation. *J. Rheol.* **1998**, *42*, 999–1019. [CrossRef]
45. Moreno, L.; Codina, R.; Baiges, J. Numerical simulation of non-isothermal viscoelastic fluid flows using a VMS stabilized finite element formulation. *J.-Non-Newton. Fluid Mech.* **2021**, *296*, 104640. [CrossRef]
46. Fernandes, C.; Araujo, M.; Ferrás, L.; Nóbrega, J. Improved both sides diffusion (iBSD): A new and straightforward stabilization approach for viscoelastic fluid flows. *J.-Non-Newton. Fluid Mech.* **2017**, *249*, 63–78. [CrossRef]
47. Rhie, C.; Chow, W. Numerical Study of Turbulent Flow Past an Isolated Airfol with Trailing Edge Separation. *AIAA J.* **1983**, *21*, 1525–1532. [CrossRef]
48. Jasak, H. Error Analysis and Estimation for the Finite Volume Method with Applications to Fluid Flows. Ph.D. Thesis, Imperial College University of London, London, UK, 1996.
49. Knechtges, P. The fully-implicit log-conformation formulation and its application to three-dimensional flows. *J.-Non-Newton. Fluid Mech.* **2015**, *223*, 209–220. [CrossRef]
50. van der Vorst, H. Bi-CGSTAB: A fast and smoothly converging variant of Bi-CG for the solution of nonsymmetric linear systems. *SIAM J. Sci. Stat. Comput.* **1992**, *13*, 631–644. [CrossRef]

polymers

MDPI

Article

Non-Isothermal Free-Surface Viscous Flow of Polymer Melts in Pipe Extrusion Using an Open-Source Interface Tracking Finite Volume Method

Célio Fernandes [1,*], Ahmad Fakhari [2] and Željko Tukovic [3]

1 Institute for Polymers and Composites (IPC), Department of Polymer Engineering, Engineering School of the University of Minho, Campus of Azurém, 4800-058 Guimarães, Portugal
2 Transport Phenomena Research Center (CEFT), Mechanical Engineering Department, Faculty of Engineering of the University of Porto, Rua Dr. Roberto Frias s/n, 4200-465 Porto, Portugal; ahmadfakhari@gmail.com
3 Faculty of Mechanical Engineering and Naval Architecture, University of Zagreb, Ulica Ivana Lučića 5, 10000 Zagreb, Croatia; Zeljko.Tukovic@fsb.hr
* Correspondence: cbpf@dep.uminho.pt

Abstract: Polymer extrudate swelling is a rheological phenomenon that occurs after polymer melt flow emerges at the die exit of extrusion equipment due to molecular stress relaxations and flow redistributions. Specifically, with the growing demand for large scale and high productivity, polymer pipes have recently been produced by extrusion. This study reports the development of a new incompressible non-isothermal finite volume method, based on the Arbitrary Lagrangian–Eulerian (ALE) formulation, to compute the viscous flow of polymer melts obeying the Herschel–Bulkley constitutive equation. The Papanastasiou-regularized version of the constitutive equation is employed. The influence of the temperature on the rheological behavior of the material is controlled by the Williams–Landel–Ferry (WLF) function. The new method is validated by comparing the extrudate swell ratio obtained for Bingham and Herschel–Bulkley flows (shear-thinning and shear-thickening) with reference data found in the scientific literature. Additionally, the essential flow characteristics including yield-stress, inertia and non-isothermal effects were investigated.

Keywords: extrudate swell; pipe die; polymer melt; Herschel–Bulkley fluids; yield stress; finite volume method; interface tracking; free-surfaces; OpenFOAM

Citation: Fernandes, C.; Fakhari, A.; Tukovic, Ž. Non-Isothermal Free-Surface Viscous Flow of Polymer Melts in Pipe Extrusion Using an Open-Source Interface Tracking Finite Volume Method. *Polymers* **2021**, *13*, 4454. https://doi.org/10.3390/polym13244454

Academic Editor: Alexander Malkin

Received: 17 November 2021
Accepted: 11 December 2021
Published: 19 December 2021

1. Introduction

The extrudate swell flow is a well-known benchmark problem in the polymer processing field, where a free-surface and a boundary stress singularity at the die exit are present. The variety of solutions for free-surface flows is usually very limited to simple cases [1]. However, with the increasing demand for large diameters and high productivity, plastic pipes have recently been produced [2].

The polymer production process is mainly non-isothermal in nature, including heat transfer phenomena such as heating and cooling, e.g., in the injection molding [3,4] or extrusion [5–8] processes. Since the flow properties are strongly dependent on rheology and temperature, it is of great interest to understand and predict this type of heat transfer phenomenon. Although there is increasing research effort on non-isothermal free-surface flows, these studies are limited to academic purposes and problems. Extrudates in general can swell up to 15% due to thermal effects [9,10]. Phuoc and Tanner [11] developed a finite element scheme based on the Galerkin method to explore the effects of thermally induced property changes in extrusion. The swelling of self-heating extruded jets was investigated, and a new phenomenon, the so-called thermal extrudate swell, was discovered. Extruded expansions of up to 70% of the die diameter have been found in a Newtonian fluid with thermal properties similar to those of low density polyethylene. Subsequently, Karagiannis et al. [12] developed a three-dimensional (3D) non-isothermal code to study viscous free-surface flows with exponential dependence of viscosity on temperature. Again, the code was based on Galerkin's finite-element formulation, and, apart from the phenomenon of thermally induced extrudate swelling, the bending and distortion of the extrudate due to temperature differences and/or geometric asymmetries were confirmed numerically and experimentally. Finally, Spanjaards et al. [13] developed a 3D transient non-isothermal finite element code to predict the extruded shape of viscoelastic fluids emerging from an asymmetric keyhole-shaped die. A systematic study was carried out to decouple the effects of shear-thinning, elasticity, and temperature in the shape of the extrudate.

Another important effect to be considered when predicting the swelling of the extrudate is the inertia, through the dimensionless Reynolds number, *Re*, defined as the ratio of inertial forces to viscous forces within a fluid that is subjected to relative internal movement due to different fluid velocities. In Newtonian flow, the extrudate swell ratio, χ, defined as the ratio of the final extrudate radius to that of the die, decreases with increasing *Re* and finally achieves a contraction rather than swelling, since χ is less than unity [14,15]. An important class of materials commonly used in extrusion is the yield-stress fluid, which does not deform when subjected to a stress below the yield stress, τ_0. Bingham [16] developed the simplest constitutive equation with a yield-stress, which was then combined with a power-law model to account for the effects of shear-thinning or shear-thickening, known as the Herschel–Bulkley model [17]. To solve fluid flows described by constitutive equations with unyielded ($\tau \leq \tau_0$) and yielded regions ($\tau > \tau_0$), a special approach to overcome the numerical difficulties created by these models (the viscosity diverges to infinity as the strain rate $\dot{\gamma}$ approaches zero) must be adopted. For that purpose, a common methodology is to do a regularization of the constitutive equation so that the same expression is uniformly valid for any value of $\dot{\gamma}$, i.e., to ensure that the viscosity is a continuous function of strain rate. Notice that here τ and $\dot{\gamma}$ represent the magnitudes of the stress tensor $\boldsymbol{\tau}$ and rate-of-strain tensor $\dot{\boldsymbol{\gamma}}$, respectively. Papanastasiou [18] developed the regularization model for the Bingham constitutive equation, which was then extended to work with the Herschel–Bulkley model by Ellwood et al. [19]. A survey of regularization models used in the scientific literature can be found in Frigaard and Nouar [20]. For creeping flow conditions ($Re \ll 1$), there are several studies where the results for the extrudate swell of materials with yield stress are presented [18,21–24]. Generally, the conclusion is that in creeping extrudate swell flow of yield-stress materials, the extrudate swell ratio decreases as yield stress increases. Additionally, it was noticed that beyond a certain value of τ_0, the swell contracts slightly ($\chi < 1$) and reaches a minimum, from which it starts to increase again to reach unity asymptotically. For moderate Reynolds numbers ($1 < Re < 200$), the scientific literature devoted to the study of the effects of inertia on the extrudate swell of yield-stress fluids is scarce [19,25]. In this situation, as the yield stress increases, swelling at lower Reynolds numbers and contraction at higher Reynolds numbers are reduced.

In recent decades, Computational Fluid Dynamics (CFD) simulation has expanded as a significant framework to help engineers understand and improve polymer processing techniques. Nonetheless, the current state of the art of numerical methods for complex problems, which is the case of non-isothermal free-surface extrudate swell flow of yield-stress fluids, is insufficient for practical needs. In addition, proprietary packages are expensive and cannot be modified to account for the engineer's needs. As the solver here presented is based in an open-source finite volume computational library, OpenFOAM [26], this allows more capabilities to be added to the code to take into account the physics involved in the phenomenon of interest, thereby making the simulation more realistic. Therefore, a general and freely available solver, based on the open-source framework OpenFOAM [26], for the simulation of the non-isothermal free-surface flow of yield-stress fluids would be of great importance. Hence, this work deals with the development of a solver to calculate non-isothermal free-surface flows of yield-stress fluids as an extension of the isothermal free-surface flow solver presented elsewhere [15,27]. Notice that in Fakhari et al. [15], we developed the predecessor of the current code, which was used to describe Newtonian isothermal fluid flows, and there the efficiency of this framework was demonstrated in terms of both considering a larger computational time-step and also the smaller CPU wall time to perform each iteration of the simulation. The objective of this work is to investigate the combined effects of inertia, yield-stress, and temperature on the extrudate swell of a pipe flow. It is worth pointing out that although the melted polymer chain strongly depends on shear and stretch [28], this work focuses on the macroscale dimensional distribution of the melted polymer in the extruded swell phenomena and does not go through the microscale changes of the material such as crystallization kinetics.

The paper is organized as follows: in Section 2, we present the description of the problem, specifically, the balance equations that govern the non-isothermal free-surface flow of generalized Newtonian fluids, the boundary and initial conditions, and the Arbitrary Lagrangian–Eulerian formulation for the mesh movement. In Section 3, the numerical method used to implement the proposed model is described. In Section 4, we present the numerical results obtained for the non-isothermal extrudate swell of the Herschel–Bulkley yield-stress fluid model, taking into account the combined effects of inertia, yield-stress and temperature. Finally, in Section 5, we summarize the main conclusions of this work.

2. Problem Description

In this section, the governing equations to solve the incompressible and non-isothermal polymer-melt extrudate flow in dies of cylindrical shape for pipe production are described.

2.1. Balance Equations

The equations of motion governing the non-isothermal flow of incompressible generalized Newtonian fluids inside an arbitrary volume V bounded by a closed moving surface S include continuity, momentum, and energy equations, which can be written as:

$$\nabla \cdot \mathbf{u} = 0, \tag{1}$$

$$\frac{\partial \mathbf{u}}{\partial t} + \nabla \cdot ((\mathbf{u} - \mathbf{u}_S)\mathbf{u}) = -\nabla P + \nabla \cdot \boldsymbol{\tau}, \tag{2}$$

$$\rho\, c_p \left(\frac{\partial T}{\partial t} + \mathbf{u} \cdot \nabla T \right) = k_p \nabla^2 T + \boldsymbol{\tau} : \mathbf{D}, \tag{3}$$

where ∇ is the Hamilton differential operator, $\mathbf{u}$ is the fluid velocity vector, $\mathbf{u}_S$ is the velocity of surface S, P is the kinematic pressure, $\boldsymbol{\tau}$ is the deviatoric stress tensor, ρ is the fluid density, c_p is the heat capacity, T is the temperature, k_p is the isotropic thermal conductivity based on the heat flux Fourier's law, and $\mathbf{D} = (\nabla \mathbf{u} + \nabla \mathbf{u}^T)/2$ is the strain rate tensor. Note that the last term on the right hand side of Equation (3) represents viscous heat dissipation and is written in a form that assumes all mechanical energy is dissipated as heat.

2.2. Constitutive Equation

The following constitutive equation is adopted to describe the relationship between stress and strain for the viscous fluid flow,

$$\boldsymbol{\tau} = 2\eta(\dot{\gamma})\mathbf{D}, \tag{4}$$

where $\dot{\gamma} = \sqrt{2\mathbf{D} : \mathbf{D}}$ is the shear rate (invariant of the strain rate tensor $\mathbf{D}$) and $\eta(\dot{\gamma})$ is the apparent viscosity, which is described by the Herschel–Bulkley model [17] with the Papanastasiou regularization [18,19],

$$\eta(\dot{\gamma}) = \min\left\{ \eta_0, \tau_0 \dot{\gamma}^{-1}[1 - \exp(-m\dot{\gamma})] + k\dot{\gamma}^{n-1} \right\}, \tag{5}$$

where k is the consistency constant of proportionality, n is the flow index exponent, which measures the degree to which the fluid is shear-thinning or shear-thickening, m is a stress growth parameter, and τ_0 is the yield shear stress. Note that the original Papanastasiou regularization does not include the artificial upper-bounding by η_0. However, this bounding is needed to avoid an infinite viscosity for $\dot{\gamma} \to 0$ and $n < 1$. The Herschel–Bulkley model is reduced to the Newtonian model when the yield-stress is zero, i.e., $\tau_0 = 0$, and the flow index exponent is one, i.e., $n = 1$.

2.3. Temperature Dependency of the Apparent Viscosity

To analyze the effect of the temperature dependency of the apparent viscosity on hydrodynamic and thermal behaviour of the flow, we keep constant the flow index n and the yield stress τ_0 and consider only a temperature-dependent consistency $k(T)$:

$$k(T) = a_T(T)k_0, \tag{6}$$

where a_T denotes the shift factor and k_0 is the consistency viscosity at the reference temperature T_0. The shift factor a_T is defined by the Williams–Landel–Ferry (WLF) function:

$$\log(a_T) = \frac{-c_1(T - T_0)}{c_2 + T - T_0}, \tag{7}$$

where c_1 and c_2 are material parameters.

2.4. Boundary and Initial Conditions

Considering that the fluid phases are immiscible, the fluid flow Equations (1)–(3) can be used for each phase individually, and at the interface, the proper boundary conditions must be used. First, the *kinematic condition* states that the fluid velocities on the two sides of the interface, $\mathbf{u}_{1f}$ and $\mathbf{u}_{2f}$, must be continuous

$$\mathbf{u}_{1f} = \mathbf{u}_{2f}. \tag{8}$$

Then, from the momentum conservation law, Equation (2), follows the *dynamic condition*, which states that forces acting on the fluid at the interface are in equilibrium,

$$[\boldsymbol{T}_2 - \boldsymbol{T}_1] \cdot \mathbf{n} = \nabla_s \sigma - \sigma\kappa\mathbf{n}, \tag{9}$$

where $\boldsymbol{T}_1$ and $\boldsymbol{T}_2$ are the stress tensors defined in terms of the local fluid pressure and velocity fields as $\boldsymbol{T}_1 = -p_1\mathbf{I} + \eta_1(\dot{\gamma})[\nabla\mathbf{u}_1 + (\nabla\mathbf{u}_1)^T]$ and $\boldsymbol{T}_2 = -p_2\mathbf{I} + \eta_2(\dot{\gamma})[\nabla\mathbf{u}_2 + (\nabla\mathbf{u}_2)^T]$, respectively, σ is the interfacial tension and $\nabla_s = [\mathbf{I} - \mathbf{nn}] \cdot \nabla = \nabla - \mathbf{n}\frac{\partial}{\partial n}$ is the tangential gradient operator, which appears because σ and $\mathbf{n}$ are defined only on the surface. From Equation (9) we derive the normal and tangential force balances [29] appropriate at a fluid–fluid interface,

$$p_2 - p_1 = \sigma\kappa - 2(\eta_2(\dot{\gamma}) - \eta_1(\dot{\gamma}))\nabla_s \cdot \mathbf{u}, \tag{10}$$

$$\begin{aligned} \eta_2(\dot{\gamma})[\mathbf{n} \cdot (\nabla\mathbf{u}_t)_2] - \eta_1(\dot{\gamma})[\mathbf{n} \cdot (\nabla\mathbf{u}_t)_1] = &- \nabla_s\sigma \\ &- \mathbf{n}(\eta_2(\dot{\gamma}) - \eta_1(\dot{\gamma}))\nabla_s \cdot \mathbf{u} \\ &- (\eta_2(\dot{\gamma}) - \eta_1(\dot{\gamma}))(\nabla_s\mathbf{u}) \cdot \mathbf{n}, \end{aligned} \tag{11}$$

where $\kappa = -\nabla_s \cdot \mathbf{n}$ is twice the mean curvature of the interface and $\mathbf{u}_t = (I - \mathbf{nn}) \cdot \mathbf{u}$ is the tangential velocity component.

Dirichlet or Neumann boundary conditions for temperature T are specified depending on the boundary wall considered. Natural convection on boundaries are treated by setting the gradient (Neumann type boundary condition) according to

$$\nabla T_b = \frac{h}{k_p}(T_b - T_\infty), \tag{12}$$

where the temperature on the boundary T_b is obtained from the previous iteration, h is the convection heat transfer coefficient, and T_∞ is the ambient temperature.

2.5. Arbitrary Lagrange–Eulerian Formulation

The above mathematical model, valid for arbitrary moving volume, is obtained from the corresponding material volume model using the Reynolds' transport theorem. For an arbitrary moving volume, the relationship between the rate of change of the volume V and the velocity $\mathbf{u}_s$ is defined by the *geometrical (space) conservation law* [30],

$$\frac{d}{dt}\int_V dV - \oint_S \mathbf{n}\cdot\mathbf{u}_s\, dS = 0. \tag{13}$$

The problem of extrudate swelling contains moving boundaries due to the movement of the free surfaces of the extrudate. Thus, the domain is described with a mesh that is moving in time in such a way that the mesh moves with the free surfaces.

3. Numerical Method

The mathematical model for the non-isothermal free-surface flow of an incompressible generalized Newtonian fluid described in Section 2 is numerically discretized using the finite volume method [27]. First, the numerical integration in time is performed using a second-order accurate implicit method [31], referred to as the backward scheme. Next, the integral forms of the fluid flow equations are discretized in space using a second-order accurate cell-centred unstructured finite volume method. The spatial domain is discretized using a mesh, which is constituted by finite volumes (the so-called cells or elements) with an arbitrary volume V bounded by a closed moving surface S, that conserve the relevant quantities, such as mass, momentum, and energy:

$$\oint_S \mathbf{n}\cdot\mathbf{u}\, dS = 0, \tag{14}$$

$$\frac{d}{dt}\int_V \mathbf{u}\, dV + \oint_S \mathbf{n}\cdot(\mathbf{u}-\mathbf{u}_S)\mathbf{u}\, dS = -\int_V \nabla P\, dV + \oint_S \mathbf{n}\cdot\boldsymbol{\tau}\, dS, \tag{15}$$

$$\rho c_p\left(\frac{d}{dt}\int_V T\, dV + \oint_S \mathbf{n}\cdot(\mathbf{u}T)\, dS\right) = \oint_S \mathbf{n}\cdot(k_p\nabla T)\, dS + \int_V (\boldsymbol{\tau}:\mathbf{D})\, dV. \tag{16}$$

Detailed information of the finite volume discretization employed in the moving mesh interface tracking algorithm can be found in Tuković and Jasak [27]. To summarize, the surface integrals of an integral conservation equation are transformed into sums of face integrals which together with the volume integrals are approximated to second-order accuracy by using the mid-point rule. Therefore, Equations (14)–(16) for each cell are written as,

$$\sum_f \mathbf{n}_f^n\cdot\mathbf{u}_f^n\, S_f^n = 0, \tag{17}$$

$$\frac{3\mathbf{u}_P^n V_P^n - 4\mathbf{u}_P^o V_P^o + \mathbf{u}_P^{oo} V_P^{oo}}{2\Delta t} + \sum_f (\dot{m}_f^n - \dot{U}_f^n)\mathbf{u}_f^n = -(\nabla P)_P^n V_P^n + \sum_f \mathbf{n}_f^n\cdot\boldsymbol{\tau}_f^n S_f^n, \tag{18}$$

$$\begin{aligned}\rho_P(c_p)_P\frac{3T_P^n V_P^n - 4T_P^o V_P^o + T_P^{oo} V_P^{oo}}{2\Delta t} + \sum_f \rho_f(c_p)_f\mathbf{n}_f^n\cdot\mathbf{u}_f^n T_f^n S_f^n = {}&\sum_f (k_p)_f\mathbf{n}_f^n\cdot(\nabla T)_f^n S_f^n \\ &+ (\boldsymbol{\tau}_P^n:\mathbf{D}_P^n)V_P^n,\end{aligned} \tag{19}$$

where Δt is the time-step, the subscripts P and f represent the cell-center and face-center values at cell with volume V_P, the superscripts n, o and oo represent values evaluated at the new time instance t^n and two previous time instances t^o and $t^{oo} = t^o - \Delta t$. Finally, the cell-face mass flux $\dot{m}_f^n = \mathbf{n}_f^n\cdot\mathbf{u}_f^n S_f^n$ must satisfy the discretized mass conservation law (Equation (17)), and the face volume flux $\dot{U}_f^n$ must satisfy the discretized geometrical conservation law (Equation (13)). Equation (17) is linearized for a pressure correction according to the consistent non-iterative PISO algorithm [32–34]. The algorithm can be summarized as follows [35]:

1. Set initial guess of the solution at time t for pressure, velocity, temperature and mass flow rate fields p^n, $\mathbf{u}^n$, T^n, and $\dot{m}_f^n$, respectively.
2. Define displacement directions for the interfacial mesh points and the control points.
3. In order to compensate the net mass flux through the interface, calculate displacement of the interface mesh points (the least-squares volume-point interpolation scheme was employed [36]).
4. Displacement of the interface mesh points is used as a boundary condition for the solution of the mesh motion problem. After mesh movement, the new face volume fluxes $\dot{U}_f^n$ are calculated.
5. Update pressure and velocity boundary conditions at the interface.
6. Assemble and solve *implicitly* the momentum equation given by Equation (18) to obtain a new velocity field $\mathbf{u}^*$.
7. Compute the mass flow rate at the cell faces $\dot{m}_f$ using the Rhie–Chow interpolation technique [37].
8. Using the new mass flow rates computed in Step 7, assemble the pressure correction equation (Equation (17)) and solve it to obtain a pressure correction field p'.
9. Update the pressure and velocity fields at the cell centroids, p^n and $\mathbf{u}^*$, respectively, and correct the mass flow rate at the cell faces $\dot{m}_f^*$, to obtain continuity-satisfying fields p^*, $\mathbf{u}^{**}$ and $\dot{m}_f^{**}$. The consistent version of the SIMPLE (Semi Implicit Method for Pressure Linked Equations) algorithm is used here by assuming that the velocity correction at point P is the weighted average of the corrections at the neighboring grid points [32–34], resulting in a better estimating for the velocity corrections, and consequently, a higher rate of convergence is obtained [15].
10. Using the latest available velocity and pressure fields, $\mathbf{u}^{**}$ and p^*, respectively, assemble and solve *explicitly* the momentum equation to obtain a new velocity field $\mathbf{u}^{***}$.
11. Update the mass flow rate at the cell faces $\dot{m}_f^{***}$ using the Rhie–Chow interpolation technique [37].
12. Using the new mass flow rates computed in Step 11, assemble the pressure correction equation and solve it to obtain a pressure correction field p''.
13. Update the pressure and velocity fields at the cell centroids, p^* and $\mathbf{u}^{***}$, respectively, and correct the mass flow rate at the cell faces $\dot{m}_f^{***}$, to obtain p^{**}, $\mathbf{u}^{****}$ and $\dot{m}_f^{****}$.
14. Go to Step 10 and repeat for a given number of corrector steps ($nCorrectors = 3$ in this work).
15. Solve the temperature equation, Equation (19), to obtain T^* and update the temperature-dependent consistency viscosity Equations (6) and (7).
16. Set the initial guess for pressure, velocity, temperature and mass flow rate as $p^n = p^{**}$, $\mathbf{u}^n = \mathbf{u}^{****}$, $T^n = T^*$ and $\dot{m}_f^n = \dot{m}_f^{****}$, respectively.
17. Repeat from Step 3 for a given number of times ($nOuterCorrectors = 10$ in this work).
18. Set the converged solution at time $t = t + \Delta t$ and advance to the next time step.
19. Return to Step 1 and repeat until the last time step is reached.

The Poisson-type equation for pressure is solved with a conjugate gradient method with Cholesky preconditioner and the velocity and temperature linear systems are solved using BiCGstab with an Incomplete Lower-Upper (ILU) preconditioning [38–40].

4. Results and Discussion

In this section, we present the validation and assessment of the newly developed moving mesh finite-volume interface tracking solver that is able to efficiently handle inelastic non-Newtonian matrix-based free-surface flows with non-isothermal effects. In this work, the fluid rheology is described by the Herschel–Bulkley constitutive equation with the Papanastasiou regularization. The newly developed solver is tested against fluid flow simulations in an axisymmetric domain geometry typical of pipe extrusion. The first case study is devoted to the extrudate swell of a Bingham fluid, and subsequently, the extrudate swell of Herschel–Bulkley shear-thinning and shear-thickening flows is presented. These studies aim to verify the solver's capabilities to accurately predict the extrudate swell of inelastic non-Newtonian matrix-based fluids. Additionally, the effects of inertia and yield stress on the extrudate swell are investigated. Finally, the non-isothermal extrudate swell effects are studied for a Bingham fluid at moderate Reynolds number ($Re = 10$), in order to test the robustness of the newly developed numerical algorithm, specifically for non-isothermal calculations.

4.1. Problem Domain and Meshes

The benchmark case study that will be discussed is the axisymmetric extrudate swell of non-Newtonian inelastic fluids. A schematic representation of the computational flow domain, the boundary faces, and the discretization mesh for the initial time step ($t = 0$) and at a steady state is shown in Figure 1. The level of mesh refinement used in the numerical studies carried out in this study corresponds to the same level employed in mesh M5 of our previous work [15], which resulted from a mesh convergence analysis. Polar coordinates are employed for the description of the axisymmetric flow domain, thus $\mathbf{x} = (r, z)$. The half width of the axisymmetric channel is denoted as R, which is considered to be the scaling length. The inlet plane is taken sufficiently far upstream from the exit so that the flow is fully developed with a mean velocity U. In the axisymmetric domain, the two lateral boundary sides are considered to be `wedge` patches (i.e., the cylinder is specified as a wedge of small angle, e.g., 5°, and a thickness cell running along the plane of symmetry, encompassing one of the coordinate planes). At the bottom, the `axis` of symmetry is considered as `empty` patch. At the solid die wall, the no-slip (tangential velocity is zero) and no-penetration (normal velocity is zero) conditions are imposed for velocity, zero-gradient pressure, and a fixed value temperature. At the free-surface, the kinematic condition, Equation (8), and the dynamic condition, Equations (10) and (11), are imposed for pressure and velocity, along with natural convection, Equation (12), for temperature. Finally, the outflow plane is taken sufficiently far downstream of the die exit so that the flow is uniform. The die exit of the axisymmetric domain is located at $x = 5R$ from the inlet, and the outflow is located at $x = 25R$ from the die exit.

The dimensionless numbers governing the flow are the Reynolds number,

$$Re = \frac{\rho U^{2-n} R^n}{k}, \tag{20}$$

which is the ratio of inertial forces to viscous forces within a fluid that is subject to relative internal movement due to different fluid velocities; the Bingham number,

$$Bn = \frac{\tau_0 R^n}{k U^n}, \tag{21}$$

which is the ratio of yield stress to viscous stress and describes the extent to which the controllable yield stress can exceed the viscous stress; the dimensionless growth exponent,

$$M = \frac{mU}{R}, \tag{22}$$

that, by using a regularized constitutive equation such as the Papanastasiou model, it will determine up to which convergent results can be obtained (in our simulations $M \geq 500$ for $0 < Bn \leq 10$); and the Prandtl number,

$$Pr = \frac{k c_p}{k_p} \left(\frac{U}{R} \right)^{n-1}, \tag{23}$$

which is the ratio of moment diffusivity (kinematic viscosity) and thermal diffusivity of a fluid, expressing the relationship between the movement quantity diffusion and the heat quantity diffusion within the fluid itself.

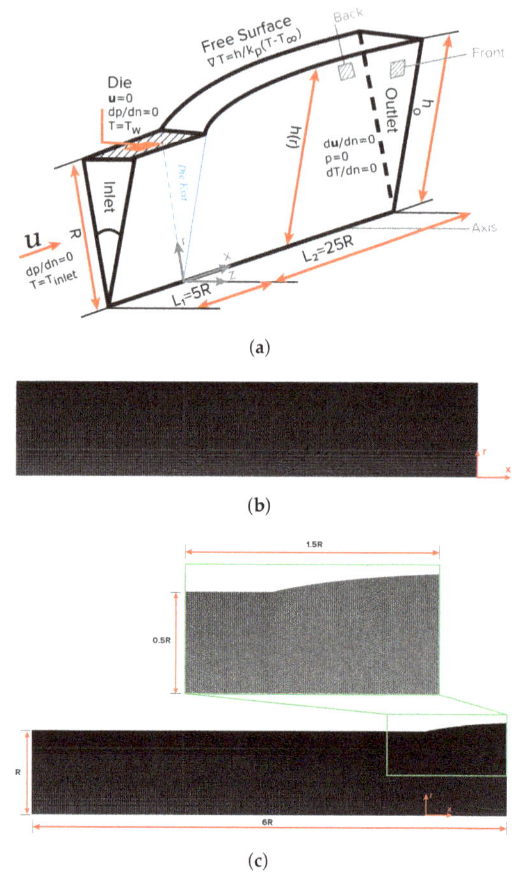

Figure 1. Schematic representation of (**a**) the axisymmetric extrudate swell domain geometry and boundary faces (**b**) of an indicative discretization mesh at the initial time step $t = 0$, and (**c**) at steady state.

4.2. The Effects of Inertia and Yield-Stress in the Extrudate Swell of Bingham Fluids

Initially, the combination of inertia and yield-stress effects on the extrudate swell ratio obtained for the isothermal flow of Bingham fluids (i.e., $n = 1$) is studied. The swell ratio is defined as the height of the free-surface away from the die exit, where the plug flow has been established, divided by the die radius, i.e., $\chi = h_0/R$. As shown in Figure 2, the extrudate swell ratio obtained by the newly developed solver is very close to the results presented by Kountouriotis et al. [41] at $Re = 1, 5$, and 10. For the lowest Reynolds number ($Re = 1$), as the yield stress effect is enhanced (i.e., increase of the Bn number), the extrudate swell shrinks steeply for $Bn > 0.1$, and above a critical value of $Bn > 5$, the swell contracts and becomes smaller than unity. At $Re = 5$ the extrudate swell ratio increases with increasing Bn, reaching a maximum at $Bn \approx 1$. Subsequently, the extrudate swell ratio decreases and becomes lower than unity for $Bn > 5$. At $Re = 10$, the extrudate swell ratio is below one for the lowest Bn numbers ($Bn < 0.5$), and then it enlarges, being higher than one as Bn enhances, reaching a maximum at $Bn \approx 2$. After that the swell ratio χ diminishes and again becomes lower than unity for $Bn > 5$. Regardless of the Re number used in the simulations, for $Bn \geq 5$, the extrudate swell ratio is very similar on each curve, which means that the yield stress effects are predominant compared to the inertial effects.

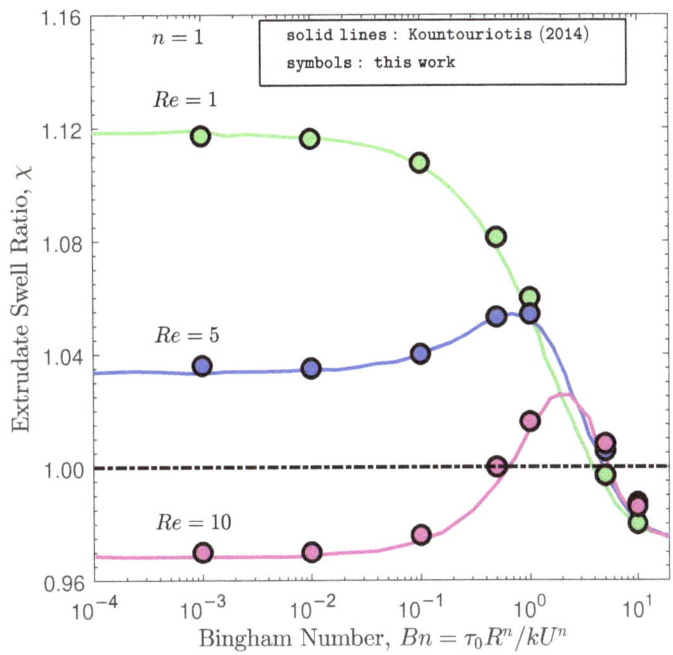

Figure 2. Steady-state extrudate swell ratio χ for the isothermal extrudate swell of Bingham fluids ($n = 1$) at $Re = \{1, 5, 10\}$ and $Bn = \{10^{-3}, 10^{-2}, 10^{-1}, 0.5, 1, 5, 10\}$. Solid lines represent the results obtained by Kountouriotis et al. [41], and the symbols represent the results obtained by the newly developed interface tracking code.

In Figure 3, the magnitude of the polymer velocity vector is shown for $Re = 1, 5$, and 10 and $Bn = 0.001$ and 10. At $Bn = 0.001$ (negligible yield-stress effects), the maximum value of the magnitude of the polymer velocity vector is twice the inlet velocity and the enhancement of inertia effects through the increase in the Re number leads to the reduction of the die swell ratio, mimicking the behavior of the extrudate swell for a Newtonian fluid, as presented by Fakhari et al. [15]. When the highest Bn number is considered ($Bn = 10$), the maximum value of the magnitude of the polymer velocity vector is 1.3 times the inlet velocity. This means that the increase in the yield stress promotes the retardation of the flow, changing from parabolic shape profile to plug-flow. The consequence is that the extrudate swell ratio decreases and, in fact, contracts. Note that the contour plots of the magnitude of the polymer velocity vector are very similar at this higher Bingham number, where increasing inertia effects do not change the shape of the flow field.

Figure 4 shows the contours of the magnitude of the polymer stress tensor at steady-state for $Re = 1, 5$, and 10 at $Bn = 0.001$ and 10. When the yield stress effects are negligible ($Bn = 0.001$), the maximum value of the stress tensor magnitude occurs for the lowest Reynolds number ($Re = 1$), corresponding to the highest extrudate swell ratio. Increasing the inertia effects decreases the localized maximum value of the magnitude of the polymer stress tensor in the top right corner of the die wall, leading to a decrease in the extrudate swell ratio. On the other hand, when the yield stress effects dominate the flow (i.e., $Bn = 10$), the maximum value of the magnitude of the polymer stress tensor is constant for all Re numbers from 1 to 10, being three orders of magnitude lower than the case with $Bn = 0.001$.

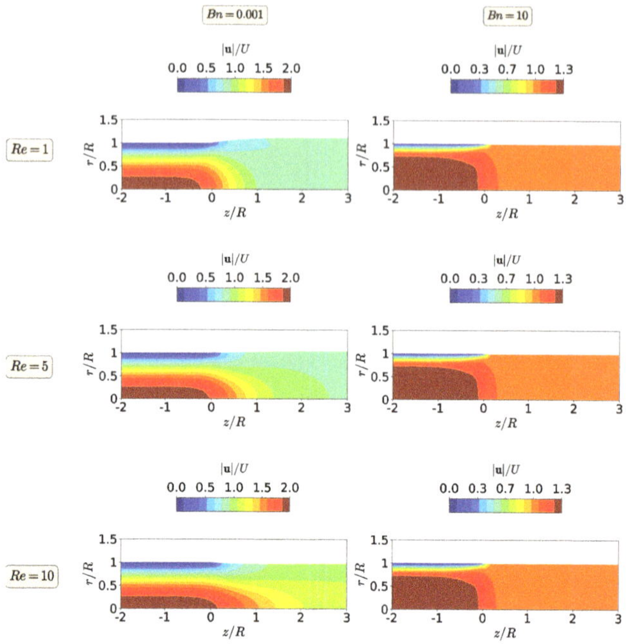

Figure 3. Contours of the magnitude of the polymer velocity vector at steady-state for the isothermal extrudate swell of Bingham fluids ($n = 1$) when $Re = 1$ (**top**), $Re = 5$ (**middle**) and $Re = 10$ (**bottom**), and $Bn = 0.001$ (**left**) and $Bn = 10$ (**right**).

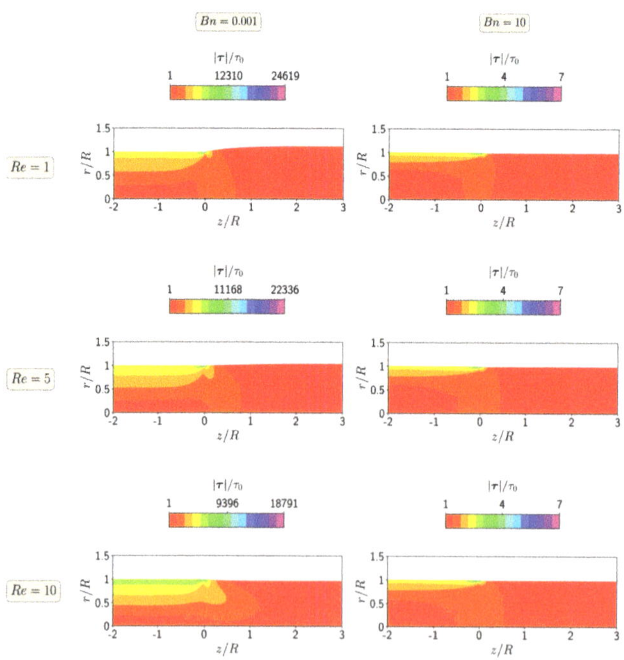

Figure 4. Contours of the magnitude of the polymer stress tensor at steady-state for the isothermal extrudate swell of Bingham fluids ($n = 1$) when $Re = 1$ (**top**), $Re = 5$ (**middle**) and $Re = 10$ (**bottom**), and $Bn = 0.001$ (**left**) and $Bn = 10$ (**right**).

4.3. The Effects of Inertia and Yield-Stress in the Extrudate Swell of Herschel–Bulkley Fluids

In this section, the effects of inertia and yield-stress are investigated on the isothermal extrudate swell of Herschel–Bulkley flows ($n \neq 1$). For this purpose, the simulations were carried out at the same Re and Bn numbers as those presented in Section 4.2, but with a flow index exponent representative of shear-thinning, $n = 0.5$, and shear-thickening, $n = 1.5$, behaviors. As shown in Figure 5, the extrudate swell ratio obtained by the newly developed solver is again very close to the results presented by Kountouriotis et al. [41] for both $n = 0.5$ and $n = 1.5$ at $Re = 1$, 5, and 10. The general trend of the shear-thinning extrudate swell ratio (see Figure 5a) at the different Re numbers is similar to the behavior shown for the Bingham flow ($n = 1$) in Figure 2. In Figure 5a, we can see that, in general, for $n = 0.5$, the extrudate swell ratio is lower than the ones obtained for $n = 1$ and $n = 1.5$. This behavior allows us to conclude that the shear-thinning nature of the Herschel–Bulkley fluid helps to reduce the swelling of the polymer. On the other hand, for the shear-thickening behavior ($n = 1.5$), the extrudate swell ratio at different Re and Bn numbers is the highest, as shown in Figure 5b. Additionally, the variation in the extrudate swell ratio with the increase in the yield stress effects is more abrupt for this shear-thickening fluid, reaching a maximum of 20% at $Re = 1$.

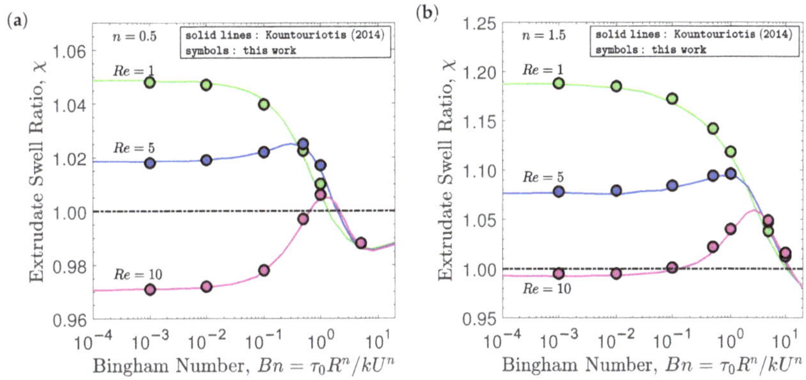

Figure 5. Steady-state extrudate swell ratio χ for the isothermal extrudate swell of Herschel–Bulkley fluids at $Re = \{1, 5, 10\}$ and $Bn = \{10^{-3}, 10^{-2}, 10^{-1}, 0.5, 1, 5, 10\}$ with (**a**) $n = 0.5$ and (**b**) $n = 1.5$. Solid lines represent the results obtained by Kountouriotis et al. [41], and the symbols represent the results obtained by the newly developed interface tracking code.

Figure 6 shows the contours of the magnitude of the polymer velocity vector for the isothermal extrudate swell of shear-thinning Herschel-Bulkley fluids ($n = 0.5$), when $Re = 1$, 5 and 10 and $Bn = 0.001$s and 10. At $Bn = 0.001$ (negligible yield stress effects), the maximum value of the magnitude of the polymer velocity vector is approximately 1.7 times the inlet velocity for $1 \leq Re \leq 10$, which is 15% smaller than the value obtained for the Bingham flow ($n = 1$), leading to variations of 7% in the extrudate swell ratio when increasing Re from 1 to 10. On the other hand, when the highest Bn number is considered ($Bn = 10$), the maximum value for the magnitude of the polymer velocity vector is 1.1 times the inlet velocity, which is again 15% smaller than the value obtained for the Bingham flow ($n = 1$). However, for this strongly dominated yield stress flow, the extrudate swell ratio is kept constant for $1 \leq Re \leq 10$ and again with a plug-flow velocity profile leading to swell contraction ($\chi < 1$).

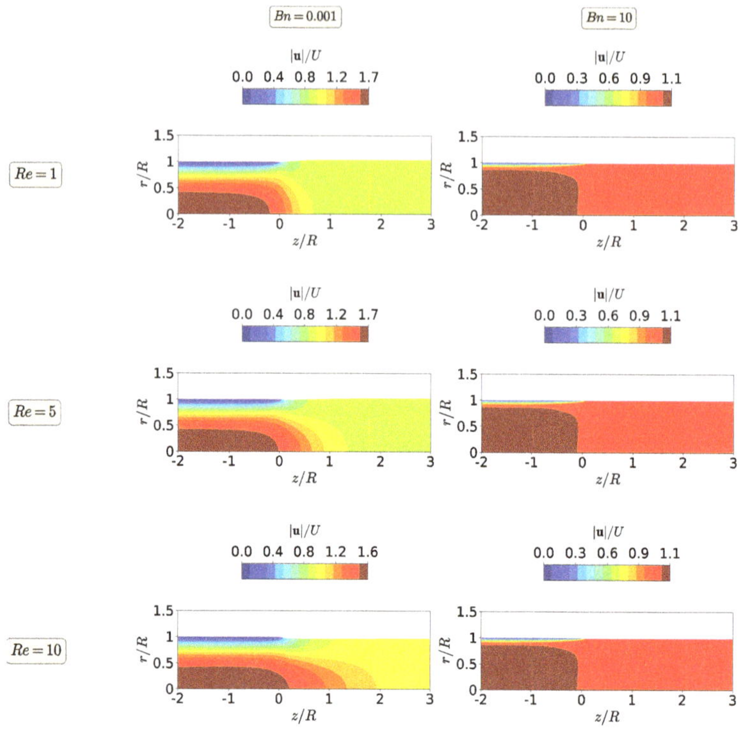

Figure 6. Contours of the magnitude of the polymer velocity vector at steady-state for the isothermal extrudate swell of shear-thinning Herschel–Bulkley fluids ($n = 0.5$) when $Re = 1$ (**top**), $Re = 5$ (**middle**) and $Re = 10$ (**bottom**), and $Bn = 0.001$ (**left**) and $Bn = 10$ (**right**).

In Figure 7, we show the contour plots of the magnitude of the polymer stress tensor at steady-state for the isothermal extrudate swell of shear-thinning Herschel–Bulkley fluids ($n = 0.5$) when $Re = 1, 5$, and 10 and $Bn = 0.001$ and 10. When the yield stress effects are negligible ($Bn = 0.001$), the maximum value of the magnitude of the polymer stress tensor decreases with the increase of inertia from $Re = 1$ to 10, while at the highest Bn number, the inertia do not have influence on the maximum value of the magnitude of the polymer stress tensor. In general, for both of the Bn numbers, the magnitudes of the polymer stress tensor are approximately one third of the ones obtained for the Bingham fluid ($n = 1$).

Figure 8 displays the contours of the magnitude of the polymer velocity vector for the isothermal extrudate swell of shear-thickening fluids ($n = 1.5$), when $Re = 1, 5$, and 10 and $Bn = 0.001$ and 10. As can be seen by the velocity contours at $Bn = 0.001$, the maximum value for the magnitude of the polymer velocity vector is 2.2 times the inlet velocity for $1 \leq Re \leq 10$, which is 10% larger than the value obtained for the Bingham flow ($n = 1$), leading to variations of 17% in the extrudate swell ratio when Re is increased from 1 to 10. At the highest Bingham number ($Bn = 10$), increasing the inertia effects does not change the extrudate swell ratio and the maximum value of the polymer velocity vector magnitude. The later remains 1.5 times the inlet velocity, which is approximately 13% larger than the one obtained for the Bingham fluid ($n = 1$).

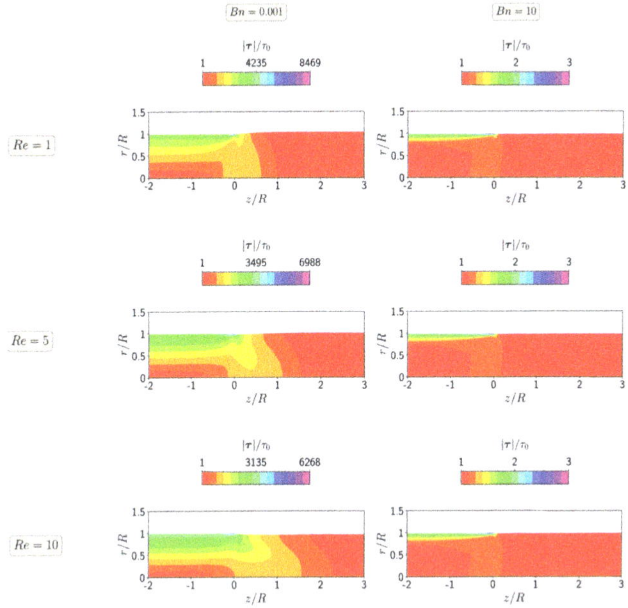

Figure 7. Contours of the magnitude of the polymer stress tensor at steady state for the isothermal extrudate swell of shear-thinning Herschel–Bulkley fluids ($n = 0.5$) when $Re = 1$ (**top**), $Re = 5$ (**middle**) and $Re = 10$ (**bottom**), and $Bn = 0.001$ (**left**) and $Bn = 10$ (**right**).

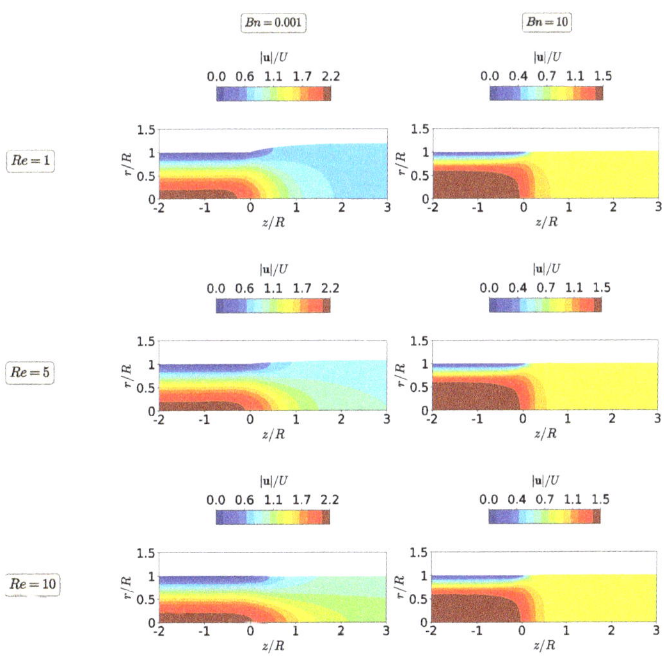

Figure 8. Contours of the magnitude of the polymer velocity vector at steady state for the isothermal extrudate swell of shear-thickening Herschel–Bulkley fluids ($n = 1.5$) when $Re = 1$ (**top**), $Re = 5$ (**middle**) and $Re = 10$ (**bottom**), and $Bn = 0.001$ (**left**) and $Bn = 10$ (**right**).

Finally, in Figure 9, we show the steady-state contour plots of the magnitude of the polymer stress tensor for the isothermal extrudate swell of shear-thickening Herschel–Bulkley fluids ($n = 1.5$) when $Re = 1, 5$, and 10 and $Bn = 0.001$ and 10. As before, the stress contours follow the same trend as in the other fluids, i.e., the maximum magnitude of the polymer stress tensor decreases with an increase in inertia at $Bn = 0.001$, while it is constant for the yield-stress-dominated flow ($Bn = 10$), regardless of the enhancement of inertia from $Re = 1$ to 10. Note also that the maximum magnitude of the polymer stress tensor for the shear-thickening Herschel–Bulkley fluid is around twice larger than the value obtained for the Bingham fluid at both Bn numbers ($Bn = 0.001$ and 10).

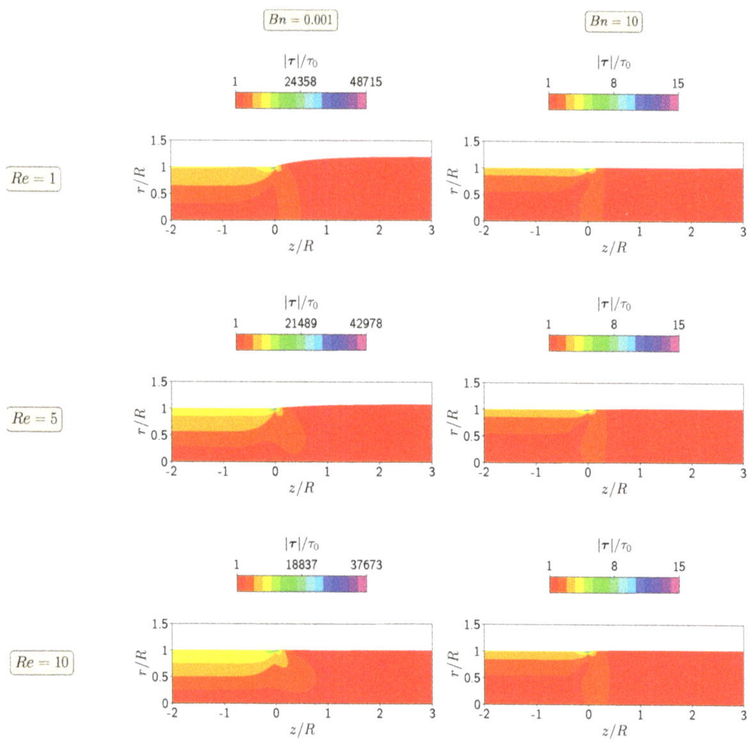

Figure 9. Contours of the magnitude of the polymer stress tensor at steady-state for the isothermal extrudate swell of shear-thickening Herschel–Bulkley fluids ($n = 1.5$) when $Re = 1$ (**top**), $Re = 5$ (**middle**) and $Re = 10$ (**bottom**), and $Bn = 0.001$ (**left**) and $Bn = 10$ (**right**).

4.4. Non-Isothermal and Yield Stress Effects in the Extrudate Swell of Bingham Fluids

In this section, we study the effects of temperature and yield stress in the extrudate swell ratio of Bingham fluids ($n = 1$) at Reynolds number $Re = 10$. We consider two different scenarios for the die wall temperature T_w, one where $T_w < T_{inlet}$ (cold wall) and another where $T_w > T_{inlet}$ (hot wall). We will then examine the thermally induced swelling behavior as the Bingham number increases for these two configurations.

The thermal and physical properties of a typical polystyrene [7] used in the simulations are listed in Table 1. An important issue for modeling the cooling of the polymer when it is extruded is the definition of the boundary condition at the polymer and air interface. In this work, we employed a defined convective heat flux as given in Equation (12). Additionally, the influence of the temperature on the rheological behavior of the material is controlled by the WLF equation for the shift factor employed in the temperature-dependent consistency parameter k (see Equations (6) and (7)). Typical extreme sets of WLF parameters (c_1, c_2) are (4.54, 150.36), leading to thermorheological coupling [42].

Table 1. General conditions used in the non-isothermal simulations of the extrudate swell of Bingham fluids ($n = 1$).

k_p (W/m °C)	0.18
ρ (kg/m^3)	1400
c_p (J/kg °C)	1000
Inlet profile temperature, T_{inlet} (°C)	180
Die wall temperature, T_w (°C)	140 or 220
Room temperature, T_∞ (°C)	20
Air convection heat transfer coefficient (free convection), h (W/m^2 °C)	5

In this case study, the ratio of moment diffusivity and thermal diffusivity of the fluid, defined by the dimensionless Prandtl number, is given by $Pr = 0.7$ (see Equation (23)). A small Pr number means that heat diffuses very easily compared to moment.

In Figure 10, we show the extrudate swell ratio obtained by the newly developed solver for the non-isothermal flow of Bingham fluids at $Re = 10$ and $0.001 \leq Bn \leq 10$, with a cold (blue symbols) and hot (red symbols) die wall. For the cold wall, the extrudate swell ratio has a decreasing monotonic behavior, but being always larger than unity, meaning that the polymer expands after the die wall. Note that the decreasing monotonic behavior is related to the enhancement of the yield stress effects. Specifically, the extrudate swell ratio χ slightly decreases from $Bn = 0.001$ to $Bn = 0.1$, and then it falls with a sharper slope, having its smallest value around 1.06 at $Bn = 10$. In contrast with the cold die wall, the extrudate swell ratio for the heating case study follows the same trend as the isothermal studies, where we first see an increase in the extrudate swell ratio followed up by a decrease due to the larger yield stress effects. Only at $Bn = 2$ and 3, the polymer expands after the die wall, but for all the other Bn numbers, the polymer melt contracts, which is a similar behavior to the shear-thinning isothermal extrudate swell at $Re = 10$ presented in Section 4.3. Note that in Labsi et al. [43], it is shown that neglecting the viscosity's temperature dependency leads to undervaluing hydrodynamic properties, especially in the cooling case. If we compare the results shown in Figures 2 and 10 at $n = 1$ and $Re = 10$, we see that the extrudate swell ratio for the case neglecting viscosity's temperature dependency (Figure 2) is smaller than the one obtained when viscosity's temperature dependence is taken into account (Figure 10), and this behavior is more noticeable for the cooling case, corroborating the conclusion presented in Labsi et al. [43].

Figure 11 shows the magnitude of the polymer velocity vector at $Re = 10$ for $Bn = 0.001$ and 10, and for the cold and hot die wall case studies. For $Bn = 0.001$ (negligible yield-stress effects), the maximum value of the magnitude of the polymer velocity vector is 30% higher in the cold die wall case study and 5% lower in the hot die wall case study, than the maximum values obtained for isothermal conditions. When the higher Bn number is considered ($Bn = 10$), the maximum value of the magnitude of the polymer velocity vector is 46% higher in the cold die wall case study and 8% lower in the hot die wall case study, than the maximum values obtained for isothermal conditions. Notice also that for the cold die wall case study at $Bn = 10$ the velocity profile has changed from a plug-flow shape in isothermal conditions to a parabolic profile shape now. This means that the non-isothermal effects are stronger than the yield stress effects for the cold die walls case study at $Re = 10$. For the hot die wall case study at $Bn = 10$, the velocity profile maintains the plug-flow shape profile, leading to a reduction in the extrudate die-swell ratio.

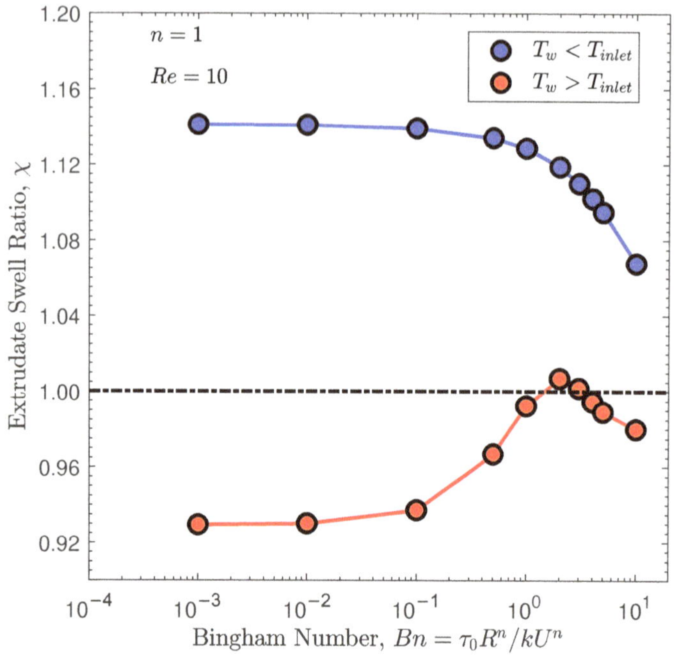

Figure 10. Steady-state extrudate swell ratio χ for the non-isothermal axisymmetric extrudate swell of Bingham fluids ($n = 1$) at $Re = 10$ and $Bn = \{10^{-3}, 10^{-2}, 10^{-1}, 0.5, 1, 2, 3, 4, 5, 10\}$.

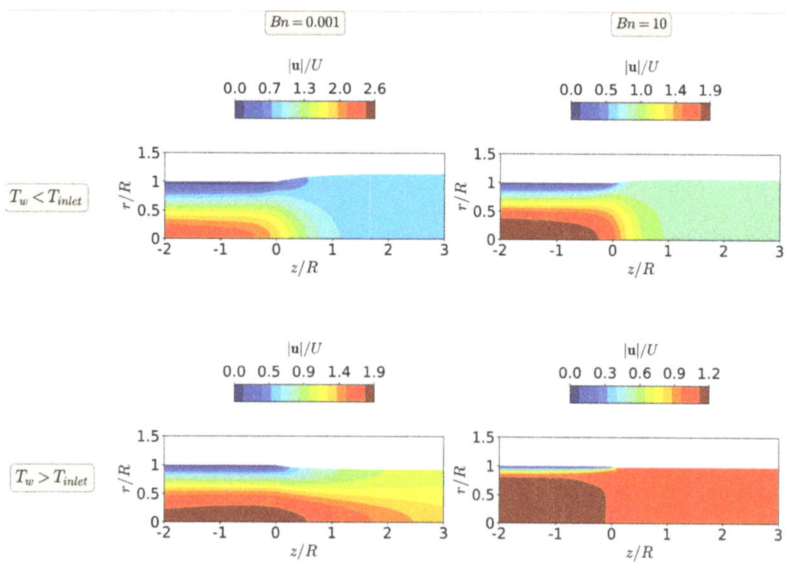

Figure 11. Contours of the magnitude of the polymer velocity vector at steady-state for the non-isothermal asymmetric extrudate swell flow of Bingham fluids ($n = 1$) at $Re = 10$ for $Bn = 0.001$ (**left**) and $Bn = 10$ (**right**), when $T_w < T_{inlet}$ (**top**) and $T_w > T_{inlet}$ (**bottom**).

In Figure 12, we show contours of the steady-state dimensionless temperature field at $Re = 10$ for $Bn = 0.001$ and 10, and for the cold and hot die wall case studies. As can be seen in Figure 12, the temperature variation $\Delta\theta$ for the case where cold die walls are used in the simulations is approximately twice the one obtained when hot die walls are employed, which will not induce a high swelling ratio for the latter case (hot die walls). This is due to the fact that the material flows toward the cooler (more viscous) region as the flow rearranges in the extrudate, which will give higher swelling ratio of the cooler side (as shown in the cold die walls).

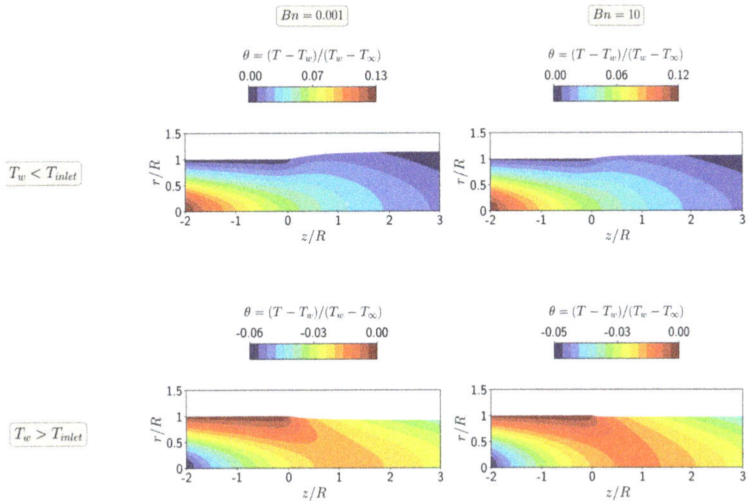

Figure 12. Contours of dimensionless temperature at steady-state for the non-isothermal asymmetric extrudate swell flow of Bingham fluids ($n = 1$) at $Re = 10$ for $Bn = 0.001$ (**left**) and $Bn = 10$ (**right**), when $T_w < T_{inlet}$ (**top**) and $T_w > T_{inlet}$ (**bottom**).

5. Conclusions

A numerical algorithm able to solve non-isothermal and inelastic non-Newtonian free-surface flows based on the Arbitrary Lagrangian–Eulerian (ALE) formulation was presented and implemented using the finite-volume method. The implementation was performed in the open-source OpenFOAM framework [26], where the interface is tracked in a semi-implicit manner, which allowed robust and stable deformations of the interface.

The newly developed algorithm was assessed in terms of accuracy for isothermal flow simulations of the axisymmetric extrudate swell of Bingham and Herschel–Bulkley fluids. The effects of inertia and yield stress on the extrudate swell ratio computed by the newly developed algorithm were studied and compared with results found in the scientific literature for the range of the dimensionless Reynolds (Re) and Bingham (Bn) numbers as follows: $1 \leq Re \leq 10$ and $0.001 \leq Bn \leq 10$. Additionally, the contours for the magnitude of the polymer velocity vector and stress tensor fields are also shown and discussed. For the isothermal Bingham flows, the extrudate swell ratio was found to vary by approximately 13% from the lowest to the highest Re number when the yield stress effects are negligible ($Bn = 0.001$). Additionally, at higher Bn numbers, the yield stress effects are dominant and the extrudate swell ratio is equal for all the Re numbers tested, being lesser than unity, which means that the swell contracts. For the isothermal Herschel–Bulkley flows of the shear-thinning fluid, the extrudate swell ratio was found to vary by approximately 8% from the lowest to the highest Re number when the yield stress effects are negligible ($Bn = 0.001$). On the other hand, at higher Bn numbers, the yield stress effects are dominant and the extrudate swell ratio is equal for all the Re numbers tested, being also lesser than unit as in the Bingham flow case, meaning that the swell contracts. In the case of using a shear-thickening fluid, the extrudate swell ratio was found to vary by approximately 17% from the lowest to the highest Re number when the yield stress effects are negligible ($Bn = 0.001$). Additionally, at higher Bn numbers, the extrudate swell ratio approaches the unitary value, which means that the polymer melt neither contracts nor expands.

Finally, the newly developed algorithm was applied to study the non-isothermal flow of axisymmetric extrudate swell using Bingham fluids at $Re = 1$. The effects of temperature and yield stress on the extrudate swell ratio computed by the newly developed algorithm were analyzed for the range of the dimensionless Bingham (Bn) numbers, $0.001 \leq Bn \leq 10$, and two different configurations, one representing a cold die wall and the other an hot die wall. Additionally, the contours for the magnitude of the polymer velocity vector and temperature fields are shown and discussed. For the cold die wall it was found that the extrudate swell ratio has a monotonic decreasing behavior but always with a value greater than unity for $1 \leq Bn \leq 10$. However, for the hot die wall, the extrudate swell ratio first increases for $0.001 \leq Bn \leq 2$ and then decreases for $2 \leq Bn \leq 10$, being greater than unity only at $Bn = 2$ and 3.

In summary, the results presented here show that the newly developed interface tracking code can accurately predict the non-isothermal extrudate swell of inelastic non-Newtonian matrix-based fluids. The code that was implemented here is being currently extended to handle viscoelastic non-Newtonian fluid flow calculations.

Author Contributions: Conceptualization, C.F.; Formal analysis, C.F.; Investigation, C.F. and A.F.; Methodology, C.F. and A.F.; Resources, C.F.; Software, C.F. and Ž.T.; Supervision, C.F. and Ž.T.; Validation, C.F. and A.F.; Writing—original draft, C.F. and A.F.; Writing—review & editing, C.F., A.F., and Ž.T. All authors have read and agreed to the published version of the manuscript.

Funding: The authors would like to acknowledge the funding by FEDER funds through the COMPETE 2020 Programme and National Funds through FCT—Portuguese Foundation for Science and Technology under the projects UIDB/05256/2020 and UIDP/05256/2020.

Acknowledgments: The authors also acknowledge the support of the computational clusters Search-ON2 (NORTE-07-0162-FEDER-000086) the HPC infrastructure of UMinho under NSRF through ERDF and FCT I.P. through the Advanced Computing Project CPCA/A00/6057/2020 using the Minho Advanced Computing Center (MACC).

Conflicts of Interest: The authors declare no conflict of interest.

References

1. Mirjalili, S.; Jain, S.S.; Dodd, M. *Interface-Capturing Methods for Two-Phase Flows: An Overview and Recent Developments;* Center for Turbulence Research Annual Research Briefs, Stanford University: Stanford, CA, USA, 2017; pp. 117–135.
2. Nie, Y.; Hao, J.; Lin, Y.-J.; Sun, W. 3D simulation and parametric analysis of polymer melt flowing through spiral mandrel die for pipe extrusion. *Adv. Polym. Technol.* **2018**, *37*, 3882–3895. [CrossRef]
3. Fernandes, C.; Pontes, A.J.; Viana, J.C.; Nóbrega, J.M.; Gaspar-Cunha, A. Modeling of Plasticating Injection Molding—Experimental Assessment. *Int. Polym. Process.* **2014**, *5*, 558–569. [CrossRef]
4. Pedro, J.; Ramôa, B.; Nóbrega, J.M.; Fernandes, C. Verification and Validation of openInjMoldSim, an Open-Source Solver to Model the Filling Stage of Thermoplastic Injection Molding. *Fluids* **2020**, *5*, 84. [CrossRef]
5. Tadmor, Z.; Klein, I. *Computer Programs for Plastic Engineers;* Rein-hold Book Corporation: New York, NY, USA, 1968.
6. Agur, E.E.; Vlachopoulos, J. Numerical Simulation of a Single-Screw Plasticating Extruder. *Polym. Eng. Sci.* **1982**, *22*, 1084–1094. [CrossRef]
7. Habla, F.; Fernandes, C.; Maier, M.; Densky, L.; Ferrás, L.L.; Rajkumar, A.; Carneiro, O.S.; Hinrichsen, O.; Nóbrega, J.M. Development and validation of a model for the temperature distribution in the extrusion calibration stage. *Appl. Therm. Eng.* **2016**, *100*, 538–552. [CrossRef]
8. Rajkumar, A.; Ferrás, L.L.; Fernandes, C.; Carneiro, O.S.; Becker, M.; Nóbrega, J.M. Design guidelines to balance the flow distribution in complex profile extrusion dies. *Int. Polym. Process.* **2017**, *32*, 58–71. [CrossRef]
9. Vlachopoulos, J.; Polychronopoulos, N.D. *Understanding Rheology and Technology of Polymer Extrusion;* Polydynamics Publisher: Hamilton, ON, Canada, 2019.
10. Vlcek, J.; Vlachopoulos, J. Effect of Die Wall Cooling or Heating on Extrudate Swell. *Polym. Eng. Sci.* **1989**, *29*, 685–689. [CrossRef]
11. Phuoc, H.B.; Tanner, R.I. Thermally-induced extrudate swell. *J. Fluid Mech.* **1980**, *98*, 253–271. [CrossRef]
12. Karagiannis, A.; Hrymak, A.N.; Vlachopoulos, J. Three-dimensional non-isothermal extrusion flows. *Rheol. Acta* **1989**, *28*, 121–133. [CrossRef]
13. Spanjaards, M.M.A.; Hulsen, M.A.; Anderson, P.D. Computational analysis of the extrudate shape of three-dimensional viscoelastic, non-isothermal extrusion flows. *J. Non–Newton. Fluid Mech.* **2020**, *282*, 104310. [CrossRef]

14. Georgiou, G.C.; Boudouvis, A.G. Converged solutions of the Newtonian extrudate-swell problem. *Int. J. Numer. Methods Fluids* **1999**, *29*, 363–371. [CrossRef]
15. Fakhari, A.; Tukovic, Z.; Carneiro, O.S.; Fernandes, C. An Effective Interface Tracking Method for Simulating the Extrudate Swell Phenomenon. *Polymers* **2021**, *13*, 1305. [CrossRef] [PubMed]
16. Bingham, E.C. *Fluidity and Plasticity*; McGraw-Hill: New York, NY, USA, 1922.
17. Herschel, W.H.; Bulkley, R. Konsistenzmessungen von Gummi-Benzollösungen. *Kolloid D* **1926**, *39*, 291–300./BF01432034. [CrossRef]
18. Papanastasiou, T.C. Flows of materials with yield. *J. Rheol.* **1987**, *31*, 385–404. [CrossRef]
19. Ellwood, K.R.J.; Georgiou, G.C.; Papanastasiou, T.C.; Wilkes, J.O. Laminar jets of Bingham plastic liquids. *J. Rheol.* **1990**, *34*, 787–812. [CrossRef]
20. Frigaard, I.A.; Nouar, C. On the usage of viscosity regularization methods for visco-plastic fluid flow computation. *J. Non-Newton. Fluid Mech.* **2005**, *127*, 1–26. [CrossRef]
21. Beverly, C.R.; Tanner, R.I. Numerical analysis of extrudate swell in viscoelastic materials with yield stress. *J. Rheol.* **1989**, *33*, 989–1009. [CrossRef]
22. Hurez, P.; Tanguy, P.A.; Bertrand, F.H. A finite element analysis of dieswell with pseudoplastic and viscoplastic fluids. *Comput. Methods Appl. Mech. Eng.* **1990**, *86*, 87–103. [CrossRef]
23. Abdali, S.S.; Mitsoulis, E.; Markatos, N.C. Entry and exit flows of Bingham fluids. *J. Rheol.* **1992**, *36*, 389–407. [CrossRef]
24. Mitsoulis, E.; Abdali, S.S.; Markatos, N.C. Flow simulation of Herschel–Bulkley fluids through extrusion dies. *J. Chem. Eng.* **1993**, *71*, 147–160. [CrossRef]
25. Kountouriotis, Z.; Georgiou, G.C.; Mitsoulis, E. Numerical study of the combined effects of inertia, slip, and compressibility in extrusion of yield stress fluids. *Rheol. Acta* **2014**, *53*, 791–804. [CrossRef]
26. OpenCFD. *OpenFOAM—The Open Source CFD Toolbox—User's Guide*; OpenCFD Ltd.: UK, Bracknell, 2007.
27. Tuković, Ž.; Jasak, H. A moving mesh finite volume interface tracking method for surface tension dominated interfacial fluid flow. *Comput. Fluids* **2012**, *55*, 70–84. [CrossRef]
28. Zhou, Y.-G.; Wu, W.-B.; Zou, J.; Turng, L.-S. Dual-scale modeling and simulation of film casting of isotactic polypropylene. *J. Plast. Film Sheeting* **2016**, *32*, 239–271. [CrossRef]
29. Tuković, Z.; Jasak, H. Simulation of free-rising bubble with soluble surfactant using moving mesh finite volume/area method. In Proceedings of the 6th International Conference on CFD in Oil & Gas, Metallurgical and Process Industries, Trondheim, Norway, 10–12 June 2008.
30. Demirdžić, I.; Perić, M. Space conservation law in finite volume calculations of fluid flow. *Int. J. Numer. Methods Fluids* **1988**, *8*, 1037–1050. [CrossRef]
31. Ferziger, J.H.; Peric, M. *Computational Methods for Fluid Dynamics*, 3rd ed.; Springer: Berlin, Germany, 2002.
32. Van Doormaal, J.P.; Raithby, G.D. Enhancement of the SIMPLE method for predicting incompressible fluid flows. *Numer. Heat Transf.* **1984**, *7*, 147–163. [CrossRef]
33. Issa, R.I. Solution of the implicitly discretised fluid flow equations by operator-splitting. *J. Comput. Phys.* **1986**, *62*, 40–65. [CrossRef]
34. Tuković, Ž.; Perić, M.; Jasak, H. Consistent second-order time-accurate non-iterative PISO-algorithm. *Comput. Fluids* **2018**, *166*, 78–85. [CrossRef]
35. Moukalled, F.; Mangani, L.; Darwish, M. *The Finite Volume Method in Computational Fluid Dynamics: An Advanced Introduction with OpenFOAM and Matlab*; Springer International Publishing: Cham, Switzerland, 2016.
36. Tuković, Ž.; Karač, A.; Cardiff, P.; Jasak, H. OpenFOAM Finite Volume Solver for Fluid-Solid Interaction. *Trans. Famena* **2018**, *42*, 1–31. [CrossRef]
37. Rhie, C.M.; Chow, W.L. Numerical study of turbulent flow past an isolated airfol with trailing edge separation. *AIAA J.* **1983**, *21*, 1525–1532. [CrossRef]
38. Jacobs, D. *Preconditioned Conjugate Gradient Methods for Solving Systems of Algebraic Equations*; Technical Report, RD/L/N193/80; Central Electricity Research Laboratories, Oxford University Press: Oxford, UK, 1980.
39. Ajiz, M.; Jennings, A. A robust incomplete Cholesky-conjugate gradient algorithm. *J. Numer. Meth. Eng.* **1984**, *20*, 949–966. [CrossRef]
40. Lee, J.; Zhang, J.; Lu, C.-C. Incomplete LU preconditioning for large scale dense complex linear systems from electromagnetic wave scattering problems. *J. Non-Newton. Fluid Mech.* **2003**, *185*, 158–175. [CrossRef]
41. Kountouriotis, Z.; Georgiou, G.C.; Mitsoulis, E. On the combined effects of slip, compressibility, and inertia on the Newtonian extrudate-swell flow problem. *Comput. Fluids* **2013**, *71*, 297–305. [CrossRef]
42. Peters, G.W.M.; Baaijens, F.P.T. Modelling of non-isothermal viscoelastic flows. *J. Non-Newton. Fluid Mech.* **1997**, *68*, 205–224. [CrossRef]
43. Labsi, N.; Benkahla, Y.K.; Boutra, A. Temperature-dependent shear-thinning Herschel–Bulkley fluid flow by taking into account viscous dissipation. *J. Braz. Soc. Mech. Sci. Eng.* **2017**, *39*, 267–277. [CrossRef]

polymers

MDPI

Article

Taguchi Optimization of Roundness and Concentricity of a Plastic Injection Molded Barrel of a Telecentric Lens

Chao-Ming Lin * and Yun-Ju Chen

Department of Mechanical and Energy Engineering, National Chiayi University, Chiayi 600355, Taiwan; yunju.chen.zoe@gmail.com
* Correspondence: cmlin@mail.ncyu.edu.tw

Abstract: Plastic is an attractive material for the fabrication of tubular optical instruments due to its light weight, high strength, and ease of processing. However, for plastic components fabricated using the injection molding technique, roundness and concentricity remain an important concern. For example, in the case of a telecentric lens, concentricity errors of the lens barrel result in optical aberrations due to the deviation of the light path, while roundness errors cause radial stress due to the mismatch of the lens geometry during assembly. Accordingly, the present study applies the Taguchi design methodology to determine the optimal injection molding parameters which simultaneously minimize both the overall roundness and the overall concentricity of the optical barrel. The results show that the geometrical errors of the optical barrel are determined mainly by the melt temperature, the packing pressure, and the cooling time. The results also show that the optimal processing parameters reduce the average volume shrinkage rate (from 4.409% to 3.465%) and the average deformations from (0.592 mm to 0.469 mm) of the optical barrel, and the corresponding standard deviation values are reduced from 1.528% to 1.297% and from 0.263 mm to 0.211 mm, respectively. In addition, the overall roundness and overall concentricity of the barrel in the four planes are positively correlated.

Keywords: plastic optical barrel; injection molding; roundness; concentricity; Taguchi method

Citation: Lin, C.-M.; Chen, Y.-J. Taguchi Optimization of Roundness and Concentricity of a Plastic Injection Molded Barrel of a Telecentric Lens. *Polymers* **2021**, *13*, 3419. https://doi.org/10.3390/polym13193419

Academic Editors: Célio Bruno Pinto Fernandes, Alexandre M. Afonso, Luís L. Ferrás and Salah Aldin Faroughi

Received: 21 September 2021
Accepted: 2 October 2021
Published: 5 October 2021

1. Introduction

Plastic injection molding is a fast and economical process for fabricating optical components with high precision, excellent performance, and good strength-to-weight properties. However, the quality of injection molded parts is highly dependent on the choice of processing parameters, including the melt temperature, mold temperature, filling time, packing time, packing pressure, injection pressure, and so on. As a result, a proper control of the processing conditions is essential [1–7].

Compared to conventional lenses, in which the magnification varies with the distance between the lens and the object, telecentric lenses have a constant field of view at all distances from the lens. As a result, they eliminate the parallax error inherent in traditional fixed-focal length lenses and, therefore, find widespread use in machine vision-based systems where precise and repeatable measurements are required, such as metrology, microlithography, semiconductor manufacturing, and so on [8–12]. Figure 1 presents a simple schematic illustration of a typical coaxial bilateral telecentric optical system consisting of a telecentric barrel, a coaxial light source, a holder, and an optical imaging system.

Among these components, the telecentric barrel plays a critical role in ensuring that the light emitted by the coaxial light source follows the preset path. In particular, the roundness and concentricity of the barrel must be strictly controlled in order to ensure a proper placement and alignment of the internal lens [13–18]. The lens barrel is generally fabricated from plastic material due to the latter's light weight and ease of manufacturing. However, as shown in Figure 1, the barrel has a varying tube diameter and an asymmetric geometry (due to the presence of the holder). Thus, controlling the injection molding parameters in such a way as to minimize the roundness and concentricity errors caused by the shrinkage and deformation of the barrel during the molding process represents a significant challenge [19,20].

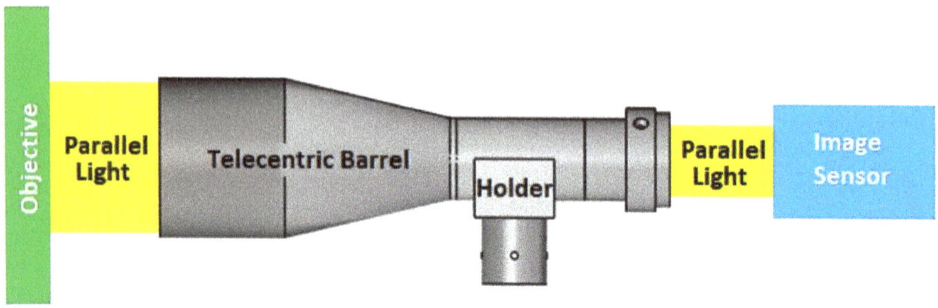

Figure 1. Coaxial bilateral telecentric optical system.

In general, the Taguchi method and mold flow analysis provide a convenient and cost-effective approach for optimizing the processing conditions employed in the injection molding process. Accordingly, the present study employs a hybrid approach consisting of the Taguchi design method and mold flow simulations to determine the optimal settings of the main injection molding parameters (i.e., the injection pressure, the packing pressure, the melt temperature, the mold temperature, and the cooling time) for simultaneously minimizing both the roundness and the concentricity errors of the telecentric barrel. Having determined the optimal processing conditions, a further investigation is performed to examine the correlation between the overall roundness and the overall concentricity of the barrel and the effects of the optimal processing conditions on the average volume shrinkage rate of the barrel following removal from the mold [21–25].

2. Theoretical Analysis

In the present study, the mold flow analysis is performed using Moldex3D computer aided engineering (CAE) simulations. The simulations assume a contact interface between the part's surface and the mold wall and separate the warpage analysis process into two parts, namely, in-mold deformation during the packing and cooling stages and free deformation following ejection from the mold. For the geometric accuracy requirements of the coaxial bilateral telecentric barrel, the final deformation analysis of the cured part from the temperature after demolding to the room temperature will be calculated based on the analysis method of the roundness and concentricity. The related theoretical calculations are explained as follows.

2.1. Flow Analysis during Filling Stage

Using the Moldex3D solid simulation, the polymer melt flow develops during the filling stage of the injection molding process where the melt flow is assumed to be incompressible. The polymer melt is assumed to be Generalized Newtonian Fluid (GNF). Therefore, the non-isothermal 3D flow motion can be mathematically described by the mass, momentum conservation, and energy conservation equations, which can be written as follows [26]:

$$\frac{\partial \rho}{\partial t} + \nabla \cdot \rho u = 0 \quad (1)$$

$$\frac{\partial (\rho u)}{\partial t} + \nabla \cdot (\rho u u - \sigma) = \rho g \quad (2)$$

$$\rho c_p \left(\frac{\partial T}{\partial t} + u \cdot \nabla T \right) = \nabla \cdot (k \nabla T) + \eta \dot{\gamma}^2 \quad (3)$$

where u is velocity, p is the pressure, ρ is the density, c_p is the heat capacity, η is the viscosity, $\dot{\gamma}$ is the shear rate, k is the heat conductivity, and σ is the stress tensor; this can be expressed as follows:

$$\sigma = -p\mathrm{I} + \eta\left(\nabla u + \nabla u^{\mathrm{T}}\right) \tag{4}$$

Considering the constitutive equation for the general polymer materials, the Modified-Cross viscosity [27] model with Arrhenius temperature is used to describe the rheological property of the polymer melt.

$$\eta(T, \dot{\gamma}) = \frac{\eta_0(T)}{1 + \left(\eta_0(T)\dot{\gamma}/\tau^*\right)^{1-n}} \tag{5}$$

$$\eta_0(T) = Be^{\left(\frac{T_b}{T}\right)}. \tag{6}$$

where η is the viscosity, η_0 is the melt viscosity under zero-shear-rate conditions, τ^* is the parameter that describes the transition region between zero shear rate and the power law region of the viscosity curve, n is the Power Law index, and B is the consistency index. The Modified-Cross viscosity model includes the Newton's fluid interval and the power-law shear thinning interval. When the shear rate approaches zero, this model predicts the zero-shear rate viscosity η_0; when the shear rate is large, it predicts the power-law behavior. The τ^* in this model is a constant, which physically represents the critical shear stress value for the transition from Newtonian fluid to power law fluid. Compared with the other model, the Modified-Cross model requires fewer parameters and can capture the dependence of viscosity on the shear rate. Therefore, the Modified-Cross model is often used in commercial simulation software, just like the Moldex3D software in this study.

A volume fraction function, f, is introduced to track the advance of the melt front. Here, $f = 0$ is defined as the air phase and $f = 1$ as the polymer melt phase. Hence, the melt front is located within cells with $0 < f < 1$. The advancement of f over time can be expressed as the following transport equation:

$$\frac{\partial f}{\partial t} + \nabla \cdot (uf) = 0 \tag{7}$$

After the part is ejected from the mold, a free thermal shrinkage happens due to the temperature and pressure difference. The warpage analysis assumes the mechanical properties are elastic. The stress–strain equilibrium equations enable us to solve the problems.

2.2. Shrinkage and Warpage

The temperature and pressure changes which occur during the injection molding process result in corresponding changes in the specific volume and density of the polymer. These changes lead in turn to a warpage of the molded component as it cools from the melt condition to the solid condition. The part additionally undergoes volume shrinkage during the molding process and following its removal from the mold. During the packing stage, the shrinkage reduces as the packing pressure and packing time increase.

For semi-crystalline polymers, the shrinkage behavior mainly depends on the degree of crystallization. If the mold temperature is low and the cooling rate is high, it is not easy to crystallize, but there is a small shrinkage; on the other hand, if the mold temperature is high and the cooling rate is low, the macromolecular chain has enough relaxation time and easily to form crystals. The amount of shrinkage will naturally increase.

For isotropic materials, the linear shrinkage is one-third the volumetric shrinkage (see Equation (8) below). However, in the injection molding process, the orientation effect of the polymer forming and the shrinkage behavior are both constrained by the mold wall.

As a result, the shrinkage deformation exhibits a non-isotropic behavior and the linear shrinkage in the part thickness is thus governed by Equation (9) [28–33].

$$S_L = 1 - (1 - S_V)^{1/3} \approx \left(\frac{1}{3}\right) S_V \tag{8}$$

$$S_L \approx 0.9 - 0.95 S_V \tag{9}$$

where S_L is the linear shrinkage rate, and S_V is the volume shrinkage rate.

2.3. Stress Analysis after the Demolding Stage

In this stage, the plastic forming part is no longer restricted to the mold after demolding and is in the free shrinking stage. The free volume shrinkage of the molded component following its removal from the mold depends mainly on the thermal stress induced by the difference between the temperature of the part and that of the environment. If the shrinkage stress exceeds the mechanical strength of the part, the part undergoes warpage. Conversely, if the plastic part is sufficiently strong to resist the thermal stress, the part retains its original geometry and dimensions. However, shrinkage voids may still be formed within the plastic component, which degrade the mechanical properties of the part and may lead to cracks and breakage under the effects of an external force. In the warpage analysis of the Moldex3D solid model, the assumptions are as follows: the material property is linear and elastic; there is a small amount of strain; the behavior is approximately steady; and the plastic part is elastically deformed. The governing equations for the material behavior in the warpage analysis can thus be expressed as follows [34]:

$$\sigma_{ij,j} + f_i = 0 \tag{10}$$

$$\sigma_{ij} = C_{ijkl}\left(\varepsilon_{kl} - \varepsilon_{kl}^0 - \alpha_{kl} \cdot \Delta T\right) + \sigma_{ij}^F \tag{11}$$

$$\varepsilon_{ij} = \left(u_{i,j} + u_{j,i}\right)/2 \tag{12}$$

where σ_{ij} is the stress tensor, σ_{ij}^F is the initial stress induced by the flow, ε_{ij} is the infinitesimal elastic strain, ε_{ij}^0 is the initial strain from the P-v-T relationship, C_{ijkl} is the elastic material stiffness, α_{kl} is the coefficient of linear thermal expansion, and ΔT is the temperature difference.

2.4. Roundness Evaluation

The most common methods for determining roundness errors include the Least Squares Circle (LSC) method, the Minimum Zone Tolerance Circle (MZC) method, the Maximum Inscribed Circle (MIC) method, and the Minimum Circumscribed Circle (MCC) method [35]. Figure 2 illustrates the LSC method, in which the center of the circle is first determined by identifying the circular contour which minimizes the sum of squared error (SSE) between the inner and outer radii of the interior surface (shown in red in Figure 2). This center point is then used to draw the circumscribed and inscribed circles of the barrel interior surface, respectively (see two black lines in Figure 2). Finally, the roundness of the circle (ΔZq) is quantified as the radial distance (R_{max}–R_{min}) between them [36]:

$$\text{Roundness} = \Delta Zq = R_{max} - R_{min} \tag{13}$$

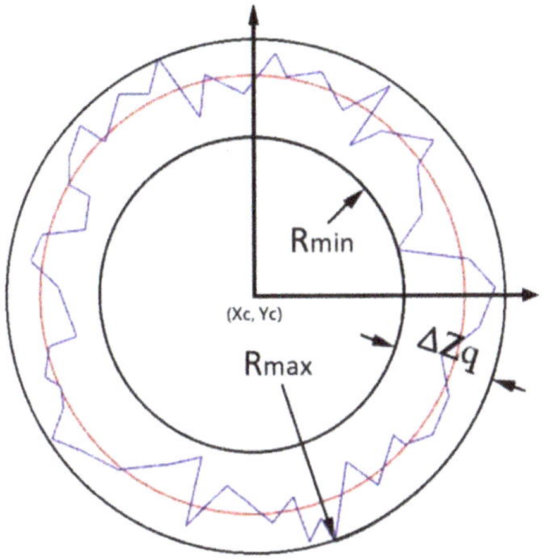

Figure 2. Least squares circle method for determination of roundness (ΔZq) at specified Z-plane.

In practice, the center point of the least square circle is unique, and its accuracy depends on the number of measurement points [37–40]. The overall roundness including n planes can be defined as

$$\text{Roundness}_{\text{overall}} = \sum_{i=1}^{n} \left[(\Delta Zq)_i \right]^2 / n \tag{14}$$

After the above calculation, one can get the center of each contour (Xc, Yc), the roundness of each contour, and the overall roundness of all contours after the injection molding processing.

2.5. Concentricity Evaluation

Concentricity refers to the deviation of the center of a circle or center of a cylinder from the center of the reference form. It is generally evaluated as either the axis concentricity tolerance (with the tolerance zone centered on the axis of the reference form) or the point concentricity tolerance (with the reference point taken as the center of the circle). For either method, the concentricity measurement process involves establishing the coordinates of the required checking plane, measuring the position of the center of the contour circle that needs to be compared after setting the datum, and calculating the distance between the original center of the circle and the center of the actual contour. As with the roundness evaluation described above, the accuracy of the concentricity tolerance process also increases with an increasing number of measurement points. Figure 3 illustrates the concentricity evaluation process for the case where the reference coordinates (X,Y) are set as (0,0) and the coordinates of the fitted circular contour are denoted as (Xc, Yc). The concentricity d at specified Z-plane is then evaluated simply as [41].

$$\text{Concentricity} = d = \sqrt{(X_c - X)^2 + (Y_C - Y)^2} \tag{15}$$

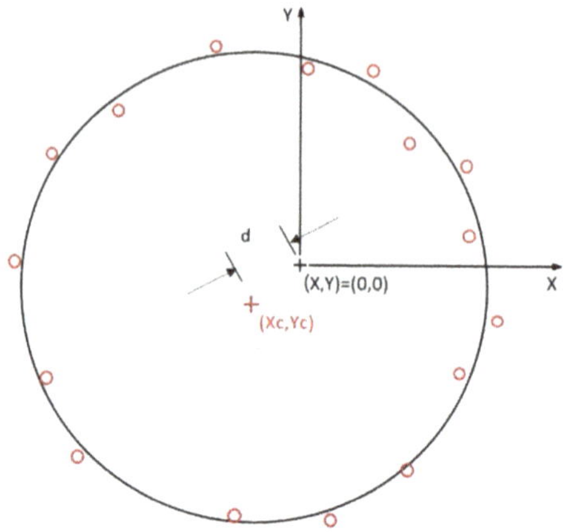

Figure 3. Calculation of concentricity (d) at specified Z-plane.

The overall concentricity including n contours can be defined as

$$\text{Concentricity}_{\text{Overall}} = \sum_{i=1}^{n} [d_i]^2/n \quad (16)$$

3. Methods and Procedures

The geometry model of the telecentric lens barrel and positions of the tracked nodes during the simulations were defined in **Rhinoceros**. The model was then imported into **Moldex3D** to design the mold cavity and the gate, runner and cooling system, as well as to perform the molding flow simulations. The simulations considered the use of **PA66** polymer material with the properties shown in Table 1 as the feedstock material. The total warpage was obtained directly from the output results of the mold flow analysis, while the roundness was calculated based on the distance between each tracked node and the offset center, and the concentricity was computed as the shortest distance between the offset circle center and the original axis.

Table 1. Material properties of PA66 (TECHNYL A 216, Solvay Engineering Plastics; Source: Moldex3D material library).

Properties	Values	Unit
Density	1140	Kg/m^3
Mold shrinkage	1.90	%
Water absorption (24 h at 23 °C)	1.30	%
Tensile modulus	3000	MPa
Tensile strength at break	55	MPa
Tensile strain at break	30	%
Flexural maximum stress	120	MPa
Melt temperature	263	°C
Heat conduction coefficient	0.25	W/(m•K)
Coefficient of linear thermal expansion (after demolding) (23 °C to 85 °C)	7	E-5/°C
Viscosity vs. shear rate under different temperature	See Figure 4a	
P-v-T	See Figure 4b	

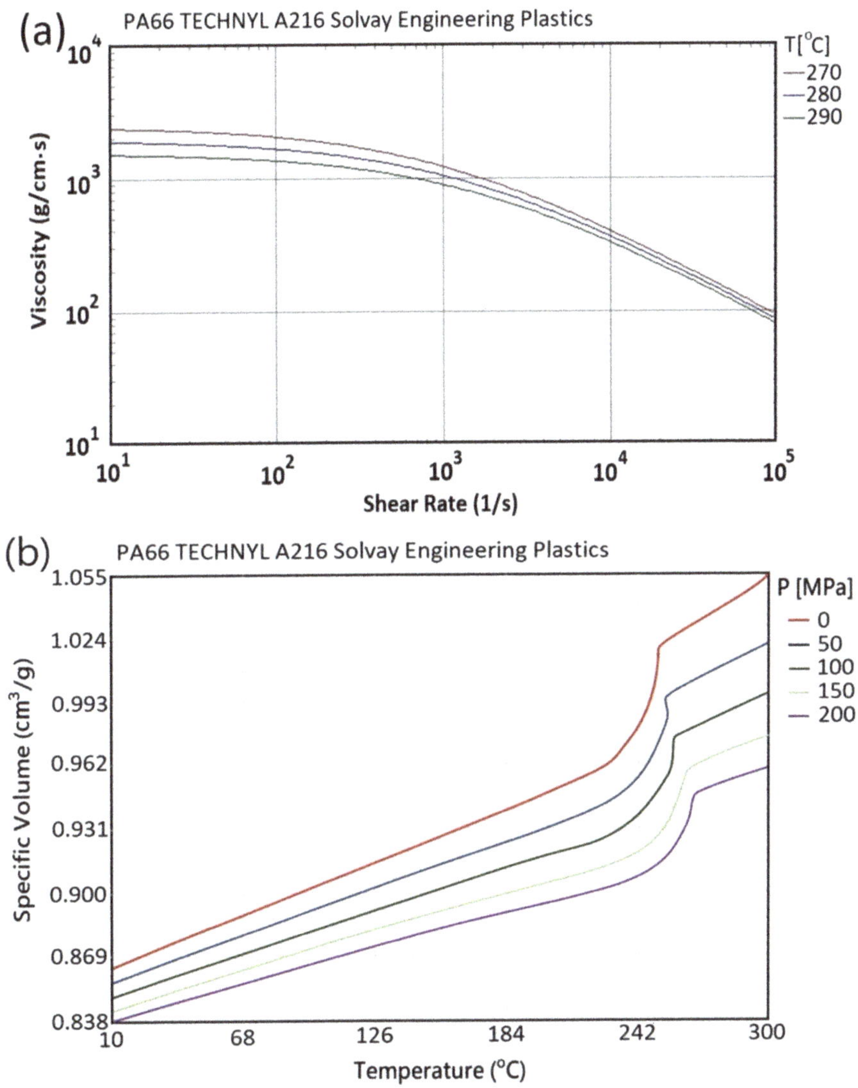

Figure 4. (**a**) Viscosity and (**b**) P-v-T properties of PA66 material (Source: Moldex3D material library).

3.1. Material Characteristics and P-v-T curves of PA66

Figure 4a shows the relationship between the viscosity of PA66 and the shear rate at different temperatures as the basis for subsequent mold flow analysis, and the Modified-Cross viscosity [27] model (See Equations (5) and (6)) with Arrhenius temperature is used to describe the rheological property of the polymer melt. Figure 4b shows the P-v-T relationship diagram of PA66. In the process of plastic processing, the plastic undergoes a very rapid cooling process under the temperature and pressure controlled by the molding process and changes from a molten state to a solid state. Usually, the volume changes greatly, and a simple comparison is no longer possible. To describe the capacity constant, the relationship between specific volume/pressure/temperature characteristics (P-v-T) is determined to calculate the degree of compression of the material in the packing stage, as well as the shrinkage rate and shrinkage warpage of the final plastic part after ejection.

The Modified Tait Model II [26] is used to describe the P-v-T relationship of semi-crystalline materials (PA66) and is also the recommended P-v-T model in Moldex3D.

$$v(T,P) = v_0(T)\left[1 - Cln\left(1 + \frac{P}{B(T)}\right)\right] + v_t(T,P) \tag{17}$$

where $v(T,P)$ is the specific volume; v_0 is the specific volume at zero gauge pressure; T is the temperature; P is the pressure; and C is the constant 0.0894.

$$\hat{v}_0(T) = \begin{cases} b_{1S} + b_{2S}\overline{T} & ,if\ T \le T_t \\ b_{1L} + b_{2L}\overline{T} & ,if\ T > T_t \end{cases} \tag{18}$$

$$B = \begin{cases} b_{3S}\exp(-b_{4S}\overline{T}) & ,if\ T \le T_t \\ b_{3L}\exp(-b_{4L}\overline{T}) & ,if\ T > T_t \end{cases} \tag{19}$$

$$v_t(T,P) = \begin{cases} b_7\exp(b_8\overline{T} - b_9P) & ,if\ T \le T_t \\ 0 & ,if\ T > T_t \end{cases} \tag{20}$$

$$\overline{T} = T - b_5 \tag{21}$$

$$T_t = b_5 + b_6P \tag{22}$$

where v_t is the value for semi-crystalline resins only applies to temperatures below the transition temperature; T_t is used to characterize the abrupt viscosity change of the material around its transition temperature; 13 parameters ($b_{1S}, b_{2S}, b_{3S}, b_{4S}, b_{1L}, b_{2L}, b_{3L}, b_{4L}, b_5, b_6, b_7, b_8, b_9$) are data-fitted coefficients. With only linear P-v-T transitions, b_7, b_8 and b_9 are for amorphous materials.

3.2. Modeling of Analyzed Product

The simulations considered a coaxial telecentric lens barrel with the dimensions and geometry shown in Figure 5.

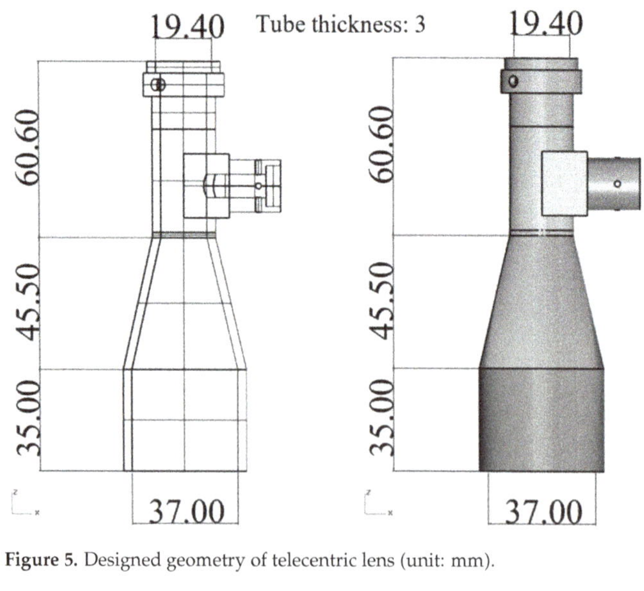

Figure 5. Designed geometry of telecentric lens (unit: mm).

Figure 6a,b show the gating and cooling system models used in the simulations. As shown in Figure 6a, the gating system was designed with four runners to accommodate the large component size and thin wall thickness (3 mm). Moreover, the cooling water runner system was numerically designed to fit snugly around the outside surface of the plastic barrel, and a baffle-type water runner was used to prevent internal heat accumulation (see Figure 6b).

(a)

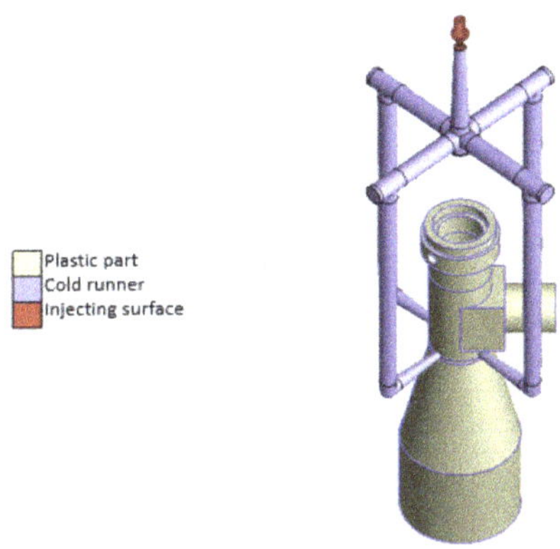

(b)

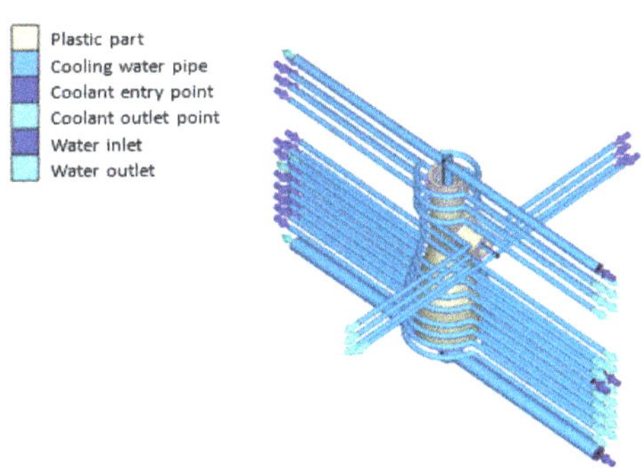

Figure 6. (**a**) Gating and (**b**) cooling systems for injection molding process.

3.3. Taguchi Design Method

Figure 7 presents a flowchart of the hybrid Taguchi/CAE optimization process performed in the present study to identify the plastic injection molding processing parameters which minimize the overall roundness and overall concentricity of the optical barrel. As shown, the process commenced by constructing the numerical model described in the previous section and establishing the build surface and solid mesh. Having chosen suitable signal-to-noise (S/N) ratios for evaluating the quality of each simulation outcome, the Taguchi design processes were defined by establishing the control factors and level settings of interest. For some specific quality requirements such as deformation, warpage, shrinkage, weld line, air trap, roundness, concentricity, etc., each of these quality characteristics will have different influential processing factors. Therefore, in order to find the best combination of parameters, the Taguchi method is usually used to screen the most influential factors. This method is to utilize the statistical operation of the orthogonal array (OA) to find the optimal parameter combination. In Taguchi method, OA is a general partial factorial design. It is based on an orthogonal design matrix, allowing users to consider selected subsets of multi-factor combinations at multiple levels. Orthogonal arrays are balanced to ensure that all levels of all factors are considered equally in statistics. Other less influential parameters adopt the recommended values of polymer materials or injection molding machines. Generally speaking, the processing factors that have an influence on deformation are related to temperature and packing. After preliminary evaluation and calculation, five control factors were chosen, namely, (A) the injection pressure, (B) the packing pressure, (C) the melt temperature, (D) the mold temperature, and (E) the cooling time. As shown in Table 2, each of the five control factors was assigned four different level settings.

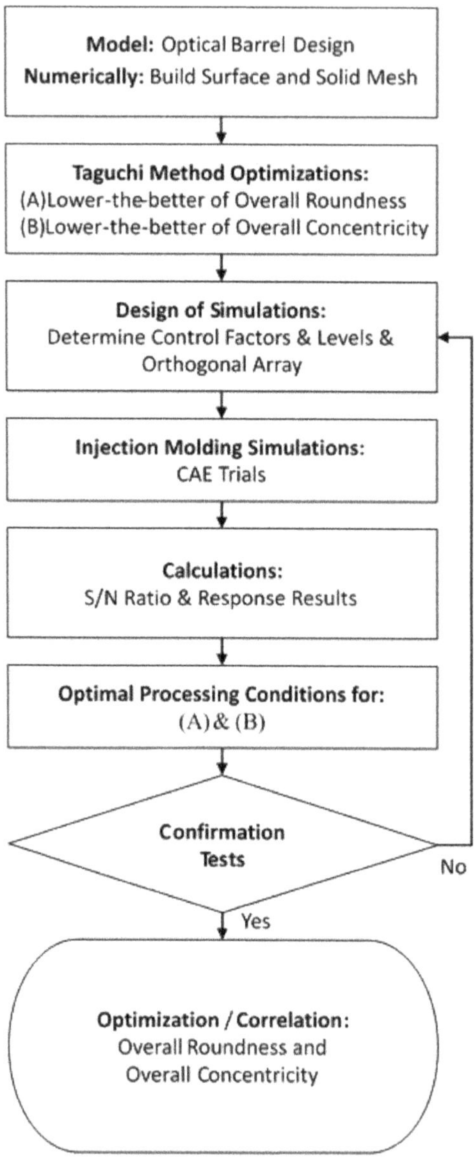

Figure 7. Flowchart showing main steps in Taguchi/CAE optimization process.

Table 2. Control factors and level settings used in Taguchi simulations.

L_{16} (4^5)	A Injection Pressure (MPa)	B Packing Pressure (MPa)	C Melt Temp. (°C)	D Mold Temp. (°C)	E Cooling Time (Sec)
Level 1	180	180	275	70	11
Level 2	200	200	280	80	13
Level 3	220	220	285	90	15
Level 4	240	240	290	100	17
Standard Parameters	200	200	280	80	13

Thus, the Taguchi simulations were configured in an $L_{16}(4^5)$ Orthogonal Array (OA), as shown in Table 3.

Table 3. Taguchi analysis results for overall roundness and overall concentricity.

Trials	Processing Factors					Overall Roundness		Overall Concentricity	
	A Injection Pressure (MPa)	B Packing Pressure (MPa)	C Melt Temp. (°C)	D Mold Temp. (°C)	E Cooling Time (Sec)	$\sum_{i=1}^{4}[(\Delta Zq)_i]^2/4$ (mm²)	S/N (dB)	$\sum_{i=1}^{4}(d_i)^2/4$ (mm²)	S/N (dB)
Standard Parameters	**200**	**200**	**280**	**80**	**13**	**0.002253**	**26.472**	**0.01487**	**18.278**
1	180	180	275	70	11	0.002259	26.460	0.01569	18.043
2	180	200	280	80	13	0.002255	26.468	0.01488	18.274
3	180	220	285	90	15	0.002236	26.504	0.01397	18.549
4	180	240	290	100	17	0.002212	26.552	0.01309	18.832
5	200	180	280	90	17	0.002213	26.551	0.01534	18.141
6	200	200	275	100	15	0.002195	26.585	0.01417	18.486
7	200	220	290	70	13	0.002282	26.418	0.01472	18.321
8	200	240	285	80	11	0.002233	26.511	0.01314	18.813
9	220	180	285	100	13	0.002315	26.355	0.01658	17.804
10	220	200	290	90	11	0.002303	26.377	0.01562	18.062
11	220	220	275	80	17	0.002123	26.730	0.01256	19.012
12	220	240	280	70	15	0.002184	26.607	0.01257	19.007
13	240	180	290	80	15	0.002324	26.338	0.01684	17.736
14	240	200	285	70	17	0.002218	26.540	0.01458	18.362
15	240	220	280	100	11	0.002231	26.516	0.01367	18.641
16	240	240	275	90	13	0.002179	26.618	0.01242	19.058
Roundness Optimized	**220**	**240**	**275**	**90**	**17**	**0.002111**	**26.755**	-	-
Concentricity Optimized	**220**	**240**	**275**	**90**	**17**	-	-	**0.01167**	**19.331**

In the present study, the aim of the optimization process was to minimize the overall roundness and overall concentricity of the selected planes in the plastic barrel. Hence, in evaluating the quality of the solutions obtained from each simulation run in the OA, the smaller-the-better S/N ratio was adopted for both quality measures, i.e.,

$$S/N = -10\log\left(\frac{1}{n}\sum_{i=1}^{n} {y_i}^2\right), \tag{23}$$

where y_i is the roundness or concentricity and n is the number of measured points in the simulation trial.

3.4. Least Squares Circle Method for Evaluation of Roundness and Concentricity

The roundness and concentricity computations were performed at four planes distributed along the barrel length, namely, Z = 0, Z = 57.75, Z = 82.3, and Z = 117.7 (mm), respectively, as shown in Figure 8. In determining the roundness using the LSC method (see Section 2.4), the center of the least squared error circle was determined using the function [42]:

$$f(x,y) = \min \sum_{i=1}^{n} [r(x,y) - R]^2 \quad (24)$$

where $r(x, y)$ is the distance between the measured point and the known center of the circle, (x, y) are the coordinates of the measured point, n is the number of measured points, and R is the radius of the least square circle [43]. For each run in the OA array, the displacements of the measurement nodes (see Figure 8) were obtained and used to obtain the center point (Xc, Yc, Zc) and radius Rc (See Table 4) of the corresponding least square circle. As described in Section 2.4, the roundness is denoted by ΔZq.

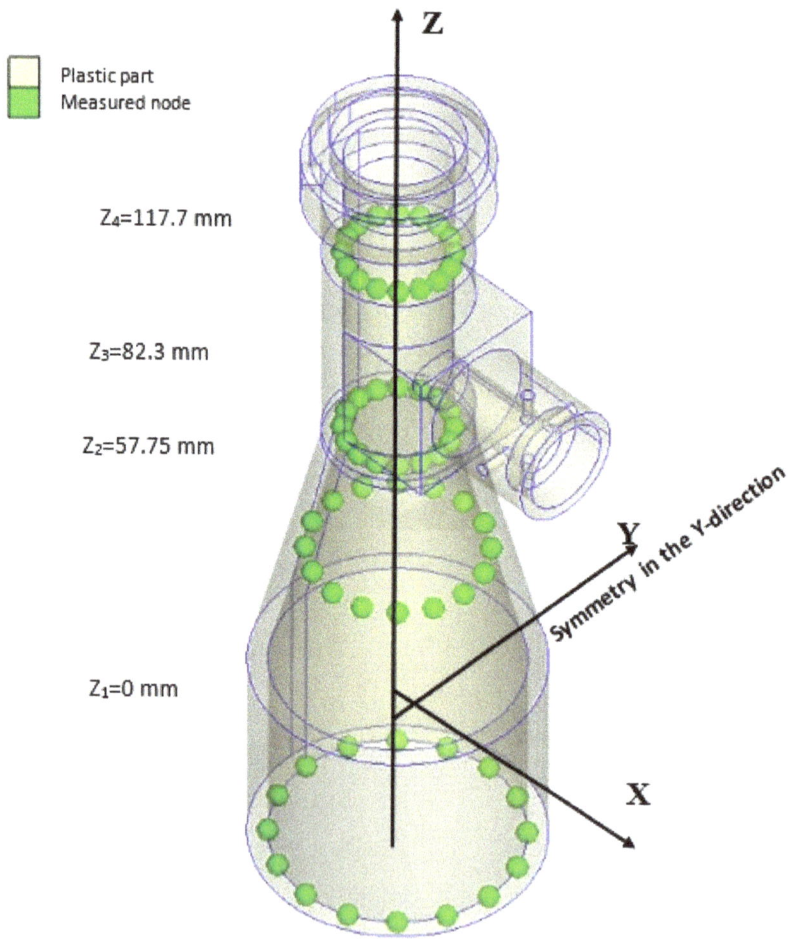

Figure 8. Measurement nodes used for roundness and concentricity evaluation at different Z-planes.

4. Results and Discussion

After Taguchi's optimization calculation, including roundness, concentricity and correlation analysis, the relevant analysis results will be confirmed and discussed. Table 3 shows the average S/N values of the overall roundness and overall concentricity obtained in each of the 16 runs in the OA over the four measurement planes (Z_1~Z_4). The table also shows the S/N values obtained under the standard injection molding conditions for the injection machine and molding material (as prescribed by the manufacturer). Finally, the table shows the S/N values for the overall roundness and overall concentricity obtained under the optimal settings of the five control factors (see Section 4.1 below).

4.1. Factor Rank Analysis and Optimal Process Parameters

Figure 9a,b show the Taguchi response graphs for the overall roundness and overall concentricity, respectively. Referring to Figure 9a, it is seen that the optimal overall roundness is obtained using factor level settings of A3, B4, C1, D3, and E4, i.e., an injection pressure of 220 MPa, a packing pressure of 240 MPa, a melt temperature of 275 °C, a mold temperature of 90 °C, and a cooling time of 17 s. Furthermore, the simulation results show that the overall roundness is dominated by the melt temperature (Rank 1), packing pressure (Rank 2), and cooling time (Rank 3) in sequence. By contrast, the injection pressure and mold temperature, with smaller S/N ranges of 0.020909 dB and 0.010688 dB, respectively, have only a relatively minor effect on the overall roundness. As shown in Table 3, the S/N value of the overall roundness under the optimal processing conditions (26.755 dB) is 0.283 dB higher than that of the barrel produced under the standard processing conditions (26.472 dB). Moreover, the S/N value is also higher than that produced in any of the simulation runs in the OA. In other words, the effectiveness of the optimized parameter design in minimizing the overall roundness of the molded plastic barrel is confirmed.

Figure 9b shows the Taguchi response graph for the overall concentricity of the molded barrel. It is seen that the optimal overall concentricity is again obtained using control factor level settings of A3, B4, C1, D3, and E4. The overall concentricity is determined mainly by the packing pressure (Rank 1), melt temperature (Rank 2), and cooling time (Rank 3). The injection pressure and mold temperature once again have only a minor effect on the overall concentricity. Referring to Table 3, it can be seen that the optimal processing conditions increase the S/N ratio (19.331 dB) by 1.053 dB compared with that obtained under the standard processing conditions (18.278 dB). In addition, the S/N ratio of the optimized design is higher than that obtained in any of the 16 runs of the OA. Thus, the effectiveness of the optimal processing conditions in improving the overall concentricity of the barrel is confirmed. Notably, the results presented in Table 3 show that the optimal values of the overall roundness and overall concentricity, respectively, are obtained using the same control factor level settings. In other words, the optimal design enables the simultaneous optimization of both the overall roundness and the overall concentricity.

The results presented in Figure 9a show that the packing pressure, melt temperature, and cooling time have similar S/N values, i.e., 0.146, 0.177, and 0.129 dB, respectively. In other words, all three factors exert a similar effect on the overall roundness of the molded barrel. However, the overall concentricity is dominated by a major factor, namely, the packing pressure (S/N = 0.997 dB) (see Figure 9b) and two moderate influence factors, namely, the melt temperature (S/N = 0.412 dB) and cooling time (S/N = 0.222 dB).

For the influence of packing pressure, when the mold cavity is completely filled with plastic melt, and the plastic melt will change from high temperature and high pressure to low temperature and low pressure. Due to changes in the temperature and pressure of the plastic melt, the final filled part may have obvious shrinkage in the mold cavity. Therefore, in order to overcome the shrinkage problem, the plastic melt in the runner will be continuously filled into the mold cavity when the filling stage is completed. This is called the packing stage of injection molding. In the packing stage, the inside of the mold cavity will reach the highest pressure, and the plastic melt will continue to solidify where it contacts the lower temperature mold wall. The packing process should continue until the injection gate is solidified. Generally speaking, increasing the packing pressure or extending the packing time will delay the curing time of the plastic melt, which will promote the dispersion of the pressure in the plastic part and reduce the volume shrinkage. Excessive packing pressure is likely to cause factors such as difficulty in demolding, high residual stress, burrs and flash. On the contrary, insufficient packing pressure will lead to larger volume shrinkage and voids and other defects.

(a)

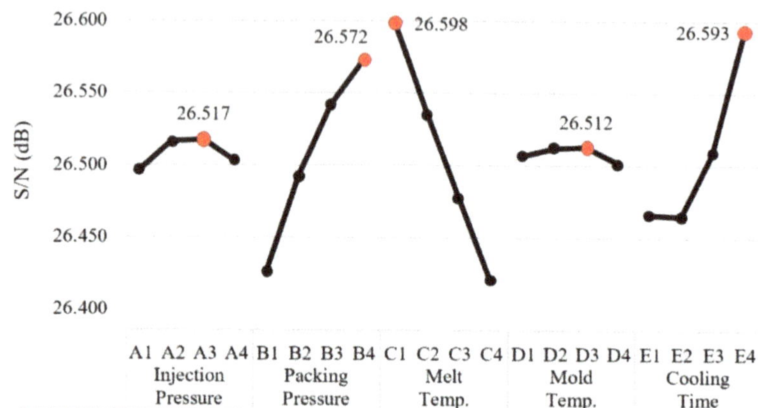

	Injection Pressure	Packing Pressure	Melt Temp.	Mold Temp.	Cooling Time
Range	0.021	0.146	0.177	0.011	0.129
Rank	**4**	**2**	**1**	**5**	**3**

(b)

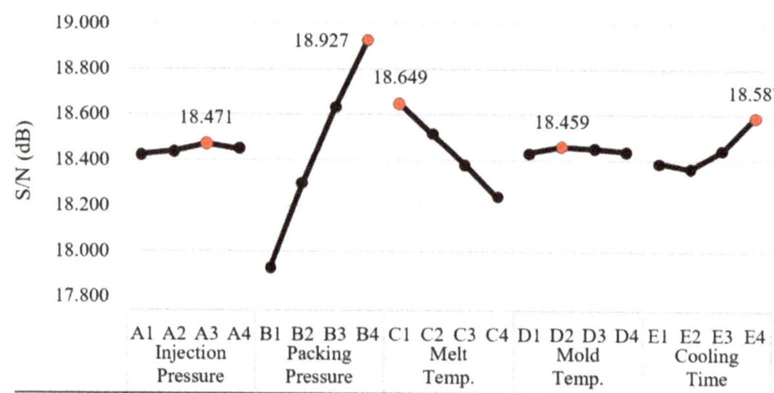

	Injection Pressure	Packing Pressure	Melt Temp.	Mold Temp.	Cooling Time
Range	0.047	0.997	0.412	0.025	0.222
Rank	**4**	**1**	**2**	**5**	**3**

Figure 9. S/N response for (**a**) Overall roundness and (**b**) Overall concentricity.

The effects of the melt temperature and cooling time mainly affect the geometric deformation of the injected part. Deformation is the most important factor that simultaneously affects the optimization of both roundness and concentricity. Generally, the control factors affecting plastic deformation are temperature and cooling. The melt temperature can determine the difference between the surface compressive stress and the internal tensile stress of the injected part during the curing process. In addition, the cooling time is mainly because the semi-crystalline polymer needs sufficient time to crystallize during the cooling process to reduce the residual stress and the shrinkage.

4.2. Correlation between Overall Roundness and Overall Concentricity

Comparing the results presented in Figure 9a,b, it can be seen that the overall roundness and overall concentricity have identical trends in terms of their dependency on the level settings of each control factor. Furthermore, for both properties, the packing pressure, melt temperature and cooling time exert the greatest effect on the simulation outcome, while the injection pressure and mold temperature have only a minor effect. Figure 10 shows the results obtained when plotting the overall roundness values in Table 3 against the corresponding overall concentricity values. Applying a regression analysis technique to the simulation data, the correlation coefficient is determined to be $R^2 = 0.7159$ (Or $R = 0.846$). Considering the general correlation evaluation, the correlation coefficient of the two variables is greater than 0.7, which can be regarded as highly correlated. In other words, the overall roundness and overall concentricity are positively related to one another, which explains why they respond in a similar manner to changes in the injection molding conditions and can be simultaneously optimized using the same control factor level settings.

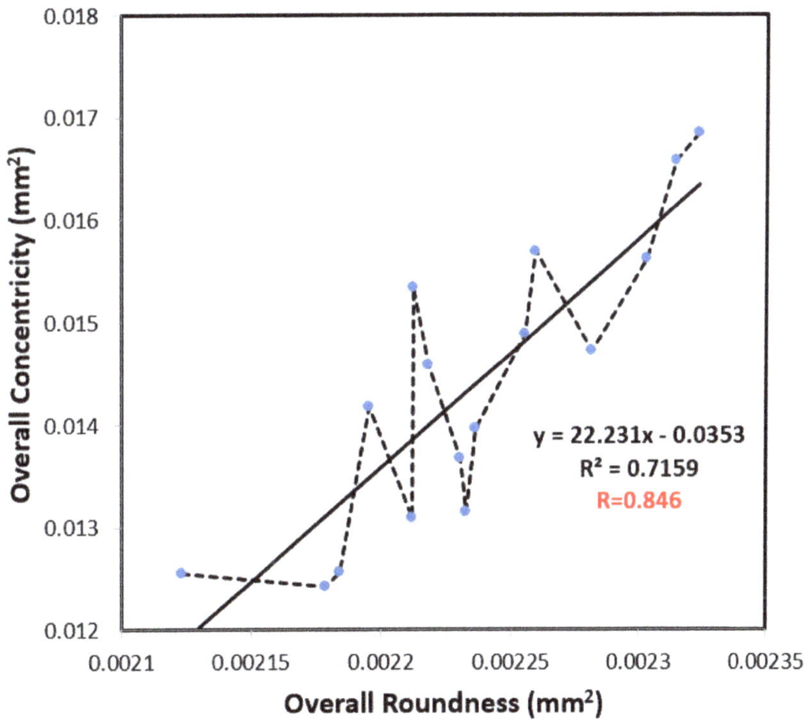

Figure 10. Analysis diagram of the correlation between overall roundness and overall concentricity.

4.3. Deformation and Shrinkage of Plastic Barrel

As shown in Figure 1, a holder structure is attached to the side of the lens barrel in order to support the barrel during use and maintain the coaxial condition of the light as it passes through the barrel. However, the addition of the holder induces a deformation of the molded barrel since the greater thickness of the holder structure relative to that of the barrel results in a corresponding reduction in the local cooling rate. Observing the left-hand schematics in Figures 11 and 12, which show the barrel produced under the standard processing conditions, the local reduction in the cooling rate results in the formation of two regions of high-volume shrinkage due to the difference in cooling rates of the outer and inner regions of the holder structure, respectively. The greater volume shrinkage rate (Figure 11) then causes the narrower portion of the barrel to deform in the direction of the holder structure (Figure 12). However, as shown in the right-hand schematics in the two figures, the optimal processing conditions suppress the local volume shrinkage effect and reduce the barrel deformation accordingly. Figure 13 compares the overall volume shrinkage rates of the barrels produced using the standard and optimized processing conditions, respectively. (Note that in the ideal case, the volume shrinkage is equal to zero.) A detailed inspection shows that the optimal processing conditions reduce the overall average shrinkage rate and standard deviation from 4.409% to 3.465% and 1.528% to 1.297%, respectively. Similarly, Figure 14 compares the overall deformations of the barrels produced using the standard and optimized processing conditions and shows that the optimal processing conditions reduce the average deformations and standard deviation from 0.592 mm to 0.469 mm and from 0.263 mm to 0.211 mm, respectively.

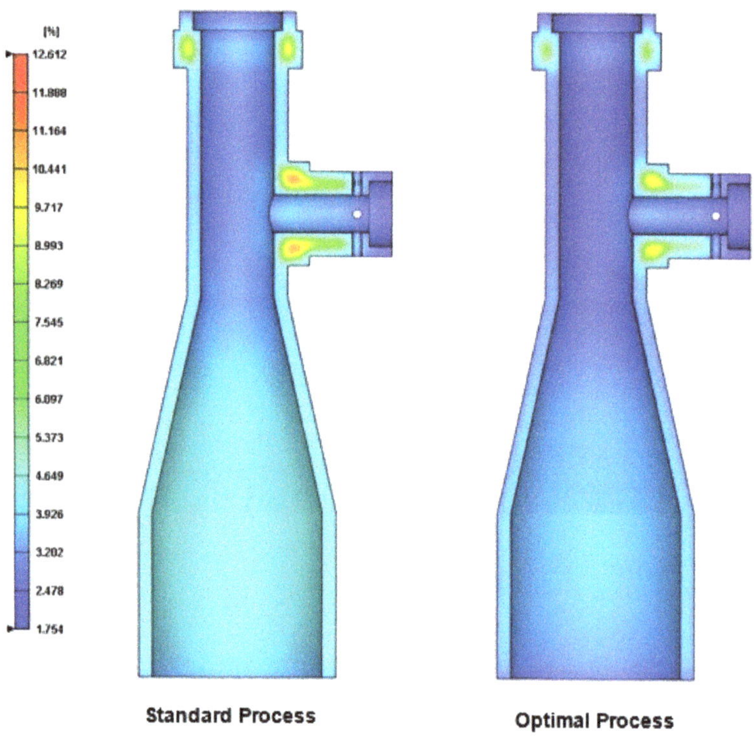

Figure 11. Cross-sectional shrinkage of final part processed using standard conditions (**left**) and optimal conditions (**right**). (Volume shrinkage: %).

4.4. Manufacturing and Processing Implications of Present Results

Table 4 shows the eccentric coordinates, least square circle radius, roundness, and concentricity values of the ideal telecentric barrel (original design) and barrels produced under the standard and optimal processing conditions, respectively.

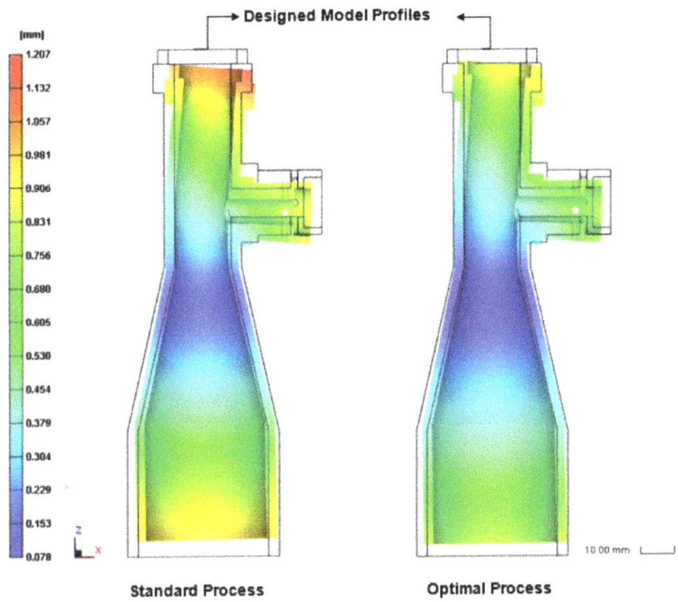

Figure 12. Side-view deformation of final part processed using standard conditions (**left**) and optimal conditions (**right**). (Displacement enlarged by factor of 10 for visualization purposes).

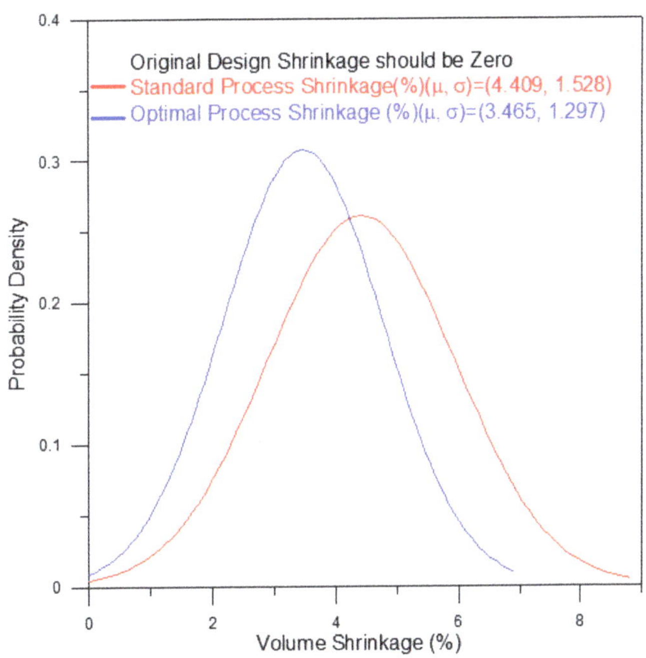

Figure 13. Shrinkage improvement of optimal process compared to standard process.

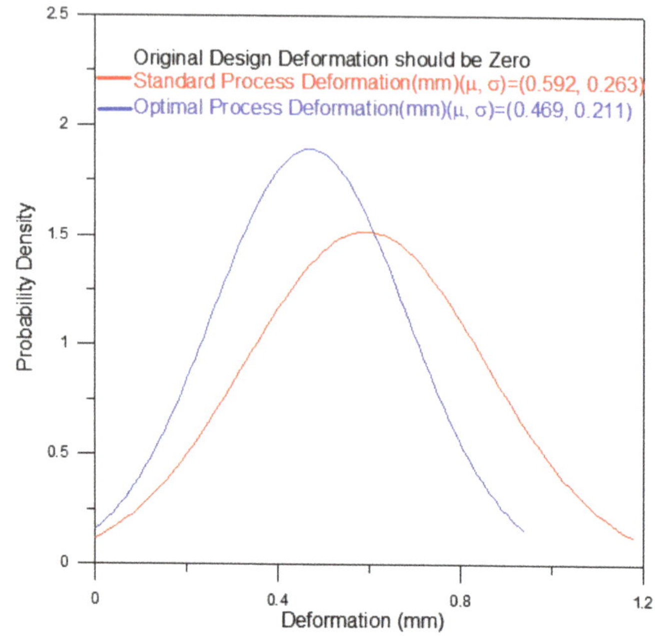

Figure 14. Deformation improvement of the optimal process compared to the standard process.

Table 4. The analyzed results of Roundness and Concentricity are on the four measured planes.

Unit: Mm	Measured Planes	Least Square Circles: Eccentric Coordinates Xc	Yc	Zc	Radius Rc	Roundness (ΔZq)	Concentricity (d)
Original Model	$Z_1 = 0$	0	0	0	18.5	0	0
	$Z_2 = 57.75$				13.25		
	$Z_3 = 82.3$				8		
	$Z_4 = 117.7$				8		
Standard Processing	$Z_1 = 0$	1.740×10^{-1}	-2.405×10^{-4}	0.9380	18.25	0.005442	0.1740
	$Z_2 = 57.75$	-6.099×10^{-3}	-2.107×10^{-4}	57.91	13.07	0.08299	0.006103
	$Z_3 = 82.3$	-5.464×10^{-2}	-1.300×10^{-4}	82.13	7.896	0.03662	0.05464
	$Z_4 = 117.7$	1.618×10^{-1}	-6.092×10^{-4}	117.0	7.916	0.02751	0.1618
Optimal Processing	$Z_1 = 0$	1.528×10^{-1}	-1.852×10^{-4}	0.7284	18.31	0.004390	0.1528
	$Z_2 = 57.75$	-1.698×10^{-2}	-1.591×10^{-4}	57.89	13.11	0.08240	0.01698
	$Z_3 = 82.3$	-5.810×10^{-2}	-9.810×10^{-5}	82.17	7.918	0.03246	0.05810
	$Z_4 = 117.7$	1.401×10^{-1}	-5.558×10^{-4}	117.2	7.934	0.02412	0.1401

Interestingly, the concentricity of the barrel produced under standard processing conditions at planes Z_2 and Z_3 (see Figure 8) is better than that of the barrel produced under the optimal process conditions at the same planes. It is speculated that this may be due to the holder. However, the overall roundness of the barrel produced using the optimal processing conditions is better than that of the barrel produced using the standard processing conditions at all of the measurement planes.

In general, the present results show that to improve the overall roundness and overall concentricity of the optical telecentric barrel, it is necessary to reduce the thermally induced residual stress to the greatest extent possible. The Taguchi optimization results suggest that this can best be achieved using an appropriate injection speed, increasing the packing pressure, extending the cooling time, reducing the product thickness difference, using an appropriate material temperature, reducing the mold temperature to improve the difference with the room temperature, and appropriately selecting the gate design.

5. Conclusions

In this study, the coaxial telecentric lens barrel was analyzed by considering the tolerances of the roundness and concentricity. Mold flow technology combined with the Taguchi design method were introduced to explore the roundness and concentricity arising from the shrinkage and deformation in the injection molding process. Five control factors and four levels of orthogonal array tables were selected for the Taguchi analysis. The results show that an appropriate selection of processing factors and levels can effectively optimize the roundness and concentricity of the lens barrel injection molding process. The simulation results support the following main conclusions.

- This study has successfully employed a hybrid Taguchi/CAE simulation approach to determine the optimal processing conditions which minimize the overall roundness and overall concentricity errors of an optical telecentric barrel produced using the plastic injection molding technique.
- The overall roundness and overall concentricity of the optical barrel are determined mainly by the packing pressure, melt temperature, and cooling time. Both properties can be improved by increasing the packing pressure, reducing the melt temperature, and extending the cooling time.
- The overall roundness and overall concentricity of the molded barrel are positively correlated with one another. Thus, an appropriate selection of the processing conditions optimizes both the overall roundness and the overall concentricity simultaneously.
- The holder structure added to the side of the lens barrel induces a local volume shrinkage effect which causes an axial deformation of the barrel. However, the optimal processing conditions reduce the overall volume shrinkage rate of the barrel from 4.409% (standard processing conditions) to 3.465% and reduce the overall deformations of the barrel from 0.592 mm (standard processing conditions) to 0.469mm. This paper successfully improves the overall roundness and overall concentricity in the vicinity of the holder structure accordingly.

Author Contributions: Conceptualization, C.-M.L., Y.-J.C.; formal analysis, C.-M.L., Y.-J.C.; writing—original draft preparation, C.-M.L., Y.-J.C.; data curation, Y.-J.C.; writing—review and editing, C.-M.L.; supervision, C.-M.L.; funding acquisition, C.-M.L. All authors have read and agreed to the published version of the manuscript.

Funding: This research was funded by the Ministry of Science and Technology of Taiwan, ROC, under Grant Numbers MOST 109-2221-E-415-001-MY3 and MOST 109-2221-E-415-002-MY3.

Institutional Review Board Statement: Not applicable.

Informed Consent Statement: Not applicable.

Acknowledgments: The authors gratefully acknowledge the financial support provided to this study by the Ministry of Science and Technology of Taiwan, ROC.

Conflicts of Interest: The authors declare no conflict of interest.

References

1. Lu, X.; Khim, L.S. A statistical experimental study of the injection molding of optical lenses. *J. Mater. Process. Technol.* **2001**, *113*, 189. [CrossRef]
2. Beaumont, J.P. *Runner and Gating Design Handbook: Tools for Successful Injection Molding*; Hanser Gardner Publications: Cincinnati, OH, USA, 2004.
3. Lin, C.M.; Hsieh, H.K. Processing optimization of Fresnel lenses manufacturing in the injection molding considering birefringence effect. *Microsyst. Technol.* **2017**, *23*, 5689–5695. [CrossRef]
4. Lin, C.M.; Chen, Y.W. Grey optimization of injection molding processing of plastic optical lens based on joint consideration of aberration and birefringence effects. *Microsyst. Technol.* **2019**, *25*, 621–631. [CrossRef]
5. Lin, C.M.; Chen, W.C. Optimization of injection-molding processing conditions for plastic double-convex Fresnel lens using grey-based Taguchi method. *Microsyst. Technol.* **2020**, *26*, 2575–2588. [CrossRef]
6. Wieme, T.; Duan, L.; Mys, N.; Cardon, L.; D'hooge, D.R. Effect of Matrix and Graphite Filler on Thermal Conductivity of Industrially Feasible Injection Molded Thermoplastic Composites. *Polymers* **2019**, *11*, 87. [CrossRef]
7. Pham, N.T.-H. Characterization of Low-Density Polyethylene and LDPE-Based/Ethylene-Vinyl Acetate with Medium Content of Vinyl Acetate. *Polymers* **2021**, *13*, 2352. [CrossRef]
8. Miks, A.; Novak, J. Design of a double-sided telecentric zoom lens. *Appl. Opt.* **2012**, *51*, 5928–5935. [CrossRef] [PubMed]
9. Tong, Y.; Guofan, J.; Junguo, Z. Design of image-side telecentric freeform imaging systems based on a point-by-point construction-iteration process. *Chin. Opt. Lett.* **2017**, *15*, 77–81. [CrossRef]
10. Peng, J.; Wang, M.; Deng, D.; Liu, X.; Yin, Y.; Peng, X. Distortion correction for microscopic fringe projection system with Scheimpflug telecentric lens. *Appl. Opt.* **2015**, *54*, 10055–10062. [CrossRef]
11. Lifu, W.; Zhu, J.; Xie, H. Single-lens 3D digital image correlation system based on a bilateral telecentric lens and a bi-prism: Validation and application. *Appl. Opt.* **2015**, *54*, 7842.
12. Liu, X.; Liu, Z.; Duan, G.; Cheng, J.; Jiang, X.; Tan, J. Precise and robust binocular camera calibration based on multiple constraints. *Appl. Opt.* **2018**, *57*, 5130. [CrossRef] [PubMed]
13. Quinn, M.; Spinosa, E.; Roberts, D. Miniaturisation of pressure sensitive paint measurement systems using low-cost, miniaturised machine vision cameras. *Sensors* **2017**, *17*, 1708. [CrossRef]
14. Ye, X.W.; Dong, C.Z.; Liu, T. A review of machine vision-based structural health monitoring: Methodologies and applications. *J. Sens.* **2016**, *2016*, 1–10. [CrossRef]
15. Li, X.; Qiao, T.; Pang, Y.; Zhang, H.; Yan, G. A new machine vision real-time detection system for liquid impurities based on dynamic morphological characteristic analysis and machine learning. *Measurement* **2018**, *124*, 130–137. [CrossRef]
16. Xu, Y.; Brownjohn, J.M. Review of machine-vision based methodologies for displacement measurement in civil structures. *J. Civ. Struct. Health Monit.* **2018**, *8*, 91–110. [CrossRef]
17. Xiaojun, W.; Gao, G. LED light design method for high contrast and uniform illumination imaging in machine vision. *Appl. Opt.* **2018**, *57*, 1694.
18. Yang, J.; Li, X.; Xu, J.; Cao, Y.; Zhang, Y.; Wang, L.; Jiang, S. Development of an optical defect inspection algorithm based on an active contour model for large steel roller surfaces. *Appl. Opt.* **2018**, *57*, 2490. [CrossRef]
19. Paul Drake, J.R. *Dimensioning & Tolerancing Handbook*; McGraw-Hill: New York, NY, USA, 1999.
20. Henzold, G. Geometrical Dimensioning and Tolerancing for Design. In *Manufacturing and Inspection*; Elsevier: Amsterdam, The Netherlands, 2006.
21. Roy, R.K. *Design of Experiments Using the Taguchi Approach*; John Wiley & Sons: Hoboken, NJ, USA, 2001; p. 538.
22. Taguchi, G. *System of Experimental Design*; Kraus International Publications: White Plains, NY, USA, 1985; Volumes 1–2.
23. Tang, S.H.; Tan, Y.J.; Sapuan, S.M.; Sulaiman, S.; Ismail, N.; Samin, R. The use of Taguchi method in the design of plastic injection mould for reducing warpage. *J. Mater Proc Technol.* **2007**, *182*, 418–426. [CrossRef]
24. Beaumont, J.P.; Nagel, R.; Sherman, R. *Successful Injection Moulding—Process, Design and Simulation*; Hanser: Gardner, NY, USA, 2002.
25. Pervez, H.; Mozumder, M.S.; Mourad, A.H.I. Optimization of injection molding parameters for HDPE/TiO2 nanocomposites fabrication with multiple performance characteristics using the Taguchi method and Grey relational analysis. *Materials* **2016**, *9*, 710. [CrossRef]
26. MOLDEX3D. Available online: http://support.moldex3d.com/2021/en/index.html (accessed on 4 October 2021).
27. Rudolph, N.; Osswald, T.A. *Polymer Rheology: Fundamentals and Applications*; Hanser: Gardner, NY, USA, 2014.
28. Malloy, R. *Plastic Part Design for Injection Molding*; Hanser: Gardner, NY, USA, 1994.
29. MOLDEX3D. Available online: https://www.moldex3d.com/moldex3d-help/ (accessed on 4 October 2021).
30. Menges, G.; Mohren, I.P. *How to Make Injection Molds*; Hanser: Gardner, NY, USA, 1986.
31. Annicchiarico, D.; Jeffrey, R.A. Review of Factors that Affect Shrinkage of Molded Part in Injection Molding. *Mater. Manuf. Process.* **2014**, *29*, 662–682. [CrossRef]
32. Fischer, J.M. *Handbook of Molded Part Shrinkage and Warpage—Determination of Shrinkage*; Beckley, B., Ed.; Plastics Design Library/William Andrew Pub.: Norwich, NY, USA, 2003.
33. Liu, F.; Zeng, S.; Zhou, H.; Li, J. A study on the distinguishing responses of shrinkage and warpage to processing conditions in injection molding. *J. Appl. Polym. Sci.* **2012**, *125*, 731–744. [CrossRef]

34. Bower, A.F. *Applied Mechanics of Solids*; CRC Press: Boca Raton, FL, USA; Taylor & Francis Group: Oxfordshire, UK, 2009.
35. Sui, W.; Zhang, D. Four Methods for Roundness Evaluation. *Phys. Procedia* **2012**, *24*, 2159–2164. [CrossRef]
36. Ahn, S.J. *Least Squares Orthogonal Distance Fitting of Curves and Surfaces in Space*; Springer: Berlin, Germany, 2004.
37. Srinivasan, V.; Shakarji, C.M.; Morse, E.P. On the enduring appeal of least-squares fitting in computational coordinate metrology. *J. Comput. Inform. Sci. Eng.* **2011**, *12*, 011008. [CrossRef]
38. Shunmugam, M.S. Criteria for computer-aided form evaluation. *J. Eng. Ind.* **1991**, *113*, 233–240. [CrossRef]
39. Kim, N.H.; Kim, S.W. Geometrical tolerances: Improved linear approximation of least-squares evaluation of circularity by minimum variance International. *J. Mach. Tools Manuf.* **1996**, *36*, 355–366. [CrossRef]
40. Hichem, N.; Pierre, B. Evaluation of roundness error using a new method based on a small displacement screw. *Meas. Sci. Technol.* **2014**, *25*, 044012.
41. GD&T BASICS. Available online: https://www.gdandtbasics.com/concentricity (accessed on 4 October 2021).
42. Berman, M.; Culpin, D. The statistical behaviour of some least squares estimators of the centre and radius of a circle. *J. R. Statist. Soc. B* **1986**, *48*, 183–196. [CrossRef]
43. Landau, U.M. Estimation of a circular arc center and its radius. *Comput. Vis. Graph. Image Process.* **1987**, *38*, 317–326. [CrossRef]

polymers MDPI

Article

A Methodology to Predict and Optimize Ease of Assembly for Injected Parts in a Family-Mold System

Chao-Tsai Huang [1,*], Tsai-Wen Lin [1], Wen-Ren Jong [2] and Shia-Chung Chen [2]

[1] Department of Chemical and Materials Engineering, Tamkang University, New Taipei City 25137, Taiwan; chdy1245@gmail.com
[2] Department of Mechanical Engineering, Chung Yung Christian University, Taoyuan City 32023, Taiwan; wenren@cycu.edu.tw (W.-R.J.); shiachun@cycu.edu.tw (S.-C.C.)
* Correspondence: cthuang@mail.tku.edu.tw or cthuang@moldex3d.com

Abstract: In this study, the assembly behavior for two injected components made by a family mold system were investigated. Specifically, a feasible method was proposed to evaluate the characteristic length of two components within a family mold system using numerical simulation and experimental validation. Results show that as the packing pressure increases, the product index (characteristic length) becomes worse. This tendency was consistent for both the simulation prediction and experimental observation. However, for the same operation condition setting through a basic test, there were some differences in the product index between the simulation prediction and experimental observation. Specifically, the product index difference of the experimental observation was 1.65 times over that of the simulation prediction. To realize that difference between simulation and experiment, a driving force index (DFI) based on the injection pressure history curve was proposed. Through the DFI investigation, the internal driving force of the experimental system was shown to be 1.59 times over that of the simulation. The DFI was further used as the basis for machine calibration. Furthermore, after finishing machine calibration, the integrated CAE and DOE (called CAE-DOE) strategy can optimize the ease of assembly up to 20%. The result was validated by experimental observation.

Keywords: injection molding; degree of assembly; a family mold system; CAE-DOE optimization

Citation: Huang, C.-T.; Lin, T.-W.; Jong, W.-R.; Chen, S.-C. A Methodology to Predict and Optimize Ease of Assembly for Injected Parts in a Family-Mold System. *Polymers* **2021**, *13*, 3065. https://doi.org/10.3390/polym13183065

Academic Editors: Célio Bruno Pinto Fernandes, Salah Aldin Faroughi, Luís L. Ferrás, Alexandre M. Afonso and Dagmar R. D'hooge

Received: 11 August 2021
Accepted: 8 September 2021
Published: 10 September 2021

1. Introduction

A family mold structure is one kind of the multi-cavity systems in the injection molding process. It has been utilized in the injection molding industry to make a series of assembly components for years. The related products are commonly in hand-held bar codes [1], automotive components [2], smartphone lenses [3], luggage [4], watercraft [5], toys [6], and so on. Due to various influencing factors and the complicated features of the components, the assembly behavior is quite sensitive to the design and to the processing during the injection molding. However, there is very little information to describe the relationship between the assembly behavior of injected components and the injection molding factors. Hence, it is very difficult to predict the assembly behavior in the design phase for those assembly components.

Moreover, the degree of assembly could be associated with the design for manufacturing and assembly (DFMA) standards. The main target is to integrate multiple components with multiple functions to minimize some indicators such as energy consumption, carbon footprint, number of parts, required amount of material, assembly time, and manufacturing costs [7–10]. The degree of assembly is the total of the ease of assembly indicators calculated by a series of manual handling and insertion analyses on the existing design based on the Boothroyd and Dewhurst (BD), Lucas Hall (LH), and Hitachi Assembly Evaluation (AEM) methods, among others. However, to the best of our knowledge, there is very little information to discuss the assembly behavior of injected components associated with the injection molding process. Meanwhile, some studies have discussed the degree of assembly of injected parts [11–13]. Unfortunately, most of them are provided as patents which are disclosed by know-how without detailed mechanism information. Hence, a determination of how to predict the sensitivity of the ease of assembly for assembling injected components from the injection molding process in the design phase has not yet been fully constructed.

Furthermore, considering the testing through the manual handling and insertion analysis on the existing design, the assembly behavior will be influenced by the final geometrical structures of the injected parts. Theoretically, if each injected component can maintain the final shape as close to the design as possible, the assembly behavior will be smooth. Therefore, the strategies developed in the literature to reduce the shrinkage and warpage of the individual injected parts might be the good solutions for the assembly behavior of family mold products. Lee and Kim [14] used the thickness of injected parts as the control factor to minimize the warpage of the product. Leo and Cuvelliez [15] focused on the gate geometry and operation parameters to modify the dimensional accuracy of the parts. Later, Yen et al. [16] selected the diameter and the length of the runner to minimize the warpage of the injected parts. Zhai et al. [17] tried to catch the balanced flow through runner size modification and then improve the product quality. They found that the target could be obtained by adjusting the runner sizes using a non-dominated sorting genetic algorithm. Othman et al. [18] conducted a study of the influence of the runner length and gate location on warpage and shrinkage to control part quality. Moreover, many researchers have moved to material modification to enhance product quality. Thomason et al. [19,20] investigated the fiber reinforced effects on the quality improvement of products experimentally. Kovacs and Solymossy [21] applied glass bead-filled PA6 to reduce the warpage and shrinkage of the injected parts. Hakimian and Sulong [22] studied different thermoplastic composites to determine the improvement on warpage. Furthermore, to enhance the quality of the injection molding parts effectively, scientists and researchers have integrated computer-aided engineering (CAE) techniques and various optimization strategies. Ozcelik and Erzurumlu [23] integrated finite element analysis, DOE, response surface methodology (RSM), and genetic algorithms to reduce warpage effectively. They claimed that after being optimized, the warpage of injected part was reduced by 51%. Zhai and Xie [24] combined CAE and sequential linear programming (SLP) to optimize the gate design to obtain a balanced flow, and then to reduce the warpage. Chiang and Chang [25] introduced RSM to optimize the shrinkage and warpage of a cell phone. They concluded the shrinkage and warpage of injected parts can be reduced by 53.9%. Fernandes et al. [26] integrated multi-objective genetic algorithms and CAE techniques to optimize the cooling channel to minimize the warpage of the injected parts. Tsai and Tang [27] utilized RSM to search for the optimal conditions to optimize the accuracy of spherical lenses. Xu and Yang [28] combined the Taguchi method, neural networks, and grey correlation analysis (GCA) to solve the multi-objective optimization problem. Kitayama et al. [29] applied a sequential approximate optimization (SAO) based on a CAE simulation to determine the optimal process parameters. They concluded that a multi-objective design optimization is effective for weld line reduction and clamping force minimization. Later, Hentati et al. [30] and Huang et al. [31] integrated CAE and the Taguchi method to optimize injection molding process parameters. In addition, Fernandes et al. [32] reviewed the studies done in the field of theoretical modeling and various optimization techniques for the injection-molding process. The strengths and weaknesses of each technique were discussed. It is noted that, in recent years, using CAE technology to perform injection molding simulations virtually can enhance efficiency in the product development and problem solving effectively. However, it is quite common to encounter some difference between simulation predictions and experimental observation. Huang et al. [31,33] proposed a feasible method to discover the cause of that difference. They concluded that to diminish that difference, both the virtual and real injection molding machines should be calibrated. Moreover, to discuss the performance of injection molding machines, Chen et al. [34,35] proposed a method to derive the correlation between product quality and machine quality indexes. They concluded that the pressure peak index, viscosity index, and energy index are strongly associated with product quality. Some quality index could be useful to discover the internal driving force in the future.

As described above, it is noted that due to the non-balanced component structure, the retention of dimensional precision for the individual components made by a family mold system is strongly affected by operation conditions. Moreover, which interaction between the individual component forming which will further influence assembly behavior is not clear. Hence, in this study, the assembly behavior of two injected components made by a family mold system was investigated. Specifically, a method to evaluate the characteristic length of two components within a family mold system is proposed using numerical simulation and experimental validation. Then, the correlation between the characteristic length and the ease of assembly is further discussed. Moreover, to enhance the simulation's accuracy of the assembly behavior, machine calibration was performed. The influence of machine calibration on the assembly behavior is then discussed. Furthermore, to optimize the ease of assembly for this complicated system, the integrated CAE and DOE (called CAE-DOE) is utilized for virtual optimization. Then the optimization efficiency in the assembly behavior is verified by the physical DOE experiment.

2. Theory and Assumption

The polymer material in this study can be assumed to be a general Newtonian fluid (GNF). During the injection molding process, the non-isothermal 3D flow motion can be mathematically described by the following equations:

$$\frac{\partial \rho}{\partial t} + \nabla \cdot \rho \mathbf{u} = 0 \tag{1}$$

$$\frac{\partial}{\partial t}(\rho \mathbf{u}) + \nabla \cdot (\rho \mathbf{u}\mathbf{u} - \boldsymbol{\sigma}) = \rho \mathbf{g} \tag{2}$$

$$\boldsymbol{\sigma} = -p\mathbf{I} + \eta\left(\nabla \mathbf{u} + \nabla \mathbf{u}^T\right) \tag{3}$$

$$\rho C_p\left(\frac{\partial T}{\partial t} + \mathbf{u} \cdot \nabla T\right) = \nabla \cdot (k \nabla T) + \eta \dot{\gamma} \tag{4}$$

where $\mathbf{u}$ is the velocity vector, T is the temperature, t is the time, p is the pressure, $\boldsymbol{\sigma}$ is the total stress tensor, ρ is the density, $\mathbf{g}$ is gravitational force, $\mathbf{I}$ is the unit matrix, η is the viscosity, k is the thermal conductivity, C_p is the specific heat, and $\dot{\gamma}$ is the generalized shear rate.

Moreover, the modified-cross model with Arrhenius temperature dependence is employed to describe the viscosity of polymer melt:

$$\eta(T, \dot{\gamma}) = \frac{\eta_o(T)}{1 + (\eta_o \dot{\gamma} / \tau^*)^{1-n}} \tag{5}$$

where

$$\eta_o(T) = BExp\left(\frac{T_b}{T}\right) \tag{6}$$

where η is the viscosity, η_o is the zero shear viscosity, n is the power law index, B is the consistency index, and τ^* is the parameter that describes the transition region between the zero shear rate and the power law region of the viscosity curve.

3. Methodology and Materials

In this study, a numerical simulation and experimental methods were utilized. The associated systems for both methods are described as follows.

3.1. Numerical Simulation System

Regarding the numerical simulation of the system, Moldex3D R16® software (supplied by CoreTech System Co. Ltd., Hsinchu County, Taiwan) was adopted. The geometrical structure of the system is shown in Figure 1. Specifically, it is a family mold system with two components of parts A and B, as shown in Figure 1a. Part A is an inner part, and part B is an outer part. The associated runner, with detailed dimensions, is listed in Figure 1b. In addition, the dimensions of part A and part B are exhibited in Figure 1c,d. Their dimensions are about 40 mm × 40 mm × 14 mm. In addition, the volumes of part A and part B were 5.9 and 6.5 cm^3, respectively. Furthermore, the moldbase and cooling channel layout are presented in Figure 2. There are four cooling channels inside the moldbase. For injection molding, the material used was acrylonitrile butadiene styrene, called ABS (PA757 supplied by Che-Mei, Tainan city, Taiwan). In order to perform the injection molding simulation, several material properties need to be measured and stored as the database. For example, the key properties to influence the flow and warpage are the temperature-shear rate-dependent viscosity and the specific volume against pressure-temperature (pvT), as presented in Figure 3. Those data were measured and provided from Moldex3D directly. Furthermore, to evaluate the assembly behavior of these components, and to find out the key practical factors for further study, a single factor test was performed with the associated factors as listed in Table 1. Specifically, each factor has five levels. The reasons we selected those factors is referred to in several studies mentioned earlier [14–18,21–25,27–31]. The key operation parameters utilized in each reference are listed in Table 2. The goal of the single factor test was to find out some practical operation parameter which can be utilized as the major control factor to evaluate the variation of the characteristic lengths. Those lengths were used to evaluate the assembly behavior and will be explained later.

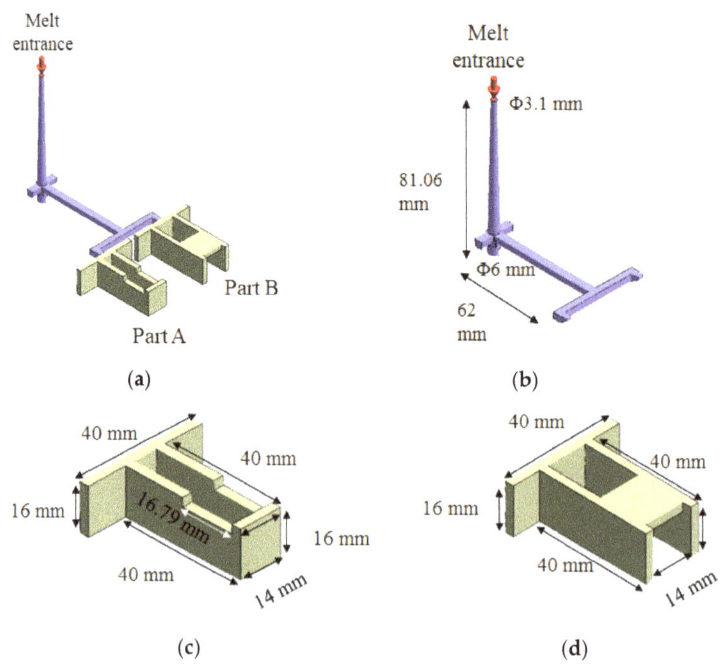

Figure 1. Geometrical structures of two components in a family mold: (**a**) the runner and cavities, (**b**) the dimensions of the runner, (**c**) the dimensions of part A, (**d**) the dimensions of part B.

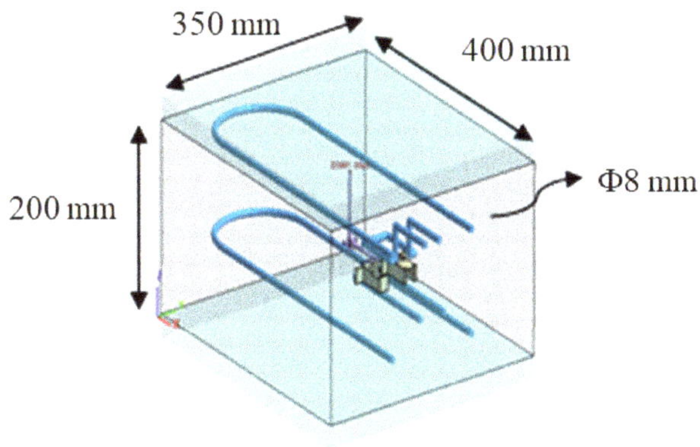

Figure 2. The moldbase and cooling channel layout.

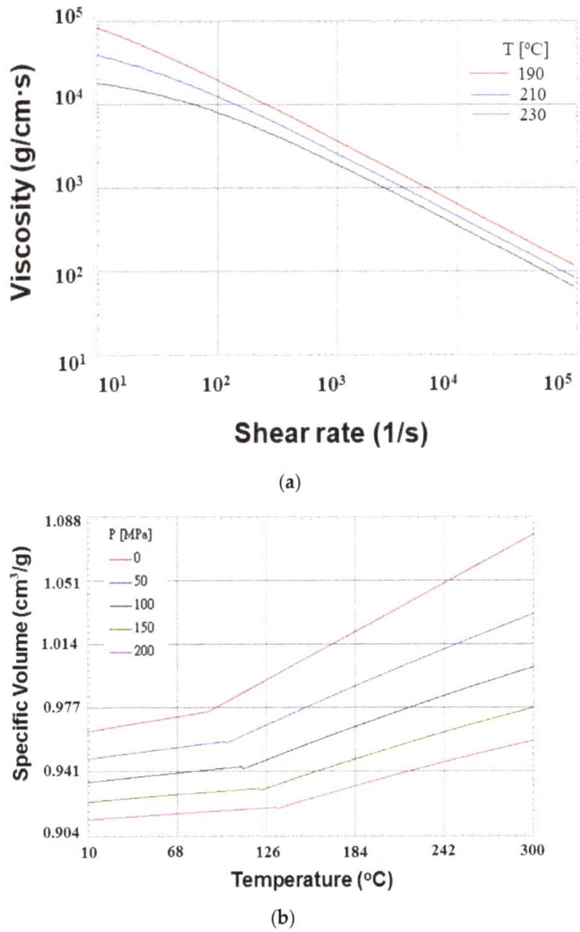

Figure 3. The material properties of ABS (PA757, supplied by Che-Mei): (a) viscosity, (b) pvT (the specific volume against pressure-temperature).

Table 1. The operation conditions for the single factor test.

Factor	Level 1	Level 2	Level 3	Level 4	Level 5
Injection speed [a] (%)	30	40	50	60	70
Melt temperature (°C)	190	200	210	220	230
Mold temperature (°C)	30	40	50	60	70
Packing time (s)	3	5	7	9	11
Packing pressure [b] (%)	50	75	100	125	150
Cooling time (s)	7	9	11	13	15

[a]: based on the maximum speed of the screw movement with 125 mm/s. [b]: based on the end of filling pressure, P_{EOF}.

Table 2. The key operation parameters utilized in the literature.

Ref.	Authors	Year	Key Operation Parameters							
			Mold Temp.	Melt Temp.	Injection Speed/Time	Injection Pressure	Packing Pressure	Packing Time	Cooling Temp.	Cooling Time
14	Lee and Kim	1995		✓	✓		✓	✓	✓	
15	Leo and Cuvelliez	1996					✓			
16	Yen et al.	2006								
17	Zhai et al.	2009	✓	✓		✓				
18	Othman et al.	2013	✓	✓	✓		✓			✓
21	Kova and Solymossy	2009	✓	✓						
22	Hakimian and Sulong	2012	✓	✓				✓		✓
23	Ozcelik and Erzurumlu	2006	✓	✓			✓	✓		✓
24	Zhai and Xie	2010								
25	Chiang and Chang	2007	✓				✓	✓		✓
27	Tsai and Tang	2014	✓	✓	✓		✓	✓		✓
28	Xu and Yang	2015	✓	✓	✓		✓	✓		✓
29	Kitayama et al.	2018		✓	✓		✓	✓	✓	✓
30	Hentati	2019	✓	✓		✓		✓		
31	Huang et al.	2020	✓	✓	✓		✓	✓		✓

Where "✓" means that parameter has been considered in that reference.

Moreover, a basic test of the injection molding simulation was performed to determine the relationship between the characteristic lengths and the assembly behavior. The operation conditions for the basic test are listed in Table 3. Specifically, the melt temperature was 210 °C. The mold temperature was 50 °C. The injection speed was setup at 50% which was based on the maximum speed of the screw movement, with 125 mm/s in the machine (afterwards called injection speed 50% setting). The packing time was 7 s. The cooling time was 11 s. The packing pressure was setup from 25% to 100%. Here the packing pressure setting was based on the end of filling pressure, P_{EOF}.

Table 3. Process conditions for the basic test.

Factor	Operation Conditions
Injection speed (%)	50
Melt temperature (°C)	210
Mold temperature (°C)	50
Packing time (s)	7
Packing pressure (%)	25; 50; 75; 100
Cooling time (s)	11

3.2. Experimental Equipment

Moreover, to discover the real variation of the assembly behavior based on the characteristic lengths and to validate the simulation predictions, an injection molding experiment was constructed as follows. Figure 4a presents the FCS 150SV injection machine supplied by Fu Chun Shin Machinery Co. Ltd., Tainan City, Taiwan. This system offers a maximum injection pressure of 140 bar, a maximum injection speed of 125 mm/s (the maximum speed of the screw movement, as mentioned previously), and a maximum movement distance of 200 mm for the screw. The screw diameter is 44 mm. In addition, the real mold structure is listed in Figure 4b. The dimensions for the cavity, runner, and cooling channels are as described in Figure 2.

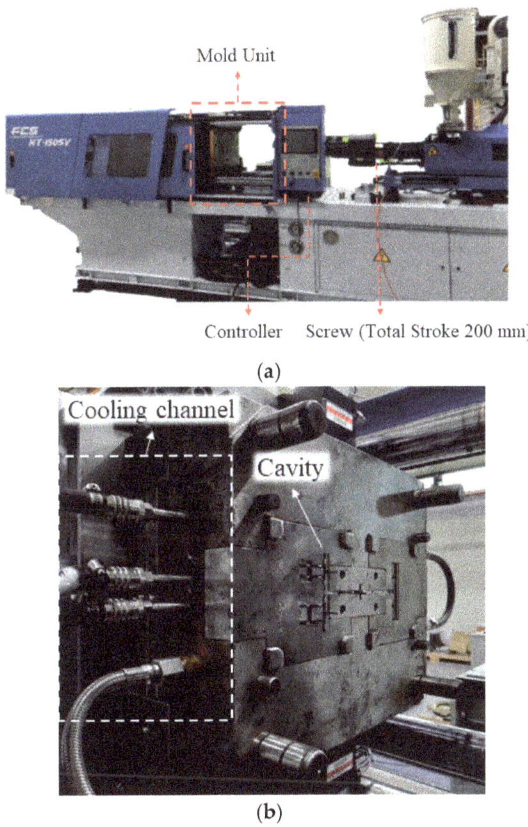

Figure 4. Machine and equipment setup: (**a**) injection machine (Fu Chun Shin Machinery Co. Ltd., Model: FCS 150-SV), and (**b**) mold layout.

3.3. Define the Characteristic Length as Product Index

To study the assembly behavior, some characteristic lengths are presented as in Figure 5. Specifically, the product index based on the characteristic length is defined as follows:

$$X_i = (X_{Bi} - X_{Ai}) \tag{7}$$

where i is from 1 to 4; X_{Ai} is the outer length of part A, and X_{Bi} is the inner length of part B.

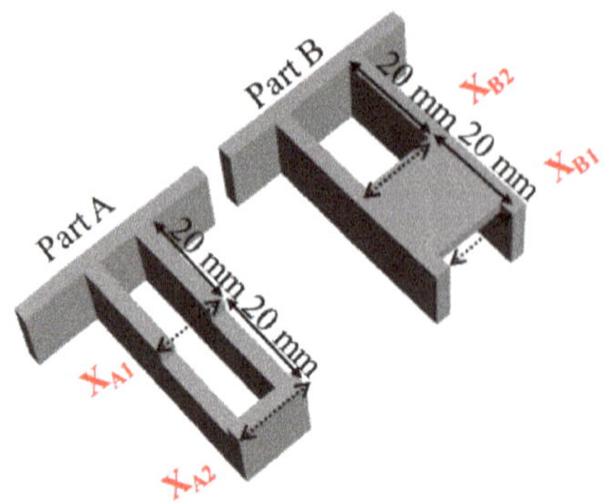

(**a**) Top view

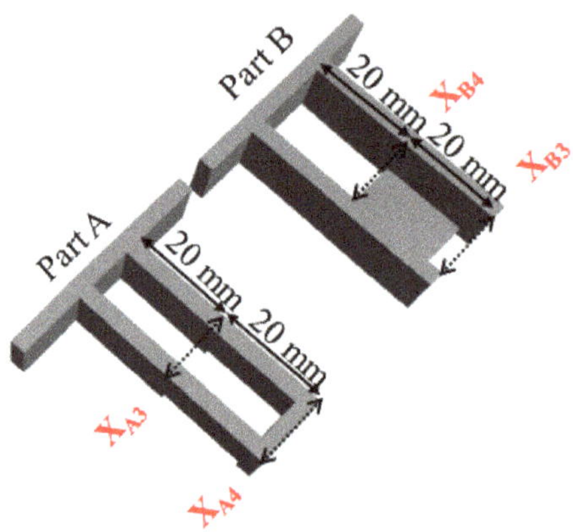

(**b**) Bottom view

Figure 5. Definition of the characteristic lengths for parts A and B: (**a**) top view, (**b**) bottom view.

For example, $X_1 = (X_{B1} - X_{A1})$ is the characteristic length at a central location (based on part A), obtained from the difference between the inner length of part B and the outer length of part A on the top plane. $X_2 = (X_{B2} - X_{A2})$ is the characteristic length at the end location of the top plane (based on part A). Similarly, X_3 and X_4 are defined on the bottom plane. Theoretically, when the characteristic length (Xi) is greater than zero where the inner length of part B is larger than the outer length of part A, the assembly should be easy. On the other hand, if the characteristic length (Xi) is smaller than zero, the assembly should be not easy.

3.4. Integrate CAE and Design of Experiments (DOE) to Optimize the Key Factors in the Assembly Behavior

The operational parameters of injection molding that influence the assembly behavior of injected parts are very complicated. To examine the influence of the key factors on the assembly behavior and to optimize them, design of experiment (DOE) optimization was introduced. Specifically, DOE methods based on CAE technology (afterwards called CAE-DOE) and physical DOE experiments were utilized. Here, to discover the optimization efficiency of the DOE method before performing the machine calibration, the key control factors include (A) injection speed, (B) mold temperature, (C) packing pressure, (D) packing time, (E) melt temperature, and (F) cooling time, as listed in Table 4. For each factor, three levels have been specified. For example, regarding the injection speed factor, level 1 to level 3 is 25 mm/s (20% injection speed setting) to 125 mm/s (100% injection speed setting). Before discussing efficiency, all the operation parameters of the level 2 column were selected as the original design setting for CAE-DOE investigation. The characteristic lengths of the injected parts based on this original design setting were used as the basis for further comparison. In addition, the corresponding orthogonal array for DOE performance using CAE (i.e., CAE-DOE) is listed in Table 5. Indeed, eighteen sets of injection molding trials was executed numerically based on the L_{18} ($2^1 \times 3^7$) orthogonal array. Since only six major factors were considered and each factor has three levels, the first and the eighth columns will be ignored for further application.

Table 4. The control factors and their levels in CAE-DOE before machine calibration.

	Control Factor	Level 1	Level 2	Level 3
A	Injection Speed (mm/s)	25 (20%)	75 (60%)	125 (100%)
B	Mold Temperature (°C)	30	50	70
C	Packing Pressure (MPa)	95	126	158
D	Packing Time (s)	5	7	9
E	Melt Temperature (°C)	200	210	220
F	Cooling Time (s)	9	11	13

Table 5. $L_{18}(2^1 \times 3^7)$ orthogonal array for CAE-DOE performance.

Exp		A	B	C	D	E	F	
		Injection Speed (mm/s)	Mold Temp. (°C)	Packing Pressure (MPa)	Packing Time (s)	Melt Temp. (°C)	Cooling Time (s)	
	1	**2**	**3**	**4**	**5**	**6**	**7**	**8**
1	1	1	1	1	1	1	1	1
2	1	1	2	2	2	2	2	2
3	1	1	3	3	3	3	3	3
4	1	2	1	1	2	2	3	3
5	1	2	2	2	3	3	1	1
6	1	2	3	3	1	1	2	2
7	1	3	1	2	1	3	2	3
8	1	3	2	3	2	1	3	1
9	1	3	3	1	3	2	1	2
10	2	1	1	3	3	2	2	1
11	2	1	2	1	1	3	3	2
12	2	1	3	2	2	1	1	3
13	2	2	1	2	3	1	3	2
14	2	2	2	3	1	2	1	3
15	2	2	3	1	2	3	2	1
16	2	3	1	3	2	3	1	2
17	2	3	2	1	3	1	2	3
18	2	3	3	2	1	2	3	1

The eighteen sets of injection molding trials were executed. The associated characteristic lengths and their average for each set were measured and will be discussed later, in the Results and Discussion section. Moreover, based on the calculated characteristic lengths for each set, the standard deviation Sn was calculated from Equation (8). Then the S/N ratio (signal-to-noise ratio) was obtained by Equation (9).

$$S_n \text{ (Standard Deviation)} = \sqrt{\frac{\Sigma(y_i - \overline{y})^2}{N - 1}} \quad (8)$$

$$S/N = -10 \times \log\left[(\overline{y} - y_0)^2 + {S_n}^2\right] \quad (9)$$

where for each injected part: y_i is the deviation between the ith characteristic length; $\overline{y}$ is the average of four characteristic length deviations; and y_0 is 0.

4. Results and Discussion

4.1. Single Factor Test and Basic Test for Assembly Behavior

4.1.1. Perform a Single Factor Test

As mentioned earlier, the purpose of the single test was to determine some practical operation parameters for further study of the assembly behavior. Figure 6 shows the results of the single factor test on the influence of the characteristic lengths. For each factor, there were five levels to test, as listed in Table 1. The deviation was estimated by the maximum value of the characteristic length minus the minimum value, at each location. Then, the average deviation was utilized to evaluate the sensitivity of each factor. For example, in Figure 6a, when the injection speed setting increased from 30% to 70%, the average deviation from the injection speed was about 0.023 mm. For the melt temperature effect, the average deviation was 0.065 mm, as seen in Figure 6b. Figure 6c presents data that the mold temperature does not provide significant influence, with only a 0.001 mm deviation. Moreover, when the packing pressure is increased, it provides a significant influence, with an average deviation of 0.121 mm, as shown in Figure 6d. Similarly, the average deviations for the packing time and cooling time effects were 0.046 mm and 0.058 mm, respectively, as shown in Figure 6e,f. Overall, the packing pressure effect had the most significant influence on the variation of the characteristic lengths. In addition, the variation tendency of the characteristic lengths was almost proportional to the changes of the packing pressures. Hence, the packing pressure effect was selected as the practical parameter for further study of the assembly behavior.

4.1.2. Perform a Basic Test

The goal for the basic test was to understand the flow behavior and the shrinkage behavior of each location for parts A and B. It was also used to realize the correlation between the characteristic length and the assembly behavior through simulation prediction and experimental verification when the injection molding simulation was performed using the operation condition of Table 3 at a 50% packing pressure setting. In Figure 7, when the volume was filled at 37.5%, the flow behavior for both parts A and B looked similar. However, from 63% to 100% volume filled, the flow imbalance phenomenon happened as expected due to the volume difference of cavities A and B. Figure 8 shows the shrinkage behavior for parts A and B. Regarding part A, X_{A1} and X_{A3} shrunk significantly because there was no constraint. The higher packing pressure, the worse the shrinkage happened, as shown in Figure 8c. However, since X_{A2} and X_{A4} were located at the end portion with strong wall constraints, the higher packing pressure provided the expansive result. On the other hand, for part B, since X_{B1} and X_{B2} were located within a concrete plane, when the packing pressure was increased, their lengths increased slightly. At the same time, X_{B3} and X_{B4} shrunk significantly because of lack of any constraint.

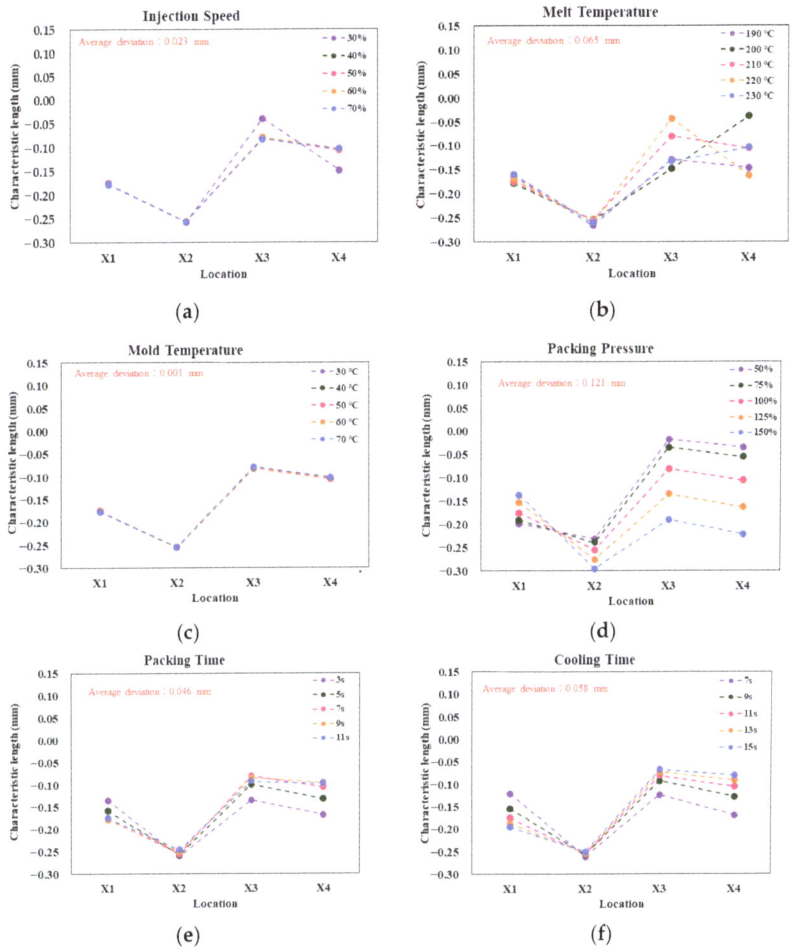

Figure 6. Characteristic lengths variation due to the single factor test: (**a**) injection speed, (**b**) melt temperature, (**c**) mold temperature, (**d**) packing pressure, (**e**) packing time, (**f**) cooling time.

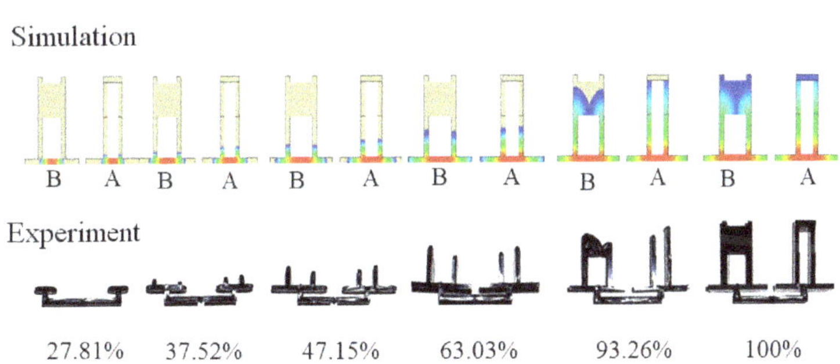

Figure 7. The flow behavior for both the simulation prediction and experimental validation.

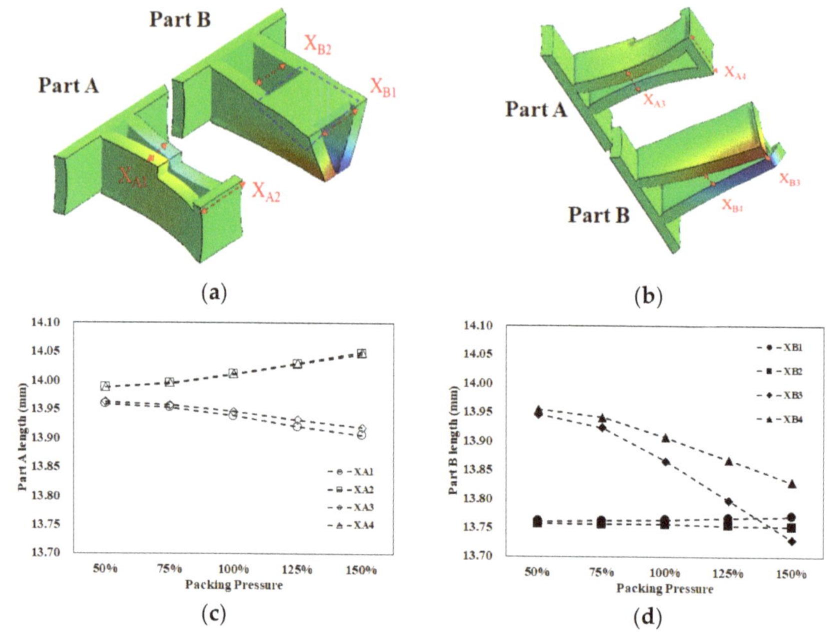

Figure 8. The shrinkage behavior for parts A and B: (**a**) top view, (**b**) bottom view, (**c**) individual length variation of part A, (**d**) individual length variation of part B.

Furthermore, since the variations of the individual lengths for parts A and B were not in the same trend, we had to define the characteristic length to measure the interaction among those individual lengths. In Figure 9a, there were four tests with different packing pressure settings for simulation prediction. As the packing pressure increased, the variation of the characteristic lengths was almost linearly changed. To give better understanding, two packing pressures with lower and higher settings were selected and presented in Figure 9b. For example, at a 25% (lower) packing pressure setting, the characteristic lengths on the top plane were less than zero ($X_1 = -0.207$ mm and $X_2 = -0.224$ mm). Also, the characteristic lengths on the bottom plane were close to or less than zero ($X_3 = 0.004$ mm and $X_4 = -0.013$ mm). This means that the inner lengths of part B were less than outer lengths of part A. Theoretically, these components are not easy to assemble. When the packing pressure was increased to the 100% (higher) packing pressure setting, the characteristic lengths on the top plane were far less than zero ($X_1 = -0.175$ mm and $X_2 = -0.255$ mm). Also, the characteristic lengths on the bottom plane were less than zero ($X_3 = -0.061$ mm and $X_4 = -0.083$ mm). When the higher packing pressure was applied, although X_1 became more positive, the others (X_2 to X_4) became more negative. This led the inner lengths of part B to become much smaller than outer lengths of part A, theoretically resulting in a more difficult assembly of parts A and B. Indeed, higher packing pressure is not a solution to manage the degree of assembly in this study.

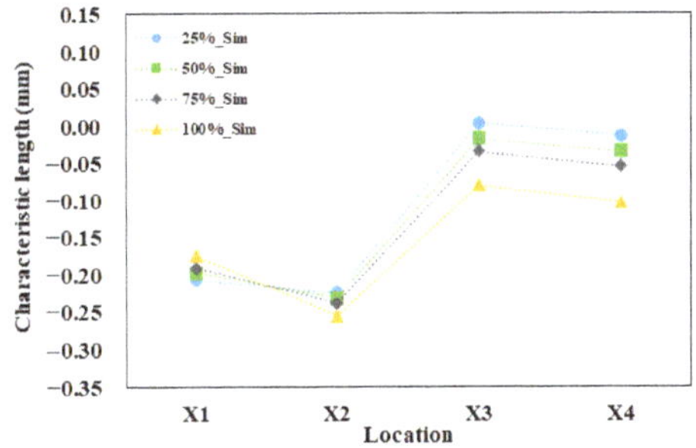

(**a**)

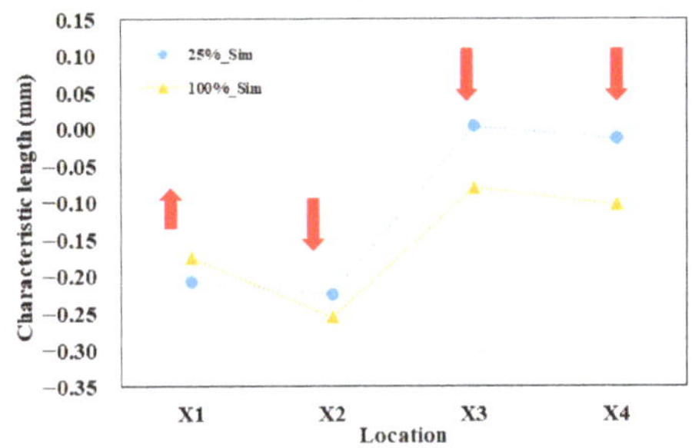

(**b**)

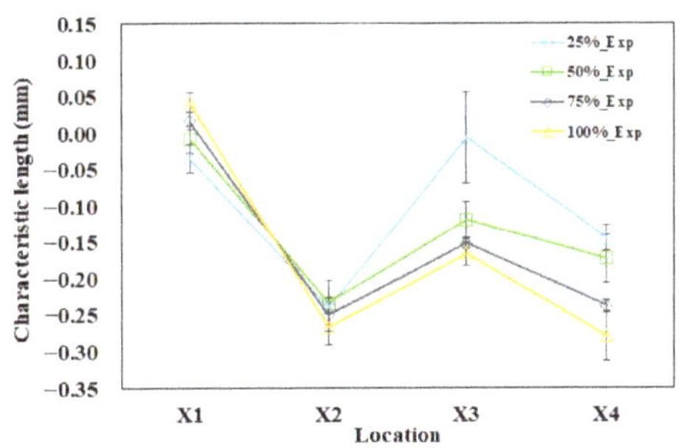

(**c**)

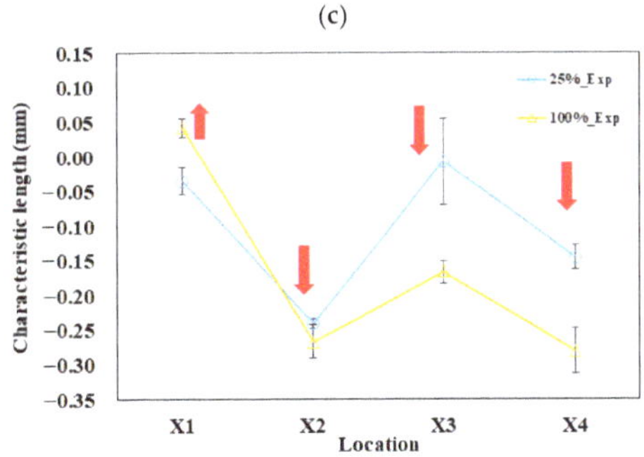

(**d**)

Figure 9. Evaluation of the ease of assembly for parts A and B: (**a**) full range of packing pressure effect for simulation prediction, (**b**) higher and lower packing pressure effect for simulation prediction, (**c**) full range of packing pressure effect for experimental measurement, (**d**) higher and lower packing pressure effect for experimental measurement.

A basic evaluation of the assembly behavior for parts A and B is also performed experimentally. The operation conditions were the same as mentioned in the basic numerical test (see Table 3). For each packing pressure operation, six samples each of part A and part B were collected to measure the inner lengths and outer lengths, as described in Figure 5 and Equation (7). Then, the associated average characteristic lengths were obtained and plotted, as in Figure 9c. As the higher packing pressure was applied, the X_1 became more positive, and the others (X_2 to X_4) became more negative. The basic tendency is similar to that of the numerical prediction. To get a better understanding, two packing pressures with lower and higher settings were selected and presented in Figure 9d. At the 25% (lower) packing pressure setting, the characteristic lengths on the top plane were less than zero ($X_1 = -0.035$ mm and $X_2 = -0.24$ mm). Also, the characteristic lengths on the bottom plane were also less than zero ($X_3 = -0.007$ mm and $X_4 = -0.145$ mm). When the packing pressure was increased to the 100% (higher) packing pressure setting, the characteristic lengths on the top plane changed. X_1 becomes more positive but X_2 became far less than zero ($X_1 = 0.042$ mm and $X_2 = -0.267$ mm). The characteristic lengths on the bottom plane were also far from zero ($X_3 = -0.167$ mm and $X_4 = -0.280$ mm). This means that the higher packing pressure will lead the inner lengths of part B to become much smaller than the outer lengths of part A, resulting in more difficulty in the assembly of parts A and B. Clearly, the tendency of the change of the characteristic lengths is in reasonable agreement for both the simulation prediction and the experimental measurement.

4.1.3. Correlation between Characteristic Length and Assembly Behavior

To realize the relationship between the characteristic lengths and the assembly behavior, a real integration test for parts A and B was performed, as shown in Figure 10. At the 25% packing pressure setting, the integration process assemble parts A and B was smooth and without difficulty. From the top view and side view, it is clearly seen that the assembly of parts A and B was completed as shown in Figure 10a. However, when the packing pressure setting was changed to 100%, the integration test for the assembly became very difficult. The integration test failed. Practically, the higher the packing pressure utilized, the more difficulty encountered in the assembly operation. The results of the integration test are consistent with the characteristic length behavior of the simulation prediction and experimental observation, as discussed previously. Obviously, this method to evaluate the assembly behavior using the characteristic lengths is feasible qualitatively so far.

4.2. *Discover the Reason for the Difference between Simulation and Experiment for the Assembly Behavior*

Figure 11 shows the comparison between the characteristic lengths of the simulation prediction and those of experimental measurement. When the packing pressure increased from 25% to 100%, X_1 increased, while the others (X_2 to X_4) decreased for both the simulation and the experiment. However, when the comparison proceeded one-by-one from X_1 via X_2 to X_4, the amounts of simulation prediction were under-predicted at the top plane for X_1 and X_2 (i.e., too much negative), and were over-predicted at bottom plane for X_3 and X_4 (i.e., too much positive). Overall, the tendency is in reasonable agreement, but the amount of characteristic length at each location was not exactly matched in both the simulation and the experiment. To discover the difference between the simulation and the experiment, the relationship between the internal driving force from the injection machine and the characteristic length difference (ΔXi) on the injected parts was investigated. Here, the characteristic length difference (ΔXi) is defined as Equation (10), as follows.

$$\Delta X_i = Xi \text{ (at 100\% Packing)} - Xi \text{ (at 25\% Packing)} \quad (10)$$

where i is from 1 to 4.

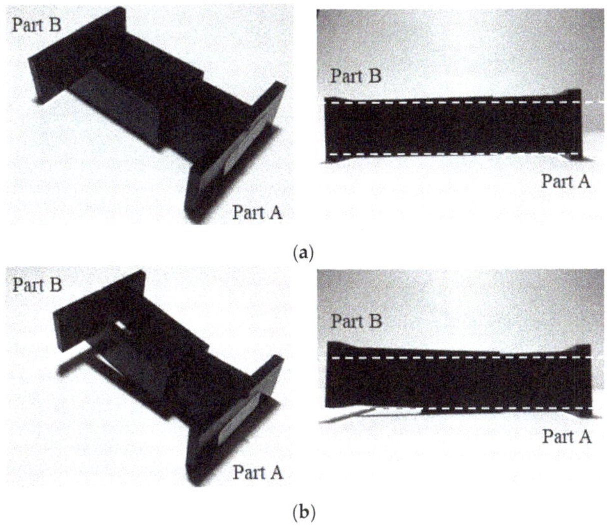

Figure 10. Evaluation on the degree of assembly difficulty through a real integration test for different packing pressure settings on the injection parts: (**a**) 25% packing: passed, (**b**) 100% packing: failed.

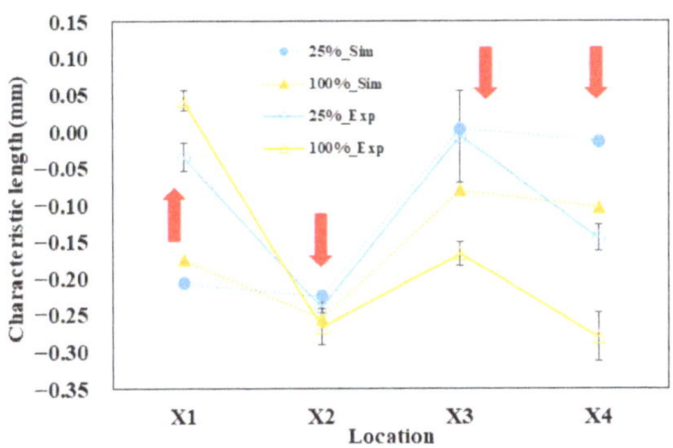

Figure 11. Comparison between simulation and experiment at 25% to 100% packing pressure settings.

For example, ΔX_1 (the characteristic length difference of the simulation) is equal to 0.032 mm (i.e., (−0.175) − (−0.207) = 0.032 mm). Other characteristic length differences of the simulation were calculated, and are listed in Table 6. The average characteristic length differences of the simulation were further calculated and are listed in the rightmost column in Table 6. The ΔX_1 of the experiment was equal to 0.077 mm (i.e., (0.042) − (−0.035) = 0.077 mm). The other characteristic length differences and their average were also calculated and are shown in Table 6. It is noted that the average of the characteristic length differences of the experiment was about 1.65 times over that of the simulation prediction (that is, 0.061/0.037 = 1.65). For the exact same operation condition settings for both simulation and experimental systems, why did the experimental system drive more dimensional variation in the final injection parts than its simulation counterpart is a very interesting question. Before we proceed to answer this question, we note that Huang et al. [33] mentioned that one of the reasons for the difference between numerical predictions and experimental observations in injection molding is because the real machine (experiment) and the virtual machine are not the same, even they have the same operation condition settings. To reduce the difference between two systems, some injection machines need to be calibrated. The details will be discussed in the following section.

Table 6. Difference of the characteristic lengths between the 100% and 25% packing pressure settings for both simulation and experiment (unit: mm).

	ΔX_1	ΔX_2	ΔX_3	ΔX_4	Ave ΔX
Simulation	0.032	−0.031	−0.065	−0.084	−0.037
Experiment	0.077	−0.027	−0.160	−0.135	−0.061

4.3. Machine Calibration Effects on Assembly Behavior

4.3.1. Perform Machine Calibration

To reduce the difference between the numerical prediction and the experimental observations, machine calibration procedures were performed based on the direction in [33]. However, the new controller was changed into the FCS injection machine and one pressure transducer was installed at gate location. Some calibration procedures needed to be modified from the the system in [33]. Specifically, the reference point to catch the injection pressure history curve was moved to the gate location; the new controller was installed, and the injection pressure history of the experiment was different. The details of the calibration procedures are as follows. Machine calibration is based on the injection pressure history curves for both the simulation and the experiment, using a circle plate system with a pressure transducer installed at the gate. The sensor location is presented in Figure 12. The data of the injection pressure can be recorded from the pressure transducer at the gate, which is further used to create the pressure history curve for machine calibration. Specifically, for the same operation condition settings, Figure 13a shows the injection pressure history curves for both the simulation and the experiment at the 50% injection speed setting. The experiment shows a higher injection pressure history curve than that of the simulation counterpart over the entire period, which means that the real injection machine has a higher driving force than that of the virtual simulation machine. To evaluate the driving force for the injection molding through the entire cycle, a driving force index (DFI) based on the total accumulated driving force was defined, as in Equation (11). It is also called the viscosity index [34,35]. This equation can be regarded as reflecting the accumulated resistance force of melt flow during injection molding.

$$\mathrm{DFI} = \langle P_{Total} \rangle_i = \int_0^t (P_{inj})_i \, dt \quad (11)$$

where i is either simulation or experiment, $\langle P_{Total} \rangle$ is the total accumulated driving force with (MPa·s) viscosity units, and P_{inj} is the injection pressure measured at the gate location at the time t.

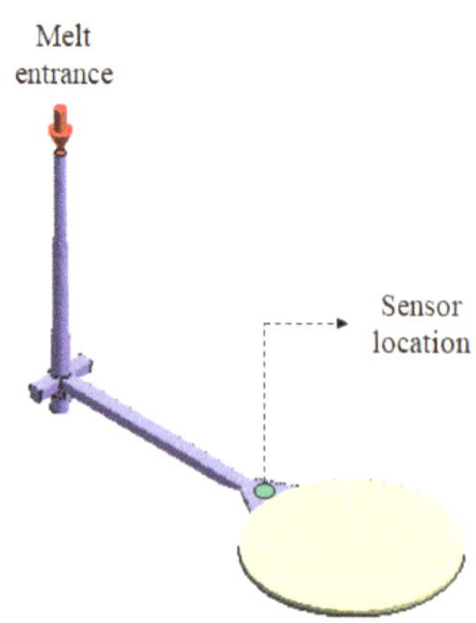

Figure 12. The calibration system and the sensor location.

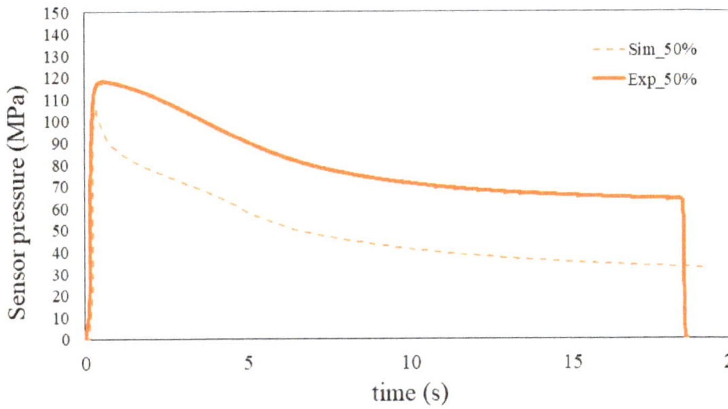

(**a**)

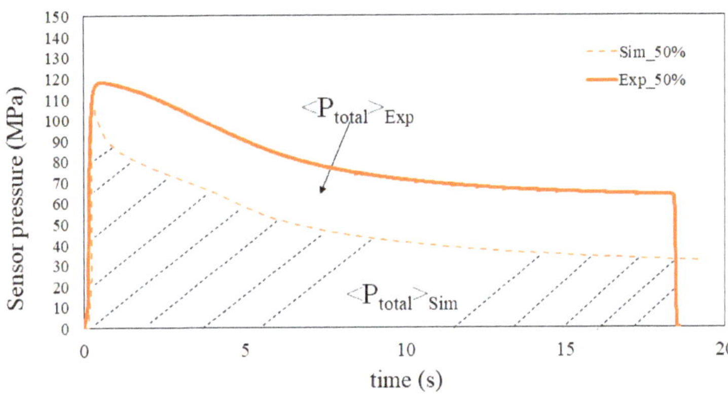

(**b**)

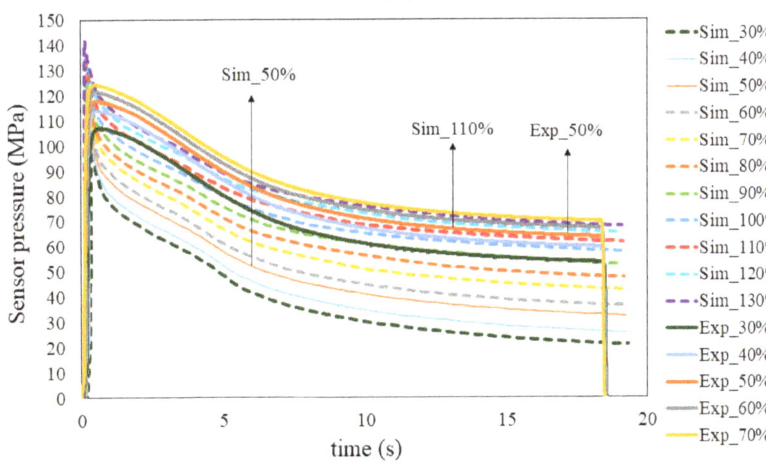

(**c**)

Figure 13. *Cont.*

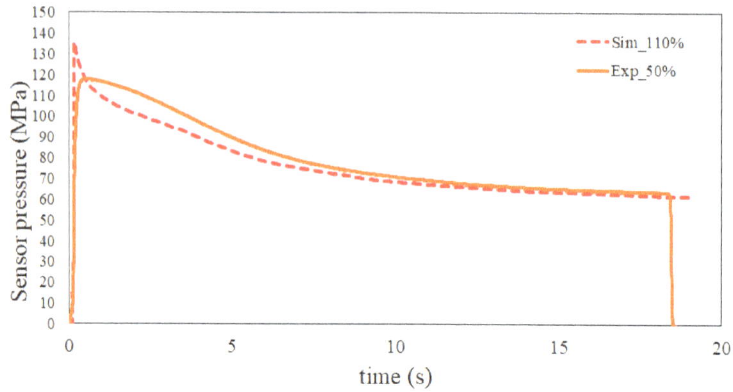

(**d**)

Figure 13. Injection pressure history curves used to perform machine calibration: (**a**) original injection pressure history curves at the 50% injection speed settings for both simulation and experiment, (**b**) schematic plots for the total driving force of the real experimental and simulation systems, (**c**) comparison of the history of injection pressure between the simulation and experiment at various injection speeds from 30% to 130%, (**d**) the matched pair for both the simulation and the experiment, where the simulation 110% injection speed setting is matched with the experimental 50% injection speed setting.

Through this equation, the total accumulated driving force is equal to the integration area under the injection pressure history curve for both the simulation and experiment systems. For instance, at the 50% injection speed setting, the DFI of the real experimental system, $\langle P_{Total} \rangle_{Exp}$, was about 1471.6 MPa·s, and the DFI of the virtual simulation system, $\langle P_{Total} \rangle_{Sim}$, was about 928.3 MPa·s, as shown in Figure 13b. Specifically, the DFI of the real experimental system was about 1.59 times over that of the virtual simulation system. This result is quite consistent with that ratio of 1.65 times for the experimental product index (characteristic lengths) difference from the simulation, as described previously. Based on this idea, the key to calibrating the machine is to determine the matched pair for both the simulation and the experimental systems that have the same DFI for the injection molding. For example, when the real injection machine keeps the 50% injection speed setting, the curve of the simulation system with the 50% injection speed setting is lower (i.e., with lower driving force). At this moment, the injection speed setting can be increased to enhance the driving force virtually. Until the injection speed setting of the simulation is increased to 110%, the injection pressure history curve is very close to that of the real injection machine with the 50% injection speed setting, as shown in Figure 13d. In this Figure, $\langle P_{Total} \rangle_{Exp}$ is about 1471.6 MPa·s, and $\langle P_{Total} \rangle_{Sim}$ is about 1449.2 MPa·s. The DFI of the real experimental system was about 1.02 times over that of the virtual simulation system. That is to say, the internal driving force of the 50% injection speed setting experimentally was matched by the 110% injection speed setting numerically. Specifically, when the injection speed setting is 50% in the experimental study, the counterpart in the simulation would be the 79.5 mm/s injection speed setting. Other matched pairs were evaluated, and are listed in Table 7.

Table 7. Matched pairs of injection speed settings for the simulation and experiment systems.

Simulation	Simulation Injection Speed Setting (mm/s)	Experiment Injection Speed Setting [a]
90%	67.0	30%
100%	73.4	40%
110%	79.5	50%
120%	85.8	60%
130%	92.0	70%

[a]: based on the maximum speed of the screw movement with 125 mm/s.

4.3.2. Evaluate Calibration Effect on Assembly Behavior

After performing the machine calibration at various injection speed settings, the machine calibration effect on the characteristic length changes could be further examined. For example, Figure 14 presents the comparison of the characteristic lengths between the simulation and experiment at the 25% to 100% packing pressure settings before and after machine calibration at the 50% injection speed setting. Since those four different characteristic lengths have different variation behavior, the calibration effect on each characteristic length was calculated individually. The calibration rate for each characteristic length is defined as in Equation (12):

$$\text{Calibration rate} = [\Delta L - (\Delta L)\text{cal}]/\Delta L * 100\% \quad (12)$$

where ΔL is the difference between the experimental characteristic length and that of the simulation before machine calibration, and (ΔL)cal is the difference between the experimental characteristic length and that of the simulation one after machine calibration.

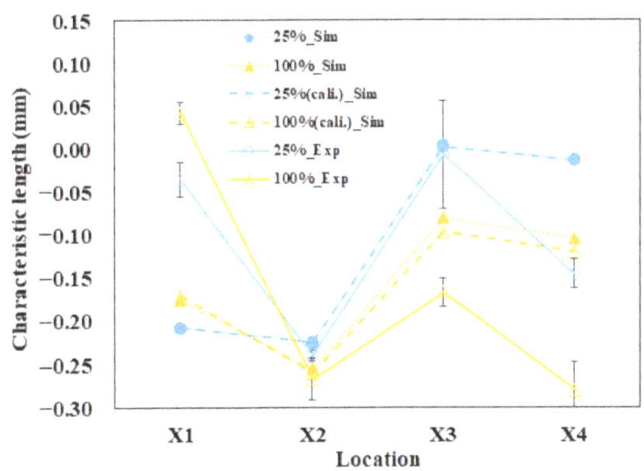

Figure 14. At the 50% injection speed setting, the comparison of the characteristic length deviation between simulation and experiment at the 25% to 100% packing pressure settings before and after machine calibration.

For example, when the 100% packing pressure setting situation is considered, the calibration rates for each characteristic length are shown as in Table 8. The average calibration rate is about 18%. This demonstrates that the difference between simulation prediction and experimental observation was reduced by 18%. In addition, the details of the machine calibration effect for the 50% injection speed setting at various packing pressure settings are listed in Table 9. It is noted that the calibration effect on the characteristic lengths improved about 10%. Moreover, after the machine calibration was completed, the relationship between the characteristic lengths and the assembly behavior was further measured using the integration test described in Figure 10. The result of the integration test is shown in Table 10. Specifically, when it was at the 25% packing pressure setting, although all X_i were smaller than zero, the real components A and B could be integrated together smoothly. Similarly, all components passed the integration test from the 25% to the 75% packing pressure settings. At the 100% packing pressure setting, the integration test failed. At this moment, it could be found that as long as one characteristic length is smaller than −0.25 mm, it would fail the integration test. Similarly, the virtual criteria to assure the good assembly of parts A and B can be obtained when one characteristic length is not smaller than −0.243 mm numerically, as listed in Table 11. Based on these results, using numerical simulation to predict the ease of assembly has been verified as a feasible and quantitative method.

Table 8. Measurements of the calibration effect for the 50% injection speed setting at the 100% packing pressure setting.

	before Calibration			after Calibration			Calibration Rate (%)
Character. Length	Sim	Exp	ΔL	(Sim)cal	(Exp)cal	(ΔL)cal	
X_1	−0.175	0.042	0.217	−0.170	0.042	0.212	2
X_2	−0.255	−0.267	−0.012	−0.260	−0.267	−0.007	42
X_3	−0.080	−0.167	−0.087	−0.096	−0.167	−0.071	18
X_4	−0.103	−0.280	−0.177	−0.119	−0.280	−0.161	9
Average calibration rate							18

Table 9. Measurement of the calibration effect for the 50% injection speed setting at various packing pressure settings.

	Machine Calibration Rate (%)				
Packing Pressure Setting (%)	X_1	X_2	X_3	X_4	Average
25	0	0	9	0	2
50	1	0	11	8	5
75	2	42	14	8	17
100	2	42	18	9	18
Total average calibration rate					10

Table 10. Quantification of the degree of assembly through the integration test for the 50% injection speed system, experimentally.

Packing Pressure (%)	X_1	X_2	X_3	X_4	Integration Test
25	−0.035	−0.240	−0.007	−0.145	passed
50	−0.008	−0.233	−0.120	−0.173	passed
75	0.017	**−0.250**	−0.152	−0.238	passed
100	0.042	−0.267	−0.167	−0.280	failed

Table 11. Quantification the degree of assembly for the 50% injection speed setting in the simulation system.

Packing Pressure (%)	X_1	X_2	X_3	X_4	Integration Test
25	−0.207	−0.224	0.003	−0.013	passed
50	−0196	−0.235	−0.028	−0.045	passed
75	−0.187	**−0.243**	−0.050	−0.069	passed
100	−0.170	−0.260	−0.096	−0.119	failed

4.4. Optimize the Assembly Behavior Using CAE-DOE

4.4.1. Optimization before the Machine Calibration

Before performing the machine calibration, the eighteen sets of injection molding trials were executed using the parameters and settings found in Tables 4 and 5. The associated characteristic lengths and their average for each set were measured are and recorded in Table 12. For example, after the first molding simulation based on Set 1 conditions, the individual characteristic lengths were −0.15, −0.25, −0.08, and −0.12 mm, respectively. Also, the average characteristic length was −0.15 mm. Based on these calculated characteristic lengths, the standard deviation Sn was 0.06 mm, which was calculated from Equation (8). Then S/N ratio (signal-to-noise ratio) was 15.82 (obtained by Equation (9)). The quality values of the remaining seventeen sets are listed in Table 12.

Table 12. Quality predictions based on the characteristic length for CAE-DOE before machine calibration.

Exp	Characteristic Lengths (mm)				Average	Sn	S/N
	X1	X2	X3	X4			
1	−0.15	−0.25	−0.08	−0.12	−0.15	0.06	15.82
2	−0.16	−0.27	−0.11	−0.14	−0.17	0.06	14.93
3	−0.13	−0.29	−0.20	−0.22	−0.21	0.06	13.27
4	−0.20	−0.24	−0.04	−0.06	−0.14	0.09	15.88
5	−0.15	−0.27	−0.13	−0.15	−0.17	0.05	14.80
6	−0.15	−0.28	−0.15	−0.19	−0.19	0.05	13.95
7	−0.15	−0.27	−0.14	−0.17	−0.18	0.05	14.52
8	−0.14	−0.28	−0.19	−0.20	−0.20	0.05	13.53
9	−0.17	−0.25	−0.09	−0.11	−0.15	0.06	15.54
10	−0.12	−0.29	−0.21	−0.23	−0.21	0.06	13.22
11	−0.18	−0.25	−0.07	−0.09	−0.15	0.07	15.74
12	−0.15	−0.26	−0.11	−0.15	−0.17	0.06	14.99
13	−0.16	−0.26	−0.14	−0.14	−0.17	0.05	14.88
14	−0.11	−0.29	−0.21	−0.25	−0.21	0.07	12.99
15	−0.18	−0.24	−0.06	−0.08	−0.14	0.08	15.93
16	−0.12	−0.29	−0.20	−0.24	−0.21	0.06	13.09
17	−0.15	−0.26	−0.16	−0.16	−0.18	0.05	14.50
18	−0.18	−0.26	−0.10	−0.13	−0.17	0.06	15.00

Moreover, before performing the machine calibration, the response for each factor was estimated and recorded into Table 13. The responses of all factors can be plotted as shown in Figure 15. From Table 13 and Figure 15, the optimized parameter set was determined as (A2, B3, C1, D2, E1, and F3). This optimized parameter set was applied in the injection molding simulation. The result is shown as "CAE-DOE (Sim)" in Figure 16. Compared to the original design, the optimized conditions reduced the average characteristic length from −0.169 mm (original) to −0.151 mm for the numerical simulation. Obviously, using the virtual DOE method (CAE-DOE), the quality can be improved about 10.7%. Moreover, to validate the efficiency of CAE-DOE optimization before performing the machine calibration, both the original design and the optimized parameter sets were utilized to execute the injection molding experimentally, and these results are also exhibited in Figure 16. The average characteristic length of the original design for the experimental system was −0.150 mm. Using the (CAE-DOE) optimized parameter set to perform the real injection molding, the average characteristic length was reduced to −0.143 mm, as demonstrated as "CAE-DOE (Exp)" in Figure 16. Clearly, the ease of assembly improved about 5%.

Table 13. Response values for various control factors before machine calibration.

	A	B	C	D	E	F
Level 1	14.66	14.57	15.57	14.67	14.61	14.54
Level 2	14.74	14.42	14.85	14.72	14.60	14.51
Level 3	14.36	14.78	13.34	14.37	14.56	14.72
E_i^{1-2}	0.08	−0.15	−0.72	0.05	−0.02	−0.03
E_i^{2-3}	−0.37	0.36	−1.51	−0.36	−0.04	0.21
Range	0.37	0.36	2.23	0.36	0.05	0.21
Rank	2	3	1	4	6	5

Where E_i^{1-2} means the influence of the "i" factor on the S/N ratio from Level 1 to Level 2; E_i^{2-3} means the influence of the "i" factor on the S/N ratio from Level 2 to Level 3.

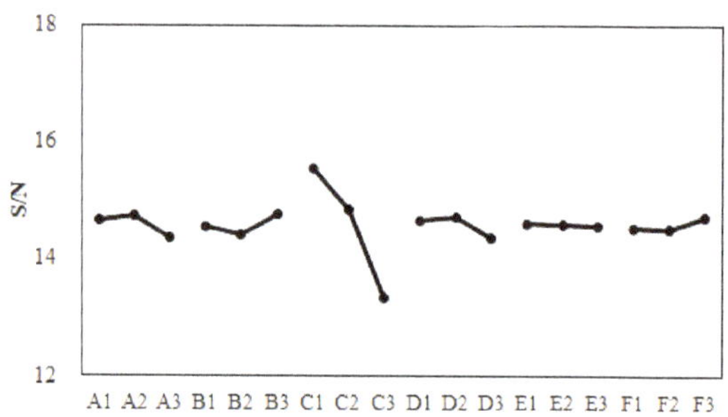

Figure 15. Response plot for various control factors before machine calibration.

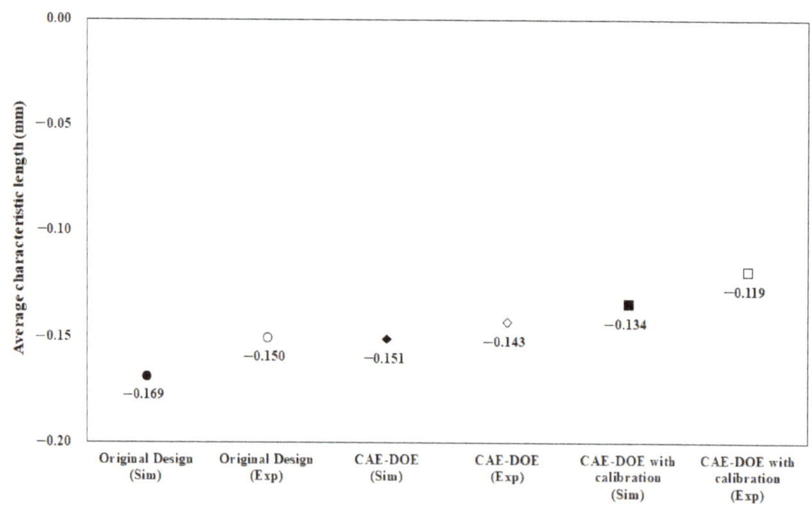

Figure 16. Average of the characteristic length quality change through DOE optimization.

4.4.2. Optimization after the Machine Calibration

After the machine was calibrated, the control factors and their levels were modified as listed in Table 14. The corresponding orthogonal array for DOE performance using CAE (i.e., CAE-DOE) is the same as that listed in Table 5. Then, eighteen sets of injection molding trials were performed. The associated characteristic lengths and their average for each set were measured and recorded into Table 15. The quality values of the eighteen sets are also listed in Table 15. Moreover, in the presence of the machine calibration effect, based on the S/N ratio, the response for each factor was estimated, as recorded in Table 16. The responses of all factors were plotted, as shown in Figure 17. From Table 16 and Figure 17, after performing the machine calibration, the optimized parameter set obtained was (A2, B3, C1, D2, E2, and F3). The optimized parameter set was used in the injection molding simulation. The result is demonstrated as "CAE-DOE with calibration (Sim)" in Figure 16. Compared to the original design, the optimized conditions reduced the characteristic length significantly from −0.169 mm (original) to −0.134 mm in the numerical simulation. After the machine calibration was performed, using CAE-DOE, the assembly behavior improved about 20.7% in the simulation system. Obviously, these results are consistent with those of the simulation prediction. Moreover, the efficiency of CAE-DOE optimization after machine calibration has been validated as well. Specifically, after the machine was calibrated, the average characteristic lengths of the injected parts, based on the optimized parameter set, was reduced significantly from −0.150 mm (original) to −0.119 mm in the experimental system. In addition, the real experimental validation through the integration test was performed, as shown in Figure 18. Obviously, after the machine was calibrated, the assembly behavior improved about 20.7% in the experimental system. Overall, the driving forces to improve the ease of assembly were quite consistent for both the simulation prediction and experimental observation. Moreover, both the simulation and the experimental systems benefitted from the machine calibration effect. The contribution of machine calibration to the ease of assembly is described in Figure 16. First, from the simulation point of view, before and after machine calibration the average characteristic by CAE-DOE went from −0.151 mm to −0.134 mm. The contribution of the machine calibration effect was to enhance the ease of assembly by 11.3% in the simulation prediction. Moreover, from the experimental point of view, before and after the machine calibration the average characteristic by real injection went from −0.143 mm to −0.119 mm. The calibration effect enhanced the ease of assembly by 16.8% in the real injected observation.

Table 14. Control factors and their levels in CAE-DOE after machine calibration.

	Control Factor	Level 1	Level 2	Level 3
A	Injection Speed (mm/s)	67	79.5	92
B	Mold Temperature (°C)	30	50	70
C	Packing Pressure (MPa)	90	120	140
D	Packing Time (s)	5	7	9
E	Melt Temperature (°C)	200	210	220
F	Cooling Time (s)	9	11	13

Table 15. Quality predictions based on the characteristic lengths for CAE-DOE after machine calibration.

Exp	Characteristic Lengths (mm)				Average	Sn	S/N
	X1	X2	X3	X4			
1	−0.14	−0.26	−0.10	−0.14	−0.16	0.06	15.41
2	−0.17	−0.26	−0.10	−0.12	−0.16	0.06	15.20
3	−0.15	−0.27	−0.16	−0.17	−0.19	0.05	14.19
4	−0.20	−0.24	−0.04	−0.05	−0.13	0.09	15.94
5	−0.15	−0.26	−0.11	−0.14	−0.17	0.06	15.08
6	−0.16	−0.27	−0.12	−0.16	−0.18	0.06	14.62
7	−0.15	−0.27	−0.13	−0.16	−0.18	0.05	14.73

Table 15. *Cont.*

Exp	Characteristic Lengths (mm)				Average	Sn	S/N
	X1	X2	X3	X4			
8	−0.16	−0.27	−0.14	−0.16	−0.18	0.05	14.46
9	−0.18	−0.24	−0.06	−0.07	−0.14	0.08	16.01
10	−0.14	−0.27	−0.16	−0.18	−0.19	0.05	14.21
11	−0.19	−0.24	−0.05	−0.08	−0.14	0.08	15.88
12	−0.17	−0.25	−0.10	−0.13	−0.16	0.06	15.33
13	−0.16	−0.26	−0.13	−0.13	−0.17	0.05	15.01
14	−0.12	−0.28	−0.18	−0.22	−0.20	0.06	13.68
15	−0.18	−0.24	−0.10	−0.02	−0.14	0.08	15.94
16	−0.12	−0.28	−0.17	−0.21	−0.19	0.06	13.85
17	−0.17	−0.25	−0.11	−0.11	−0.16	0.06	15.30
18	−0.18	−0.26	−0.09	−0.11	−0.16	0.07	15.25

Table 16. Response values for various control factors after machine calibration.

	A	B	C	D	E	F
Level 1	15.04	14.86	15.75	14.93	15.02	14.89
Level 2	15.04	14.93	15.10	15.12	15.05	15.00
Level 3	14.94	15.22	14.17	14.97	14.95	15.12
E_i^{1-2}	0.00	0.08	−0.65	0.19	0.03	0.10
E_i^{2-3}	−0.11	0.29	−0.93	−0.15	−0.10	0.12
Range	0.11	0.36	1.58	0.19	0.10	0.23
Rank	5	2	1	4	6	3

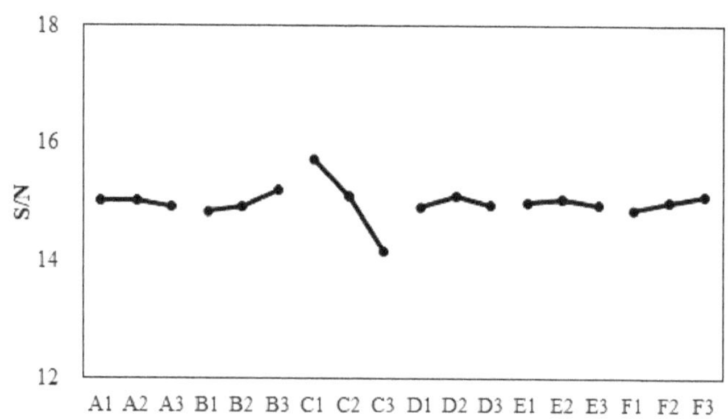

Figure 17. Response plot for various control factors after machine calibration.

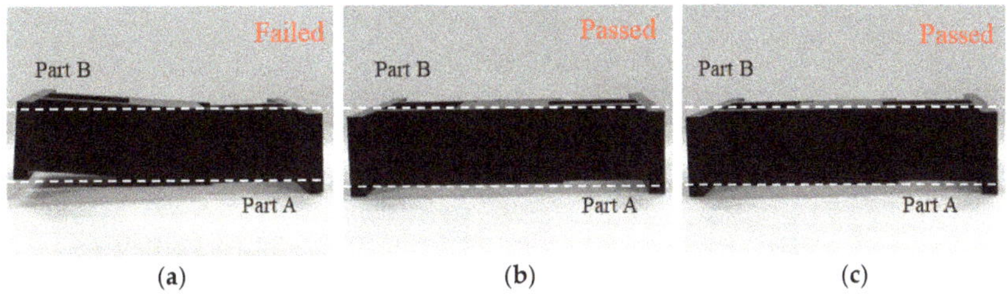

Figure 18. Experimental validation for the degree of assembly through the real integration test for CAE-DOE: (**a**) original design, (**b**) optimization before machine calibration, (**c**) optimization before machine calibration.

5. Conclusions

In this study, we proposed a feasible method to predict assembly behavior using the characteristic length as the product index for two components within a family mold system, using numerical simulation and experimental observation. Several key points can be obtained, as follows:

(1) For the same operation condition settings of simulation and experimental systems, as the packing pressure is higher, the assembly behavior based on the characteristic lengths becomes poorer. The trend is consistent for both simulations and experiments, but there is some difference between the simulation and experimental results.

(2) Based on the characteristic length variation (product index difference) investigation, under the same operation condition setting, the product index difference of the experimental observation was 1.65 times over that of the simulation prediction. Through the DFI investigation, the internal driving force of the experimental system was 1.59 times over that of the simulation one. This shows the internal driving force is quite matched with the product quality index. It also demonstrates that the simulation and experimental systems are not the same. Hence, the injection machine needs to be calibrated.

(3) After the injection machine was calibrated, the criteria for good assembly based on the integration test could be constructed. Specifically, the individual characteristic lengths should be not smaller than −0.250 mm in the real system (or not smaller than −0.243 mm in the virtual simulation system). The consistency was good.

(4) To handle complex injection molding processing, the CAE-DOE optimization method was verified with high efficiency in ease of assembly improvement. Moreover, after finishing the machine calibration, the improvement of the CAE-DOE optimization method could approach 20%. In addition, the driving forces to improve the assembly behavior were quite consistent for both the simulation prediction and the experimental observation. To handle the huge parameter operation window and optimize the assembly behavior, the CAE-DOE optimization strategy was applied. After finishing the machine calibration, the CAE-DOE strategy could optimize the ease of assembly up to 20%. The result is validated by experimental observation.

Author Contributions: Conceptualization, C.-T.H. and T.-W.L.; methodology, C.-T.H. and T.-W.L.; software, T.-W.L.; validation, C.-T.H. and T.-W.L.; formal analysis, C.-T.H. and T.-W.L.; investigation, C.-T.H. and T.-W.L.; resources, W.-R.J. and S.-C.C.; data curation, T.-W.L.; writing—original draft preparation, C.-T.H.; writing—review and editing, C.-T.H.; visualization, C.-T.H., W.-R.J. and S.-C.C.; supervision, W.-R.J. and S.-C.C.; project administration, C.-T.H., W.-R.J. and S.-C.C.; funding acquisition, W.-R.J. and S.-C.C. All authors have read and agreed to the published version of the manuscript.

Funding: The authors would like to thank Ministry of Science and Technology of Taiwan. (Project number: MOST 109-2218-E-033-001-) for partly financially supporting for this research.

Institutional Review Board Statement: Not applicable.

Informed Consent Statement: Not applicable.

Data Availability Statement: The data presented in this study are available on request from the corresponding author.

Conflicts of Interest: The authors declare no conflict of interest.

References

1. Barkan, E.; Drzymala, M.; Li, Y. Hand-Held Bar Code Reader with Single Printed Circuit Board. US Patent No. 6,817,529, 16 November 2004.
2. Horton, T.; Saroken, J.; Sparks, D.; Brant, A.; Mally, T. Vehicle Console Assembly. US Patent No. 8,915,531, 23 December 2014.
3. Springer, G. Smart Phone Lens Assembly. US Patent No. 9,001,176, 7 April 2015.
4. Oh, S. Luggage Having an Integrated Structure and Method of Manufacturing the Same. US Patent No. 9,854,884, 2 January 2018.
5. Darling, J. Personal Watercraft Fabrication Using Thermoforming. US Patent No.10,391,702, 27 August 2019.
6. Tokyo Tower and Osaka Temple Toys. Available online: https://www.toy-people.com/?p=44377 (accessed on 1 June 2021).
7. Boothroyd, G.; Alting, L. Design for Assembly and Disassembly. *CIRP Annals* **1992**, *41*, 625–636. [CrossRef]
8. Azri, N.M.; Effendi, M.S.M.; Rosli, M.F. Preliminary studies on DFMA and sustainable design approach: A case study on a cordless drill. *AIP Conf. Proc.* **2018**, *2030*, 020069. [CrossRef]
9. Zhafri, Z.M.; Effendi, M.S.M.; Rosli, M.F. A Review on Sustainable Design and Optimum Assembly Process: A Case Study on a Drone. *AIP Conf. Proc.* **2018**, *2030*, 020071. [CrossRef]
10. Hamzah, N.S.M.; Rosli, F.; Effendi, M.S.M. Analysis on Laser jet Printer Using Design for Manufacture and Assembly. *AIP Conf. Proc.* **2018**, *2030*, 020137. [CrossRef]
11. Matthias, W. Rechargeable Battery for Hand-Guided Electromechanical Tools. US Patent 8,652,677, 18 February 2014.
12. Peirsman, D.; Vandekerckhove, S. Beverage Dispensing Apparatus Comprising an Integrated Pressure Reducing Channel. US Patent 9,016,528, 28 April 2015.
13. Schlittmeier, T.; Sauer, G.; Brunhofer, A. Optical Waveguide System for a Display Device and a Display Device. US Patent 9,071,369 B2, 30 June 2015.
14. Lee, B.H.; Kim, B.H. Optimization of Part Wall Thicknesses to Reduce Warpage of Injection-Molded Parts Based on The Modified Complex Method. *Polym.-Plast. Technol. Eng.* **1995**, *34*, 793–811. [CrossRef]
15. Leo, V.; Cuvelliez, C.H. The effect of the packing parameters, gate geometry, and mold elasticity on the final dimensions of a molded part. *Polym. Eng. Sci.* **1996**, *35*, 1961–1971. [CrossRef]
16. Yen, C.; Lin, J.C.; Li, W.J.; Huang, M.F. An abductive neural network approach to the design of runner dimensions for the minimization of warpage in injection mouldings. *J. Mater. Process. Technol.* **2006**, *174*, 22–28. [CrossRef]
17. Zhai, M.; Lam, Y.C.; Au, C.K. Runner sizing in multiple cavity injection mould by non-dominated sorting genetic algorithm. *Eng. Comput.* **2009**, *25*, 237–245. [CrossRef]
18. Othman, M.H.; Hassan, S.; Ibrahim, M.Z.; Li, L. The Effects of Mould Design on Shrinkage and Warpage of Polypropylene Integral Hinges. In Proceedings of the 3rd International Conference on Trends in Mechanical and Industrial Engineering (3rd ICTMIE), Kuala Lumpur, Malaysia, 8–9 January 2013.
19. Thomason, J.L.; Vlug, M.A. Influence of fiber length and concentration on the properties of glass fiber-reinforced polypropylene: Part 1-Tensile and flexural modulus. *Composites* **1996**, *27A*, 477–484. [CrossRef]
20. Thomason, J.L. The influence of fibre length and concentration on the properties of glass fibre reinforced polypropylene: Interface strength and fibre strain in injection moulded long fibre PP at high fibre content. *Compost. Part A Appl. Sci. Manuf.* **2007**, *38*, 210–216. [CrossRef]
21. Kovacs, J.G.; Solymossy, B. Effect of glass bead content and diameter on shrinkage and warpage of injection-molded PA6. *Polym. Eng. Sci.* **2009**, *49*, 2218–2224. [CrossRef]
22. Hakimian, E.; Sulong, A.B. Analysis of warpage and shrinkage properties of injection-molded micro gears polymer composites using numerical simulations assisted by the Taguchi method. *Materials Design* **2012**, *42*, 62–71. [CrossRef]
23. Ozcelik, B.; Erzurumlu, T. Comparison of the warpage optimization in the plastic injection molding using ANOVA, neural network model and genetic algorithm. *J. Mater. Process. Technol.* **2006**, *171*, 437–445. [CrossRef]
24. Zhai, M.; Xie, Y. A study of gate location optimization of plastic injection molding using sequential linear programming. *Int. J. Adv. Manuf. Technol.* **2010**, *49*, 97–103. [CrossRef]
25. Chiang, K.T.; Chang, F.P. Analysis of shrinkage and warpage in an injection-molded part with a thin shell feature using the response surface methodology. *Int. J. Adv. Manuf. Technol.* **2007**, *35*, 468–479. [CrossRef]
26. Fernandes, C.; Pontes, A.J.; Viana, J.C.; Gaspar-Cunha, A. Using Multi-objective Evolutionary Algorithms for Optimization of the Cooling System in Polymer Injection Molding. *Int. Polym. Process.* **2012**, *27*, 213–223. [CrossRef]
27. Tsai, K.M.; Tang, B.H. Determination of injection molding process window based on form accuracy of lens using response surface methodology. *Int. J. Adv. Manuf. Technol.* **2014**, *75*, 947–958. [CrossRef]
28. Xu, G.; Yang, Z. Multiobjective optimization of process parameters for plastic injection molding via soft computing and grey correlation analysis. *Int. J. Adv. Manuf. Technol.* **2015**, *78*, 525–536. [CrossRef]

29. Kitayama, S.; Tamada, K.; Takano, M.; Aiba, S. Numerical and experimental investigation on process parameters optimization in plastic injection molding for weldlines reduction and clamping force minimization. *Int. J. Adv. Manuf. Technol.* **2018**, *97*, 2087–2098. [CrossRef]
30. Hentati, F.; Hadriche, I.; Masmoudi, N.; Bradai, C. Optimization of the injection molding process for the PC/ABS parts by integrating Taguchi approach and CAE simulation. *Int. J. Adv. Manuf. Technol.* **2019**, *104*, 4353–4363. [CrossRef]
31. Huang, C.T.; Xu, R.T.; Chen, P.H.; Jong, W.R.; Chen, S.C. Investigation on the Machine Calibration Effect on the Optimization through Design of experiments (DOE) in Injection Molding Parts. *Polym. Test.* **2020**, *90*, 106703. [CrossRef]
32. Fernandes, C.; Pontes, A.J.; Viana, J.C.; Gaspar-Cunha, A. Modeling and Optimization of the Injection-Molding Process: A Review. *Adv. Polym. Technol.* **2018**, *37*, 429–449. [CrossRef]
33. Huang, C.T.; Hsu, Y.H.; Chen, B.H. Investigation on the internal mechanism of the deviation between numerical simulation and experiments in injection molding product development. *Polym. Test.* **2019**, *75*, 327–336. [CrossRef]
34. Chen, J.Y.; Yang, K.Y.; Huang, M.S. Online Quality Monitoring of Molten Resin in Injection Molding. *Int. J. Heat Mass Transf.* **2018**, *122*, 681–693. [CrossRef]
35. Chen, J.Y.; Tseng, C.C.; Huang, M.S. Quality Indexes Design for Online Monitoring Polymer Injection Molding. *Adv. Polym. Technol.* **2019**, *2019*, 3720127. [CrossRef]

polymers

MDPI

Article

Effects of Different Mold Materials and Coolant Media on the Cooling Performance of Epoxy-Based Injection Molds

Chil-Chyuan Kuo [1,2,*], Jing-Yan Xu [1], Yi-Jun Zhu [1] and Chong-Hao Lee [1]

1 Department of Mechanical Engineering, Ming Chi University of Technology, New Taipei City 24301, Taiwan; U09217119@mail.mcut.edu.tw (J.-Y.X.); M09118003@mail.mcut.edu.tw (Y.-J.Z.); U09117014@mail.mcut.edu.tw (C.-H.L.)

2 Research Center for Intelligent Medical Devices, Ming Chi University of Technology, No. 84, Gungjuan Road, New Taipei City 24301, Taiwan

* Correspondence: jacksonk@mail.mcut.edu.tw

Abstract: Metal additive manufacturing techniques are frequently applied to the manufacturing of injection molds with a conformal cooling channel (CCC) in order to shorten the cooling time in the injection molding process. Reducing the cooling time in the cooling stage is essential to reducing the energy consumption in mass production. However, the distinct disadvantages include higher manufacturing costs and longer processing time in the fabrication of injection mold with CCC. Rapid tooling technology (RTT) is a widely utilized technology to shorten mold development time in the mold industry. In principle, the cooling time of injection molded products is affected by both injection mold material and coolant medium. However, little work has been carried out to investigate the effects of different mold materials and coolant media on the cooling performance of epoxy-based injection molds quantitatively. In this study, the effects of four different coolant media on the cooling performance of ten sets of injection molds fabricated with different mixtures were investigated experimentally. It was found that cooling water with ultrafine bubble is the best cooling medium based on the cooling efficiency of the injection molded parts (since the cooling efficiency is increased further by about 12.4% compared to the conventional cooling water). Mold material has a greater influence on the cooling efficiency than the cooling medium, since cooling time range of different mold materials is 99 s while the cooling time range for different cooling media is 92 s. Based on the total production cost of injection mold and cooling efficiency, the epoxy resin filled with 41 vol.% aluminum powder is the optimal formula for making an injection mold since saving in the total production cost about 24% is obtained compared to injection mold made with commercially available materials.

Keywords: conformal cooling channel; rapid tooling technology; mold material; cooling medium

Citation: Kuo, C.-C.; Xu, J.-Y.; Zhu, Y.-J.; Lee, C.-H. Effects of Different Mold Materials and Coolant Media on the Cooling Performance of Epoxy-Based Injection Molds. *Polymers* **2022**, *14*, 280. https://doi.org/10.3390/polym14020280

Academic Editor: Célio Bruno Pinto Fernandes

Received: 25 December 2021
Accepted: 10 January 2022
Published: 11 January 2022

1. Introduction

Wax injection molding is a frequently applied manufacturing technique for the production of investment casting wax patterns due to time efficiency. The cooling time of the wax patterns constitutes a large portion in cycle time. Nowadays, the conformal cooling channel (CCC) was incorporated in the wax injection mold to improve the cooling efficiency of the wax patterns in terms of uniform cooling during the cooling stage [1]. Low-pressure wax injection molding is one of the most widely used approaches for producing wax patterns, which can be used for investment casting [2] to manufacture metal components with intricate shapes [3]. Kitayama et al. [4] optimized the process parameters of plastic injection molding to reduce warpage of the plastics products using CCC. This work confirmed that the CCC provides the reduction on the warpage of the injection molded part. Rajaguru et al. [5] employed commercially available aluminum (Al)-filled epoxy resin to make a cavity insert for low-volume production of plastic parts. The tool life is acceptable under general process parameters. Lim et al. [6] proposed a method for designing the cooling channel by means of the energy balance principle and arrangement method. It was found that both the tensile strengths and hardness of the roof side products were improved. Chen et al. [7] showed segmented finite element models to optimize the geometries of the cooling system. It was found that the computational time saves up to 92.6% compared with the entire model of the U-shape component. Additive manufacturing (AM) techniques are frequently used in various industries to create end-use parts or physical prototypes. Maji et al. [8] developed patient-specific craniofacial implant through the AM, rapid tooling (RT), and investment casting technologies. Gill and Kaplas [9] proposed rapid casting technologies to fabricate a tool. It was found that the proposed rapid casting technologies are effective for the production of cast technological prototypes in very short times avoiding any tooling phase and with dimensional tolerances that are completely consistent with metal casting processes. Nandi et al. [10] developed a new composite material to enhance the solidification efficiency in the RT process. It was found that the solidification time was minimized appreciably keeping the same advantages of RT process. Wang et al. [11] optimized the molds with spherical spiral conformal cooling system and product structure to reduce service stress of the injection molded parts. This work suggested that injection molding defects such as warpage and residual stress cannot be ignored, especially for the assembly edge. Mercado-Colmenero et al. [12] proposed a new method for the automatic design of CCC based on the discrete geometry of the plastic part. The CCC follows the profile of the mold core or cavity to perform a uniform cooling process for injection molding process. This work demonstrated that the proposed algorithm is independent of the computer-aided design (CAD) modeler used to create the part since it performs a recognition analysis of the part surfaces, being able to be implemented in any CAD system. Additionally, the design data can be utilized in later applications including the automated design of the injection mold.

The wax injection mold with highly complex-shaped CCC capable of reduction cooling times can be fabricated by the metal powder AM, such as laser fusing, direct metal laser sintering, electron beam melting, selective laser melting, selective laser sintering, diffusion bonding, or direct metal deposition. However, operating consumables of the metal powder AM systems are costly. Thus, those approaches are not suitable for developing a new product since higher initial cost of the capital equipment and maintenance. To overcome this obstacle, RT technology was developed to meet this requirement since it is a cost-effective method that builds tools without the need for complex conventional machining operations. Typically, the RT technology is divided into two categories, namely direct tooling and indirect tooling. Indirect tooling is widely used to manufacture an injection mold for batch production of a new product in the research and development stage using commercial Al-filled epoxy resin. According to practical experience, both mechanical and physical properties of the wax injection mold fabricated by Al-filled epoxy resin are limited due to intrinsic material properties. Tomori et al. [13] proposed a silicon carbide filled epoxy casting system intended for prototype molds in plastic injection molding applications. Results revealed that the flexural strength of the molded parts increased with silicon carbide concentration. Kovacs and Bercsey [14] investigated the influence of mold properties on the quality of injection molded parts. Results demonstrated that the rapid tool inserts are useful in the injection molding technology, although the warpage of the molded parts could be more significant. In addition, injection molding simulation programmers can analyze the cooling time of the molded part or minimize the warpage of the molded part using CCC. Khushairi et al. [15] proposed metal filled epoxy as an alternative material used in RT application for injection molding. Results showed that the compressive strength is increased from 76.8 to 93.2 MPa with 20% of brass fillers. In addition, metal fillers also enhance the thermal properties and density of Al-filled epoxy resin. Cost-effectiveness depends on the cooling time of the molded part greatly in the injection molding process. Metal additive manufacturing (MAM) [16] technology is applied to speed up manufacturing for the production of injection mold with CCC. However, the distinct disadvantage is expensive spending [17] and longtime taking [18] in the fabrication of injection mold with CCC. Under certain condition of less expenditure is possible to manufacture injection molds by RT technology with conformal cooling channel.

In general, the cooling time of injection molded products is affected by both injection mold material and coolant medium. However, little work has been conducted to investigate the effects of different mold materials and coolant media on the cooling performance of epoxy-based injection molds. The goal of this study is to investigate the cooling performance of epoxy-based injection molds fabricated from different mold materials using different coolant media. In this study, ten sets of injection molds were fabricated using commercially available Al-filled epoxy resin and epoxy resin added with three different particle sizes of stainless steel, Al, and copper (Cu) powders. Four different kinds of cooling media, including cooling water, compressed gas, cooling water with ultrafine bubble, and cold stream are used to study the cooling performance using a low-pressure wax injection molding.

2. Experimental Details

Figure 1 shows the research process of this study. The three different particle sizes of stainless steel (SUS) 316 powder, Al powder, and Cu powder were used in this study. Figure 2 shows the three different kinds of fillers with the purity of the fillers is approximately 96–99%. Average particle sizes of the three different kinds of fillers were examined by field-emission scanning electron microscopy (FE-SEM) (JEC3000-FC, JEOL Inc., Tokyo, Japan). The Al powder and Cu powder have three average particle sizes of 45 μm, 75 μm, and 150 μm, respectively. 316 stainless steel powder has three average particle sizes of 13 μm, 75 μm and 125 μm. The epoxy resin (EP-2N1, Ruixin Inc., Taipei, Taiwan) with the viscosity about 13,000 mPa-s was used as matrix material. The mixture is composed of epoxy resin, hardener, and filler. The curing agent and epoxy resin were mixed in a weight ratio of 1:2 first and then filler particle was added. The mixture was stirred manually about 10 min until the mixture is well blended. The mixture was then degassed by a vacuum pump (F-600, Feiling Inc., Taipei, Taiwan). Finally, the fabricated wax injection molds were post-cured in a thermal oven at 60 °C for obtaining desired strength to withstand injection and hold pressures. Note that the wax injection molds have very limited shrinkage after post-cured process. The thermal conduction of the wax injection mold fabricated by different mixtures was examined using a hot plate (YS-300S, Yong-Xin Inc., Taipei, Taiwan) with the heating temperature of 40. The chemical composition of the filler was examined using an energy-dispersive X-ray spectroscopy (EDS). The X-ray phase analysis was carried out using an X-ray diffractometer (XRD) (D8 ADVANCE, Bruker Inc., Taipei, Taiwan) to study the structures of the fabricated wax injection molds. To assess the benefits of the optimum material formulation, commercially available Al powder-filled epoxy resin (TE-375, Jasdi Chemicals, Inc., Taipei, Taiwan) was also employed to fabricate injection molds.

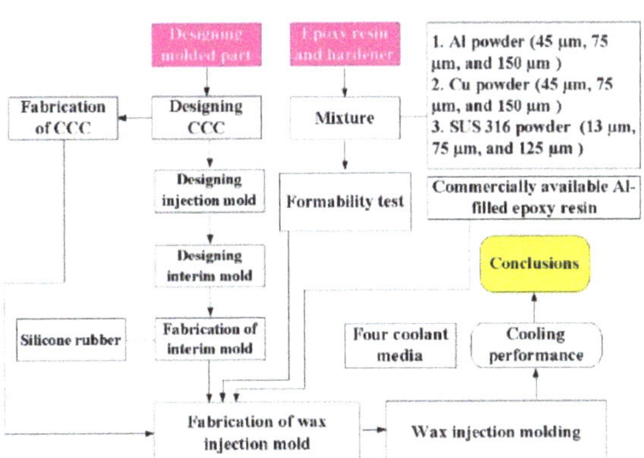

Figure 1. Research process of this study.

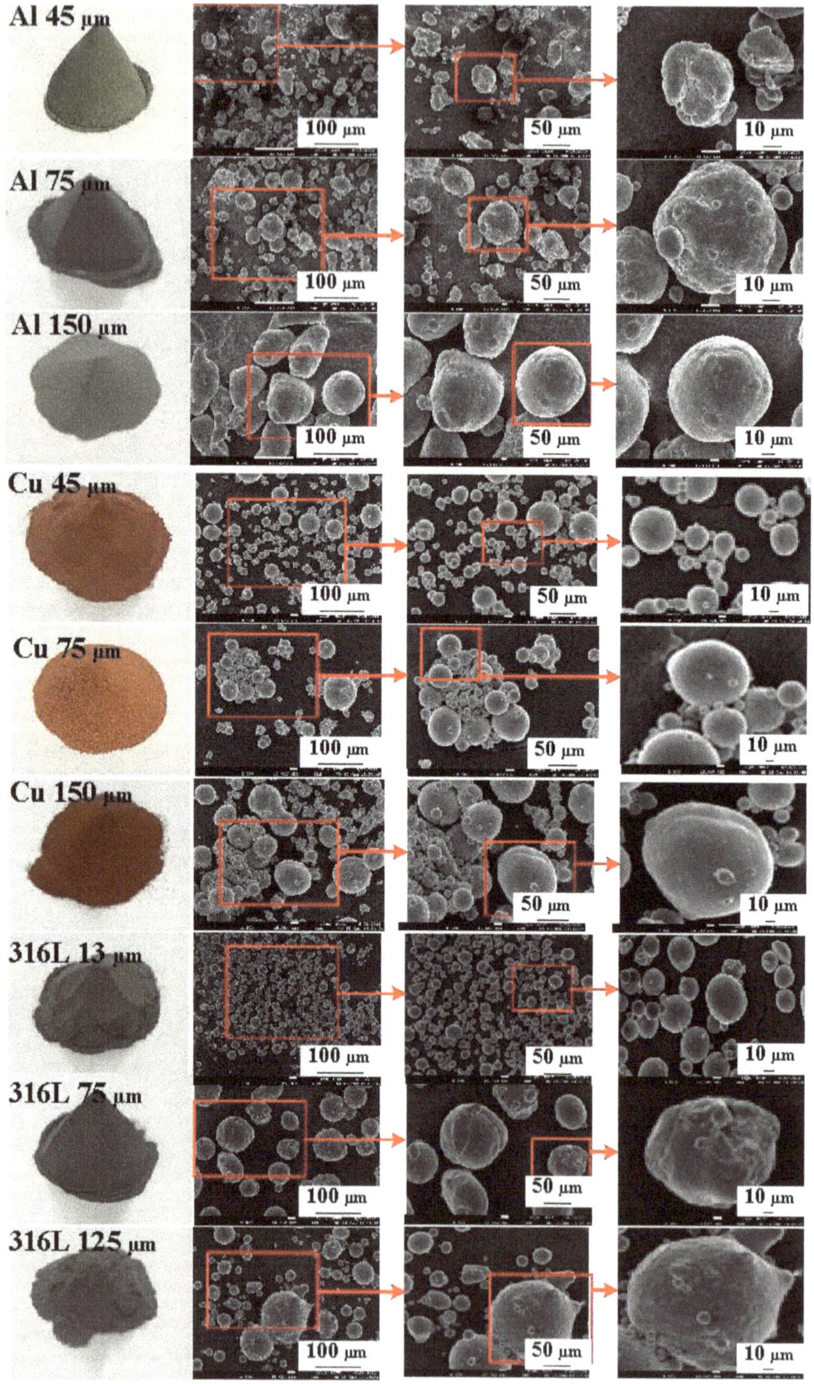

Figure 2. Three different kinds of fillers used in this study.

Figure 3 shows the 3D CAD model and dimensions of the CCC, core and cavity inserts. In this study, a pipe cover with the dimensions of 32 mm in outer diameter, 17.5 mm in height, and 2.5 mm in thickness was selected as the master pattern. The circular CCC was used in this study since it has the smallest pressure loss during the cooling stage [19]. According to the design guideline of the CCC, the distance between the wall of cooling channel to the mold surface, and the pitch distance between central lines of cooling channels are 4 mm, 7.5 mm, and 17 mm, respectively. Generally, there are various different types of CC (i.e., longitudinal, zigzag, spiral, parallel, scaffold, or Voronoi diagram). In this study, the spiral-shaped CCC was used in the wax injection mold. The core and cavity inserts have a length of 55 mm, a width of 55 mm, a height of 35 mm. Thanks to the distinct advantage of AM technique, the AM-based cooling channel was used to fabricate wax injection mold via fused filament fabrication using polyvinyl butyral (PVB) filament since the industrial alcohol solution is capable of dissolving the PVB plastic.

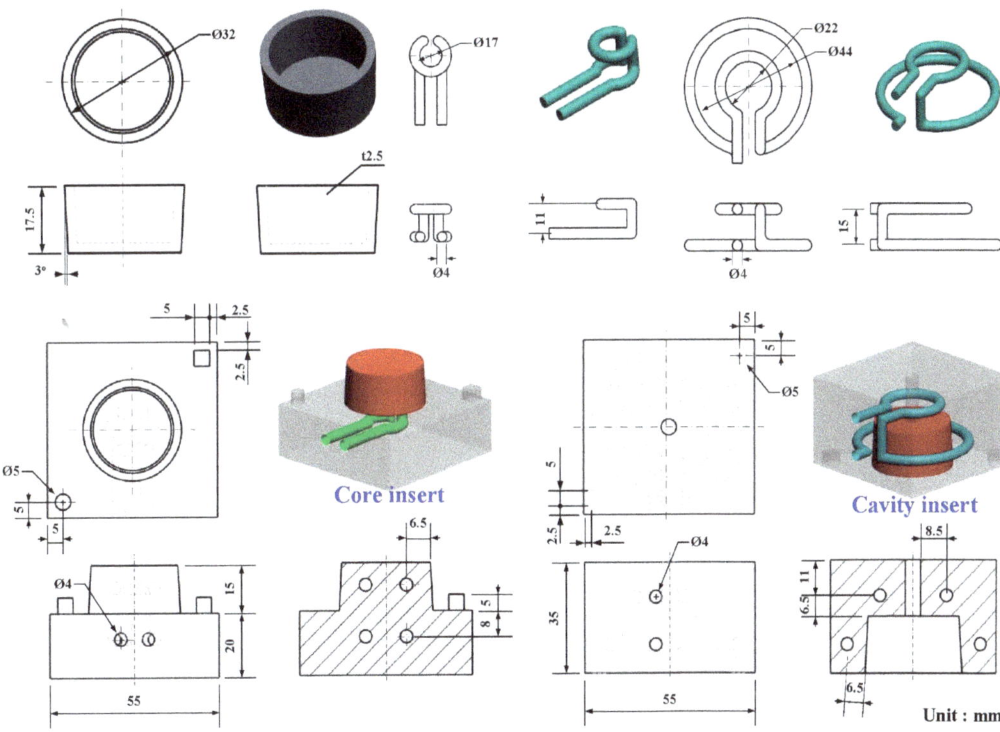

Figure 3. 3D CAD model and dimensions of the CCC, core, and cavity inserts.

Cooling media is a substance, which is widely applied to regulate or reduce temperature of injection mold during the injection molding process. An ideal cooling medium involves low cost, low viscosity, high specific heat capacity, non-toxic, and chemically inert of the cooling system in the injection mold. Air is a common form of a cooling media, which uses forced circulation to reduce temperature of injection mold during the injection molding process. Thus, four different kinds of cooling media (i.e., compressed gas, cooling water, cooling water with ultrafine bubble, and cold stream) were applied to study the cooling efficiency of injection molds by low-pressure wax injection molding. Cold air comes from the vortex tube. To evaluate cooling performance of the injection molds, a simple and effective cooling time measurement system was designed and implemented. Figure 4 shows an in-house cooling time measurement system. This measurement system comprises a low-pressure wax injection molding machine (0660, W&W Inc., Taoyuan, Taiwan), three k-type thermocouples (C071009-079, Cheng Tay Inc., Taipei, Taiwan) with a measurement sensitivity of ±1 °C, a mold temperature controller (JCM-33A, Shinko Inc., Taipei, Taiwan), and a water reservoir with a thermo-electric cooler (TEC12706AJ, Caijia Inc., Taipei, Taiwan). The process parameters for low-pressure wax injection molding involve injection temperature of 95 °C, injection pressure of 0.06 MPa, and an ejection temperature of 30 °C. The temperature sensors were installed in the mold cavity. The other end of the temperature sensor was connected to a data acquisition system (MRD-8002L, IDEA System Inc., Taipei, Taiwan). With the help of the temperature sensors, the temperature histories of the injection molded wax patterns after low-pressure wax injection molding were recorded in time interval of 1 s for determining the cooling time of the injection molded wax patterns. The room temperature was kept at 27 °C for the entire study. The injection mold was held in horizontal position and the molten wax at 82 °C was injected into the mold cavity at 27 °C. The ejection temperature of the molded wax patterns was set at 30 °C via a series of test runs. The inlet coolant temperature was kept at room temperature.

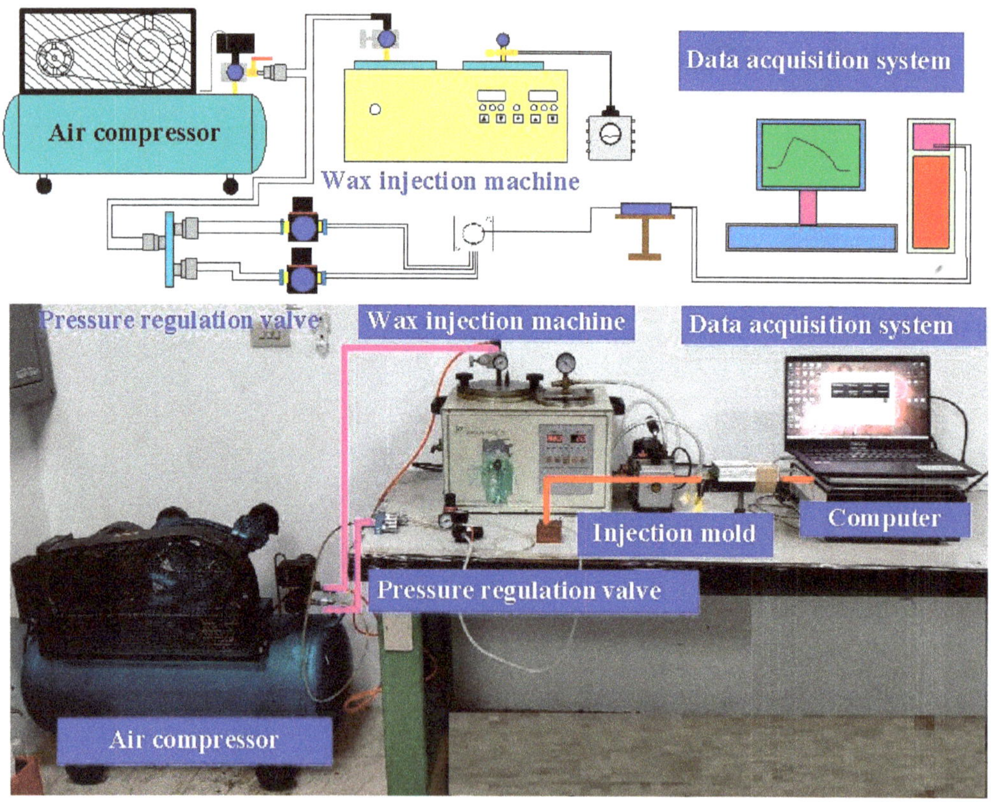

Figure 4. An in-house cooling time measurement system.

3. Results and Discussion

To ensure the molding materials can fill the cavity completely and smoothly, a direct gate was employed in this study. Figure 5 shows the mesh of the injection molded part and cooling channels. The evolution of the flow front during the filling stage of wax injection molding is also illustrated in this figure. It shows that the filling time is approximately 2 s for the full filing of the molded part, showing the designed filling system is appropriate.

In addition, no weld lines or air traps of the injection molded part were found at the end of filling.

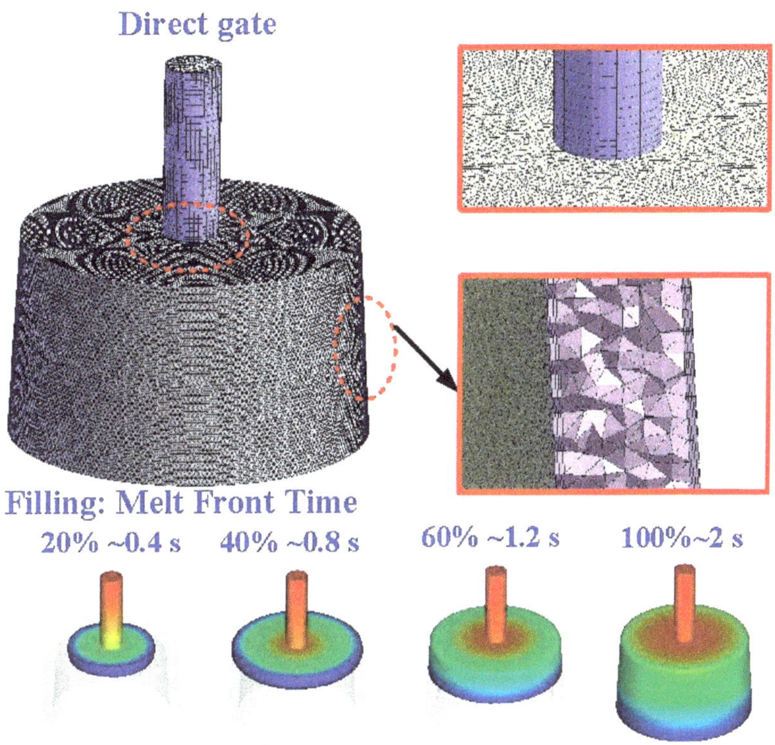

Figure 5. Mesh of the injection molded part and cooling channels.

To investigate the upper limit of the volume percentage (vol.%) of the filler particles that can be added to the matrix material of epoxy resin, the vol.% of these three powders was varied from 15 to 65 vol.%. Figure 6 shows the formability results of the epoxy resin filled with three different powders with three different particle sizes. Based on the formability and manufacturability of the specimens, the vol.% for 316 stainless steel powder with an average particle size of 13 μm, 75 μm, and 125 μm can be added to the epoxy resin are about 44 vol.%, 41 vol.% and 38 vol.%, respectively. The vol.% for Al powder with an average particle size of 45 μm, 75 μm, and 125 μm can be added to the epoxy resin are about 41 vol.%, 35 vol.% and 33 vol.%, respectively. The vol.% for Cu powder with an average particle size of 45 μm, 75 μm, and 125 μm can be added to the epoxy resin are about 41 vol.%, 39 vol.% and 36 vol.%, respectively.

Figure 6. *Cont.*

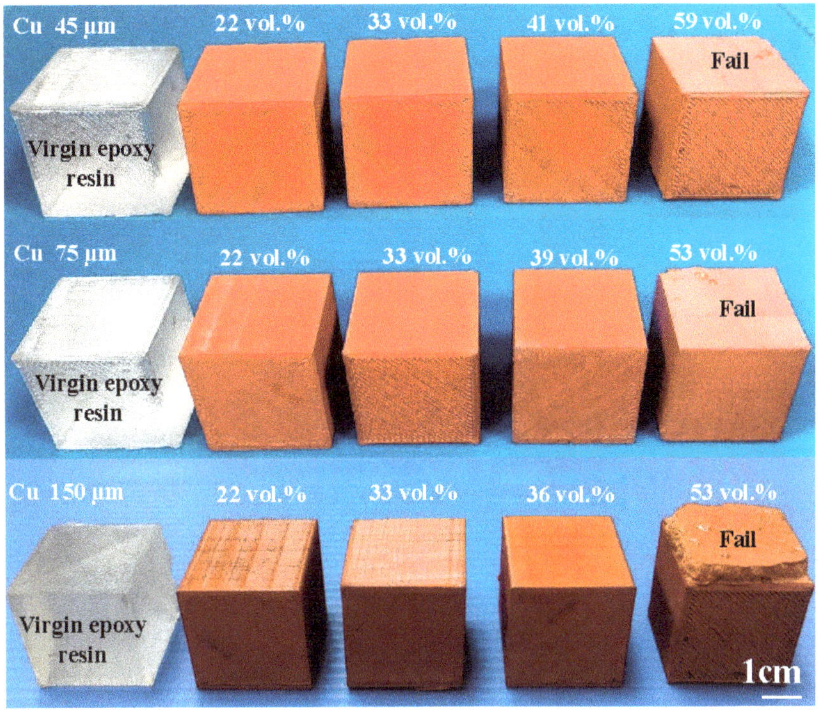

Figure 6. Formability results of the epoxy resin filled with three different powders with three different particle sizes.

To investigate the precipitation of fillers in the injection mold, EDS was carried out in this study. Figure 7 shows the precipitation analysis of the injection mold fabricated by epoxy resin filled with SUS 316 powder. As can be seen, the colors of the top, middle, and bottom of the injection mold are similar, showing that there is no any powder precipitation. This inference is supported by the EDS measurements of the wax injection mold. It was found that elements of Cr and Ni at the top, middle and bottom of the injection mold are similar. This means the filler of SUS 316 powder can be dispersed uniformly in the injection mold. Figure 8 shows the ten pairs of injection molds for low-pressure wax injection molding. Figure 9 shows the results of before and after removing CCC. The weight of the CCC is 27 g. The weight of the injection mold with CCC is 482 g. After removing the CCC, the weight of the injection mold without CCC is only 455 g. This result revealed that the CCC inside the injection mold was removed completely.

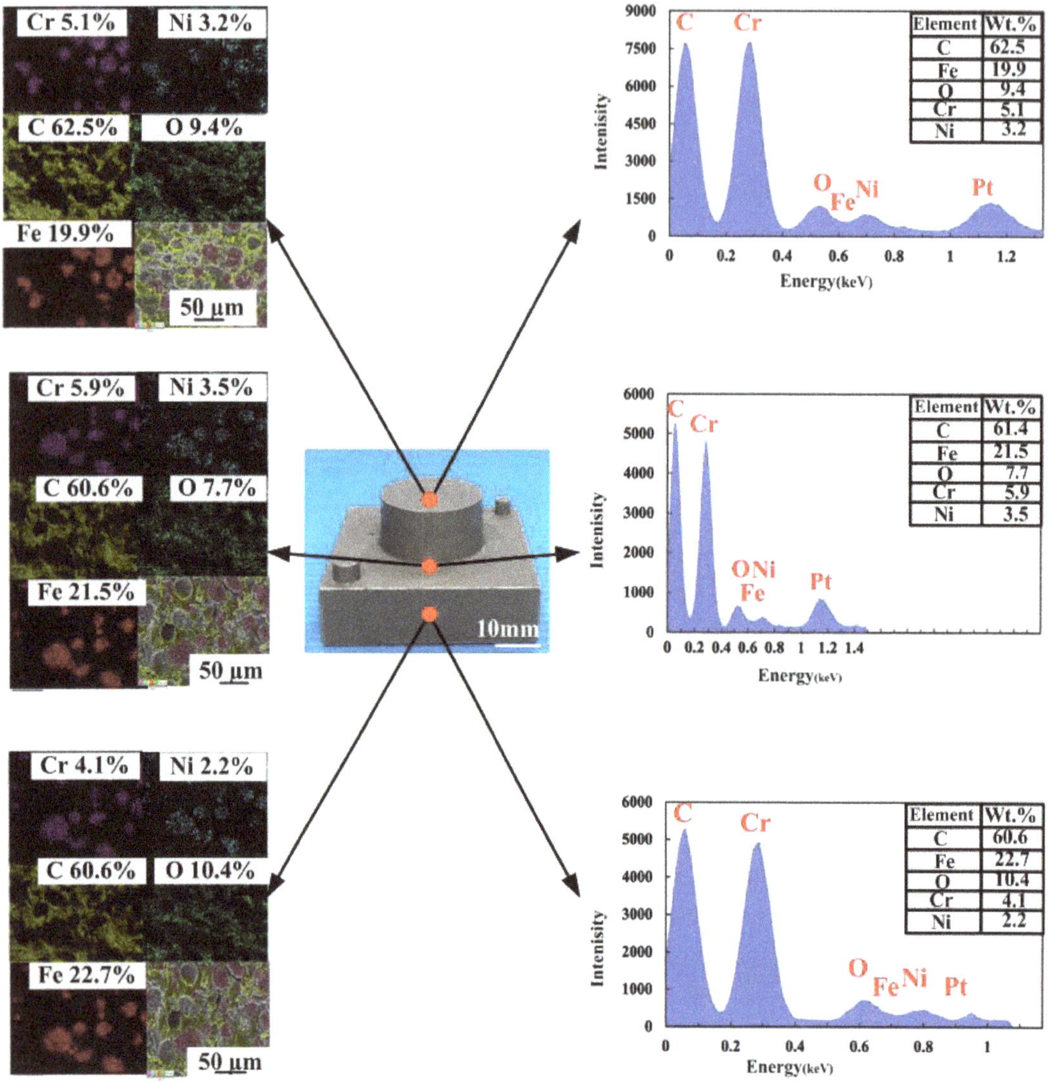

Figure 7. Precipitation analysis of the injection mold fabricated by epoxy resin filled with SUS 316 powder.

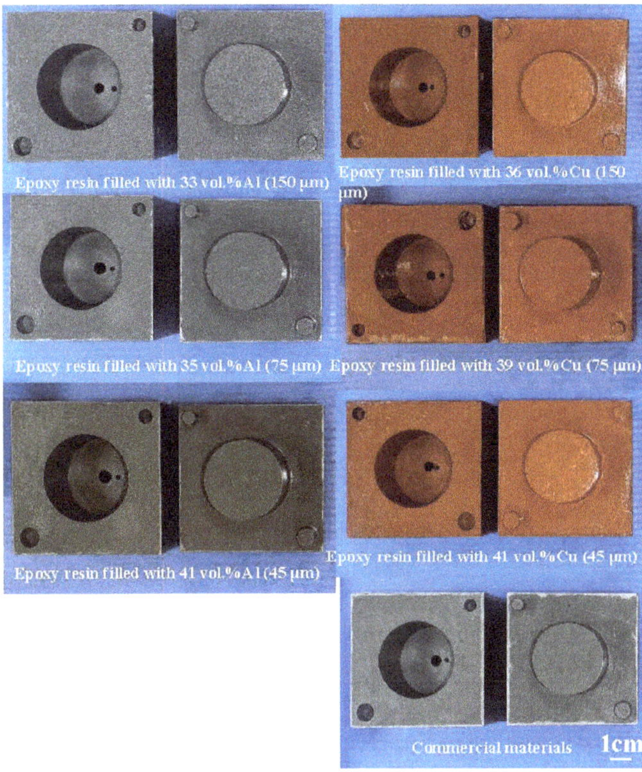

Figure 8. Ten pairs of injection molds for low-pressure wax injection molding.

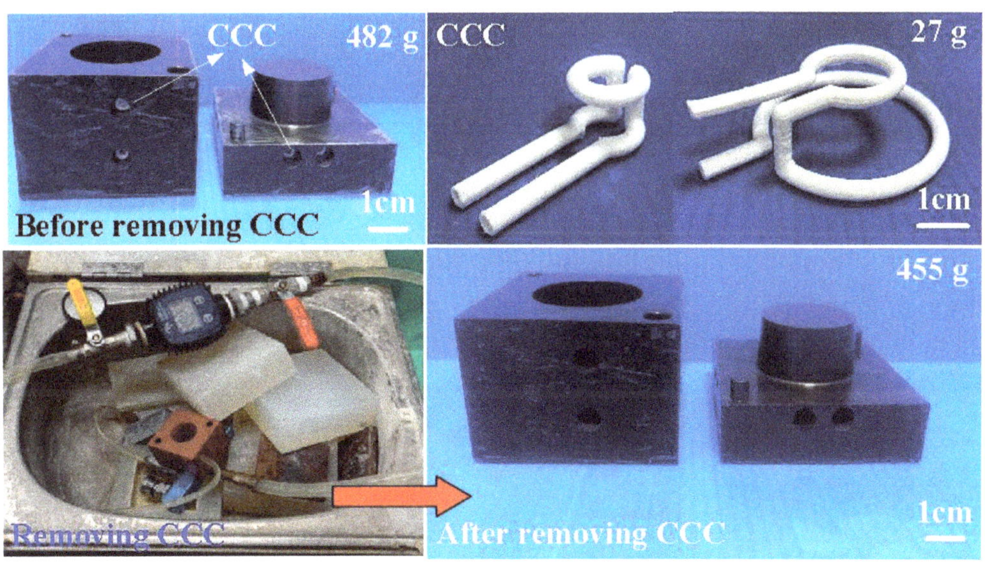

Figure 9. Results of before and after removing CCC.

Figure 10 shows the results of the heat conduction experiment for core and cavity of the wax injection mold fabricated by epoxy resin filled with three different particle sizes of Cu powder. The results showed that the thermal conductivity of the wax injection mold fabricated by epoxy resin filled with average particle size of 45 μm Cu powder is the highest, followed by the wax injection mold fabricated by epoxy resin filled with average particle size 75 μm Cu powder. The wax injection mold fabricated by epoxy resin filled with average particle size 150 μm Cu powder has the lowest thermal conductivity. To further verify the heat transfer characteristics of the wax injection mold fabricated by epoxy resin filled with three different particle sizes of Cu powder, a series of experiment of low-pressure wax injection molding were carried out. Figure 11 shows the temperature of wax pattern as a function of the cooling time. The ejection temperature of the wax pattern was set as 30 °C. Similarly, the cooling time of the wax pattern is affected by the particle size of Cu powder filled into the epoxy resin. The cooling time of the wax pattern fabricated by epoxy resin filled with 45 μm Cu powder is the shortest. The cooling times of the wax pattern duplicated by wax injection mold fabricated by epoxy resin filled with average particle size of 45 μm, 75 μm, and 150 μm Cu powder are around 471 s, 511 s, and 608 s, respectively. Thus, the wax injection mold fabricated by epoxy resin filled with Cu powder with a smaller average particle size has better heat conduction effect. The schematic illustrations of heat conduction of the wax injection mold fabricated by epoxy resin filled with Cu powder with larger and smaller average particle sizes are shown in Figure 12. It should be noted that the same phenomena were found when wax injection mold fabricated by epoxy resin filled with SUS or Al powder with a smaller average particle size.

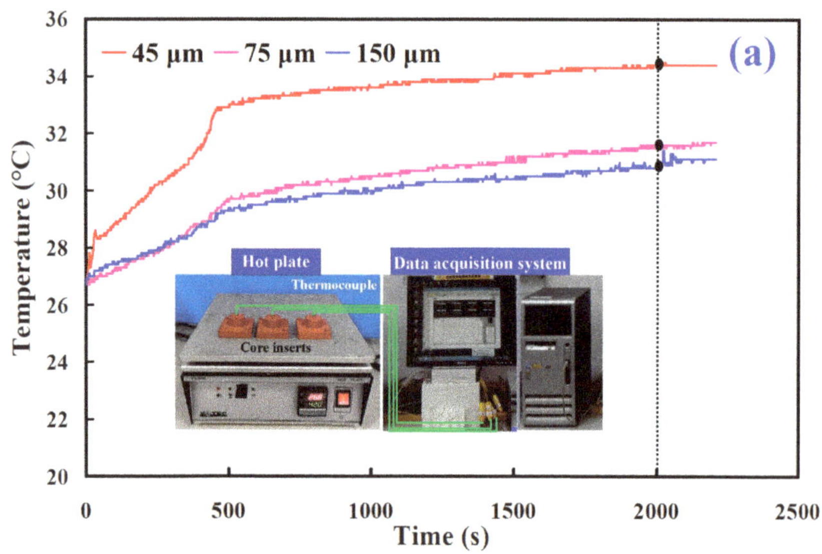

Figure 10. *Cont.*

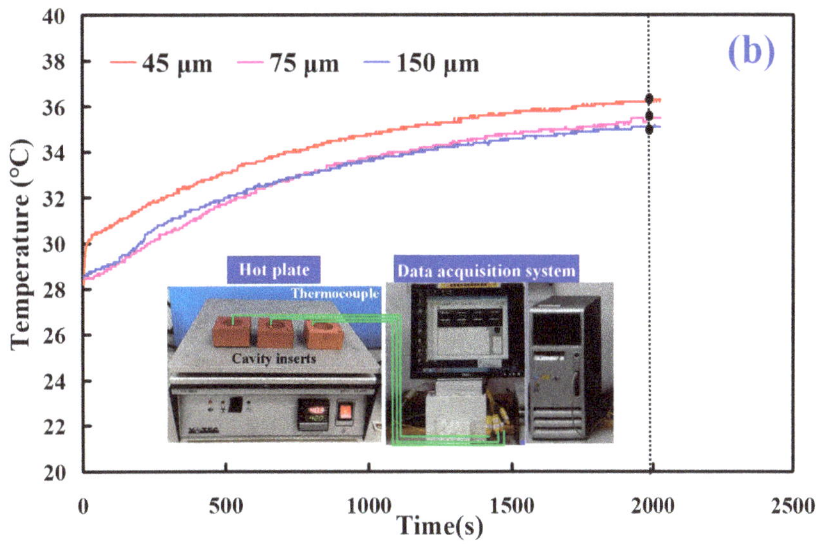

Figure 10. Results of the heat conduction experiment for (**a**) core and (**b**) cavity of the wax injection mold fabricated by epoxy resin filled with three different particle sizes of Cu powder.

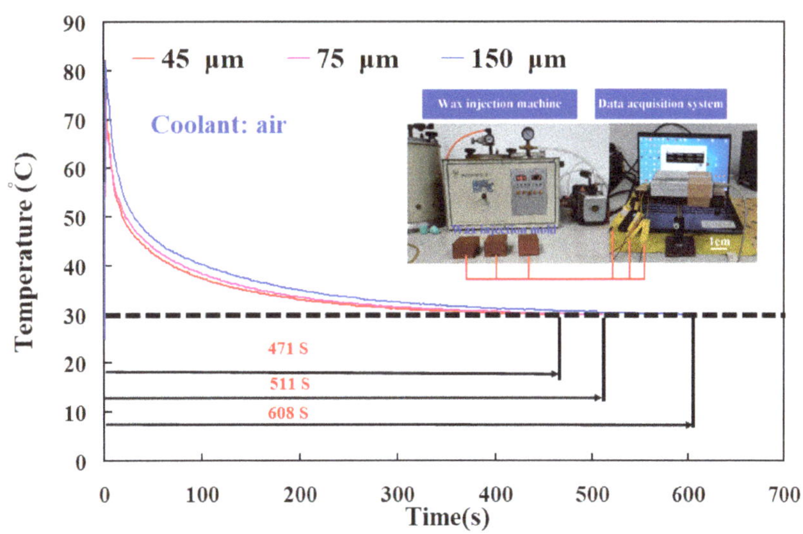

Figure 11. Temperature of wax pattern as a function of the cooling time.

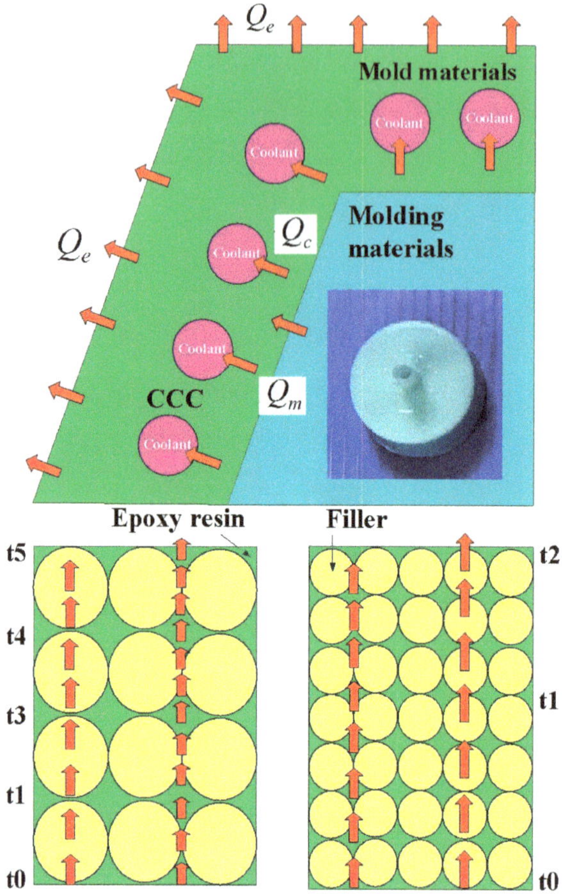

Figure 12. Schematic illustrations of heat conduction of the wax injection mold fabricated by epoxy resin filled with Cu powder with larger and smaller average particle sizes.

Figure 13 shows typical temperature history of the injection molded part after low-pressure wax injection molding. Temperature history involves four stages, including filling, early cooling, middle cooling, and late cooling stages. The cooling time of the wax patterns was defined the duration form the filling stage to the temperature of wax pattern reaching the ejection temperature of 30 °C. The heat of the molten wax was removed by the coolant in the mold, which is recommended based on the solidification of the wax patterns. The cooling time (t_c) of the injection molded parts can be calculated by the following equation [20–22]:

$$t_c = \frac{s^2}{\pi^2 \alpha} \ln\left[\frac{4}{\pi}\left(\frac{T_m - T_w}{T_e - T_w}\right)\right] \tag{1}$$

α is the thermal diffusivity;
T_w is the mold cavity surface temperature;
s is the part thickness;
T_m is the melt temperature;
T_e is the average ejection temperature.

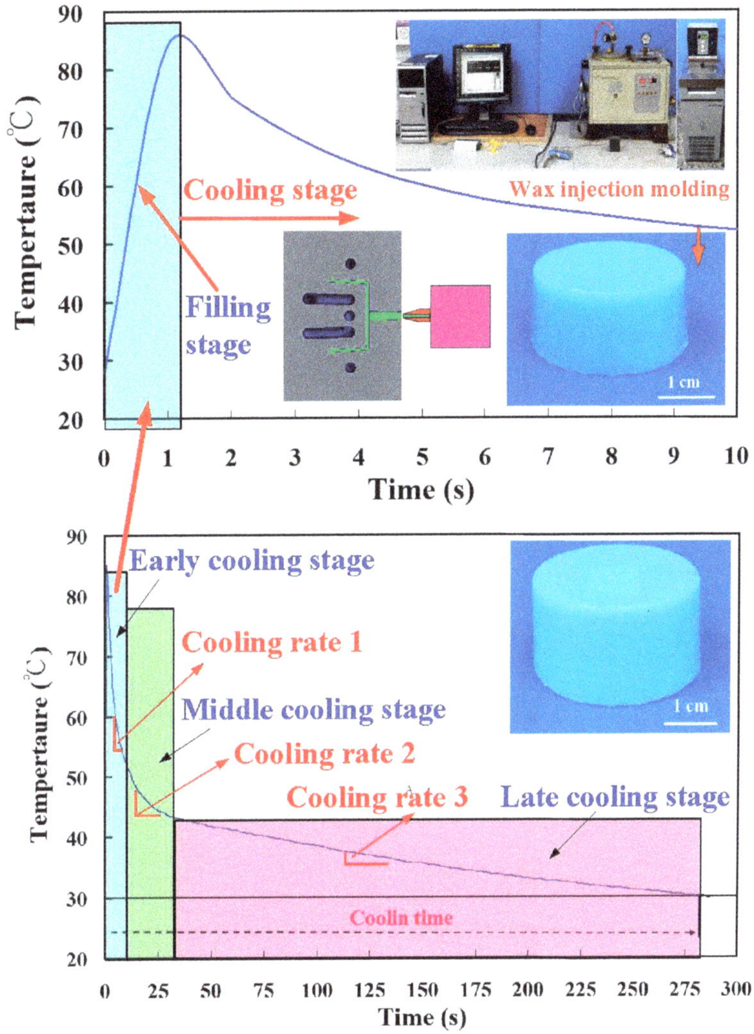

Figure 13. Typical temperature history of the injection molded part.

Figure 14 shows cooling time of injection molded parts fabricated by ten sets of injection molds cooled with different cooling water temperatures. Based on these results, four phenomena were found: (a) cooling time of the wax pattern is shorter when the cooling system uses a low temperature coolant; (b) cooling time of the wax pattern is not sensitive to injection mold with different thermal conductivity when the cooling system uses a higher temperature coolant; (c) cooling time of the wax pattern is sensitive to injection mold with different thermal conductivity when the cooling system uses a lower temperature coolant; and (d) cooling time of the wax pattern fabricated by injection mold made with Cu-filled epoxy resin is not shorter than that fabricated by injection mold made with commercially available Al-filled epoxy resin. Figure 15 shows the SEM micrograph of injection mold made with Cu-filled epoxy resin. The underlying reason is that the mixture is easier to generate porosity [23] when the Cu powder was added in the epoxy resin, which will affect the heat transfer effect of the fabricated injection mold.

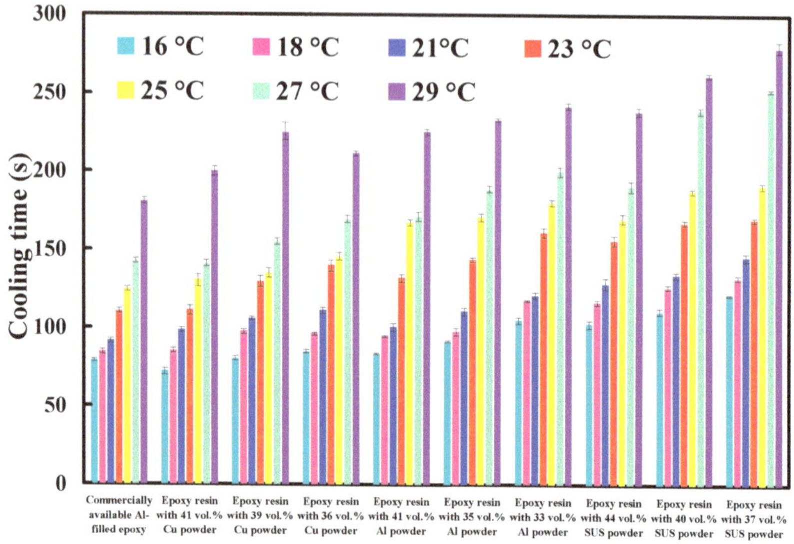

Figure 14. Cooling time of injection molded parts fabricated by ten sets of injection molds cooled with different cooling water temperatures.

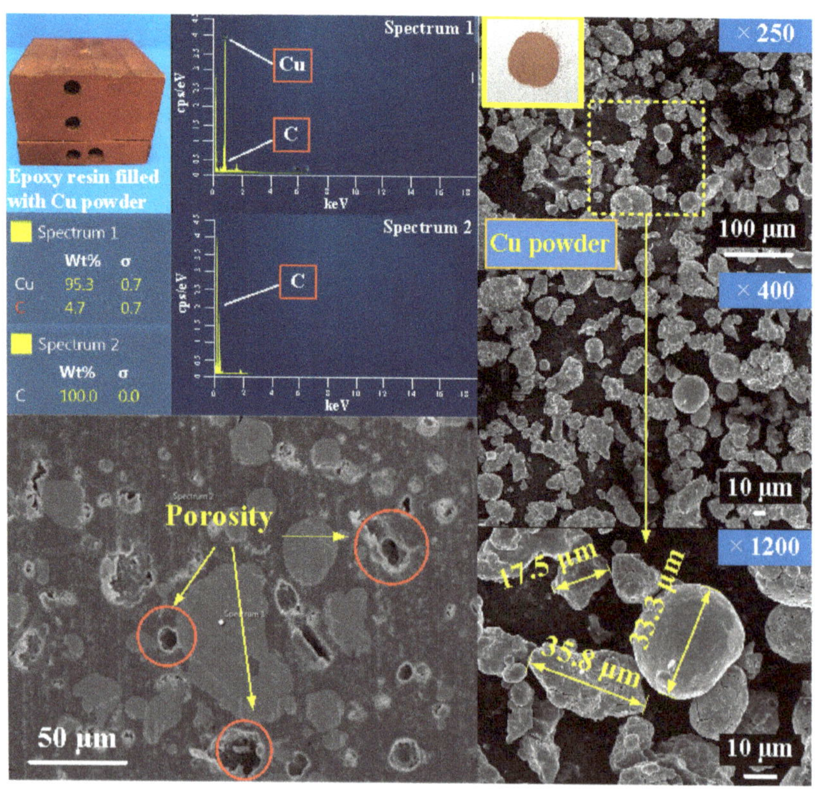

Figure 15. SEM micrograph of injection mold made with Cu-filled epoxy resin.

Generally, the cooling water flow rate is an important issue in the cooling performance of injection mold with CCC. The turbulent flow [24] having a Reynolds number greater than 4000 provides three to five times as much heat transfer as laminar flow having a Reynolds number less than 2100 [25]. Figure 16 shows cooling time of injection molded parts fabricated by ten sets of injection molds cooled with different cooling water flow rates. The Reynolds number for four different coolant flow rates is about 4927, 5913, 6897, and 7883, respectively. As can be seen, the cooling time of the wax patterns was not affected by the different coolant flow rates while cooling water reaches turbulence fully. The Reynolds number can be calculated by the following Equation (2) [26].

$$Re = \frac{4Q\rho}{\pi D\eta} \tag{2}$$

Q denotes the flow rate of the coolant;
D denotes the diameter of the cooling channel;
ρ denotes the density of the coolant;
η denotes the viscosity of the coolant.

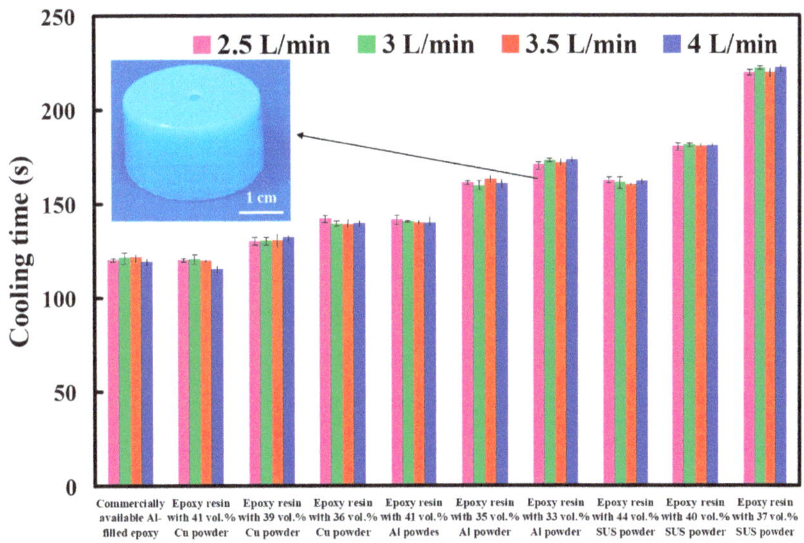

Figure 16. Cooling time of injection molded parts fabricated by ten sets of injection molds cooled with different cooling water flow rates.

Figure 17 shows cooling time of injection molded parts fabricated by ten sets of injection molds cooled with different compressed gas pressure. The results showed that the cooling time of injection molded parts become shorter when compressed gas pressure is higher. The main reason is that better cooling effect derived from faster gas flowing caused by high pressure in the mold. Figure 18 shows cooling time of injection molded parts fabricated by ten sets of injection molds cooled with different cold air temperatures. It is interesting that two results were found. One is that the cooling time of injection molded parts become shorter when cold air temperature is lower. The other is that the cooling performance for cooling water is better than cold air under the same temperature.

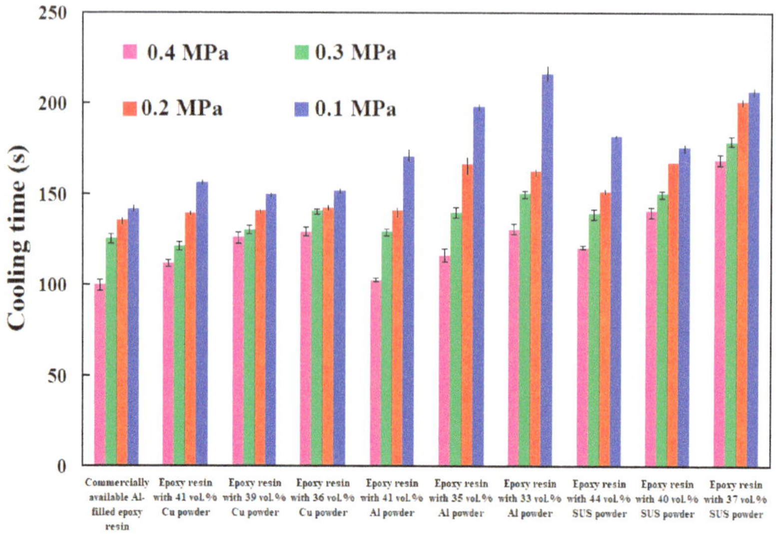

Figure 17. Cooling time of injection molded parts fabricated by ten sets of injection molds cooled with different compressed gas pressure.

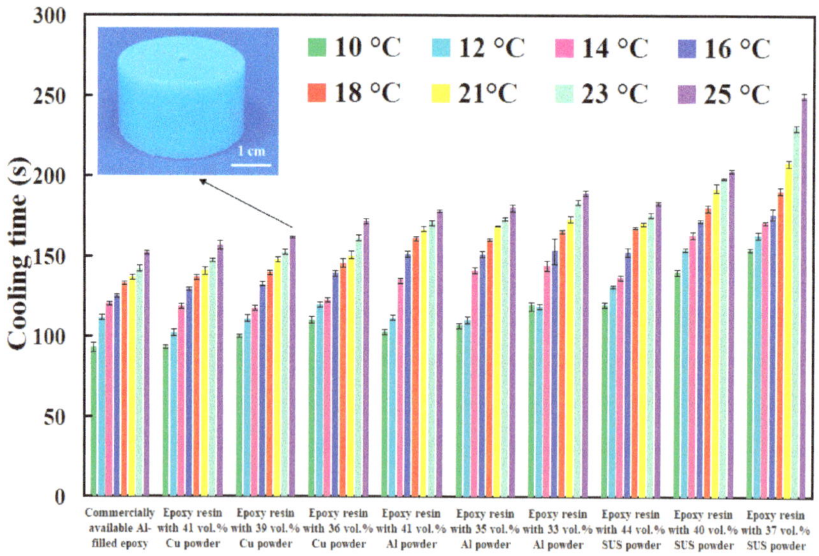

Figure 18. Cooling time of injection molded parts fabricated by ten sets of injection molds cooled with different cold air temperatures.

The viscosity of water and air is about 0.8 m Pa-s and 18 μ Pa-s, respectively. The viscosity of water is about 44.5 times that of air. Therefore, the gas is mixed into the water to form cooling water with ultrafine bubbles is efficient to improve cooling efficiency and shorten the cooling time of injection molded parts. Figure 19 shows the cooling time of injection molded parts fabricated by ten sets of injection molds cooled with different temperatures of cooling water with ultrafine bubbles. As can be seen, the cooling water with ultrafine bubbles is the best candidate of cooling medium. Cooling time of the injection molded parts fabricated by commercially available Al-filled epoxy resin cooled with cooling water with ultrafine bubbles of 16 °C, 18 °C, 21 °C, 23 °C and 25 °C are 66 s, 72 s, 84 s, 93 s, and 114 s, respectively. However, cooling time of the injection molded parts fabricated by commercially available Al-filled epoxy resin cooled with conventional cooling water of 16 °C, 18 °C, 21 °C, 23 °C and 25 °C are 79 s, 85 s, 92 s, 110 s, 24 s, respectively. Cooling efficiency of the injection molded parts can be increased by approximately 15.6%, 14.6%, 8.4%, 15.4% and 8%, respectively. Thus, average cooling efficiency about 12.4% can be further increased compared with commonly used cooling water.

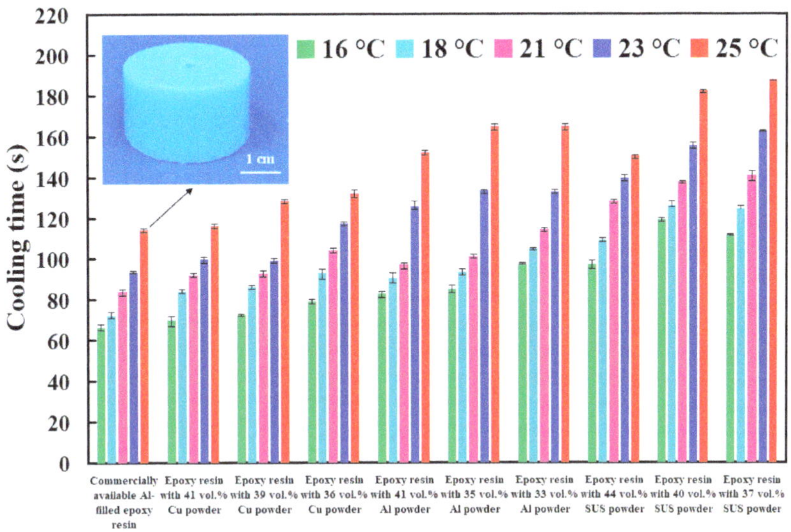

Figure 19. Cooling time of injection molded parts fabricated by ten sets of injection molds cooled with different temperature of cooling water with ultrafine bubbles.

To investigate cooling efficiency of cooling water with ultrafine bubbles at a temperature of 25 °C, four different coolant flow rates of cooling water with ultrafine bubbles are carried out. Figure 20 shows the cooling time of injection molded parts fabricated by ten sets of injection molds cooled with different flow rates of cooling water with ultrafine bubbles. The Reynolds number for four different coolant flow rates is about 4927, 5913, 6897, and 7883, respectively. It should be noted that the cooling time of the injection molded parts was not affected by the different coolant flow rates while cooling water reaches turbulence completely. Figure 21 shows the shortest cooling time of injection molded parts fabricated by ten sets of injection molds cooled with different four different cooling media. As can be seen, the cooling water with ultrafine bubbles is the best candidate of the cooling medium. It should be noted that the cooling performance will decrease significantly when the diameter of the cooling channel is less than 1.5 mm. According to practical experience, the time to cool the air is significantly lower than that of the water. Thus, the cooling medium can be replaced with cold air when the diameter of the cooling channel is less than 1.5 mm, since the cooling performance of cold air is acceptable. In addition, the mold material has a significant effect on the cooling time of the injection molded part since cooling time range of different mold materials is 99 s since the maximum and minimum cooling time is 279 s and 180 s, respectively. However, the cooling time range for different cooling media is only 92 s since the maximum and minimum cooling time is 279 s and 187 s, respectively. Figure 22 shows the longest and shortest cooling time of injection molded parts fabricated by ten sets of injection molds cooled with different four different cooling media. The total production cost of an injection mold fabricated with commercially available Al-filled epoxy resin is new Taiwan dollar (NTD) 3,046, while the total production cost of a injection mold fabricated by epoxy resin filled with 41 vol.% Al powder is only NTD 3999. Based on total production cost of injection mold and cooling efficiency, epoxy resin filled with 41 vol.% Al powder is the optimal formula for making an injection mold since saving in the total production cost about 24% is obtained compared to injection mold made with commercially available materials.

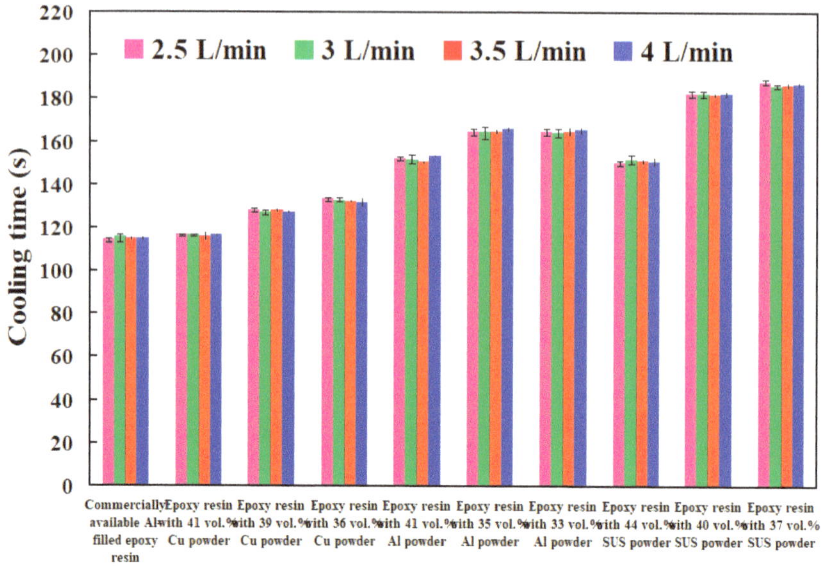

Figure 20. Cooling time of injection molded parts fabricated by ten sets of injection molds cooled with different flow rates of cooling water with ultrafine bubbles.

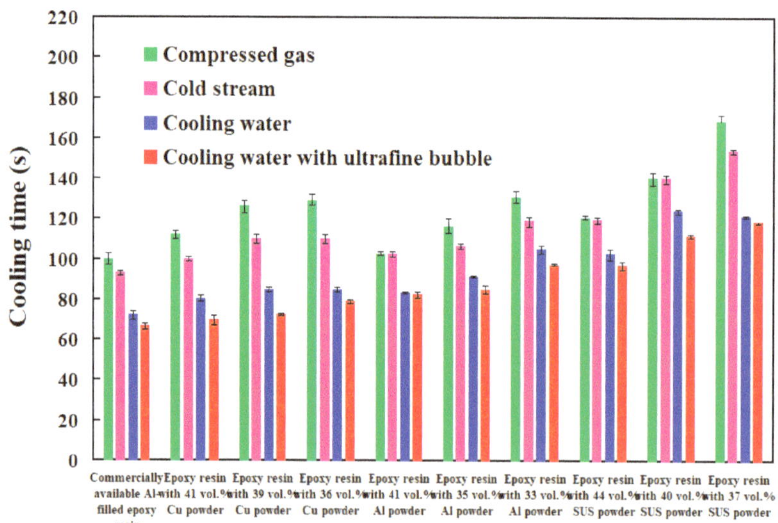

Figure 21. Shortest cooling time of injection molded parts fabricated by ten sets of injection molds cooled with different four different cooling media.

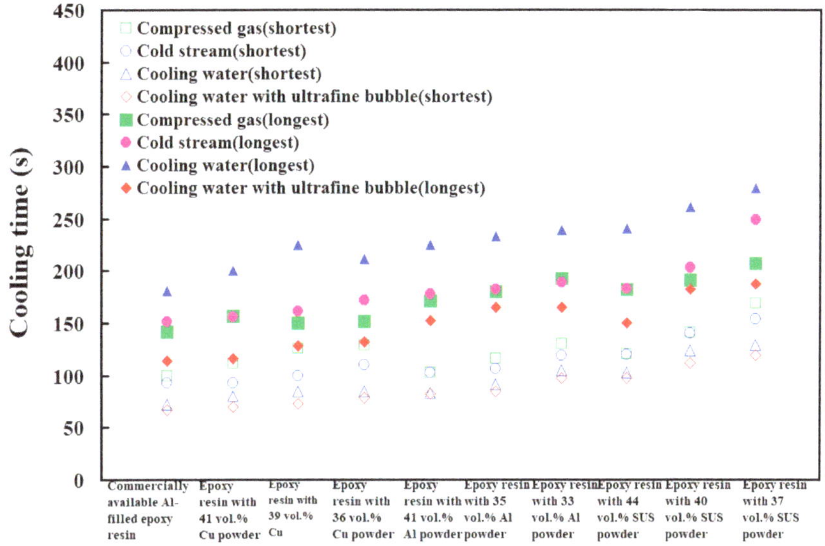

Figure 22. The longest and shortest cooling time of injection molded parts fabricated by ten sets of injection molds cooled with different four different cooling media.

According to the results described above, the findings of this study are very practical and provide the greatest application potential in the precision mold industry, especially in the mold design stage. In this study, the internal surface of the CCC fabricated by additive manufacturing technology possesses high surface roughness compared with conventional cooling channel made by computer-numerical control machining [27]. The service life of the mold will be affected significantly due to stress concentration. The main reason is that the mold is repeatedly heated [28] and cooled during the wax injection process. Ongoing research is focused on improving the surface roughness of the CCC by abrasive blasting, abrasive flow machining [29], electrochemical polishing, chemical polishing, laser polishing, or ultrasonic cavitation abrasive finishing. In various industries, hydrogen [30,31] is widely applied as a high-performance gaseous cooling medium since its thermal conductivity is higher than all other gases. In addition, helium gas [32] or hydrogen [33] can also be applied as cooling media in gas-cooled nuclear reactors since its low tendency to absorb neutrons. Sulfur hexafluoride [34,35] is also widely applied to cool some high-voltage power systems, such as circuit breakers, transformer, or switches. These issues are currently being investigated and the results will be presented in a later study.

4. Conclusions

Generally, the cost-effectiveness depends on the cycle time of the injection molded products significantly in the mass production process of new products. The MAM method is frequently applied to speed up manufacturing for the production of injection mold with CCC. However, the major issue is expensive spending and longtime taking in the mold making process. The objective of this study is to study the cooling performance of epoxy-based injection molds fabricated by different mold materials using different coolant media. Based on the experimental results obtained from this work, the contributions and findings of this study can be summarized as follows:

1. A low-cost recipe that can produce cooling efficiency better than that made of commercial materials was demonstrated. The epoxy resin with 41 vol.% Al powder seems to be the optimum recipe for making injection mold with high cooling efficiency.
2. The cooling time of the injection molded parts is extremely sensitive to both injection mold material and the cooling medium. Cooling water with ultrafine bubbles is the

best candidate of the cooling medium. Average cooling efficiency approximately 12.4% can be further increased compared with commonly used cooling water.

3. Mold material has a significant effect on the cooling time of the injection molded part, which has a greater influence on the cooling efficiency than the cooling medium since cooling time range of different mold materials is 99 s while the cooling time range for different cooling media is only 92 s. In addition, The cooling medium can be replaced with cold air while the diameter of the cooling channel is less than 1.5 mm and the coolant medium is water.

Author Contributions: C.-C.K. wrote the paper/conceived and designed the analysis/performed the analysis/conceptualization; Y.-J.Z., J.-Y.X., C.-H.L. collected the data/contributed data or analysis tools. All authors have read and agreed to the published version of the manuscript.

Funding: This study received financial support by the Ministry of Science and Technology of Taiwan under contract nos. MOST 110-2221-E-131-023 and MOST 109-2637-E-131-004.

Institutional Review Board Statement: Not applicable.

Informed Consent Statement: Not applicable.

Data Availability Statement: Not applicable.

Conflicts of Interest: The authors declare no conflict of interest.

Nomenclature

Acronyms	
CCC	Conformal cooling channel
Al	Aluminum
AM	Additive manufacturing
RT	Rapid tooling
CAD	Computer-aided design
MAM	Metal additive manufacturing
Cu	Copper
FE-SEM	Field-emission scanning electron microscopy
EDS	Energy-dispersive X-ray spectroscopy
XRD	X-ray diffractometer
SUS	Stainless steel
PVB	Polyvinyl butyral
NTD	New Taiwan dollar
List of symbols	
T_w	Mold cavity surface temperature
s	Part thickness
T_m	Melt temperature
T_e	Average ejection temperature
Q	Flow rate of the coolant
D	Diameter of the cooling channel
Greek symbols	
α	Thermal diffusivity
ρ	Density of the coolant
η	Viscosity of the coolant

References

1. Torres-Alba, A.; Mercado-Colmenero, J.M.; Caballero-Garcia, J.D.D.; Martin-Doñate, C. A Hybrid Cooling Model Based on the Use of Newly Designed Fluted Conformal Cooling Channels and Fastcool Inserts for Green Molds. *Polymers* **2021**, *13*, 3115. [CrossRef]
2. Mukhtarkhanov, M.; Perveen, A.; Talamona, D. Application of Stereolithography Based 3D Printing Technology in Investment Casting. *Micromachines* **2020**, *11*, 946. [CrossRef] [PubMed]
3. Dong, Y.W.; Li, X.L.; Zhao, Q.; Yang, J.; Dao, M. Modeling of shrinkage during investment casting of thin-walled hollow turbine blades. *J. Mater. Process. Technol.* **2017**, *244*, 190–203. [CrossRef]

4. Kitayama, S.; Miyakawa, H.; Takano, M.; Aiba, S. Multi-objective optimization of injection molding process parameters for short cycle time and warpage reduction using conformal cooling channel. *Int. J. Adv. Manuf. Technol.* **2014**, *88*, 1735–1744. [CrossRef]
5. Rajaguru, J.; Duke, M.; Au, C. Development of rapid tooling by rapid prototyping technology and electroless nickel plating for low-volume production of plastic parts. *Int. J. Adv. Manuf. Technol.* **2015**, *78*, 31–40. [CrossRef]
6. Lim, W.S.; Choi, H.S.; Ahn, S.Y.; Kim, B.M. Cooling channel design of hot stamping tools for uniform high-strength components in hot stamping process. *Int. J. Adv. Manuf. Technol.* **2014**, *70*, 1189–1203. [CrossRef]
7. Chen, J.; Gong, P.; Liu, Y.; Zheng, X.; Ren, F. Optimization of hot stamping cooling system using segmented model. *Int. J. Adv. Manuf. Technol.* **2017**, *93*, 1357–1365. [CrossRef]
8. Maji, P.K.; Banerjee, P.S.; Sinha, A. Application of rapid prototyping and rapid tooling for development of patient-specific craniofacial implant: An investigative study. *Int. J. Adv. Manuf. Technol.* **2008**, *36*, 510–515. [CrossRef]
9. Gill, S.S.; Kaplas, M. Comparative Study of 3D Printing Technologies for Rapid Casting of Aluminium Alloy. *Mater. Manuf. Process.* **2009**, *24*, 1405–1411. [CrossRef]
10. Nandi, A.K.; Deb, K.; Datta, S. Genetic Algorithm–Based Design and Development of Particle-Reinforced Silicone Rubber for Soft Tooling Process. *Mater. Manuf. Process.* **2013**, *28*, 753–760. [CrossRef]
11. Wang, X.; Li, Z.; Gu, J.; Ruan, S.; Shen, C.; Wang, X. Reducing service stress of the injection-molded polycarbonate window by optimizing mold construction and product structure. *Int. J. Adv. Manuf. Technol.* **2016**, *86*, 1691–1704. [CrossRef]
12. Mercado-Colmenero, J.M.; Rubio-Paramio, M.A.; Marquez-Sevillano, J.d.J.; Martin-Doñate, C. A new method for the automated design of cooling systems in injection molds. *Comput.-Aided Des.* **2018**, *104*, 60–86. [CrossRef]
13. Tomori, T.; Melkote, S.; Kotnis, M. Injection mold performance of machined ceramic filled epoxy tooling boards. *J. Mater. Process. Technol.* **2004**, *145*, 126–133. [CrossRef]
14. Kovacs, J.G.; Bercsey, T. Influence of mold properties on the quality of injection molded parts. *Period. Polytech. Mech. Eng.* **2005**, *49*, 115–122.
15. Khushairi, M.T.M.; Sharif, S.; Jamaludin, K.R.; Mohrun, A.S. Effects of Metal Fillers on Properties of Epoxy for Rapid Tooling Inserts. *Int. J. Adv. Sci. Eng. Inf. Technol.* **2017**, *7*, 1155–1161. [CrossRef]
16. Waalkes, L.; Längerich, J.; Imgrund, P.; Emmelmann, C. Piston-Based Material Extrusion of Ti-6Al-4V Feedstock for Complementary Use in Metal Injection Molding. *Materials* **2022**, *15*, 351. [CrossRef]
17. Armen, J.; Bruck, H.A. Improving contact resistance in metal–ceramic heat exchangers running liquid metal by additive manufacturing and ceramic tubes with electroplated films. *Int. J. Adv. Manuf. Technol.* **2021**, *113*, 2101–2119. [CrossRef]
18. Hossain, M.S.; Taheri, H. In-situ process monitoring for metal additive manufacturing through acoustic techniques using wavelet and convolutional neural network (CNN). *Int. J. Adv. Manuf. Technol.* **2021**, *116*, 3473–3488. [CrossRef]
19. Kuo, C.-C.; Qiu, S.-X. A Simple Method of Reducing Coolant Leakage for Direct Metal Printed Injection Mold with Conformal Cooling Channels Using General Process Parameters and Heat Treatment. *Materials* **2021**, *14*, 7258. [CrossRef] [PubMed]
20. Çalışkan, C.İ.; Coşkun, M.; Özer, G.; Koç, E.; Vurkır, T.A.; Yöndem, G. Investigation of manufacturability and efficiency of micro channels with different geometries produced by direct metal laser sintering. *Int. J. Adv. Manuf. Technol.* **2021**, *117*, 3805–3817. [CrossRef]
21. Muvunzi, R.; Dimitrov, D.M.; Matope, S.; Harms, T. A case study on the design of a hot stamping tool with conformal cooling channels. *Int. J. Adv. Manuf. Technol.* **2021**, *114*, 1833–1846. [CrossRef]
22. Yamaguchi, M.; Furumoto, T.; Inagaki, S.; Tsuji, M.; Ochiai, Y.; Hashimoto, Y.; Koyano, T.; Hosokawa, A. Internal face finishing for a cooling channel using a fluid containing free abrasive grains. *Int. J. Adv. Manuf. Technol.* **2021**, *114*, 497–507. [CrossRef]
23. Hailan, S.M.; Ponnamma, D.; Krupa, I. The Separation of Oil/Water Mixtures by Modified Melamine and Polyurethane Foams: A Review. *Polymers* **2021**, *13*, 4142. [CrossRef] [PubMed]
24. Franzelli, B.; Scouflaire, P.; Darabiha, N. Using In Situ Measurements to Experimentally Characterize TiO2 Nanoparticle Synthesis in a Turbulent Isopropyl Alcohol Flame. *Materials* **2021**, *14*, 7083. [CrossRef] [PubMed]
25. Kim, Y.; Moon, C.; Nematollahi, O.; Kim, H.D.; Kim, K.C. Time-Resolved PIV Measurements and Turbulence Characteristics of Flow Inside an Open-Cell Metal Foam. *Materials* **2021**, *14*, 3566. [CrossRef] [PubMed]
26. Abbasnezhad, N.; Kebdani, M.; Shirinbayan, M.; Champmartin, S.; Tcharkhtchi, A.; Kouidri, S.; Bakir, F. Development of a Model Based on Physical Mechanisms for the Explanation of Drug Release: Application to Diclofenac Release from Polyurethane Films. *Polymers* **2021**, *13*, 1230. [CrossRef] [PubMed]
27. Kuo, C.-C.; Chen, W.-H. Improving Cooling Performance of Injection Molding Tool with Conformal Cooling Channel by Adding Hybrid Fillers. *Polymers* **2021**, *13*, 1224. [CrossRef] [PubMed]
28. Asif, A.; Manzoore, E.M.S.; Ali, B.; Mohammed, K. Thermal Performance of Compression Ignition Engine Using High Content Biodiesels: A Comparative Study with Diesel Fuel. *Sustainability* **2021**, *13*, 7688.
29. Dong, Z.; Zhang, Y.; Lei, H. Cutting factors and testing of highly viscoelastic fluid abrasive flow machining. *Int. J. Adv. Manuf. Technol.* **2021**, *112*, 3459–3470. [CrossRef]
30. Chai, M.; Li, Z.; Yan, H.; Huang, Z. Flow field characteristics analysis of interelectrode gap in electrochemical machining of film cooling holes. *Int. J. Adv. Manuf. Technol.* **2021**, *112*, 525–536. [CrossRef]
31. da Silva, L.J.; Souza, D.M.; de Araújo, D.B.; Reis, R.P.; Scotti, A. Concept and validation of an active cooling technique to mitigate heat accumulation in WAAM. *Int. J. Adv. Manuf. Technol.* **2020**, *107*, 2513–2523. [CrossRef]

32. Abdelrazek, A.H.; Choudhury, I.A.; Nukman, Y.; Kazi, S.N. Metal cutting lubricants and cutting tools: A review on the performance improvement and sustainability assessment. *Int. J. Adv. Manuf. Technol.* **2020**, *106*, 4221–4245. [CrossRef]
33. Fan, J.; Peng, B.; Zhao, W. Study on hydrogen behaviors around micropores within heavy forging during heating process. *Int. J. Adv. Manuf. Technol.* **2021**, *113*, 523–533. [CrossRef]
34. Ricciardi, M.R.; Papa, I.; Coppola, G.; Lopresto, V.; Sansone, L.; Antonucci, V. Effect of Plasma Treatment on the Impact Behavior of Epoxy/Basalt Fiber-Reinforced Composites: A Preliminary Study. *Polymers* **2021**, *13*, 1293. [CrossRef] [PubMed]
35. Khettabi, I.; Benyoucef, L.; Boutiche, M.A. Sustainable reconfigurable manufacturing system design using adapted multi-objective evolutionary-based approaches. *Int. J. Adv. Manuf. Technol.* **2021**, *115*, 3741–3759. [CrossRef]

polymers

MDPI

Article

Effect of Local Heat Pipe Cooling on Throughput Distribution and Thermal Homogeneity in a Binary Melt Pre-Distributor for Polyolefin Extrusion

Christian Hopmann, Lisa Leuchtenberger *, Malte Schön and Lena Wallhorn

Institute for Plastics Processing (IKV) in Industry and Craft at RWTH Aachen University, Seffenter Weg 201, 52074 Aachen, Germany; sekretariat@ikv.rwth-aachen.de (C.H.); malte.schoen@ikv.rwth-aachen.de (M.S.); lena.wallhorn@rwth-aachen.de (L.W.)

* Correspondence: lisa.leuchtenberger@ikv.rwth-aachen.de

Abstract: In polymer blown film extrusion, inhomogeneous die temperature distributions lead to an inhomogeneous temperature and cause film thickness variations. To avoid an inhomogeneous film thickness and to achieve good film qualities, thermal homogenisation of the melt is necessary. Therefore, a new approach for cooling hot spots with heat pipes is investigated. CFD Simulations in OpenFOAM show that heat pipes can be used to influence melt temperatures locally in the places in which a temperature reduction is required. Since the outlets interact in a pre-distribution die, one heat pipe is not sufficient to homogenise the temperature at every outlet to similar temperatures. Two heat pipes show much better results with lower average temperature deviations between the distributor outlets. In order to equalise the temperature at all outlets, at least one heat pipe per outlet will be required.

Keywords: thermal homogenisation; pre-distribution; heat pipe; blown film extrusion; CFD

Citation: Hopmann, C.; Leuchtenberger, L.; Schön, M.; Wallhorn, L. Effect of Local Heat Pipe Cooling on Throughput Distribution and Thermal Homogeneity in a Binary Melt Pre-Distributor for Polyolefin Extrusion. *Polymers* **2022**, *14*, 2271. https://doi.org/10.3390/polym14112271

Academic Editors: Célio Bruno Pinto Fernandes, Salah Aldin Faroughi, Luís L. Ferrás and Alexandre M. Afonso

Received: 27 April 2022
Accepted: 30 May 2022
Published: 2 June 2022

1. Introduction

About 39% of all plastics are processed into packaging [1]. In turn, 49% of all packaging is produced using extrusion processes, such as blown film extrusion [2]. In blown film extrusion, plastics granules are melted and homogenised in an extruder. Afterwards, the melt is transformed into a tubular cross-section in an extrusion die consisting of a pre-distributor and a main distributor, normally a spiral die (Figure 1). At the die outlet, the melt is hauled-off and inflated. Air cooling then initiates the cooling process via convection. At the same time, the melt is stretched in the circumferential direction by the blow up process and in the extrusion direction by the haul-off. A great advantage of this process is the variable film width, which can be easily adjusted by the blow-up ratio without changing the die. Moreover, the induced biaxial stretching renders a stronger and less permeable film [3,4].

The most important quality aspects of films are the homogeneous film thickness and film flatness. Packaging films, in particular, often consist of multilayer films, where each layer performs a specific function. If one of the layers contains thin spots, functions such as keeping oxygen or water out can no longer be provided, which can lead to, e.g., spoiled food [4]. To ensure the reliable function of every layer, the respective thinnest spot has to reach a specified minimum thickness, which leads to unnecessary material usage in case of inhomogeneous film thickness distribution. Whether the film has a homogeneous layer thickness depends on various influences. One influence is the throughput distribution at the die outlet [5]. Circumferential segments of the die outlet that carry more of the melt cause an increased wall thickness in the film. Another influence is the temperature and the viscosity distribution at the die outlet [6]. Areas with higher melt temperatures are more stretched in the blow-up-process due to the lower extensional viscosity. Therefore, these areas appear as thin sections in the film.

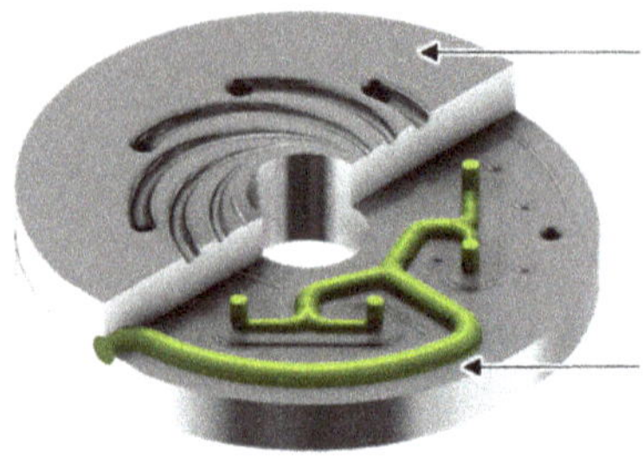

Figure 1. The 2^n-pre-distributor in combination with a spiral die as main distributor in a blown film process.

To reach a perfectly homogeneous blown film, a large number of empirical and simulative studies from research and industry has been carried out. They find a temperature influence on the melt distribution and temperature distribution in extrusion dies. Astarita et al., pointed out, using the example of the mandrel holder die, that non-uniform dissipation and asymmetric heat conduction in the die material can lead to an inhomogeneous temperature distribution in the die [6]. The fact that inhomogeneous temperature distributions at the die outlet can influence the throughput distribution was shown by Vergnes et al., using the example of a flat die [7]. This also includes the influence of shear heating in the extrusion die [8–11]. Zatloukal et al., performed a non-isothermal flow analysis on the flow channel in a spiral die [12]. In their work, the influence of the temperature distribution calculations was investigated. A comparison of the non-isothermal and isothermal calculations show that the simulation results of the non-isothermal calculations agree better with experimental data. This observation explained the viscous dissipation causing a temperature increase and viscosity decrease. Another study on spiral distributor dies was carried out by Skabrahova et al. [13]. They simulated the flow channel of an axial spiral distributor to investigate whether the assumption of a homogeneous throughput and temperature distribution at the inlet to the spiral distributor is permissible. By simulating different asymmetric inlet conditions at the spiral distributor inlets, they showed that some equalisation processes for thermal and rheological homogenisation take place along the spirals. Nevertheless, inhomogeneous inlet conditions lead to a significant deterioration of the melt distribution profile at the die exit. However, these considerations on non-uniform inlet conditions are not based on real measurements, since only different hypothetical scenarios were run [13]. The general importance of thermal influences on the throughput distribution in extrusion dies is also emphasized by Pittman [14]. In a review of the state-of-the-art extrusion die simulation, he explains that both dissipative shear heating and temperature distribution in the die can negatively affect the velocity distribution at the die exit. Pittman recommends introducing differences between the melt and flow channel wall temperatures in a targeted manner in order to influence and equalize the flow in this way. Burmann et al., also discussed the temperature influence in spiral mandrels. They give general advice on how spiral mandrel dies can be optimized with regard to processing temperatures [15,16]. The installation of internal temperature control systems (water or oil temperature control), thermal separation by insulating layers or improvement of heat exchange (e.g., by ceramic fins) are recommended as possible solutions to improve thermal conditions. In later contributions, it was pointed out that the plastic melt is subjected to locally different dwell time and temperature-related stresses in the overall extrusion die system, depending on the design of the main and pre-distributor [17]. Saul investigated the automated die design of radial spiral dies in his dissertation [18]. He suggested for future investigations to consider non-isothermal calculations, due to local melt shear heating. Moreover, he pointed out the expansion potential of his investigations, which optimise the temperature distribution in the die. The temperature effect in a pre-distributor was analysed four years later by te Heesen, using the example of a centrally fed star pre-distributor [19]. The non-isothermal flow simulations, taking into account both shear heating of the melt and heat conduction in the die material, show that the shear stresses and the pressure drop are influenced by the dissipative shear heating. However, no influence on the throughput distribution was observed, which can be attributed to the rotationally symmetric geometry of a star distributor. Furthermore, te Heesen notes that the temperature influence can affect the throughput distribution in more complex pre-distributors, such as a 2^n-distributor. The analysation and homogenisation of the thermal effects in a 2^3-pre-distributor was carried out by Yesildag [5,20]. The different melt temperatures cause viscosity differences between the outlets, which lead to an inhomogeneous mass throughput. Yesildag's approach to homogenising the throughput distribution was to integrate heating cartridges and thermally conductive inserts to homogenise the temperature distribution in the die. Prior to practical trials, the effect of heating cartridges and thermally conductive inserts was investigated using CFD Simulations. He succeeded in homogenising the melt distribution at the die outlets, but to optimise the temperature distribution for one process point, 12 heating cartridges are necessary. Each of them has a different temperature, which makes this approach highly process point dependent and requires a separate control circuit for each cartridge. Another disadvantage is the occurrence of overheated spots in the die, which rules out the usage of thermally sensitive materials such as EVOH or PVDC. In addition, heating cartridges require high maintenance due to their short service life. Therefore, a simpler approach in the form of adjustable nozzles was developed [21]. Through adjustable nozzles at every pre-distributor outlet, an equal distribution of the melt flows can be achieved. However, the temperature distribution remained outlet dependent. Therefore, film thickness variations will appear, despite a melt distribution homogenisation in the pre-distributor.

In addition to the described direct heating methods for extrusion dies, which are based on electricity, another method is indirect heating by liquids. Indirect heating is preferred in the processing of elastomers [22]. The key feature that distinguishes heating with liquids from electrical heating is the ability to provide heat regardless of ambient conditions. This proves to be an advantage in the typical, relatively low temperatures that are common in the processing of elastomers [22]. Fluid heating is also occasionally used in relatively small dies for thermoplastics. To add or remove heat from a specific area, such as from the exit zone of the mandrel of blown film dies, oil heating is used. The reassembly of the dies, when they are changed or serviced, is usually more complicated than with direct electric heating, because of the heating flow channel cleaning [22]. A further disadvantage of liquid heating is the technical complexity and the cost in cases where different die zones must be maintained at different temperatures. Current technology requires a separate heating unit for each temperature zone.

Since the temperature has a major influence on polymer viscosity, it affects not only the flow resistances and thus the mass flow distribution, but also the extensional viscosity beyond the outlet of the main distributor. During inflation, the melt parts of lower extensional viscosity can be stretched more easily, resulting in thin spots in the film. To avoid thickness variations in the film, this paper's approach is to develop an alternative thermal die control with heat pipes in order to avoid producing temperature hot spots in the die, which is not only less process point dependent, but also requires low maintenance and can be easily retrofitted.

Heat pipes are already used in many areas such as electronics cooling and aerospace [23–25]. They consist of a closed outer body, usually cylindrical in shape, with a capillary structure on the inner walls and are partially filled with a fluid [25,26]. This structure is shown in Figure 2.

When the temperature at one end of a heat pipe is increased, the fluid starts to evaporate causing localised cooling. A pressure gradient is established inside the cylinder, and the gas flows to the colder end. Here the evaporated fluid cools down and condenses. Through the capillary structures, the fluid is transported back to the warmer end. The process repeats as long as there exists a temperature gradient between the heat pipe's ends. Due to the high enthalpies of evaporation and condensation, heat pipes are able to transport large amounts of heat. If, for example, a solid copper rod with a diameter of 8 mm and a length of 300 mm is used to transfer a heat quantity of 100 W, a driving temperature gradient of theoretically 1493 °C would be required. According to heat pipe manufacturers, a heat pipe achieves the same heat transfer with a driving temperature gradient of approx. 0.5 K [24]. Through the choice of cylinder material and working fluid, it is possible to create heat pipes for nearly every application temperature range. According to Stephan, heat pipes with a copper cylinder and water as working fluid are suitable for most polymer extrusion application temperature ranges [25]. The heat pipe performance can be calculated with Equation (1) and is dependent on the temperature difference between the heat pipe ends, the chosen material combination, the cross section and the capillary structure [26].

$$\Delta T = \dot{Q} R_{tot}, \quad (1)$$

R_{tot} is the total thermal resistance, including the radial resistance between external heat source and evaporator wall and the axial resistance of the steam flow. The heat pipe performance or heat flux density is restricted by several physical boundaries shown in Figure 3 [26]:

- Viscosity limit This limits the heat flux density at working temperatures just above the fluid's melting point. Here, the pressure difference driving the steam between evaporator and condenser is small, and the steam flow is determined by high viscous forces and can even be interrupted.
- Sound velocity limit If the vapour reaches the sonic velocity at a certain evaporator temperature, this velocity cannot be exceeded even by reducing the steam pressure in the condenser. This limits the maximum heat transfer.
- Interaction limit At high heat flux density, liquid is entrained by the vapour, and a partial drying out of the capillary leads to a breakdown in the liquid flow.
- Capillary force limit The capillary force limit is reached when the flow losses of the liquid heat transfer medium are greater than the existing capillary pressure. This scenario is dependent on the temperature difference between evaporator and condenser.
- Boiling limit Bubbles boiling in the capillary restrict the flow of the liquid or stop it.

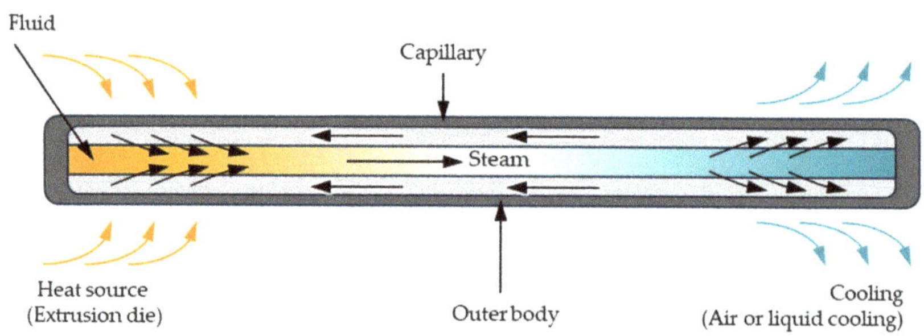

Figure 2. Schematic of a heat pipe.

If one of these five limits is exceeded, the capillary structure in the heating zone dries out. This significantly increases the thermal resistance in the evaporator. Since this thermal resistance has a significant influence on the overall thermal resistance, this results in a significant temperature increase in the heat source.

Heat pipes have short reaction times, a compact design and enable operation without an external power source [27,28]. Lakemeyer et al., already investigated the use of heat pipes or the heat pipe principle for temperature control of extruder screws [29]. In this case, the extruder screw acts as a heat pipe, the temperature peaks in the screw are reduced and the thermal homogenisation of the screw takes place through heat equalisation processes. Furthermore, Kartelmeyer et al., used heat pipes to control the temperature of injection moulds, as this approach is a low-maintenance and efficient way of transporting heat [27, 28].

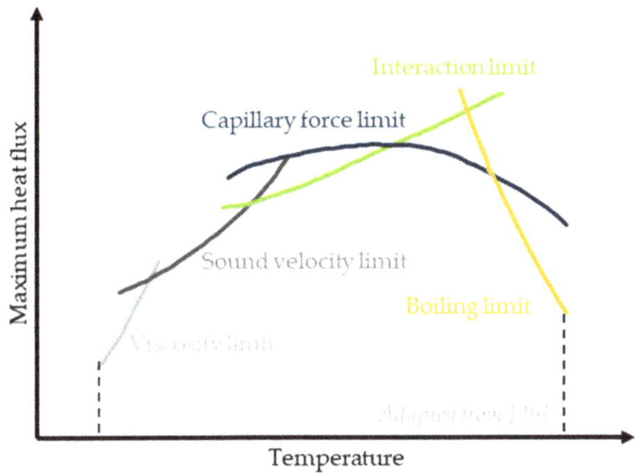

Figure 3. Performance restrictions of heat pipes [26].

To compare different heat pipes, a specific performance rating can be useful. One possibility is to calculate the merit number M, which is composed of the surface tension σ, the evaporation enthalpy Δh_e and the kinematic viscosity ν_l (Equation (2)).

$$M = \frac{\sigma \Delta h_e}{\nu_l}, \tag{2}$$

The performance number M should be as large as possible in the application-specific temperature range.

Considering the high performance of heat pipes, in this paper it is investigated whether a maximum of two heat pipes per pre-distribution half is sufficient to achieve thermal homogenisation. For this purpose, CFD simulations are performed to determine suitable heat pipe positions. Starting from an initial baseline simulation without heat pipes, four suitable positions are defined. To assess the suitability of the heat pipes, the average temperature deviations in each outlet, the thermal melt homogeneity and the melt distribution are considered. Two promising positions are further investigated by varying the heat pipe power up to 80 W for one heat pipe or 210 W in the case of two heat pipes. The effects of heat pipes with different cooling capacities need to be investigated to ensure that an appropriate heat pipe capacity is considered when a heat pipe is specified for a practical application. The aim is to achieve temperature and melt distribution deviations of less than 1% between the outlets.

2. Materials and Methods

In order to investigate local cooling through the operating behaviour of the heat pipes and their effect on melt temperature and melt distribution, CFD simulations were performed using OpenFOAM software (OpenFOAM Foundation Ltd., London, UK). In Section 2, the authors present the material and methods used for the CFD-Simulations.

2.1. Materials

As low-density Polyethylen (PE-LD) is the most processed material in the blown film extrusion, a PE-LD type 2101N0W from SABIC, Riyadh, Saudi Arabia is investigated in the simulations. In order to describe the shear-thinning, temperature-dependent material behaviour, the combined approach of Carreau and Williams, Landel and Ferry (WLF) is suitable [18,19,21,26]. The Carreau and WLF parameters were determined from high-pressure capillary measurements at IKV. The shear viscosity η in dependence of the shear rate $\dot{\gamma}$ and temperature T is described in Equations (3) and (4).

$$H(\dot{\gamma}) = \frac{18{,}757.01}{(0.86\,\dot{\gamma})^{0.703}}, \tag{3}$$

$$\lg(a_T) = -\frac{8.86(T - 257.481\ K)}{101.6\ K + T - 257.481\ K}, \tag{4}$$

Another relevant material parameter is the specific melt density. The pressure and temperature dependent density was determined using p-v-T measurements. As the density is almost constant at 200 bar and temperatures between 180 °C and 230 °C, incompressibility of the melt is assumed. For the simulations, an average density value of 769 kg/m^3 is set. To perform a non-isothermal simulation, thermal properties are required as well. The thermal conductivity of 0.282 W/(Km) at 473.15 K was determined in a laser flash analysis (LFA). The specific heat capacity of 2.722 J/(gK) at 473.15 K was determined by differential scanning calorimetry (DSC).

2.2. Start and Boundary Conditions

In accordance with common practice, a steady state with laminar, perfectly wall-adherent and incompressible flow is considered. Moreover, a homogeneous temperature and velocity at the die inlet are assumed (Figure 4) [5,20,21].

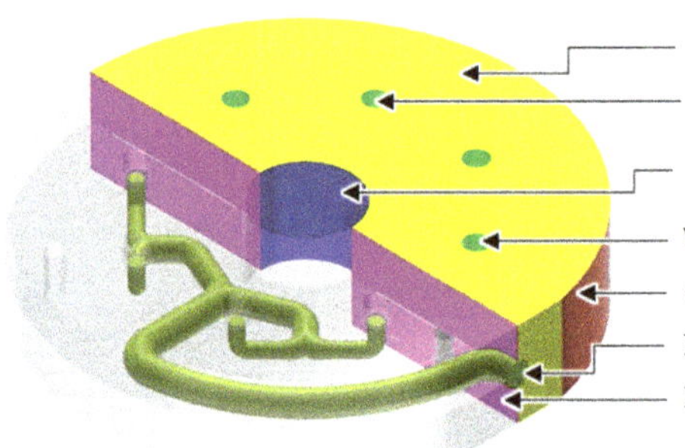

Figure 4. Start and boundary conditions.

2.3. Meshing

The die geometry analysed is based on the work of Yesildag [5]. In the pre-processing, all chamfers, holes and threads are neglected. Furthermore, the die geometry is reduced to one half of the symmetrical die. Since a future comparison of heat pipes and fluid oil heating is intended, an easy mesh adjustment is advantageous. Therefore, the use of the immersed boundary method is appropriate at this point [30]. The defining feature of this approach is the possibility to perform the entire simulation on a fixed Cartesian grid.

To discretise the pre-distribution geometry, the mesh generator snappyHexMesh is used, which performs a representation of the complete computational domain by hexahedron-shaped control volumes. All elements are first assigned a very high-flow resistance. For this purpose, according to Khadra et al., at the moment the conservation equation is extended with a term in which a proportionality factor for the velocity vector is integrated. The proportionality factor represents the inverse Darcy coefficient, i.e., the permeability of the material [30]. If this value is very high, the flow of the fluid through this geometry volume is suppressed, representing the steel body. The flow channels are then taken into account by means of an adjustment of the local flow resistances. In this process, the flow resistance is reduced in all elements of the die grid that are located within the flow channel geometry. Although a coarser partitioning is sufficient outside the flow channel, since only the heat conduction is calculated, the mesh in the flow channel is refined to represent flow effects sufficiently accurately. Mesh independence checks have shown that a one-level refinement is sufficient. Furthermore, areas where an integration of heat pipes would potentially be useful are refined. For reasons of symmetry, it is sufficient to perform the calculations for only one half of the mould. This results in a mesh with 620,141 volumes, which is shown in Figure 5 and leads to a computation time of about 3.5 h.

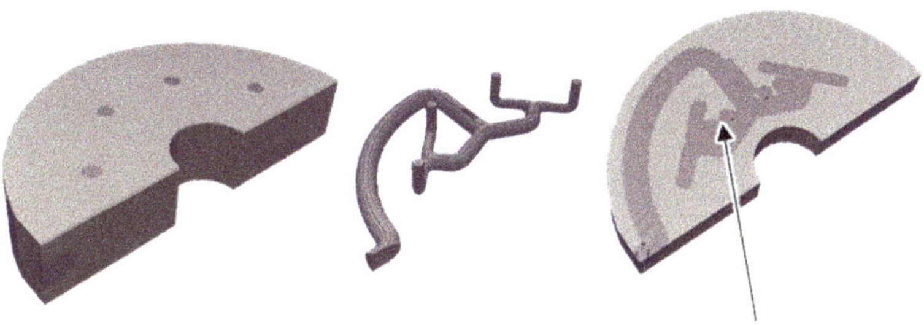

Figure 5. Meshed simulation space and flow channel geometry.

2.4. Solver

A solver based on the SIMPLEC (Semi-Implicit Method for Pressure Linked Equations-Consistent) method is used [31,32]. The solver is implemented in the open-source finite volume software OpenFOAM (OpenFOAM Foundation, London, UK). The differential equations of mass, momentum and energy conservation are solved for the variables pressure, flow velocity and temperature, whereat the shear heating is additionally taken into account for the latter. This flow solver was extended in [33] by an immersed boundary approach according to Khadra et al. [30].

$$\nabla \cdot \mathbf{u} = 0, \tag{5}$$

$$\nabla \mathbf{p_{norm}} = \nabla \cdot (\mathbf{u} \times \mathbf{u}) - \nabla \cdot \boldsymbol{\tau}_{\mathbf{norm}} - \delta \mathbf{u}, \tag{6}$$

$$0 = \nabla \cdot (\mathbf{uT}) + \nabla \cdot (\alpha \nabla \mathbf{T}) + \frac{\boldsymbol{\tau}_{\mathbf{norm}}}{c_p} : (\nabla \times \mathbf{u}), \tag{7}$$

In the conservation equations, $\mathbf{u}$ is the vector of velocities and $\mathbf{p_{norm}}$ is the pressure normalized to the density ρ. Moreover, $\mathbf{T}$ is the temperature and α is the local thermal diffusivity. The zero-velocity Dirichlet condition is used as a penalty condition. The inverse Darcy coefficient δ in the momentum equation is 0 for the fluid region and becomes $>>0$ for the die region, which corresponds to a complete suppression of flow. In [33], the temperature distribution could not yet be described. Therefore, a heat equation, including the shear heating calculated from the shear stresses in the energy equation, is implemented as follows [34].

$$0 = \nabla \cdot (\mathbf{uT}) + \nabla \cdot (\alpha \nabla \mathbf{T}) + \frac{\boldsymbol{\tau}_{\mathbf{norm}}}{c_p} : (\nabla \times \mathbf{u}), \tag{8}$$

$$\boldsymbol{\tau}_{\mathbf{Norm}} = \boldsymbol{\nu}((\nabla \times \mathbf{u}) + (\nabla \times \mathbf{u})^T), \tag{9}$$

$\boldsymbol{\tau}_{\mathbf{norm}}$ is the tensor of the shear stress normalized to the density ρ, c_p is the local specific heat capacity and $\boldsymbol{\nu}$ is the kinematic viscosity (nu).

To take into account the highly variable placement of the heat pipes a similar approach is applied. For simplicity, the heat pipes are considered as simple cylindric geometries, functioning as a constant heat sink. In those cells defined as part of the heat pipes, a constant cooling rate in K/s is assumed. The cooling rate is freely adjustable. Therefore, the energy equation is extended as follows:

$$0 = \nabla\cdot(\mathbf{uT}) + \nabla\cdot(\alpha\nabla\mathbf{T}) + \frac{\tau_{norm}}{c_p} : (\nabla\times\mathbf{u}) + \varepsilon C, \tag{10}$$

For die regions without heat pipes ε is 0. Cells overlapping with the heat pipe geometry (ε = 1) act as a heat sink for C < 0 and as heat source for C > 0. The immersed boundary method is reasonable at this point, laminar material behaviour is assumed and no turbulent wall conditions are required. In addition, the heat fluxes at the boundary between solid steel and liquid melt are expected to be low so that the assumption of perfect heat transfer between solid and liquid is reasonable.

3. Simulation Results and Discussion

In order to determine the temperature distribution without thermal homogenisation, a first basic simulation is performed without heat pipes. According to Yesildag, the temperature and throughput variation between the outlets increases as the difference between the melt and die temperature increases. Moreover, the shear heating effect is affected by the total mass throughput. A homogeneous inlet mass flow of 100 kg/h with a constant melt temperature of 453.15 K and an outer die temperature of 473.15 K is assumed, which is in the range of realistic processing temperatures [35]. The temperature of the cool air in the centre of the die is assumed 293.15 K; this causes a heat transfer coefficient of 18 W/(m^2*K) [5]. Since the top and bottom of the die are in contact with other heated components of the extrusion line, e.g., the spiral die, it can be assumed that the heat loss is comparatively low. Therefore, the top and bottom are considered adiabatic. Figure 6 shows the resulting temperature distribution both in the die and at the outlets. The outlets are numbered from 1 to 4. It becomes clear that the temperature in the outlet is not homogeneous. In addition, the temperature at outlets 1 and 4 seems to be lower than at outlets 2 and 3, which is in good agreement with the work of Yesildag and Hopmann and shows that the immersed boundary approach works [5,20,21].

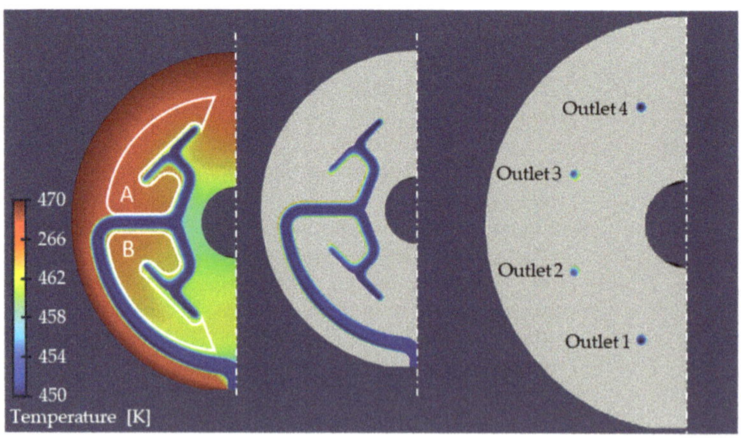

Figure 6. Temperature distribution in a 2^n-pre-distributor without heat pipes.

A consideration of the mass flow distribution in dependence of the outlets (Figure 7) as well as the representation of the temperature across the outlet diameter shown in Figure 8 supports this thesis. Due to a shielding effect of the flow channel, the die region A is warmer than the die region B (Figure 6), which leads to a slightly warmer melt at outlet 3 where a temperature maximum of 460.63 K on average is present. Outlet 2 shows a temperature of 459.63 K. The minimum temperature of 456.32 K is found at outlet 1.

The temperature differences between the minimum and maximum temperature of 4.31 K meaningfully impacts the shear and extensional viscosity. The shear viscosity drops by 8.9% (Equation (4)).

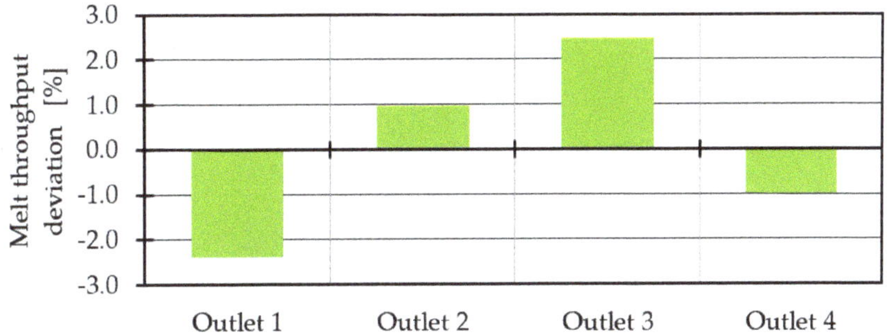

Figure 7. Melt throughput deviation without heat pipes.

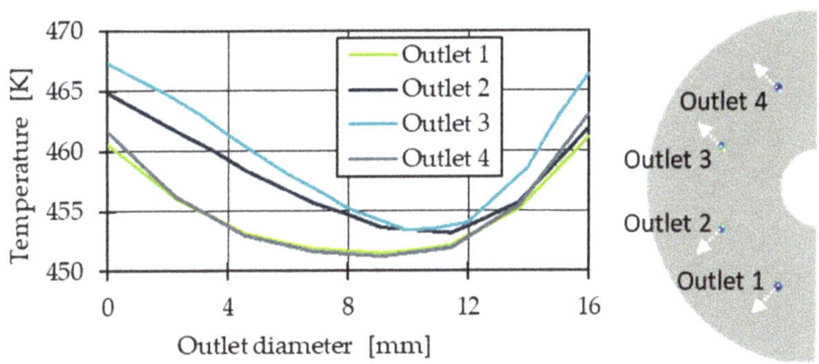

Figure 8. Temperature distribution in the cross section of each individual outlet.

To homogenise the temperature and the mass throughput, the third outlet was cooled. Therefore, four different heat pipe positions are investigated, which are shown in Figure 9. In each simulation, one heat pipe with a radius of 3 mm and a length of 30 mm was integrated. Since the pre-distributor is an interacting system where the effect on one outlet leads to deviations at all other outlets, the efficiency of the heat pipes and the average of the temperature deviation across all four outlets (Equation (11)) were compared. The resulting average temperature deviations for a cooling rate of −25 K/s, which corresponds to a heat transfer capacity of 44 W, are shown in Figure 10. According to a manufacturer's data sheet, a heat pipe with a diameter of 6 mm and a length of 100 mm can reach up to 137 W [36].

$$\Delta T_{\text{average, x}} = \frac{\frac{\sum_{j=1}^{n_x} T_j}{n_x} - \frac{\sum_{x=1}^{4} \sum_{j=1}^{n_x} T_j}{\sum_{x=1}^{4} n_x}}{\frac{\sum_{x=1}^{4} \sum_{j=1}^{n_x} T_j}{\sum_{x=1}^{4} n_x}}, \tag{11}$$

The average temperatures at the third outlet are decreasing the closer the heat pipe is positioned to the outlet. This is caused by the large surrounding melt-carrying surface. Position 4 is nearly surrounded by melt flow channel walls, which is resulting in a larger heat flow at constant heat flow density, and thus a smaller temperature difference between outlet 1 (minimum temperature) and outlet 3 (maximum temperature). Without heat pipes the temperature difference is 4.31 K. With one heat pipe at position 4 with a cooling rate of −25 K/s it decreases to 3.51 K.

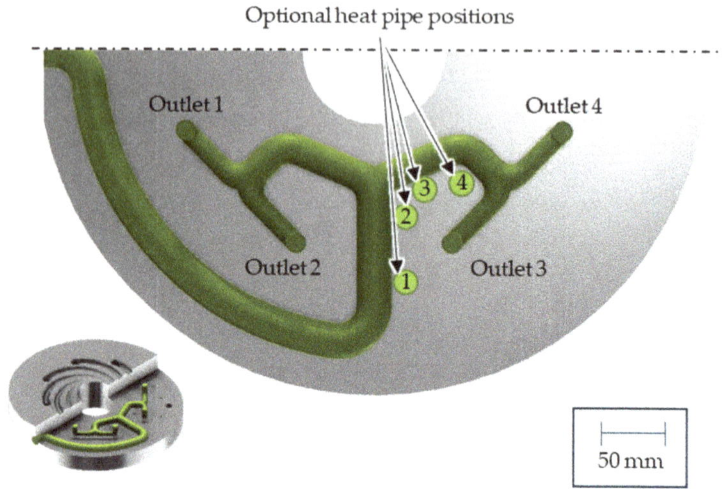

Figure 9. Heat pipe position 1–4.

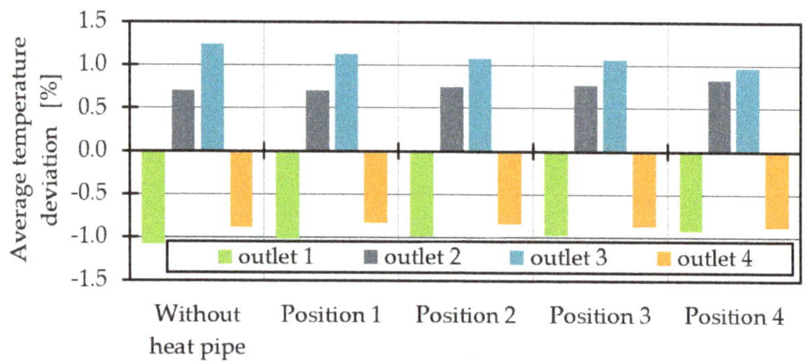

Figure 10. Average temperature deviation for different heat pipe positions.

As the temperature distribution is not homogeneous over the outlet surface (Figures 6 and 8), $e_{thermal}$ was used to evaluate the thermal homogenisation at all outlet's cross sections. The indicator $e_{thermal}$ is often used to evaluate mixing devices and allows a comparison of the temperature distribution in one cross section compared to the temperature distribution in another cross section [37,38]. The calculation of $e_{thermal}$ is given in Equations (12) and (13), where σ describes the thermal homogeneity at the outlet without heat pipe influence and the outlet with heat pipe influence with A = total area, A_i = cell area, υ_i = cell temperature, $\bar{\upsilon}$ = average temperature in the outlet plane.

$$e_{thermal} = \frac{\sigma_{without\_heat\_pipe} - \sigma_{with\_heat\_pipe}}{\sigma_{without\_heat\_pipe}}, \quad (12)$$

$$\sigma = \sum_{i=1}^{n} \frac{A_i(\upsilon_i - \bar{\upsilon})^2}{A}, \quad (13)$$

The sum of the squared temperature differences of each cell to the mean temperature normalised to the cell area is calculated. In the case of a constant temperature in a cross section, $\sigma = 0$, in all other cases $\sigma > 0$. If $e_{thermal}$ is negative, there is a less homogeneous temperature distribution at the outlet compared to the original temperature profile without heat pipe cooling. If no change in temperature homogeneity is calculated, $e_{thermal} = 0$. If, on the other hand, $0 < e_{thermal} < 1$, the thermal homogeneity at the outlet with the heat pipe is better than without the heat pipe. In the special case $e_{thermal} = 1$, the melt has a perfectly homogeneous temperature at the outlet with the heat pipe. When comparing the thermal homogeneity (Figure 11) it is noticeable that the addition of a heat pipe improves the thermal homogeneity at all outlets regardless of the position. As expected, the influence is greatest for any heat pipe position at outlet 3. In agreement with the average temperatures, the homogeneity increases with the melt-carrying surface influenced by a heat pipe. This leads to an $e_{thermal}$ of 0.53 for outlet 3 at heat pipe position 4. Moreover, the homogeneity for outlet 4 at position 4 is increasing. For outlet 1 and 2 the maximum homogeneity is reached at position 2.

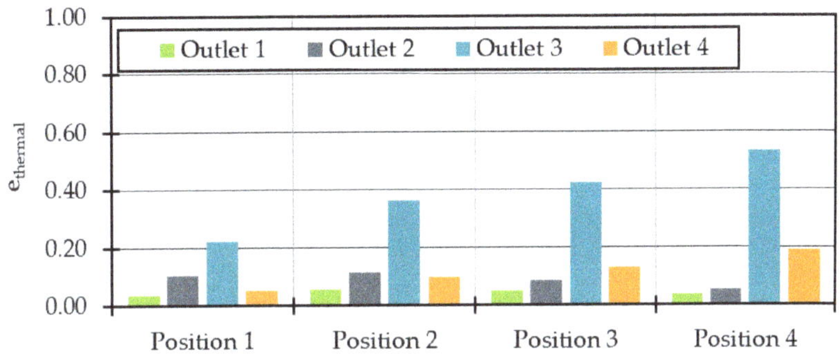

Figure 11. Thermal homogenisation indicator $e_{thermal}$ at each outlet for different heat pipe positions.

Finally, the temperature changes will influence the melt throughput distribution. Therefore, the mass throughput deviations with dependence on the heat pipe positions are shown in Figure 12. For a perfectly homogenous distribution, the deviation should be 0% at all outlets. It is obvious that as the influence of the heat pipe on the melt channels increase, the throughput at outlet 3 decreases. This was to be expected since the temperature and thus the shear viscosity drop. Since the viscosity describes the flow resistance, throughput decreases at lower viscosities or higher flow resistances. However, the heat pipe influence at position 4 is so great that the system is overcompensating. The mass flow deviation at outlet 3 drops into the negative.

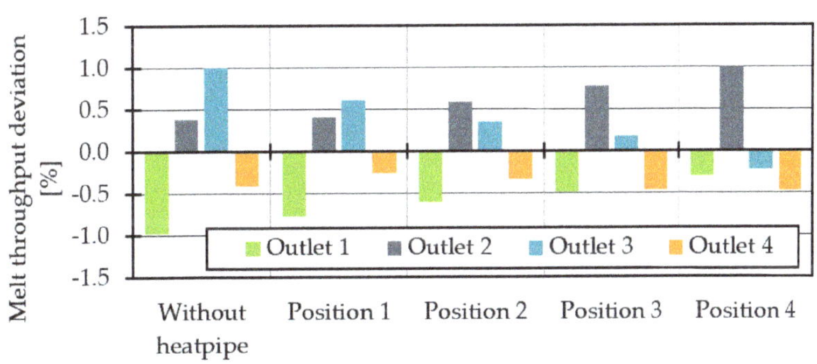

Figure 12. Melt throughput deviation at each outlet for different heat pipe positions.

All in all, the first simulations show that it is possible to influence the temperature and melt distribution with heat pipes. In terms of thermal homogeneity, positions 3 and 4 seem to be particularly promising. In the following, the heat transfer capacity of the heat pipes at position 3 and 4 is varied, starting with an increasing cooling rate at position 3 from −25 K/s to −35 K/s, which corresponds to the heat transfer capacity of 44.4 W and 62.1 W. The increasing heat transfer capacity has the effect of a further temperature drop at outlet 3 (Figure 13). At the same time, the temperature at outlet 2 rises. One explanation is a mass throughput redistribution (Figure 14). Due to the lower temperature at outlet 3, the flow resistance at that outlet increases, caused by higher viscous forces. Therefore, the output at outlet 2 increases, which is leading to an increasing viscous shear heating at that very same outlet. As shown in Figure 14, the increasing cooling rate leads to a linearly lower mass flow at outlet 3. At the same time, the melt is redistributed to outlets 1 and 2 where the melt flow also increases linearly. The mass flow at outlet 4 remains nearly unaffected. The parameters of the linearisation in Equation (14) are given in Table 1 and could be further investigated in the future to predict the required heat pipe performances to achieve optimal mass flow distributions for different process points.

$$\text{Melt throughput deviation} = \text{Cooling rate} \times x + y, \tag{14}$$

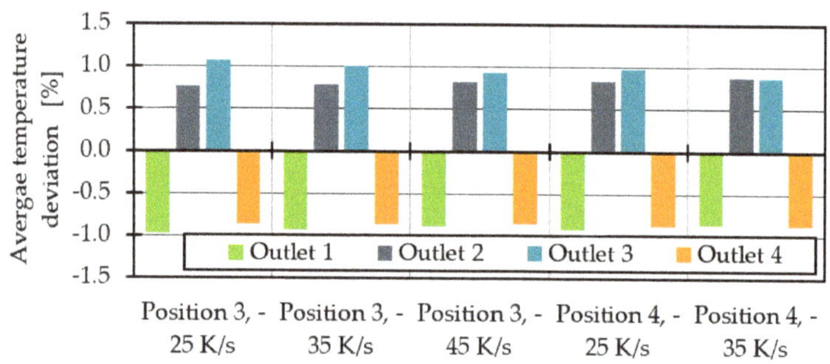

Figure 13. Average temperature deviation for different heat pipe performances and positions.

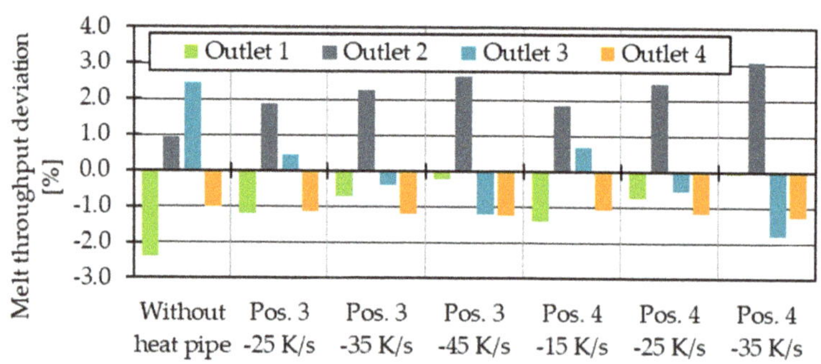

Figure 14. Melt throughput deviation for different heat pipe performances and positions.

Table 1. Linearisation parameters to predict the melt throughput deviation in case of one heat pipe at position 3.

Outlet	x	y	R^2
1	0.49	−1.69	1
2	0.39	1.5	1
3	−0.82	1.25	1
4	−0.06	−1.06	0.9996

A similar, slightly weaker effect can be observed for the average temperatures (Figure 13). The temperature change dependent on the cooling rate can also be linearised (Equation (15)). The corresponding parameters are given in Table 2. Due to the small number of process points on which the linearisation is based, this correlation should be confirmed by further investigation.

$$\Delta T_{average} = \text{Cooling rate} \times x + y, \tag{15}$$

Table 2. Linearisation parameters to predict the average outlet temperatures in case of one heat pipe at position 3.

Outlet	x	y	R^2
1	0.041	−1.02	1
2	0.023	0.75	0.9999
3	−0.070	1.14	1
4	0.005	−0.87	0.9998

The analysis of the simulation results with a heat pipe at position 4 shows again the larger heat flux surface at this heat pipe position. The melt flows are increasing and decreasing more strongly, but the same effects tend to be observed. With a cooling rate of −35 K/s, it is possible to equalise the average temperatures in outlets 2 and 3. To achieve similar average temperatures at heat pipe position 3, the cooling rate was raised to −45 K/s (79.9 W). The remaining temperature difference between outlet 2 and 3 is 0.2 K. In addition to mass flow redistribution, the decreasing average temperatures at outlet 3 result in increasing thermal homogeneity (Figure 15). The heat pipes at position 3 seem to result in a more homogeneous temperature distribution at all outlets compared to position 4.

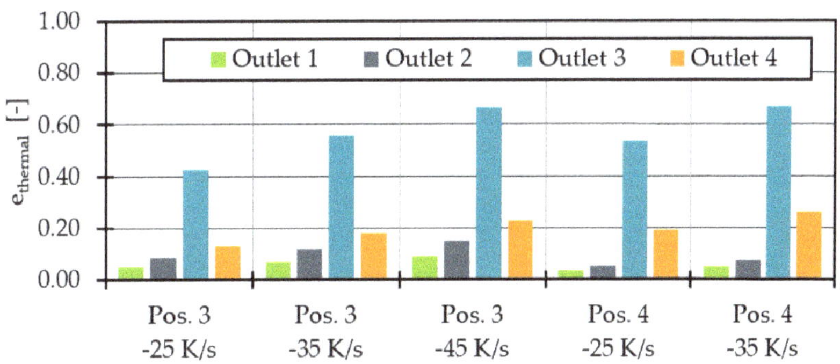

Figure 15. Thermal homogenisation indicator for different heat pipe performances and positions.

As the output at position 2 represents a new, excessively high mass flow, the integration of another heat pipe could be helpful to homogenise the temperature and mass flow distribution. Therefore, the integration of another heat pipe is tested by integrating two heat pipes symmetrically, as shown in Figure 16. To cool down the temperature at outlets 2 and 3, and since the average temperature at outlet 3 is higher than at outlet 2 (Figure 6), the cooling rates of the heat pipes at position 3′ are set lower than at position 3. The difference between the two positions is 10 K/s and 20 K/s, respectively. Maximum cooling rates of up to −70 K/s are simulated, with all cooling rates listed in Table 3. The extrusion operating point remains the same.

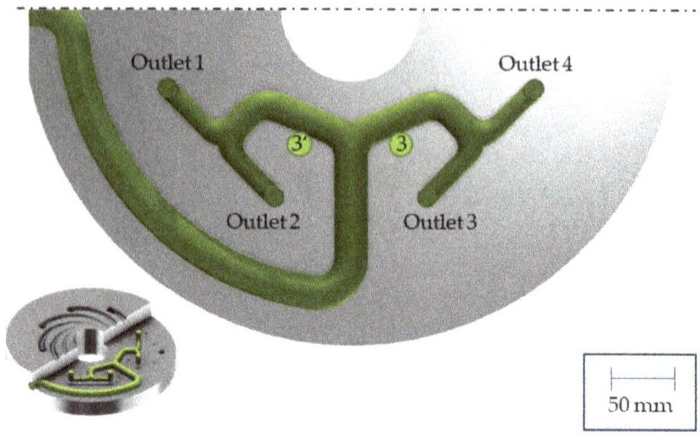

Figure 16. Heat pipe position for two heat pipes.

Table 3. Heat pipe performances and process point for simulations with two heat pipes.

Material	Throughput	Melt Temperature	Die Temperature	Support Air Temperature	Cooling Rate Position 3	Cooling Rate Position 3′
LDPE 2101N0W	100 kg/h	453.15 K	473.15 K	298.15 K	−25 K/s −35 K/s −45 K/s −50 K/s −60 K/s −70 K/s	−15 K/s −25 K/s −35 K/s −30 K/s −40 K/s −50 K/s

By integrating a second heat pipe, the temperature at outlet 2 decreases as expected (Figure 17). By increasing the cooling rates, the average temperatures converge further. A maximum cooling rate at which the average outlet temperatures fall below the optimum temperature is not yet reached in the process points investigated. For all selected cooling rates, the difference between the average temperature at outlet 2 and 3 remains between 0.8 K and 1 K. The temperature differences seem to tend to be slightly higher for the cooling rate difference of 20 K/s.

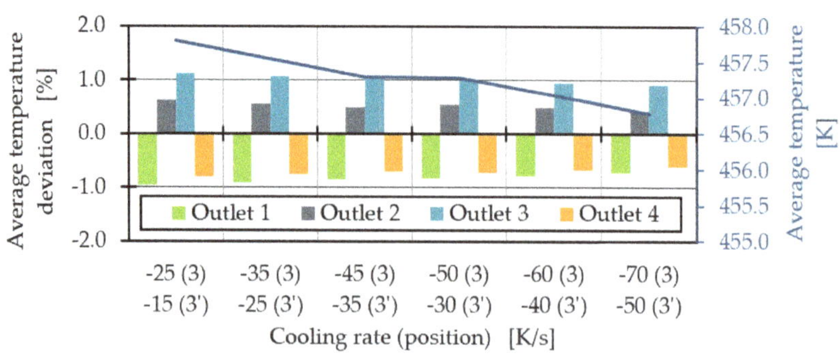

Figure 17. Average temperature deviation for two heat pipes with different performances at position 3 and 3′.

By increasing the cooling rate difference between position 3 and 3′, the average temperatures converge further. The average temperature deviation for the cooling rate combination of −60 K/s and −40 K/s is less than 1%.

As shown in Figure 18, thermal homogenisation is improving for all outlets when increasing the cooling rates. Moreover, higher cooling rate differences between position 3 and 3′ seem to result in more homogenous temperatures at all outlets. The most homogenous process point is reached with a cooling rate of −70 K/s at position 3 and −50 K/s at position 3′. The mass flow deviation (Figure 19) shows that the mass flow distribution is significantly improved by the heat extraction of the two heat pipes at position 3 and 3′ compared to the state without heat pipes. The mass flow deviation for all test points with two heat pipes is less than 1%. Compared to one heat pipe (Figure 14), the mass flow distribution is no longer overcompensated and significantly improved. Especially at cooling rates of −45 K/s at position 3 and −35 K/s at position 3′, to then −50 K/s at position 3 and −30 K/s at position 3′, respectively, the mass flow distribution is excellent. At higher cooling rates, the throughput deviation becomes worse again for the process point investigated but is still acceptable compared to the distribution without heat pipes.

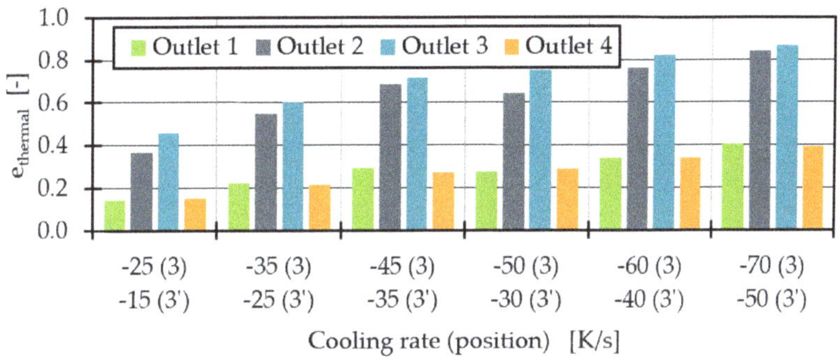

Figure 18. Thermal homogenisation indicator for two heat pipes with different performances at position 3 and 3′.

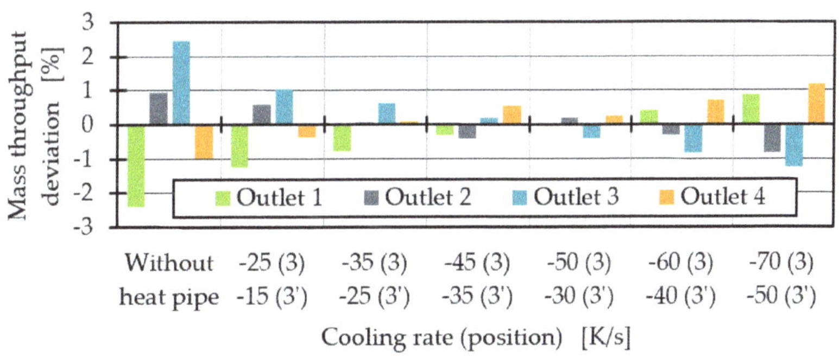

Figure 19. Average throughput deviation for two heat pipes with different performances at position 3 and 3′.

4. Conclusions

Inhomogeneous die temperatures lead to an inhomogeneous temperature distribution at pre-distributor outlets, and thus to an inhomogeneous distribution of mass flow over the pre-distributor outlets. This is caused by the externally applied heating and simultaneous flow of cold blowing air through the die centre as well as local shear heating of the melt. Since the temperature has a major influence on the polymer viscosity, it affects not only the flow resistances and thus the mass flow distribution, but also the extensional viscosity in the film bubble. During inflation, the melt parts of lower extensional viscosity can be stretched more easily, resulting in thin spots in the film. To avoid an inhomogeneous film thickness distribution and to achieve good film qualities, thermal homogenisation of the melt is necessary in the pre-distributor. Therefore, a new approach for cooling hot spots with heat pipes was investigated. Our simulations show that heat pipes can be used to influence melt temperatures at those locations where a temperature reduction is required. Two suitable heat pipe positions were identified that improve the homogeneity of the melt temperature by using only one heat pipe per die half. In addition to the heat pipe position, the heat pipe performance has a major influence on the temperature distribution and the melt flow distribution. With one heat pipe with a performance of 79.9 W the average temperature drops, but no thermal homogenization is achieved. Instead, the originally maximum flow at one of the outlets falls below the optimal average flow per outlet, and the melt flow deviation is overcompensated. Furthermore, the simulations show that both the average temperatures and the melt throughput deviation can be predicted with the cooling rates. Since the outlets of a pre-distribution die interact, one heat pipe is not sufficient to fully homogenise the temperature at each outlet to similar temperatures. Cooling the melt stream with the highest temperature creates a new maximum temperature and melt flow at another outlet. This changes as soon as a second heat pipe is integrated to cool the new maximum melt flow outlet. With heat pipe cooling of −60 K/s and −40 K/s, a maximum average temperature deviation and melt flow deviation of less than 1% can be achieved for all outlets. Whether our assumption that two heat pipes are sufficient for thermal homogenisation and also applies to other process points must be investigated in further simulation studies. For other process points, at least one heat pipe per outlet could be required in perspective in order to equalise the temperature at all outlets independently of the process point. Of course, the heat pipe performances would still have to be adjustable via cooling and set depending on the process point.

Due to the small number of process points investigated, the observed linear correlation between the cooling rate and the resulting average temperature and melt flow deviation should be confirmed by further CFD simulations, but also in practical tests. For future practical tests, further investigations are necessary to also develop a prediction of the average temperature and melt flow deviation for systems with more than one heat pipe per die half. For the practical tests, a 2^3 pre-distributor will be retrofitted with heat pipes at the simulated suitable positions. An external air cooling system will be developed to cool the heat pipes. For this purpose, bent heat pipes will be used coming out of the side of the die where a pipe system with cool air flow will cool them. One challenge will be the isolation against die surfaces where cooling through the heat pipes is not desired. Another limitation is the available space around the pre-distributor in industrial blown film extrusion lines. In addition, the spiral die also has an influence on the temperature deviation. Another simulation setup with a spiral die could help to estimate the thermal influence in the downstream flow channel.

Author Contributions: Conceptualization, L.L.; methodology, L.L. and L.W.; software, M.S.; formal analysis, L.L.; investigation, L.L. and L.W.; data curation, L.L. and L.W.; writing—original draft preparation, L.L.; writing—review and editing, M.S. and C.H.; visualization, L.L.; supervision, M.S. and C.H.; project administration, M.S.; funding acquisition, C.H. All authors have read and agreed to the published version of the manuscript.

Funding: This research was funded as part of the "Industrielle Gemeinschaftsforschung und -entwicklung (IGF)" by the German Bundesministerium für Wirtschaft und Klimaschutz (BMWK) due to an enactment of the German Bundestag through the AiF, grant number 21776 N.

Institutional Review Board Statement: Not applicable.

Informed Consent Statement: Not applicable.

Data Availability Statement: The data that support the findings of this study are available from the corresponding author, L.L., upon reasonable request.

Acknowledgments: The authors acknowledge the generous donation of raw materials by SABIC Europe B.V. and would like to extend their thanks to these parties.

Conflicts of Interest: The authors declare no conflict of interest.

References

1. PlasticsEurope. Plastics—The Facts. 2020. Available online: https://plasticseurope.org/de/wp-content/uploads/sites/3/2021/11/Plastics_the_facts-WEB-2020_versionJun21_final-1.pdf (accessed on 1 March 2022).
2. IK Industrievereinigung Kunststoffverpackungen e.V. IK Jahresbericht 2021, Bad Homburg. 2021. Available online: https://kunststoffverpackungen.de/wp-content/uploads/2021/10/IK_Jahresbericht_2021_online.pdf (accessed on 7 March 2022).
3. Osswald, T.A. *Polymer Processing: Modeling and Simulation*; Hanser Publishers: Munich, Germany, 2006; ISBN 3-446-40381-7.
4. Schrenk, W.J.; Alfrey, T. Coextruded Multilayer Polymer Films and Sheets. In *Polymer Blends*; Paul, D.R., Newman, S., Eds.; Academic Press: New York, NY, USA, 1978; pp. 129–165. ISBN 9780125468022.
5. Yesildag, N. *Simulative Auslegung von Thermisch-Rheologisch Homogenen Vorverteilern in Wendelverteilerwerkzeugen*; RWTH Aachen University, Diss.; Günter Mainz: Aachen, Germany, 2017; ISBN 9783958861718.
6. Sidiropoulos, V.; Vlachopoulos, J. Temperature gradients in blown film bubbles. *Adv. Polym. Technol.* **2005**, *24*, 83–90. [CrossRef]
7. Vergnes, B.; Saillard, P.; Agassant, J.F. Non-isothermal flow of a molten polymer in a coat-hanger die. *Polym. Eng. Sci.* **1984**, *24*, 980–987. [CrossRef]
8. Carneiro, O.; Nóbrega, J.M.; Pinho, F.T.; Oliveira, P.J. Computer aided rheological design of extrusion dies for profiles. *J. Mater. Process. Technol.* **2001**, *114*, 75–86. [CrossRef]
9. Seibel, S. *Konzeptionelle Aspekte bei der Gestaltung und Auslegung von Extrusionswerkzeugen*; Paderborn, University, Diss.; Shaker: Aachen, Germany, 2007; ISBN 9783832265526.
10. Blömer, P. *Untersuchung und Weiterentwicklung von Berechnungsverfahren für Strömungen in Wendelverteilern: = Investigation and Further Development of Calculation Methods for Flow in Spiral Mandrel Dies*; RWTH Aachen University, Diss.; Mainz: Aachen, Germany, 2006; ISBN 3861308517.
11. Sun, Y.; Gupta, M. An analysis of the effect of elongational viscosity on the flow in a spiral-mandrel die. *Adv. Polym. Technol.* **2006**, *25*, 90–97. [CrossRef]
12. Zatloukal, M.; Tzoganakis, C.; Perdikoulias, J.; Saha, P. Numerical simulations of polymer flow in flat spiral dies. *Polym. Eng. Sci.* **2001**, *41*, 1683–1694. [CrossRef]
13. Skabrahova, P.; Svabik, J.; Perdikoulias, J. A Non-Isothermal 3D Fem Study of Spiral Mandrel Dies with Non-Symmetrical Input. In *ANTEC-CONFERENCE PROCEEDINGS-*; 2003; Volume 1, pp. 305–309.
14. Pittman, J.F.T. Computer-aided design and optimization of profile extrusion dies for thermoplastics and rubber: A review. *Part E J. Process Mech. Eng.* **2001**, *225*, 280–321. [CrossRef]
15. Burmann, G.; Fischer, P.; Wortberg, J. Schmelzefluss im Circularverteiler. *Kunststoffe* **2010**, *100*, 150–154.
16. Burmann, G.; Wortberg, J. (Eds.) Grundlagen und Strategie bei der Werkzeugauslegung. In Proceedings of the IKV Seminar—Auslegung von Schnecke und Werkzeugen in der Extrusion, Institute for Plastics Processing in Industry and Craft at RWTH Aachen University, Aachen, Germany, 26–27 April 2004.
17. Burmann, G.; Te Heesen, O. (Eds.) Verweilzeitoptimierte Extrusionswerkzeuge.: Auslegung Unter Berücksichtigung Verfahrens- und Konstruktionstechnischer Einflüsse. In Proceedings of the 10th Duisburger Extrusionstagung, Duisburg, Germany, 19–20 March 2013.
18. Saul, K. Automatisierte Auslegung von Extrusionswerkzeugen. Doctoral Disertation, Universität Duisburg-Essen, Duisburg, Germany, 2011.
19. te Heesen, O. Ganzheitlicher Ansatz zur Dimensionierung und Optimierung von Extrusionswerkzeugen am Beispiel von Wendelverteilern. Ph.D. Thesis, Universität Duisburg-Essen, Duisburg, Germany, 2015.
20. Hopmann, C. *Integrative Simulation des Temperatureinflusses in Vorverteilern von Wendelverteilerwerkzeugen*; Final Report on the AiF Project 17645 N; Institute for Plastics Processing in Industry and Craft at RWTH Aachen University: Aachen, Germany, 2015.
21. Hopmann, C. *Homogenisierung der Vorverteilung in Wendelverteilerwerkzeugen Mittels Verstellbarer Fließwiderstände*; Final Report on the AiF Project 19777 N; Institute for Plastics Processing in Industry and Craft at RWTH Aachen University: Aachen, Germany, 2019.
22. Michaeli, W.; Hopmann, C. *Extrusion Dies for Plastics and Rubber*, 4th ed.; Hanser Publishers: Munich, Germany, 2016.
23. Groll, M.; Zimmermann, P. Kenngrößen zum Beurteilen von Wärmeträgern fur Wärmerohre. *Chemie-Ing.-Techn.* **1969**, 1294–1300. [CrossRef]
24. SITUS Technicals GmbH. Heatpipe OderWärmerohr. Available online: https://www.situs-tec.de/app/download/5781352533/W%C3%A4rmerohr.pdf (accessed on 1 March 2022).
25. Skrabek, E.A. *Heat Pipe Design Handbook, Part 1*; NASA Technical Reports Server (NTRS): Washington, DC, USA, 1972.
26. VDI EVE. *VDI-Wärmeatlas: 11., Bearbeitete und Erweiterte Auflage*; Verein Deutscher Ingenieure, Ed.; Springer: Berlin/Heidelberg, Germany, 2013; ISBN 978-3-642-19980-6.
27. Kartelmeyer, S.; Hüttemann, M.; Moritzer, E.; Jaroschek, C. Werkzeugkonstruktion mit Heatpipes. In *Plastverarbeiter*; Hanser: München, Germany; pp. 20–25.
28. Kartelmeyer, S.; Hüttemann, V.; Jaroschek, C. "No Cool"-Werkzeuge-richtig-gerechnet: Simulation von Wärmeströmen in Spritzgießwerkzeugen mit Heatpipes. In *Kunststoffe*; Hanser: München, Germany; pp. 44–48.

29. Lakemeyer, C.; Schöppner, V. Temperature Calculation for Extruder Screws with Internal Heat pipes. In Proceedings of the 29th International Conference of the Polymer Processing Society, Nuremberg, Germany, 15–19 July 2013; AIP Publishing: New York, NY, USA, 2014.
30. Khadra, K.; Angot, P.; Parneix, S.; Caltagirone, J.-P. Fictitious domain approach for numerical modelling of Navier-Stokes equations. *Int. J. Numer. Methods Fluids* **2000**, *34*, 651–684. [CrossRef]
31. Patankar, S.; Spalding, D. A calculation procedure for heat, mass and momentum transfer in three-dimensional parabolic flows. *Int. J. Heat Mass Transf.* **1972**, *15*, 1787–1806. [CrossRef]
32. Versteeg, H.K.; Malalasekera, W. *An Introduction to Computational Fluid Dynamics: The Finite Volume Method*, 2nd ed.; Pearson/Prentice Hall: Harlow, UK, 2007; ISBN 9780131274983.
33. *Evaluation of Mesh Interface and Immersed Boundary Models for the Optimisation of Mixing Elements*; Hopmann, C.; Schön, M. (Eds.) ANTEC: Fremont, CA, USA, 2020.
34. Becker, M. *ViscousHeatingSolver: Adding Temperature Transport and Viscous Heating to SimpleFoam*; DHCAE Tools GmbH: Krefeld, Germany, 2012.
35. *Saechtling Kunststoff Taschenbuch, 31. Auflage*; Hanser: München, Germany, 2013; ISBN 9783446434424.
36. Rodgers, A. Boyd-Copper-Water Heat Pipes. Available online: https://info.boydcorp.com/hubfs/Thermal/Two-Phase-Cooling/Boyd-Copper-Water-Heat-Pipes.pdf (accessed on 2 March 2022).
37. Hopmann, C.; Schön, M.; Theunissen, M.; Meiners, W.; Shen, L. Simulative evaluation of the Optimization Potential of Additively Manufactured Static Mixing Elements for Extrusion. In Proceedings of the 33rd International Conference of the Polymer Processing Society, Cancun, Mexico, 10–14 December 2017.
38. Evaluation of the Thermal Homogenising Effect of Mixing Elements for Single-Screw Extrusion. In Proceedings of the 56th Annual Technical Conference of the Society of Plastic Engineers (ANTEC), Atlanta, GA, USA, 10–12 April 1998.

polymers

MDPI

Article

Leakage-Flow Models for Screw Extruders

Christian Marschik [1,2,*], Wolfgang Roland [1], Marius Dörner [3], Georg Steinbichler [1] and Volker Schöppner [3]

[1] Institute of Polymer Extrusion and Compounding, Johannes Kepler University Linz, 4040 Linz, Austria; wolfgang.roland@jku.at (W.R.); georg.steinbichler@jku.at (G.S.)
[2] Competence Center CHASE GmbH, 4040 Linz, Austria
[3] Kunststofftechnik Paderborn, Paderborn University, 33098 Paderborn, Germany; marius.doerner@ktp.uni-paderborn.de (M.D.); Volker.Schoeppner@ktp.upb.de (V.S.)
* Correspondence: christian.marschik@jku.at; Tel.: +43-732-2468-6574

Abstract: Many theoretical analyses of extrusion ignore the effect of the flight clearance when predicting the pumping capability of a screw. This might be reasonable for conventional extruder screws with "normal" clearances but leads to errors when more advanced screw designs are considered. We present new leakage-flow models that allow the effect of the flight clearance to be included in the analysis of melt-conveying zones. Rather than directly correcting the drag and pressure flows, we derived regression models to predict locally the shear-thinning flow through the flight clearance. Using a hybrid modeling approach that includes analytical, numerical, and data-based modeling techniques enabled us to construct fast and accurate regressions for calculating flow rate and dissipation rate in the leakage gap. Using the novel regression models in combination with network theory, the new approximations consider the effect of the flight clearance in the predictions of pumping capability, power consumption and temperature development without modifying the equations for the down-channel flow. Unlike other approaches, our method is not limited to any specific screw designs or processing conditions.

Keywords: leakage flow; extrusion; modeling and simulation; polymer processing

Citation: Marschik, C.; Roland, W.; Dörner, M.; Steinbichler, G.; Schöppner, V. Leakage-Flow Models for Screw Extruders. *Polymers* **2021**, *13*, 1919. https://doi.org/10.3390/polym13121919

Academic Editors: Salah Aldin Faroughi, Luís L. Ferrás, Alexandre M. Afonso and Célio Bruno Pinto Fernandes

Received: 4 May 2021
Accepted: 7 June 2021
Published: 9 June 2021

1. Introduction

Plasticating extruders abound in the polymer industry. Due to their great versatility, they are used in many polymer-shaping operations, producing semi-finished plastic products such as films, pipes, profiles, sheets, and fibers. In addition, they are frequently found in compounding and recycling operations. Although plasticating extruders come in various designs, the elementary processing steps are generally the same: (i) transport and (ii) melting of particulate solids, followed by (iii) mixing and (iv) pumping of the polymer melt.

Numerous theoretical studies have modeled the extrusion process to increase the understanding of the transport mechanisms governing physical operation. The fundamentals of extrusion modeling were summarized in various books: Tadmor and Klein [1], White and Potente [2], Tadmor and Gogos [3], Campbell and Spalding [4], Rauwendaal [5], Agassant et al. [6], and others. Recently, Wilczyński et al. [7] presented a thorough review of global extrusion modeling.

The assessment of leakage flow and its prediction has been of interest since the earliest theories for the metering zone. Over the past decades, many analyses have modeled polymer-melt flows in single-screw extruders. Most of these, however, ignored the effect of flight clearance.

1.1. Analysis of Flow in Metering Channels

Early theoretical analyses that investigated melt conveying in single-screw extruders dealt with one- and two-dimensional flows of temperature-independent Newtonian fluids. These problems have exact analytical solutions for the drag and pressure flows in the cross- and down-channel directions, respectively. The first model of screw viscosity pumps was published anonymously [8] and later extended by Rowell and Finlayson [9]. Similarly, Carley et al. [10] proposed a simplified flow theory for screw extruders. Mohr et al. [11,12] investigated the characteristics of the transverse flow in the screw channel. Both the complexity and the accuracy of the analysis increase when the non-Newtonian flow behavior of polymer melts is included. It is well known that polymer melts are shear-thinning fluids whose viscosity decreases with increasing shear rate. Pseudo-plastic behavior complicates the mathematical model such that the governing flow equations must be solved numerically, and exact closed-form analytical solutions are no longer possible. The viscosity being dependent on the shear rate, the drag and pressure flows are coupled. For multidimensional flows, complexity is increased further by the combined effect of shear in the down- and cross-channel directions. To gain insights into how a material's shear-thinning nature affects melt conveying in single-screw extruders, several authors presented numerical solutions for one- and two-dimensional flows of power-law fluids in metering channels. Rotem and Shinnar [13] obtained numerical results for a one-dimensional flow under isothermal conditions. Including the effect of transverse flow, Griffith [14], Zamodits and Pearson [15], and Karwe and Jaluria [16] presented numerical solutions for shear-thinning flows in infinitely-wide screw channels.

With the development of more advanced computers, a trend emerged towards a numerical analysis of three-dimensional flows in metering channels. Including non-Newtonian and non-isothermal effects, a number of studies provided detailed insights into the recirculating extruder channel flow. Spalding et al. [17] applied the finite-element method to examine the flow in a helical metering section. Ghoreishy et al. [18] presented a non-isothermal analysis of elastomer melt flow in unwound screw channels. Polychronopoulos and Vlachopoulos [19] analyzed the development of Moffat Eddies in the flight root corners of the screw channel. Vachagina et al. [20] investigated the non-isothermal conveying behavior of a Giesekus fluid in a helical screw section.

To avoid numerical procedures, various approximate solutions for shear-thinning flows were developed. Booy [21] employed the classical Newtonian pumping model to derive effective viscosity values and approximate shear-thinning flow behavior. Rauwendaal [22] proposed correction factors to the drag- and pressure flows in the tra-ditional model to include the shear-rate-dependent flow behavior of polymer melts. Similarly, Spalding et al. [23] introduced a correction factor for the drag flow.

Rather than presenting corrections to the traditional Newtonian theory, we pro-posed an alternative approach for power-law fluids. Using a hybrid modeling approach that combines analytical, numerical, and data-based modeling techniques, we approximated a large number of numerical solutions of scaled flow equations obtained for two- and three-dimensional formulations. Assuming two-dimensional flows of power-law fluids in infinitely-wide screw channels, Pachner et al. [24] and Roland et al. [25,26] proposed regression models for the prediction of throughput and viscous dissipation. To include the influence of the flight flanks, we [27–30] extended the theories to three-dimensional flows. These widely applicable models increase prediction accuracy by including the combined effects of shear-thinning flow behavior, transverse flow, and screw flights.

1.2. Analysis of Leakage Flow

The influence of leakage flow on the pumping capability of extruder screws has been investigated since the earliest theories for the metering zone. Rowell and Finlayson [9] approximated the annular clearance between flight land and barrel surface by two parallel plates and assumed the leakage flow to be a pressure flow through an infinitely wide slit. Gore and McKelvey [31] extended the analysis by adding the effect of flight clearance on the drag flow. Mohr and Mallouk [12] presented a more sophisticated theory by considering the pressure flow caused by the cross-channel pressure gradient in the screw channel and the effect of flight clearance on the drag flow. A detailed review of the early Newtonian theories was provided by Tadmor and Klein [1]. For extruder screws with "normal" clearances, the velocity profile in the flight clearance was shown to be determined mainly by the drag-flow component and influenced only marginally by the transverse pressure gradient.

Taking the shear-thinning flow behavior of polymer melts into account, Rauwendaal and Housz [32] used a numerical approach to readdress the flow characteristics in the flight clearance. For large flight clearances and small power-law indices, a significant influence of the transverse pressure flow was identified. Further, the total power consumption of the screw was shown to be affected considerably by the power consumption in the clearance. To account for the effect of leakage on throughput, corrections were made to (i) the transverse flow in the screw channel and (ii) the net flow rate. Focusing on pressure-generating metering zones, their analysis omitted overridden functional zones that are subject to a negative axial pressure gradient.

Including non-isothermal effects, Meyer et al. [33] numerically evaluated the temperature development in the flight clearance. Taking a pure drag flow into account, the results indicated that the thermal development length is generally smaller than the available gap length. When predicting the velocity and temperature distribution at the exit of the flight clearance, the flow can therefore be considered fully developed. To investigate the effect of leakage flow on the temperature distribution in the screw channel, Pittman and Rashid [34] and Rauwendaal [35] carried out numerical analyses of two- and three-dimensional power-law-model-based flows.

1.3. Research Approach

This work presents a new modeling approach to including leakage effects in the analysis of melt-conveying zones. Rather than directly correcting the drag and pressure flows, we developed new analytical regressions to predict the flow rate and dissipation locally in the leakage gap. In the construction of the leakage-flow models, particular attention was paid to their usability in our screw-simulation routine for predicting the conveying behavior of melt-conveying zones. For detailed information, the reader is referred to the following articles and PhD-theses: Aigner [36], Marschik et al. [37–39], Roland et al. [40], and Luger [41].

The main idea of the simulation approach is based on network theory: To reduce the complexity of the flow situation, the screw channel is subdivided into very small interconnected segments of constant cross-section and physical parameters. These sections are represented by network elements that consist of a source and a resistance connected in parallel. Using melt-conveying models from the literature, the network elements describe the flow locally in the screw channel. Analogously to electrical circuits, the network elements are connected via nodal points to form an equivalent flow network, whose flow rates and pressure differences are evaluated by means of nodal analysis. In contrast to time-consuming numerical simulations based on computational fluid dynamics, the method iteratively solves a linearized set of network equations and is therefore significantly faster.

To illustrate the objective of this research, Figure 1 shows a flow network of a conventional metering zone, which consists of down- and cross-channel elements. Due to the changes in channel depth in the transverse direction, the latter is initialized with three network elements in series: (a) one from the channel center to the pushing flight, (b) one over the flight clearance, and (c) one from the trailing flight to the channel center one turn behind. The extrusion literature provides numerous theories for predicting the characteristics of the down-channel flow. Accurate models for the analysis of leakage effects, in contrast, remain elusive.

A flow chart of our work is given in Figure 2. First, we described mathematically the flow under consideration, converted the model into a dimensionless form, and carried out a detailed analysis of the governing equations. The dimensionless model was then solved numerically to evaluate flow rate and dissipation for a large number of physically independent setups. To remove the need for numerical methods, we derived approximate equations for flow rate and dissipation, using symbolic regression based on genetic programming. The structure of the regression models and their parameters were optimized iteratively until we obtained satisfying prediction accuracies. The new regressions were designed to be able to describe the properties of the network elements positioned in the flight gap.

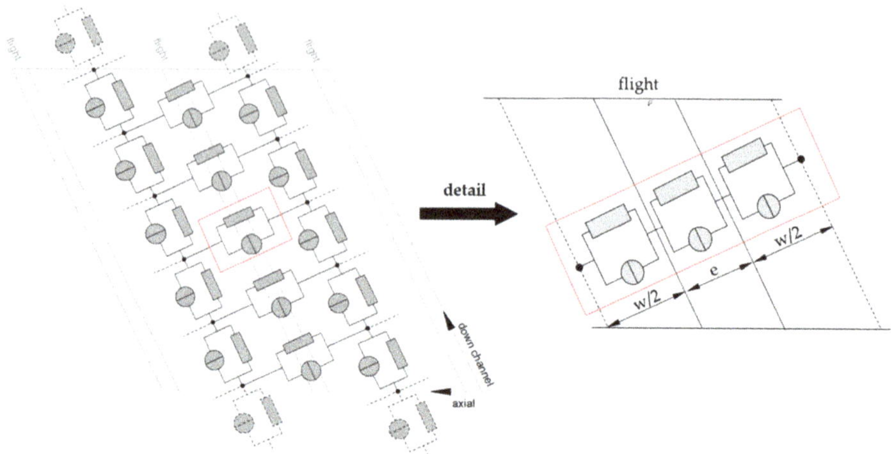

Figure 1. The development of leakage-flow models to predict flow rate and dissipation in the leakage gap of a discretized screw channel.

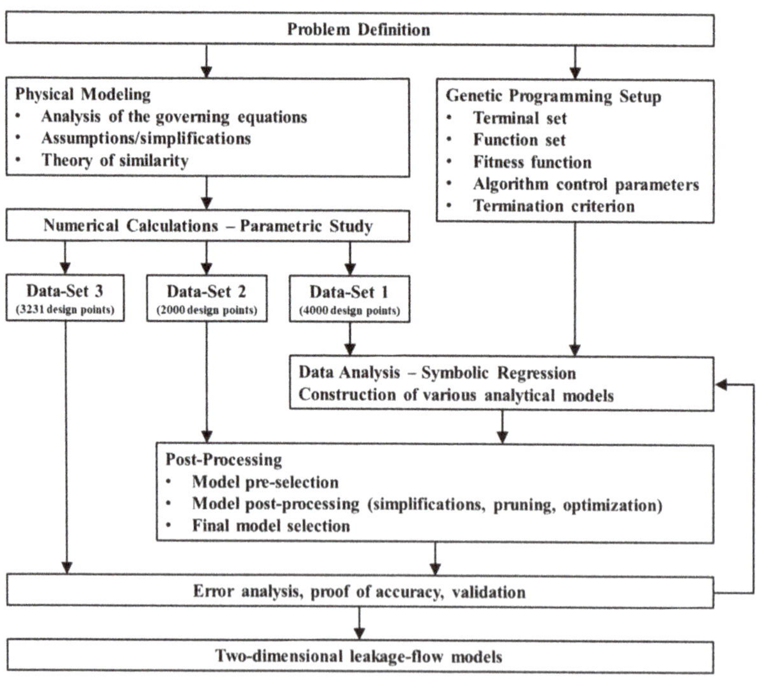

Figure 2. Schematic work-flow chart of the hybrid modeling approach, including analytical, numerical, and data-based modeling.

2. Model Development

We used a hybrid modeling approach to derive approximate equations for the flow rate and the dissipation in the flight clearance. The method incorporates (i) analytical, (ii) numerical, and (iii) data-based modeling techniques. A detailed description of the main characteristics of our hybrid modeling approach was provided in [42].

2.1. Analytical Modeling

2.1.1. Problem Definition

We applied the flat-channel approximation to simplify the helical reference frame of the screw. The screw channel is unwound and considered as a rectangular flow channel covered by an infinite flat plate. The top view of the unwound screw channel is shown in Figure 3. The simplified channel representation is based on a Cartesian coordinate system: x is the cross-channel direction and z the down-channel direction. Avoiding cylindrical coordinates, the effect of channel curvature is ignored. The validity of the flat-plate model was discussed critically by Sun and Rauwendaal [43] and Roland et al. [44] for Newtonian and non-Newtonian polymer melts, respectively. Both studies confirmed the correctness of the approximation for small channel-depth-to-diameter ratios. This geometrical precondition is particularly fulfilled in the flight clearance. For conventional extruder screws with standard clearances, the ratio is typically in the range of $\delta / D_b = 0.001$. As a result, the error introduced by unwinding the screw channel at the radial position of the barrel is small. The influence of channel curvature can hence be ignored. The following parameters were used to describe the geometry of the screw: D_b is the screw diameter, t the screw pitch, e the flight width, w the channel width, and φ_b the pitch angle.

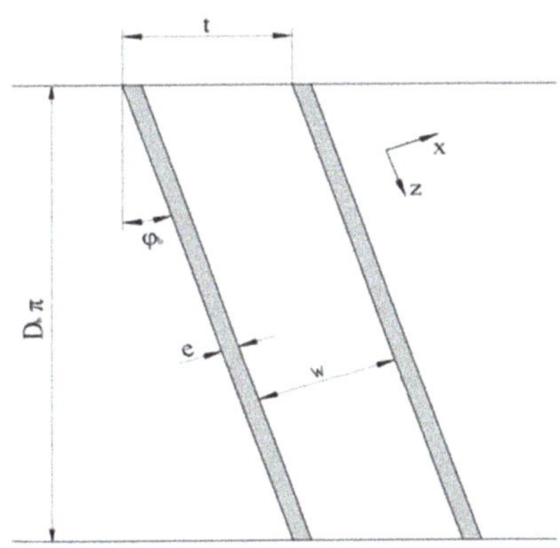

Figure 3. The top view of the unwound screw channel.

Drawing on the traditional pumping model, we reversed the kinematics of the process; that is, the screw is stationary, and the barrel moves at circumferential speed v_b:

$$v_b = D_b \, \pi \, N, \tag{1}$$

where N is the screw speed. The barrel velocity can be decomposed into a cross- and a down-channel component, $v_{b,x}$ and $v_{b,z}$:

$$v_{b,x} = v_b \, sin(\varphi_b); \qquad v_{b,z} = v_b \, cos(\varphi_b); \tag{2}$$

$$tan(\varphi_b) = \frac{t}{D_b \, \pi}. \tag{3}$$

Unlike previous studies that investigated the effect of the flight clearance [9,12,31,32], we locally analyzed the flow between flight land and barrel surface. Figure 4 illustrates the flat-plate model of the screw channel and the flight clearance.

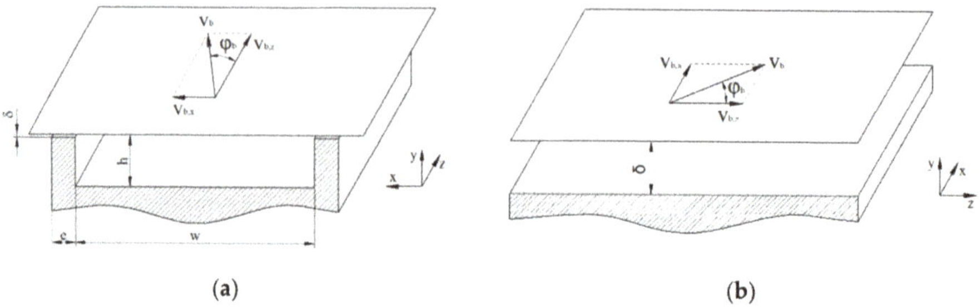

Figure 4. Flat-plate approximation of the unwound screw channel (**a**) and the flight clearance (**b**). In the representation of the flight clearance, the flow channel was rotated clockwise by 90°: x is the direction across and z the direction along the clearance.

The space through which the leakage flow occurs can be considered infinitely wide, resulting in a two-dimensional flow. Note that the coordinate system is the same in both cases. For convenience, however, the flow channel of the flight clearance was rotated clockwise by 90°: (i) x is the direction across the clearance, and (ii) z is the direction along the clearance. The channel gap between barrel surface and flight land is given by the clearance δ.

2.1.2. Governing Equations

In the first step, we derived the governing equations to mathematically describe the flow through the flight clearance. To this end, we made the following assumptions: (i) the flow is independent of time, fully developed, and isothermal, (ii) the fluid is incompressible and wall-adhering, and (iii) gravitational forces are ignored. A detailed discussion on the validity of these commonly-applied assumptions was provided by Tadmor and Klein [1]. The velocity field was reduced to a two-dimensional flow:

$$\mathbf{v} = \begin{pmatrix} v_x(y) \\ 0 \\ v_z(y) \end{pmatrix}. \tag{4}$$

Assuming a fully developed flow, the velocity components are functions of the channel-height coordinate y only; that is, the two-dimensional formulation is not capable of predicting any flow rate variations in the x- and z-directions. Meyer et al. [33] confirmed the validity of the assumption for both Newtonian and power-law fluids. Due to the low velocities and high viscosities of extruder flows, the Reynolds number is usually small (Re≪1), and the flow can be considered laminar [28,30]. Viscous forces being dominant over inertial forces, we reduced the problem to Stokes flow. By omitting time derivates, we ended up with the following simplified momentum equations:

$$\frac{\partial p}{\partial x} = \frac{\partial \tau_{yx}}{\partial y}. \tag{5}$$

$$\frac{\partial p}{\partial z} = \frac{\partial \tau_{yz}}{\partial y}. \tag{6}$$

In our isothermal theory, the momentum equations are uncoupled from the energy equation, which allowed us to evaluate viscous dissipation from the velocity field subject to the following boundary conditions:

$$\mathrm{v}_x(y=0)=0. \qquad \mathrm{v}_x(y=\delta)=\mathrm{v}_{b,x}. \tag{7}$$

$$\mathrm{v}_z(y=0)=0. \qquad \mathrm{v}_z(y=\delta)=\mathrm{v}_{b,z}. \tag{8}$$

The rheological behavior of the polymer melt was expressed by a nonlinear constitutive equation that relates the stress tensor $\boldsymbol{\tau}$ to the rate-of-deformation tensor $\boldsymbol{D}$, which is given by the symmetric part of the velocity-gradient tensor $\boldsymbol{L}$:

$$\boldsymbol{\tau}=2\,\eta\,\boldsymbol{D}. \tag{9}$$

$$\boldsymbol{D}=\frac{1}{2}\left(\boldsymbol{L}+\boldsymbol{L}^{\mathrm{T}}\right). \qquad \boldsymbol{L}=\nabla\mathbf{v}. \tag{10}$$

For most polymer melts, the viscosity is strongly dependent on the velocity gradients in the flow field, rendering them shear thinning. To describe the shear-rate dependency of the viscosity, we applied a power-law model, where K is the consistency and n the power-law index. The magnitude of shear rate $|\dot{\gamma}|$ is related to the second invariant of the rate-of-deformation tensor:

$$\eta=K|\dot{\gamma}|^{n-1}. \tag{11}$$

$$|\dot{\gamma}|=\sqrt{2(\boldsymbol{D}:\boldsymbol{D})}=\left[\left(\frac{\partial \mathrm{v}_x}{\partial y}\right)^2+\left(\frac{\partial \mathrm{v}_z}{\partial y}\right)^2\right]^{\frac{1}{2}}. \tag{12}$$

With these definitions, the shear stresses were obtained from:

$$\tau_{yx}=\eta\frac{\partial \mathrm{v}_x}{\partial y}=K\left[\left(\frac{\partial \mathrm{v}_x}{\partial y}\right)^2+\left(\frac{\partial \mathrm{v}_z}{\partial y}\right)^2\right]^{\frac{n-1}{2}}\frac{\partial \mathrm{v}_x}{\partial y}. \tag{13}$$

$$\tau_{yz}=\eta\frac{\partial \mathrm{v}_z}{\partial y}=K\left[\left(\frac{\partial \mathrm{v}_x}{\partial y}\right)^2+\left(\frac{\partial \mathrm{v}_z}{\partial y}\right)^2\right]^{\frac{n-1}{2}}\frac{\partial \mathrm{v}_z}{\partial y}. \tag{14}$$

Finally, the momentum equations were rewritten as:

$$\frac{\partial p}{\partial x}=\frac{\partial}{\partial y}\left\{K\left[\left(\frac{\partial \mathrm{v}_x}{\partial y}\right)^2+\left(\frac{\partial \mathrm{v}_z}{\partial y}\right)^2\right]^{\frac{n-1}{2}}\frac{\partial \mathrm{v}_x}{\partial y}\right\}. \tag{15}$$

$$\frac{\partial p}{\partial z}=\frac{\partial}{\partial y}\left\{K\left[\left(\frac{\partial \mathrm{v}_x}{\partial y}\right)^2+\left(\frac{\partial \mathrm{v}_z}{\partial y}\right)^2\right]^{\frac{n-1}{2}}\frac{\partial \mathrm{v}_z}{\partial y}\right\}. \tag{16}$$

These coupled nonlinear partial differential equations are the governing equations of our model. In combination with the boundary conditions in Equations (7) and (8), the equation system describes the velocity field of the two-dimensional flow. Since, generally, valid closed-form analytical solutions are unknown, the shooting method was used to calculate the velocity field and the volume flow rate $\dot{V}$ per unit width and total viscous dissipation per unit area ($\dot{q}_{diss}=\boldsymbol{\tau}:\boldsymbol{L}$):

$$\dot{V}=\int_0^\delta \mathrm{v}_x(y)dy. \tag{17}$$

$$\dot{q}_{diss} = \eta \left[\left(\frac{\partial \mathrm{v}_x}{\partial y} \right)^2 + \left(\frac{\partial \mathrm{v}_z}{\partial y} \right)^2 \right]. \tag{18}$$

$$\dot{Q}_{diss} = \int_0^\delta \dot{q}_{diss}(y) dy. \tag{19}$$

2.1.3. Theory of Similarity

Before the flow equations were solved numerically, the model was converted into a dimensionless form, using the theory of similarity, for the following reasons [26–30]: (i) Two systems described by the same dimensionless quantities are similar; (ii) varying any one of the independent input parameters changes the physical conditions of the flow; (iii) solutions obtained for a specific set of dimensionless input parameters apply to all dimensional variations that yield the same dimensionless input parameters.

We introduced the following dimensionless spatial coordinate and fluid velocities:

$$\xi = \frac{y}{\delta}. \qquad v_x = \frac{\mathrm{v}_x}{\mathrm{v}_{b,x}} \qquad v_z = \frac{\mathrm{v}_z}{\mathrm{v}_{b,x}}. \tag{20}$$

These definitions were used to derive dimensionless parameters for shear rate and viscosity in the flight clearance:

$$\eta_\delta^* = \frac{\eta\ \delta^{n-1}}{K\ \mathrm{v}_{b,x}^{n-1}} = \left| \dot{\gamma}^* \right|_\delta^{n-1} = \left[\left(\frac{\partial v_x}{\partial \xi} \right)^2 + \left(\frac{\partial v_z}{\partial \xi} \right)^2 \right]^{\frac{n-1}{2}}. \tag{21}$$

Similarly, the momentum equations were transformed into dimensionless form:

$$6\ \Pi_{p,x}^\delta = \frac{\partial}{\partial \xi} \left\{ \left[\left(\frac{\partial v_x}{\partial \xi} \right)^2 + \left(\frac{\partial v_z}{\partial \xi} \right)^2 \right]^{\frac{n-1}{2}} \frac{\partial v_x}{\partial \xi} \right\}; \tag{22}$$

$$6\ \Pi_{p,z}^\delta = \frac{\partial}{\partial \xi} \left\{ \left[\left(\frac{\partial v_x}{\partial \xi} \right)^2 + \left(\frac{\partial v_z}{\partial \xi} \right)^2 \right]^{\frac{n-1}{2}} \frac{\partial v_z}{\partial \xi} \right\}, \tag{23}$$

where the parameters $\Pi_{p,x}^\delta$ and $\Pi_{p,z}^\delta$ are dimensionless pressure gradients:

$$\Pi_{p,i}^\delta = \frac{\frac{\partial p}{\partial i}\ \delta^{n+1}}{6\ K\ \mathrm{v}_{b,x}^n}, \qquad i = x, z. \tag{24}$$

In addition, the boundary conditions were rewritten to:

$$v_x(\xi = 0) = 0. \qquad v_x(\xi = 1) = 1. \tag{25}$$

$$v_z(\xi = 0) = 0. \qquad v_z(\xi = 1) = tan(90^\circ - \varphi_b). \tag{26}$$

Solving the dimensionless momentum Equations (22) and (23) in combination with the boundary conditions (25) and (26) required an additional mathematical constraint to be defined. Previous theories that modeled two-dimensional flows of polymer melts in infinitely wide screw channels considered the cross-channel net flow to be zero [22,26]. Physically, this means that leakage flow across the screw flights was ignored. Since this assumption is not reasonable in the analysis presented here, we instead assumed the pressure gradient in the z-direction to be zero. In our local formulation, this implies that the corresponding flow component is governed exclusively by the rotation of the screw:

$$\frac{\partial p}{\partial z} = 0 \quad \rightarrow \quad \Pi_{p,z}^\delta = 0. \tag{27}$$

Our assumption is based on the results of a similar study [26], in which the authors examined the flow of polymer melts in infinitely wide screw channels. For a variety of screw designs and processing conditions, the dimensionless down-channel pressure gradient along the screw channel $\Pi_{p,z}$ was shown to be within the range of $(-1.0; 1.0)$. Rather than using the channel depth h, our theory requires the flight clearance δ to be predefined for the calculation of the dimensionless pressure gradient in the z-direction. Since the latter is significantly smaller, the dimensionless down-channel pressure gradient along the flight clearance $\Pi^{\delta}_{p,z}$ converges to zero.

A comparison of typical parameter values of $\Pi_{p,z}$ and $\Pi^{\delta}_{p,z}$ is given in Table 1. The results indicate the different orders of magnitude of the parameters.

Table 1. Dimensionless pressure gradients in the screw channel $\Pi_{p,z}$ and in the flight clearance $\Pi^{\delta}_{p,z}$.

Parameter	Unit	Screw Channel	Flight Clearance
K	Pas^n	1000	1000
n	-	0.3	0.3
D	mm	45	45
N	rpm	100	100
h or δ	mm	3.5	0.075
$\partial p/\partial z$	Pa/m	$50{\cdot}10^5$	$50{\cdot}10^5$
$\Pi^{\delta}_{p,z}$ or $\Pi^{\delta}_{p,z}$	-	0.82	0.006

Finally, we derived dimensionless parameters for the volume flow rate Π^{δ}_{V} and the total viscous dissipation Π^{δ}_{Q}:

$$\Pi^{\delta}_{V} = \frac{2\,\dot{V}}{\delta\,v_{b,x}} = 2\int_0^1 v_x(\xi)\,d\xi. \tag{28}$$

$$\Pi^{\delta}_{Q} = \frac{\dot{Q}_{diss}\,\delta^n}{K\,v_{b,x}^{n+1}} = \int_0^1 \left[\left(\frac{\partial v_x}{\partial \xi}\right)^2 + \left(\frac{\partial v_z}{\partial \xi}\right)^2\right]^{\frac{n+1}{2}} d\xi. \tag{29}$$

Our dimensionless model has three independent input parameters that completely describe the physics of the flow: (i) the screw-pitch ratio t/D_b, indicated by $tan(\varphi_b)$, (ii) the power-law index n, and (iii) a dimensionless pressure gradient across the flight clearance $\Pi^{\delta}_{p,x}$. The first is part of the boundary conditions (25), while the second and third are included in the momentum Equations (22) and (23).

2.1.4. Set-Up of Parametric Study

We created four sets of physically independent design points by varying the input parameters t/D_b, n, and $\Pi^{\delta}_{p,x}$ (Tables 2–4). For all sets, the screw-pitch ratio and the power-law index were varied within the ranges (0.5–2.47) and (0.2–1.0), respectively. These ranges include almost all extruder screws and polymer melts in industrial use. The variation of the dimensionless pressure gradient across the flight clearance $\Pi^{\delta}_{p,x}$ was adjusted case by case. To determine the industrially relevant parameter ranges, we carried out screw calculations for various types of extruder screws, using our network-based routine [1–40].

Table 2. The ranges of variation for t/D_b, n, and $\Pi^{\delta}_{p,x}$ for Data Set 1.

Variable	Min	Max	Delta
t/D_b	0.50	2.47	variable
n	0.2	1.0	0.1
$\Pi^{\delta}_{p,x}$	-1.0	1.0	0.1

Table 3. The ranges of variation for t/D_b and n for Data Sets 2 to 4.

Variable	Min	Max	Delta
t/D_b	0.50	2.47	variable
n	0.2	1.0	0.1

Table 4. The ranges of variation for $\Pi_{p,x}^{\delta}$ for Data Sets 2 to 4.

n	Data Set 2			Data Set 3			Data Set 4		
	Min	Max	Delta	Min	Max	Delta	Min	Max	Delta
1.0	−2000	2000	200	−40	40	4	−20	20	2
0.9	−2000	2000	200	−20	20	2	−10	10	1
0.8	−1000	1000	100	−20	20	2	−10	10	1
0.7	−400	400	40	−20	20	2	−10	10	1
0.6	−200	200	20	−10	10	1	−5	5	0.5
0.5	−80	80	8	−10	10	1	−5	5	0.5
0.4	−32	32	3.2	−5	5	0.5	−3	3	0.3
0.3	−12.8	12.8	1.28	−3	3	0.3	-	-	-
0.2	−6.4	6.4	0.64	-	-	-	-	-	-

For conventional extruder screws with normal clearances ($\delta \approx 0.001\ D_b$), the dimensionless pressure gradient is relatively small. Its magnitude, however, increases if the extruder screw is constructed with an undercut flight. This situation can be found, for example, in barrier screws. High dimensionless pressure gradients arise in the context of high-performance screws, such as energy-transfer screws, in which screw flights are, by design, undercut to promote cross-channel mixing. In addition, the direction of leakage flow depends on the conveying characteristics of the screw. For pressure-generating functional zones, leakage flow typically reduces the throughput. For strongly overridden functional zones, in contrast, polymer melt passing through the clearance is forced towards the screw tip to increase the net flow rate of the processing machine. This effect can be increased if the flight clearance is undercut.

For Data Set 1, the dimensionless pressure gradient was varied within the range $(-1.0; 1.0)$. Since pressure development also depends on the shear-thinning nature of the polymer melt, the minima, maxima, and increments for Data Sets 2–4 were based on the power-law index. In total, we created 11,781 design points. Finally, Data Sets 1–4 were merged, and design points with multiple occurrences were deleted, which yielded 9231 physically independent setups.

2.2. Numerical Modeling

2.2.1. Numerical Solution Procedure

In the next step, we numerically evaluated the target variables (dimensionless volume flow rate Π_V^{δ} and dimensionless dissipation Π_Q^{δ}) of our model for all 9,231 physically independent setups, using the shooting method. To this end, we calculated the velocity field in the flight clearance by solving the governing equations of our dimensionless model. A detailed description of the numerical solution procedure was given in [26]. Transforming the boundary value into an initial-value problem, we derived explicit forms of the dimensionless momentum equations:

$$\frac{\partial v_x}{\partial \xi} = \frac{1}{\eta_\delta^*}\left(6\,\Pi_{p,x}^{\delta}\xi + C_1\right); \quad (30)$$

$$\frac{\partial v_z}{\partial \xi} = \frac{C_2}{\eta_\delta^*}, \quad (31)$$

where the dimensionless viscosity was rewritten as:

$$\eta_\delta^* = \left[\left(6\,\Pi_{p,x}^{\delta}\xi + C_1\right)^2 + C_2^2\right]^{\frac{n-1}{2n}}. \tag{32}$$

The initial estimates of the integration constants C_1 and C_2 were taken from the Newtonian solution. Applying the Simpsons rule yielded the following equations for the velocity profiles:

$$v_x(\xi) = \underbrace{v_x(\xi = 0)}_{0} + \int_0^1 \frac{\partial v_x}{\partial \xi} d\xi. \tag{33}$$

$$v_z(\xi) = \underbrace{v_z(\xi = 0)}_{0} + \int_0^1 \frac{\partial v_z}{\partial \xi} d\xi. \tag{34}$$

To iteratively solve the unknowns, we used a Newton-Raphson scheme:

$$\boldsymbol{x}_{n+1} = \boldsymbol{x}_n - \boldsymbol{J}(\boldsymbol{x}_n)^{-1}[\boldsymbol{f}(\boldsymbol{x}_n) - \boldsymbol{f}(\boldsymbol{x})], \tag{35}$$

where $\boldsymbol{x}$ is the vector of unknowns, $\boldsymbol{J}$ the Jacobian matrix, and $\boldsymbol{f}$ is the vector of boundary conditions:

$$\begin{pmatrix} C_{1,n+1} \\ C_{2,n+1} \end{pmatrix} = \begin{pmatrix} C_{1,n} \\ C_{2,n} \end{pmatrix} - \begin{pmatrix} \frac{\partial v_{x,1}}{\partial C_1} & \frac{\partial v_{x,1}}{\partial C_2} \\ \frac{\partial v_{z,1}}{\partial C_1} & \frac{\partial v_{z,1}}{\partial C_2} \end{pmatrix}^{-1} \left[\begin{pmatrix} v_{x,1,n} \\ v_{z,1,n} \end{pmatrix} - \begin{pmatrix} tan(\varphi_b) \\ 1 \end{pmatrix}\right]. \tag{36}$$

The velocity boundary conditions at the barrel surface will not be met unless the initial values are perfect. The converged solutions for the velocity profiles were then used to determine the dimensionless target variables Π_V^{δ} and Π_Q^{δ}.

For all calculations, the dimensionless channel height was divided into 1000 equidistant segments. A solution was considered converged if the difference in dimensionless volume flow rate Π_V^{δ} between two iterations was smaller than 10^{-8}. Previous analyses have shown that these settings are sufficient to obtain mesh-independent results for our target variables [26].

2.2.2. Numerical Results

Our parametric design study encompassing 9,231 independent setups provided numerical solutions for the dimensionless volume flow rate Π_V^{δ} and the dissipation Π_Q^{δ} in the flight clearance as functions of the dimensionless input parameters t/D_b, n, and $\Pi_{p,x}^{\delta}$.

Figure 5 shows the dimensionless volume flow rate as a function of the dimensionless pressure gradient across the flight clearance for various power-law indices. For all setups, the curves are symmetrical about the point of pure drag flow ($\Pi_{p,x}^{\delta} = 0$); that is, an equidistant increase in the pressure gradient (positive or negative) affects the magnitude of the dimensionless volume flow rate equally.

The power-law index is a measure of the shear-thinning behavior of the polymer melt: the lower the power-law index, the more shear-thinning is the fluid. Assuming a Newtonian fluid, the widely-known linear behavior is evident: For $\Pi_{p,x}^{\delta} = 0$ (pure drag flow), the curve satisfies $\Pi_V^{\delta} = 1$, while for $\Pi_{p,x}^{\delta} = 1$, the zero-throughput condition is fulfilled $\Pi_V^{\delta} = 0$.

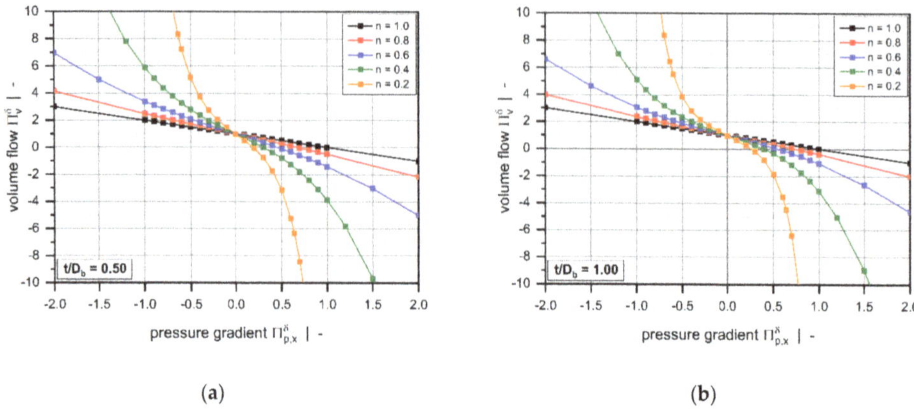

Figure 5. The volume flow rate Π_V^δ as a function of pressure gradient $\Pi_{p,x}^\delta$: Influence of power-law index for $t/D_b = 0.5$ (**a**) and $t/D_b = 1.0$ (**b**).

Generally, the volume flow rates become negative if a critical pressure gradient is reached. From a mathematical viewpoint, this means that the direction of flow changes to the negative x-direction. Physically, it implies that the pressure flow caused by the pressure build-up across the clearance exceeds the drag flow, which yields a negative net throughput. According to our model definition, this behavior is subject to strongly overridden melt-conveying zones. Positive flow rates, in contrast, are found in pressure-generating metering zones, in which the leakage flow reduces the net throughput. With decreasing power-law index, the critical pressure gradient shifts to lower values.

The curves become increasingly nonlinear and pressure-sensitive with decreasing power-law index. Two effects are evident: (i) For a given dimensionless volume flow rate, the more the dimensionless pressure gradient (positive and negative) increases, the less shear-thinning the fluid. (ii) For highly shear-thinning polymer melts, small variations in the pressure gradient can lead to pronounced variations in the volume flow rate.

The influence of the screw-pitch ratio on the dimensionless volume flow rate is less pronounced (Figure 6). For positive pressure gradients, the target variable increases with increasing screw-pitch ratio, while the opposite behavior is evident for negative pressure gradients. This effect is caused by the influence of the flow along the flight clearance (in the z-direction) on the deformation rates, which becomes more pronounced the lower the screw-pitch ratio.

For markedly positive or negative dimensionless pressure gradients (Figure 6b), the effect of the screw-pitch ratio decreases significantly since the flow is governed mainly by the pressure gradient across the flight clearance.

Figure 7 illustrates the influence of the power-law index on the dimensionless dissipation for a square-pitched screw with $t/D_b = 1$. Viscous dissipation is mainly responsible for the temperature development in the channel. Due to inner friction, mechanical energy is converted into heat, causing a rise in melt temperature.

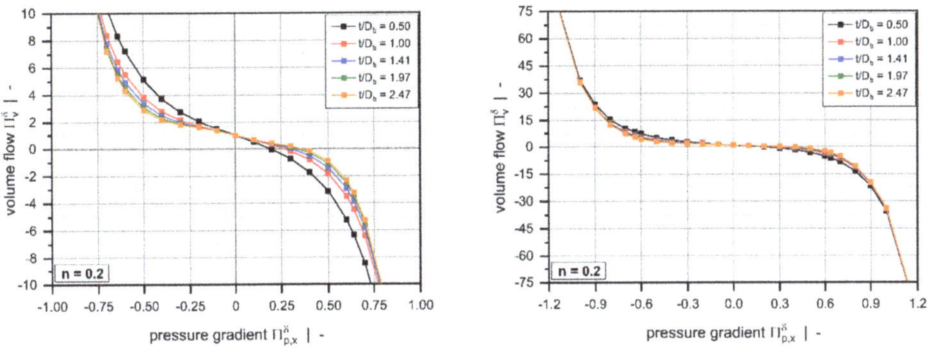

(a) (b)

Figure 6. The volume flow rate Π_V^δ as a function of pressure gradient $\Pi_{p,x}^\delta$: Influence of screw-pitch ratio for $n = 0.2$. The scaling of the diagrams was adjusted to better visualize the influence for smaller pressure gradients (**a**) and larger pressure gradients (**b**).

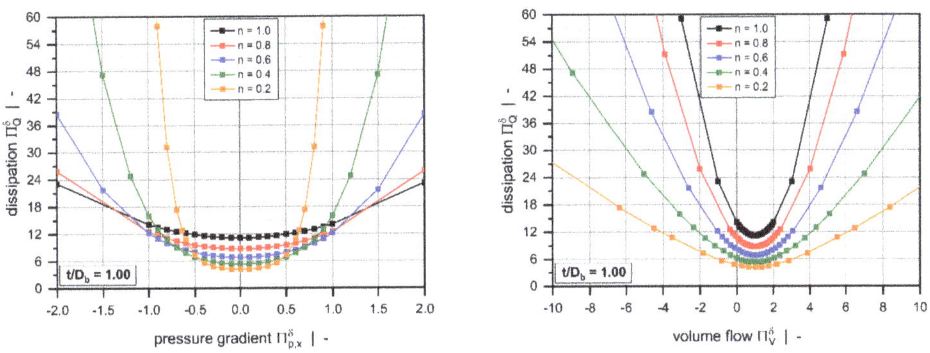

(a) (b)

Figure 7. Dimensionless dissipation $\boldsymbol{\Pi_Q^\delta}$ as a function of the dimensionless pressure gradient $\boldsymbol{\Pi_{p,x}^\delta}$ (**a**) and as a function of the dimensionless volume flow rate $\boldsymbol{\Pi_V^\delta}$. (**b**) The influence of the power-law index for a square-pitched screw with $\boldsymbol{t/D_b = 1.0}$.

The dimensionless dissipation can be plotted as a function of either the dimensionless pressure gradient (Figure 7a) or the dimensionless volume flow rate (Figure 7b). For both representations, the curves are again symmetrical about the point of pure drag flow ($\Pi_{p,x}^\delta = 0$ or $\Pi_V^\delta = 1$), where the target variable reaches a minimum. In general, the dimensionless dissipation increases if the pressure flow contributes to the flow characteristics; that is, the higher the dimensionless pressure gradient, the more pronounced is the frictional heat generation. Similarly, dissipation becomes highly dependent on the power-law index for strongly pressure-generating or pressure-consuming flows.

Three effects are observed: (i) For moderate pressure gradients, the more dimensionless dissipation decreases, the more shear-thinning the polymer melt. (ii) For higher magnitudes, in contrast, the more frictional heat generation decreases, the more Newtonian the fluid. (iii) For constant dimensionless volume flow rates, viscous heating increases with the increasing power-law index.

Figure 8 shows the influence of the screw-pitch ratio on the dimensionless dissipation for a polymer melt with power-law index n = 0.2. For both constant dimensionless pressure gradients (Figure 8a) and constant dimensionless flow rates (Figure 8b), viscous dissipation increases with decreasing screw-pitch ratio. This result is again caused by the effect of transverse flow in the leakage gap. The influence of the screw-pitch ratio on dimensionless dissipation vanishes almost completely in strongly pressure-generating and pressure-consuming flows.

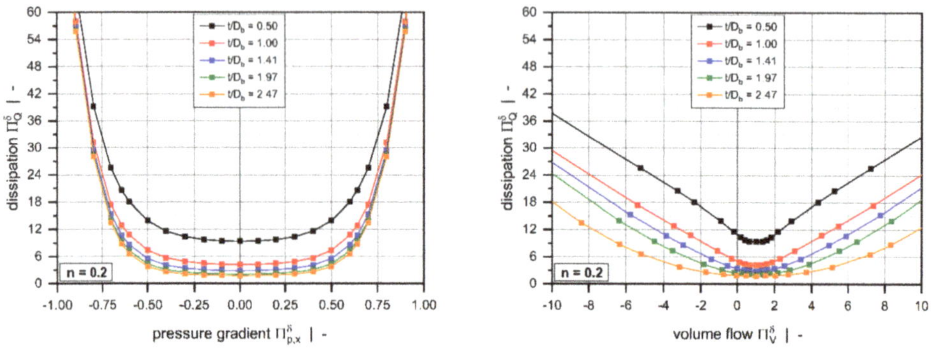

(**a**) (**b**)

Figure 8. Dimensionless dissipation $\boldsymbol{\Pi_Q^\delta}$ as a function of the dimensionless pressure gradient $\boldsymbol{\Pi_{p,x}^\delta}$ (**a**) and as a function of the dimensionless volume flow rate $\boldsymbol{\Pi_V^\delta}$. (**b**) The influence of the screw-pitch ratio for a polymer melt with power-law index $\boldsymbol{n = 0.2}$.

To represent the diverse characteristics of the leakage flow, we considered a wide range of dimensionless pressure gradients. Especially for highly shear-thinning polymer melts with low power-law indices, our extended dataset caused a significant nonlinear increase in the target variables, yielding values higher than $\left|\Pi_V^\delta\right| > 10^5$ and $\left|\Pi_Q^\delta\right| > 10^6$, as illustrated in Figure 9. Recently, we have shown that the parameters are limited to $\left|\Pi_V\right| < 40$ and $\left|\Pi_Q\right| < 140$ when analyzing the flow in metering channels [26]. This comparison illustrates the increased complexity of the following symbolic regression analysis.

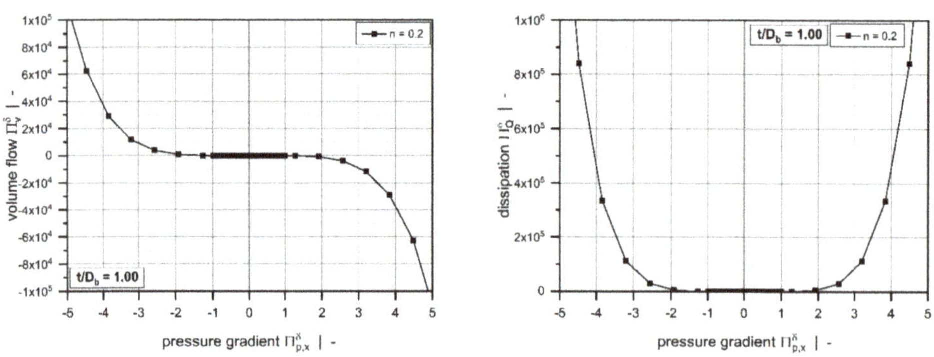

(**a**) (**b**)

Figure 9. The dimensionless volume flow rate $\boldsymbol{\Pi_V^\delta}$ (**a**) and dimensionless dissipation $\boldsymbol{\Pi_Q^\delta}$ (**b**) as functions of the dimensionless pressure gradient $\boldsymbol{\Pi_{p,x}^\delta}$ for a square-pitched screw with $\boldsymbol{t/D_b = 1.0}$ and a polymer melt with the power-law index $n = 0.2$.

2.3. Data-Based Modeling

2.3.1. Symbolic Regression Analysis

Removing the need for numerical simulations required two analytical regressions that accurately predict the numerical solutions of our parametric design study. For this reason, we approximated the numerical solutions of our parametric design study analytically by using symbolic regression based on genetic programming implemented in the open-source software HeuristicLab (Hagenberg, Austria) [45]. This data-based modeling approach searches the space of mathematical expressions to find regressions that relate sets of input and output data. For detailed information, the reader is referred to [46,47].

To reduce the ranges of our target variables and the nonlinearities in our dataset, we took advantage of the following phenomenon: For large dimensionless pressure gradients, our numerical results showed that the influence of the screw-pitch ratio converges to zero, causing the pressure flow across the flight clearance to dominate the overall flow characteristics. Ignoring the effect of the flow in the z-direction, we used the following analytical relationships to approximate the dimensionless volume flow rate and the dissipation [26]:

$$\Pi^{\delta}_{V,app} = 1 - sign\left(\Pi^{\delta}_{p,x}\right)\frac{3^{\frac{1}{n}}\, n}{2\,n+1}\left|\Pi^{\delta}_{p,x}\right|^{\frac{1}{n}}. \tag{37}$$

$$\Pi^{\delta}_{Q,app} = \frac{(2n+1)^{n}}{n^{n}}\left|\Pi^{\delta}_{V} - 1\right|^{(1+n)}. \tag{38}$$

Considering the flow of a power-law fluid, the first relationship describes the dimensionless volume flow rate as a function of the pressure gradient by a linear superposition of a one-dimensional drag and pressure flow. The second relationship, in contrast, approximates the dimensionless dissipation as a function of the volume flow rate by a one-dimensional pressure flow. These parameters were used to create new target variables by correcting the numerical results:

$$\Delta\Pi^{\delta}_{V} = \Pi^{\delta}_{V} - \Pi^{\delta}_{V,app}. \tag{39}$$

$$\Delta\Pi^{\delta}_{Q} = \frac{\Pi^{\delta}_{Q}}{\Pi^{\delta}_{Q,app}+1}. \tag{40}$$

Figure 10 illustrates the characteristics of the modified target variables for a square-pitched screw and various power-law indices. Rather than showing a strong nonlinear increase for increased dimensionless pressure gradients, both parameters reach a plateau, thereby decreasing the value range of the target variables. Again, the curves are symmetrical about the point of pure drag flow ($\Pi^{\delta}_{p,x} = 0$ or $\Pi^{\delta}_{V} = 1$).

In the construction of the regression models, we randomly divided our dataset into three subsets: (i) a training set, (ii) a test set, and (iii) a validation set, including 4000, 2000, and 3231 design points, respectively. The first two subsets were employed to develop and optimize the symbolic regression solutions, while the third subset was used to validate the models against unseen numerical results not used in model development.

We applied the NSGA-II algorithm [48] to construct two regressions of the form:

$$\Delta\Pi^{\delta}_{V} = f\left(\Pi^{\delta}_{p,x},\ n,\ t/D_b\right); \tag{41}$$

$$\Delta\Pi^{\delta}_{Q} = f\left(\Pi^{\delta}_{V},\ n,\ t/D_b\right). \tag{42}$$

This multi-objective non-dominant genetic algorithm simultaneously optimizes model quality and complexity. The latter was adjusted by restricting (a) the model size to a maximum tree length of 100, and (b) the function set, which defines the functions used to generate symbolic regression solutions, to (i) constant, (ii) state variable, (iii) addition, (iv) multiplication and division, and (v) cosine and sine, thus limiting the search space of the regression analysis. Model optimization was driven by a constant optimization evaluator, which calculates Pearson R^2 of a symbolic regression solution and optimizes the constants used:

$$R^2 = 1 - \frac{\sum_{i=1}^{n} (y_i - \hat{y}_i)^2}{\sum_{i=1}^{n} (y_i - \overline{y})^2}, \tag{43}$$

where y_i and $\hat{y}_i$ are the numerical and approximated results, respectively, and $\overline{y}$ is the mean of the numerical solutions.

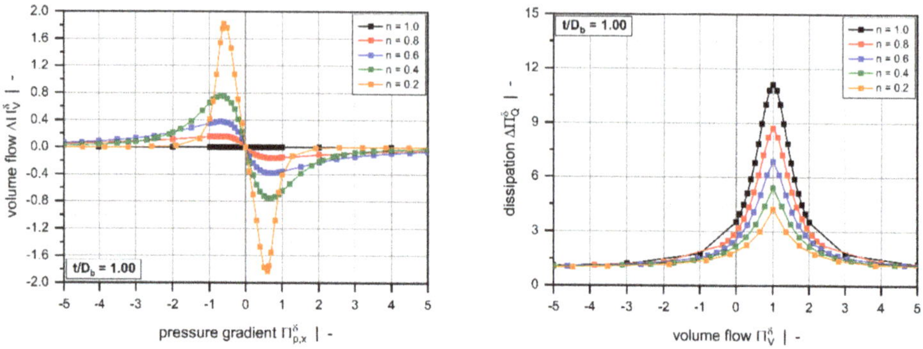

(a) (b)

Figure 10. The corrected dimensionless volume flow rate $\boldsymbol{\Delta\Pi_V^\delta}$ (**a**) and corrected dimensionless dissipation $\boldsymbol{\Delta\Pi_Q^\delta}$ (**b**) as functions of the dimensionless volume flow rate $\boldsymbol{\Pi_V^\delta}$ for a square-pitched screw with $\boldsymbol{t/D_b = 1.0}$ and a polymer melt with power-law index $\boldsymbol{n = 0.2}$.

For each model, we first performed 20 runs to generate a set of symbolic regression solutions, using the training and test data and then selected the most accurate approximation. To evaluate model quality, we carried out an error analysis for all subsets.

2.3.2. Symbolic Regression Results

Our hybrid modeling approach provided two analytical regression for the corrected dimensionless volume flow rate $\Delta\Pi_V^\delta$ and the corrected dimensionless dissipation $\Delta\Pi_Q^\delta$:

$$\Delta\Pi_V^\delta\left(\Pi_{p,x}^\delta,\ n,\ t/D_b\right) = a_{00} + \frac{A_1 + A_2\ A_3}{A_4 + A_5 + A_6}; \tag{44}$$

$$\Delta\Pi_Q^\delta\left(\Pi_V^\delta,\ n,\ t/D_b\right) = b_{00} + \frac{B_1 + \frac{B_2}{B_3} + B_4}{B_5 + B_6}, \tag{45}$$

where A_1–A_6 and B_1–B_6 are the subfunctions, which contain 25 (a_{00}–a_{24}) and 31 (b_{00}–b_{30}) coefficients, respectively. The subfunctions and their coefficients are given in the Appendix A. Note that the regressions predict the corrected target variables, while the actual parameters result from:

$$\Pi_V^\delta = \Delta\Pi_V^\delta + \Pi_{V,app}^\delta. \qquad \Pi_Q^\delta = \Delta\Pi_Q^\delta\left(\Pi_{Q,app}^\delta + 1\right). \tag{46}$$

Avoiding complex nested analytical functions, the models include only basic arithmetic operations and analytical functions. Due to their relatively simple mathematical structure, the approximations allow fast computation of the target variables without the need for further numerical simulations.

To evaluate the accuracy of the regression solutions, we carried out an error analysis for all subsets, including 9,231 design points in total. For this reason, we determined the mean absolute error (AE_{mean}), the maximum absolute error (AE_{max}), and the coefficient of determination (R^2). The accuracy of the dissipation model was additionally investigated by comparing the mean relative error (RE_{mean}) and the maximum relative error (RE_{max}):

$$AE_{mean} = \frac{1}{N}\sum_{i=1}^{n}|y_i - \hat{y}_i|. \qquad AE_{max} = \max\left(|y_i - \hat{y}_i|\right). \tag{47}$$

$$RE_{mean} = \frac{1}{N}\sum_{i=1}^{n}\frac{|y_i - \hat{y}_i|}{y_i}. \qquad RE_{max} = \max\left(\frac{|y_i - \hat{y}_i|}{y_i}\right). \tag{48}$$

Overviews of the results of the error measures are given in Tables 5 and 6. Both models achieved a coefficient of determination $R^2 > 0.999$ for all subsets, which indicates excellent accuracy of the approximations. Since the results of the error measures of all subsets fall within a similar range, we conclude that overfitting was avoided. The low mean absolute and mean relative errors confirm the validity of the new models over the entire range of input parameters. Note that the design points of the validation set were not used for model development and therefore enabled an unbiased estimation of model quality. To demonstrate the performance of the regression models on the original numerical data, including 9231 design points, Table 7 shows the error measures of the final models in Equation (46). As the corrected target variables were constructed by using addition and multiplication, the models for the flow rate and dissipation exhibit the same absolute and relative errors, respectively. Comparisons of numerical and approximated results are illustrated in Figures 11 and 12 for various setups. The points indicate numerical results, and the continuous lines approximated solutions. Further, Figures 13 and 14 show a normalized representation of all design points in the form of scatter plots that compare numerical and approximated results for all subsets. Removing the need for numerical simulations, the new leakage-flow models can be used to include the effect of the flight clearance in the analysis of melt-conveying zones.

Table 5. The error measures of $\Delta\Pi_V^\delta = f\left(t/D_b,\ n,\ \Pi_{p,x}^\delta\right)$.

Quality Measure	Unit	Training Set	Test Set	Validation Set
R^2	-	0.99985	0.99983	0.99986
AE_{mean}	-	0.00536	0.00543	0.00545
AE_{max}	-	0.04511	0.06068	0.06735

Table 6. The error measures of $\Delta\Pi_Q^\delta = f\left(t/D_b,\ n,\ \Pi_V^\delta\right)$.

Quality Measure	Unit	Training Set	Test Set	Validation Set
R^2	-	0.99999	0.99999	0.99999
AE_{mean}	-	0.00638	0.00698	0.00634
AE_{max}	-	00.15041	0.17306	0.20197
RE_{mean}	%	0.31	0.33	0.3
RE_{max}	%	6.65	6.29	6.64

Table 7. The error measures of $\Pi_V^\delta = f\left(t/D_b,\ n,\ \Pi_{p,x}^\delta\right)$ and $\Pi_Q^\delta = f\left(t/D_b,\ n,\ \Pi_{p,x}^\delta\right)$.

Quality Measure	Unit	Π_V^δ	Π_Q^δ
R^2	-	0.99985	0.99999
AE_{mean}	-	0.00513	1047.77
AE_{max}	-	0.06735	30270.9
RE_{mean}	%	-	0.31
RE_{max}	%	-	6.64

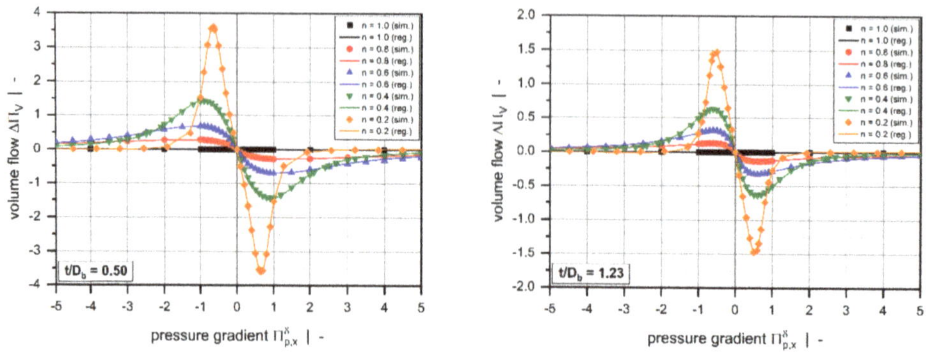

(a) (b)

Figure 11. A comparison of approximated results obtained from $\Delta\Pi_V^\delta = f\left(t/D_b,\ n,\ \Pi_{p,x}^\delta\right)$ and numerical solutions for $t/D_b = 0.50$ (**a**) and $t/D_b = 1.23$ (**b**). The points indicate numerical results, and the continuous lines approximated solutions.

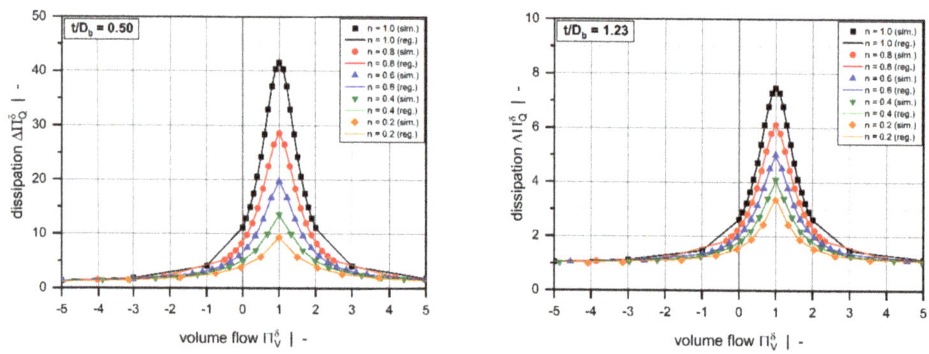

(a) (b)

Figure 12. A comparison of approximated results obtained from $\Delta\Pi_Q^\delta = f\left(t/D_b,\ n,\ \Pi_V^\delta\right)$ and numerical solutions for $t/D_b = 0.50$ (**a**) and $t/D_b = 1.23$ (**b**). The points indicate numerical results, and the continuous lines approximated solutions.

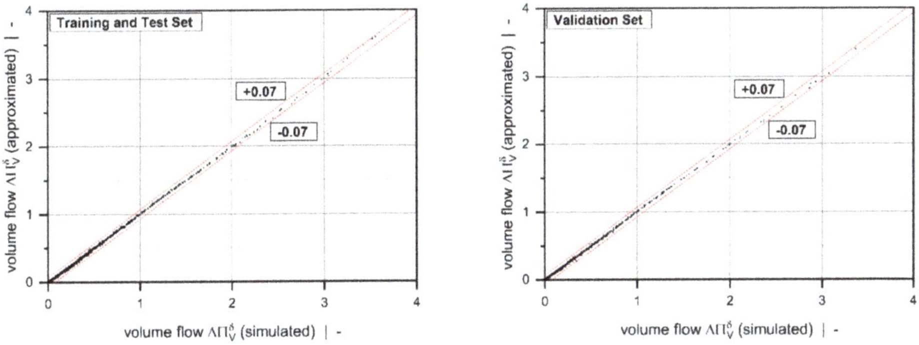

(a) (b)

Figure 13. A scatter plot of $\Delta\Pi_V^\delta = f\left(t/D_b,\ n,\ \Pi_{p,x}^\delta\right)$: training and test set (**a**) and validation set (**b**). The dashed lines indicate an absolute error of 0.07.

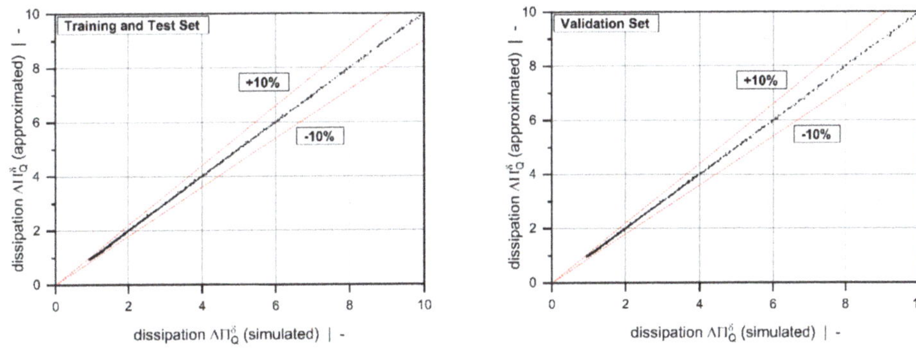

(a) (b)

Figure 14. A scatter plot of $\Delta\Pi_Q^\delta = f\left(t/D_b,\ n,\ \Pi_V^\delta\right)$: training and test set (**a**) and validation set (**b**). The dashed lines indicate a relative error of 10%.

3. Conclusions

We have proposed novel analytical regression models for predicting the volume flow rate and the viscous dissipation rate in the flight clearance of single-screw extruders. Unlike theories that correct the drag and pressure flows in the traditional pumping model, these approximations were designed to locally evaluate the characteristics of the leakage flow. Using a local reference system of the flight clearance allowed us to relax a variety of modeling assumptions and, therefore, to increase the accuracy of the models.

The approach presented here combines analytical, numerical, and data-based modeling techniques. The governing flow equations were rewritten in a dimensionless form by applying the theory of similarity. Three physically independent dimensionless input parameters of the flow were identified: (i) the screw-pitch ratio t/D_b, (ii) the power-law index n, and (iii) a dimensionless pressure gradient across the flight clearance $\Pi_{p,x}^\delta$. These dimensionless parameters were varied to create 9,231 physically independent setups, whose volume flow rates and dissipations were evaluated numerically with the shooting method. The numerical solutions of the design study were then approximated by means of symbolic regression based on genetic programming. The hybrid modeling approach yielded two analytical relationships for volume flow rate and dissipation, the accuracy of which we have demonstrated in an error analysis, yielding a Pearson $R^2 > 0.999$ for both models. The first model showed a maximum absolute error of $AE_{max} = 0.06735$, while the second model produces a maximum relative error of $RE_{max} = 6.64\%$.

A novel feature of our theory is the flat-plate representation of the flight clearance. This local reference system gives rise to a two-dimensional flow in which the drag- and pressure flows in the cross- and down-channel directions are coupled due to the shear-rate-dependent viscosity of the power-law fluid. To be able to solve the governing flow equations, we introduced a mathematical constraint by omitting the pressure gradient along the flight clearance (in the z-direction). From a physical viewpoint, this means that the corresponding flow component is governed exclusively by the rotation of the screw. This assumption was shown to be valid if the flight clearance is significantly smaller than the channel depth ($\delta/h \ll 1$).

To extend the applicability of the new leakage-flow models to a variety of screw designs and processing conditions in industrial use, our approach considered a significant range of dimensionless pressure gradients. Preliminary analyses indicated that these parameter ranges cover several orders of magnitude, depending on the depth of the leakage gap. Low values were observed for conventional screw designs with standard clearances ($\delta \approx 0.001\ D_b$), while high values were detected for high-performance screws with pronounced undercut distances between main and secondary flight (e.g., wave-dispersion screws). In addition, the calculations illustrated that the direction of leakage flow is governed mainly by the conveying characteristics of the extruder screw. For pressure-generating melt-conveying zones, leakage flow reduces the net flow of the processing machine. For strongly overridden functional zones, in contrast, leakage flow increases the net flow.

Since the diverse characteristics of the leakage flow were considered in the construction of the design points, the numerical results of our parametric study extend over several orders of magnitude. To decrease the parameter ranges for the symbolic regression analyses, two modified target variables were constructed by assuming that, for large dimensionless pressure gradients, the flow can be approximated mathematically by linear superposition of one-dimensional drag and pressure flows. The corrected parameters were designed to reach a plateau for increased dimensionless pressure gradients, thereby decreasing the nonlinearities in the dataset.

The novel leakage-flow models will be implemented in our screw-simulation routine based on network theory (presented in [1–40]). The aim is to accurately describe the properties of the network elements positioned in the leakage gap to include the effect of the flight clearance in the prediction of pumping capability and power consumption. Using the new leakage-flow models in the network calculation, the modeling approach is no longer limited to any specific geometrical design and allows fast and accurate analysis of various types of extruder screws, including both conventional and high-performance screws.

Author Contributions: Conceptualization, C.M. and W.R.; Methodology, C.M. and W.R.; Software, C.M. and W.R.; Validation, C.M.; Formal Analysis, C.M.; Investigation, C.M. and W.R.; Resources, C.M., W.R., M.D., and G.S.; Data Curation, C.M.; Writing—Original Draft Preparation, C.M.; Writing—Review and Editing, W.R. and M.D.; Visualization, C.M. and W.R.; Supervision, W.R., M.D., and G.S.; Project Administration, C.M., W.R., M.D., G.S., and V.S.; Funding Acquisition, C.M, W.R., M.D., G.S, and V.S. All authors have read and agreed to the published version of the manuscript.

Funding: This work was funded by the Austrian Science Fund (FWF): I 4872-N. This work was performed within the Competence Center CHASE GmbH, funded by the Austrian Research and Promotion Agency (grant number 868615). The authors acknowledge financial support by the COMET Centre CHASE (project No 868615), which is funded within the framework of COMET—Competence Centers for Excellent Technologies—by BMVIT, BMDW, and the Federal Provinces of Upper Austria and Vienna. The COMET program is run by the Austrian Research Promotion Agency (FFG).

Institutional Review Board Statement: Not applicable.

Informed Consent Statement: Not applicable.

Data Availability Statement: The data presented in this study are available on request from the corresponding author.

Acknowledgments: Open Access Funding by the Austrian Science Fund (FWF).

Conflicts of Interest: The authors declare no conflict of interest.

Appendix A

Appendix A.1. Subfunctions of Equations (44) and (45)

$$\left(\frac{t}{D_b}\right)_{eff} = \pi \tan\left(90 - \arctan\left(\frac{t}{D_b \pi}\right)\right) \tag{A1}$$

$$A_1 = a_{01}\, \Pi^{\delta}_{p,x} \left(\frac{t}{D_b}\right)_{eff} \tag{A2}$$

$$A_2 = a_{02}\,(a_{03}+n)\,(a_{04}+n)\,\Pi^{\delta}_{p,x}\left(a_{05} + (a_{06}+n)\,{\Pi^{\delta}_{p,x}}^2 + (a_{07}+a_{08}\,n)\left(\frac{t}{D_b}\right)_{eff}\right). \tag{A3}$$

$$A_3 = n + \left(a_{09} + a_{10}\,{\Pi^{\delta}_{p,x}}^2\right)\left(\frac{t}{D_b}\right)_{eff}. \tag{A4}$$

$$A_4 = a_{11} + a_{12}\,{\Pi^{\delta}_{p,x}}^2. \tag{A5}$$

$$A_5 = a_{13}\left(a_{14} + {\Pi^{\delta}_{p,x}}^2 + a_{15}\left(\frac{t}{D_b}\right)_{eff}\right)\left(a_{16}\,{\Pi^{\delta}_{p,x}}^2 + {\Pi^{\delta}_{p,x}}^4 + a_{17}\left(\frac{t}{D_b}\right)_{eff}\right). \tag{A6}$$

$$A_6 = a_{18}\,n^3\left(a_{19} + n + a_{20}\left(\frac{t}{D_b}\right)_{eff}\right) + a_{21}\,n^2\,(a_{22}+n)\left({\Pi^{\delta}_{p,x}}^2 + a_{23}\left(\frac{t}{D_b}\right)_{eff}\right)\left({\Pi^{\delta}_{p,x}}^2 + a_{24}\left(\frac{t}{D_b}\right)_{eff}\right). \tag{A7}$$

$$B_1 = b_{01} + b_{02}\,n + b_{03}\left(\frac{t}{D_b}\right)_{eff} + b_{04}\,n\left(\frac{t}{D_b}\right)_{eff} + n\left(\frac{t}{D_b}\right)_{eff}\left(b_{05}n + b_{06}\left(\frac{t}{D_b}\right)_{eff}\right) + b_{07}\left(\frac{t}{D_b}\right)^2_{eff}. \tag{A8}$$

$$B_2 = b_{08} + b_{09}\,\Pi^{\delta}_{V} + \left(\frac{t}{D_b}\right)_{eff}\left(b_{10} + b_{11}\,n^3\right) + \left(\frac{t}{D_b}\right)^2_{eff}\left(b_{12} + b_{13}\,n^2\right). \tag{A9}$$

$$B_3 = b_{14} + b_{15}\,n + b_{16}\,\Pi^{\delta}_{V} + b_{17}\left(\Pi^{\delta}_{V}\right)^2 + \frac{1}{b_{18} + b_{19}\left(\Pi^{\delta}_{V}\right)^2} + b_{20}\left(\frac{t}{D_b}\right)_{eff}. \tag{A10}$$

$$B_4 = b_{21}\,\sin\left(b_{22}\,\Pi^{\delta}_{V}\right). \tag{A11}$$

$$B_5 = b_{23} + b_{24}\,\Pi^{\delta}_{V} + b_{25}\left(\Pi^{\delta}_{V}\right)^2 + b_{26}\left(\frac{t}{D_b}\right)_{eff}. \tag{A12}$$

$$B_6 = n\left(b_{27} + \frac{b_{28}}{b_{29} - \left(\Pi^{\delta}_{V}\right)^2} + b_{30}\left(\frac{t}{D_b}\right)_{eff}\right). \tag{A13}$$

A.2. Model Coefficients

Table A1. Model coefficients.

a_{00}	0.0000413479	b_{00}	0.998487
a_{01}	−2.97497	b_{01}	−11.4595
a_{02}	390.644	b_{02}	9.75142
a_{03}	−1.00056	b_{03}	1.69836
a_{04}	−0.121986	b_{04}	−0.775636
a_{05}	−0.394801	b_{05}	−0.533892
a_{06}	−0.317506	b_{06}	0.1505
a_{07}	−0.093065	b_{07}	0.063818
a_{08}	0.762303	b_{08}	0.910203
a_{09}	0.11879	b_{09}	−0.148087
a_{10}	0.121586	b_{10}	0.169683
a_{11}	4.04094	b_{11}	−0.471171
a_{12}	11.6959	b_{12}	0.015469
a_{13}	89.2206	b_{13}	0.30448
a_{14}	−0.147822	b_{14}	1.74533
a_{15}	0.0113796	b_{15}	0.343839
a_{16}	−0.0381859	b_{16}	−3.1969
a_{17}	0.0171605	b_{17}	1.597370
a_{18}	791.442	b_{18}	−67815.8
a_{19}	−1.01937	b_{19}	67789.9
a_{20}	0.209565	b_{20}	0.00156
a_{21}	4118.38	b_{21}	1.01936
a_{22}	−0.26066	b_{22}	1.45018
a_{23}	0.00593881	b_{23}	17.2059
a_{24}	0.26419	b_{24}	−13.3529
		b_{25}	6.68431
		b_{26}	0.35355
		b_{27}	−2.15945
		b_{28}	−0.00005
		b_{29}	0.99991
		b_{30}	−0.368

Table A2. Nomenclature.

a_{ij}	coefficients of regression model	$\dot{V}$	volume flow rate
A_i	subfunctions of regression model	w	channel width
AE_{max}	maximum absolute error	x	cross-channel coordinate
AE_{mean}	mean absolute error	$\boldsymbol{x}$	vector of unknowns
b_{ij}	coefficients of regression model	y	up-channel coordinate
B_i	subfunctions of regression model	y_i	numerical result
$C_{1,2}$	constants	$\hat{y}_i$	approximated result
D_b	barrel diameter	$\overline{y}$	mean value of numerical results
$\mathbf{D}$	rate-of-deformation tensor	z	down-channel coordinate
e	flight width	α	temperature coefficient
$\boldsymbol{f}$	vector of boundary conditions	δ	flight clearance
i	number of parallel screw flights	$\dot{\gamma}$	shear rate
$\boldsymbol{J}$	Jacobian	$\dot{\gamma}_\delta^*$	dimensionless shear rate
K	consistency	η	viscosity
$\mathbf{L}$	velocity gradient tensor	η_δ^*	dimensionless viscosity
$\dot{m}$	mass-flow rate	v_i	dimensionless velocities
n	power-law index	ξ	dimensionless height direction
N	screw speed	$\Pi_{p,i}^\delta$	dimensionless pressure gradients
R^2	coefficient of correlation	Π_Q^δ	dimensionless dissipation
RE_{max}	maximum relative error	$\Pi_{Q,app}^\delta$	approximated dimensionless dissipation
RE_{mean}	mean relative error	$\Delta\Pi_Q^\delta$	corrected dimensionless dissipation
p	pressure	Π_V^δ	dimensionless flow rate
t	screw pitch	$\Pi_{V,app}^\delta$	approximated dimensionless flow rate
v_i	velocities	$\Delta\Pi_V^\delta$	corrected dimensionless flow rate
v_b	barrel velocity	ρ_m	melt density
$\mathrm{v}_{b,x}$	barrel velocity in x-direction	τ_{ij}	shear stresses
$\mathrm{v}_{b,z}$	barrel velocity in z-direction	τ	stress tensor
$\mathbf{v}$	velocity vector	φ_b	pitch angle

References

1. Tadmor, Z.; Klein, I. *Engineering Principles of Plasticating Extrusion*; Van Nostrand Reinhold, Co.: New York, NY, USA; ISBN 9780442156350.
2. White, J.L.; Potente, H. *Screw Extrusion*; Hanser Publishers: Munich, Germany, 2001; ISBN 9783446196247.

3. Tadmor, Z.; Gogos, Z.G. *Principles of Polymer Processing*, 2nd ed.; Wiley and Sons: Hoboken, NJ, USA, 2002; ISBN 9780471387701.
4. Campbell, G.A.; Spalding, M. *Analyzing and Troubleshooting Single-Screw Extruders*; Hanser Publishers: Munich, Germany, 2013; ISBN 9783446413719.
5. Rauwendaal, C. *Polymer Extrusion*, 5th ed.; Hanser Publishers: Munich, Germany, 2014; ISBN 9783446217744.
6. Agassant, J.F.; Avenas, P.; Carreau, P.J.; Vergnes, B.; Vincent, M. *Polymer Processing. Principles and Modelling*, 2nd ed.; Hanser Publishers: Munich, Germany, 2017; ISBN 9781569906057.
7. Wilczyński, K.; Nastaj, A.; Lewandowski, A.; Wilczyński, K.J.; Buziak, K. Fundamentals of global modeling for polymer extrusion. *Polymers* **2019**, *11*, 2106. [CrossRef] [PubMed]
8. Anonymous. Screw Viscosity Pumps. *Engineering* **1922**, *114*, 606–607.
9. Rowell, H.S.; Finlayson, D. Screw viscosity pumps. *Engineering* **1928**, *126*, 249–250, 385–387.
10. Carley, J.F.; Mallouk, R.S.; McKelvey, J.M. Simplified flow theory for screw extruder. *Ind. Eng. Chem.* **1953**, *45*, 974–978. [CrossRef]
11. Mohr, W.D.; Saxton, R.L.; Jepson, C.H. Theory of mixing in the single-screw extruder. *Ind. Eng. Chem.* **1957**, *49*, 1857–1862. [CrossRef]
12. Mohr, W.D.; Mallouk, R.S. Flow, power requirement, and pressure distribution of fluid in a screw extruder. *Ind. Eng. Chem.* **1959**, *51*, 765–770. [CrossRef]
13. Rotem, Z.; Shinnar, R. Non-newtonian flow between parallel boundaries in linear movements. *Chem. Eng. Sci.* **1961**, *15*, 130–149. [CrossRef]
14. Griffith, R.M. Fully developed flow in screw extruders. *Ind. Eng. Chem. Fundam.* **1962**, *1*, 180–187. [CrossRef]
15. Zamodits, H.J.; Pearson, J.R.A. Flow of polymer melts in extruders. Part I. The effect of transverse flow and of a superpose steady temperature profile. *Trans. Soc. Rheol.* **1969**, *13*, 357–385. [CrossRef]
16. Karwe, M.V.; Jaluria, Y. Numerical simulation of fluid flow and heat transfer in a single-screw extruder for non-newtonian fluids. *Numer. Heat Transf. Part. A* **1990**, *17*, 167–190. [CrossRef]
17. Spalding, M.A.; Dooley, J.; Hyun, K.S.; Strand, S.R. *Three-Dimensional Numerical Analysis of a Single-Screw Extruder*; SPE ANTEC Technical Papers; Society of Plastics Engineers: New Orleans, LA, USA, 1993; pp. 1533–1541.
18. Ghoreishy, M.H.R.; Razavi-Nouri, M.; Naderi, G. Finite element analysis of a thermoplastic elastomer melt flow in the metering region of a single screw extruder. *Comp. Mater. Sci.* **2005**, *34*, 389–396. [CrossRef]
19. Polychronopoulos, N.D.; Vlachopoulos, J. Computer flow simulation of moffat eddies in single screw extrusion. *Int. Polym. Process.* **2018**, *5*, 662–668. [CrossRef]
20. Vachagina, E.K.; Kadyirov, A.; Karaeva, J. Simulation of Giesekus fluid flow in extruder using helical coordinate system. *IOP Conf. Ser. Mater. Sci. Eng.* **2020**, *733*, 1–5. [CrossRef]
21. Booy, M.L. The influence of non-newtonian flow on effective viscosity and channel efficiency in screw pumps. *Polym. Eng. Sci.* **1981**, *21*, 93–99. [CrossRef]
22. Rauwendaal, C. Throughput-pressure relationships for power law fluids in single screw extruders. *Polym. Eng. Sci.* **1986**, *26*, 1240. [CrossRef]
23. Spalding, M.A.; Campbell, G.A. *An Engineering Approach to the Correction of Rotational Flow Calculations for Single-Screw Extruders-Equation Correction*; SPE ANTEC Technical Paper; Society of Plastics Engineers: New Orleans, LA, USA, 2011; pp. 1211–1216.
24. Pachner, S.; Löw-Baselli, B.; Affenzeller, M.; Miethlinger, J. A Generalized 2d output model of polymer melt flow in single-screw extrusion. *Int. Polym. Process.* **2017**, *32*, 209. [CrossRef]
25. Roland, W.; Miethlinger, J. Heuristic analysis of viscous dissipation in single-screw extruders. *Polym. Eng. Sci.* **2018**, *58*, 2055–2070. [CrossRef]
26. Roland, W.; Kommenda, M.; Marschik, C.; Miethlinger, J. extended regression models for predicting the pumping capability and viscous dissipation of two-dimensional flows in single-screw extrusion. *Polymers* **2019**, *11*, 334. [CrossRef]
27. Marschik, C.; Roland, W.; Loew-Baselli, B.; Miethlinger, J. *Modeling Three-Dimensional Non-Newtonian Flows in Single-Screw Extruders*; SPE ANTEC Technical Papers; Society of Plastics Engineers: New Orleans, LA, USA, 2017; pp. 1125–1130.
28. Marschik, C.; Roland, W.; Loew-Baselli, B.; Miethlinger, J. A heuristic method for modeling three-dimensional non-Newtonian flows of polymer melts in single-screw extruders. *J. Non Newt. Fluid Mech.* **2017**, *248*, 27–39. [CrossRef]
29. Marschik, C.; Osswald, T.; Roland, W.; Loew-Baselli, B.; Miethlinger, J. *A Heuristic Model for Predicting Three-Dimensional Non-Newtonian Flows in Metering Channels*; SPE ANTEC Technical Papers; Society of Plastics Engineers: New Orleans, LA, USA, 2018.
30. Roland, W.; Marschik, C.; Krieger, M.; Löw-Baselli, B.; Miethlinger, J. Symbolic regression models for predicting viscous dissipation of three-dimensional non-Newtonian flows in single-screw extruders. *J. Non Newt. Fluid Mech.* **2019**, *268*, 12–19. [CrossRef]
31. Gore, W.L.; McKelvey, J.M. Theory of Screw Extruders. In *Rheology: Theory and Applications*, 3rd ed.; Eirich, F.R., Ed.; Academic Press: New York, NY, USA, 1959.
32. Rauwendaal, C.; Ingen Housz, J.F. Leakage flow of an isothermal power law fluid. *Adv. Polym. Tech.* **1988**, *8*, 290. [CrossRef]
33. Meyer, H.E.H.; Ingen Housz, J.F.; Gorissen, W.C.M. temperature development in the leakage flow of screw extruders. *Polym. Eng. Sci.* **1978**, *18*, 288. [CrossRef]
34. Pittman, J.F.T.; Rashid, K. Heat transfer in recirculating extruder channel flow. *J. Polym. Eng.* **1985**, *5*, 1. [CrossRef]

35. Rauwendaal, C. Finite element studies of flow and temperature evolution in single screw extruder. *Plast. Rubber Comp.* **2004**, *33*, 390. [CrossRef]
36. Aigner, M. Computational and Experimental Modelling of Transport Phenomena in Single Screw Plasticating Units under Consideration of the Melt Quality. Ph.D. Thesis, Johannes Kepler University, Linz, Austria, 2014.
37. Marschik, C.; Roland, W.; Miethlinger, J. A network-theory-based comparative study of melt-conveying models in single-screw extrusion: A. Isothermal flow. *Polymers* **2018**, *10*, 929. [CrossRef]
38. Marschik, C.; Dörner, M.; Roland, W.; Miethlinger, J.; Schöppner, V.; Steinbichler, G. Application of network analysis to flow systems with alternating wave channels: Part A (pressure flows). *Polymers* **2019**, *11*, 1488. [CrossRef]
39. Marschik, C.; Roland, W.; Dörner, M.; Schaufler, S.; Schöppner, V.; Steinbichler, G. Application of network analysis to flow systems with alternating wave channels: Part B (superimposed drag-pressure flows in extrusion). *Polymers* **2020**, *12*, 1900. [CrossRef]
40. Roland, W.; Marschik, C.; Hammer, A.; Steinbichler, G. *Modeling The Non-Isothermal Conveying Characteristics in Single-Screw Extrusion by Application of Network Analysis*; SPE ANTEC Technical Papers; Society of Plastics Engineers: New Orleans, LA, USA, 2020.
41. Luger, H.J. High-Performance Extrusion under Consideration of Extensional and Wedge Flows. Ph.D. Thesis, Johannes Kepler University, Linz, Austria, 2020.
42. Marschik, C.; Roland, W.; Loew-Baselli, B.; Steinbichler, G. *Application of Hybrid Modeling in Polymer Processing*; SPE ANTEC Technical Papers; Society of Plastics Engineers: New Orleans, LA, USA, 2020.
43. Sun, J.; Rauwendaal, C. *Analysis of Flow in Single Screw Extruders*; SPE ANTEC Technical Papers; Society of Plastics Engineers: New Orleans, LA, USA, 2002.
44. Roland, W.; Marschik, C.; Loew-Baselli, B.; Miethlinger, J. *The Effect of Channel Curvature on the Flow Rate and Viscous Dissipation of Power-Law Fluids*; SPE ANTEC Technical Papers; Society of Plastics Engineers: New Orleans, LA, USA, 2018.
45. Wagner, S.; Kronberger, G.; Beham, A.; Kommenda, M.; Scheibenpflug, A.; Pitzer, E.; Vonolefen, S.; Kofler, M.; Winkler, S.; Dorfer, V.; et al. Architecture and Design of the Heuristic Lab Optimization Environment. In *Advanced Methods and Applications in Computational Intelligence*; Klempous, R., Nikodem, J., Jacak, W., Chaczko, Z., Eds.; Springer: Heidelberg, Germany, 2014; Volume 6, pp. 197–261. ISBN 9783319014364.
46. Poli, R.; Langdon, W.B.; McPhee, N.F. *A Field Guide to Genetic Programming*; Lulu Press: Morrisville, NC, USA, 2008; ISBN 9781409200734.
47. Affenzeller, M.; Wagner, S.; Winkler, S.; Beham, A. *Genetic Algorithms and Genetic Programming: Modern Concepts and Practical Applications*; Chapman & Hall/CRC Press: New York, NY, USA, 2018; ISBN 978-1-58488-629-7.
48. Deb, K.; Pratap, A.; Agarwal, S.; Meyarivan, T. A Fast and Elitist Multiobjective Genetic Algorithm: NSGA-II. *IEEE Trans. Evol. Comp.* **2002**, *6*, 182–197. [CrossRef]

polymers

MDPI

Article

Optimization of the Warpage of Fused Deposition Modeling Parts Using Finite Element Method

Daniyar Syrlybayev, Beibit Zharylkassyn, Aidana Seisekulova, Asma Perveen and Didier Talamona *

Department of Mechanical and Aerospace Engineering, School of Engineering & Digital Sciences, Nazarbayev University, Nur-Sultan 010000, Kazakhstan; daniyar.syrlybayev@nu.edu.kz (D.S.); beibit.zharylkassyn@nu.edu.kz (B.Z.); aidana.seisekulova@nu.edu.kz (A.S.); asma.perveen@nu.edu.kz (A.P.)
* Correspondence: didier.talamona@nu.edu.kz

Abstract: Fused deposition modeling (FDM) is one of the most affordable and widespread additive manufacturing (AM) technologies. Despite its simplistic implementation, the physics behind this FDM process is very complex and involves rapid heating and cooling of the polymer feedstock. As a result, highly non-uniform internal stresses develop within the part, which can cause warpage deformation. The severity of the warpage is highly dependent on the process parameters involved, and therefore, currently extensive experimental studies are ongoing to assess their influence on the final accuracy of the part. In this study, a thermomechanical Finite Element model of the 3D printing process was developed using ANSYS. This model was compared against experimental results and several other analytical models available in the literature. The developed Finite Element Analysis (FEA) model demonstrated a good qualitative and quantitative correlation with the experimental results. An L9 orthogonal array, from Taguchi Design of Experiments, was used for the optimization of the warpage based on experimental results and numerical simulations. The optimum process parameters were identified for each objective and parts were printed using these process parameters. Both parts showed an approximately equal warpage value of 320 μm, which was the lowest among all 10 runs of the L9 array. Additionally, this model is extended to predict the warpage of FDM printed multi-material parts. The relative percentage error between the numerical and experimental warpage results for alternating and sandwich specimens are found to be 1.4% and 9.5%, respectively.

Keywords: warpage; finite element analysis (fem); FDM; Taguchi; multilateral

Citation: Syrlybayev, D.; Zharylkassyn, B.; Seisekulova, A.; Perveen, A.; Talamona, D. Optimization of the Warpage of Fused Deposition Modeling Parts Using Finite Element Method. *Polymers* **2021**, *13*, 3849. https://doi.org/10.3390/polym13213849

Academic Editors: Célio Bruno Pinto Fernandes, Salah Aldin Faroughi, Luís L. Ferrás and Alexandre M. Afonso

Received: 21 October 2021
Accepted: 2 November 2021
Published: 8 November 2021

1. Introduction

Fused deposition modeling is one of the Additive Manufacturing (AM) processes in which the part is manufactured layer by layer from the thermoplastic polymers extruded through a heated nozzle, which moves along the programmed path. It was originally developed by Stratasys Inc., and nowadays has become one of the most popular and affordable AM processes [1]. One of the most significant advantages of the FDM process is its ability to produce parts of complex shapes [2]. In addition, the FDM process requires no tooling [3] and offers a high degree of customization, as the cost per part produced by AM is lower for small batches [4]. Nonetheless, several drawbacks limit its use in the industry, and the most important among them are build speed, mechanical properties, and part dimensional accuracy [5,6].

The accuracy of the parts produced by FDM is highly dependent on the process parameters employed. For this reason, recently there have been several studies conducted to optimize the quality of the end-product produced by FDM. An approach involving benchmark artifacts was also used in several studies to compare the accuracy of the FDM with other popular AM processes [7,8]. In addition, Mahmood et al. (2018) [9] performed Taguchi optimization of the 13 common printing parameters to achieve the highest accuracy of the features of the benchmark artifact. Anitha et al. [10] used Taguchi optimization to study how surface roughness is affected by printing parameters. It was found that layer thickness is the most significant parameter while printing speed is the least significant one. Similar conclusions were obtained by [11,12]. Multi-objective Grey Taguchi optimization of the FDM was performed by Sood et al. [3] to minimize length, width, and thickness deviations. It was found that shrinkage is predominant along the length and width. This occurs during the cooling from the glass transition temperature of the plastic to room temperature. However, thickness deviations are always positive. One of the reasons for this is the positive round-off error of the printer.

Apart from the dimensional deviations, the warpage of the parts is also a serious issue in FDM. Due to rapid cooling and heating during the deposition process, non-uniform shrinkage occurs within a part, and it starts to warp. Several studies were performed to investigate how the warpage is affected by printing parameters. Experiments show that warpage is highly affected by the layer thickness, and the lower the layer thickness, the higher the warpage [13–16]. On the other hand, several analytical models in which an elastic material behavior was assumed [15,17–19] showed the direct correlation between the layer thickness and the warpage. Similarly, Armilota et al. [20] developed an analytical model, which considers reheating of the layers and yielding. This model showed a greater accuracy compared to the simple models based on the theory of elasticity.

The reasons for the discrepancy between the analytical models and experiments are still unknown and under investigation. Finite Element Analysis (FEA) of the FDM process can be used to understand these discrepancies. Recent studies [21–24] have shown that coupled thermomechanical studies using FEA can be successfully employed to recreate the thermal history of the part and model its residual stresses and deformations. It was found that residual stresses are highly affected by the rate of cooling and increasing the convection will increase the development of residual stresses, which leads to excessive distortions and delamination. Cattenone et al. [25] studied how the Finite Element setup affects the result for distortions and residual stresses. Distortions of the semi-crystalline polymers were studied by Samy et al. [26], who found a direct correlation among warpage, residual stresses, and relative crystallization. However, works mentioned above consider simple-shaped bodies, which can be meshed by the structured grid. Several studies were also employed to model complex shapes [27–30]. An approach used in these studies is to approximate geometry around the boundary by voxelization.

As discussed, warpage was studied in numerous works previously, however, for now, the results are inconsistent. This is especially true for the layer thickness. In addition, although the effect of the cooling rate on the warpage is known, to the authors' knowledge no study attempted to consider the effects of the nozzle and build-plate temperatures. Furthermore, FEA has already shown its reliability in modeling the FDM printing process. It allows obtaining and assessing data that cannot be measured during the experiments and provides a better insight into the warpage occurrence. However, such nonlinear, transient simulations require large computational power. Hence, the main objective of this work was to develop a transient thermomechanical, simple material, model using FEM. This model was used to optimize and study the effect of three parameters such as layer thickness, bed temperate and nozzle temperature. FEA results were validated against several analytical models and experimental results. In a second phase, this FEA model was extended to multi-material FDM printing.

2. Methodology

The Finite Element Model of the FDM process was built and used to predict and optimize warpage. The results were verified against experimental results and analytical models available in the literature. The following sections described the design of experiment (DOE) of the procedure involved in every step of the investigation.

2.1. Finite Element Model

The model selected for Finite Element Analysis and 3D printing is the standard sample for tensile testing along with the build platform, as shown in Figure 1a. The build platform having dimensions equal to those in the actual printer was added to the analysis to represent the heat transfer through the bottom layer more accurately. The part was selected as it is long and thin, which allows obtaining larger warpage and facilitates the measurements.

To simplify the analysis, the following assumptions were used:

(1) The phase change and creep effects at high temperatures were neglected. This is a common assumption, which was employed in several previous studies [21,25] and did not show any significant deviations.
(2) It was assumed that the entire layer is deposited at once (or instantaneously). This assumption is also commonly used in analytical models [15,17–20]. The results from El Moumen et al. [23] also show that this assumption does not cause significant deviations when deformations are modeled using FEA.
(3) Plastic was assumed to have isotropic material properties with flawless microstructure.
(4) Chamber and plate temperatures were assumed to have constant temperatures, and natural convection effects were neglected.

The assumptions (3) and (4) were also successfully employed in previous studies [21,25] and did not lead to significant errors between experimental and numerical results.

Due to the second assumption, the printed part is symmetrical and only one-quarter of the part needs to be modeled, with proper symmetry conditions to be applied at the boundaries. This reduced the computational time of the analysis significantly. The final domain used for Finite Element simulations is shown in Figure 1b.

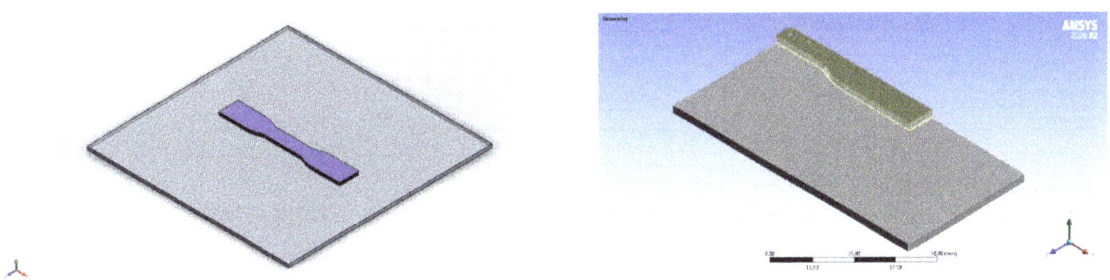

(a) (b)

Figure 1. Problem domain (**a**) tensile test sample (**b**) quarter symmetry.

During the FDM printing process, the plastic filaments are heated and extruded through a nozzle. Upon cooling, the strains and internal stresses start to develop within the part. For this reason, in the following study, the thermal history of the built part was re-created. The equation governing the thermal analysis is given as follows:

$$c\frac{\partial T}{\partial t} = \frac{\partial}{\partial x_i}\left(k\frac{\partial T}{\partial x_j}\right) + q \tag{1}$$

with boundary conditions $T = T_b$ on Γ_u and $-\frac{\partial T}{\partial x_i}n_i = q_n$ on Γ_j. The initial condition is given by $T(x, y, z, 0) = T_0$. Here, T is temperature, k is heat conductivity, c—specific heat, and q—body heat per unit volume (zero in this study), q_n—heat transfer rate at the boundary per unit area, T_b—bed temperature, Γ_u—Dirichlet boundary, Γ_j—Neumann boundary, and T_0—initial temperature. The initial temperature was assumed to be the temperature of the nozzle used in real printing. The Dirichlet boundary in the following analysis was imposed on the whole build plate, as shown in Figure 2a. The Neumann boundary was set on the whole surface of the part, including the top surface of the platform.

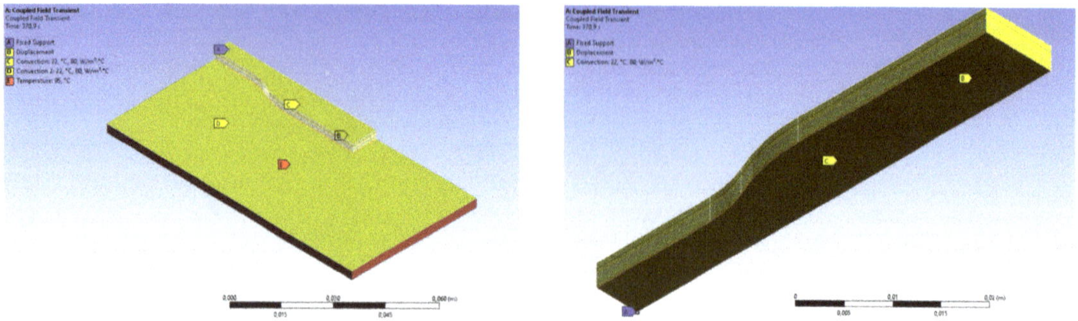

(**a**) (**b**)

Figure 2. (**a**) Thermal boundary conditions. (**b**) Structural boundary conditions (zero translation along the normal direction and rotation on each node of symmetry surfaces).

In the following study, the heat transfer at the boundary might occur due to convection and radiation. Convective heat transfer q_c can be found by the following equation.

$$q_c = h(T - T_c) \tag{2}$$

where h is the convective heat transfer coefficient and T_c is the temperature of the surroundings. Because the printer used for the experiments is open, it was assumed that T_c is constant and equal to 22 °C (room temperature). The convective heat transfer coefficient can be found using an empirical relation given as follows:

$$Nu_L = \frac{hL}{k_{air}} = (0.037Re_L - 871)\sqrt[3]{Pr} \tag{3}$$

where Re_L and Pr are the Reynolds and Prandtl numbers of the air around the part [31]. Using this relation, convective heat transfer was calculated, and it is equal to 80 W/m^{2}C. This is consistent with the values commonly found in the literature [21,22,27]. Usually, heat radiation from the part surface is very small and was ignored in the previous studies [21,25]. As it was suggested by Costa et al. [32], radiation heat transfer can be neglected when the convective loss is large (larger than 60 W/m^{2}C). Hence in this work, radiation was ignored.

The solution of Equation (1) was used to find the strain field using the following equation.

$$\varepsilon_{ij}^t = \alpha I(T - T_c) \tag{4}$$

where ε_{ij}^t is a thermal strain and α is the linear heat expansion coefficient, I—identity matrix. The result of the thermal strain was used as a boundary condition for the structural analysis. This is governed by the equilibrium equation given by

$$\frac{\partial}{\partial x_j}\left(C_{ij}\frac{\partial u_j}{\partial x_i}\right) + f_i = \rho\frac{\partial^2 u_i}{\partial t^2} \tag{5}$$

with the boundary conditions, $u = u_g$ on Γ_u and $\frac{\partial u}{\partial x_i} = f_s$ on Γ_j. Here, u is the displacements field, f_i—the body force per volume (zero in this study), f_s—the surface traction per area, C_{ij}—material stiffness matrix, ρ—the material density, u_g—prescribed displacements. Moreover, the strains are given as the sum of elastic, thermal, and plastic stresses ε_{ij}^e, ε_{ij}^t, ε_{ij}^p, respectively.

$$\varepsilon_{ij} = \frac{\partial u_i}{\partial x_j} = \varepsilon_{ij}^e + \varepsilon_{ij}^t + \varepsilon_{ij}^p \tag{6}$$

For the structural analysis, no surface traction was imposed on the part. However, a fully constrained displacement boundary condition was imposed, as shown in Figure 2b. These supports will be deleted during the spring-back phase of the simulation and will be discussed later. To avoid rigid body translation at this stage, one vertex at the center of the full part was also fully constrained for the duration of the whole simulation.

To discretize the model, the structured hex mesh was used, as shown in Figure 3a. The order of the mesh is two, meaning that there is a mid-side node on each edge of an element, as shown in Figure 3b. This allows using second-order shape functions, alleviates shear locking, and increases the accuracy of the solution for a given number of elements. The maximum size of the mesh is 1 mm along the x and z axes. A second-order interpolation was used and therefore, there were three nodes per element edge and the distance between two nodes is comparable with one road-width of the deposited filament. Along the y-axis, the size of a mesh is equal to the height of the layer.

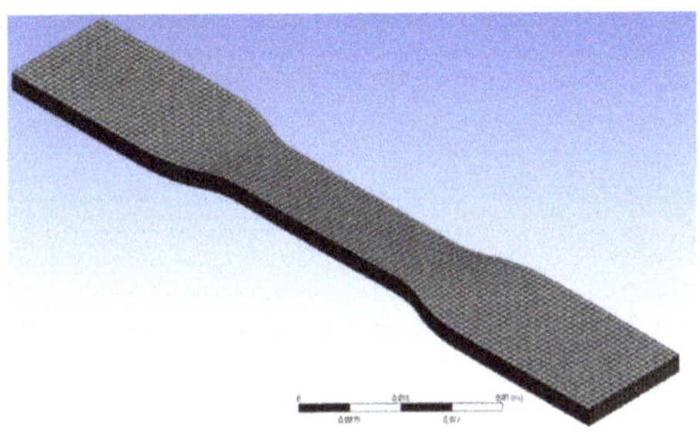

(a)

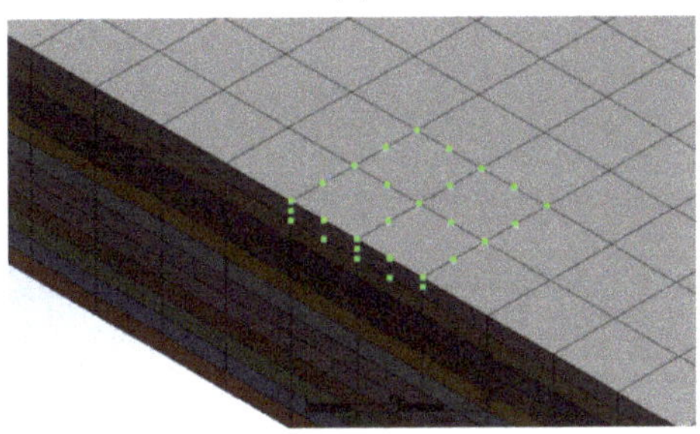

(b)

Figure 3. (**a**) General mesh of a body. (**b**) Second-order elements.

The deposition of the molten plastic was modeled using the element birth and death method. In the following method, different elements can be activated at different time steps, and the part topology is updated according to the activation algorithm [21].

The approach utilized for our development is represented in Figure 4. First, the elements were deactivated except for the platform. Starting from the time step 1, the layers were activated one by one and left for cooling. Simultaneously, after each time sub-step within a time step, the thermal analysis was conducted first according to Equations (1)–(3). Then the thermal result was used to calculate thermal loading using Equation (4), and it was used as input for equilibrium Equations (5)–(6). The time sub-step incremented, and the equations were solved again until the whole step was resolved. After one layer was resolved entirely, the next layer was activated. After all the layers are activated, the platform's temperature boundary condition is turned off, and it is left for cooling until it reaches equilibrium with the environment. Afterward, the part detachment is performed, and due to the thermal loading and constrained shrinkage, the part warps and deformations are obtained.

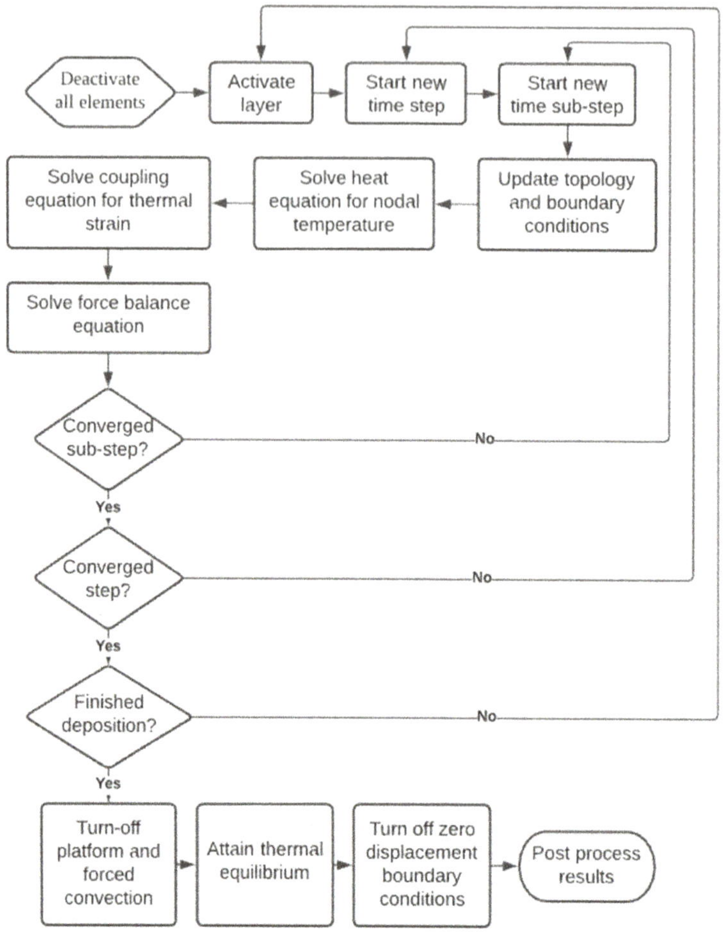

Figure 4. Algorithm used to perform simulations.

The material used in these simulations is Acrylonitrile Butadiene Styrene (ABS), and the material constants from Equations (1)–(6) are listed in Table 1. Other properties: heat capacity, heat transfer coefficient, Young's modulus, and yield stress were set as temperature-dependent, and their variation was obtained from [25]. For this simulation, an elastic perfectly plastic hardening model was assumed. This assumption is in agreement with the findings of [33]. To avoid the convergence problem, the secant modulus in the plastic region was set to 10% of young's modulus at the corresponding temperature.

Table 1. Constant material properties.

Density (ρ)	1040 kg/m^3	Poisson Ratio (v)	0.38
Glass transition temperature (T_g)	105 °C	Thermal expansion (α)	9×10^{-5}/°C

2.2. Experimental Setup

The samples were printed using the Ultimaker S3 printer (Ultimaker B.V., Utrecht, The Netherlands), which has a dual extrusion print head and a nozzle of 0.4 mm diameter. The geometry of the samples is shown in Figure 5. It was sliced in Ultimaker Cura software where extrusion temperature, bed temperature, and layer thickness were set individually for each experimental run according to the Taguchi orthogonal array (Tables 2 and 3), while other parameters were not changed throughout the experiments and are shown in Table 4. The ABS filaments (Bestfilament", Tomsk, Russia) were 2.85 mm in diameter. Three samples were printed for each experimental run resulting in 27 samples in total. Depending on the position, three samples were labeled such that the sample in the middle was denoted as "0" and the samples to the left and right of it were labeled "−1" and "1", respectively (Figure 5b).

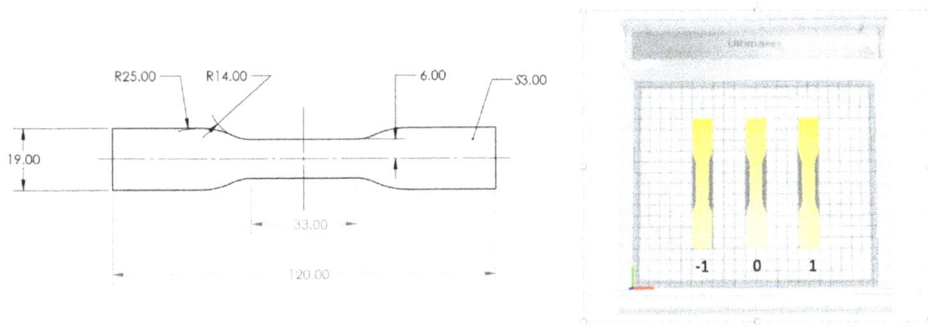

(a) (b)

Figure 5. (**a**) The geometry of the sample. (**b**) The position of the samples during printing.

Table 2. Factors and levels.

Factor	Symbol	Level			Unit
		1	2	3	
Layer thickness	A	0.1	0.2	0.3	mm
Bed temperature	B	85	95	105	°C
Nozzle temperature	C	220	230	240	°C

Table 3. L9 orthogonal array.

Experimental Run №	A (mm)	B (°C)	C (°C)
1	0.2	85	220
2	0.2	95	230
3	0.2	105	240
4	0.25	85	230
5	0.25	95	240
6	0.25	105	220
7	0.3	85	240
8	0.3	95	220
9	0.3	105	230

Table 4. Default Printing factors.

Factor	Value	Unit
Wall thickness	1.3	mm
Infill density	100	%
Infill pattern	Rectilinear	-
Print speed	55	mm/s
Fan speed	2	%

Before printing, the surface of the building platform was cleaned with the ethanol solution. To avoid excessive adhesion to the platform surface and subsequent damaging removal, no glue was applied. However, without glue, the samples were displaced from the specified positions by the movement of the nozzle and severely warped, which caused the nozzle to scratch the surface of the samples. Moreover, this scraping could damage the nozzle. Hence, brims were added to samples. After printing was completed, the samples were allowed to cool, then they were carefully removed from the platform and the brims were cut off. The platform was cleaned for the next experimental run and the procedure described above was repeated. The samples were then measured using a digital caliper.

Each sample was measured three times. Then the values were averaged. The parameter that denotes warpage was labeled as "H" and is shown in Figure 6.

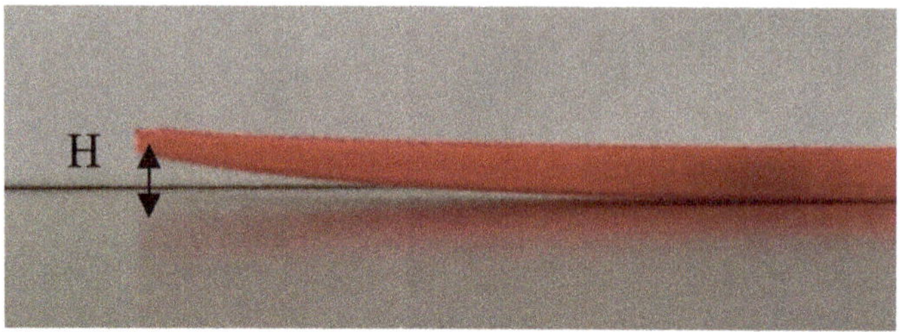

Figure 6. The warped edge of a sample.

3. Simulations and Experimental Results

The simulations were run, parts were manufactured, and warpage was measured according to the procedure. Figure 7 shows the deformation of the part just after the removal of the supports. It is seen that the part warps and the maximum deformation is expected at the corners of the part. There is also a shrinkage center at the geometrical center of a part. The deformations close to it are low, which was also observed by [15]. Thus, the part attains the shape of the bowl. The reason for this pattern is the shrinkage of the part during cooling. Due to the shrinkage, internal forces are generated within a constrained part. These forces cause internal moments, and after the removal of the part, they cause warping [18,20].

Figures 8 and 9 show the warpage deformation along the central half-length and utmost half-width. It is seen that the warpage progresses along the length and width. At the center of the part, zero warpage is expected, while close to the end it attains maximum value. In addition, warpage along the length increases more compared to the width dimension. Thus, for longer dimensions, the warpage is larger. Similar findings were also observed by [20].

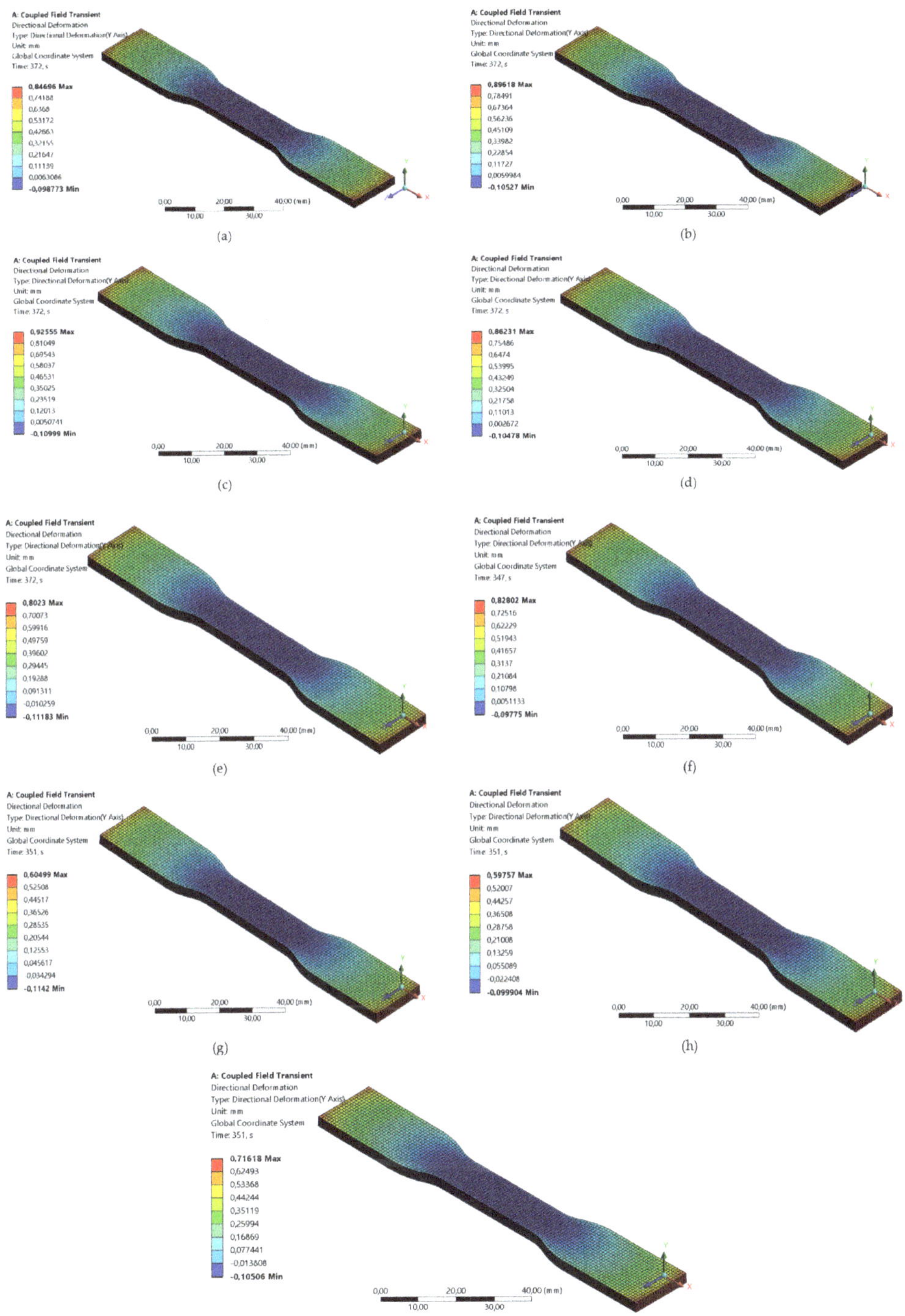

Figure 7. Directional deformation along the *y*-axis for run: (**a**) 1; (**b**) 2; (**c**) 3; (**d**) 4; (**e**) 5; (**f**) 6; (**g**) 7; (**h**) 8; (**i**) 9.

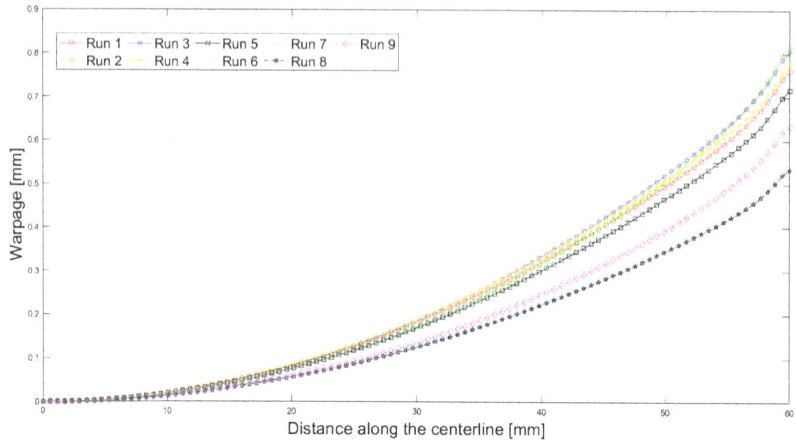

Figure 8. Warpage along the central half-length.

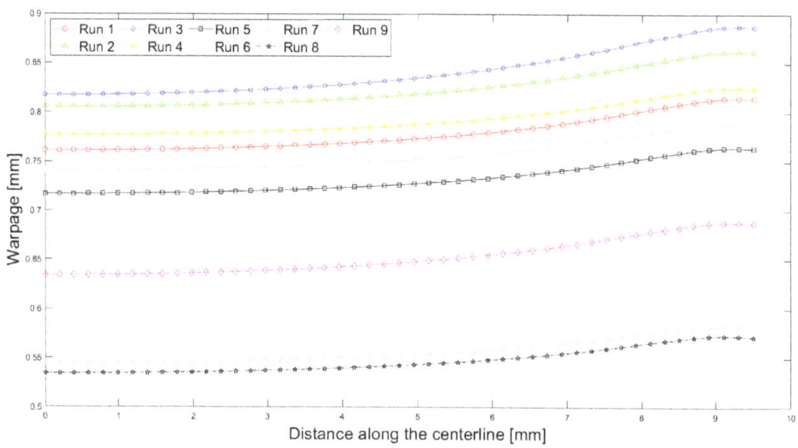

Figure 9. Warpage along the central half-width.

In order to compare the experimental results, numerical and several published analytical models, all the results were plotted, as shown in Figure 10. In addition, a comparison of experimental results and numerical predictions are listed in Table 5. Several analytical models were used to calculate warpage as the function of printing conditions. The results obtained using models developed by [17,19] were identical, as seen from the figure. This happened because they used similar principles (equilibrium) and assumptions (elastic loading at room temperature) in the derivations of their models. Note that warpage was predicted twice by Armillota et al. [20] using material properties at the room temperature (RT) conditions and the average temperature (AT) of the range. The latter model offered the best predictions for warpage among the given analytical models so far. The reason for this might be the inclusion of the yielding and layer reheating [21].

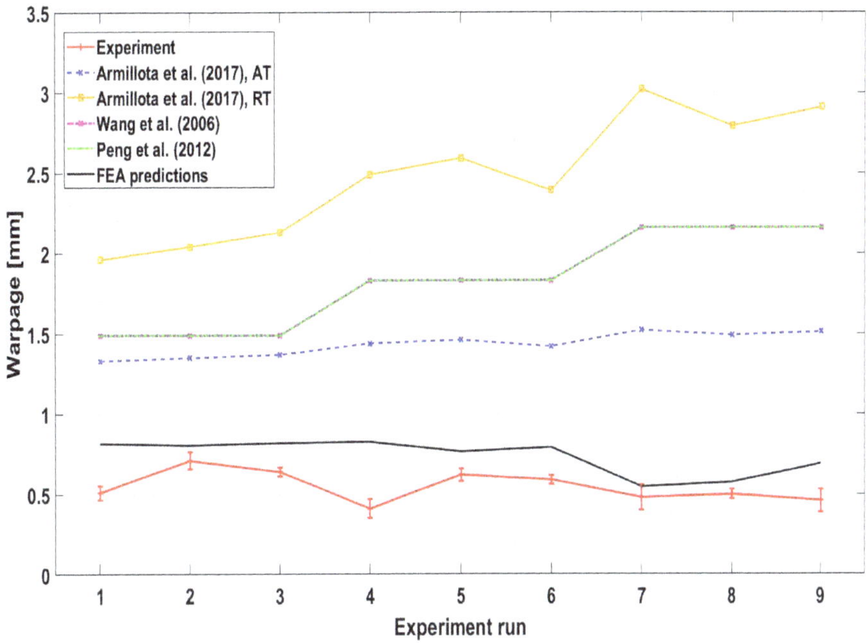

Figure 10. Comparison of FEA and analytical predictions with experimental results for warpage.

Table 5. Comparison of simulation predictions with experimental results for warpage.

Run №	Layer Thickness (mm)	Bed Temperature (°C)	Extrusion Temperature (°C)	Experimental H (μm)	Predicted H (μm)	Error %
1	0.2	85	220	506	814	60.9
2	0.2	95	230	709	805	12.8
3	0.2	105	240	639	818	28.0
4	0.25	85	230	414	825	99.3
5	0.25	95	240	617	764	23.8
6	0.25	105	220	588	790	34.4
7	0.3	85	240	483	545	12.8
8	0.3	95	220	501	573	14.3
9	0.3	105	230	457	689	50.8

From Table 5, the predictions of the finite element method were found to be close to the experimental results in some simulations but diverged in others. The model fits the results well for Runs 2, 5, 7 and 8. However, for Run 4, the discrepancy between Finite Element prediction and experimental measurement is high. The reasons for this might be the assumptions employed in Finite Element modeling and human errors during the measurements. Indeed, from Figure 10, it can be noticed that 410 μm of warpage measured during the experiment is abnormally low compared to other printings with similar process parameters. It can be noted that the predictive capability of the model becomes better at higher levels of the layer thickness. These findings may be supported with the aid of surface chemistry and roughness. It is reported that decreased coating weight generates higher hydrophobicity and surface roughness while thick layers come with fewer empty spaces between the layers, resulting in a reduced hydrophobic effect. In addition, thin layers of filament are most likely to retain the intrinsic unevenness of the surface [34,35]. Some studies reported on the extensive hydrophobic nature of thin coating related to the higher surface roughness [36], while other studies [37] suggested decreased surface roughness as well as hydrophobicity due to filled up voids and formation of large aggregate in the case of multiple layers. However, this hypothesis needs to be further investigated.

It can also be noticed that the values obtained by the proposed model are consistently higher than the experimental results. Similar overprediction was obtained by [25]. Normally, the strain energy of the approximate FE solution is always not greater than that of the exact solution [38], and hence predicted deformations should be lower than measured. This discrepancy might be explained by the fact that in the current model, creeping of the part was not included in the calculation. Due to heating from the printing bed and nozzle, the printed part is always heated during the building process. The creep rate of ABS is significantly higher at elevated temperatures [39]. Hence, because of the action of the adhesion of part to the platform, which acts in the opposite direction of the warpage, the part experiences severe deformation and straightens. Because of neglecting this effect in the Finite Element model, results obtained using FEM are larger than those in the actual experiments.

4. Taguchi Optimization

The Taguchi method can help to design experiments to study the effects of process parameters on response parameters. In addition, it allows reducing the number of experimental runs without resorting to complicated calculations. In this study, the process parameters are layer thickness, extrusion, and bed temperature, while warpage was chosen as a response parameter. It is desired to reduce the warping of the samples. Hence, the smaller-is-better approach was used. To analyze the effects of the process parameters on the warpage, the S/N ratios need to be calculated. Equation (7) is used to calculate η (S/N ratio) for the smaller-is-better approach in Taguchi analyses, where σ, Y_{avg}, and Y_0 are variance, average, and target value, respectively. In this study, the target value is 0.

$$\eta = -10\log\left(\sigma^2 + \left(Y_{avg} - Y_0\right)^2\right) \tag{7}$$

$$SST = \sum_{i=1}^{N}(\eta_i - \overline{\eta}) \tag{8}$$

$$SS_j = \sum_{i=1}^{L}(\eta_{ji} - \overline{\eta}) \tag{9}$$

$$MS_j = \frac{SS_j}{DOF_j} \tag{10}$$

$$F_j = \frac{MS_j}{MS_e} \tag{11}$$

For ANOVA analysis, SST, SS_j, MS_j, F_j values were calculated using Equations (8)–(11), where SST is the total sum of squares and $\overline{\eta}$ is called the average of S/N ratios of N number of experiments. SS_j is called the sum of squared deviations of the jth factor and L is the level of that factor. MS_j and DOF_j are called the variance and the degree of freedom of the jth factor, respectively. F_j is the F-value of the jth factor and is calculated by dividing MS_j by the error's variance (MS_e).

The results of calculations can be seen in Tables 6 and 7 for experimental and FEM predicted values of H, respectively. The larger values of S/N ratios indicate the optimum level of the parameter. Taguchi optimization from experimental results showed that for minimum warpage deviation layer thickness, bed temperature and extrusion temperature should be at levels 3, 1, and 2, respectively (Figure 11). ANOVA analysis can show the statistical significance of factors if the p-value is less than 0.05. The p-values from experimental H analyses were 0.272, 0.243, 0.607 for layer thickness, bed temperature, and extrusion temperature, respectively (Table 8).

Table 6. S/N ratios response table for experimental H.

Level	Layer Thickness (mm)	Bed Temperature (°C)	Nozzle Temperature (°C)
1	−55.74	−53.37	−54.49
2	−54.51	−55.61	−54.18
3	−53.62	−54.90	−55.20
Delta	2.11	2.24	1.01
Rank	2	1	3

Table 7. S/N ratios response table for FEM predicted H.

Level	Layer Thickness	Bed Temperature	Nozzle Temperature
1	−58.19	−57.09	−57.11
2	−57.98	−56.98	−57.74
3	−55.55	−57.65	−56.88
Delta	2.64	0.67	0.86
Rank	1	3	2

Table 8. ANOVA table for the warpage optimization based on experimental results.

Source	DOF	SS	MS	F	*p*	Contribution
A	2	6.713	3.356	2.675	0.272	35.961
B	2	7.818	3.909	3.116	0.243	41.885
C	2	1.626	0.813	0.648	0.607	8.711
Error	2	2.509	1.255	-	-	13.443
Total	8	18.666	-	-	-	100

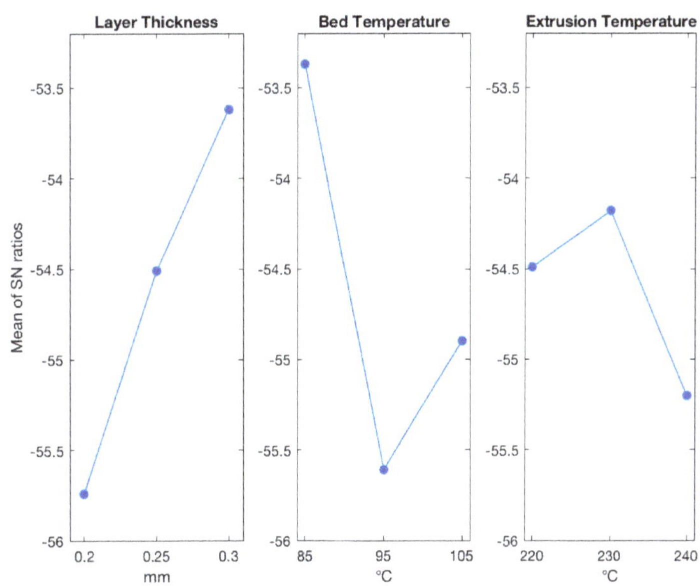

Figure 11. Experimentally derived results for warpage.

Taguchi optimization of warpage deviations using FEM results are shown for minimum warpage when levels of input parameters are as follows, 3, 2, and 3 (Figure 12). The *p*-values from ANOVA analyses were 0.042, 0.419, 0.343 for layer thickness, bed temperature, and extrusion temperature, respectively (Table 9).

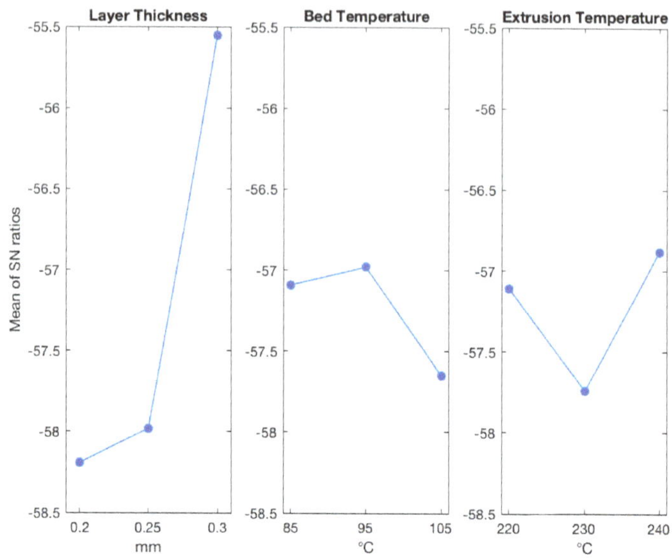

Figure 12. FEM results for warpage.

Table 9. ANOVA table for FEM predicted H.

Source	DOF	SS	MS	F	*p*	Contribution
A	2	12.911	6.455	22.819	0.042	83.60
B	2	0.784	0.392	1.385	0.419	5.07
C	2	1.184	0.592	2.092	0.343	7.66
Error	2	0.566	0.283	-	-	3.66
Total	8	15.444	-	-	-	100

According to the optimization based on experimental results (Figure 11 and Table 7), layer thickness and bed temperature are the most significant factors affecting the warpage of the part. Layer thickness and bed temperature have a contribution of approximately 36% and 42% to the final warpage. Additionally, the dependence of the warpage on the layer thickness is monotonic, and with an increase in the layer thickness, the warpage is minimized. On the other hand, the dependence of the warpage on the bed temperature is not monotonic.

Similarly, according to the optimization based on simulation results (Figure 12 and Table 9), layer thickness solely has the largest impact on the warpage. Its contribution is about 83.6% Again, simulation results show the inverse monotonic correlation between warpage and layer thickness, which agrees with experimental results. Similar results were also observed in other works [13–16].

To verify the results, the optimum levels of the process parameters were set, and the samples were printed using those parameters. The measured values of the warpage can be seen in Table 10. Optimum process parameters based on the results of the FE simulations lead to a part with a slightly smaller warpage value of 310 microns. At the same time, optimum process parameters based on the experimental results produce a part with a warpage equal to 320 microns. Nevertheless, both samples yield to lower warpage compared to the results from nine runs shown in Table 5.

Table 10. Warpage at optimum parameters.

	Factor and Levels	Measured H (μm)
Experiment	$A_3B_1C_2$	320
Simulation	$A_3B_2C_3$	310

5. Model Validation for Multilaterals

The application of FEM using ANSYS ® (ANSYS 2020R2, Canonsburg, Pennsylvania) was further extended to predict the warpage of FDM printed multi-material parts. In this study, HIPS (High Impact Polystyrene, Bestfilament", Tomsk, Russia) thermoplastic was used in different combinations with ABS (Acrylonitrile Butadiene Styrene (Bestfilament", Tomsk, Russia) material because of their better compatibility and uniformity when printed on top of each other [40]. As in the case of pure ABS part, a bilinear plastic model was used for HIPS Material. Both constant and transient material properties for HIPS material were based on the secondary findings, as shown in Table 11.

Table 11. Material properties for HIPS material.

Property	Value	Source
Glass transition temperature (°C)	100	[40]
Density (kg/m^3)	1048	[41]
CTE (Coefficient of thermal expansion) (1/°C)	6.7×10^{-5}	[42]
Thermal conductivity (W/mK)	Transient	[41]
Specific Heat (J/kgK)	Transient	[41]
Elastic Modulus (MPa)	Transient	[43]
Yield Strength (MPa)	Transient	[43]

The effect of material combinations on the warpage of printed multi-material parts was studied using a numerical study. The following material combinations were studied both numerically and experimentally:

Alternating specimen (AA HH AA HH AA HH)

Sandwich specimen (AAA HHHH AAA)

Note that HH stands for the two layers of the HIPS material, whereas AAA denotes the three layers of the ABS plastic (see Figure 13). Figure 14 shows the illustration of a printed multi-material sandwich specimen (AAA HHHH AAA).

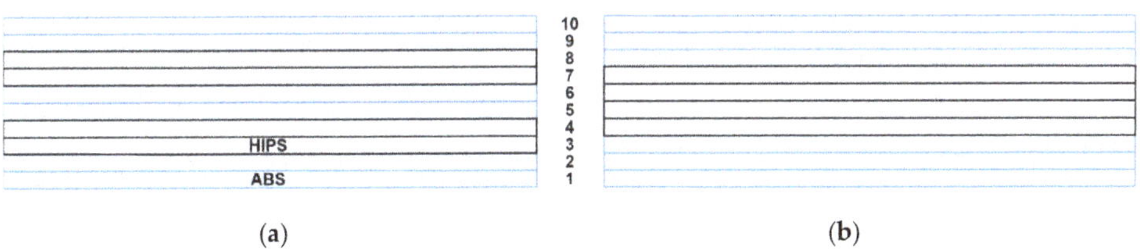

Figure 13. Multi-material combination schematics: (**a**) alternating specimen, (**b**) sandwich specimen.

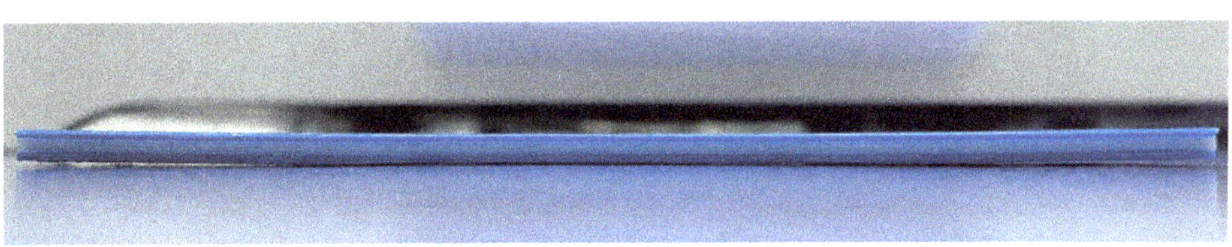

Figure 14. Printed multi-material sandwich specimen.

The same process parameters (0.3 mm layer thickness, 95 °C platform temperature, and 240 °C nozzle temperature) were used for both numerical and experimental studies. The numerical simulation result for the part warpage is presented in Figure 15.

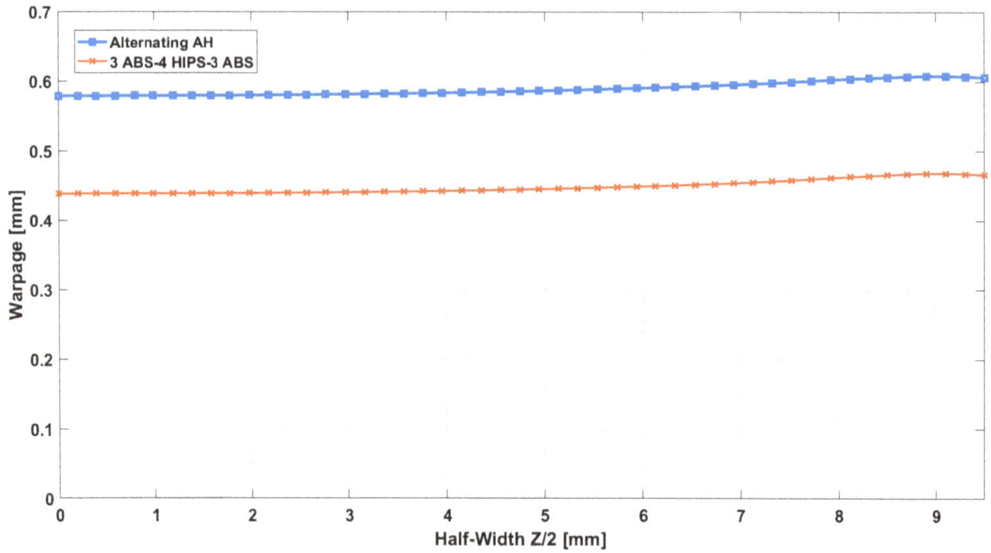

Figure 15. Warpage of multi-material parts along the half-width.

Table 12 provides detailed values of the numerical and experimental findings in terms of printed part warpage. The same material combinations were printed using a commercial Ultimaker S3 FDM printer. HIPS and ABS thermoplastics were obtained from the "Best filaments" manufacturer.

Table 12. Numerical simulation and experimental results.

Material Combination	Warpage (μm)		Error (%)
	FEM	Experimental	
Alternating specimen (AA HH AA HH AA HH)	607.93	616.67	1.4
Sandwich specimen (AAA HHHH AAA)	467.63	516.67	9.5

It can be noted that the FEM predicted values for the part warpage are bigger than the corresponding experimental findings. This implies that FEM overestimates the dimensional deviation of FDM printed parts. The same finding was stated in other literature [25]. The relative percentage error between the numerical and experimental warpage results for alternating and sandwich specimens are 1.4% and 9.5%, respectively (see Table 12). In this study, all the material properties were obtained from the existing literature and therefore might not be the same as the utilized thermoplastics. This can be considered as a feasible reason for the discrepancy between the numerical and experimental results. For example, the warpage prediction using FEM was shown to be linearly dependent on the CTE of the assigned material [42]. Therefore, the accuracy of numerical simulation in predicting the warpage of FDM printed parts can be enhanced by implementing the exact material properties as an input.

6. Conclusions

In this study, the FDM printing process was simulated to predict the warping deformation of the printed samples made from ABS only and from altering ABS-HIPS combinations (multi-material parts). The results were compared with analytical models from the literature and with the experimental results. The FEA model showed that samples warp in a bowl-like shape, which was also observed on experimentally printed parts. The predictions of the FEA model are closer to the actual warpage at higher values of the layer thickness. From this investigation, the following conclusions were observed:

- Both simulated and experimental results showed that the warpage decreases with increasing layer thickness.
- With regards to the analytical models, all models predicted much higher warping deformation compared to the experimental values and their respective numerical approximations. It was observed that using a model developed by Armillota et al. [20], calculated warpage values became more in line with experimental data when the average temperature was used instead of room temperature.
- In all analytical models and the developed FEA model, the warpage was overestimated. On the other hand, the FE results for displacement should be lower because the stiffness matrix obtained through the Finite Element solution is stiffer. This might happen because the assumptions employed in the FE modeling for the simplicity effect of the creep were not ignored.
- Regarding simulations of multi-material parts, the relative percentage error between the numerical and experimental warpage results for alternating and sandwich specimens are 1.4% and 9.5%, respectively.

Author Contributions: Conceptualization, A.P. and D.T.; methodology; validation, D.T.; investigation, D.S., B.Z., A.S.; writing and editing, D.S., B.Z., A.S., A.P., D.T.; supervision, A.P. and D.T.; project administration, D.T.; funding acquisition, D.T. All authors have read and agreed to the published version of the manuscript.

Funding: This research study was funded by Nazarbayev University under the project "Cost Effective Hybrid Casting Methods for Cellular Structures", grant No.: 240919FD3923.

Institutional Review Board Statement: Not applicable.

Informed Consent Statement: Not applicable.

Data Availability Statement: The data presented in this study are available on request from the corresponding author.

Conflicts of Interest: The authors declare no conflict of interest.

Nomenclature

Symbol	Meaning
c	Specific Heat
T	Temperature
t	Time
L	Length
A	Coefficient of Thermal Expansion
ρ	Material Density
k	Coefficient of Thermal Conductivity
q	Internal Heat Generation per unit Volume
Γ_u	Dirichlet Boundary
Γ_j	Neuman Boundary
n_i	Unit normal vector
T_0	Initial Temperature
q_c	Convective Heat Flux

h	Convective Heat Transfer Temperature
T_c	Temperature of the surrounding medium
Nu_L	Nusselt Number
ε_{ij}^{t}	Thermal Strain
ε_{ij}^{e}	Elastic Strain
ε_{ij}^{p}	Plastic Strain
u_j	Displacement Vector
C_{ij}	Material Stiffness Matrix
f_i	Body Force
f_s	Surface Force
η	S/N Ratio
σ	Variance
Y_{avg}	Average Response
Y_0	Target Response
SST	Total Sum of the Squares
SS_j	Sum of Squared Deviations of the *j*th Factor
MS_j	Variance of the *j*th factor
DOF_j	Degree of Freedom of the *j*th factor
F_j	F-value of the *j*th factor

References

1. Syrlybayev, D.; Zharylkassyn, B.; Seisekulova, A.; Akhmetov, M.; Perveen, A.; Talamona, D. Optimisation of Strength Properties of FDM Printed Parts—A Critical Review. *Polymers* **2021**, *13*, 1587. [CrossRef] [PubMed]
2. De Laurentis, K.J.; Mavroidis, C. Rapid fabrication of a non-assembly robotic hand with embedded components. *Assem. Autom.* **2004**, *24*, 394–405. [CrossRef]
3. Sood, A.K.; Ohdar, R.; Mahapatra, S.S. Improving dimensional accuracy of fused deposition modelling processed part using grey Taguchi method. *Mater. Des.* **2009**, *30*, 4243–4252. [CrossRef]
4. Thomas, D.S.; Gilbert, S.W. Costs and cost effectiveness of additive manufacturing. *NIST Spec. Publ.* **2014**, *1176*, 12.
5. Gibson, I.; Rosen, D.; Stucker, B.; Khorasani, M. *Additive Manufacturing Technologies*; Springer: Cham, Switzerland, 2021; Volume 17.
6. Hopkinson, N.; Hague, R.; Dickens, P. *Rapid Manufacturing: An Industrial Revolution for the Digital Age*; John Wiley & Sons: Chichester, UK, 2006.
7. Moylan, S.; Slotwinski, J.; Cooke, A.; Jurrens, K.; Donmez, A.M. An additive manufacturing test artifact. *J. Res. Natl. Inst. Stand. Technol.* **2014**, *119*, 429. [CrossRef] [PubMed]
8. Al-Ahmari, A.; Ashfaq, M.; Mian, S.H.; Ameen, W. Evaluation of additive manufacturing technologies for dimensional and geometric accuracy. *Int. J. Mater. Prod. Technol.* **2019**, *58*, 129–154. [CrossRef]
9. Mahmood, S.; Qureshi, A.; Talamona, D. Taguchi based process optimization for dimension and tolerance control for fused deposition modelling. *Addit. Manuf.* **2018**, *21*, 183–190. [CrossRef]
10. Anitha, R.; Arunachalam, S.; Radhakrishnan, P. Critical parameters influencing the quality of prototypes in fused deposition modelling. *J. Mater. Process. Technol.* **2001**, *118*, 385–388. [CrossRef]
11. Nancharaiah, T.; Raju, D.R.; Raju, V.R. An experimental investigation on surface quality and dimensional accuracy of FDM components. *Int. J. Emerg. Technol.* **2010**, *1*, 106–111.
12. Kumar, Y.R. An application of Taguchi's technique to improve the accuracy of rapid prototyped FDM parts. *Int. J. Mater. Eng. Innov.* **2012**, *3*, 228–246. [CrossRef]
13. Kuo, C.-C.; Wu, Y.-R.; Li, M.-H.; Wu, H.-W. Minimizing warpage of ABS prototypes built with low-cost fused deposition modeling machine using developed closed-chamber and optimal process parameters. *Int. J. Adv. Manuf. Technol.* **2019**, *101*, 593–602. [CrossRef]
14. Liu, L.; Wan, C.; Li, K.; Wu, J. Research on warpage deformation mechanism & control method of fused deposition parts. In *Advanced Graphic Communications and Media Technologies*; PPMT 2016: Lecture Notes in Electrical Engineering; Zhao, P., Ouyang, Y., Xu, M., Yang, L., Ouyang, Y., Eds.; Springer: Singapore, 2017; Volume 417. [CrossRef]
15. Xinhua, L.; Shengpeng, L.; Zhou, L.; Xianhua, Z.; Xiaohu, C.; Zhongbin, W. An investigation on distortion of PLA thin-plate part in the FDM process. *Int. J. Adv. Manuf. Technol.* **2015**, *79*, 1117–1126. [CrossRef]
16. Panda, B.N.; Shankhwar, K.; Garg, A.; Jian, Z. Performance evaluation of warping characteristic of fused deposition modelling process. *Int. J. Adv. Manuf. Technol.* **2017**, *88*, 1799–1811. [CrossRef]
17. Wang, T.-M.; Xi, J.-T.; Jin, Y. A model research for prototype warp deformation in the FDM process. *Int. J. Adv. Manuf. Technol.* **2007**, *33*, 1087–1096. [CrossRef]
18. Huynh, T.T.; Nguyen, T.V.T.; Nguyen, Q.M.; Nguyen, T.K. Minimizing warpage for macro-size fused deposition modeling parts. *CMC Comput. Mater. Contin.* **2021**, *68*, 2913–2923.

19. Peng, A.H. Research on the interlayer stress and warpage deformation in FDM. *Adv. Mater. Res.* **2012**, *538–541*, 1564–1567. [CrossRef]
20. Armillotta, A.; Bellotti, M.; Cavallaro, M. Warpage of FDM parts: Experimental tests and analytic model. *Robot. Comput. Integr. Manuf.* **2018**, *50*, 140–152. [CrossRef]
21. Zhang, Y.; Chou, Y. Three-dimensional finite element analysis simulations of the fused deposition modelling process. *Part B Proc. Inst. Mech. Eng. J. Eng. Manuf.* **2006**, *220*, 1663–1671. [CrossRef]
22. Chockalingam, K.; Karthic, M.; Jawahar, N. Development of thermal and structural deformation model to predict the part build dimensional error in fused deposition modeling. *ADMT J.* **2018**, *11*, 27–37.
23. El Moumen, A.; Tarfaoui, M.; Lafdi, K. Modelling of the temperature and residual stress fields during 3D printing of polymer composites. *Int. J. Adv. Manuf. Technol.* **2019**, *104*, 1661–1676. [CrossRef]
24. De Macedo, R.Q.; Ferreira, R.T.L. COBEM-2017-0124 residual thermal stress in fused deposition modelling. In Proceedings of the 24th ABCM International Congress of Mechanical Engineering, Curitiba, Brazil, 3–8 December 2017.
25. Cattenone, A.; Morganti, S.; Alaimo, G.; Auricchio, F. Finite element analysis of additive manufacturing based on fused deposition modeling: Distortions prediction and comparison with experimental data. *J. Manuf. Sci. Eng.* **2019**, *141*, 011010. [CrossRef]
26. Samy, A.A.; Golbang, A.; Harkin-Jones, E.; Archer, E.; McIlhagger, A. Prediction of part distortion in Fused Deposition Modelling (FDM) of semi-crystalline polymers via COMSOL: Effect of printing conditions. *CIRP J. Manuf. Sci. Technol.* **2021**, *33*, 443–453. [CrossRef]
27. Zhou, Y.; Lu, H.; Wang, G.; Wang, J.; Li, W. Voxelization modelling based finite element simulation and process parameter optimization for Fused Filament Fabrication. *Mater. Des.* **2020**, *187*, 108409. [CrossRef]
28. Naranjo-Lozada, J.; Ahuett-Garza, H.; Orta-Castañón, P.; Verbeeten, W.M.H.; Sáiz-González, D. Tensile properties and failure behavior of chopped and continuous carbon fiber composites produced by additive manufacturing. *Addit. Manuf.* **2019**, *26*, 227–241. [CrossRef]
29. Favaloro, A.J.; Brenken, B.; Barocio, E.; Pipes, B.R. Simulation of polymeric composites additive manufacturing using Abaqus. In Proceedings of the SIMULIA Science in the Age of Experience, Chicago, IL, USA, 15–18 May 2017; pp. 103–114.
30. Wang, J.; Papadopoulos, P. Coupled thermomechanical analysis of fused deposition using the finite element method. *Finite Elem. Anal. Des.* **2021**, *197*, 103607. [CrossRef]
31. Rathore, M.M. *Engineering Heat and Mass Transfer*; Laxmi Publications (P) Ltd.: New Delhi, India, 2006.
32. Costa, S.; Duarte, F.; Covas, J. Thermal conditions affecting heat transfer in FDM/FFE: A contribution towards the numerical modelling of the process: This paper investigates convection, conduction and radiation phenomena in the filament deposition process. *Virtual Phys. Prototyp.* **2015**, *10*, 35–46. [CrossRef]
33. Rodríguez-Panes, A.; Claver, J.; Camacho, A.M. The influence of manufacturing parameters on the mechanical behaviour of PLA and ABS pieces manufactured by FDM: A comparative analysis. *Materials* **2018**, *11*, 1333. [CrossRef]
34. Henn, K.A.; Forsman, N.; Zou, T.; Österberg, M. Colloidal lignin particles and epoxies for bio-based, durable, and multiresistant nanostructured coatings. *ACS Appl. Mater. Interfaces* **2021**, *13*, 34793–34806. [CrossRef]
35. Eslami, H.; Gharibi, A.; Müller-Plathe, F. Mechanisms of nucleation and solid–solid-phase transitions in Triblock Janus Assemblies. *J. Chem. Theory Comput.* **2021**, *17*, 1742–1754. [CrossRef] [PubMed]
36. Forsman, N.; Lozhechnikova, A.; Khakalo, A.; Johansson, L.-S.; Vartiainen, J.; Österberg, M. Layer-by-layer assembled hydrophobic coatings for cellulose nanofibril films and textiles, made of polylysine and natural wax particles. *Carbohydr. Polym.* **2017**, *173*, 392–402. [CrossRef]
37. Söz, C.K.; Yilgör, E.; Yilgör, I. Influence of the average surface roughness on the formation of superhydrophobic polymer surfaces through spin-coating with hydrophobic fumed silica. *Polymer* **2015**, *62*, 118–128.
38. Bathe, K.-J. *Finite Element Procedures*; Klaus-Jurgen Bathe: Cambridge, MA, USA, 2006.
39. An, T.; Selvaraj, R.; Hong, S.; Kim, N. Creep Behavior of ABS Polymer in Temperature–Humidity Conditions. *J. Mater. Eng. Perform.* **2017**, *26*, 2754–2762. [CrossRef]
40. Singh, R.; Kumar, R.; Farina, I.; Colangelo, F.; Feo, L.; Fraternali, F. Multi-material additive manufacturing of sustainable innovative materials and structures. *Polymers* **2019**, *11*, 62. [CrossRef] [PubMed]
41. Dos Santos, W.N. Thermal properties of melt polymers by the hot wire technique. *Polym. Test.* **2005**, *24*, 932–941. [CrossRef]
42. Alsewailem, F.D. On the thermal expansion behavior of polystyrene/polyethyleneterephthalate blend systems: Experimental study. *J. Thermoplast. Compos. Mater.* **2009**, *22*, 585–604. [CrossRef]
43. Momanyi, J.; Herzog, M.; Muchiri, P. Analysis of thermomechanical properties of selected class of recycled thermoplastic materials based on their applications. *Recycling* **2019**, *4*, 33. [CrossRef]

polymers

MDPI

Article

Exact Solution for Viscoelastic Flow in Pipe and Experimental Validation

Ekaterina Vachagina, Nikolay Dushin, Elvira Kutuzova and Aidar Kadyirov *

Institute of Power Engineering and Advanced Technologies, FRC Kazan Scientific Center, Russian Academy of Sciences, 420111 Kazan, Russia; vachaginae@mail.ru (E.V.); ndushin@bk.ru (N.D.); elvira.kutuzova@list.ru (E.K.)
* Correspondence: aidarik@rambler.ru; Tel.: +7-843-292-7597

Abstract: The development of analytical methods for viscoelastic fluid flows is challenging. Currently, this problem has been solved for particular cases of multimode differential rheological equations of media state (Giesekus, the exponential form of Phan-Tien-Tanner, eXtended Pom-Pom). We propose a parametric method that yields solutions without additional assumptions. The method is based on the parametric representation of the unknown velocity functions and the stress tensor components as a function of coordinate. Experimental flow visualization based on the SIV (smoke image velocimetry) method was carried out to confirm the obtained results. Compared to the Giesekus model, the experimental data are best predicted by the eXtended Pom-Pom model.

Keywords: polymer solution; Giesekus; eXtended Pom-Pom; visualization; analytical solution

Citation: Vachagina, E.; Dushin, N.; Kutuzova, E.; Kadyirov, A. Exact Solution for Viscoelastic Flow in Pipe and Experimental Validation. *Polymers* **2022**, *14*, 334. https://doi.org/10.3390/polym14020334

Academic Editors: Célio Bruno Pinto Fernandes, Salah Aldin Faroughi, Luís L. Ferrás and Alexandre M. Afonso

Received: 10 December 2021
Accepted: 12 January 2022
Published: 15 January 2022

1. Introduction

Nonlinear differential rheological equations of state are being used increasingly often to describe the rheological properties of viscoelastic fluids and to solve fluid mechanics problems related to polymer melts and solutions. First of all, the features of viscoelastic polymer-based materials behavior include stress-relaxation phenomena, the presence of elastic properties, nonlinear dependence of effective viscosity on shear rate, the occurrence of normal stresses in shear flows, and the ability to swell.

The rheological models based on relaxation equations of state and the structure of polymer molecules, which describe all the above phenomena that occur in flows of viscoelastic polymer materials in the channels of production equipment, are well known in the literature. Analytical solutions can be obtained using simple equations or under rather rough simplifying conditions. As a consequence, most of the studies deal with unimodal viscoelastic rheological models, e.g., Gruz and Pinho [1,2] (Poiseuille-Couette flows of PTT fluids), Oliveira [3] (Fene-P and Giesekus fluid flows in round pipes and flat channels), Schleiniger and Weinacht [4], or a number of works by Oliveira et al. [5] and Coelho et al. [6,7] (isothermal and non-isothermal Fene-P and PTT fluid flows in round pipes and flat channels [5,8]), Hashemabadi [9,10]. Nevertheless, the rheological properties of polymeric melts and concentrated polymer solutions are very often more complex than predicted by such unimodal models, and hence multimode models providing more adequate description are required. In [11], an analytical solution was obtained for a simplified multimode rheological model of a PTT fluid flow. In most cases, calculations of multimode viscoelastic flows are carried out by numerical methods, e.g., [12,13]. The disadvantage of the numerical approach to solution of the considered problems is that for each particular case, it is necessary to perform a full complex of numerical studies, which, due to strong nonlinearity of the considered equations, requires significant computational resources. In contrast, the parametric representation of solution yields distributions of velocities and stresses for a wide variety of problems with any degree of accuracy.

Analytical methods for viscoelastic fluid flows described in [4–6] employed simplified rheological models which led to relations connecting the nonzero components of the elastic strain tensor and the shear-rate gradient. Using the solution to the momentum transfer equation in projection onto the direction of a viscoelastic medium motion for such simplified models, the authors [4–6] then obtained a correlation between a specific selected unknown function and a transverse coordinate. In this case, the remaining unknown functions are expressed in terms of the selected one using the obtained relations. Our solution method does not have the drawback of using simplified rheological models and can be adapted to an arbitrary number of the rheological model modes. Some experimental studies were carried out to visualize the viscoelastic medium flow in a round pipe to check the obtained results of theoretical studies. The experimental data were compared with the results obtained using the Giesekus model and Extended Pom-Pom.

2. Problem Statement

2.1. Mathematical Model

A steady laminar isothermal flow of viscoelastic incompressible fluid in around pipes was considered. We assumed that the velocity vector had a single axial velocity component V_z, which is a function of the only variable r of a cylindrical coordinate system r, φ, z with the z-axis directed along the pipe axis. Under the assumptions made, the system of equations for the momentum transfer and continuity in the selected coordinate system can be written as

$$-\frac{\partial P}{\partial z}+\frac{1}{r}\frac{d(r\sigma_{rz})}{dr}=0, -\frac{\partial P}{\partial r}+\frac{1}{r}\frac{d(r\sigma_{rr})}{dr}-\frac{\sigma_{\varphi\varphi}}{r}=0, -\frac{\partial P}{\partial \varphi}=0, \frac{dV_z}{dz}=0 \quad (1)$$

where P is the pressure; σ_{rr}, $\sigma_{\varphi\varphi}$, σ_{rz} are physical components of the stress tensor in the cylindrical coordinate system. Let us assume that the liquid adheres to the pipe wall. It follows from (1) that

$$\sigma_{rz}=-|C_0|r/2, \ \partial P/\partial z=C_0=const=-|C_0| \quad (2)$$

The rheological equation of state for a multimode viscoelastic fluid can be written as

$$\boldsymbol{\sigma}=\sum_{k=1}^{n}\boldsymbol{\sigma}_k+\boldsymbol{\sigma}_N, \boldsymbol{\sigma}_N=2\eta_N\mathbf{D}$$

where n is the total number of modes; $\boldsymbol{\sigma}$ is the extra stress tensor; $\boldsymbol{\sigma}_N$ is the Newtonian part of extra stress tensor; η_N is the solvent viscosity; $\boldsymbol{\sigma}_k$ is the elastic part of extra stress tensor corresponding to each mode.

Two rheological models of viscoelastic behavior were used to determine the elastic components of each mode:

Giesekus model [14]:

$$\boldsymbol{\sigma}_k+\lambda_k\overset{\nabla}{\boldsymbol{\sigma}}_k+\frac{\alpha_k\lambda_k}{\eta_k}\boldsymbol{\sigma}_k\cdot\boldsymbol{\sigma}_k=2\eta_k\mathbf{D}, (k=1,\ldots,n) \quad (3)$$

Single-equation eXtended Pom-Pom [15] (hereinafter mentioned as eXt. Pom-Pom):

$$f(\boldsymbol{\sigma}_k)\boldsymbol{\sigma}_k+\lambda_k\left(\overset{\nabla}{\boldsymbol{\sigma}}_k\right)+\frac{\lambda_k\alpha_k}{\eta_k}(\boldsymbol{\sigma}_k\cdot\boldsymbol{\sigma}_k)+\frac{\eta_k}{\lambda_k}(f(\boldsymbol{\sigma}_k)-1)\mathbf{I}=2\eta_k\mathbf{D} \quad (4)$$

where $f(\boldsymbol{\sigma}_k)=\frac{2}{\varepsilon_k}\exp\left[\frac{2(\Lambda_k-1)}{q_k}\right]\left(1-\frac{1}{\Lambda_k}\right)+\frac{1}{\Lambda_k{}^2}\left(1-\frac{\alpha_k\lambda_k{}^2 tr(\boldsymbol{\sigma}_k\cdot\boldsymbol{\sigma}_k)}{3\eta_k{}^2}\right)$, $\Lambda_k=\sqrt{1+\frac{\lambda_k tr\boldsymbol{\sigma}_k}{3\eta_k}}$ is the backbone stretch, $\varepsilon_k=\frac{\lambda_{s0k}}{\lambda_k}$, $k=1,\ldots,n$; k is the current mode number, $\overset{\nabla}{\boldsymbol{\sigma}}=\frac{d\boldsymbol{\sigma}}{dt}-\boldsymbol{\sigma}\cdot\nabla\mathbf{V}^T-\nabla\mathbf{V}\cdot\boldsymbol{\sigma}$ is the upper convective derivative of the tensor $\boldsymbol{\sigma}$; $\mathbf{I}$ is the unit tensor; p is the pressure; $\mathbf{D}=\left(\nabla\mathbf{V}+(\nabla\mathbf{V})^T\right)/2$ is the rate of deformation tensor; $\mathbf{V}$ is the velocity; λ_k is the relaxation time; η_k is the polymeric viscosity; q_k is the number of arms at the backbone extremity of the Pom-Pom molecule; α_k and ε_k are the rheological parameters.

2.2. Parametric Method

In the present paper, we suggest that a solution method based on the search for an unknown functional dependency $V_z(r)$ should be used in a parametric form. Let us consider a solution method for multimode Giesekus fluid flows in round pipes. For this purpose, we rewrite (3) in a cylindrical coordinate system according to [16]. After some transformations and introduction of the parameter

$$\rho_k = \frac{\lambda_k}{\eta_k}\left|\sigma_{rr(k)}\right| \quad (5)$$

we obtain the following expression for nonzero components of the elastic stress tensor and the shear rate gradient:

$$\sigma_{rz(k)} = -\frac{\eta_k}{\lambda_k}A_k,\ \sigma_{zz(k)} = \frac{\eta_k\rho_k}{\alpha_k\lambda_k}\left(\frac{2-\alpha_k(\rho_k+1)}{(1-\rho_k)}\right),\ \sigma_{rr(k)} = -\frac{\eta_k}{\lambda_k}\rho_k \quad (6)$$

$$\frac{dV_z}{dr} = -\frac{1+\rho_k(1-2\alpha_k)}{(1-\rho_k)^2}\frac{A_k}{\lambda_k}(k=1,\ldots,n) \quad (7)$$

Hereinafter $A_k = \frac{\sqrt{\rho_k(1-\alpha_k\rho_k)}}{\sqrt{\alpha_k}}$.

Introducing the parameter ρ_k into (6), we get

$$\sigma_{rz} =-\frac{\eta_N(1-2\alpha_k\rho_k+\rho_k)}{\lambda_k(1-\rho_k)^2}A_k - \sum_{j=1}^{n}\frac{\eta_j}{\lambda_j}A_j\ (k=1,\ldots,n) \quad (8)$$

$$\widetilde{r} = \frac{2R}{|C_0|}\left(\frac{\eta_N(1-2\alpha_k\rho_k+\rho_k)}{\lambda_k(1-\rho_k)^2}A_k + \sum_{j=1}^{n}\frac{\eta_j}{\lambda_j}A_j\right),\ (\widetilde{r} = r/R) \quad (9)$$

The remaining components of the stress tensor can be obtained similarly: σ_{zz}, σ_{rr}.

The parameter ρ_k was introduced for each of the modes; therefore, to obtain the parametric dependences $dV_z(r)/dr$, $V_z(r)$, and $\sigma_{ij}(r)$ we select the parameter ρ_1, corresponding to the first mode. The remaining parameters $\rho_k(k \neq 1)$ must be expressed in terms of the main one ρ_1 using equations (7).

The parametric dependence of the dimensionless shear rate gradient will have the form:

$$\begin{cases} \frac{dv_z(\rho_1)}{dr} = -\frac{1+\rho_1(1-2\alpha_1)}{(1-\rho_1)^2}\frac{A_1}{\widetilde{\lambda}_1 Wi} \\ \widetilde{r}(\rho_1) = \frac{1}{Kr}\left(\frac{\widetilde{\eta}_N(1-2\alpha_1\rho_1+\rho_1)A_1}{\widetilde{\lambda}_1(1-\rho_1)^2} + \sum_{j=1}^{n}\frac{\widetilde{\eta}_j\sqrt{\rho_j(\rho_1)(1-\alpha_j\rho_j(\rho_1))}}{\sqrt{\alpha_j}\widetilde{\lambda}_j}\right) \end{cases} \quad (10)$$

where $\rho_j(\rho_1)$ is determined with any degree of accuracy from (7) using numerical methods, for example, by dichotomy method; $\widetilde{\lambda}_j = \lambda_j/\lambda_a$, $\widetilde{\eta}_j = \eta_j/\eta_0$ and $\widetilde{\eta}_N = \eta_N/\eta_0$ are dimensionless simplexes; $\lambda_a = \sum_{i=1}^{n}\lambda_i\eta_i/\sum_{i=1}^{n}\eta_i$ is the average relaxation time; $\eta_0 = \eta_N + \sum_{i=1}^{n}\eta_i$ is the maximum possible fluid viscosity; $v_z = V_z/V_a$ is the dimensionless velocity; $V_a = Q/(\pi R^2)$ is the average bulk velocity; Q is the fluid flow rate through the cross-section of the circular pipe; R is the radius of the circular pipe. Here $Wi = (\lambda_a R/V_a)$ is the Weissenberg number, $Kr = Wi \cdot Eu \cdot \mathrm{Re}_0 \cdot \Gamma$ is the dimensionless complex; $Eu = |C_0 l|/(\rho {V_a}^2)$ is the Euler number; $\mathrm{Re}_0 = (\rho V_a(2R))/\eta_0$ is the Reynolds number; l is the pipe length; $\Delta P = |C_0 l|$ is the pressure drop in a round pipe across the length l.

To obtain the dependence of the dimensionless velocity v_z on the dimensionless variable in a parametric form, it is necessary to use the condition

$$\int_0^1 v_z \tilde{r} d\tilde{r} = \frac{1}{2} \tag{11}$$

The parametric dependence of the axial component of the dimensionless velocity on the dimensionless coordinate can be obtained from (10) by integrating these relations

$$\begin{cases} v_z(\rho_1) = \frac{\tilde{\eta}_N}{2KrWi\tilde{\lambda}_1{}^2\alpha_1} \frac{\left(1+(1-2\alpha_1)\rho_1^w\right)^2 \rho_1^w \left(1-\alpha_1\rho_1^w\right)}{\left(1-\rho_1^w\right)^4} \\ -\frac{\tilde{\eta}_N}{2KrWi\tilde{\lambda}_1{}^2\alpha_1} \frac{(1+(1-2\alpha_1)\rho_1)^2\rho_1(1-\alpha_1\rho_1)}{(1-\rho_1)^4} + \frac{1}{2KrWi}\sum\limits_{j=1}^{n} \frac{\tilde{\eta}_j}{\tilde{\lambda}_j{}^2\alpha_j} B(\rho_j) \\ \tilde{r}(\rho_1) = \frac{1}{Kr}\left(\frac{\tilde{\eta}_N(1-2\alpha_1\rho_1+\rho_1)A_1}{\lambda_1(1-\rho_1)^2} + \sum\limits_{j=1}^{n} \frac{\tilde{\eta}_j\sqrt{\rho_j(\rho_1)(1-\alpha_j\rho_j(\rho_1))}}{\sqrt{\alpha_j}\lambda_j} \right) \end{cases} \tag{12}$$

where $B(\rho_j) = \int\limits_{\rho_j}^{\rho_1^w} \frac{\left(1+(1-2\alpha_j)\rho_j\right)\left(1-2\alpha_j\rho_j\right)}{\left(1-\rho_j\right)^2} d\rho_j$ is calculated analytically; ρ_1^w is the parameter value on the pipe wall.

The direct use of analytical expressions (12) is hampered by the fact that there is a functional dependence between the dimensionless complexes Wi and Kr, which can be determined using (11).

Similarly, a parametric problem solution for multimode fluid flows can be obtained using eXt. Pom-Pom can be obtained. In this case, it is convenient to use the following expression as a parameter (ρ_k)

$$\sigma_{\varphi\varphi(k)} = \frac{\tilde{\eta}_k}{\tilde{\lambda}_k} \frac{\eta_0}{\lambda_a r^2}(-\rho_k),\ \tilde{\eta}_k = \eta_k/\eta_0,\ \tilde{\lambda}_k = \lambda_k/\lambda_a.\ (0 \le \rho_k \le 1) \tag{13}$$

Then

$$\begin{cases} v_z(\rho_1) = \int\limits_{\rho_1}^{\rho_1^w} \dot{\gamma}_k(\rho_1)(d\tilde{r}/d\rho_1)/\tilde{\lambda}_k d\rho_1 \\ \tilde{r}(\rho_1) = -\left(\frac{\tilde{\eta}_N}{\tilde{\lambda}_m}\dot{\gamma}_m(\rho_m) + \sum\limits_{k=1}^{n} \frac{\tilde{\eta}_k}{\tilde{\lambda}_k}\frac{\lambda_k}{\eta_k}\sigma_{rz(k)}(\rho_k) \right)(Kr)^{-1} \end{cases} \tag{14}$$

where dependences $\rho_k(\rho_1)$ can be obtained from the definition of the shear rate gradient

$$\dot{\gamma} = \frac{dv_z}{d\tilde{r}} = \frac{\dot{\gamma}_1}{Wi\tilde{\lambda}_1} = \ldots = \frac{\dot{\gamma}_n}{Wi\tilde{\lambda}_n} \tag{15}$$

A more detailed description of the method is provided in [17].

3. Materials and Methods

3.1. Experimental Setup

Experimental studies to investigate the structure of the viscoelastic fluid flow were carried out on a test bench (FRC KazSC of RAS, Kazan, Russia) shown in Figure 1. The test sections were made of transparent plexiglass with the diameter D1 = 39 (inner)/D1 = 45 (outer) mm (Figure 1c). The inlet and outlet sections had the length of 43 D_1 sufficient to exclude the inflow and outflow effects. The test section was placed in a rectangular duct (Figure 1d) to compensate for optical distortions. The space between the test section and the duct was filled with a fluid with a refractive index close to that of the test section and the duct. The actual fluid temperature was controlled using a resistance temperature device DTS034-RT100.A3.25/1.5 (OVEN, Moscow, Russia) with the uncertainty of 0.1 °C

(Figure 1f,g). The fluid temperature was kept constant using a KENTATSU (KENTATSU DENKI, Guangdong, China) air conditioner installed in the room.

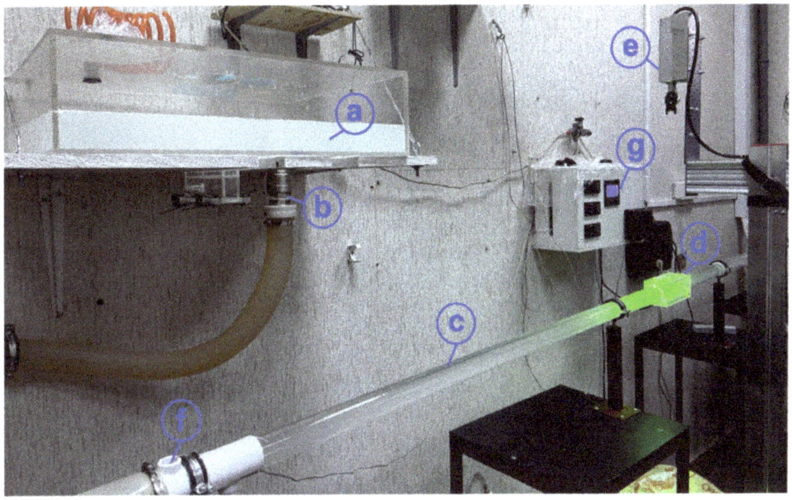

Figure 1. Photo of the test bench. (**a**) tank with polymer solution in distilled water; (**b**) valve; (**c**) Plexiglass pipe; (**d**) rectangular duct; (**e**) laser; (**f**) resistance thermocouple; (**g**) temperature monitor.

3.2. Visualization

Velocity profiles were estimated using the Smoke Image Velocimetry (SIV, Kazan, Russia) method developed at our institute. It is an optical measurement method. Articles [18,19] give a detailed description of the SIV method, estimation of its accuracy, and applicability to the measurement of various flow characteristics. The SIV method was chosen due to the peculiarities of the image-processing algorithm. In particular, the SIV method requires neither uniform seeding of the flow with tracers nor smoothing the maximum of the cross-correlation function when refining the displacement of tracers with subpixel accuracy. These aspects simplify the preparation of the working fluid and increase the measurement accuracy. Figure 2 illustrates the tracer's motion at different times. For example, the motion of one particle is highlighted with a red rectangle. The detailed description of the SIV method is presented in [18,19]. The used frequency value was sufficient to fulfill the condition for the permissible displacement of the tracer images with respect to the size of the compared interrogation windows at the highest flow rates and provided the convenience of a virtual decrease in the shooting frequency at lower flow rates. The virtual decrease in the shooting frequency was used to increase the displacement of tracers in the compared pairs of frames to the optimal values from the standpoint of measurement accuracy (about half the longitudinal size of the interrogation window). The size of the compared interrogation windows in all cases was 28×20 pix^2 and was selected based on the patterns of the distribution of tracers in the flow visualization frames. The scaling factor was 24.4 pix/mm. The spacing between grid nodes was 20 pix (0.82 mm). The shooting time and the number of captured frames were sufficient for time averaging of the data. For most of the investigated modes, 1000 pairs of frames were compared.

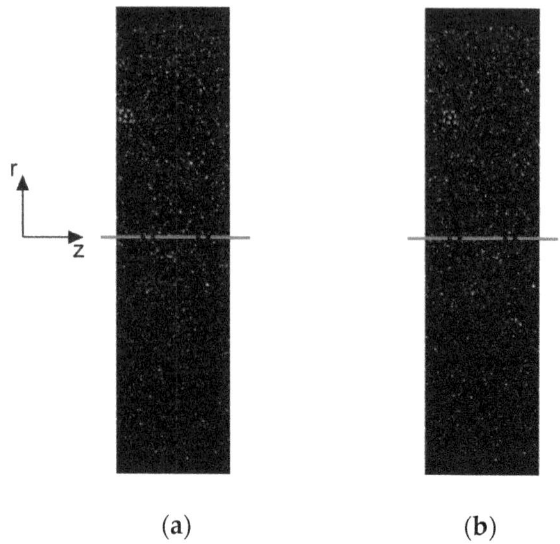

Figure 2. The tracer's motion at different time: (**a**) $t = t_0$; (**b**) $t = t_0 + \Delta t$.

The laser light sheet in the experiments was created by a continuous diode laser SSP-ST-532-NB-LED (Changchun New Industries Optoelectronics Tech. Co. Ltd, Changchun, China) (Figure 1e). The laser light sheet was no more than 0.6 mm thick. Filming was carried out with a Phantom Miro C110 digital video camera (Vision Research, Inc., Wayne, New Jersey, USA) with a frequency of 60 fps in steady flow regimes. The image processing procedure included elimination of static objects and noise filtering. The procedures for searching and filtering erroneous velocity vectors were applied to the primary quantitative data. The average number of erroneous vectors in the entire volume of the obtained data was no more than 0.56%.

3.3. *Polymer Solution Preparation*

An aqueous solution of polyacrylamide with polyamide particles (average diameter 20 μm) was used as a viscoelastic liquid. The influence of gravity and inertia forces on the position of tracers relative to the carrier medium during the measurements was negligible. The fluid was prepared in the following order. Polyamide particles with a concentration of 0.013% of the total mass were added to warm distilled water and slightly (manually) mixed. Then, the mixing process was carried out using an overhead stirrer Heidolph Hei-TORQUE Value 100 (Heidolph Instruments GmbH & Co. KG, Schwabach, Germany) with a ViscoJet stirrer (Heidolph Instruments GmbH & Co. KG, Schwabach, Germany). After two minutes of mixing, the polyacrylamide powder was gradually added. It took 15 min to mix 700 mL volume. The uncertainty in weighing of distilled water, polyacrylamide powder, and polyamide particles was less than 0.1%. Prior to the experiment, the finished liquid was stored in 5 L canisters in a dark cabinet.

3.4. *Rheological Measurements*

The viscosity curve, as well as dynamic moduli of two concentration of polymer solutions (2500 and 7500 ppm weight), were measured using MCR 102 rheometer (Anton Paar, Graz, Austria) equipped with a Peltier (H-PTD200) (Anton Paar, Graz, Austria) temperature control system with an accuracy of 0.01 °C; parallel-plate geometry was employed with a plate diameter of 50 mm with 1 mm gap.

4. Results and Discussion

Figure 3 shows the experimental viscosity curve and dynamic moduli of tested samples of polymer solutions (the temperature value was taken from experimental studies of flow visualization). Intermediate calculations showed that four modes are sufficient to approximate the dynamic modulus curves and the viscosity curve, which is consistent with the literature [11] (Figure 3). The literature data [20] were used to find the linear spectrum of the relaxation time for the four-modal rheological equations of state of a viscoelastic medium. The search for a set of nonlinear parameters of rheological models (3) and (4) is based on the flow curve approximation; e.g., for the Giesekus model, a detailed algorithm was presented in our earlier published work [21]. For the reader's convenience, Table 1 presents the parameters of the Giesekus and eXt. Pom-Pom models that we found, which characterize the rheological behavior of the tested samples (2500 and 7500 ppm). From the figure, we can see that the eXt. Pom-Pom model, in comparison with the Giesekus model, best approximates the experimental viscosity curves for both concentrations, especially in the shear rate range $0.1 < \dot{\gamma} < 60$. A slight deviation in the storage modulus approximation in the interval $\omega < 0.02$ was caused by the use of four modes; however, as is shown below, this deviation does not significantly affect the final result.

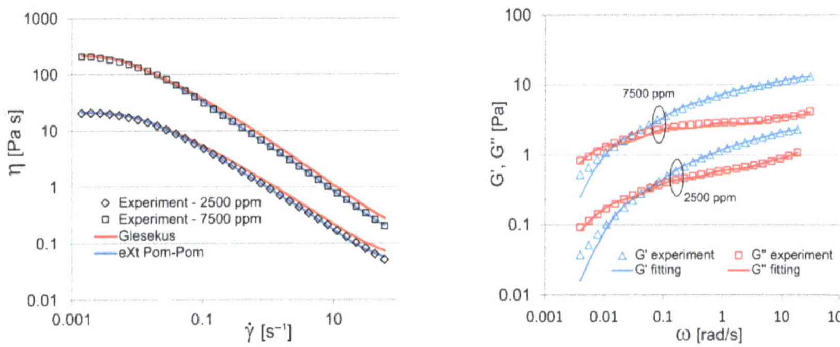

(a) (b)

Figure 3. Viscosity curve (**a**) and dynamic moduli (**b**); experiment and fitting with four-mode Giesekus and eXt. Pom-Pom models (2500 ppm at T = 22.8 °C and 7500 ppm at T = 20.0 °C).

Table 1. Parameters of Giesekus and eXt. Pom-Pom models (2500 ppm at T = 22.8 °C, 7500 ppm at T = 20.0 °C).

				Giesekus	eXt. Pom-Pom		
Concentration [ppm]	λ_i **[s]**	η_i **[Pa·s]**	η_N **[Pa·s]**	α_i	q_i	ε_k	α_i
2500	0.1406	0.1253	0.0411	0.495	1	0.1	0.9
	0.991	0.71		0.495	2	0.4	0.5
	7.1911	4.064		0.495	1	0.1	0.7
	62.6357	17.1691		0.495	1	0.1	0.6
7500	0.1106	0.4699	0.0962	0.495	1	0.1	0.6
	1.0032	3.8742		0.495	2	0.4	0.1
	9.0587	31.7676		0.495	1	0.1	0.6
	98.5914	191.802		0.495	1	0.1	0.4

Since the specific values of the average bulk velocity (V_a) used in flow visualization were calculated during post-processing by integrating the velocity profile over the pipe section according to the formula

$$V_a = 2\int_0^R V_z r dr / R^2 \tag{16}$$

then analytical solutions were obtained using the initial experimental data.

For 2500 ppm, we visualized four variants of the flow with the values Va = 1.866; 4.579; 7.511; 9.588 [mm/s], which corresponded to the following Weissenberg numbers: Wi = 4.80; Wi = 11.76; Wi = 19.29; Wi = 24.63. For 7500 ppm, two variants with the values Va = 0.993; 1.561 [mm/s] were investigated, which corresponded to the following Wi = 4.29; 6.74.

The visualization method was preliminarily tested on the flow of 40% glycerol and 60% propylene glycol (Figure 4). Good agreement of the experimental data with the parabolic velocity profile characterizing the laminar flow of Newtonian fluid in a round pipe was obtained. The relative uncertainty did not exceed 0.6%.

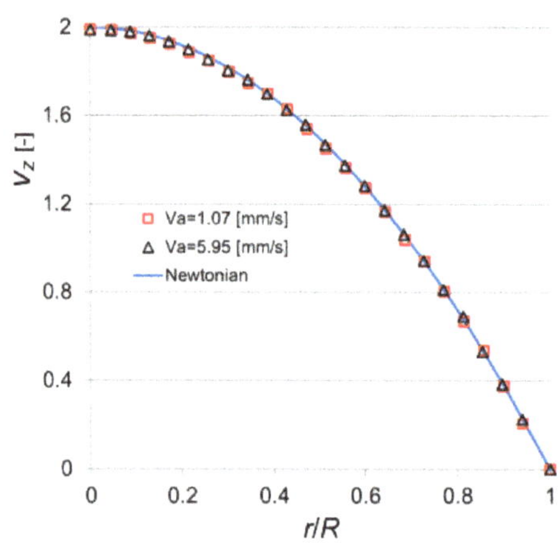

Figure 4. Dimensionless axial velocity profiles in the pipe for various v_a [mm/s] (Newtonian flow, the mixture of 40% glycerol and 60% propylene glycol with dynamic viscosity $\mu = 166 \times 10^{-3}$ Pa·s; T = 22.6 °C).

Figures 5 and 6 show the profiles of dimensionless axial velocity in the longitudinal section of the channel, which characterize the laminar flow of tested polymer solutions in the circular pipe. The illustration of the obtained data shows that the experimental data are best predicted using the eXt. Pom-Pom model, while the relative uncertainty consequently does not exceed 0.9 and 0.96% for 2500 and 7500 ppm. For the Giesekus model, there is a tendency to overestimate the axial velocity values on the channel axis with a decrease in the Weissenberg number. For 2500 ppm, the relative uncertainty on the channel axis does not exceed 1.2% at Wi = 24.63 and it is 4.66% at Wi = 4.80. For 7500 ppm the discrepancy is more pronounced, so the relative uncertainty equals 8.77% at Wi = 4.29 and 7.49% at Wi = 6.74. Additional calculation was performed for a particular case, i.e., $Wi \to 0$ (tending to Newtonian flow), to check the correctness of the developed method. At Wi = 0.01, the calculated velocity profiles predicted by both the Giesekus and eXt. Pom-Pom models coincided with the parabolic profile. The latter indicates that even for the simplest case of viscoelastic fluid flow in a round pipe, the Giesekus model overestimates the value of axial velocity in the central region of the channel in the range of Weissenberg numbers $0.1 < Wi < 25$. The obtained results agree with the results of other authors who investigated more complex geometries [22].

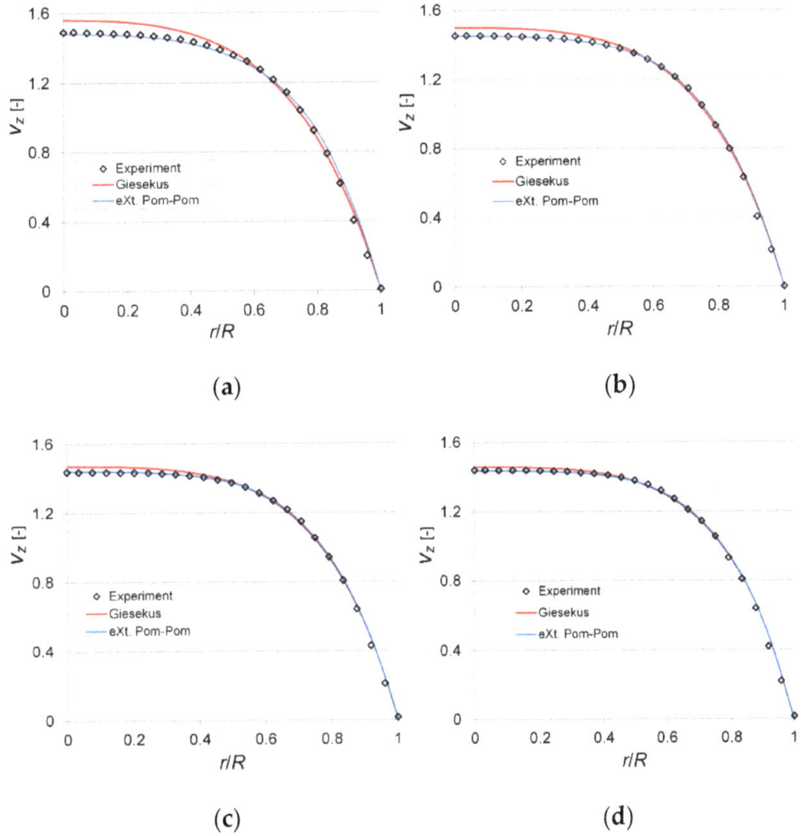

Figure 5. Dimensionless axial velocity profiles in the pipe for various *Wi* ($\lambda_a = 50.087\ \mathrm{s}^{-1}$): (**a**) *Wi* = 4.80; (**b**) *Wi* = 11.76; (**c**) *Wi* = 19.29; (**d**) *Wi* = 24.63.

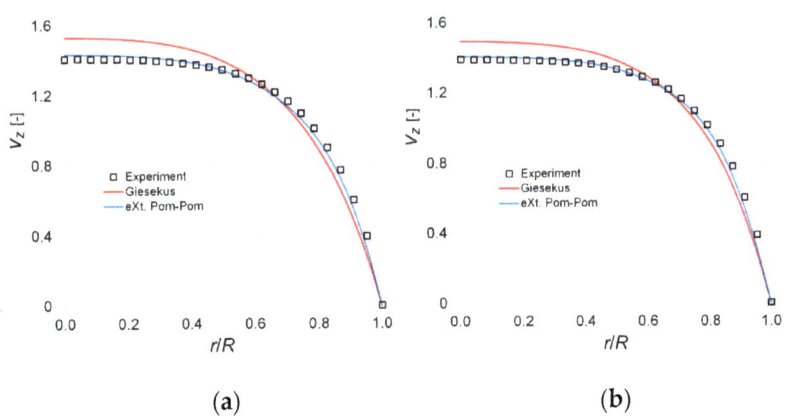

Figure 6. Dimensionless axial velocity profiles in the pipe for various *Wi* ($\lambda_a = 84.25\ \mathrm{s}^{-1}$): (**a**) *Wi* = 4.29; (**b**) *Wi* = 6.74.

5. Conclusions

Our proposed parametric method for multimode Giesekus fluid flows remains stable in a wide range of Weissenberg numbers, but the velocity profiles predicted by this model are slightly higher than the experimental data in the central region of the channel. The best agreement with the experimental data was obtained using the eXt. Pom-Pom model for example of pipe flows of aqueous polyacrylamide solutions with 2500 and 7500 ppm concentrations. The SIV method used to visualize turbulent flows demonstrated its efficiency for visualization of viscoelastic fluid flows and can be successfully applied to the analysis of more complex flows.

Author Contributions: Supervision E.V.; visualization N.D.; data curation E.K.; methodology A.K. All authors have read and agreed to the published version of the manuscript.

Funding: The experimental investigation was carried out with the financial support of the Russian Science Foundation (Project no. 19-11-00220). The adaptation of the SIV method was carried out with the financial support of the state assignment for the FRC KazSC of RAS.

Institutional Review Board Statement: Not applicable.

Informed Consent Statement: Not applicable.

Data Availability Statement: The data presented in this study are available on request from the corresponding author.

Conflicts of Interest: The authors declare no conflict of interest.

References

1. Cruz, D.O.A.; Pinho, F.T. Skewed Poiseuille-Couette Flows of sPTT Fluids in Concentric Annuli and Channels. *J. Non-Newt. Fluid Mech.* **2004**, *121*, 1–14. [CrossRef]
2. Cruz, D.O.A.; Pinho, F.T.; Oliveira, P.J. Analytical solutions for fully developed laminar flow of some viscoelastic liquids with a Newtonian solvent contribution. *J. Non-Newt. Fluid Mech.* **2005**, *132*, 28–35. [CrossRef]
3. Oliveira, P.J. An exact solution for tube and slit flow of a FENE-P fluid. *Acta Mech.* **2002**, *158*, 157–167. [CrossRef]
4. Schleiniger, G.; Weinacht, R.J. Steady Poiseuille flows for a Giesekus fluid. *J. Non-Newt. Fluid Mech.* **1991**, *40*, 79–102. [CrossRef]
5. Oliveira, P.J.; Coelho, P.M.; Pinho, F.T. The Graetz problem with viscous dissipation for FENE-P fluids. *J. Non-Newt. Fluid Mech.* **2004**, *121*, 69–72. [CrossRef]
6. Coelho, P.M.; Pinho, F.T.; Oliveira, P.J. Thermal entry flow for a viscoelastic fluid: The Graetz problem for the PTT model. *Int. J. Heat Mass Transf.* **2003**, *46*, 3865–3880. [CrossRef]
7. Coelho, P.M.; Pinho, F.T.; Oliveira, P.J. Fully-developed forced convection of the Phan-Thien-Tanner fluid in ducts with a constant wall temperature. *Int. J. Heat Mass Transf.* **2002**, *45*, 1413–1423. [CrossRef]
8. Oliveira, P.J.; Pinho, F.T. Analysis of forced convection in pipes and channels with the simplified Phan-Thien-Tanner fluid. *Int. J. Heat Mass Transf.* **2000**, *43*, 2273–2287. [CrossRef]
9. Hashemabadi, S.H.; Etemad, S.G.; Thibault, J.; Golkar-Naranji, M.R. Analytical solution for dynamic pressurization of viscoelastic fluids. *Int. J. Heat Fluid Flow* **2003**, *24*, 137–144. [CrossRef]
10. Hashemabadi, S.H.; Etemad, S.G.; Thibault, J.; Golkar-Naranji, M.R. Mathematical modeling of laminar forced convection of simplified Phan-Thien-Tanner (SPTT) fluid between moving parallel plates. *Int. Com. Heat Mass Transf.* **2003**, *30*, 197–205. [CrossRef]
11. Cruz, D.O.A.; Pinho, F.T. Fully-developed pipe and planar flows of multimode viscoelastic fluids. *J. Non-Newt. Fluid Mech.* **2007**, *141*, 85–98. [CrossRef]
12. Oliveira, P.J. On the numerical implementation of nonlinear viscoelastic models in a finite-volume method. *Numer. Heat Transf. Part B* **2001**, *40*, 283–301. [CrossRef]
13. Oishia, C.M.; Martinsa, F.P.; Toméb, M.F.; Cuminatob, J.A.; McKee, S. Numerical solution of the eXtended Pom-Pom model for viscoelastic free surface flows. *J. Non-Newton. Fluid Mech.* **2011**, *166*, 169–179. [CrossRef]
14. Giesekus, H. A simple constitutive equation for polymer fluids based on the concept of deformation-dependent tensorial mobility. *J. Non-Newton. Fluid Mech.* **1982**, *11*, 69–109. [CrossRef]
15. Verbeeten, W.M.H.; Peters, G.W.M.; Baaijens, F.P.T. Differential constitutive equations for polymer melts: The extended Pom-Pom Model. *J. Rheol.* **2001**, *45*, 823–844. [CrossRef]
16. Astarita, G.; Marrucci, G. *Principles of Non-Newtonian Fluid Mechanics*; McGraw-Hill: New York, NY, USA, 1974; p. 289.
17. Kadyirov, A.I.; Vachagina, E.K. Semi-analytical solution for the problemof extended Pom-Pom fluid flow in a round pipe. *J. Phys. Conf. Ser.* **2021**, *2057*, 012007. [CrossRef]

18. Mikheev, N.I.; Dushin, N.S. A method for measuring the dynamics of velocity vector fields in a turbulent flow using smoke image-visualization videos. *Instrum. Exp. Tech.* **2016**, *59*, 882–889. [CrossRef]
19. Molochnikov, V.; Mikheev, N.; Paereliy, A.; Dushin, N.; Dushina, O. SIV measurements of flow structure in the near wake of a circular cylinder at Re = 3900. *Fluid Dyn. Res.* **2019**, *51*, 055505. [CrossRef]
20. Calin, A.; Wilhelm, M.; Balan, C. Determination of the non-linear parameter (mobility factor) of the Giesekus constitutive model using LAOS procedure. *J. Non-Newton. Fluid Mech.* **2010**, *165*, 1564–1577. [CrossRef]
21. Kadyirov, A.; Zaripov, R.; Karaeva, J.; Vachagina, E. The viscoelastic swirled flow in the confusor. *Polymers* **2021**, *13*, 630. [CrossRef] [PubMed]
22. Peters, G.W.M.; Schoonen, J.F.M.; Baaijens, F.P.T.; Meijer, H.E.H. On the performance of enhanced constitutive models for polymer melts in a cross-slot flow. *J. Non-Newton. Fluid Mech.* **1999**, *82*, 2–3. [CrossRef]

polymers

Article

Data-Driven Modelling of Polyethylene Recycling under High-Temperature Extrusion

Fanny Castéran [1], Karim Delage [1], Nicolas Hascoët [2], Amine Ammar [3], Francisco Chinesta [2] and Philippe Cassagnau [1,*]

1 Centre National de la Recherche Scientifique, Ingénierie des Matériaux Polymères, Université Claude Bernard Lyon 1, 15 Boulevard André Latarjet, 69622 Villeurbanne, France; fanny.casteran@univ-lyon1.fr (F.C.); karim.delage@univ-lyon1.fr (K.D.)
2 ESI Group Chair@PIMM, Arts et Métiers Institute of Technology, 151 Boulevard de l'Hôpital, 75013 Paris, France; nicolas.hascoet@ensam.eu (N.H.); francisco.chinesta@ensam.eu (F.C.)
3 ESI Group Chair@LAMPA, Arts et Métiers Institute of Technology, 2 Boulevard du Ronceray, 49035 Angers, France; amine.ammar@ensam.eu
* Correspondence: philippe.cassagnau@univ-lyon1.fr

Abstract: Two main problems are studied in this article. The first one is the use of the extrusion process for controlled thermo-mechanical degradation of polyethylene for recycling applications. The second is the data-based modelling of such reactive extrusion processes. Polyethylenes (high density polyethylene (HDPE) and ultra-high molecular weight polyethylene (UHMWPE)) were extruded in a corotating twin-screw extruder under high temperatures (350 °C < T < 420 °C) for various process conditions (flow rate and screw rotation speed). These process conditions involved a decrease in the molecular weight due to degradation reactions. A numerical method based on the Carreau-Yasuda model was developed to predict the rheological behaviour (variation of the viscosity versus shear rate) from the in-line measurement of the die pressure. The results were successfully compared to the viscosity measured from offline measurement assuming the Cox-Merz law. Weight average molecular weights were estimated from the resulting zero-shear rate viscosity. Furthermore, the linear viscoelastic behaviours (Frequency dependence of the complex shear modulus) were also used to predict the molecular weight distributions of final products by an inverse rheological method. Size exclusion chromatography (SEC) was performed on five samples, and the resulting molecular weight distributions were compared to the values obtained with the two aforementioned techniques. The values of weight average molecular weights were similar for the three techniques. The complete molecular weight distributions obtained by inverse rheology were similar to the SEC ones for extruded HDPE samples, but some inaccuracies were observed for extruded UHMWPE samples. The Ludovic® (SC-Consultants, Saint-Etienne, France) corotating twin-screw extrusion simulation software was used as a classical process simulation. However, as the rheo-kinetic laws of this process were unknown, the software could not predict all the flow characteristics successfully. Finally, machine learning techniques, able to operate in the low-data limit, were tested to build predicting models of the process outputs and material characteristics. Support Vector Machine Regression (SVR) and sparsed Proper Generalized Decomposition (sPGD) techniques were chosen to predict the process outputs successfully. These methods were also applied to material characteristics data, and both were found to be effective in predicting molecular weights. More precisely, the sPGD gave better results than the SVR for the zero-shear viscosity prediction. Stochastic methods were also tested on some of the data and showed promising results.

Keywords: polyethylene recycling; artificial engineering; polymer extrusion; machine learning

Citation: Castéran, F.; Delage, K.; Hascoët, N.; Ammar, A.; Chinesta, F.; Cassagnau, P. Data-Driven Modelling of Polyethylene Recycling under High-Temperature Extrusion. *Polymers* **2022**, *14*, 800. https://doi.org/10.3390/polym14040800

Academic Editor: Célio Bruno Pinto Fernandes

Received: 11 November 2021
Accepted: 15 February 2022
Published: 18 February 2022

1. Introduction

Considering the current situation of plastic consumption worldwide, the issue of end-of-life of polymer materials has become a significant problem. Polyethylene (PE) accounts for most plastic packaging and, consequently, plastic waste [1]. As PE is a thermoplastic, the most common method for its recycling is mechanical recycling, which involves reprocessing the materials [2–4]. These processes can induce the formation of radicals by homolytic cleavage of the polymers, inducing degradation, branching or even crosslinking of the materials leading to different final properties [5–9]. Consequently, to these properties changes, the applications of the mechanically recycled polymers have to be adapted [10–12]. Whereas the majority of recycled high and low-density polyethylenes (HDPE and LDPE) are however produced that way, Ultra High Molecular Weight Polyethylenes (UHMWPE), mostly used for high-performance applications due to their superior mechanical properties, are more difficult or impossible to process due to their high viscosity [13]. The other principal way of recycling polymers is by chemical recycling, which consists of a chemical transformation leading to new raw materials. Whereas polyethylene terephthalate (PET) can be depolymerized into dimethyl terephthalate and ethylene glycol by mathanolysis [14], no such reactions are possible for PE. The main way of PE chemical recycling is then pyrolysis, leading to smaller carbonated molecules, which can, in theory, be reinjected into the chemical industry [2,6,15–21].

Whilst thermal degradation of polymers that include heteroatoms in their structure (PMMA, for instance) leads to simple products and mechanisms, PE degradation is more complex [22]. A simplification of PE degradation mechanisms is presented in Figure 1.

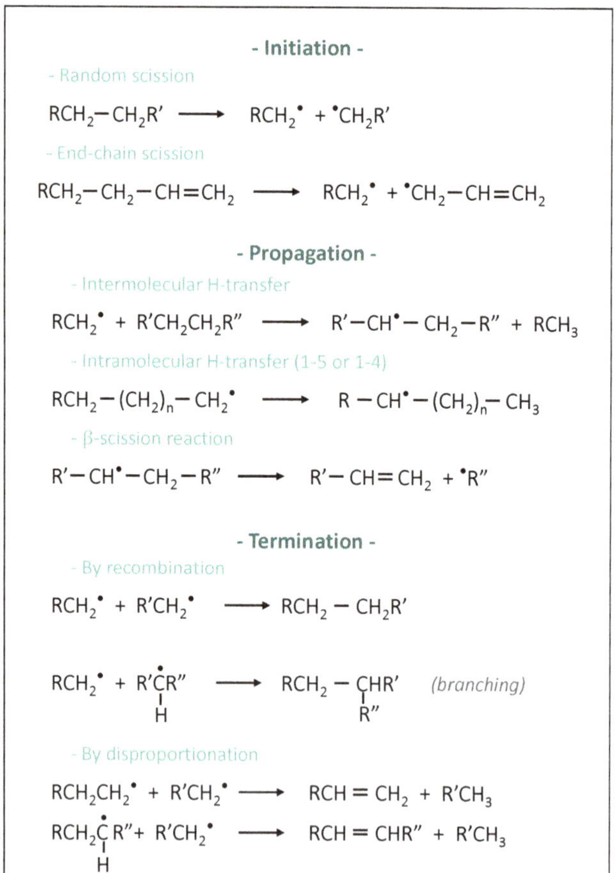

Figure 1. Simplified mechanisms of thermal degradation of polyethylene [23,24].

As shown in this figure, different degradation mechanisms are possible for PE thermal degradation. The process conditions then define their probability. Whereas for polymers with a more complex structure, the end-chain scission mechanism would be preponderant, in the case of polyolefins, random scissions are more significant [24]. Therefore, the higher the molecular weight of the polyolefin, the more random scissions occur, resulting in a narrowing of the molecular weight distribution [25]. As schematized in Figure 2, the molecular weight decreases with the increase of the temperature of the reaction, leading first to oligomers and then to smaller molecules whose nature depends on propagation and termination mechanisms. A short reaction time at high temperatures would favour β-scissions, leading to the formation of a certain yield of the ethylene monomer. However, longer reaction times tend to favour the production of cyclic compounds due to intramolecular and intermolecular transfers, which are less easy to valorize afterwards [25].

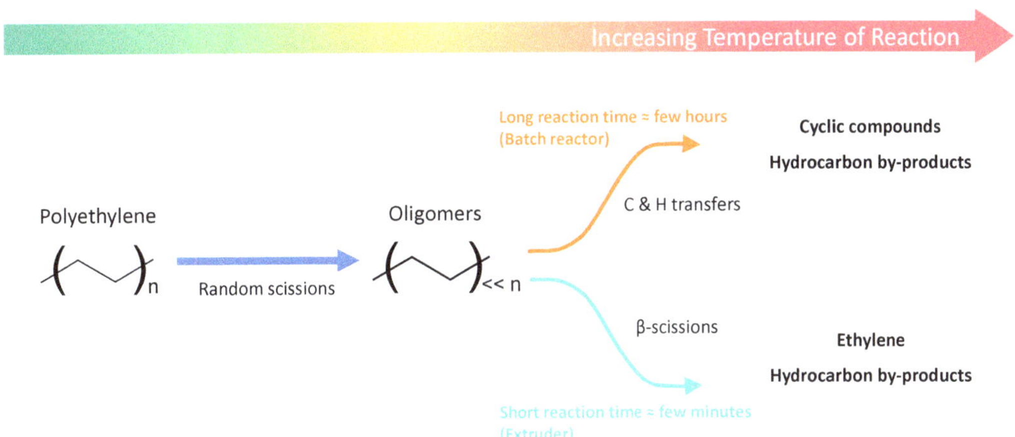

Figure 2. Representation of the different products of polyethylene degradation depending on temperature and residence time, inspired from Vollmer et al. [25].

The final products are then highly dependent on the processing conditions. Moreover, classic pyrolysis processes induce heat and mass transfer problems, leading to highly heterogeneous products without the possibility of controlling the degradation [25].

Other innovative ways of PE recycling were recently studied. Manas et al. [26,27], for instance, studied the recycling of PE crosslinked by irradiations by using it as a filler for virgin LDPE, Elmanowich et al. [28] studied the use of supercritical fluids for PE recycling, and recent promising studies are about the enzymatic degradation of polymers [29,30].

The present work aims at controlling the thermo-mechanical degradation of PE by carrying out a twin-screw extrusion process at high temperatures ($320 < T$ °C < 420). Whereas polyethylene extrusion usually leads to branching and crosslinking, the extrusion thermal conditions in this work favour degradation mechanisms closer to pyrolysis conditions.

With the question of the process control comes the issue of its simulation. Extrusion simulation has been widely studied in the last decades. However, the complexity of the physical phenomena involved in the extrusion process involves either a long time and a significant computing power or a lot of hypotheses and simplifications. Furthermore, in this study, the viscoelastic properties of the materials evolve with their degradation. Due to the temperatures reached in the extruder, the materials are at the limit of pyrolysis. Moreover, the presence of oxygen involves additional chemical reactions due to high-temperature oxidation. Thus, the complexity of the degradation mechanisms increases the imprecision of the simulation.

On the other hand, machine learning does not need to understand these complex physics, only to have accurate experimental data to predict the results of new experiments.

This approach often leads to faster computing and sometimes to more precise results, the imprecision coming from unknown phenomena or simplifications unavoidable in classical simulations that are not necessary with machine learning. Such methodologies were successfully employed in previous works on various reactive extrusion systems [31,32].

The following is a study of the degradation of HDPE and UHMWPE by twin-screw extrusion at high temperatures. Then, several methods will be applied and compared to determine the extruded polyethylenes' final viscosities and molecular weights. Finally, the modelling of this process with the twin-screw extrusion Ludovic® simulation software (SC-Consultants, Saint-Etienne, France) will be compared to the results obtained with Machine-Learning methodologies.

2. Experimental Section

2.1. Materials and Extrusion

HDPE (HDPE XRT70, TOTAL, Melt Flow Index (MFI: 190 °C/5 kg) = 0.7 g/10 min) and UHMWPE (GUR 4130, Celanese, Melt Flow Index (MFI: 190 °C/21.6 kg) < 0.1 g/10 min) were extruded in an intermeshing corotating twin-screw extruder ZSE 18 MAXX/HPe (Leistritz, Nuremberg, Germany) for various temperatures, screw rotation speeds and exit flow rates, reaching 29 different configurations. These processing conditions are summarized in Table 1.

Table 1. Processing conditions used in this study.

Materials	T_{max} *	Flow Rate	Screw Rotation Speed
HDPE XRT70 TOTAL, (MFI = 0.7 g/10 min (190 °C, 5 kg))	350 °C 390 °C 420 °C	1 to 6 kg/h	300 to 1000 rpm
UHMWPE GUR 4130, Celanese, (MFI < 0.1 g/10 min (190 °C, 21.6 kg))	390 °C 420 °C	1 and 3 kg/h	300 to 1000 rpm

* T_{max} correspond to the maximal setpoint temperature cf. Figure 3.

As the Polyethylene samples are highly viscous, and in order to have a maximum of possibilities concerning flow rate and rotation speed combinations, the screw profile was designed with quite a low shear and few restrictive elements. The temperature was increased progressively along with the barrel blocks and progressively decreased before the die for security issues and to limit the degradation at the die exit. The polyethylenes were introduced at the extruder entrance with a gravimetric feeder. The die had a 3 mm diameter. At the exit, the materials were cooled by air and then pelletized. The extruder configuration is described in Figure 3.

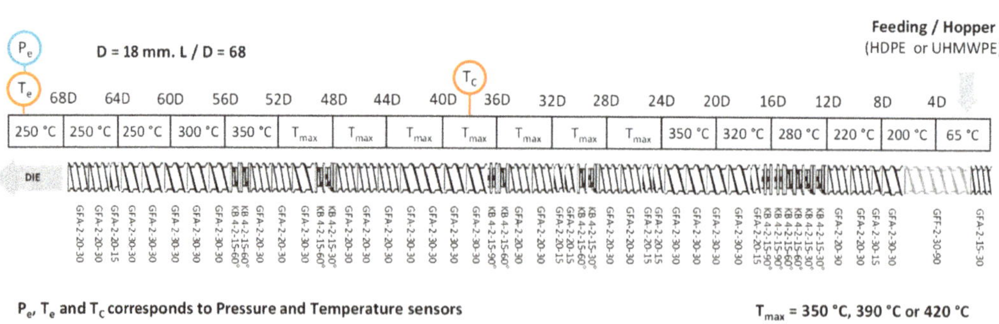

Figure 3. Scheme of the screw and temperature profiles used.

The details about the different extrusion configurations tested and the in-line measures are presented in the Appendix A of this article, in Table A1.

The torque, the Engine Power and the die pressure were measured for each experiment. Thermocouples placed around the centre of the extruder (T_c at L = 38D) and in the die permitted to measure the melt temperature. However, it appears that the temperature measured is the one of the inner surface of the barrel and not precisely the bulk material temperature. Therefore, a manual thermocouple was used to measure it at the die exit of the extruder.

2.2. Characterizations

The rheological behaviour of the extruded and raw materials was studied using a DHR-2 (TA Instruments, New Castle, DE, USA), a stress-controlled rheometer. Frequency sweeps have been carried out at 190 °C from 100 to 0.01 rad·s^{-1} under nitrogen and with 1% deformation. The geometry used was 25 mm diameter parallel plates with a 1 mm gap.

The molecular weight distributions of five samples (two degraded HDPE, two degraded UHMWPE and raw HDPE XRT70) were measured using high-temperature steric exclusion chromatography (HT-SEC) using a Viscositek (Malvern Panalytical Ltd., Malvern, UK) device. The samples were previously dissolved in toluene at 100 °C for 30 min.

2.3. Theoretical Methodologies

2.3.1. Determination of *Mw* from Viscoelastic Behaviour

The molecular weight distributions have been calculated using the TA instrument tool implemented in the Trios software (TA Instruments, New Castle, DE, USA). This tool, based on the double reptation theories [33,34], uses a model linking the molecular weight distribution to the relaxation modulus via the following relationship:

$$G(t) = G_N^0 \cdot \left[\int_{\ln(Me)}^{\infty} F^{\frac{1}{\beta}}\, w(M) \mathrm{d}ln(M) \right]^{\beta} \tag{1}$$

where $G(t)$ is the linear viscoelastic relaxation modulus, G_N^0 is the plateau modulus, $w(M)$ is the molecular weight distribution, $F(M,t)$ is the monodisperse relaxation function, and β takes the value 1 for simple reptation and 2 for double reptation. Several models exist to define the F function. In this study, as in most cases, a single exponential form described by Equation (2) is applied. Table 2 presents the definition of the constants used by the model and their values in this work, which correspond to classical values for polyethylene.

$$\begin{gathered} F^{\frac{1}{2}}(M,t) = exp\left(\frac{-t}{2\lambda(M)}\right) \\ \text{with } \lambda(M) = K_\lambda(T) M^{\alpha}; \\ K_\lambda(T) = K_\lambda(T_0)\, exp\left(\frac{E_a}{RT}\right) \end{gathered} \tag{2}$$

Table 2. Parameters used for the molecular weight distribution calculation from inverse rheology method.

Symbol	Parameter	Value
T	Test Temperature	190 °C
α	Relaxation time exponent	3.6
G_N^0	Plateau modulus	2.3×10^6 Pa
E_a	Activation Energy	30 kJ/mol
K_λ	Front Factor	2.5×10^{-21} s·(mol/g)$^{3.6}$
Me	Entanglement Molecular weight	1250 g/mol
Mr	Reptation Molecular weight	2500 g/mol

2.3.2. Determination of *Mw* from Measured Die Pressure

As the pressure measured at the extruder die depends on the viscosity of the viscous polymer, which can be related to the molecular weight, it is then possible to estimate the viscosity and subsequently the molecular weight without the need for post-process characterizations. Figure 4 summarises the different steps required to measure *Mw* from the die pressure measurement.

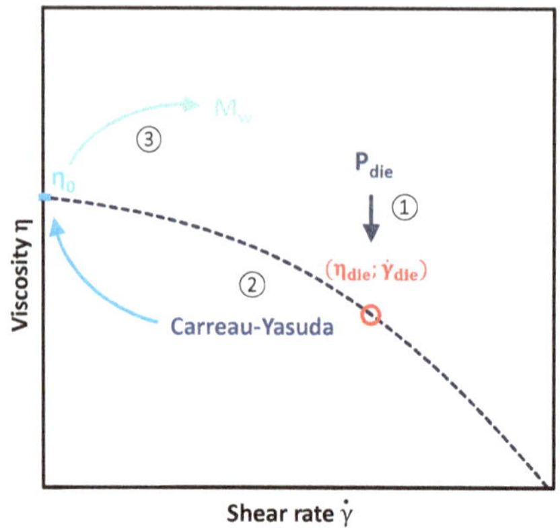

Figure 4. Scheme of the *Mw* determination from die pressure measurement.

The first step is to determine the value of the viscosity and the shear rate in the die. For this purpose, the die is considered a capillary rheometer. The experimentally measured temperatures and pressures are then approximated to those of the narrowest section of the die. As in a capillary rheometer, an apparent shear rate $\dot{\gamma}_{app}$ and an apparent viscosity η_{app} are calculated as a function of the flow rate Q, the die cross-section and length r and L and the difference between die and outside drop pressure ΔP according to Equation (3).

$$\dot{\gamma}_{app} = \frac{4\,Q}{\pi\,r^3} \quad ; \quad \eta_{app} = \frac{\Delta P\,r}{2\,L\dot{\gamma}_{app}} \tag{3}$$

These values are valid at the temperature of the die T_{die}, it is then necessary to calculate them at a reference temperature T_0 which is set at 190 °C in this study. This correction is then done assuming an Arrhenius dependence of the viscosity (Equation (4)).

$$\eta_{app}(T_{die}) = a_T \times \eta_{app}(T_0) \quad ; \quad a_T = exp\left(\frac{E_a}{R} \times \left(\frac{1}{T_0} - \frac{1}{T_{die}}\right)\right) \tag{4}$$

E_a = activation energy;
R = ideal gases constant.

Then, the Rabinowitsch correction (Equation (5)) has to be applied to calculate real shear rate and viscosity $\dot{\gamma}_{die}$ and η_{die}.

$$\dot{\gamma}_{die} = \frac{3m+1}{4m}\,\dot{\gamma}_{app} \;;\; \eta_{die} = \frac{4m}{3m+1}\,\eta_{app}\,; \quad \text{with } m - 1 = \frac{\partial\,\log(\eta)}{\partial \log(\dot{\gamma})} \tag{5}$$

Usually, the parameter m is obtained by measuring several values of $\dot{\gamma}_{app}$ and η_{app} and calculating the slope of the resulting line by plotting $log(\eta_{app}) = f(log(\dot{\gamma}_{app}))$. However, in the present study, it is impossible to obtain more than one point of the curve because measuring another pressure value means changing the process and thus changing the material itself. A first regression is then performed considering a Carreau-Yasuda model (Equation (6)) passing through this single point; it is thus considered that all degraded

PE still follow a Carreau-Yasuda law. m and a parameters are set as the ones of the raw HDPE XRT70 (m = 0.058 and a = 0.248). They have been determined by a frequency sweep with the rheometer and a fitting with the Carreau-Yasuda model. The Cox-Merz law was assumed to compare the actual shear viscosity and the complex viscosity measured by oscillatory rheometry tests. According to the literature [35], the viscosity at high shear rates does not depend on molecular weight. The viscosity curves of different molecular weight polyethylene samples converge then on an identical power-law behaviour. Making this assumption, λ and η_0 have been determined for each sample through classical regression methods, and the equation is thus completely described.

$$\eta(\dot{\gamma}, T) = \eta_0 \cdot \left(1 + (\lambda \times \dot{\gamma})^a\right)^{\frac{m-1}{a}} \tag{6}$$

m and a = dimensionless indexes;
λ [s] = characteristic time;
η_0 [Pa·s] = zero shear viscosity.

The m parameter can then be calculated with the resulting curve, which enables the calculation of real shear rate $\dot{\gamma}_{die}$ and viscosity η_{die} in the die thanks to Rabinowitsch correction. Finally, the real viscosity curve passing through this point is estimated by a second regression with a Carreau-Yasuda model in the same way as the first regression.

It is then possible to link the average molecular weight M_w to η_0 considering Equation (7). Often estimated at 3.4 for entangled polymers, the α exponent has been fixed at 3.6 in this study to be homogeneous with previous calculations and other HDPE studies [36–38] (the results being similar with the two values). The K constant can be determined from the values of η_0 and M_w of the reference HDPE.

$$\eta_0 = K \times M_w^{\alpha} \tag{7}$$

3. Modelling and Machine Learning

3.1. Simulation

Ludovic® software (SC-Consultants, Saint-Etienne, France) is a well-known twin-screw extrusion simulation software. Due to its computing speed, ease of use and flexibility, it has proven its reliability in many different fields and applications. Initially developed for starch extrusion, it is now developed and used for plastics compounding and the pharmaceutical, cosmetics, food, and construction industries. This broad use makes it an interesting choice for this study and for comparison with new approaches such as machine learning.

Its efficiency is due to many simplifications and hypotheses and its adaptability to each situation. First, the melting process is considered instantaneous, but the user can also implement a melting model. Then, specific geometries are used depending on the element, allowing the flow calculation in only one dimension. Elements are divided into several sections in which the fluid is considered as Newtonian and isothermal. Specific viscosities depending on shear rate and temperature are however defined and chosen by the user. Dedicated articles were published for a complete description of the method [39].

Here, the software was configured to match screw profile, extruder, die and temperature profile with the experiments. Transfer coefficients were fixed at 50 $W \cdot m^{-2} \cdot K^{-2}$ for the die and 350 $W \cdot m^{-2} \cdot K^{-2}$ for the barrel, corresponding to similar and previous simulations [31,40].

Concerning the viscosity of the polyethylene, several options are available: Choosing a rheological model between Power Law and Carreau-Yasuda and indicating the corresponding parameters, implementing a new model, or entering a set of points (SoP) corresponding to rheological measurements. It is also possible to couple the viscosity with some simulation results, such as reaction rate (requires entering a description of the kinetics), total residence time, cumulated strain or total dissipated energy. Nevertheless, it requires knowing the relation between these parameters and viscosity. In our case, polyethylene viscosity evolves during the extrusion due to the degradation under high temperature.

Berzin et al. [41] developed a method for coupling the starch viscosity variation with the SME on Ludovic® (SC-Consultants, Saint-Etienne, France) However, here, the dependence between degradation and the process parameters is unknown, and the objective is to use the simulation classically.

HDPE viscosity has been defined as following a Carreau-Yasuda model, in which parameters have been determined from rheological measurements on the raw polyethylenes.

Concerning UHMWPE, it appears that the available rheological measurement methods underestimate the viscosity. Several hypotheses and simplifications had then to be made in order to approximate it for our simulation.

As a first simplification, the elastic modulus (G') has been considered as constant and equal to its rubbery plateau value G_N^0, which can be calculated the following way:

$$G_N^0 = \frac{\rho R\,T}{M_e} \quad (8)$$

where ρ is the density of the polymer estimated at 930 kg/m^3 according to the supplier, R is the perfect gas constant, T is the reference temperature (T = 190 °C) and M_e is the molecular weight between entanglements (1.25 kg/mol for polyethylene).

Then, as the loss modulus is significantly lower than the value of the elastic modulus, its contribution to viscosity calculation has been neglected.

Finally, the Cox-Merz hypothesis has been made, allowing to assimilate the actual viscosity η and shear rates $\dot{\gamma}$ of the extruder to the complex viscosity η^* and angular frequency ω, thus obtaining the following approximation:

$$\eta(\dot{\gamma}) = \eta^*(\omega) = \frac{G\prime(\omega)}{\omega} = \frac{G_N^0}{\omega} \quad (9)$$

All the polymer characteristics used have been summarised in Table 3.

Table 3. Thermal Properties and viscosity laws used in the simulation.

Thermal Properties	HDPE XRT70	UHMWPE GUR 4130
Heat Capacity [J kg^{-1} K^{-1}]	1550	1840
Density [kg m^{-3}]	947	930
Thermal Conductivity [W mK^{-1}]	0.35	0.41
Melting Temperature [°C]	129	135
Melting enthalpy [kJ kg^{-1}]	190	122
Viscosity Law	Carreau-Yasuda: $\eta(\dot{\gamma}) = \eta_0 \cdot (1 + (\lambda \times \dot{\gamma})^a)^{\frac{m-1}{a}}$ η_0 = 2.5 × 10^6 Pa·s λ = 0.33 s a = 0.25 m = 0.058 T_{ref} = 190 °C E_a = 30 kJ·mol^{-1}	Power Law: $\eta(\dot{\gamma}) = K\,\dot{\gamma}^{\,m-1}$ m = 0 K = 2.86 × 10^6 Pa·s T_{ref} = 190 °C E_a = 30 kJ·mol^{-1}

3.2. *Machine-Learning*

Since a data-driven model is only fed by data, the more data there is, the more accurate the prediction will be. While today, in the era of "big data", one of the main concerns is to classify this huge amount of data successfully, this study is in the opposite situation dealing with the low data limit imposed by the number of extrusions and the available hardware. The dataset used in our study includes four types of inputs (HDPE or UHMWPE, flow rate, screw rotation speed and maximal imposed temperature), and only around 27 data, each corresponding to an extrusion configuration. To find a law linking the outputs to these inputs, it is consequently necessary to use algorithms able to perform with few data. As regression methods are well adapted to this case, Support Vector Machine Regression (SVR) and Sparsed Proper Generalized Decomposition (sPGD), two regression methods, have been tested for those data. Their specific modes of operation are described hereafter.

The process is as follows: the dataset is first randomly divided into training and test groups. Then, the training inputs and outputs are implemented in the algorithm, which, depending on the method, will "learn" from these data, creating a model linking the inputs to the outputs. The model is then tested with the remaining data, and the outputs predicted by it are compared to the real ones, allowing qualifying its accuracy.

3.2.1. Support Vector Machine Regression—SVR

SVR is a derivate of the Support Machine Vector—SVM—classification method. It is a classical method known for its effectiveness in high dimensional spaces and is widely described in the literature, such as Smola et al. in 2004 [42]. As with all classification methods, SVM aims to find the equation of the limit between two classes. As it can be tricky or impossible to find this limit in the original space of the data, the strategy here is to represent the data in a higher-dimensional space where the equation of the limit would be more simple. This principle is represented in Figure 5. This transformation is carried out via a transformation function called "kernel" (noted φ in the figure). This kernel is defined by the user, depending on the system and its complexity. Here the "RBF" kernel was chosen as it is adapted to nonlinear systems.

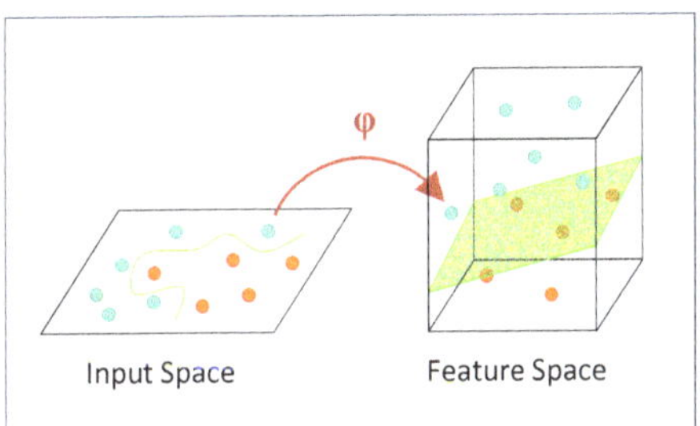

Figure 5. Scheme of SVM principle.

The decision surface defining the different class areas is defined according to two parameters C and ε defined by the user. ε represents the maximum error between the decision surface and the experimental points, and C characterizes the smoothness of this surface. To be more specific, a high value of C would make the decision surface fit exactly all the experimental points but can lead to overfitting. It would hardly represent reality and would not fit new points that are not part of the training points. On the contrary, a low value of C would smooth the decision surface, enabling more errors and bringing more realism. This principle can also be used for regression purposes, and this is how the SVR method used here works. Then, the decision surface does not represent a border between classes but a hyper-surface approximating the points, predicting numerical values for new testing points.

3.2.2. Sparsed Proper Generalized Decomposition—sPGD

This regression method has been developed and fully described by Ibáñez et al. [43]. To give an idea of the principle of this method, let us consider an output y that depends on two input parameters x_1 and x_2. The simplest regression method, particularly in the case of few experimental data, is a linear regression such as described by Equation (10).

$$y(x_1, x_2) = a + b \times x_1 + c \times x_2 \tag{10}$$

where three experimental data, i.e., values of (y, x_1, x_2), are necessary to find a, b and c parameters of Equation (10).

However, the dependence between the inputs and outputs is often nonlinear, and calculating richer regressions would necessitate more data. The principle of sPGD is considering instead of (10) the following equation:

$$y(x_1, x_2) = X_1(x_1) \times X_2(x_2) \tag{11}$$

The dependence relation between the output y and the inputs x_1 and x_2 would thus be a product of functions depending each on x_1 and x_2. To determine these functions without needing more data, the following method is applied. First, $X_1(x_1)$ is fixed and $X_2(x_2)$ is estimated thanks to the available data. Only two data are then necessary to have a linear function in each coordinate and then a bilinear regression with only three data (some regularisation being needed to avoid unphysical behaviours). However, a more significant amount of data would lead to richer approximations. Then, $X_1(x_1)$ can be estimated by fixing $X_2(x_2)$ to its just calculated value using the same data. With this method, three experimental data could lead to a quadratic dependence instead of a linear one with Equation (10).

3.2.3. Stochastic Methods

Classically, the way of estimating the equation matching a maximum with data is to apply the classical least-squares procedure. However, another method called "stochastic" based on statistics is possible.

Let consider a set of two variables x_i and y_i that we attempt to describe by a linear function $y_i = y_i(x_i) = ax_i + b$. The classical lest squares procedure consists of minimizing S.

$$S = \frac{1}{2}\sum_i (y_i - f(x_i))^2 = \frac{1}{2}\sum_i (y_i - (ax_i + b))^2 \tag{12}$$

The number of samples is n ($i = 1..n$).

The linear regression applied at each data point reads:

$$\begin{aligned} b + ax_1 &= y_1 \\ &\dots \\ b + ax_n &= y_n \end{aligned} \tag{13}$$

whose matric form reads

$$\begin{pmatrix} x_1 & 1 \\ \dots & \dots \\ x_n & 1 \end{pmatrix} \begin{pmatrix} a \\ b \end{pmatrix} = \begin{pmatrix} y_1 \\ \dots \\ y_n \end{pmatrix} \tag{14}$$

or using a more compact form

$$\mathbb{X} \begin{pmatrix} a \\ b \end{pmatrix} = \mathbb{Y} \tag{15}$$

The least-squares procedure used for solving the just overdetermined system consists of pre-multiplying the previous system by $\mathbb{X}^T$ that results in

$$\mathbb{X}^T \mathbb{X} \begin{pmatrix} a \\ b \end{pmatrix} = \mathbb{X}^T \mathbb{Y} \tag{16}$$

or more explicitly

$$\begin{pmatrix} \sum_i x_i^2 & \sum_i x_i \\ \sum_i x_i & n \end{pmatrix} \begin{pmatrix} a \\ b \end{pmatrix} = \begin{pmatrix} \sum_i x_i y_i \\ \sum_i y_i \end{pmatrix} \tag{17}$$

It is then possible to prove that the solution of this system results

$$a = \frac{\text{Cov}(x,y)}{\text{Var}(x)}$$
$$b = \overline{y} - a\overline{x} \tag{18}$$

with $\overline{y}$ and $\overline{x}$ the mean value in the sample of variables y and x, respectively.

We suppose now that for each input x_i the response y_i follow a Gaussian distribution with a standard deviation σ_i. Let denotes by $N_\sigma(.)$ the Gaussian distribution around the zero value with standard deviation σ. The least-squares procedure previously described is here modified. The idea consists in saying that each y_i must follow a normal distribution centred at $ax_i + b$ and with a standard deviation σ_i.

The optimization problem results then in the maximization of the following sum

$$S = \frac{1}{2}\sum_i N_{\sigma_i}(y_i - f(x_i)) = \frac{1}{2}\sum_i N_{\sigma_i}(y_i - (ax_i + b)) \tag{19}$$

To get back on a classical optimization problem expressed in terms of minimization, we just have to add a negative sign on the sum. The resolution cannot be made analytically, as was done in the case of the least-squares' procedure. We can however use a descent scheme.

Assuming that the standard deviation does not depend on x, one could look for the statistical regression with a minimum standard deviation. For that, the most straightforward procedure consists of, for a tentative a and b coefficients of the regression, compute the standard deviation of the sampling, and then apply the previous rationale for computing the regression, that is, for updating a and b coefficients and iterate until reaching convergence.

The procedure described above is easily generalized to handle multiparametric regressions with richer nonlinear regressions. This method was tried on 10 data corresponding only to HDPE extruded at 190 °C and coupled to SVR previously described methodology.

4. Results and Discussion

4.1. Comparison of Estimated and Measured Viscosities

The viscosities of the extruded materials deduced from the pressure measured at the die were compared to the rheometer measurements. The viscosity curves obtained from die pressure measurements are compared to data calculated from rheometer experiments in Figure 6A only for the samples resulting from HDPE 390 °C extrusion. Figure 6B compares the zero shear viscosities obtained for all samples with the two different methods. A Carreau-Yasuda model was thus used to fit the experimental viscosity curves, which validate the use of such a model for the degraded PE. With this type of representation, the closer the points on the $x = y$ line are, the closer the values of the two methods for the same sample are. The samples from extrusions performed at different temperatures or with the UHMWPE give similar results. The results are fully detailed in Table A3 of the Appendix A.

Despite the assumptions made, the estimated viscosity curves are close to the rheometer experiments. However, the region of the Newtonian plateau at low frequency seems to be more pronounced for the rheometer data, leading to a higher value of the parameter a of the Carreau-Yasuda law. But the interest of this method is to avoid offline characterizations. Given the lack of information to model the curve correctly, it seems unlikely that better results can be obtained. Besides, the comparison of zero-shear viscosities gives satisfactory results and proves that this method can rapidly approximate the final viscosity of extruded materials.

4.2. Molecular Weight Distribution

Figure 7A shows the comparison of the values of weight average molecular mass (Mw) obtained with the three methods previously described, i.e., (i) viscoelastic behaviour measurements, (ii) from die pressure and (iii) from size exclusion chromatography (SEC). Only the five samples analyzed by SEC are in the 3D figure, the other ones being in the die pressure versus viscoelastic behaviour plane. Figure 7B compares Mw values obtained by the three techniques and *Mn* values obtained by SEC and viscoelastic measurements. The results are fully detailed in Table A3 of the Appendix A. It appears that despite the simplifications they induce, the three methodologies give similar results for the determination of *Mw* for all samples, with a factor of less than 2 between values in most cases. Thus, it appears that, concerning the determination of *Mw*, the viscoelastic measurement method has few advantages over the die pressure method, which does not require offline characterizations. However, its advantages actually lie in the fact that this technique allows obtaining the complete distribution of molecular weights, unlike the die pressure method, which only allows obtaining the *Mw*.

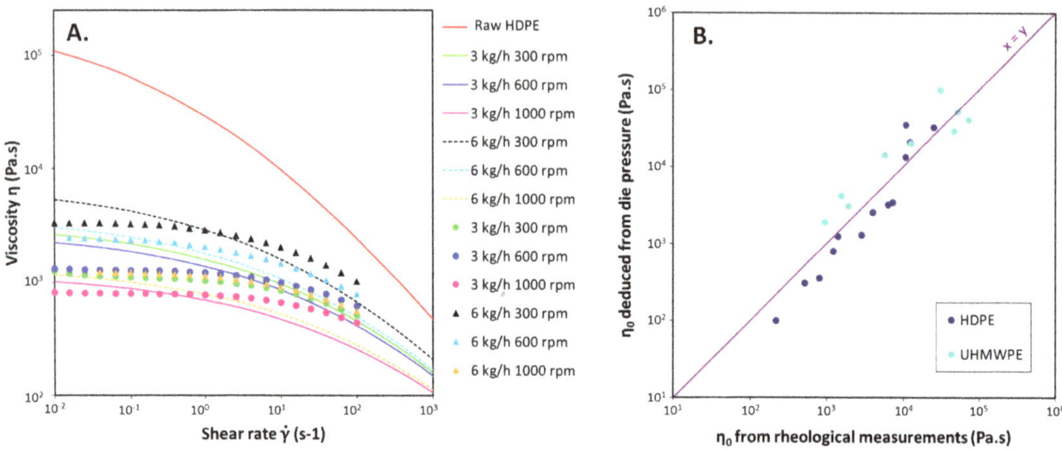

Figure 6. (**A**) Representation of calculated (lines) and measured (dots) viscosity curves for some HDPE extrusions at 390 °C for the indicated flow rates and screw rotation speeds; (**B**) Zero shear viscosities deduced from die pressure compared to the one obtained from rheological measurements.

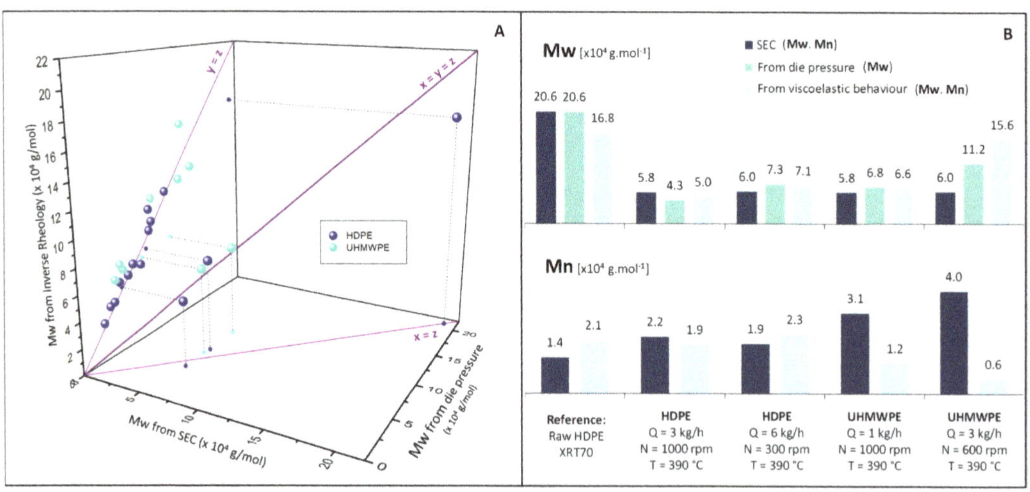

Figure 7. (**A**) Comparison of weight average molecular weight (*Mw*) of degraded polyethylenes obtained from the estimated zero shear rate viscosity versus the one obtained from viscoelastic behaviour measurements. (**B**) Comparison of weight average (*Mw*) and number average (*Mn*) molecular weight values obtained with different specified methodologies.

The *Mn* values obtained are compared to SEC values in Figure 7B. Figure 8 compares complete molecular weight distributions obtained by SEC and calculated from viscoelastic measurements for an HDPE and a UHMWPE. These figures point out that concerning HDPE samples, the distributions calculated are close to the ones obtained by SEC, which is the more precise methodology. It is consequently possible for this type of sample to determine the molecular weight distribution only from frequency sweep experiments, avoiding using SEC, which is a more complex and less accessible process involving the use of hot CMR solvents. However, for UHMWPE samples, *Mn* values are not as close to the SEC values. It can be due to structural differences with HDPE samples, inducing different parameters for the reptation model (relaxation time exponent, entanglement and reptation molecular weights, etc.).

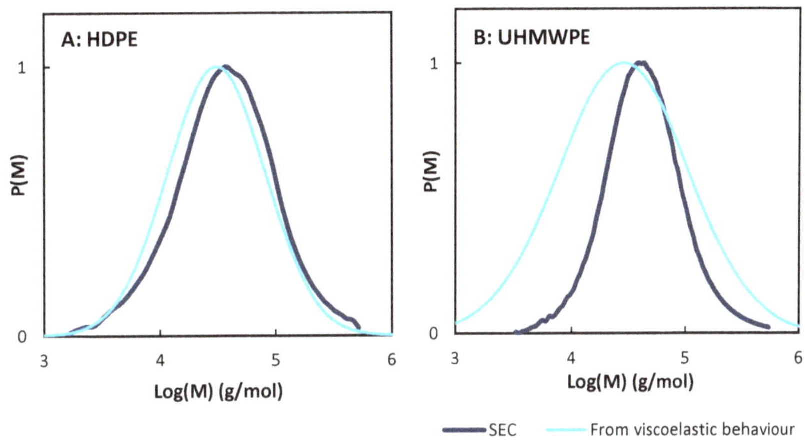

Figure 8. Molecular weight distribution obtained by SEC and viscoelastic measurements for (**A**) HDPE extruded at 390 °C, 3 kg/h and 1000 rpm and (**B**) UHMWPE extruded at 390 °C, 1 kg/h and 1000 rpm.

4.3. *Ludovic® Simulation*

Figure 9 compares the values obtained with Ludovic® simulation (x-axis) to the experimental measures (y-axis) for different process parameters. This representation involves that the closer the values are to the $x = y$ line, the closer the predictions of Ludovic® are to the measures. These results are fully detailed in Table A2 of the Appendix A.

The die temperatures measured by the extruder thermocouple are all around 200 °C, while the output temperatures measured with a manual thermocouple are much higher and more scattered. It is often a problem with extruder thermocouples which, being placed on the walls of the die, are influenced by its temperature and do not measure the actual melt temperature.

Concerning the simulation results, the first thing to notice is that all temperatures seem to be overestimated by the software, which matches the fact that polyethylene degradation is not considered. The actual viscosity decreases along with the screws and causes less self-heating than what could be expected without degradation. The temperature is, in fact, closer to the one imposed by the extruder.

This viscosity error also causes an overvaluation of the pressure in the die, more accentuated for UHMWPE because of its high viscosity. Concerning the torque and the engine power, and in the case of HDPE, the experimental values match pretty well with the simulation. It can be surprising considering the error between simulated and experimental viscosities caused by degradation. However, as viscosity decreases along the extruder, we can think that the torque value is mainly ruled by the most viscous part, which is the raw polyethylene present in the first screw elements and not yet degraded. The torque is then ruled by the viscosity of raw polyethylene, which is the one implemented in Ludovic® (SC-Consultants, Saint-Etienne, France) The ability of machine learning algorithms to make better predictions than classic simulation is studied in what follows.

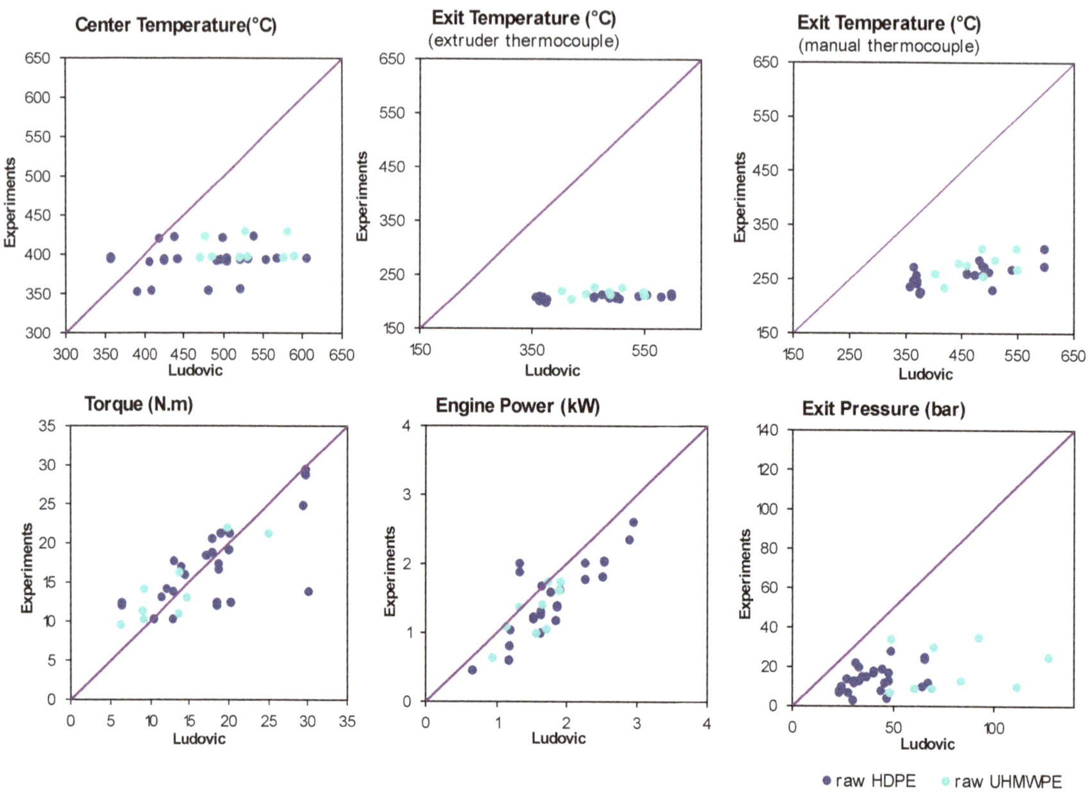

Figure 9. Comparison of Ludovic® Simulation versus Experimental data for in-line measured parameters.

4.4. Data-Based Modelling

4.4.1. Modelling of In-Line Measures with Machine-Learning Methods

This part presents the results obtained by Machine-Learning (ML) method on in-line measured parameters, which correspond to the parameters measured directly during the extrusion without needing additional experiments.

Figures 10 and 11 show the results obtained for centre and exit temperatures (manual thermocouple), torque, engine power and exit pressure for SVR and sPGD methods, respectively. To obtain these results, the model obtained after training with SVR regression has been applied on inputs and these figures represent the resulting outputs compared with the measured ones. Blue dots correspond to the data used for training and constructing the model, whereas the red star ones represent data that are new for the model as they have not been used for the training. Regarding these results, it appears that both methodologies give acceptable results as the dots are well distributed along the $x = y$ line and relatively close to it.

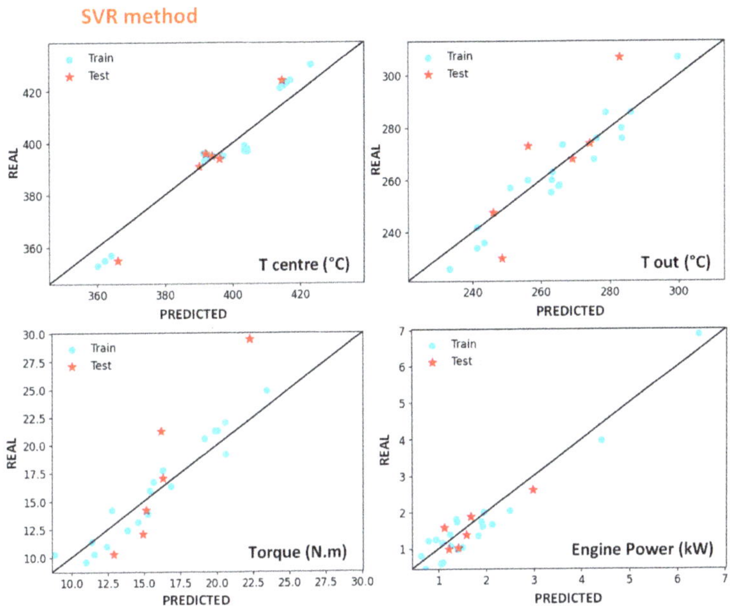

Figure 10. Results of the outputs predicted by the SVR method versus the experimental ones. The blue dots correspond to training data, and the red stars correspond to test data for HDPE and UHMWPE indiscriminately.

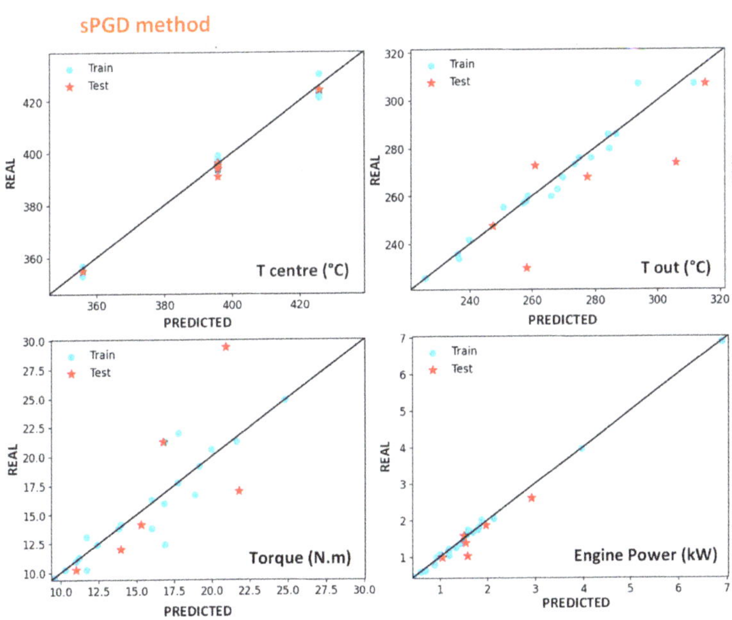

Figure 11. Results of the outputs predicted by the sPGD method versus the experimental ones. The blue dots correspond to training data, and the red stars correspond to test data for HDPE and UHMWPE indiscriminately.

In order to have more precise comparison tools, R^2 scores were calculated on the results and presented in Table 4. The closer the score is to 1, the closer the model is to measures. Whereas obtaining a good score for training data is accessible, obtaining it for both training and test data is trickier. This table shows that both methods give acceptable errors but that the sPGD can be more precise for most parameters, particularly for the die pressure. On the contrary, exit temperature is a little bit more precisely modelled with SVR.

Table 4. R^2 scores errors for SVR and sPGD methods.

R^2 Error	Centre Temperature	Exit Temperature	Torque	Engine Power	Die Pressure
SVR train	0.93	0.93	0.91	0.93	0.92
SVR global	0.92	0.88	0.8	0.92	0.75
sPGD train	0.99	0.91	0.82	1	0.98
sPGD global	0.99	0.88	0.71	0.99	0.84

Finally, either of these methods gives better results than the classical Ludovic® (SC-Consultants, Saint-Etienne, France) software model and good predictions without the need to understand the physical phenomena involved in the extrusion process. However, if this last point appears to be an advantage in favour of this method, it should be noted that one should be wary of it because the algorithm can model data that are false in the absolute. For example, the measured exit and centre temperature are very probably underestimated. Both algorithms, however, succeed to predict them, which proves that there is a logic between input parameters and these values. Nevertheless, they do not prevent eventual systematic errors in the measurements.

4.4.2. Modelling of Viscosity and Molecular Weight

One of the main interesting aspects of data-based simulations is that there is no need to understand the physics behind the measurements to obtain predictive models of these results. Consequently, it is a less time and power-consuming way to obtain predictions for any measurements, as long as the correct inputs are given. ML methods can also succeed in predicting values depending on unknown phenomena. However, they work as black boxes and cannot help to understand these phenomena.

For example, in this work, understanding the degradation mechanisms sufficiently to predict the final molecular weights of the material seems out of reach. On the contrary, predicting these values with ML methods seems entirely feasible. Figure 12 presents the results obtained with SVR and sPGD methods when predicting zero-shear viscosity η_0 and weight and number average molecular weights Mw and Mn. Table 5 presents the R^2 scores obtained, indicating the precision of these methods.

Both methods successfully model Mw values with reasonably high precision, but Mn values predictions are less accurate. Mw and Mn values implemented in the software are deduced from the viscoelastic behaviours of the melt samples. It has previously been shown that this method is more accurate for predicting Mw than Mn. This fact can explain the significant error noticeable for some of the data. Concerning the viscosity, the results obtained with the SVR method are pretty bad, and the predicted values seems shifted from the real ones. The cause of this phenomenon is unclear since the algorithm succeeds to predict Mw which, as seen previously, can directly be related to viscosity. sPGD algorithm presents similar R^2 scores for viscosity, but less shifted values, which shows that the choice of the regression method is crucial.

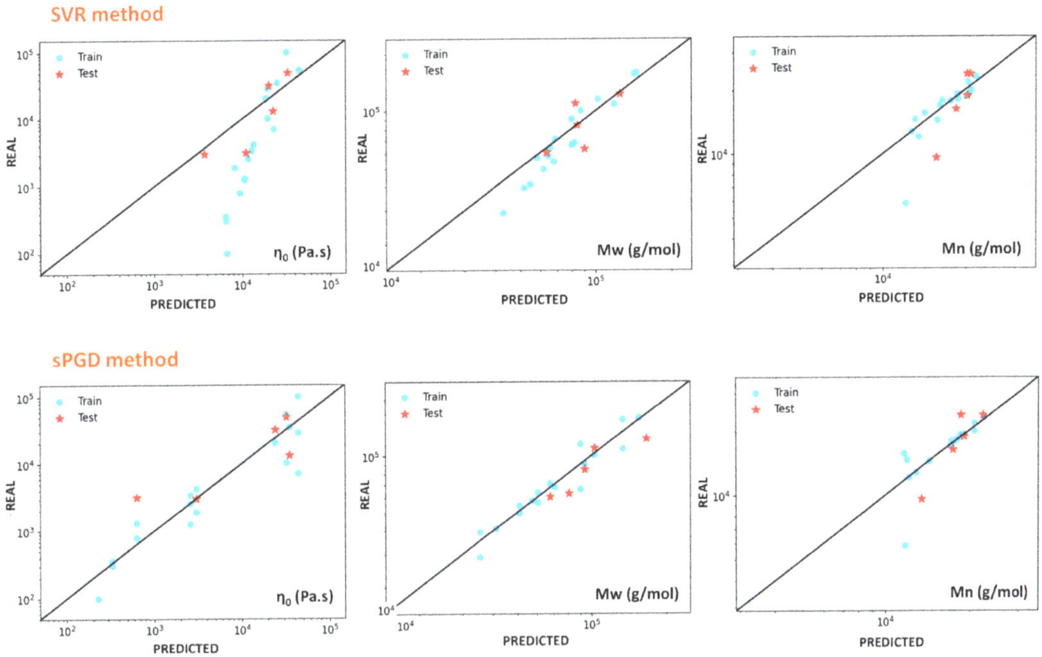

Figure 12. Results of zero-shear viscosity and weight and number average molecular weights predicted by SVR and sPGD methods versus the experimental ones. The blue dots correspond to training data, and the red stars correspond to test data for HDPE and UHMWPE indiscriminately.

Table 5. R^2 scores errors for the determination of η_0, Mw and Mn with SVR and sPGD methods.

R^2 Error	η_0	Mw	Mn
SVR train	0.48	0.91	0.77
SVR global	0.50	0.86	0.70
sPGD train	0.49	0.90	0.77
sPGD global	0.50	0.82	0.75

4.4.3. Stochastic Models

One problem with data-driven models is that their accuracy depends on the accuracy of the data, which necessarily includes inaccuracies due to measurement techniques. Stochastic models allow considering probability curves instead of points as data, thus smoothing out these inaccuracies and generally simplifying the model. The SVR method coupled with the stochastic approach was tested on ten data corresponding to HDPE extruded at T_{max} = 390 °C. The results are shown in Figure 13.

As there are only two inputs for these data (Screw rotation speed and Flow rate), the results can be plotted in 3D graphs. The middle surface corresponds to the predictions, and the translucent ones correspond to the superior and inferior acceptation boundaries. Despite the limited amount of data, the method gave satisfactory results. Therefore, this method is promising and needs to be tested with the different polymers and temperatures as inputs and other outputs such as viscosities or molecular weights to see if the stochastic approach can make an improvement.

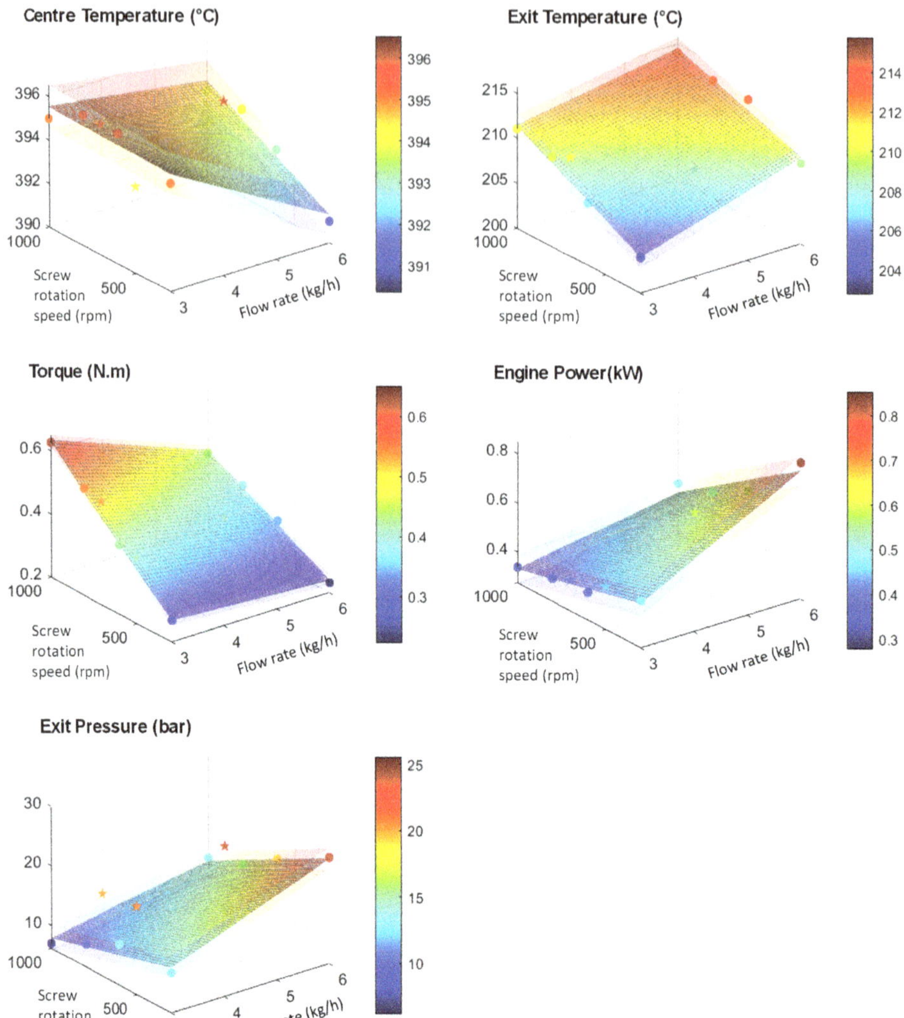

Figure 13. Results obtained with the stochastic method coupled to SVR. The dots correspond to training values, and the stars correspond to test ones.

5. Conclusions

HDPE and UHMWPE were degraded by twin-screw extrusion under different high temperatures ($320 < T$ °C < 420) and for different process conditions (flow rate and screw rotation speed), leading to numerous different extrusion configurations. Several parameters were measured for each configuration, creating a dataset with four different inputs, five outputs, and thirty-eight data.

The shear viscosity curves of the extruded materials were estimated from the measured die pressure and temperature. Their comparison with frequency sweep measurements showed that despite the numerous simplifications, the results were accurate. This fact shows that this method can be used to rapidly obtain an approximation of the final zero-shear viscosity of extruded materials.

Two methods were tested to estimate the molecular weight of extruded polyethylene. One was based on the viscoelastic behaviour of the material, and the other was deduced from die pressures and temperatures. The results showed that the average molecular weight *Mw* values were similar for both methods and similar to those obtained by SEC for the five samples tested. The method determining *Mw* only from measured die pressure and temperature thus seems more advantageous because it does not involve offline characterizations. However, this method is not sufficient to obtain the complete molecular weight distribution. In contrast, the other method based on viscoelastic measurements determined the complete molecular weight distribution. The results were good for HDPE but with some inaccuracies for UHMWPE samples. Although, as with the SEC, it requires offline characterizations, it is faster and is an interesting alternative.

The Ludovic® (SC-Consultants, Saint-Etienne, France) twin-screw extrusion simulation software was used as a classical model of the extrusion experiments. Since the degradation mechanisms occurring in the extruder are unknown, the simulation was performed considering the viscosities of the raw materials, which led to overestimated pressures and temperatures. Consequently, SVR and sPGD Machine-Learning methods were applied to the dataset and succeeded in modelling the extrusions' torque, engine power, die pressure, and die and centre temperatures. They also gave good results for the predictions of *Mw*. *Mn* has also been successfully predicted but with more inaccuracies, probably caused by its method of determination. Besides, whereas the SVR method gave inaccurate results for zero-shear viscosity modelling, sPGD's results were more acceptable. Finally, stochastic methods were tested on ten of the data giving promising results.

Machine-Learning seems to be a valuable tool for extrusion simulation as it is possible to obtain quickly accurate models. However, it is essential to keep in mind that ML methods cannot be used as predictive tools and also that the accuracy of the results depends on the accuracy of the data. In perspective, it could be interesting to think about the scale-up of this process and about how machine-learning could be a helpful tool for this purpose. Few experiments on a larger scale could then be necessary to adapt the whole model to it.

Author Contributions: Conceptualization, P.C. and F.C. (Francisco Chinesta); methodology, F.C. (Fanny Castéran) and K.D., software, F.C. (Fanny Castéran), N.H. and A.A.; validation, F.C. (Fanny Castéran). All authors have read and agreed to the published version of the manuscript.

Funding: This research received no external funding.

Data Availability Statement: Data is contained within the article.

Acknowledgments: Authors acknowledge ESI Group by its research chair at Arts et Metiers Institute of Technology and the French ANR through the DataBEST project.

Conflicts of Interest: The authors declare no conflict of interest.

Appendix A. Detail of Data's

Table A1. In-line measurements for the different tested extrusion parameters.

Process Inputs				Extrusion in-Line Measurements				
PE Type	*Q* (kg/h)	N (rpm)	T_{max} (°C)	Torque (N·m)	Pe (Bar)	*Tc* (°C)	Engine Power (kW)	Te (°C)
HDPE	3	300	350	21.3	28	355	1.04	248
HDPE	3	600	350	17.7	22	357	1.68	273
HDPE	6	300	350	13.8	12	353	3.97	236
HDPE	6	600	350	12.4	12	355	6.89	258
HDPE	1	300	390	10.3	7	395	0.46	226

Table A1. *Cont.*

	Process Inputs			Extrusion in-Line Measurements				
PE Type	Q (kg/h)	N (rpm)	T_{max} (°C)	Torque (N·m)	Pe (Bar)	*Tc* (°C)	Engine Power (kW)	Te (°C)
HDPE	3	300	390	16.7	13	395	0.81	242
HDPE	3	300	390	17.4	17	393	0.81	247
HDPE	3	500	390	16.0	15	394	1.23	260
HDPE	3	500	390	16.0	15	394	1.20	261
HDPE	3	600	390	13.8	12	395	1.31	263
HDPE	3	700	390	14.2	14	396	1.59	268
HDPE	3	1000	390	12.4	8	397	2.01	274
HDPE	5	600	390	20.6	18	396	2.01	276
HDPE	6	300	390	28.7	25	391	0.69	273
HDPE	6	600	390	21.3	19	393	1.02	286
HDPE	6	1000	390	17.0	20	396	0.38	307
HDPE	3	300	420	12.4	4	423	0.60	226
HDPE	3	600	420	10.3	3	424	1.00	230
HDPE	6	300	420	24.8	10	421	0.59	257
HDPE	6	600	420	19.2	8	422	1.81	258
UHMWPE	1	300	390	13.1	25	397	0.64	260
UHMWPE	1	600	390	11.4	13	398	1.09	276
UHMWPE	1	1000	390	9.6	9	399	1.38	286
UHMWPE	3	400	390	22.0	35	398	1.41	280
UHMWPE	3	600	390	16.3	30	397	1.74	307
UHMWPE	3	1000	390	14.2	34	397	1.74	307
UHMWPE	3	300	420	21.3	10	424	1.00	234
UHMWPE	3	600	420	11.0	9	430	1.06	256
UHMWPE	3	1000	420	10.3	7	430	1.61	268

Table A2. Ludovic® results obtained for the different tested extrusion parameters.

	Process Inputs			Ludovic® Simulation				
Polymer	Q (kg/h)	N (rpm)	T_{max} (°C)	Torque (N·m)	Te (°C)	*Tc* (°C)	Pe (bar)	Engine Power (kW)
HDPE	3	300	350	19.0	363	409	49	1.19
HDPE	3	600	350	13.1	491	520	31	1.65
HDPE	6	300	350	30.2	357	390	67	1.90
HDPE	6	600	350	20.3	473	481	45	2.55
HDPE	1	300	390	10.4	376	443	27	0.65
HDPE	3	300	390	18.7	369	426	47	1.18
HDPE	3	500	390	14.6	459	496	34	1.53
HDPE	3	500	390	14.6	459	496	34	1.53

Table A2. *Cont.*

Polymer	Process Inputs			Ludovic® Simulation				
	Q (kg/h)	N (rpm)	T_{max} (°C)	Torque (N·m)	Te (°C)	*Tc* (°C)	Pe (bar)	Engine Power (kW)
HDPE	3	600	390	13.0	499	530	30	1.64
HDPE	3	700	390	12.1	539	568	27	1.78
HDPE	3	800	390	11.4	579	606	24	1.91
HDPE	3	1000	390	6.4	597	356	23	1.33
HDPE	5	600	390	17.9	487	503	40	2.26
HDPE	6	300	390	29.8	363	407	65	1.87
HDPE	6	600	390	20.1	481	491	44	2.53
HDPE	6	800	390	17.2	553	554	36	2.89
HDPE	6	1000	390	14.1	597	520	32	2.95
HDPE	3	300	420	18.5	374	439	47	1.17
HDPE	3	600	420	12.9	505	538	30	1.63
HDPE	6	300	420	29.5	368	419	64	1.85
HDPE	6	600	420	20.0	488	498	43	2.51
UHMWPE	1	300	390	14.8	402	471	127	0.93
UHMWPE	1	600	390	9.0	459	530	83	1.13
UHMWPE	1	1000	390	6.3	510	590	60	1.32
UHMWPE	3	400	390	19.8	444	485	92	1.66
UHMWPE	3	600	390	13.8	486	519	70	1.74
UHMWPE	3	1000	390	9.2	547	577	49	1.92
UHMWPE	3	300	420	25.0	419	477	112	1.57
UHMWPE	3	600	420	13.7	488	527	69	1.72
UHMWPE	3	1000	420	9.1	550	581	48	1.90

Table A3. Viscosities and molecular weights for the different tested extrusion parameters.

Polymer	Process Inputs			Zero Shear-Rate Viscosities		Molecular Weights (Inverse Rheology)			*Mw* (From Die Pressure)
	Q (kg/h)	N (rpm)	*T* (°C)	η_0 from Die Pressure (Pa·s)	η_0 Rheometer (Pa·s)	*Mw* (g/mol)	Mz (g/mol)	*Mn* (g/mol)	*Mw* (g/mol)
HDPE	3	300	350	2.6×10^4	3.2×10^4	1.1×10^5	5.0×10^5	2.4×10^4	1.1×10^5
HDPE	3	600	350	1.3×10^4	2.1×10^4	9.1×10^4	4.3×10^5	1.9×10^4	8.6×10^4
HDPE	6	300	350	1.1×10^4	3.5×10^4	1.0×10^5	5.2×10^5	1.9×10^4	8.3×10^4
HDPE	6	600	350	1.1×10^4	1.3×10^4	8.4×10^4	3.7×10^5	1.9×10^4	8.2×10^4
HDPE	3	300	390	6.4×10^3	3.2×10^3	6.0×10^4	1.5×10^5	2.4×10^4	7.0×10^4
HDPE	6	300	390	7.4×10^3	3.4×10^3	7.1×10^4	2.2×10^5	2.3×10^4	7.3×10^4
HDPE	6	600	390	4.0×10^3	2.5×10^3	6.4×10^4	2.0×10^5	2.0×10^4	6.1×10^4
HDPE	6	1000	390	1.4×10^3	1.2×10^3	5.4×10^4	1.5×10^5	1.9×10^4	4.5×10^4
HDPE	3	300	420	3.5×10^2	9.0×10^1	2.8×10^4	5.9×10^4	1.3×10^4	3.0×10^4
HDPE	3	600	420	2.2×10^2	1.0×10^2	2.9×10^4	5.9×10^4	1.4×10^4	2.6×10^4
HDPE	6	300	420	8.1×10^2	3.5×10^2	4.1×10^4	9.3×10^4	1.8×10^4	3.8×10^4
HDPE	6	600	420	5.2×10^2	3.1×10^2	3.9×10^4	8.7×10^4	1.8×10^4	3.4×10^4

Table A3. *Cont.*

Process Inputs				Zero Shear-Rate Viscosities		Molecular Weights (Inverse Rheology)			*Mw* (From Die Pressure)
Polymer	Q (kg/h)	N (rpm)	*T* (°C)	η_0 from Die Pressure (Pa·s)	η_0 Rheometer (Pa·s)	*Mw* (g/mol)	Mz (g/mol)	*Mn* (g/mol)	*Mw* (g/mol)
UHMWPE	1	300	390	7.4×10^4	5.0×10^4	1.2×10^5	1.4×10^6	9.6×10^3	1.4×10^5
UHMWPE	1	600	390	1.3×10^4	5.4×10^4	1.1×10^5	8.8×10^5	1.3×10^4	8.6×10^4
UHMWPE	1	1000	390	5.8×10^3	1.0×10^4	6.6×10^4	3.5×10^5	1.2×10^4	6.8×10^4
UHMWPE	3	400	390	5.3×10^4	7.0×10^3	1.6×10^5	1.5×10^6	1.6×10^4	1.3×10^5
UHMWPE	3	600	390	3.1×10^4	1.0×10^5	1.6×10^5	3.4×10^6	5.8×10^3	1.1×10^5
UHMWPE	3	1000	390	4.8×10^4	2.9×10^4	1.2×10^5	8.7×10^5	1.5×10^4	1.3×10^5
UHMWPE	3	300	420	1.9×10^3	3.1×10^3	6.3×10^4	2.4×10^5	1.6×10^4	4.9×10^4
UHMWPE	3	600	420	1.5×10^3	4.2×10^3	6.8×10^4	2.6×10^5	1.8×10^4	4.6×10^4
UHMWPE	3	1000	420	9.4×10^2	1.9×10^3	5.8×10^4	2.0×10^5	1.7×10^4	4.6×10^4

References

1. PlascticEurope-Association of Plastics Manufactures, Plastics—The Facts 2020. *PlasticEurope* **2020**, 1–64. Available online: https://www.plasticseurope.org/en/resources/publications/4312-plastics-facts-2020 (accessed on 2 June 2021).
2. Al-Salem, S.M.; Lettieri, P.; Baeyens, J. Recycling and recovery routes of plastic solid waste (PSW): A review. *Waste Manag.* **2009**, *29*, 2625–2643. [CrossRef] [PubMed]
3. Meran, C.; Ozturk, O.; Yuksel, M. Examination of the possibility of recycling and utilizing recycled polyethylene and polypropylene. *Mater. Des.* **2008**, *29*, 701–705. [CrossRef]
4. Gu, F.; Guo, J.; Zhang, W.; Summers, P.A.; Hall, P. From waste plastics to industrial raw materials: A life cycle assessment of mechanical plastic recycling practice based on a real-world case study. *Sci. Total Environ.* **2017**, *601–602*, 1192–1207. [CrossRef]
5. Pinheiro, L.A.; Chinelatto, M.A.; Canevarolo, S.V. The role of chain scission and chain branching in high density polyethylene during thermo-mechanical degradation. *Polym. Degrad. Stab.* **2004**, *86*, 445–453. [CrossRef]
6. Mastellone, M.L.; Perugini, F.; Ponte, M.; Arena, U. Fluidized bed pyrolysis of a recycled polyethylene. *Polym. Degrad. Stab.* **2002**, *76*, 479–487. [CrossRef]
7. Basfar, A.A.; Idriss Ali, K.M. Natural weathering test for films of various formulations of low density polyethylene (LDPE) and linear low density polyethylene (LLDPE). *Polym. Degrad. Stab.* **2006**, *91*, 437–443. [CrossRef]
8. Al-Salem, S.M. Influence of natural and accelerated weathering on various formulations of linear low density polyethylene (LLDPE) films. *Mater. Des.* **2009**, *30*, 1729–1736. [CrossRef]
9. Benoit, N.; González-Núñez, R.; Rodrigue, D. High density polyethylene degradation followed by closedloop recycling. *Prog. Rubber, Plast. Recycl. Technol.* **2017**, *33*, 17–37. [CrossRef]
10. Salomão, R.; Pandolfelli, V.C. Anti-spalling fibers for refractory castables: A potential application for recycling drinking straws. *Ceram. Int.* **2020**, *46*, 14262–14268. [CrossRef]
11. Adhikary, K.B.; Pang, S.; Staiger, M.P. Dimensional stability and mechanical behaviour of wood-plastic composites based on recycled and virgin high-density polyethylene (HDPE). *Compos. Part B Eng.* **2008**, *39*, 807–815. [CrossRef]
12. Lei, Y.; Wu, Q.; Yao, F.; Xu, Y. Preparation and properties of recycled HDPE/natural fiber composites. *Compos. Part A Appl. Sci. Manuf.* **2007**, *38*, 1664–1674. [CrossRef]
13. Christakopoulos, F.; Troisi, E.; Friederichs, N.; Vermant, J.; Tervoort, T.A. "Tying the Knot": Enhanced Recycling through Ultrafast Entangling across Ultrahigh Molecular Weight Polyethylene Interfaces. *Macromolecules* **2021**, *54*, 9452–9460. [CrossRef]
14. Kulkarni, G.S. *Introduction to Polymer and Their Recycling Techniques*; Elsevier Inc.: Amsterdam, The Netherlands, 2018; ISBN 9780323511346.
15. Kaminsky, W.; Schlesselmann, B.; Simon, C. Olefins from polyolefins and mixed plastics by pyrolysis. *J. Anal. Appl. Pyrolysis* **1995**, *32*, 19–27. [CrossRef]
16. Mccaffrey, C.; Cooper, C.; Kamal, M.R. Thermolysis of Polyethylene/Polystyrene Mixtures. *J. Appl. Polym. Sci.* **1996**, *60*, 2133–2140. [CrossRef]
17. Bockhorn, H.; Hornung, A.; Hornung, U. Stepwise pyrolysis for raw material recovery from plastic waste. *J. Anal. Appl. Pyrolysis* **1998**, *46*, 1–13. [CrossRef]
18. Horvat, N.; Ng, F.T.T. Tertiary polymer recycling: Study of polyethylene thermolysis as a first step to synthetic diesel fuel. *Fuel* **1999**, *78*, 459–470. [CrossRef]
19. Nishino, J.; Itoh, M.; Fujiyoshi, H.; Uemichi, Y. Catalytic degradation of plastic waste into petrochemicals using Ga-ZSM-5. *Fuel* **2008**, *87*, 3681–3686. [CrossRef]
20. Del Remedio Hernández, M.; García, Á.N.; Marcilla, A. Catalytic flash pyrolysis of HDPE in a fluidized bed reactor for recovery of fuel-like hydrocarbons. *J. Anal. Appl. Pyrolysis* **2007**, *78*, 272–281. [CrossRef]

21. Achilias, D.S.; Roupakias, C.; Megalokonomos, P.; Lappas, A.A.; Antonakou, V. Chemical recycling of plastic wastes made from polyethylene (LDPE and HDPE) and polypropylene (PP). *J. Hazard. Mater.* **2007**, *149*, 536–542. [CrossRef]
22. Buekens, A.G.; Huang, H. Catalytic plastics cracking for recovery of gasoline-range hydrocarbons from municipal plastic wastes. *Resour. Conserv. Recycl.* **1998**, *23*, 163–181. [CrossRef]
23. Guy, L.; Fixari, B. Waxy polyethylenes from solution thermolysis of high density polyethylene: Inert and H-donor solvent dilution effect. *Polymer* **1999**, *40*, 2845–2857. [CrossRef]
24. Scheirs, J.; Kaminsky, W. *Recycling and Pyrolysis of Waste Plastics: Converting Waste Plastics into Diesel and Other Fuels*; Scheirs, J., Kaminsky, W., Eds.; John Wiley & Sons, Ltd.: Hoboken, NJ, USA, 2006; Volume 3, ISBN 0-470-02152-7.
25. Vollmer, I.; Jenks, M.J.F.; Roelands, M.C.P.; White, R.J.; van Harmelen, T.; de Wild, P.; van der Laan, G.P.; Meirer, F.; Keurentjes, J.T.F.; Weckhuysen, B.M. Beyond Mechanical Recycling: Giving New Life to Plastic Waste. *Angew. Chem.-Int. Ed.* **2020**, *59*, 15402–15423. [CrossRef] [PubMed]
26. Manas, D.; Manas, M.; Mizera, A.; Navratil, J.; Ovsik, M.; Tomanova, K.; Sehnalek, S. Use of irradiated polymers after their lifetime period. *Polymers* **2018**, *10*, 641. [CrossRef] [PubMed]
27. Manas, D.; Manas, M.; Mizera, A.; Stoklasek, P.; Navratil, J.; Sehnalek, S.; Drabek, P. The high density polyethylene composite with recycled radiation cross-linked filler of rHDPEx. *Polymers* **2018**, *10*, 1361. [CrossRef]
28. Elmanovich, I.V.; Stakhanov, A.I.; Kravchenko, E.I.; Stakhanova, S.V.; Pavlov, A.A.; Ilyin, M.M.; Kharitonova, E.P.; Gallyamov, M.O.; Khokhlov, A.R. Chemical recycling of polyethylene in oxygen-enriched supercritical CO_2. *J. Supercrit. Fluids* **2021**, *181*, 105503. [CrossRef]
29. Inderthal, H.; Tai, S.L.; Harrison, S.T.L. Non-Hydrolyzable Plastics—An Interdisciplinary Look at Plastic Bio-Oxidation. *Trends Biotechnol.* **2021**, *39*, 12–23. [CrossRef]
30. Zhu, B.; Wang, D.; Wei, N. Enzyme discovery and engineering for sustainable plastic recycling. *Trends Biotechnol.* **2022**, *40*, 22–37. [CrossRef]
31. Castéran, F.; Ibanez, R.; Argerich, C.; Delage, K.; Chinesta, F.; Cassagnau, P. Application of Machine Learning Tools for the Improvement of Reactive Extrusion Simulation. *Macromol. Mater. Eng.* **2020**, *305*, 2000375. [CrossRef]
32. Ibañez, R.; Casteran, F.; Argerich, C.; Ghnatios, C.; Hascoet, N.; Ammar, A.; Cassagnau, P.; Chinesta, F. On the Data-Driven Modeling of Reactive Extrusion. *Fluids* **2020**, *5*, 94. [CrossRef]
33. Cassagnau, P.; Montfort, J.P.; Marin, G.; Monge, P. Rheology of polydisperse polymers: Relationship between intermolecular interactions and molecular weight distribution. *Rheol. Acta* **1993**, *32*, 156–167. [CrossRef]
34. Des Cloizeaux, J.; Jannik, G. *Polymers in Solution Their Modelling and Structure*; Oxford University Press: Oxford, UK, 1990; ISBN 0-19-852036-0.
35. Vega, J.F.; Rastogi, S.S.; Peters, G.G.; Meijer, H.E.H. Rheology and reptation of linear polymers. Ultrahigh molecular weight chain dynamics in the melt. *J. Rheol.* **2004**, *48*, 663–678. [CrossRef]
36. Wasserman, S.H.; Graessley, W.W. Prediction of linear viscoelastic response for entangled polyolefin melts from molecular weight distribution. *Polym. Eng. Sci.* **1996**, *36*, 852–861. [CrossRef]
37. Rochefort, W.E.; Smith, G.G.; Rachapudy, H.; Raju, V.R.; Graessley, W.W. Properties of amorphous and crystallizable hydrocarbon polymers. II. Rheology of linear and star-branched polybutadiene. *J. Polym. Sci. Polym. Phys. Ed.* **1979**, *17*, 1197–1210. [CrossRef]
38. Wood-Adams, P.M.; Dealy, J.M.; DeGroot, A.W.; Redwine, O.D. Effect of molecular structure on the viscoelastic behavior of epoxy resin. *Kobunshi Ronbunshu* **1976**, *33*, 19–27. [CrossRef]
39. Vergnes, B.; Della Valle, G.; Delamare, L. A global computer software for polymer flows in corotating twin screw extruders. *Polym. Eng. Sci.* **1998**, *38*, 1781–1792. [CrossRef]
40. Fel, E.; Massardier, V.; Mélis, F.; Vergnes, B.; Cassagnau, P. Residence time distribution in a high shear twin screw extruder. *Int. Polym. Process.* **2014**, *29*, 71–80. [CrossRef]
41. Berzin, F.; Tara, A.; Tighzert, L.; Vergnes, B. Importance of Coupling Between Specific Energy and Viscosity in the Modeling of Twin Screw Extrusion of Starchy Products. *Polym. Eng. Sci.* **2010**, *50*, 1758–1766. [CrossRef]
42. Smola, A.J.; Scholkopf, B. A tutorial on support vector regression. *Stat. Comput.* **2004**, *14*, 199–222. [CrossRef]
43. Ibáñez, R.; Abisset-Chavanne, E.; Ammar, A.; González, D.; Cueto, E.; Huerta, A.; Duval, J.L.; Chinesta, F. A multidimensional data-driven sparse identification technique: The sparse proper generalized decomposition. *Complexity* **2018**, *2018*, 5608286. [CrossRef]

polymers

MDPI

Article

Lime/Sodium Carbonate Treated Seawater to Improve Flocculation and Sedimentation of Clay-Based Tailings

Francisco Pulgar [1], Luis Ayala [2], Matías Jeldres [1], Pedro Robles [3], Pedro G. Toledo [4,*], Iván Salazar [5] and Ricardo I. Jeldres [1,*]

1 Departamento de Ingeniería Química y Procesos de Minerales, Facultad de Ingeniería, Universidad de Antofagasta, P.O. Box 170, Antofagasta 1240000, Chile; francisco.pulgar.gonzalez@ua.cl (F.P.); hugo.jeldres.valenzuela@ua.cl (M.J.)

2 Faculty of Engineering and Architecture, Universidad Arturo Prat, Almirante Juan José Latorre 2901, Antofagasta 1240000, Chile; luisayala01@unap.cl

3 Escuela de Ingeniería Química, Pontificia Universidad Católica de Valparaíso, Valparaíso 2340000, Chile; pedro.robles@pucv.cl

4 Departamento de Ingeniería Química, Universidad de Concepción, Concepción 4030000, Chile

5 Departamento de Ingeniería Civil, Universidad Católica del Norte, Antofagasta 1270709, Chile; isalazar@ucn.cl

* Correspondence: petoledo@udec.cl (P.G.T.); ricardo.jeldres@uantof.cl (R.I.J.)

Citation: Pulgar, F.; Ayala, L.; Jeldres, M.; Robles, P.; Toledo, P.G.; Salazar, I.; Jeldres, R.I. Lime/Sodium Carbonate Treated Seawater to Improve Flocculation and Sedimentation of Clay-Based Tailings. *Polymers* **2021**, *13*, 4108. https://doi.org/10.3390/polym13234108

Academic Editors: Célio Bruno Pinto Fernandes, Salah Aldin Faroughi, Luís L. Ferrás and Alexandre M. Afonso

Received: 25 October 2021
Accepted: 9 November 2021
Published: 25 November 2021

Abstract: Seawater treated with lime and sodium carbonate in different proportions to reduce magnesium and calcium contents is used in flocculation and sedimentation tests of artificial quartz and kaolin tailings. Solid complexes were separated from water by vacuum filtration, and factors such as lime/sodium carbonate ratio, kaolin content, flocculation time, and flocculant dose are evaluated. The growth of the aggregates was captured in situ by a focused beam reflectance measurement (FBRM) probe. Solid magnesium and calcium complexes are formed in raw seawater at pH 11, impairing the performance of flocculant polymers based on polyacrylamides. The results show that the settling rate improved when the treatment's lime/sodium carbonate ratio increased. That is, when a greater removal of magnesium is prioritized over calcium. The amount of magnesium required to be removed depends on the mineralogy of the system: more clay will require more significant removal of magnesium. These results respond to the structural changes of the flocs, achieving that the more magnesium is removed, the greater the size and density of the aggregates. In contrast, calcium removal does not significantly influence flocculant performance. The study suggests the necessary conditions for each type of tailing to maximize water recovery, contributing to the effective closure of the water cycle in processes that use seawater with magnesium control.

Keywords: tailings flocculation; seawater; calcium and magnesium removal; lime; sodium carbonate

1. Introduction

An important number of mineral deposits are located in arid or semi-arid regions, where any water consumption has significant economic and socio-environmental repercussions [1,2]. Thus, reducing water consumption in the concentration stages is an urgent task [3,4]. In this sense, solid–liquid separation has a fundamental role because it is the stage in which the greatest amount of water is recovered [5]. This process separates solid particles by flocculating them in water and settling by gravity [6,7]. As a result, a highly concentrated sediment is obtained that is transported through a pipe to a disposal area for further dewatering and consolidation. At the same time, the clarified overflow is recycled in the upstream operations. A common flocculant is a high molecular weight anionic polyacrylamide (A-PAM), which generates high sedimentation rates when applied at relatively low doses [8–10].

Currently, several mining operations use seawater in their processes, either directly or after desalination by reverse osmosis [2,11]. For example, by 2030, the consumption of seawater in Chile would increase by 156% compared to 2019, which represents 47% of the water required by the copper mining at the national level [2]. The use of desalinated seawater ensures an adequate water supply. However, desalination plants pose numerous environmental challenges, mainly related to the wastewater concentrated in salt and chemicals. Discarding can affect coastal water quality and dramatically change the marine environment [12,13]. Furthermore, this process requires energy supplied by fuels, which implies high emissions of atmospheric pollutants [14]. The direct use of seawater offers the advantage of avoiding the construction of desalination plants, reducing the associated economic and environmental costs. However, this implementation should be directed to new plants since older facilities are not prepared to the withstand high salinities [2].

Several studies have been dedicated to evaluating the consequences of a saline environment in the mineral flocculation processes and its impact on the thickening stages. For example, Ji et al. [15] used different flocculants to sediment quartz and albite in a highly saline solution, finding high sedimentation rates and a clearer supernatant. Liu et al. [16] suggested that salts enhance the aggregation of bentonite and illite, but hinder the aggregation of kaolinite. Recently, Jeldres et al. [17] and Quezada et al. [18] have used NaCl brines, showing that a saline medium can favor the flocculation of mineral particles by increasing the adsorption of the flocculant on the surface of the particles. These results are interesting because they challenge the idea that flocculants roll up in the presence of salts, losing the adsorptive capacity of their mineral particles. Thus, salinity can favor or harm flocculation, depending on the type of flocculant and mineral considered. The process is more complicated in seawater, especially when the concentration circuit is carried out in alkaline conditions that are characteristic of froth flotation operations. It is typical for the copper minerals processing to operate at a pH higher than 10.5, with the aim of depressing non-valuable minerals such as pyrite, which tend to float due to their hydrophobic nature, thus contaminating the concentrate [19,20]. However, this strategy cannot be implemented when using seawater since solid Ca/Mg complexes can arise, producing a buffering effect at a pH above 10 and a considerable reduction in the recovery of molybdenite [21,22]. Furthermore, the consequences for thickening operations are challenging. The few studies that have been performed in this regard show that the presence of precipitates leads to considerably lower sedimentation rates. For example, Ramos et al. [23] analyzed the flocculation of mine tailings using seawater in a wide range of pH in the pulp, and reported a severe detriment to the sedimentation rate once the pH of the slurry reaches the value at which solid precipitates (pH > 10.3). In their main results, Ramos et al. found that magnesium causes the greatest detriment. They also found that calcium does not affect the process within the pH range considered (pH < 11.1). Ramos et al. suggested that magnesium precipitates would have a higher affinity for the flocculant, which impairs the unique interaction of the polymer with the mineral. This forced the use of high doses, first to saturate the surface of the precipitates, and then to flocculate the mineral particles.

To reduce the impact caused by precipitates and solid complexes on foam flotation, Castro [24] proposed reducing the magnesium content of seawater by treating it with lime. Later, Jeldres et al. [25] supplemented the strategy by using a mixture of lime and sodium carbonate to reduce magnesium and calcium content. The authors achieved good recovery of molybdenite and chalcopyrite at an alkaline pH, promoting pyrite depression. Recently, Jeldres et al. [26] studied thickening operations, finding that a lime removal treatment can improve the performance of the thickeners. The authors considerably increased the sedimentation rate of mineral tailings at pH 11, obtaining higher values than at natural pH. More recently, Arias et al. [27] proposed a biotechnological treatment to remove divalent ions from seawater, using a fluidized bed bioreactor completed with the halotolerant ureolytic strain Bacillus subtilis LN8B. These latest studies [26,27] have yielded promising results in reducing magnesium in seawater, opening up a new line of research that needs to be explored systematically.

In this work, different mixtures of lime with sodium carbonate are used to produce seawater with varying magnesium contents. These treated waters are evaluated in flocculation and sedimentation tests of synthetic quartz and kaolin tailings in terms of lime/sodium carbonate ratio, kaolin content, flocculation time, and flocculant dose.

2. Materials and Methods

2.1. Materials

Seawater from San Jorge Bay in Antofagasta (Chile) was filtered through a U.V. purification system to eliminate bacterial activity. The electrical conductivity was 50.2 mS/cm at 25 °C, and the pH was natural (pH 7.5). The concentration of the primary ions was determined by the methods indicated in Table 1.

Table 1. Ionic concentration of seawater and analytical methods.

Ion	Concentration [g/L]	Analytical Method
Na^+	10.8	Atomic absorption spectrometry
Mg^+	1.42	Atomic absorption spectrometry
Ca^{2+}	0.42	Atomic absorption spectrometry
K^+	0.39	Atomic absorption spectrometry
Cl^-	18.9	Argentometry
$HCO_3{}^-$	0.14	Volumetric acid-base titration

Synthetic tailings were prepared from mixtures of quartz and kaolin in different proportions. The quartz particles were available in our lab (XRD in Figure 1a) and the kaolin was procured from Ward's Science (Rochester, NY, USA). The kaolin composition included mainly kaolinite and halite and a small proportion of SiO_2 (XRD in Figure 1b). The XRD analyses were performed on a Bruker D8 Advance X-ray diffractometer (Bruker, Karlsruhe, Germany). Once collected, diffraction data were processed and analyzed using the latest versions of the International Centre for Diffraction (ICDD) databases.

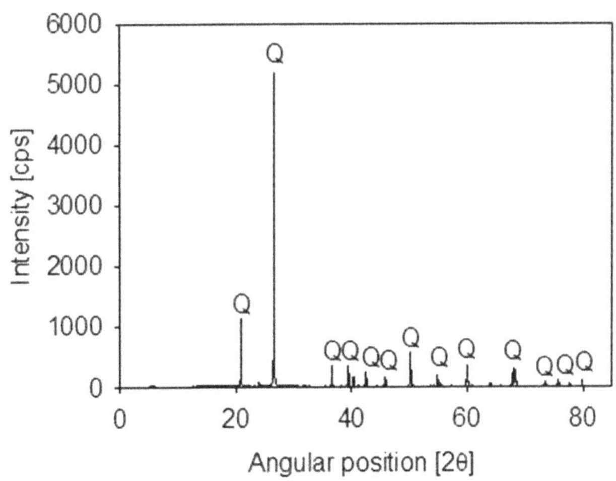

(a)

Figure 1. *Cont.*

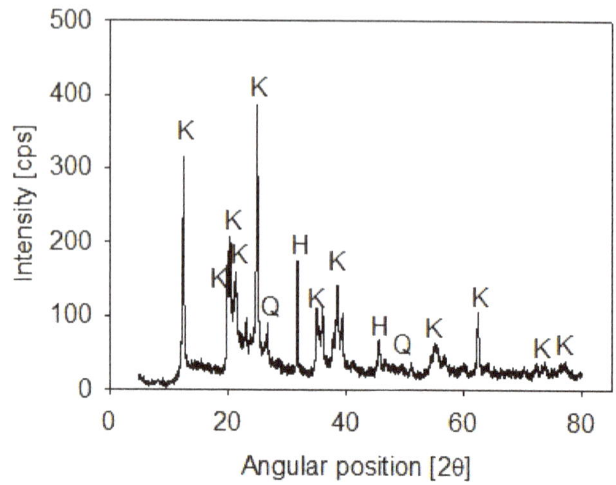

(**b**)

Figure 1. X-ray diffraction of quartz (**a**) and kaolin (**b**) powder, showing kaolinite (K), halite (H) and quartz (Q).

Volume-weighted particle size distribution (PSD) was measured by laser diffraction using a Microtrac S3500 instrument (Verder Scientific, Newtown, PA, USA). As shown in Figure 2, 10% of the particles were smaller than d10 = 1.8 and 3.8 μm in the kaolin and quartz samples, respectively. SNF 704, provided by SNF Chile S.A., was used as a flocculant. This reagent has a molecular weight of 18×10^6 and a medium charge density (30–50% anionic functionalities). The alkalizing agents used to precipitate magnesium at alkaline pH were analytical-grade lime and sodium carbonate, obtained from Sigma-Aldrich (Santiago, Chile). The pH of the suspensions was controlled with sodium hydroxide, obtained from Sigma-Aldrich, Chile.

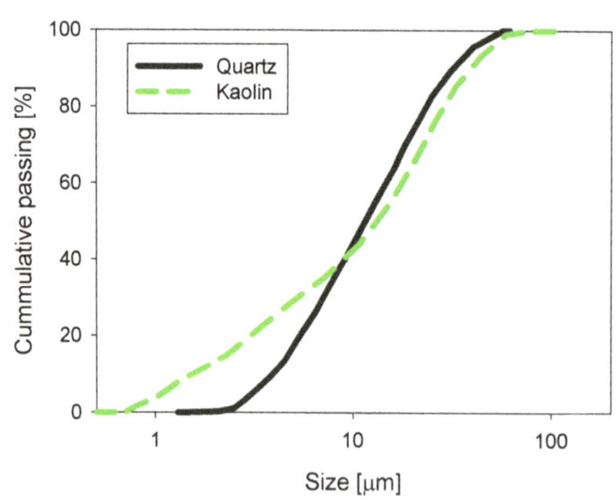

Figure 2. Size distribution for quartz and kaolin particles in distilled water at natural pH.

2.2. *Magnesium Removal Treatment*

Magnesium ions were partially removed from seawater by precipitation with lime and sodium carbonate in different proportions and overall concentration of 0.07 M. The resulting solution was mixed for 30 min at room temperature, generating a highly alkaline environment that favored the formation of the calcium and magnesium precipitates that were finally separated by vacuum filtration. This filtrate, referred to as treated seawater throughout this document, was used in flocculation and sedimentation tests of artificial quartz and kaolin tailings. Ca and Mg concentrations were measured by inductively coupled plasma mass spectrometry (ICP-MS, Varian 220 FS Atomic Absorption Spectrophotometer, Varian, Palo Alto, CA, USA).

2.3. Flocculation Kinetic

In a 1 L beaker, 24 g of the quartz–kaolin-based mineral was mixed with 246 g of water, and the pH was adjusted to 11 with sodium hydroxide. For this, a stirring rate (600 rpm) was applied for 5 min using a 30 mm diameter turbine impeller at the end of a vertical axis (4 mm diameter) placed 20 mm above the bottom of the container. Subsequently, the stirring was reduced to 220 rpm, then the flocculant solution and the remaining volume of water were added to reach 300 g of suspension. This methodology ensured the same solids concentration (8% by weight) for all the flocculation experiments. The calcium and magnesium that were not precipitated in the lime/sodium carbonate treatment did so during the pH control with sodium hydroxide in the conditioning for the flocculation tests.

The evolution of the aggregate size of each pulp was determined using the focused beam reflectance measurement (FBRM) system (Particle Track E25, Mettler Toledo, Columbus, OH, USA), which consists of a processing unit and a probe with a 19 mm diameter tip and a sapphire window (14 mm diameter) at the measuring tip. The probe was inserted vertically into the container with the pulp, 10mm above the stirrer and 20 mm off-axis. The FBRM probe features a laser that is focused through the sapphire window and scans a circular path at a tangential velocity of 2 m/s. Once the beam encounters suspended solids in the focal plane, backscattered light is generated. A chord length is obtained from the persistence of any high backscattered light intensity and the speed of the laser. The software processes the recorded data into histograms of the counts corresponding to chord lengths in selected channel sizes ranging from 1 μm to 1 mm as quickly as every 2 s. In this case, the chord length distributions (CLD) represent 100 channels in the full range, but the histograms are presented as line graphs for easy comparison. The FBRM system offers two types of particle size distributions, the unweighted CLD, which is more sensitive to finer particles, and the squared-weighted CLD, which is more susceptible to coarse particle aggregates. The raw data were processed in this work without any weighing to detect dispersed fine particles. However, the average size was obtained as a function of the square-weighted size distribution.

2.4. Sedimentation Tests

The settling rate was determined by interrupting the flocculation tests at specific preset times of 20, 40, 60, and 80 s. Then, the suspension was poured from the bottom of the flocculation cell into a cylindrical tube (35 mm inner diameter). The cylinder, with its contents, was slowly inverted three times and then placed on a surface to determine the sedimentation rate classically.

2.5. Fractal Dimension

The fractal dimension of the aggregates was determined according to the methodology of Heath et al. [28], expressed in the equation

$$U_h = \frac{\overline{d_{agg}^2} g(\rho_s - \rho_l)\left(\frac{\overleftarrow{d_{agg}}}{\overleftarrow{d_p}}\right)^{D_f-3}}{18\mu} \left(1 - \phi_s\left(\frac{\overleftarrow{d_{agg}}}{\overleftarrow{d_p}}\right)^{3-D_f}\right)^{4.65} \quad (1)$$

where U_h is the hindered settling rate in m/s, d_p is the average size of the primitive particles, d_{agg} is the average size of the aggregates after some flocculation time, ρ_s and ρ_l are, respectively, the densities of the solid and liquid phases, g is the acceleration of gravity, μ is the fluid viscosity, ϕ_s is the solid fraction, and D_f is the mass fractal dimension. To determine the fractal dimension, the hindered settling rate of the previous section was used, and the squared weighted mean chord length was used for the average size of the aggregates. All the other parameters in the equation were constant for all the systems studied.

3. Results and Discussion

Water recovery in mineral concentration processes using seawater requires fine control of magnesium. In this study, seawater was treated with lime and sodium carbonate to precipitate magnesium, and then filtered. Various proportions of lime and sodium carbonate were used. Treated seawater was used in flocculation and sedimentation tests of artificial tailings of quartz and kaolin to evaluate the lime/sodium carbonate ratio, kaolin content, flocculation time, and flocculant dose. The growth of the aggregates was captured by the FBRM probe.

3.1. Seawater Treatment

The main reactions when seawater is treated with various proportions of lime and sodium carbonate are as follows.

$$Ca(OH)_2 + Mg^{2+} \longleftrightarrow Mg(OH)_2 + Ca^{2+} \tag{2}$$

$$Ca^{2+} + 2OH^- \longleftrightarrow Ca(OH)_2 \tag{3}$$

$$Ca(OH)_2 + SO_4^{2-} + 2H_2O \longleftrightarrow CaSO_4{\cdot}2H_2O + 2OH^- \tag{4}$$

$$Mg(OH)_2 + SO_4^{2-} \longleftrightarrow MgSO_4 + 2OH^- \tag{5}$$

$$Ca(OH)_2 + CO_3^{2-} \longleftrightarrow CaCO_3 + 2OH^- \tag{6}$$

$$Mg(OH)_2 + CO_3^{2-} \longleftrightarrow MgCO_3 + 2OH^- \tag{7}$$

Magnesium combines to give rise to various complexes with varying solubilities. The solubility constants are summarized in Table 2.

Table 2. Ionic concentration of seawater and analytical methods.

Product	Solubility Product (Ksp)
$Mg(OH)_2$	5.61×10^{-12}
$MgCO_3$	6.82×10^{-6}
$Ca(OH)_2$	5.02×10^{-6}
$CaCO_3$	3.36×10^{-9}

Figure 3 shows the concentration of calcium and magnesium in seawater after being treated with different proportions of lime and sodium carbonate, always maintaining a total concentration of 0.07 M of these reagents. When using 100% sodium carbonate, the magnesium concentration decreases from 1420 to 920 mg/L because it precipitates as magnesium carbonate. However, as the lime/sodium carbonate ratio increases, the amount of magnesium in the solution decreases significantly, with almost complete removal when 100% lime is used. This result shows that lime precipitates more magnesium as magnesium hydroxide than sodium carbonate as magnesium carbonate. The explanation lies in the higher solubility of magnesium carbonate, 2.61×10^{-3} at 25 °C, versus the solubility of magnesium hydroxide, which is only 1.77×10^{-4} at 25 °C. Regarding calcium, the concentration decreases from 400 to 113 mg/L when using 100% sodium carbonate and increases when the lime content increases, reaching 2412 mg/L when using 100% lime. Seawater treatment with lime/sodium carbonate ratios below 50% is ineffective for magnesium reduction, as shown in Figure 3.

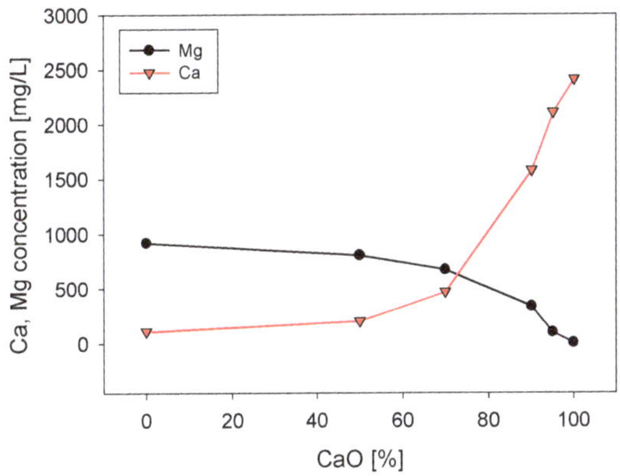

Figure 3. Calcium and magnesium concentrations in seawater treated with different lime/sodium carbonate ratios with a concentration of 0.07 M. Initial Mg and Ca are, respectively, 1420 mg/L and 420 mg/L.

3.2. Implications of Magnesium Content

The magnesium content in tailings prepared with seawater treated with different lime/sodium carbonate ratios has great implications for flocculation kinetics and settling rates. Figure 4 shows the flocculation kinetics of artificial clay-based tailings in treated and raw seawater obtained using the FBRM technique. It is important to recall that treated seawater has a reduced magnesium content according to the lime and sodium carbonate ratio used. Figure 5 shows the tailing settling rates measured when the flocculation test was interrupted at 20, 40, 60, and 80 s, which we call flocculation times. Every flocculation test began from scratch. In general, the kinetics led to large aggregates in the short term, but prolonged agitation produced fragmentation of the aggregates and thus a decrease in settling rates. This fragmentation was practically irreversible when the flocculant chains ere torn apart [29,30]. Agitation not only deteriorated the flocculation but also the quality of the seawater with respect to its magnesium content. When seawater treatment was performed with lime/sodium carbonate ratios below 50%, the flocculation kinetics were not much different from the kinetics in raw seawater, according to Figure 4.

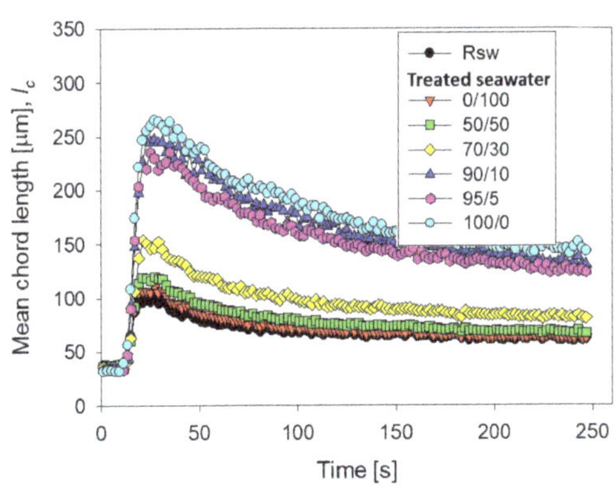

Figure 4. Clay-based tailing flocculation kinetics in raw seawater (Rsw) and seawater treated with different ratios of lime/sodium carbonate. Quartz/kaolin = 80/20, pH 11, stirring rate 220 rpm, and flocculant dose 17 g/t.

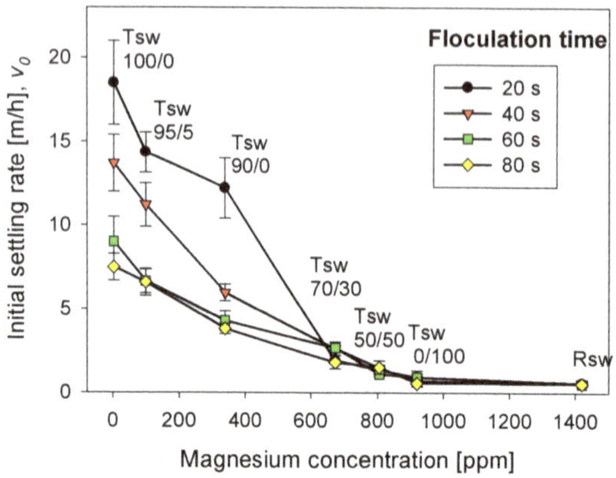

Figure 5. Effect of seawater treatment and flocculation time on the settling rate of flocculated tailings. Quartz/kaolin = 80/20, pH 11, stirring rate 220 rpm, and flocculant dose 17 g/t. Seawater treated with lime/sodium carbonate in different ratios (Tsw). Raw seawater is included for comparison.

Small aggregates, no more than ca. 120 microns, were formed, that rapidly disaggregated after brief stirring. When the treatment involved lime/sodium carbonate ratios over 50%, the amount of magnesium cations in the liquor decreased rapidly after the massive formation and separation of insoluble magnesium hydroxides. In the subsequent flocculation tests, large aggregates formed, which disaggregated more slowly, but significantly, with prolonged agitation. For example, when the content of lime in seawater was over 90%, the aggregates reached ca. 250 microns in a short time, and after stirring for 250 s, the average size reached ca. 150 microns. Such fragmentation as flocculation time increases was due to the sustained hydrodynamic shear and more frequent contact of magnesium complexes which invariably form at alkaline pH with tailings particles. These complexes are expected to eventually coat the tailings particles and flocculant chains, inhibiting their interactions. Therefore, flocculation improves with water treated with high lime/sodium carbonate ratios and deteriorates markedly with low ratios. In the latter case, the formation of magnesium complexes due to the strong alkalinity of the medium and the ion-pairing of magnesium ions and anionic groups in the flocculant suppresses the action of the flocculant in the same way as in raw seawater. In Figure 4, the size reached by the aggregates in treated and raw seawater can be determined when flocculation is interrupted after 20, 40, 60, and 80 s, and settling begins. Figure 5 summarizes the corresponding settling rate data.

The settling rates increased to more than acceptable values for the industry as the lime/sodium carbonate increased from 70 to 100% at all flocculation times, although the higher the flocculation time, the lower the settling rate. As the ratio increases toward 100%, almost all magnesium is precipitated in complexes, including magnesium hydroxide, and was removed prior to flocculation tests. However, the magnesium that does not precipitate during the treatment does so in the alkaline environment in which flocculation occurs. If the flocculation time is short (20 s in Figures 4 and 5), the complexes do not interfere with the flocculation of the tailings particles; rather, they are incorporated into the process, forming large agglomerates that settle easily. However, if the flocculation under agitation lasts a long time (60 and 80 s in Figures 4 and 5) before being interrupted, sedimentation is notoriously impeded due to fragmented aggregates and a degraded flocculant. The high concentration of calcium cations under high lime conditions does not appear to affect the flocculation of the tailings particles. Motta et al. [31] observed a synergistic effect between Ca^{2+} concentration and polyacrylamide dose leading to a beneficial effect on particle flocculation. However, there is no consensus; other works suggest that a high calcium concentration would affect the performance of flocculants due to coiling [32,33].

In seawater treated with lime/sodium carbonate ratios lower than 70%, the remaining magnesium content in the water is very high, and the sedimentation rates are unacceptably low, in the order of 1 m/h, only comparable to the rates obtained in raw seawater, as shown in Figure 5. The fate of the sedimentation in these cases and in raw seawater is decided in the flocculation stage regardless, of the flocculation time. In seawater with low lime (less than 50%), the concentration of magnesium that remains in solution after the treatment is over 920 mg/L, so it is ineffective for magnesium reduction, as shown in Figure 3. Then, in the strongly alkaline medium of flocculation, this remaining magnesium forms hydroxylated complexes that, being deposited on the particles and the flocculant chains, prevent their interaction. Additionally, the residual magnesium cations interact with the acrylate groups in the flocculant chains by ion-pairing, thus preventing the particle–flocculant interaction even more.

For Ramos et al. [23], the presence of magnesium precipitates distracts the flocculant from the tailings particles, generating selectivity problems. According to these authors, the magnesium precipitates effectively interact with the flocculant chains occupying their functional groups. In recent work, Quezada et al. [29] used molecular dynamics simulations to study the interaction between flocculants and brucite, the crystalline form of magnesium hydroxide. The interaction between the deprotonated oxygen of the acrylic group of the polymer and the oxygen on the brucite surface dominates. A minor but significant contribution is that of hydrogen bonds between the nitrogen of the acrylamide group and the oxygen on the surface of the brucite.

3.3. *Effect of Flocculant Dose*

Figure 6 shows the flocculation kinetics of artificial clay-based tailings in seawater treated with lime and different flocculant doses. The FBRM technique provided the average size of the aggregates. Figure 7 shows the corresponding settling rates of the tailings. Settling rates were measured once the flocculation tests were interrupted after 20 s, that is, when the aggregates reached their largest size. Thus, each flocculation test began from scratch.

The flocculation kinetics in Figure 6 show a continuous growth of the tailing aggregates with increasing flocculant doses. For example, the size peaks at doses of 8 to 21 g/t ranged from 165 to 310 μm. It is also true that the larger the aggregates, the greater the fragmentation with the flocculation time. For example, at 21 g/t, after 250 s of flocculation, the aggregate size decreased from ca. 300 microns to less than 200 microns, a 66% relative decrease from the initial size. On the other hand, at 8 g/t, after 250 s of flocculation, the size decreased from ca. 175 microns to ca. 100 microns, that is, 57% of the initial size. Flocculation enhancement with flocculant dose also occurs in seawater treated with different lime/sodium carbonate ratios, but this amount was less than when the ratio was 100% lime. These data are not shown here but are available on request.

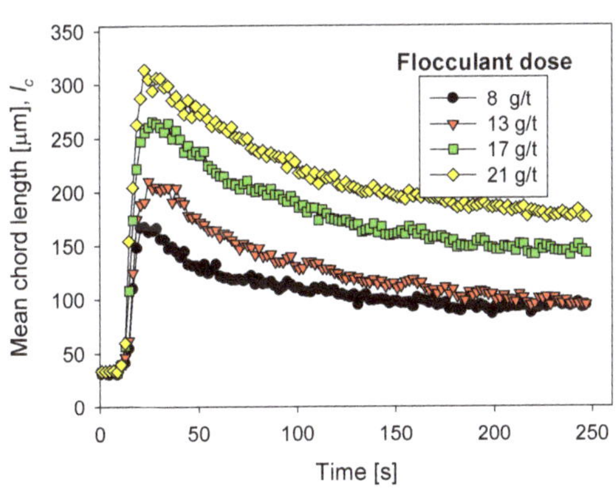

Figure 6. Clay-based tailings flocculation kinetics in seawater treated with lime and different flocculant doses. Quartz/kaolin = 80/20, pH 11, stirring rate 220 rpm.

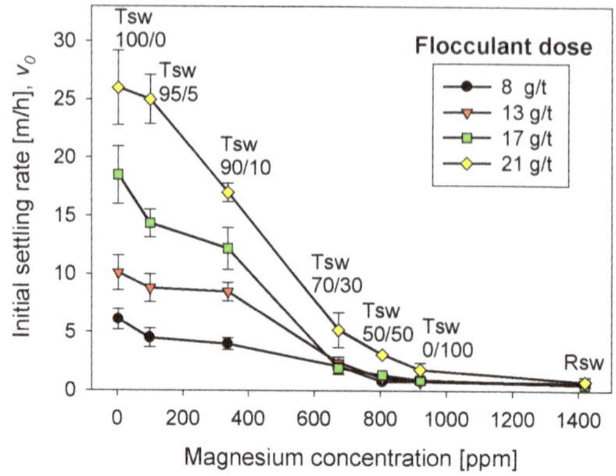

Figure 7. Effect of flocculant dose and seawater treatment on the settling rate of flocculated tailings. Quartz/kaolin = 80/20, pH 11, stirring rate 220 rpm, and flocculation time 20 s.

The sedimentation behavior of the tailings particles, once the agitation in the flocculation cell was stopped at 20 s, was similar for different doses of the flocculant (Figure 7). If the dose of flocculant was high (21 g/t) and the lime/sodium carbonate ratio was high (>90%), the settling rate was also high. As the lime/sodium carbonate ratio decreased, the settling rate also decreased, regardless of how high the flocculant dose was; this result was not different from that of raw seawater. If the flocculant dose was very low, such as 8 g/t, it did not matter much that the seawater treatment was only with lime, the settling rates were unacceptably low, almost as low as in seawater. Thus, seawater treatment for magnesium abatement and flocculant type/dose are critical for adequate solid–liquid separation. Both must be optimal to correctly control the rupture of the aggregates and the flocculant's degradation, especially the flocculant's interaction with the magnesium complexes formed in each seawater treatment and the ion pairing between the magnesium cations and the anionic sites of the flocculant chains. According to Ramos et al. [1], magnesium complexes monopolize the flocculant up to a critical dose. Above this dose, the flocculation of the mineral begins. The sedimentation data from seawater treated with a 70% lime/sodium carbonate ratio support this thesis. Settling was poor for flocculant doses of 8 to 17 g/t; however, at 21 g/t, a settling rate greater than 5 m/h was obtained. When the ratio was 100% lime and the flocculant dose was as low as 8 g/t, the settling rate was already acceptable, about 5 m/h, but doubling the flocculant dose quadrupled the settling rate. It is important to remember that Figure 5 corresponds to 20 s after stopping the flocculation test. This short time minimizes the interaction between the magnesium complexes and the flocculant chains.

3.4. Effect of the Kaolin Content

Clays are generally tricky to flocculate unless they are the flocculating agents [34]. Therefore, separating clays from water in large thickeners requires extremely well-defined conditions and well-designed flocculating polymers, especially if the ionic strength is high such as in raw seawater or partially desalinated or treated seawater.

Figure 8 shows the flocculation kinetics of tailings prepared with different quartz/kaolin ratios in seawater treated with lime to abate magnesium at a fixed flocculant dose of 17 g/t. The largest aggregates were formed in the absence of clay particles. The aggregates of quartz tailings reached a maximum size of 330 microns very quickly (ca. 20 s), while the aggregates of quartz and kaolin tailings in a 60/40 ratio reached less than 250 microns at longer times (ca. 40 s). Therefore, flocculation time requires special control. Large quartz aggregates break under prolonged shear, so the sedimentation operation should be carried out in short times, close to 20 s. Clay-rich aggregates are smaller. In this case, the settling operation must occur immediately after a sufficient particle–flocculant contact time, between 40 and 50 s. The flocculation kinetics curves for tailings in seawater treated with lime and sodium carbonate in different proportions are not shown here, but are available on request.

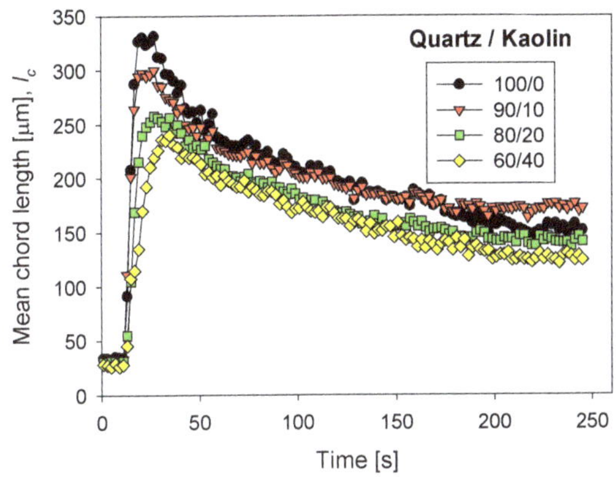

Figure 8. Clay-based tailings flocculation kinetics in seawater treated with lime with different quartz/kaolin ratios. pH 11, stirring rate 220 rpm, and flocculant dose 17 g/t.

Figure 9 shows the settling rates of tailings of quartz/kaolin for treated and raw seawater at a flocculant dose of 17 g/t. Settling rates were measured once the flocculation tests are interrupted after 20 s, that is, when the aggregates reached large sizes, although not exactly their maximum size, as was the case with the kaolin-rich tailings. The tests for each particle system were independent and began from scratch. Figure 9 shows that kaolin deteriorates sedimentation, reducing the settling rate, which was a known impact [16]. When the seawater treatment was with lime alone, practically all the magnesium formed insoluble complexes that were removed and discarded before the flocculation tests. The settling rate for quartz tailings exceeded 30 m/h, and the settling rate for quartz/kaolin tailings in a 60/40 ratio was only 5 m/h. These low settling rates were due to the smaller size of the aggregates in the presence of the kaolin particles, due to the fine size of the clay particles and their plate-like shape with a high specific surface that consumes flocculant. Interestingly, when using water treated with sodium carbonate only, almost all the magnesium remained in the ionic state in the solution. As a result, the settling rate was practically zero for tailings with kaolin content less than 20%, and 2 m/h for tailings with more than 40% kaolin. At first, it seems that these settling results are better for systems with higher clay content; however, when they are compared with results in raw seawater, it is seen that they are not different. The latter suggests that the treatment only with carbonate sodium (without lime), is unsuitable for systems with high clay content, from the point of view of sedimentation.

The thickening stages generally do not require sedimentation rates as high as those observed with seawater treated 100% with lime; in fact, discharge from underflows is hampered at high rates. Rates between 10 and 20 m/h can be considered acceptable in the copper mining industry. Therefore, removing all magnesium from seawater is unnecessary to achieve optimal sedimentation conditions. However, this depends on the clay content. The higher the kaolin content, the greater the amount of magnesium to abate.

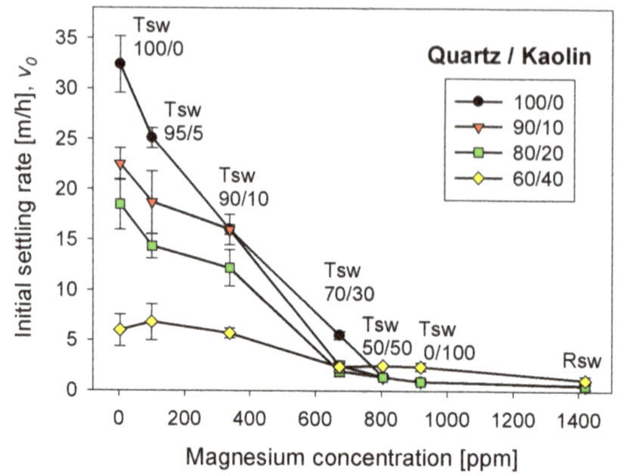

Figure 9. Effect of quartz/kaolin ratio and seawater treatment on the settling rate of flocculated tailings. pH 11, stirring rate 220 rpm, flocculant dose 17 g/t, and flocculation time 20 s.

3.5. Structure of Aggregates

The fractal dimension can be a handy indicator of the structural characteristics of tailings aggregates formed under different magnesium concentration conditions, which depend on the lime/sodium carbonate ratio, flocculant dosage, flocculation time, and kaolin content. The structural characteristics include compactness, resistance to hydrodynamic shear, the effectiveness of the flocculant at the applied dose, the effect of insoluble complexes formed during the process, and even permeability. In addition, depending on the type of system, flocculation conditions define the structure of the aggregates and the settling behavior [28,35]. Figure 10 shows the fractal dimensions and densities of aggregates of quartz and kaolin tailings at various conditions. Figure 10a shows that the fractal dimension and density of the aggregates change little if the flocculation time is extended. Thus, the size of the aggregates decreases with flocculation time, but their structure and density remain relatively constant. For example, the fractal dimension starts with a value of 2.36 at 20 s of flocculation and ends at 2.21 after 80 s of flocculation. The density begins at a value of at 1480 g/cm^3 at 20 s and ends at 1410 g/cm^3 at 80 s. These results are in agreement with those of Quezada et al. [36], who observed that aggregates from mine tailings flocculated with anionic polyacrylamides maintain their structure when the shear rate is less than 200 s^{-1}. The effect of the magnesium content is different when seawater is treated with varying proportions of lime and sodium carbonate. Figure 10b shows that the fractal dimension of the aggregates at 400 ppm magnesium were ca. 2.3 but at a high magnesium concentration at the limit of raw seawater, the fractal dimension reached values close to 1. The implications for the water recovery operation are dramatic: sedimentation is very slow because the aggregates are structurally very loose and light, and water clarification may require prohibitive times. The aggregates are so light that the density of the aggregates decreases from 1480 g/cm^3 in liquor with 400 ppm magnesium to 1320 g/cm^3 in raw seawater. Regarding the dose of the flocculant, Figure 10c shows that the higher the dose, the greater the fractal dimension, although the increase is slight, from 2.2 to 8 g/t to 2.4 to 21 g/t. This result is consistent with the results of a previous study by Jeldres et al. [36]. In contrast, the density of the aggregates remained relatively constant at 1450 g/cm3 in the dose range used. Finally, Figure 10d shows the impact of the clay content of the tailings on the fractal dimension and the density of the aggregates that are formed, both of which decrease. The fractal dimension decreased slightly, maintaining values between 2.4 without kaolin and 2.3 with 40% kaolin. The density of the aggregates decreased dramatically from 1500 g/cm^3 without kaolin to 1330 g/cm^3 with 40% kaolin. Here, the difficulty for the water recovery operation is the lightness of the aggregates in tailings with high clay content, the clarification times in the latter case would also be prohibitive.

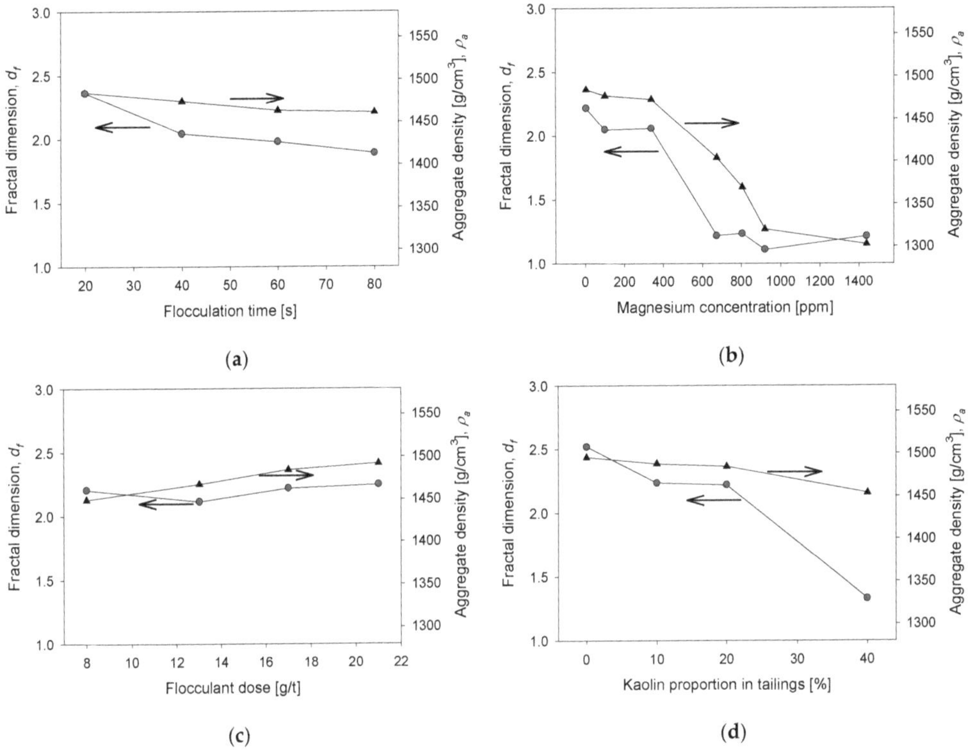

Figure 10. Fractal dimensions and density of aggregates as functions of (**a**) flocculation time (quartz/kaolin 80/20, flocculant dose 17 g/t, magnesium concentration 3 ppm), (**b**) magnesium concentration (quartz/kaolin ratio 80/20, flocculant dose 17 g/t, flocculation time 20 s), (**c**) flocculant dose (quartz/kaolin ratio 80/20, flocculation time 20 s, magnesium concentration 3 ppm), (**d**) and kaolin content in quartz tailings (flocculant dose 17 g/t, flocculation time 20 s, magnesium concentration 3 ppm). All tests were performed at pH 11.

4. Conclusions

Seawater treated with different lime and sodium carbonate proportions was used to improve the flocculation and sedimentation of clay-based tailings under highly alkaline conditions. The precipitated solids, mainly magnesium and some calcium, were separated from the water by vacuum filtration. When operating with raw seawater at pH 11, solid magnesium complexes impair the selectivity of the flocculating polymer, drastically reducing its performance. However, promising results were obtained when using magnesium control-treated seawater, including a significant increase in the sedimentation rate caused by structural changes in the aggregates, including size, fractal dimension, and density. A low fractal dimension revealed open and porous structures, intensifying at high magnesium concentrations and clay contents. If, in addition, the density of the aggregates was low, then the sedimentation of the tailings and the clarification of water were of little practical use. The results suggest that both the flocculation and the sedimentation of low-clay content tailings occur with industrially attractive yields even in the presence of magnesium. However, the flocculation and sedimentation of high-clay tailings requires a significant lowering of the magnesium concentration to be of industrial interest. The latter can be achieved by using seawater treated with lime.

Author Contributions: Conceptualization, M.J. and R.I.J.; methodology F.P.; formal analysis, P.G.T., P.R., M.J. and R.I.J.; data curation, F.P., L.A. and I.S.; writing—original draft preparation, M.J.; writing—review and editing, F.P., L.A., M.J., P.R., I.S., P.G.T. and R.I.J.; project administration, R.I.J.; funding acquisition, P.G.T. and R.I.J. All authors have read and agreed to the published version of the manuscript.

Funding: Please add: This research was funded by ANID Fondecyt 1211606 and Centro CRHIAM Project Anid/Fondap/15130015.

Data Availability Statement: The data presented in this study are available on request from the corresponding author.

Acknowledgments: Ricardo I. Jeldres is grateful for the support of ANID FONDECYT 1211606. Francisco Pulgar, Ri-cardo I. Jeldres and Pedro G. Toledo are grateful for the financial support of Centro CRHIAM Project Anid/Fondap/15130015. Pedro Robles thanks the Pontificia Universidad Católica de Val-paraíso for the support provided. Matías Jeldres and Francisco Pulgar acknowledge the infra-structure and support of the Programa de Doctorado en Ingeniería de Procesos de Minerales of the Universidad de Antofagasta.

Conflicts of Interest: The authors declare no conflict of interest.

References

1. Northey, S.; Mudd, G.; Werner, T.T.; Jowitt, S.M.; Haque, N.; Yellishetty, M.; Weng, Z. The exposure of global base metal resources to water criticality, scarcity and climate change. *Glob. Environ. Chang.* **2017**, *44*, 109–124. [CrossRef]
2. Cisternas, L.A.; Gálvez, E.D. The use of seawater in mining. *Miner. Process. Extr. Met. Rev.* **2018**, *39*, 18–33. [CrossRef]
3. Gunson, A.J.; Klein, B.; Veiga, M.; Dunbar, S. Reducing mine water requirements. *J. Clean. Prod.* **2012**, *21*, 71–82. [CrossRef]
4. Northey, S.; Mudd, G.M.; Saarivuori, E.; Wessman-Jääskeläinen, H.; Haque, N. Water footprinting and mining: Where are the limitations and opportunities? *J. Clean. Prod.* **2016**, *135*, 1098–1116. [CrossRef]
5. Scott, K.J. Continuous Thickening of Flocculated Suspensions. Comparison with Batch Settling Tests and Effects of Floc Compression Using Pyrophyllite Pulp. *Ind. Eng. Chem. Fundam.* **1970**, *9*, 422–427. [CrossRef]
6. Gladman, B.; de Kretser, R.; Rudman, M.; Scales, P. Effect of Shear on Particulate Suspension Dewatering. *Chem. Eng. Res. Des.* **2005**, *83*, 933–936. [CrossRef]
7. de Kretser, R.; Scales, P.J.; Boger, D.V. Improving clay-based tailings disposal: Case study on coal tailings. *AIChE J.* **1997**, *43*, 1894–1903. [CrossRef]
8. Nasser, M.; James, A. Effect of polyacrylamide polymers on floc size and rheological behaviour of kaolinite suspensions. *Colloids Surf. A Physicochem. Eng. Asp.* **2007**, *301*, 311–322. [CrossRef]
9. Lee, L.T.; Rahbari, R.; Lecourtier, J.; Chauveteau, G. Adsorption of polyacrylamides on the different faces of kaolinites. *J. Colloid Interface Sci.* **1991**, *147*, 351–357. [CrossRef]
10. Shaikh, S.M.; Nasser, M.; Hussein, I.A.; Benamor, A. Investigation of the effect of polyelectrolyte structure and type on the electrokinetics and flocculation behavior of bentonite dispersions. *Chem. Eng. J.* **2017**, *311*, 265–276. [CrossRef]
11. Wang, B.; Peng, Y. The effect of saline water on mineral flotation—A critical review. *Miner. Eng.* **2014**, *66*, 13–24. [CrossRef]
12. Lattemann, S.; Höpner, T. Environmental impact and impact assessment of seawater desalination. *Desalination* **2008**, *220*, 1–15. [CrossRef]
13. Panagopoulos, A.; Haralambous, K.-J.; Loizidou, M. Desalination brine disposal methods and treatment technologies—A review. *Sci. Total Environ.* **2019**, *693*, 133545. [CrossRef]
14. Sadhwani, J.J.; Veza, J.M.; Santana, C. Case studies on environmental impact of seawater desalination. *Desalination* **2005**, *185*, 1–8. [CrossRef]
15. Ji, Y.; Lu, Q.; Liu, Q.; Zeng, H. Effect of solution salinity on settling of mineral tailings by polymer flocculants. *Colloids Surf. A Physicochem. Eng. Asp.* **2013**, *430*, 29–38. [CrossRef]
16. Liu, D.; Edraki, M.; Berry, L. Investigating the settling behaviour of saline tailing suspensions using kaolinite, bentonite, and illite clay minerals. *Powder Technol.* **2018**, *326*, 228–236. [CrossRef]
17. Jeldres, R.I.; Piceros, E.C.; Leiva, W.H.; Toledo, P.G.; Herrera, N. Viscoelasticity and yielding properties of flocculated kaolinite sediments in saline water. *Colloids Surf. A Physicochem. Eng. Asp.* **2017**, *529*, 1009–1015. [CrossRef]
18. Quezada, G.R.; Jeldres, R.I.; Fawell, P.D.; Toledo, P.G. Use of molecular dynamics to study the conformation of an anionic polyelectrolyte in saline medium and its adsorption on a quartz surface. *Miner. Eng.* **2018**, *129*, 102–105. [CrossRef]
19. Zanin, M.; Lambert, H.; du Plessis, C. Lime use and functionality in sulphide mineral flotation: A review. *Miner. Eng.* **2019**, *143*, 105922. [CrossRef]
20. Mu, Y.; Peng, Y.; Lauten, R.A. The depression of pyrite in selective flotation by different reagent systems—A Literature review. *Miner. Eng.* **2016**, *96–97*, 143–156. [CrossRef]
21. Castro, S.; Lopez-Valdivieso, A.; Laskowski, J. Review of the flotation of molybdenite. Part I: Surface properties and floatability. *Int. J. Miner. Process.* **2016**, *148*, 48–58. [CrossRef]

22. Jeldres, R.I.; Forbes, L.; Cisternas, L.A. Effect of Seawater on Sulfide Ore Flotation: A Review. *Miner. Process. Extr. Met. Rev.* **2016**, *37*, 369–384. [CrossRef]
23. Ramos, J.J.; Leiva, W.H.; Castillo, C.N.; Ihle, C.F.; Fawell, P.D.; Jeldres, R.I. Seawater flocculation of clay-based mining tailings: Impact of calcium and magnesium precipitation. *Miner. Eng.* **2020**, *154*, 106417. [CrossRef]
24. Castro, S. Use of Seawater and Other Saline Water (Salinity between 1–7%) Pre-Treated with Lime or Other Alkalinizing Reagents, on Mineral Flotation Process, Where the Pre-Treatment Removes the Ions of Seawater Hardness without Significantly Altering Its Salinity. *Chil. Pat.* **2010**, *52248*, 2010.
25. Jeldres, R.I.; Arancibia-Bravo, M.P.; Reyes, A.; Aguirre, C.E.; Cortes, L.; Cisternas, L.A. The impact of seawater with calcium and magnesium removal for the flotation of copper-molybdenum sulphide ores. *Miner. Eng.* **2017**, *109*, 10–13. [CrossRef]
26. Jeldres, M.; Piceros, E.C.; Toro, N.; Robles, P.; Nieto, S.; Quezada, G.R.; Jeldres, R.I. Enhancing the sedimentation of clay-based tailings in seawater by magnesium removal treatment. *Sep. Purif. Technol.* **2020**, *242*, 116762. [CrossRef]
27. Arias, D.; Villca, G.; Pánico, A.; Cisternas, L.A.; Jeldres, R.I.; González-Benito, G.; Rivas, M. Partial desalination of seawater for mining processes through a fluidized bed bioreactor filled with immobilized cells of Bacillus subtilis LN8B. *Desalination* **2020**, *482*, 114388. [CrossRef]
28. Heath, A.R.; Bahri, P.A.; Fawell, P.D.; Farrow, J.B. Polymer flocculation of calcite: Relating the aggregate size to the settling rate. *AIChE J.* **2006**, *52*, 1987–1994. [CrossRef]
29. Quezada, G.R.; Jeldres, M.; Toro, N.; Robles, P.; Toledo, P.G.; Jeldres, R.I. Understanding the flocculation mechanism of quartz and kaolinite with polyacrylamide in seawater: A molecular dynamics approach. *Colloids Surf. A Physicochem. Eng. Asp.* **2021**, *608*, 125576. [CrossRef]
30. Healy, T. Flocculation-dispersion behavior of quartz in the presence of a polyacrylamide flocculant. *J. Colloid Sci.* **1961**, *16*, 609–617. [CrossRef]
31. Motta, F.L.; Gaikwad, R.; Botha, L.; Soares, J.B. Quantifying the effect of polyacrylamide dosage, Na^+ and Ca^{2+} concentrations, and clay particle size on the flocculation of mature fine tailings with robust statistical methods. *Chemosphere* **2018**, *208*, 263–272. [CrossRef]
32. Witham, M.; Grabsch, A.; Owen, A.; Fawell, P. The effect of cations on the activity of anionic polyacrylamide flocculant solutions. *Int. J. Miner. Process.* **2012**, *114–117*, 51–62. [CrossRef]
33. Costine, A.; Cox, J.; Travaglini, S.; Lubansky, A.; Fawell, P.; Misslitz, H. Variations in the molecular weight response of anionic polyacrylamides under different flocculation conditions. *Chem. Eng. Sci.* **2018**, *176*, 127–138. [CrossRef]
34. García, K.I.; Quezada, G.R.; Arumí, J.L.; Urrutia, R.; Toledo, P.G. Adsorption of Phosphate Ions on the Basal and Edge Surfaces of Kaolinite in Low Salt Aqueous Solutions Using Molecular Dynamics Simulations. *J. Phys. Chem. C* **2021**, *125*, 21179–21190. [CrossRef]
35. Jeldres, M.; Piceros, E.C.; Toro, N.; Torres, D.; Robles, P.; Leiva, W.H.; Jeldres, R.I. Copper Tailing Flocculation in Seawater: Relating the Yield Stress with Fractal Aggregates at Varied Mixing Conditions. *Metals* **2019**, *9*, 1295. [CrossRef]
36. Quezada, G.R.; Ayala, L.; Leiva, W.H.; Toro, N.; Toledo, P.G.; Robles, P.; Jeldres, R.I. Describing Mining Tailing Flocculation in Seawater by Population Balance Models: Effect of Mixing Intensity. *Metals* **2020**, *10*, 240. [CrossRef]

polymers

MDPI

Article

Feasibility Study of the Flatness of a Plastic Injection Molded Pallet by a Newly Proposed Sequential Valve Gate System

Hsi Hsun Tsai [1,2,*] and Yi Lin Liao [1]

[1] Department of Mechanical Engineering, Ming Chi University of Technology, New Taipei City 24301, Taiwan; M09118018@mail2.mcut.edu.tw
[2] Research Center for Intelligent Medical Devices, Ming Chi University of Technology, New Taipei City 24301, Taiwan
* Correspondence: hhtsai@mail.mcut.edu.tw

Abstract: The investigation of plastic pallet molding, assisted by a sequential valve gate system, has not yet been performed due to the limitations of the pallet scale. Furthermore, at present, the application of recycled plastics by chemical industries has become extremely popular around the world. This study aimed to determine pallet flatness experimentally and numerically using recycled polypropylene with a large-scale pallet. Short-shot testing on injection molding was performed to obtain short-shot samples for confirmation of the flow front during simulated filling. The real injected pallet profile, which was measured by an ATOS, was compared after confirmation to the numerical profile of the pallet. The pallet's flatness was accurately compared to the real experimental and numerical results. By adjusting the temperature of the cooling channel within the cavity plate to 55 °C, the flatness of the pallet achieved by the newly proposed sequential valve gate-opening scheme was about 7 mm, which meets the height directional warpage standard determined by the pre-set sequential scheme. The numerical flatness is in line with existing flatness values for pallets. Furthermore, the proposed cooling temperature gives the highest yield in terms of pallet molding from the perspective of the stakeholders.

Keywords: plastic pallet; injection molding; flatness; sequential valve gate system; molding flow analysis

Citation: Tsai, H.H.; Liao, Y.L. Feasibility Study of the Flatness of a Plastic Injection Molded Pallet by a Newly Proposed Sequential Valve Gate System. *Polymers* **2022**, *14*, 616. https://doi.org/10.3390/polym14030616

Academic Editors: Célio Bruno Pinto Fernandes, Salah Aldin Faroughi, Luís L. Ferrás and Alexandre M. Afonso

Received: 7 January 2022
Accepted: 2 February 2022
Published: 4 February 2022

1. Introduction

As defined in the SFS-EN ISO 445 standard, a pallet is a "rigid horizontal platform of minimum height, compatible with being handled by pallet trucks, forklift trucks and/or other appropriate handling equipment" and can be used "as a base for assembling, loading, storing, handling, stacking, transporting, or displaying goods and loads" [1]. Pallets can be made of wood, plastic, aluminum and composites, and used under the three pallet management strategies of single use, buy/sell, and pooled. A pooled pallet is leased to customers without transfer of ownership. A standardized pallet is designed to last several trips, under a scheme called the "buy/sell" strategy. However, single-use pallets are the simplest strategy, as they are discarded after one trip [2], after having been loaded with goods and transported by container ships to places all over the world. Recycled plastic pallets were found to be superior to conventional plastic pallets by an impact category analysis of the results per trip. The recycled plastic pallets also performed better in terms of environmental impact compared of wooden pallets [3]. Single-use pallets made of recycled plastic via injection molding are the subject of this study.

Recycled plastics are based on recycled PE (rPE) and recycled PP (rPP), and they have outstanding potential. They have the potential to significantly contribute to new markets with more demanding and critical applications. One of the applications with the highest demand is found in the lower-quality end of the agricultural and building sectors, serving as a structural part [4]. Gall et al. [5] revealed the properties of the recycled plastics, and found that the density, melt flow rate (MFR) and Charpy impact strength of the rPP materials varied from 0.904 to 0.924 g/cm^3 (by ISO 1183), from 13 to 22 g/10 min (by ISO 1133), and from 5.9 to 6.8 kJ/m^2 (by ISO 179). The content of calcium carbonate in the rPP, analyzed by a thermos-gravimetric analyzer (TGA), ranged from 0.29 to 1.92% in mass. As they contain diverse contaminants, recyclables should be blended with legacy calcium carbonate and a polymeric cross-contaminant to modify the MFR for later applications in plastic engineering processes. One of the more common industrial plastic-processing technologies is plastic injection molding.

When using rPP in micro-injection molding, a uniaxial extension test showed that the increases in Young's modulus, yield stress and ultimate stress values were 3.07%, 10.97% and 27.33%, respectively. A 1.29% reduction was found in the breakage strain compared to virgin PP samples [6]. The variations in the recycled plastic's properties may disturb the final quality of the injection part, due to the injection parameter setting remaining constant throughout the whole process. An important quality of the injection part is the warpage, which can face problems related to a combination of poor material characterization and inadequate control of the processing parameters [7–11]. The temperature differences between the two mold surfaces significantly affect the morphology distributions of the molded parts. The cooling rate may affect the injection parts in terms of relaxation/reorganization levels and give rise to an asymmetric distribution of mechanical properties [12]. An imbalance in the mold-filling is one of the factors affecting the asymmetric temperature distribution of the injection part [13,14]. Plastic injection molding involves four major stages: filling, packing, cooling, and ejection. The injection pressure and rate, packing pressure and time, and cooling temperature and time may affect the quality of the injection part. The mold temperature seems to be one of the main process parameters that affects the properties of molded parts [15,16]. Nevertheless, the mold temperature is actually unstable during the process. The temperature of the mold has never been controlled individually, as it is affected by the cooling channel and cooling time, as well as the mold opening time.

Filling the mold cavity with melted material via a gate is essential for the small injection part. Considering the limitation of the flow length from the gate, multi-gate filling can reduce a machine's required injection parameters, and the filling time, during injection molding. Not using multi-gate filling injection is associated with a higher number of weld lines. A weld line is formed when two separate melt fronts join into one flow. Moreover, while weld lines are not appropriate for parts, it is impossible to avoid all of them due to the filling efficiency of large molding components. It is well known that the strength of weld lines is lower than the strength of the general plastic-molding material, since a lower temperature, along with air bubbles, occurs between the fronts. The structural parts molded by rPP, such as the pallet and water cage, endure the external load. The weld lines within the molded structure parts face an increased risk due to the injection molding of plastic pallets via multiple gates, whereby the melted materials flow into the mold cavity from a molding machine.

The concurrent filling of multiple gates increases the filling pressure, meaning that a larger injection molding machine is required to clamp and pack the mold during the filling and packing processes. A sequential valve gate-opening system can be used to decrease the clamping and packing forces, which divides the gates into several groups during the filling stage. Via this sequential valve gate-opening system, the flow front from the initial gates spreads to the lateral gates. The lateral gates are activated to pass on the melt material when they come into contact with the front. This approach could eliminate many weld lines from the molded part. Moreover, the scale of the injection machine can be decreased. During the injection molding of ASTM-D638 standard specimens (200 mm in length, 20 mm in width, and 2 mm in thickness) made of acrylonitrile butadiene styrene (ABS), sequentially setting the on/off times of these filling gates enabled the sequential valve gate system to eliminate the welding lines and increase the tensile strength [17]. In addition, the different temperature levels of the melted material's flow fronts tensile strengths, and melt polyamide flow front temperature (PA6) were correlated with the strength of the welding line [18].

The sequential valve gate-opening system could be implemented to create large-scale plastic parts with a moderately sized injection machine. Knowing the flow front of the melt material in the mold cavity is essential for setting the switching time of all the gates in the sequential valve gate system—a theoretical model or a numerical approach could be used to predict the flow front of the injection in the mold cavity. Iwko et al. [19] derived numerical results to verify the experimental results by constructing a comprehensive model of the plasticization process in a screw-barrel system injection molding machine. They found that the output pressure and temperature of the plasticization process, determined numerically by the model, fit the experimental results with an average error of less than 10%, but the flow front in the mold cavity was never assessed.

Cardozo [20] reviewed the numerical approaches to filling via injection molding, and indicated that the Moldex3D software, a commercial software available for injection molding, could provide an understanding of the physical effects occurring in the mold cavity. Moldex3D was applied to investigate the molding process, while the prediction of the flow front during filling was derived from the Hele–Shaw model [21]. The finite-volume approach was used to determine the multi-physical quantities involved in the packing and cooling processes. Furthermore, the equation governing the jetting behavior of the filling from the gate was discretized by a control volume-based, finite-volume method [22]. By using each of the commercialized software, including ANSYS Fluent [23,24], Moldflow [25], Open FOAM [26] and Moldex3D [10,27–30], one could analyze the multiple physical parameters of the output molding parts in relation to the operational parameters of the injection molding process. Notably, the experimental investigator could easily make a comparison between the simulation results and the real operational results in the laboratory.

Using Moldex3D, the authors undertook a numerical feasibility study of a single-use pallet created via a sequential valve gate system [10]. However, the experimental results of this rib-structured pallet have not been determined. Differing from previous rib-structured pallets, the target pallet has a flat top surface. The aim of this study is to investigate the temperature, pressure, stress and warpage of the injection-molded rPP flat surface pallet with size dimensions of 1 m × 1 m × 0.13 m by CAE simulation and experimental methods. The results for the numerical warpage of a plastic pallet produced via a sequential valve gate system, derived by Moldex3D 2020, are compared with the real profile of the rPP pallet measured by the ATOS scan box 5120 system. Using the injection parameters of the plastic pallet, the specifications of the injection machine, and a polymer database, a fundamental investigation can be conducted to understand the injection molding process. By using a sequential valve gate-opening system to mold a plastic pallet, the flow fronts during the filling stage can be derived for an evaluation of the actuating time of the gates. A detailed understanding of the pre-setting sequence applied to the valve gates may help to determine the ideal sequence. A pre-set valve gate-opening sequence, provided by the cooperating company, would enable a comparison of the flatness of the pallet. A new proposed valve gate-opening sequence can then be used to derive an improved pallet flatness, which will help to evaluate the accuracy of the numerical predictions.

2. Experimental Setup and Software

Figure 1 shows the dimensions of the plastic pallet that were used in this study. It has a flat top surface and a reinforced structure with ribs at the base. The ribs are 3 mm in thickness and were changed according to the draft angle used for the injection molding. A general rule when arranging the positions of the filling gates is to ensure that the ratio of flow length to rib thickness is below 150. When the ratio is lower than 150, the running length of the melted polymer within the mold during injection is sufficient to maintain the melting state. In this application, the maximum filling length from gates #1, #4, #13 and #16 to the four corner legs of the pallet was 394 mm. The ratio of the flow length to rib thickness in this study was 131.3, which is smaller than the general standard when arranging the positions of the filling gates. As such, there were 16 filling gates (7.0 mm in diameter and 20 mm in height), as shown in Figure 1a. rPP was used as the injection molding plastic material to simulate real-world plastic waste reduction. The pallet's dimensions were 1000 mm in length and width and 130 mm in height, as shown in Figure 1b. An isometric view of the pallet's base (Figure 1c) shows the complex features that should be cooled during the molding process by the baffle cooling flow system, as shown in Figure 1d. The molding experiments included sequential gate-opening and concurrent valve gate-opening systems for the melt-filling process, in order to numerically analyze the mold flow. The gates' opening and closing were sequentially controlled by a pneumatic system, so that the previous gates were closed by relay during melt filling, and the following gates were concurrently opened—this assumes that the melted material starts to fill the mold cavity once the valve gate is completely open.

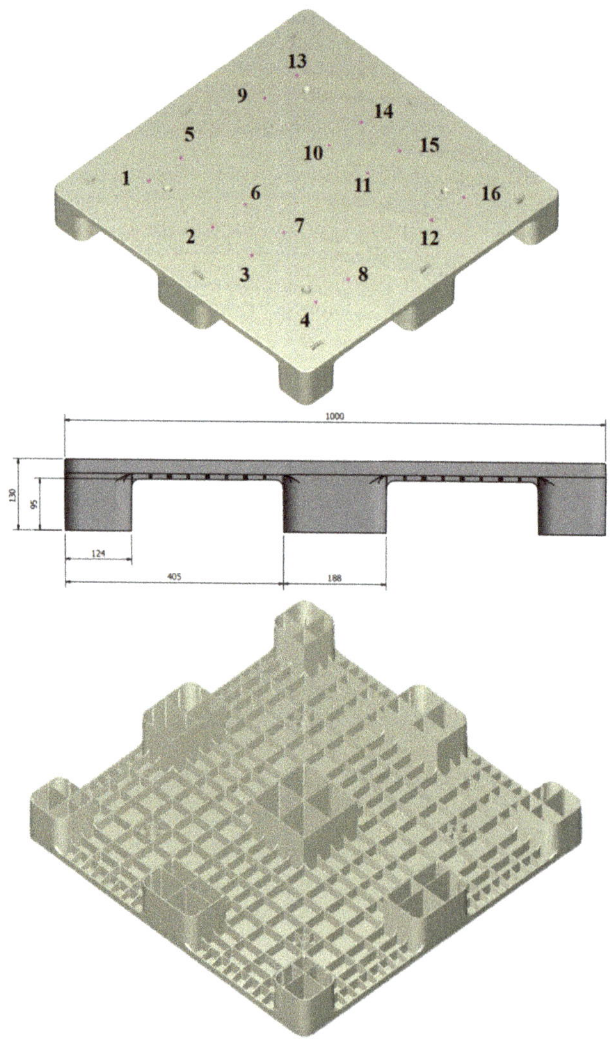

(a)

(b)

(c)

Figure 1. *Cont.*

(**d**)

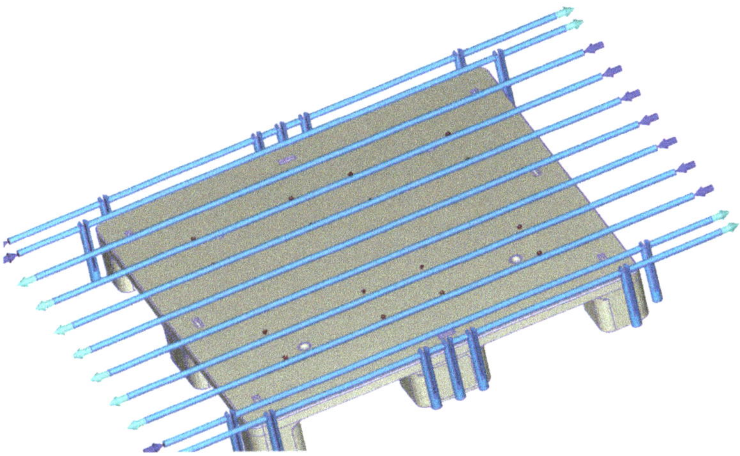

(**e**)

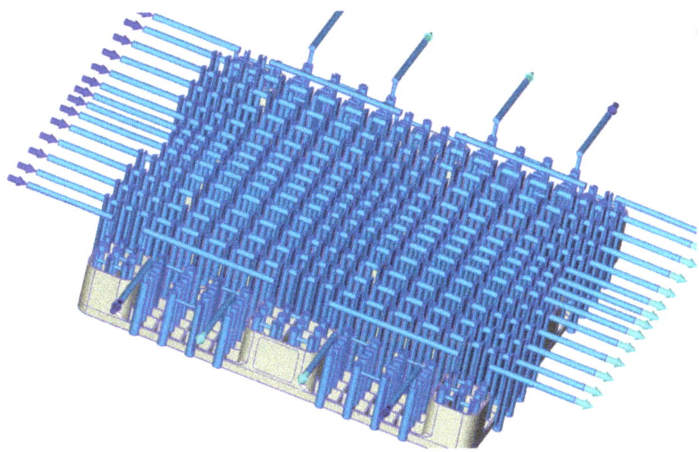

Figure 1. Injection-molded pallet (1000 mm × 1000 mm × 130 mm). (**a**) Sixteen filling gates in the pallet; (**b**) dimensions of the pallet (front view); (**c**) isometric view of the pallet (bottom side); (**d**) isometric view of the water flow system on the top side of the pallet; (**e**) isometric view of the baffle-cooling flow system on the bottom side of the pallet.

Within the simulation, the 3600-ton injection, performed by the Supermaster 3600E1 molding machine (https://chenhsong.com/, Taoyuan Taiwan, 3 February 2022), is modeled. This molding machine has a screw diameter of 225 mm, a maximum screw stroke of 4400 mm, a maximum injection pressure of 159.7 MPa, and a maximum injection volume of 49,278 cm^3. The simulation analysis was performed using Moldex3D software. In this software, both the skin and core materials are considered to be compressible, generalized Newtonian fluids. The surface tension at the melt front is neglected. The modified Cross model with Arrhenius temperature dependence was employed to describe the viscosity of the polymer melt. During the polymer melt-filling phase, the velocity and temperature were specified at the mold inlet. While the core material was injected, the flow rate was specified at the mold inlet. On the mold wall, the non-slip boundary condition was applied, and a fixed mold wall temperature was assumed.

In Moldex3D, the finite volume method was used to discretize the Navier–Stokes equation based on the pressure-based decoupled procedure and solve the transient flow field in a complex three-dimensional geometry. A compressive, bounded, high-order differencing scheme was also utilized to directly solve the hyperbolic advection equation of the fractional volume function to track the melt front during the filling process [21]. Modeling the flow field in Moldex3D is an iterative decoupled procedure for coupling velocity and pressure, in which the three linearized momentum equations are solved for an estimated pressure field, then sequentially followed by the solution of the pressure correction equation. The mass fluxes and pressure are then corrected. This will satisfy both local and global continuity, but can cause the momentum to deviate. Hence, a new outer iteration is activated. The process is then repeated until the prescribed tolerance for each equation is achieved [21].

All sixteen of the valve gates are concurrently opened to allow the molten material to enter, with a hot runner used to determine the filling flow front, the clamping force, the temperature distribution, the thermal stress, and the deformation; then, the 16 gates are opened and closed in a controlled sequence. The pre-set sequence of the sequential gate-opening scheme is depicted in Table 1. Gates 3, 5, 9, and 13–16 were initially opened to fill the mold cavity with the melted material. Gates 8 and 11 were then actuated within the first second. After this filling, gates 1, 10, and 12 were relayed. Gate 4 was turned on in the third second, and then gate 7 was activated after five seconds. Gates 6 and 2 were opened in the sixth and eighth seconds, respectively, until the end of filling. The total filling time was 9.3 s.

Table 1. Sequence of the filling gates.

Time (s)	Start	1	2	3	5	6	8
Gate (#)	#3, #5, #9, #13~16	#8, #11	#1, #10, #12	#4	#7	#6	#2

Using the previous setting sequence, the Moldex 3D 2020 software was used to simulate the sequential valve gate system used for the melt-filling process in the mold flow analysis. During melt-filling, the gate-opening and -closing times are sequentially controlled by the pneumatic system, so that the previous gates are closed by relay, with subsequent gates opening at the same time. This software package was also used to simultaneously open all 16 gates and direct the melted material into the mold cavity, which is the concurrent valve gate-opening scheme. The distribution of the weld lines, the filling pressure, and the estimated clamping force can also be derived. The molding analysis was then conducted for the same injection filling time via the sequential gate-opening scheme. The molding pressure within the mold cavity was measured numerically, using the same injection parameters used to set the timing control of the 16 gates. Under the sequential gate-opening scheme, the hot runners were opened at different times. Within the pallet, the temperature distributions, filling pressures, deformations and thermal stresses, as well as the shrinkages, were compared under the sequential gate-opening and concurrent valve gate-opening schemes.

During the filling stage, the fill flow front of the molten material is closely dependent on the viscosity, the material temperature, and the runner and gate of the mold. The pressure of the molten material is usually a consequence of viscosity. During the filling stage of the injection molding process, the flow front is controlled by the flow rate and pressure gradient. Adding mica powder to the rPP gives the material properties [10] such as those shown in Table 2, with the density of rPP being 1.026 g/cm^3, which is more than that of raw PP. In the design phase, the volumes of the plastic pallet and mold were 10,039 and 1,179.92 cm^3, respectively. The associated solid mesh of the plastic pallet, mold, and cooling channel contained 1,562,598, 4,724,368, and 4,189,448 elements, respectively. The volume and the mesh number of the 16 hot runners were 784.06 cm^3 and 839,450 elements, respectively. The surface mesh of the pallet contained 311,368 elements. The experimental viscosity with respect to the shear rate and temperature, the specific heat with respect to temperature, and the mechanical properties of the melt rPP resin were derived from

Cheng [10]. The injection molding parameters of the rPP that were used for this plastic pallet are shown in Table 1.

Table 2. Injection molding parameters of recycled polypropylene [10].

Melt temperature (°C)	250
Curing temperature (°C)	117
Mold temperature (°C)	50
Fill rate (%)	72
Filling pressure (max) [MPa]	40
Filling time (s)	9.3
Packing time (s)	3.0
Packing pressure (%)	70
Cooling time (s)	45
Mold opening time (s)	10

It was assumed that the melted material would begin filling the mold cavity once the valve gate was completely open, and the 16 gates were opened sequentially to direct the melted material into the mold cavity. The flow fronts that occurred during short-shot testing in the injection molding of rPP pallets were compared to ensure correct simulation modeling. Then, the most suitable sequential control scheme of the gates with the same injection parameters was investigated. Via the appropriate sequential control of the gates, the filling pressures, temperature distribution, warpage deformation, thermal stresses and shrinkages were assessed to determine the advantages of this analysis.

The rPP pallet molded by the Supermaster 3600E1 molding machine is 1 m in length and width and 0.13 m in height. This pallet scale is too large to measure its three-dimensional profiles using a general coordinate measuring machine (CMM). The ATOS scan box 5120 system (GOM, Swiss, www.gom.com, 3 February 2022) is a non-contact three-dimensional measuring system that operates via high-speed sensors used to scan all the parts. A 3D graph could be generated from these scans for 3D printing, reverse engineering or part inspection. The injection-molded pallet was measured by RATC in Taiwan (https://www.ratc.com.tw/, 3 February 2022) with ATOS scan box 5120. The authors imported the 3D graph generated by the ATOS scanning system into Creo Parametric CAD to calculate the flatness of each surface on the pallet. The measured flatness and the profile of the pallet's surface were compared with the numerical results derived by Moldex3D.

3. Results and Discussion

3.1. Flow Front Comparison of Short Shot Testing

When producing a fully molded part via the plastic injection molding process, the screw-back position refers to the screw within the injection machine barrel being brought back to the starting position before the start of the next cycle. Incomplete filling of a part, called short-shot molding, can be achieved in plastic injection molding by manually shortening the screw-back position to check the real flow fronts that propagated from the filling gates during the filling stage. The implementation of short-shot testing may help investigators to estimate the propagating times of the initial flow fronts from the filling gates to the next gate. Using previously recorded times, one could propose an actuating sequence for the sequential valve gate-opening system. In addition, the real shapes of the short-shot molding samples and the simulated flow fronts can be verified to complete the numerical validation.

The concurrent opening of all the filling gates produces a higher injection pressure and, thus, a higher clamping force in general plastic injection molding. The concurrent opening scheme is different from the sequential valve gate-opening scheme. Short-shot simulations of 30% and 60% filling are shown in Figure 2. Figure 2a,c show the flow fronts when the 16 gates are opened at the same time. The fill flow fronts that form around gates 2 and 3, 6 and 7, 10 and 11, and 14 and 15 initially interact, and are stitched at the half center of the plastic pallet. The fronts are expanded to the four corners, which then become the portions with the longest flow lengths during injection mold filling. The color distribution is almost uniform, since all gates are open. Weld lines are located on the borders of each of the pairs of gates, and there are 17 borders on the top surface of the pallet. The other borders are located across the pallet.

In terms of weld lines, variant filling can be improved by the one-gate or sequential valve gate-opening schemes. In this study, for a large-scale plastic pallet, a pre-set sequence of opening the valve gates was used to expand the filling fronts. The pre-set sequence was provided by the industrial company we are working with. Figure 2b,d show that the 16 gates are controlled in a seven-stage sequence, as also shown in Table 2. The comparison of the filling fronts produced under the concurrent opening and sequential valve gate-opening schemes is shown in the form of iso-surfaces. An iso-surface is a surface produced over an equal time within a volume of space. Changes to the iso-surfaces' shapes show differences close to the gates—that is, the flow fronts of the melt are increased close to these gates. The iso-surface shows a higher velocity of melt material in the middle of the pallet. During the sequential valve gate-opening scheme, the gates are not ideal due to the number of weld lines, while this is slightly decreased under the concurrent opening scheme. Furthermore, the initial gates that were opened under this sequential pre-set valve gate-opening scheme are inappropriate for real-world applications.

(**a**)

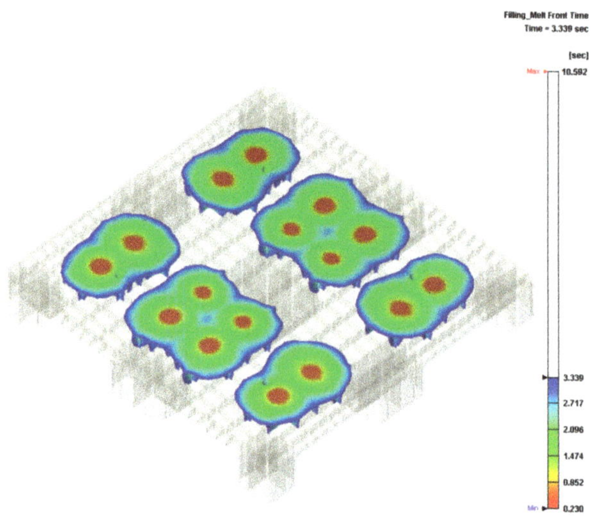

Figure 2. *Cont.*

(b)

(c)

(d)

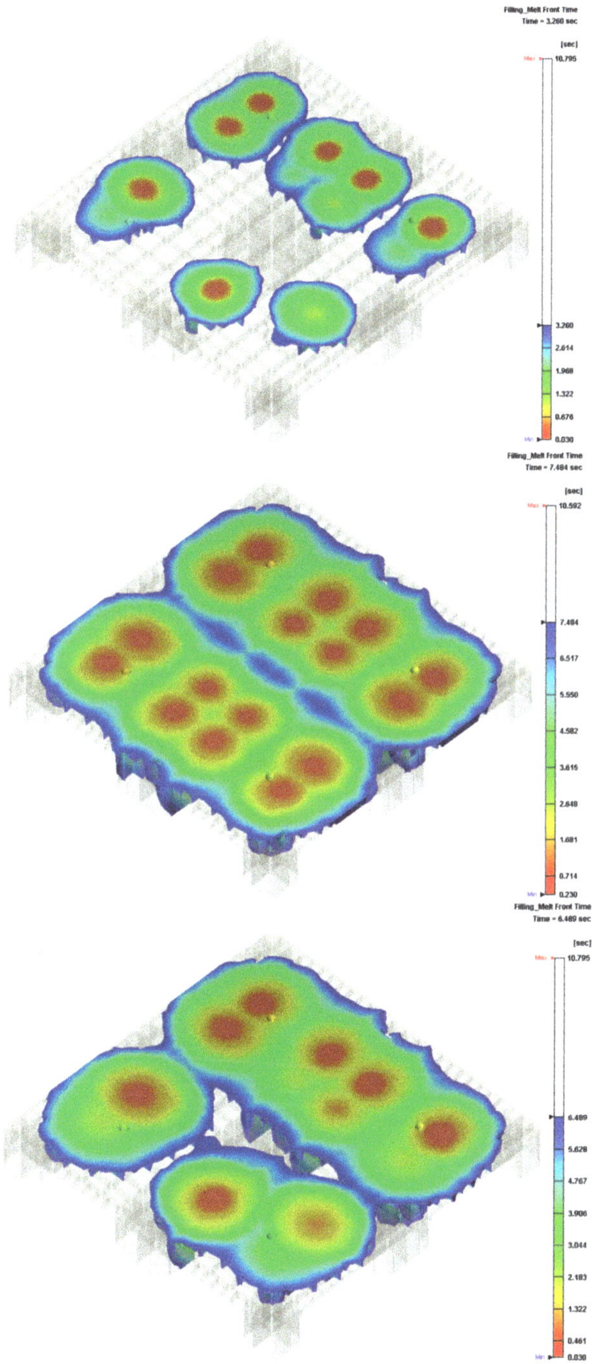

Figure 2. Flow fronts in short-shot testing: (**a**) 30% flow front under the concurrent valve gate-opening scheme; (**b**) 30% flow front under sequential valve gate-opening scheme; (**c**) 60% flow front under concurrent valve gat- opening scheme; (**d**) 60% flow front under sequential valve gate-opening scheme.

The real temperatures of the core and cavity plates of the mold were unknown before the filling stage of the injection process. The mold temperature was increased with a shortened cycle time. As the residual heat within the mold, which would not be transferred into the cooling channel, increased, the mold temperature is unknown, although the mold temperature can be detected at the ejection stage. However, the detected mold temperature cooled down during the ejection stage. A preliminary analysis of the flatness can be performed by assuming the mold temperature. A further serial numerical analysis would help us to determine the exact mold temperature by adjusting the cooling channel mold temperature to fit the experimental surface flatness of the pallet. By pre-setting the sequential valve gate-opening scheme (Table 2), a serial short-shot test can be performed experimentally. A 70% short-shot sample was produced by the Supermaster injection machine (Figure 3a,c), showing the top and bottom surfaces of the rPP injected pallets). The 70% short-shot of the real injected pallet was compared to the fronts on the top and bottom sides of the pallet via a numerical approach (Figure 3b,d). The profile of the real pallet qualitatively agreed with the simulated flow front. In Figure 3e, the profile of the top surface of the pallet produced by 90% short-shot testing indicates a close fit with the numerical flow front, as shown in Figure 3f. The bottom side of the 90% short-shot pallet provides significant evidence of the prediction accuracy of the experimental and numerical results shown in Figure 3g,h. Figure 3b,f show that the imbalanced flow fronts spread to the entire pallet due to the inaccurate setting of the valve gate-opening sequence. This mis-setting may induce a fluctuation in the spread front. Through the simulation results, a suitable sequence of the valve gates' switching times can be predicted. The flow front thus spreads continuously once it meets the next gate.

(a)

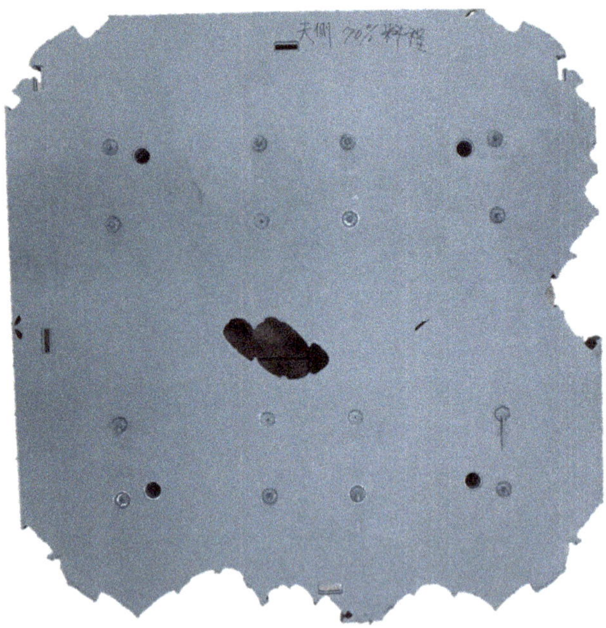

Figure 3. *Cont.*

(b)

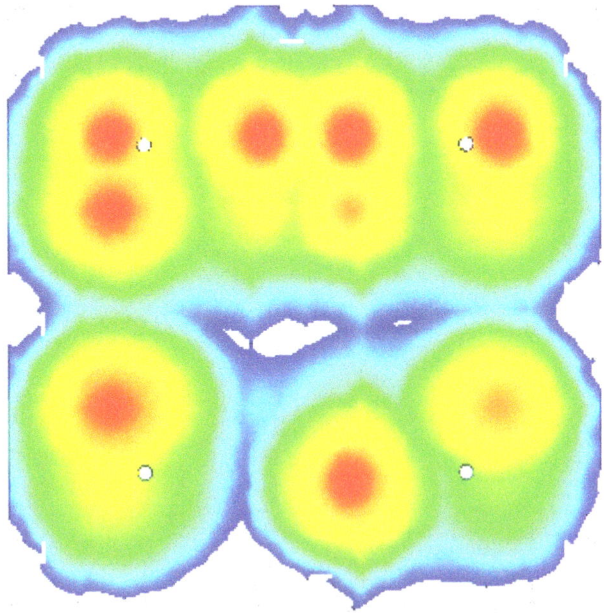

(c)

Figure 3. *Cont.*

(d)

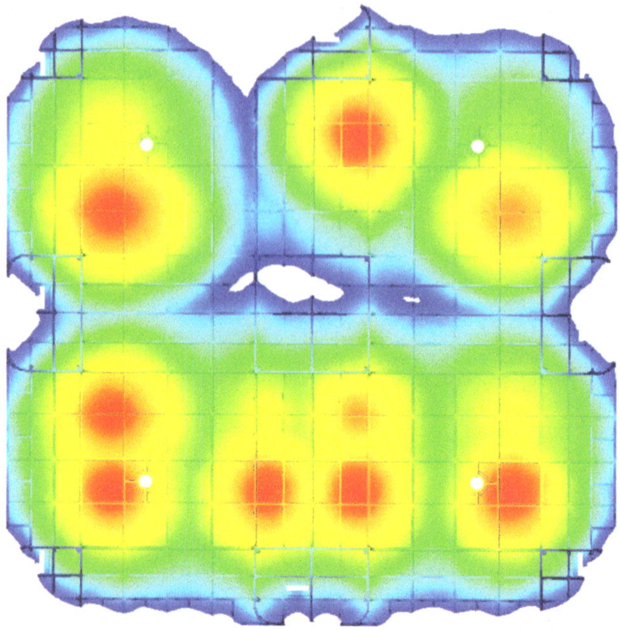

(e)

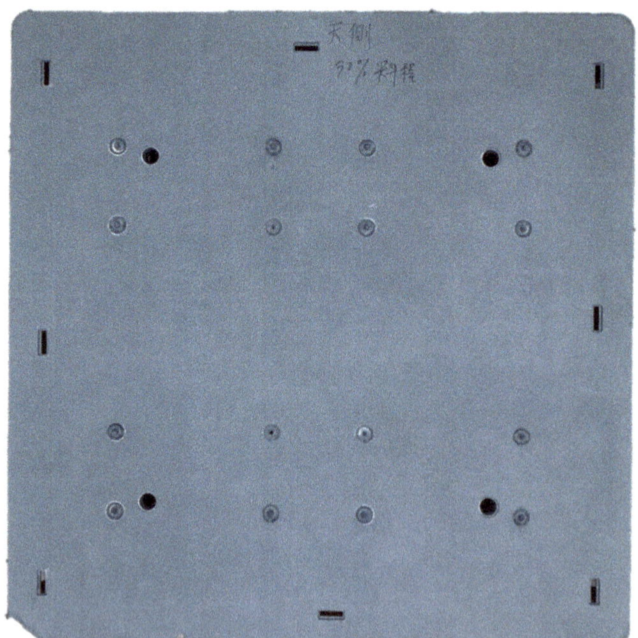

Figure 3. *Cont.*

(f)

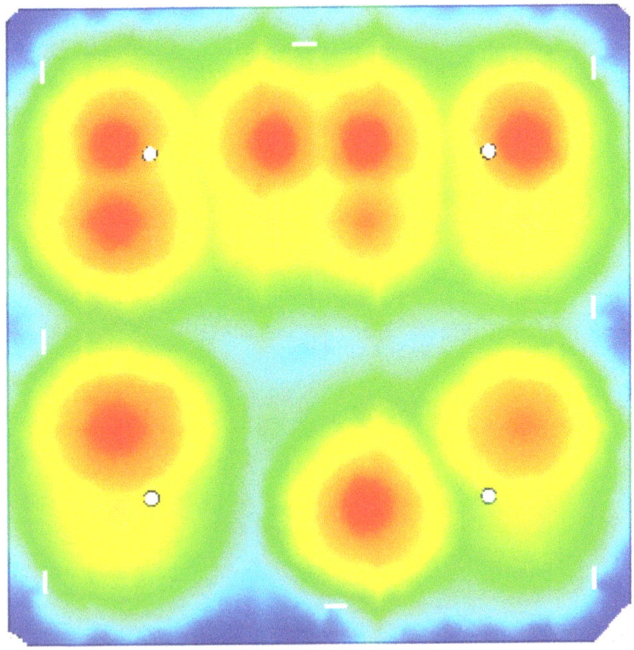

(g)

Figure 3. *Cont.*

(**h**)

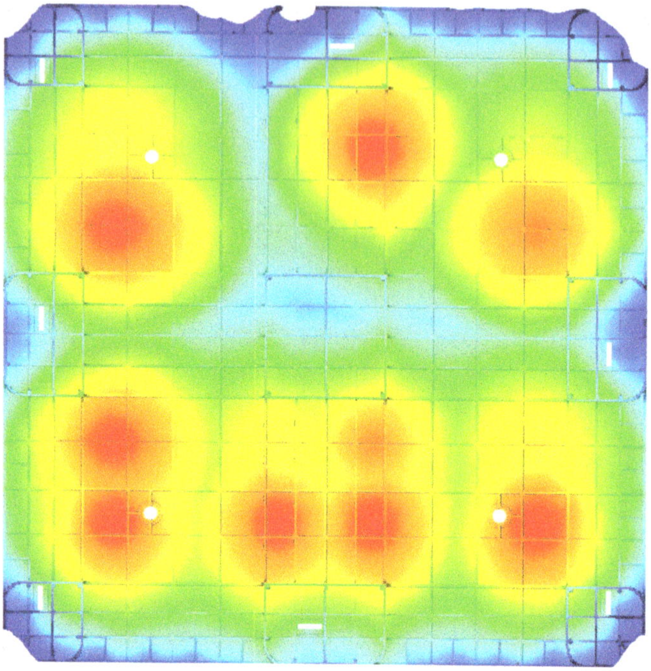

Figure 3. Comparisons of the experimental and numerical flow fronts in short-shot testing under the pre-set sequential valve gate-opening scheme. (**a**) Top view of the 70% short-shot pallet; (**b**) top view of the 70% numerical flow front; (**c**) bottom view of the 70% short shot pallet; (**d**) bottom view of the 70% numerical flow front; (**e**) top view of the 90% short-shot pallet; (**f**) top view of the 90% numerical flow front; (**g**) bottom view of the 90% short-shot pallet; (**h**) bottom view of the 90% numerical flow front.

3.2. Flatness

The temperature of the cooling channel within the mold plate, including the core and cavity plates, was set to 20 °C. The injection molding samples were acquired after 20 cycles to ensure that the temperature of the mold plate was in a steady state. Then, the temperatures of the mold's core and cavity were measured by K type thermocouple after the pallet was ejected from the core plate. These two measured temperatures were higher than the cooling channel temperature since their heat is never effectively transferred to the cooling channel. The exact temperatures of the mold's core and cavity could not be measured during the packing and cooling stages of injection molding. To fit the numerical warpage to the molded one using the trial-and-error method, an empirical approach was used to adjust the temperatures of the cooling channels within the cavity and core plates of the mold, which induce different temperature levels in the two plates compared to the predicted temperature for the core and cavity of the mold. Hence, the entire pallet was measured by way of the ATOS scan box 5120 system, and this was then compared with the numerical results.

The difference between the highest and lowest positions of a surface profile in the height direction can be called the flatness. Figure 4a depicts the measured flatness of the top surface of the molded pallet by the previous ATOS system. The entire profile of the pallet's top surface is convex and has an excellent flatness value of 6.721 mm. By adjusting the cooling channel temperature of the mold's cavity plate to 55 °C, and the core plate temperature to 20 °C, for simulation on Moldex3D, the pallet's numerical top profile is symmetrically convex and the surface flatness of this top surface is 6.044 mm, as shown in Figure 4b. This molded pallet has nine legs to support the entire top surface and the loads.

The flatness of the pallet's nine-leg plane is, therefore, important for stably supporting goods. The flatness of the nine-leg plane of the pallet's bottom surface is acquired by an ATOS of 6.772 mm, as shown in Figure 4c. The predicted flatness of the bottom side of the pallet is 5.960 mm in Figure 4d. The deviation of 0.812 mm, with regard to the 130 mm pallet height, is relatively accurate.

(**a**)

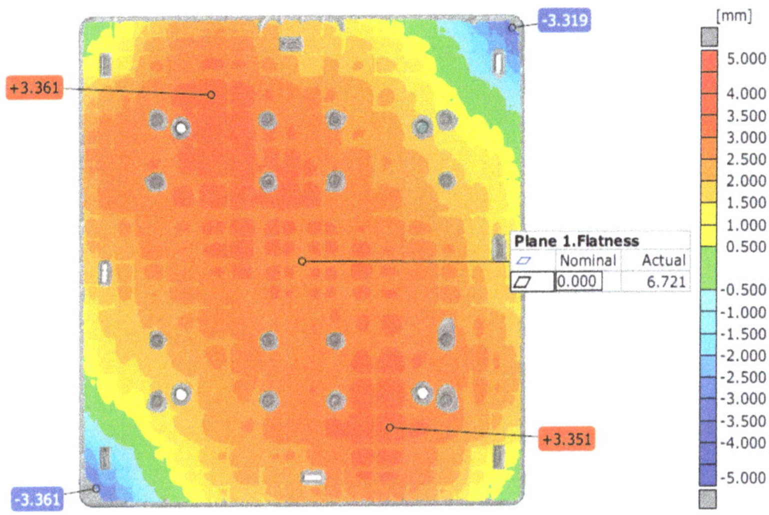

(**b**)

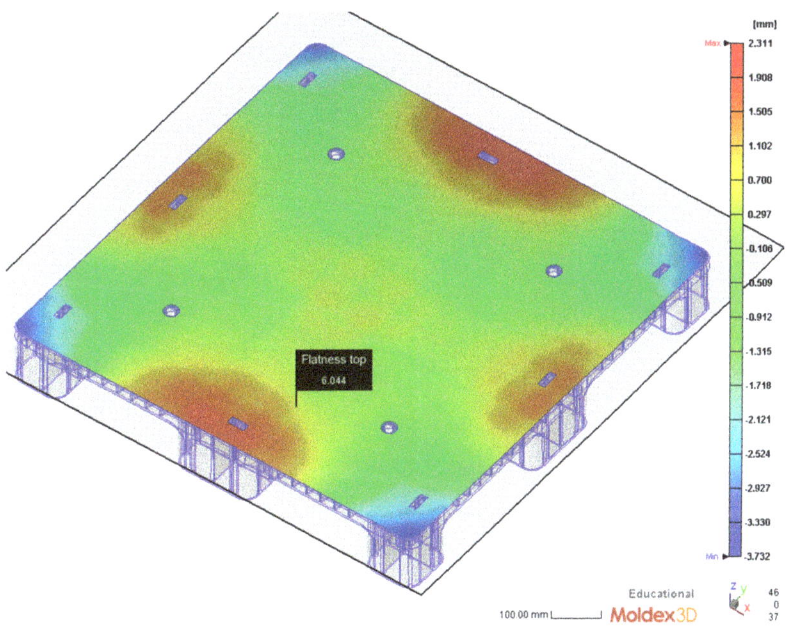

Figure 4. *Cont.*

(**c**)

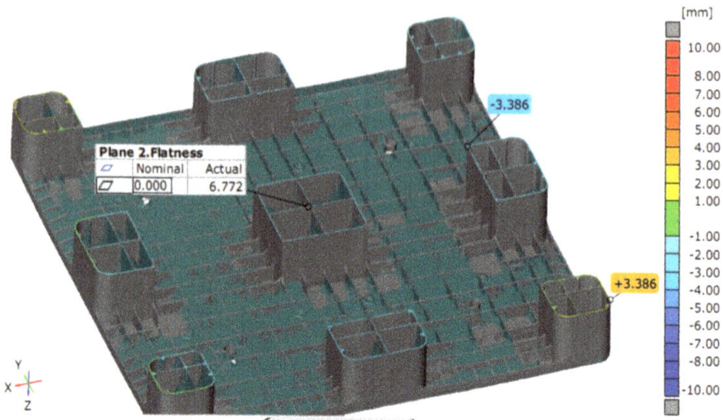

(**d**)

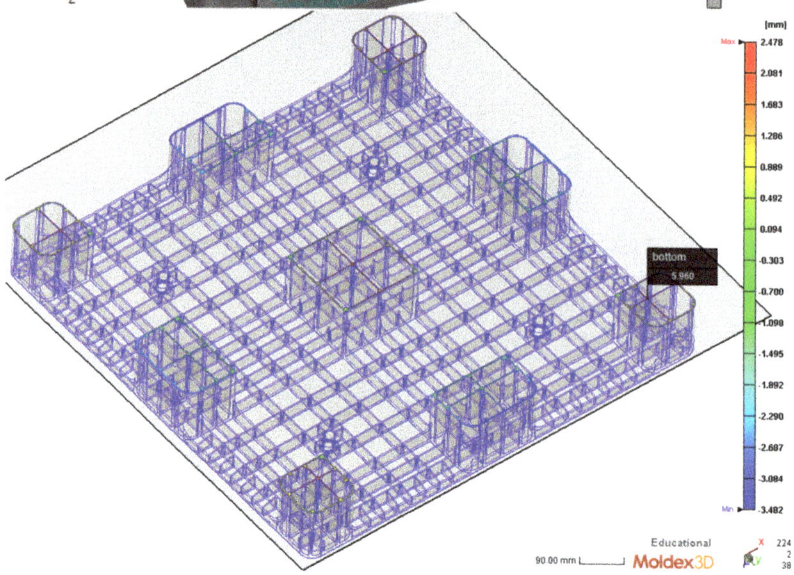

Figure 4. *Cont.*

(e)

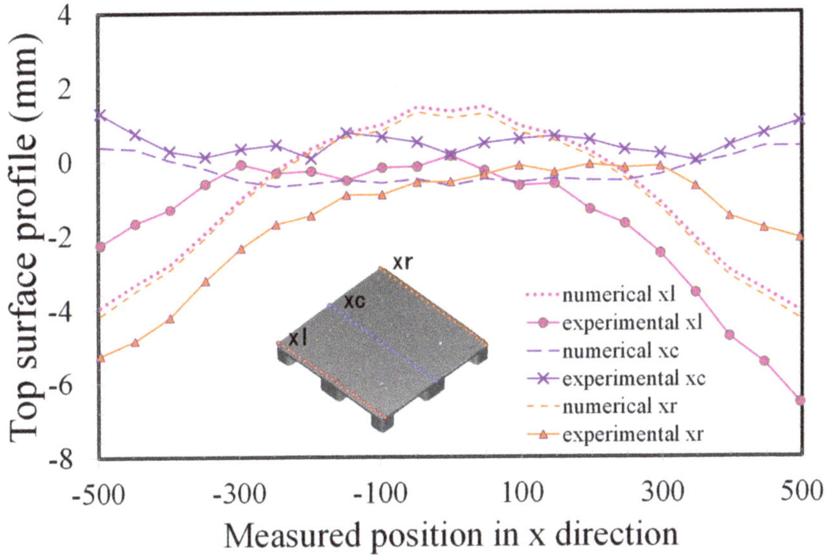

(f)

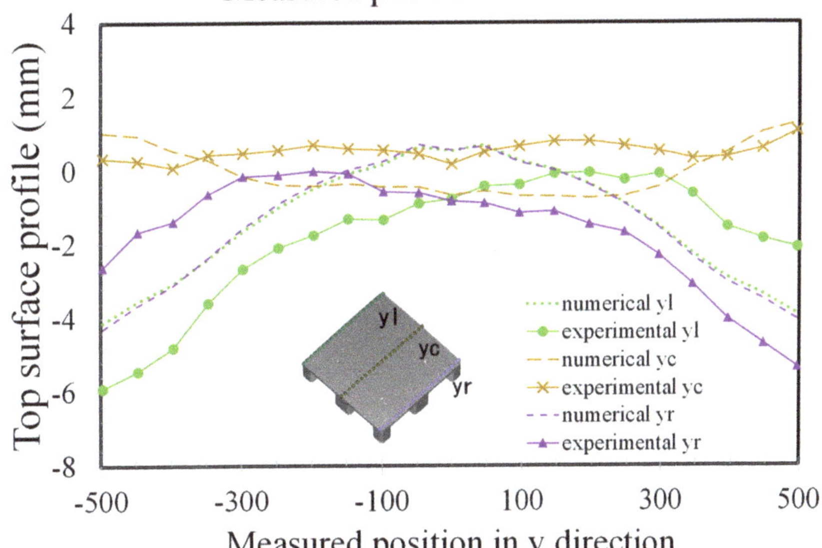

Figure 4. Comparisons of the numerical and experimental flatness under the pre-set sequential valve gate-opening scheme (20 °C cooling channel). (**a**) Isometric view of ATOS-measured top flatness of pallet; (**b**) Isometric view of numerical top flatness of pallet; (**c**) isometric view of ATOS-measured bottom flatness of pallet; (**d**) Isometric view of numerical bottom flatness of pallet; (**e**) x-axial top surface profiles; (**f**) y-axial top surface profiles.

We compared the results, showing the true measured pallet height in the ideal coordinate system by ATOS with the reconstructed deformations in Figure 4a and the numerical heights produced by Moldex3D in Figure 4b. A comparison of experimental and numerical profiles of the molded pallet along the x axis is shown in Figure 4e, where the purple crosses denote the heights measured every 50 mm along the middle (x_c) of the top profile of the pallet, and the purple dashed line denotes the simulated height of the same profile.

In the middle of the x axis, the simulated profile shows good conformity with the measured profile, with a 0.84 mm deviation between the experimental and numerical profiles. The warpage in the middle profile along the x axis is very small, as the heights of the profile are all below 1.29 mm and the average height of the profile is 0.49 mm.

The red circles denote the measured height on the left-hand side (x_l) of the pallet and the red dotted line denotes the numerical profile. The orange triangles denote the measured height on the right-hand side (x_r) of the pallet and the orange dashed line denotes the numerical profile. Numerical profiles on both the left- and right-hand sides show the convex symmetric heights of the pallet, but the experimental profiles on both sides of pallet show either positive skewness or negative skewness. The experimental deviations between the maximum and minimum heights of both profiles on the left- and right-hand sides of the pallet are 6.3 and 5.2 mm, respectively. Along the y axis, the predicted heights of the profiles also showed good conformity with the measured profile of the pallet, with a 0.92 mm deviation between the experimental and numerical profiles, as shown in Figure 4f. The middle profile (y_c) along the y axis was experimentally shown to be deformed by under 1.07 mm. The experimental profiles on both sides of the pallet are characterized by positive or negative skewness. The experimental deviations between the maximum and minimum heights of both profiles, on the left-hand (y_l) and right-hand (y_r) sides of the pallet, are 5.9 and 4.2 mm, respectively. The numerical predictions of the profiles by Moldex3D perfectly fit the true profiles of the pallet due to the large-scale geometrical sizes of 1000 mm × 1000 mm × 130 mm.

To derive the warpage of the large-scale pallet, an injection molding process was applied, using a numerical approach. The structure of plastic pallet is composed of ribs, beams and rods, making the construction of a close-form solution inaccessible. Using Moldex3D software, one may acquire the pallet's temperature distribution under each stage of injection molding. However, after demolding, the pallet is free to deform, and the asymmetric internal stress of the molded pallet may cause the warpage of the pallet to develop. Since the temperature in the pallet is still high and the pallet is not stiff enough to endure this, it immediately warps towards the cold side upon ejection. The hot side of the pallet has a higher temperature than the cold side; it is less stiff and experiences more cooling than the cold side as the whole pallet eventually cools down to room temperature. Hence, the hot side will deform more than the cold side during the free quench of the pallet, and this will cause the warpage to decreasingly incline toward the cold side [31–33].

3.3. *Proposed Sequence of Valve Gate-Opening*

An unbalanced filling resulting from the pre-set sequence is shown in Figure 2b,d, as well as in Figure 3b,d,f,h. The pre-set sequence shown in Table 2 is improper due to the well-maintained weld lines between the 16 gates. An essential feature of the sequential valve gate scheme is that the valve gate is actuated once the flow front spreads towards it. This arrangement may drastically reduce the number of weld lines in the injection molding parts. A newly proposed sequential valve gate scheme is shown in Table 3. The total filling time is 10.591 s, which is slightly greater than the pre-set sequence of 10.795 s.

Table 3. Proposed sequence of filling valve gates.

Time (s)	Start	2.2	5.42	5.46	5.5	5.85	5.88	6.04
Gate (#)	#6, #7, #10, #11	#2, #3, #14, #15	#5	#8, #12	#9	#1	#4, #16	#13

The cooling channel temperatures within the core and cavity plates of the mold were 20 °C and 55 °C, respectively. The injection molding parameters are shown in Table 1. In the Moldex3D simulation, the flow front initially spreads from gates #6, #7, #10, and #11. Gates #2, #3, #14, and #15 are activated when the initial flow front meets the previous gates. The flow front profile is shown in Figure 5a, in which there are only three weld lines on the top surface of the pallet. The front continuously spreads to the four corners of the pallet. The gates begin filling the melt material into the cavity when the flow front meets the lateral gates, as shown in Figure 5b,c. Weld lines appear at four corner legs of the pallet, which cannot be avoided due to the two filling approaches used during the molding of the corner legs. However, this proposed filling valve gate scheme sequence eliminates a high number of weld lines on the top surface.

(a)

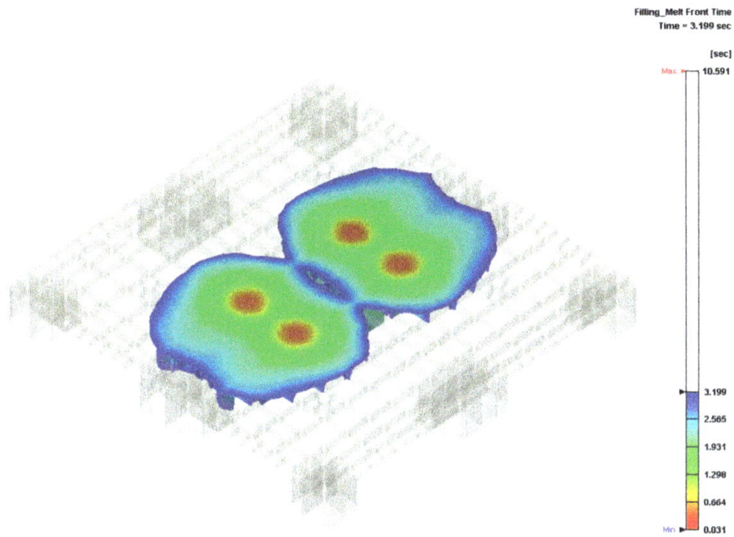

Figure 5. *Cont.*

(**b**)

(**c**)

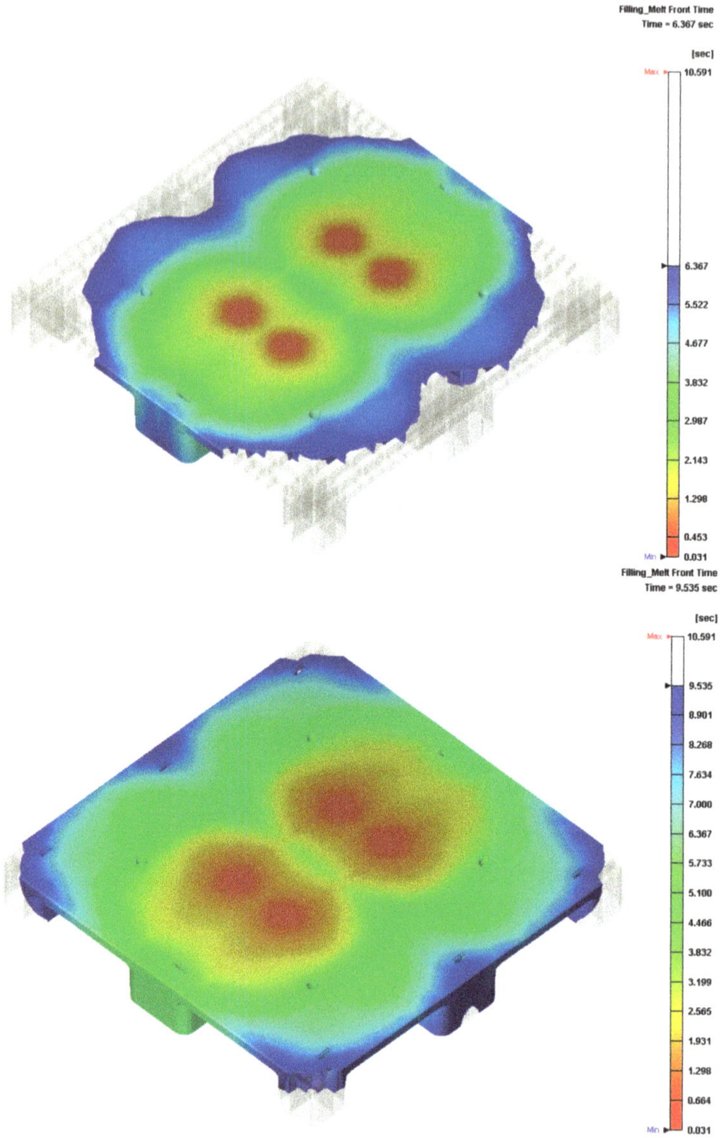

Figure 5. Flow front under the proposed sequential valve gate scheme: (**a**) 30% of total filling; (**b**) 60% of total filling; (**c**) 90% of total filling.

Figure 6a depicts the numerical flatness of the top surface of the pallet under the newly proposed sequential valve gate scheme. The top flatness of the pallet, 7.76 mm, is drastically decreased with respect to the 64.052 mm flatness by pre-setting the sequence of the valve gate-opening scheme. The profile of the top surface differs from the convex profile (Figure 4a). This reduced flatness demonstrates that the proposed sequence scheme is a successful approach, and that this flatness excellently meets the specifications of the commercial plastic pallet. Figure 6b shows the bottom surface of the pallet, with a 7.027 mm flatness. This flatness of 7.027 mm can be compared to the numerical flatness of 64.072 mm of the bottom surface of the pallet, achieved by pre-setting the valve gate system system.

The proposed valve gate scheme sequence may numerically decrease due to the warpage of the rPP pallet.

(**a**)

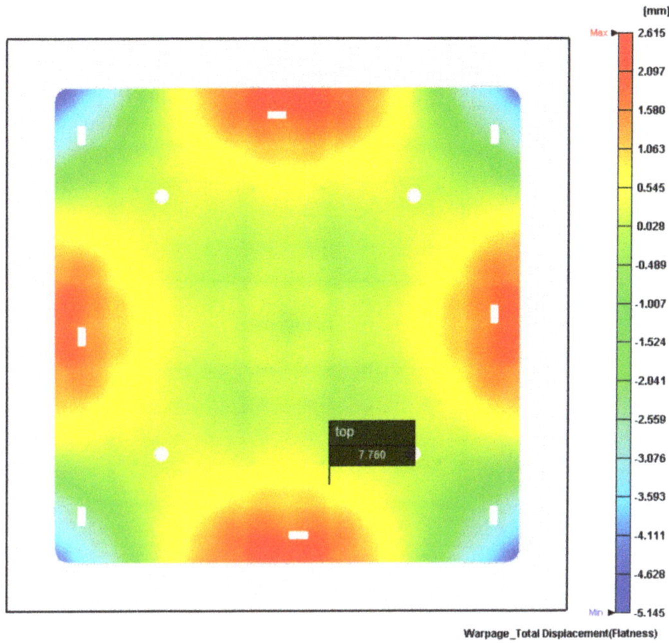

(**b**)

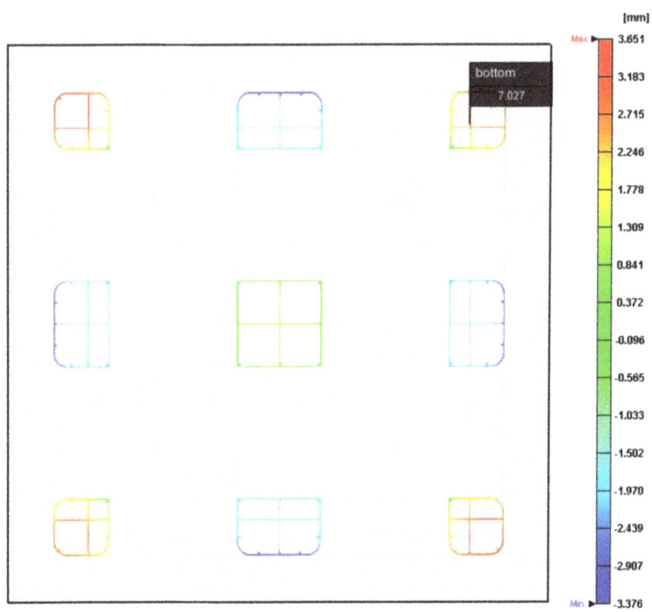

Figure 6. Flow front under the proposed sequential valve gate scheme: (**a**) Top flatness of pallet; (**b**) bottom flatness of pallet.

In this study, the mold temperatures of the core and cavity plates are the same, which causes a large warpage in this thin, rib-reinforced pallet due to the total thin-rib structure of the core plate. The top profiles of the pallet are, therefore, shown to be numerically in convex Figure 4a and experimentally convex in Figure 4b. Under the pre-set sequential valve gate scheme, the flatness of the pallet in Figure 7 shows that the temperature of the cavity plate of the mold has a negative proportional relationship at temperatures below 60 °C. This trend may help us to set the local mold temperature. Consequently, the simulation verifies the feasibility of using the proposed sequential valve gate system. A further experimental investigation is essential to verify the previously determined numerical flatness of the pallet. The temperatures of the core and cavity plates of the mold has to be experimentally determined via the pallet's warpage.

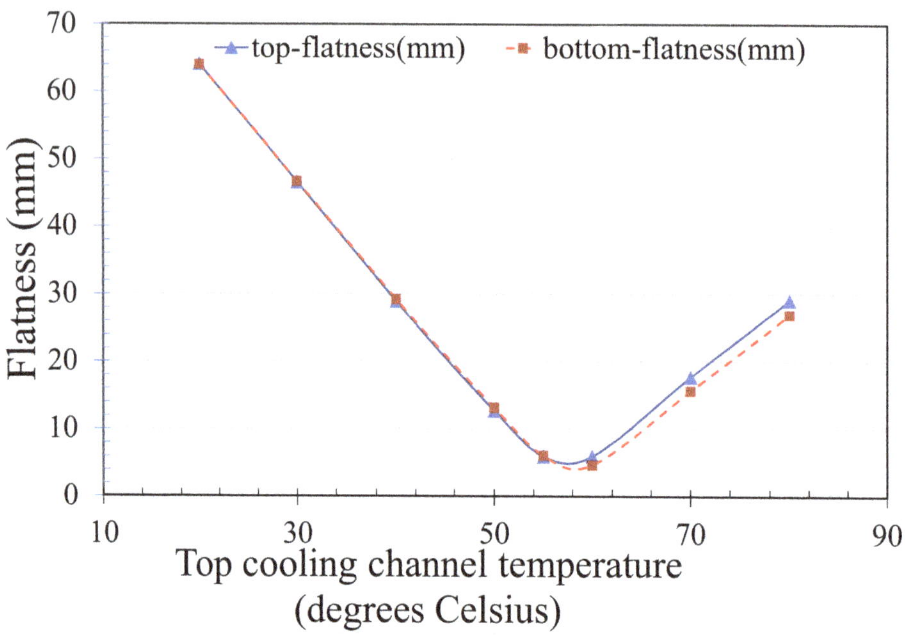

Figure 7. Numerical flatness with respect to the cooling channel of the cavity plate under the pre-set sequential valve gate-opening scheme (20 °C cooling channel within the core plate).

4. Conclusions

The flatness of an rPP pallet, created using a pre-set and newly proposed sequential valve gate-opening injection molding scheme, was investigated and found to be successful. The pallet's required specifications were achieved. The pre-set sequence of valve gate-opening produced a poor pallet product due to the weld lines present on the top surface of the pallet as the flatness neared 7 mm. The authors proposed a new sequential valve gate-opening scheme to ensure the smooth spreading of the flow front from the middle four gates. The newly proposed, eight-stage, sequential gate-opening, compared to the pre-set valve gate-opening scheme, propagates the flow front from the central gate to the four corners of the pallet, with fewer welding lines being formed between each set of four legs, located in the rib portions. The weld lines on the top surface of the pallet are, therefore, dramatically decreased. The flatness of the pallet was accurately demonstrated via experimental and numerical results. The flatness of the pallet produced by the newly proposed sequential valve gate-opening scheme is about 7 mm, which coheres with the height directional warpage achieved under the pre-set sequential scheme. By adjusting the temperature of the cooling channel within the cavity plate to 55 °C, the flatness measured by ATOS is in line with the numerical flatness of the pallet.

Author Contributions: Conceptualization, methodology, original draft preparation, and writing—review and editing, H.H.T.; software and data curation, Y.L.L. All authors have read and agreed to the published version of the manuscript.

Funding: The authors would like to thank Nan-Ya Plastics Inc., Taiwan for their financial support of 990,000 N.T.D. (grant no. I01109E021).

Institutional Review Board Statement: Not applicable.

Informed Consent Statement: Not applicable.

Data Availability Statement: All the data generated or analyzed during this study are included in the published article.

Acknowledgments: The authors acknowledge the supports from C. D. Cheng, Lecture in Department. of Mechnical Engineering, Ming Chi University of Technology and RATC in Taiwan.

Conflicts of Interest: The authors declare no conflict of interest with this work.

References

1. SFS-EN ISO 445 Pallets for Materials Handling. Vocabulary. 2013. Available online: https://www.iso.org/standard/61915.html (accessed on 18 September 2021).
2. Deviatkin, I.; Khan, M.; Ernst, E.; Horttanainen, M. Wooden and Plastic Pallets: A Review of Life Cycle Assessment (LCA) Studies. *Sustainability* **2019**, *11*, 5750. [CrossRef]
3. Edge Environment Pty Ltd Pallet Life Cycle Assessment and Benchmark. 2017. Available online: https://re-pal.com/wp-content/uploads/2019/03/Edge-Environment-Pallet-Life-Cycle-Assessment-and-Benchmark-Report.pdf (accessed on 18 September 2021).
4. PlasticsEurope. Plastics–the Facts 2020: An Analysis of European Plastics Production, Demand and Waste Data. 2020. Available online: https://www.plasticseurope.org/en/resources/publications/4312-plastics-facts-2020 (accessed on 17 September 2021).
5. Gall, M.; Freudenthaler, P.J.; Fischer, J.; Lang, R.W. Characterization of Composition and Structure–Property Relationships of Commercial Post-Consumer Polyethylene and Polypropylene Recyclates. *Polymers* **2021**, *13*, 1574. [CrossRef]
6. Gaxiola-Cockburn, R.; Martínez-Romero, O.; Elías-Zúñiga, A.; Olvera-Trejo, D.; Reséndiz-Hernández, J.E.; Soria-Hernández, C.G. Investigation of the Mechanical Properties of Parts Fabricated with Ultrasonic Micro Injection Molding Process Using Polypropylene Recycled Material. *Polymers* **2020**, *12*, 2033. [CrossRef] [PubMed]
7. Huang, M.-C.; Tai, C.-C. The effective factors in the warpage problem of an injection-molded part with a thin shell feature. *J. Mater. Process. Technol.* **2001**, *110*, 1–9. [CrossRef]
8. Ozcelik, B.; Sonat, I. Warpage and structural analysis of thin shell plastic in the plastic injection molding. *Mater. Des.* **2009**, *30*, 367–375. [CrossRef]
9. Wang, C.; Huang, M.; Shen, C.; Zhao, Z. Warpage prediction of the injection-molded strip-like plastic parts. *Chin. J. Chem. Eng.* **2016**, *24*, 665–670. [CrossRef]
10. Cheng, C.D.; Tsai, H.H.; Liao, Y.L. Investigation of the Large-Scale Pallet by Recycled Polypropylene and the Sequential Valve Gate System during the Injection Molding. *Adv. Polym. Technol.* **2021**, *2021*, 19. [CrossRef]
11. Kurt, M.; Saban Kamber, O.; Kaynak, Y.; Atakok, G.; Girit, O. Experimental investigation of plastic injection molding: Assessment of the effects of cavity pressure and mold temperature on the quality of the final products. *Mater. Des.* **2009**, *30*, 3217–3224. [CrossRef]
12. Liparoti, S.; Speranza, V.; Sorrentino, A.; Titomanlio, G. Mechanical Properties Distribution within Polypropylene Injection Molded Samples: Effect of Mold Temperature under Uneven Thermal Conditions. *Polymers* **2017**, *9*, 585. [CrossRef]
13. Wilczyński, K.; Narowski, P. A Strategy for Problem Solving of Filling Imbalance in Geometrically Balanced Injection Molds. *Polymers* **2020**, *12*, 805. [CrossRef]
14. Wilczyński, K.; Narowski, P. Simulation Studies on the Effect of Material Characteristics and Runners Layout Geometry on the Filling Imbalance in Geometrically Balanced Injection Molds. *Polymers* **2019**, *11*, 639. [CrossRef]
15. Liparoti, S.; Speranza, V.; Titomanlio, G.; Pantani, R. Effect of Rapid Mold Heating on the Structure and Performance of Injection-Molded Polypropylene. *Polymers* **2020**, *12*, 341. [CrossRef] [PubMed]
16. Park, K.; Kim, Y.-S. Effect of Mold Temperature on Mechanical Properties of an Injection-Molded Part with Microfeatures. *J. Polym. Eng.* **2009**, *29*, 135–154. [CrossRef]
17. Fang, M.L. Study of Eliminating Welding Line by Sequential Valve Gate in Hot Runner System. Master's Thesis, Department of Mold and Die Engineering, National Kaohsiung University of Applied Science, Kaohsiung, Taiwan, 2006.
18. Wu, Y.Y. Study on Strength of Welding Line Using Sequential Valve Gates. Master's Thesis, Department of Mold and Die Engineering, National Kaohsiung University of Applied Science, Kaohsiung, Taiwan, 2009.

19. Iwko, J.; Steller, R.; Wróblewski, R. Experimentally Verified Mathematical Model of Polymer Plasticization Process in Injection Molding. *Polymers* **2018**, *10*, 968. [CrossRef]
20. Cardozo, D. Three models of the 3D filling simulation for injection molding: A brief review. *J. Reinforced Plas. Compos.* **2008**, *27*, 1963–1974. [CrossRef]
21. Chang, R.Y.; Yang, W.H. Numerical simulation of mold filling in injection molding using a three-dimensional finite volume approach. *Int. J. Numer. Meth. Fluids* **2001**, *37*, 125–148. [CrossRef]
22. Hua, S.; Zhang, S.; Cao, W.; Wang, Y.; Shao, C.; Liu, C.; Dong, B.; Shen, C. Simulation of Jetting in Injection Molding Using a Finite Volume Method. *Polymers* **2016**, *8*, 172. [CrossRef]
23. Baldi-Boleda, T.; Sadeghi, E.; Colominas, C.; García-Granada, A. Simulation Approach for Hydrophobicity Replication via Injection Molding. *Polymers* **2021**, *13*, 2069. [CrossRef] [PubMed]
24. Jiang, B.; Peng, H.; Wu, W.; Jia, Y.; Zhang, Y. Numerical Simulation and Experimental Investigation of the Viscoelastic Heating Mechanism in Ultrasonic Plasticizing of Amorphous Polymers for Micro Injection Molding. *Polymers* **2016**, *8*, 199. [CrossRef]
25. Zink, B.; Szabó, F.; Hatos, I.; Suplicz, A.; Kovács, N.K.; Hargitai, H.; Tábi, T.; Kovács, J.G. Enhanced Injection Molding Simulation of Advanced Injection Molds. *Polymers* **2017**, *9*, 77. [CrossRef]
26. Krebelj, K.; Krebelj, A.; Halilovič, M.; Mole, N. Modeling Injection Molding of High-Density Polyethylene with Crystallization in Open-Source Software. *Polymers* **2021**, *13*, 138. [CrossRef]
27. Huang, C.-T.; Chen, X.-W.; Fu, W.-W. Investigation on the Fiber Orientation Distributions and Their Influence on the Mechanical Property of the Co-Injection Molding Products. *Polymers* **2020**, *12*, 24. [CrossRef] [PubMed]
28. Chung, C.-Y.; Hwang, S.-S.; Chen, S.-C.; Lai, M.-C. Effects of Injection Molding Process Parameters on the Chemical Foaming Behavior of Polypropylene and Polystyrene. *Polymers* **2021**, *13*, 2331. [CrossRef]
29. Guerrier, P.; Tosello, G.; Kirstein Nielsen, K.; Hattel, J.H. Three-dimensional numerical modeling of an induction heated injection molding tool with flow visualization. *Int. J. Adv. Manuf. Technol.* **2016**, *85*, 643–660. [CrossRef]
30. Guerrier, P.; Tosello, G.; Hattel, J.H. Flow visualization and simulation of the filling process during injection moulding. *CIRP J. Manuf. Sci. Technol.* **2017**, *16*, 12–20. [CrossRef]
31. Moayyedian, M.; Mamedov, A. Multi-objective optimization of injection molding process for determination of feasible moldability index. *Procedia CIRP* **2019**, *84*, 769–773. [CrossRef]
32. Nian, S.C.; Wu, C.Y.; Huang, M.S. Warpage control of thin-walled injection molding using local mold temperatures. *Int. Commun. Heat Mass Transf.* **2015**, *61*, 102–110. [CrossRef]
33. Fan, B.F.; Kazmer, D.O.; Bushko, W.C.; Theriault, R.P.; Poslinski, A.J. Warpage prediction of optical media. *J. Polym. Sci. Part B Polym. Phys.* **2003**, *41*, 859–872. [CrossRef]

polymers

Article

Pyrolysis Study of Mixed Polymers for Non-Isothermal TGA: Artificial Neural Networks Application

Ibrahim Dubdub

Department of Chemical Engineering, King Faisal University, Al-Hassa 31982, Saudi Arabia; idubdub@kfu.edu.sa

Abstract: Pure polymers of polystyrene (PS), low-density polyethylene (LDPE) and polypropylene (PP), are the main representative of plastic wastes. Thermal cracking of mixed polymers, consisting of PS, LDPE, and PP, was implemented by thermal analysis technique "thermogravimetric analyzer (TGA)" with heating rate range (5–40 K/min), with two groups of sets: (ratio 1:1) mixture of PS and PP, and (ratio 1:1:1) mixture of PS, LDPE, and PP. TGA data were utilized to implement one of the machine learning methods, "artificial neural network (ANN)". A feed-forward ANN with Levenberg-Marquardt (LM) as learning algorithm in the backpropagation model was performed in both sets in order to predict the weight fraction of the mixed polymers. Temperature and the heating rate are the two input variables applied in the current ANN model. For both sets, 10-10 neurons in logsig-tansig transfer functions two hidden layers was concluded as the best architecture, with almost (R > 0.99999). Results approved a good coincidence between the actual with the predicted values. The model foresees very efficiently when it is simulated with new data.

Keywords: pyrolysis; mixed polymers; thermogravimetric analyzer (TGA); artificial neural networks (ANN)

Citation: Dubdub, I. Pyrolysis Study of Mixed Polymers for Non-Isothermal TGA: Artificial Neural Networks Application. *Polymers* **2022**, *14*, 2638. https://doi.org/10.3390/polym14132638

Academic Editors: Salah Aldin Faroughi, Luís L. Ferrás, Alexandre M. Afonso and Célio Bruno Pinto Fernandes

Received: 11 June 2022
Accepted: 27 June 2022
Published: 28 June 2022

1. Introduction

Recently, most of the researchers are aiming to deal with machine learning methods "ANN" for the forecasting of different data since it approved that it has a strong performance to deal with non-linearity relationships. Therefore, ANN is considered as another option to deal with the TGA datum.

The literatures surveyed listed below will be limited only for the papers handling ANN for TGA data [1–18].

Conesa et al. [1] was the first to explore ANN in the thermal analysis by initiating a way to treat with the pyrolysis kinetics at different samples for non-isothermal runs. Bezerra et al. [2] applied the ANN model to the thermal cracking of carbon fiber/phenolic resin composite laminate. Yıldız et al. [3] examined the oxidation of mixtures of different ratio by enforcing ANN. Çepelioĝullar et al. [4] extended an ANN to foresee the pyrolysis of waste fuel. Ahmad et al. [5] established ANN for the pyrolysis of Typha latifolia. They collected 1021 data for the feed-forward Levenberg–Marquardt back-propagation algorithm. Çepelioĝullar et al. [6] performed the ANN models for Lignocellulosic forest residue (LFR) and olive oil residue (OOR) in two different sets: (i) two separate networks for each sample, and (ii) one network for both samples. Later, Chen et al. [7] studied the co-combustion characteristics of sewage sludge and coffee grounds (CG) mixtures. Naqvi et al. [8] suggested an ANN to tip the thermal cracking of one type of sludge and offered a strong harmonization for the predicted with experimental figures. In this paper, a richly powerful promoted ANN model ($R \approx 1.0$) predicted a pyrolytic behavior of mixed polymers. Ahmad et al. [9] validated the pyrolysis of Staghorn Sumac by ANN model.

Bi et al. [10] investigated the co-combustion co-pyrolysis of sewage sludge and peanut shell by ANN model. Bong et al. [11] applied the ANN model for the catalytic pyrolysis of pure microalgae, peanut shell wastes, and their binary mixtures with the microalgae ash as a catalyst. In addition, Bi et al. [12] repeated the study for the co-pyrolysis of coal gangue and peanut shell. In both papers, they found there was consistency between the experimental and the ANN model results. Liew et al. [13] predicted the co-pyrolysis of corn cob and high-density polyethylene (HDPE) mixtures, with chicken and duck egg shells as catalysts. Zaker et al. [14] investigated the effects of two catalysis (HZSM5 and sludge-derived activated char) on the pyrolysis of sewage sludge. Dubdub and Al-Yaari [15,16] and Al-Yaari and Dubdub [17,18] tried to use the ANN to predict the performance for different samples. They used a feed-forward LM optimization technique for backpropagation process in the ANN model, in two hidden layers. In the first paper, they applied two input variables, temperature and heating rate, and one output variable, weight left %, while in the second paper, catalyst/polymer weight ratio was added as third input.

Almost all of the above-mentioned studies have good agreement between the experimental collected data and the ANN predicted results efficiently in common. The architecture details of all the papers above are similar to this work (non-isothermal TGA data) are summarized in Table 1. Most of these papers used the temperature and the heating rate for the input variables with weight left % as the only output. This table showed and approved that the application of ANN to predict TGA data is feasible and promising research. In this work, the novelty of this work is in applying the ANN for new two mixture of polymers (PS, LDPE, and PP), and using the final best architecture efficiently in the simulation of new input data.

Table 1. Literature summary of ANN applications for non-isothermal TGA data.

Author	Input Variables			Output Variables	Architecture Model	No. of Hidden Layers	Transfer Function for Hidden Layers	Data Points
Bezerra et al. [2]	temperature	heating rate	-	mass retained	2-21-21-1	2		1941
Yıldız et al. [3]	temperature	heating rate	blend ratio	Mass loss %	3-5-15-1	2	tangsig-tansig	
Ahmad et al. [5]	temperature	Heating rate	-	weight loss		2		1021
Çepelioĝullar et al. [6] Individual	temperature	heating rate	-	weight loss	2-20–20-1 (LFR)2-19–16-1 (OOR)	2	tangsig-logsig	4000
Çepelioĝullar et al. [6] Combined					2-7–6-1	2		8000
Chen et al. [7]	temperature	heating rate	mixing ratio	mass loss %	3-3-19-1	2	tansig-tansig	
Naqvi et al. [8]	temperature	heating rate	-	weight loss	2-5-1	1	tansig	1400
Ahmad et al. [9]	temperature	Heating rate	-	weight loss	2-10-1	1		1155
Bi et al. [10] (combustion), (pyrolysis)	temperature	mixing ratio	-	residual mass	2-3-18-1 2-3-15-1	2	tangsig-tangsig	
Bong et al. [11]	temperature	heating rate	-	weight loss %	2-(9-12)-(9-12)-1	2	tansig-tansig and logsig-tansig	
Bi et al. [12]	temperature	heating rate	mixing ratio	remaining mass %	3-5-10-1	2	tangsig-tangsig	5000
Zaker et al. [14]	temperature	heating rate	-	weight loss (%)	2-7-1	1	tansig	
Al-Yaari and Dubdub [17]	temperature	heating rate	mass ratio	mass left %	3-10-10-1	2	tansig-logsig	900

2. Materials and Methods

2.1. Thermal Decomposition

Pyrolysis experiments were conducted under nitrogen with different compositions of three polymers: PP, PS, and LDPE. Table 2 shows six tests of two sets: tests 1–3 (ratio 1:1) binary of PS and PP, and tests 4–6 (ratio 1:1:1) of PS, LDPE, and PP. 10 mg of each powder sample was used throughout the study. Proximate and ultimate analysis that was performed to characterize the polymer samples can be found in reference [16]. Thermal decomposition experiments were conducted under N2 (99.999%) gas flowing at 100 cm^3/min using the thermogravimetric analyzer (TGA-7), manufactured by PerkinElmer, Shelton, CT, USA [16].

Table 2. List of six runs of different PS, LDPE, and PP polymers compositions.

Set No.	Test No.	Heating Rate (K/min)	Weight %			Comment
			PP	PS	LDPE	
1	1	5	50	50	0	mixture of PS, and PP
	2	20	50	50	0	
	3	40	50	50	0	
2	4	5	33.3	33.3	33.3	mixture of PS, LDPE, and PP
	5	20	33.3	33.3	33.3	
	6	40	33.3	33.3	33.3	

2.2. Structure of ANNs

The common procedure for modelling engineering units is to develop a model depending on the basic principles of physics and chemistry and then the values of the model parameters are estimated from some experimental data by some numerical techniques. However, formulating any model and finding the values of the parameters are the most difficult works in most of the cases, especially when the final model is very complicated with non-linear relations among the variables. In these cases, the ANN may become the alternative option. One of the strengths of ANN is its ability to model the non-linear functions and complex process by mapping these relations by some approximation functions. Moreover, ANN can deal with the noisy data.

ANN architecture is ordered in three consecutive layers: input, hidden/s, and output. Every layer possesses a number of neurons, a weight, a bias, and output [19]. Initially, one must figure out all the variables, with the effect on the main process being variable. The data collection, normally established before the ANN steps, becomes the mirror of the problem area. The best ANN architecture is subjected to learning quality and generalization ability, which relies on whether the collected data fall within the variation margin of the variables and are big enough in size [8].

The type of the task to be handled by the ANN is crucial in finding the best architecture. For better performance of ANNs, the parameters such as the number of neurons in the hidden layer(s), number of the hidden layers, the momentum, and the learning rates should be optimized.

The performance of an ANN model in portending the output can be checked and assessed by five statistical correlations [3,5,7,10,20,21]:

$$\text{Average correlation factor } \left(R^2\right) = 1 - \frac{\sum\left((W\,\%)_{est} - (W\,\%)_{\exp}\right)^2}{\sum\left((W\,\%)_{est} - \overline{(W\,\%)_{\exp}}\right)^2} \quad (1)$$

$$\text{Root mean square error (RMSE)} = \sqrt{\frac{1}{N}\sum\left((W\,\%)_{est} - (W\,\%)_{\exp}\right)^2} \quad (2)$$

$$\text{Mean absolute error (MAE)} = \frac{1}{N}\sum\left|(W\,\%)_{est} - (W\,\%)_{\exp}\right| \quad (3)$$

$$\text{Mean bias error (MBE)} = \frac{1}{N}\sum\left((W\,\%)_{est} - (W\,\%)_{\exp}\right) \quad (4)$$

$$\text{Correlation coefficient (R)} = \frac{\sum_{m=1}^{n}\left((W\,\%)_{\exp,m} - \overline{(W\,\%)_{\exp,m}}\right)\left((W\,\%)_{est,m} - \overline{(W\,\%)_{est,m}}\right)}{\sqrt{\sum_{m=1}^{n}\left((W\,\%)_{\exp,m} - \overline{(W\,\%)_{\exp,m}}\right)^2 \sum_{m=1}^{n}\left((W\,\%)_{est,m} - \overline{(W\,\%)_{est,m}}\right)^2}} \quad (5)$$

where

$(W\,\%)_{est}$: is the estimated value of the weight left % by ANN model;
$(W\,\%)_{exp}$, is the experimental value of the weight left %; and
$\overline{(W\,\%)}$: is the average values of weight left %.

In order to get the best ANN model, it should be targeted to get the lowest error with (RMSE, MAE, MBE), and the highest with (R^2, R) correlations [10]. In this investigation, weight left % of mixed polymers has been predicted by an ANN model. There are some advantages and some disadvantages for using ANN. Some of these advantages can be summarized as being easy to work with linear and non-linear relationships and learning these relationships directly from the data used, while a disadvantages is that doing the fitting needs big memory and computational efforts [22].

3. Results and Discussion

3.1. TGA of Mixed Polymers

TGA provides us with the thermogravimetric (TG), and the derivative thermogravimetric (DTG) at different heating rates of the pyrolysis of two sets at different polymers compositions, which are shown in Figures 1 and 2, respectively [16].

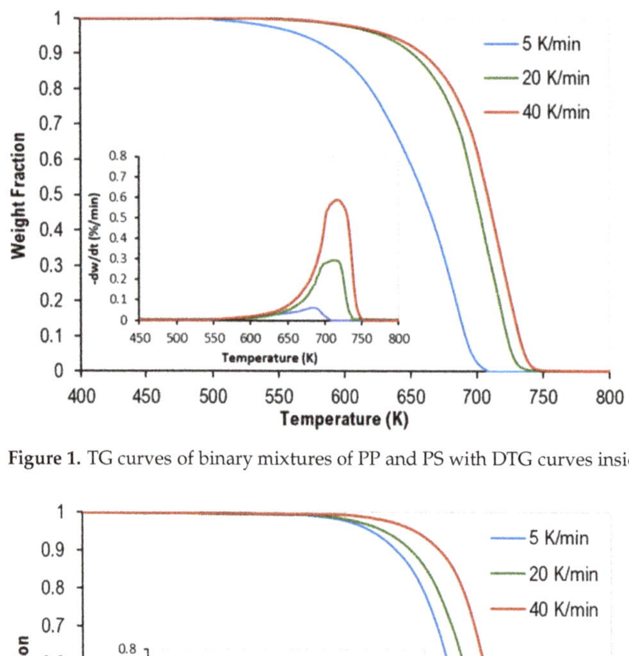

Figure 1. TG curves of binary mixtures of PP and PS with DTG curves inside.

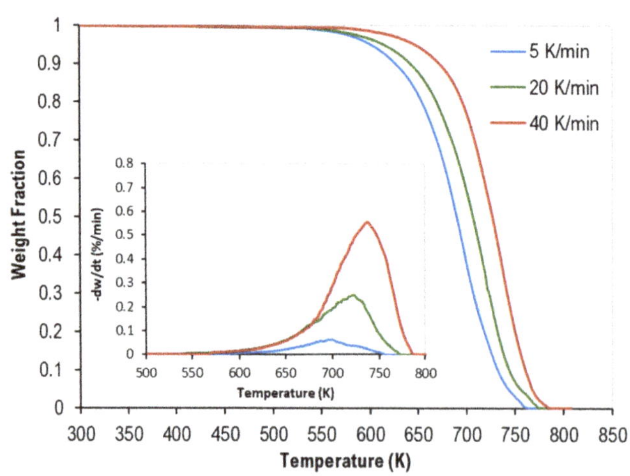

Figure 2. TG curves of ternary mixtures of PP, PS, and LDPE with DTG curves inside.

3.2. Pyrolysis Prediction by ANN Model

Neural Network with "Feed-Forward, Back-Propagation" (FFBPNN) was established in "nntool" function in MATLAB® R2020a based on 358, 752 data for the two sets. This type of ANN model is widely used because it is very efficient and simple [3]. Usually, in the thermal analysis instrument TGA, the raw signal (weight left %) will be the output of the ANN model and the independent variables (temperature and heating rate in the non-isothermal TGA data) could be the inputs of the ANN model.

The collected data will be divided by three subsets: training set will be used to establish the network learning and correct the weights by minimizing the error function; the validation set checks the performance of the network; and finally, the test set will test the generalization of the network [23].

The whole data comprising 358, 752 sets are shown in Table 3, and randomly divided into three sets as follows: 70% for training, 15% for validation and testing. Osman and Aggour [24] mentioned that collecting large sets of data could help the model with high accuracy.

Table 3. Data set number of six tests.

Set No.	Test No.	Heating Rate (K/min)	Data Set Number	Total
1	1	5	126	358
	2	20	101	
	3	40	131	
2	4	5	251	752
	5	20	251	
	6	40	250	

Table 4 listed the parameters of the ANN "nntool" model and Table 5 shows a comparison of different ANN structure performance with different numbers of hidden layers and different numbers of neuron and transfer functions in each hidden layer. Usually, the best architecture is found by a trial and error process [8]. The value of R is examined as the criteria in judging the most efficient network architecture for finding the percentage weight loss %. Values of four statistical correlations will be tabulated only for the last best-selected architecture.

Table 4. Main parameters of the ANN "nntool" model.

Number of inputs	2 (Temperature (K), Heating rate (K/min)
Number of output	1 (Mass left %)
Number of hidden layers	1-2
Transfer function of hidden layers	logsig-tansig
Number of neurons of hidden layers	10-10
Transfer function of out layer	purelin
Data division function	Dividerand
Learning algorithm	Levenberg-Marquardt (TRAINLM)
Data division (Training-Validation-Testing)	70%-15%-15%
Data number (Training-Validation-Testing)	250-54-54 = 358 526-113-113 = 752
Data number (Simulation)	9-9
Performance function	MSE
Validation checks	6

Table 5. Comparison between different ANN structures for the two sets: (i) mixtures of PS and PP, (ii) mixtures of PS, LDPE, and PP.

Model	Network Topology (no. of Neurons) 2 Input-Hidden Layers (1 or 2 Layers)-1 Output	Hidden Layers		R
		1st Transfer Function	2nd Transfer Function	
		i		
AN1-A	2-5-1	*tansig*	-	0.99881
AN2-A	2-5-1	*logsig*	-	0.99972
AN3-A	2-10-1	*tansig*	-	0.99995
AN4-A	2-10-1	*logsig*	-	0.99997
AN5-A	2-15-1	*tansig*	-	0.99997
AN6-A	2-15-1	*logsig*	-	0.99999
AN7-A	**2-10-10-1**	***logsig***	***tansig***	**1.00000**
		ii		
AN1-B	2-5-1	*tansig*	-	0.99976
AN2-B	2-5-1	*logsig*	-	0.99997
AN3-B	2-10-1	*tansig*	-	0.99999
AN4-B	2-10-1	*logsig*	-	0.99999
AN5-B	2-15-1	*tansig*	-	0.99999
AN6-B	2-15-1	*logsig*	-	0.99999
AN7-B	**2-10-10-1**	***logsig***	***tansig***	**1.00000**

The final and best ANN architecture is AN7-A and AN7-B, as shown in Figure 3 for both sets. This network is utilized for the next simulation step. This architecture constitutes 10 neurons with logsig-tansig functions in the two hidden layers with linear transfer function for the output layer. Hidden layers with non-linear functions were used to deal with complex functions [2]. Usually, linear function is not recommended in the hidden layers in order to avoid a linearly separable prediction, while tansig is more preferable since it has larger range of output [11]. Most of the researchers mentioned in Table 1 implied more than one hidden layer [11]. The number of neurons in the hidden layer is a crucial parameter in the efficiency and the accuracy of the ANN output. To avoid the underfitting and the overfitting (too many neurons), one should select the number of neurons in such a way that the performance function will get eventually the optimum value [6,23,25]. There are different supervised learning algorithms such as Levenberg–Marquardt (LM), Bayesian Regularization, and Scaled Conjugate Gradient, but LM is used due its best performance and relevance for this data number [8,10,26]. This optimization LM algorithm technique will update the values of the weighted and biases factors in order to get the calculated output close to the target [5,10].

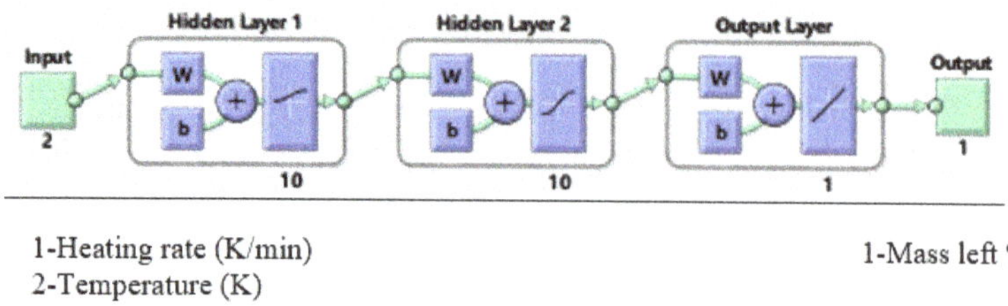

Figure 3. Topology of the selected AN7-A and AN7-B network.

Figure 4 shows all the results fall close to the diagonal, which confirms a strong agreement and good correlation for ANN prediction with experimental values at minimum mean square error (MSE) values of 2.1275×10^{-7} and 4.58×10^{-8} of the two sets, respectively (Figure 5). This MSE's values are too small ($<2.1275 \times 10^{-7}$), which shows that the prediction of the system is very reliable [8]. Naqvi et al. [8] also pointed out that for a good prediction ANN, output values should be close to the target values, and ANN model is a good fit for TGA data.

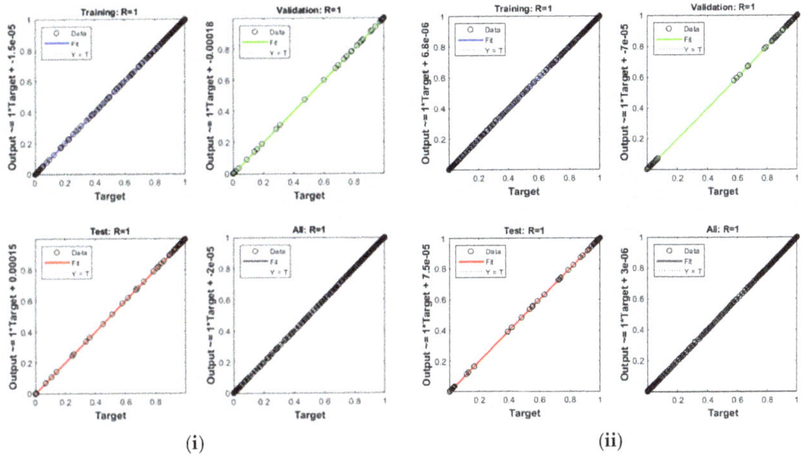

Figure 4. Regression of training, validation, and test plots for the selected (**i**) AN7-A, (**ii**) AN7-B.

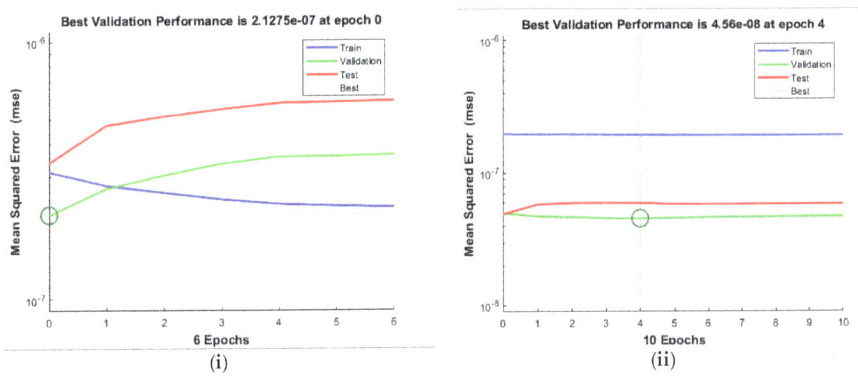

Figure 5. Mean square error for training, validation, and test plots for the selected (**i**) AN7-A, (**ii**) AN7-B.

The performance of the current AN7-A and AN7-B model in predicting the weight left % was measured by calculating these four statistical correlations. Table 6 shows all these four statistical correlations. Notice that values of RMSE, MAE, and MBE are significantly low. Consequently, this model can powerfully predict the output within an acceptable limit of error.

Table 6. Statistical parameters of the (A) AN7-A, (B) AN7-B model.

Set	AN7-A				AN7-B			
	Statistical Parameters				Statistical Parameters			
	R^2	RMSE	MAE	MBE	R^2	RMSE	MAE	MBE
Training	1.0	0.00055	0.00030	−0.00001	1.0	0.00044	0.00016	1.49×10^{-6}
Validation	1.0	0.00046	0.00029	−0.00001	1.0	0.00021	0.00012	-1.74×10^{-6}
Test	1.0	0.00058	0.00032	0.000018	1.0	0.00024	0.00014	0.000034
All	**1.0**	**0.00054**	**0.00030**	**−0.000012**	**1.0**	**0.000389**	**0.000154**	$\mathbf{6.018 \times 10^{-6}}$

Once checking the ANN for the two sets, the final architecture will be simulated by new input data. Table 7 presented the simulation stage with nine datasets for each

AN7-A and AN7-B for only new input data, and the final network will produce the simulated output according to the final architecture AN7-A and AN7-B. Figure 6 shows the comparison between the simulated network with the actual output and indicates very high performance of the selected network. In addition, Table 8 lists all statistical parameters for each set: AN7-A and AN7-B. As presented, the value of R is slightly high (>0.9900) and RMSE, MAE, and MBE have reasonably low values. Finally, Figure 7 shows the error histogram for the two sets, which is distributed across the zero error normally [11]. The error lies in a very small value range (−0.00085 to 0.002678) for the first set and (−0.00123 to 0.000489) for the second set, which indicates very good performance of the proposed ANN model.

Table 7. Simulation input data and output data: mixtures of PS and PP mixtures of PS, LDPE, and PP.

No.	Mixture of PS and PP for AN7-A			Mixture of PS, LDPE, and PP for AN7-B		
	Input Data		Output Data	Input Data		Output Data
	Heating Rate (K/min)	Temperature (K)	Weight Fraction	Heating Rate (K/min)	Temperature (K)	Weight Fraction
1	5	690	0.11471	5	731	0.10335
2	5	668	0.41012	5	697	0.40892
3	5	634	0.70892	5	669	0.70090
4	20	716	0.21154	20	731	0.20736
5	20	698	0.51639	20	705	0.51387
6	20	672	0.80757	20	669	0.80014
7	40	718	0.32648	40	741	0.30962
8	40	700	0.62535	40	717	0.60931
9	40	658	0.90289	40	671	0.90323

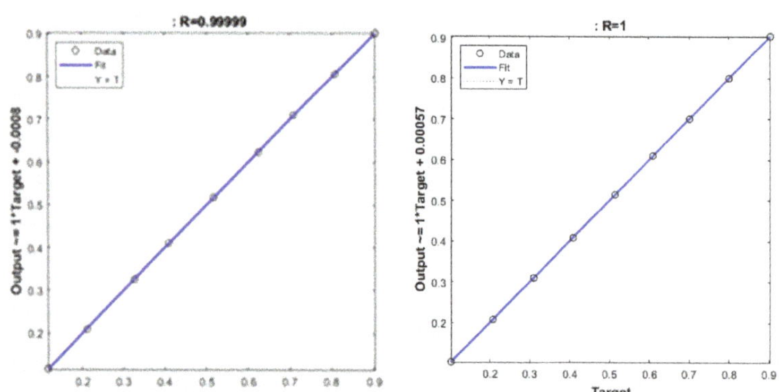

(i) (ii)

Figure 6. Regression of simulated data for (**i**) AN7-A, (**ii**) AN7-B.

Table 8. Statistical parameters for the simulated data of AN7-A and AN7-B.

AN7-A				AN7-B			
Statistical Parameters				Statistical Parameters			
R^2	RMSE	MAE	MBE	R^2	RMSE	MAE	MBE
0.99999	0.00144	0.00123	−0.00052	0.99999	0.00062	0.00049	0.00026

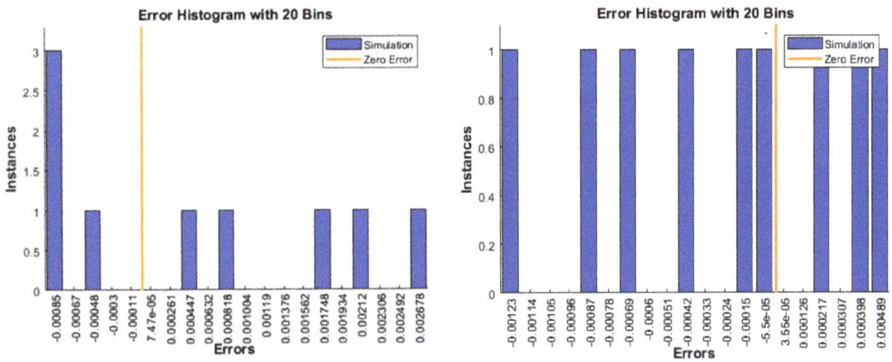

(i) (ii)

Figure 7. Error histogram of simulated data for (**i**) AN7-A, (**ii**) AN7-B.

4. Conclusions

Thermal cracking of polymers, consisting of PS, LDPE, and PP, was implemented using TGA at heating rate range (5–40 K/min), with two groups of sets: (ratio 1:1) a mixture of PS and PP, and (ratio 1:1:1) a mixture of PS, LDPE, and PP. TGA data are used in modeling ANN for two sets of PS, LDPE, and PP polymers in order to predict the weight left %.

However, an efficient ANN model has been created to predict the thermal decomposition of these two sets separately. The best architecture of 2-10-10-1 (*logsig-tansig-purelin*) transfer functions has been adopted as the highest efficient model. This could foresee the output very precisely with high regression coefficient value. After that, the best model has been simulated with untrained input data, and its behavior (calculated output) shows a close agreement with the actual values (high $R > 0.9999$).

Funding: This research and APC was funded by Deanship of Scientific Research at King Faisal University (Saudi Arabia).

Institutional Review Board Statement: Not applicable.

Informed Consent Statement: Not applicable.

Data Availability Statement: Not applicable.

Acknowledgments: The author gratefully thank the Deanship of Scientific Research at King Faisal University (Saudi Arabia) for supporting this research as a part of the Research Grants Program (Old No.: NA000169, New No.:GRANT963).

Conflicts of Interest: The authors declare no conflict of interest.

References

1. Conesa, J.A.; Caballero, J.A.; Labarta, J.A. Artificial neural network for modelling thermal decompositions. *J. Anal. Appl. Pyrolysis* **2004**, *71*, 343–352. [CrossRef]
2. Bezerra, E.; Bento, M.; Rocco, J.; Iha, K.; Lourenço, V.; Pardini, L. Artificial neural network (ANN) prediction of kinetic parameters of (CRFC) composites. *Comput. Mater. Sci.* **2008**, *44*, 656–663. [CrossRef]
3. Yıldız, Z.; Uzun, H.; Ceylan, S.; Topcu, Y. Application of artificial neural networks to co-combustion of hazelnut husk–lignite coal blends. *Bioresour. Technol.* **2016**, *200*, 42–47. [CrossRef] [PubMed]
4. Çepelioğullar, Ö.; Mutlu, I.; Yaman, S.; Haykiri-Acma, H. A study to predict pyrolytic behaviors of refuse-derived fuel (RDF): Artificial neural network application. *J. Anal. Appl. Pyrolysis* **2016**, *122*, 84–94. [CrossRef]
5. Ahmad, M.S.; Mehmood, M.A.; Taqvi, S.T.H.; Elkamel, A.; Liu, C.-G.; Xu, J.; Rahimuddin, S.A.; Gull, M. Pyrolysis, kinetics analysis, thermodynamics parameters and reaction mechanism of Typha latifolia to evaluate its bioenergy potential. *Bioresour. Technol.* **2017**, *245*, 491–501. [CrossRef]
6. Çepelioğullar, Ö.; Mutlu, I.; Yaman, S.; Haykiri-Acma, H. Activation energy prediction of biomass wastes based on different neural network topologies. *Fuel* **2018**, *220*, 535–545. [CrossRef]

7. Chen, J.; Xie, C.; Liu, J.; He, Y.; Xie, W.; Zhang, X.; Chang, K.; Kuo, J.; Sun, J.; Zheng, L.; et al. Co-combustion of sewage sludge and coffee grounds under increased O_2/CO_2 atmospheres: Thermodynamic characteristics, kinetics and artificial neural network modeling. *Bioresour. Technol.* **2018**, *250*, 230–238. [CrossRef]
8. Naqvi, S.R.; Tariq, R.; Hameed, Z.; Ali, I.; Taqvi, S.A.; Naqvi, M.; Niazi, M.B.; Noor, T.; Farooq, W. Pyrolysis of high-ash sewage sludge: Thermo-kinetic study using TGA and artificial neural networks. *Fuel* **2018**, *233*, 529–538. [CrossRef]
9. Ahmad, M.S.; Liu, H.; Alhumade, H.; Tahir, M.H.; Çakman, G.; Yıldız, A.; Ceylan, S.; Elkamel, A.; Shen, B. A modified DAEM: To study the bioenergy potential of invasive Staghorn Sumac through pyrolysis, ANN, TGA, kinetic modeling, FTIR and GC–MS analysis. *Energy Convers. Manag.* **2020**, *221*, 113173. [CrossRef]
10. Bi, H.; Wang, C.; Lin, Q.; Jiang, X.; Jiang, C.; Bao, L. Combustion behavior, kinetics, gas emission characteristics and artificial neural network modeling of coal gangue and biomass via TG-FTIR. *Energy* **2020**, *213*, 118790. [CrossRef]
11. Bong, J.T.; Loy, A.C.M.; Chin, B.L.F.; Lam, M.K.; Tang, D.K.H.; Lim, H.Y.; Chai, Y.H.; Yusup, S. Artificial neural network approach for co-pyrolysis of Chlorella vulgaris and peanut shell binary mixtures using microalgae ash catalyst. *Energy* **2020**, *207*, 118289. [CrossRef]
12. Bi, H.; Wang, C.; Lin, Q.; Jiang, X.; Jiang, C.; Bao, L. Pyrolysis characteristics, artificial neural network modeling and environmental impact of coal gangue and biomass by TG-FTIR. *Sci. Total Environ.* **2020**, *751*, 142293. [CrossRef] [PubMed]
13. Liew, J.X.; Loy, A.C.M.; Chin, B.L.F.; AlNouss, A.; Shahbaz, M.; Al-Ansari, T.; Govindan, R.; Chai, Y.H. Synergistic effects of catalytic co-pyrolysis of corn cob and HDPE waste mixtures using weight average global process model. *Renew. Energy* **2021**, *170*, 948–963. [CrossRef]
14. Zaker, A.; Chen, Z.; Zaheer-Uddin, M. Catalytic pyrolysis of sewage sludge with HZSM5 and sludge-derived activated char: A comparative study using TGA-MS and artificial neural networks. *J. Environ. Chem. Eng.* **2021**, *9*, 105891. [CrossRef]
15. Dubdub, I.; Al-Yaari, M. Pyrolysis of Low Density Polyethylene: Kinetic Study Using TGA Data and ANN Prediction. *Polymers* **2020**, *12*, 891. [CrossRef] [PubMed]
16. Dubdub, I.; Al-Yaari, M. Thermal Behavior of Mixed Plastics at Different Heating Rates: I. Pyrolysis Kinetics. *Polymers* **2021**, *13*, 3413. [CrossRef]
17. Al-Yaari, M.; Dubdub, I. Application of Artificial Neural Networks to Predict the Catalytic Pyrolysis of HDPE Using Non-Isothermal TGA Data. *Polymers* **2020**, *12*, 1813. [CrossRef]
18. Al-Yaari, M.; Dubdub, I. Pyrolytic Behavior of Polyvinyl Chloride: Kinetics, Mechanisms, Thermodynamics, and Artificial Neural Network Application. *Polymers* **2021**, *13*, 4359. [CrossRef]
19. Quantrille, T.E.; Liu, Y.A. *Artificial Intelligence in Chemical Engineering*; Elsevier Science: Amsterdam, The Netherlands, 1992.
20. Halali, M.A.; Azari, V.; Arabloo, M.; Mohammadi, A.H.; Bahadori, A. Application of a radial basis function neural network to estimate pressure gradient in water–oil pipelines. *J. Taiwan Inst. Chem. Eng.* **2016**, *58*, 189–202. [CrossRef]
21. Govindan, B.; Jakka, S.C.B.; Radhakrishnan, T.K.; Tiwari, A.K.; Sudhakar, T.M.; Shanmugavelu, P.; Kalburgi, A.K.; Sanyal, A.; Sarkar, S. Investigation on Kinetic Parameters of Combustion and Oxy-Combustion of Calcined Pet Coke Employing Thermogravimetric Analysis Coupled to Artificial Neural Network Modeling. *Energy Fuels* **2018**, *32*, 3995–4007. [CrossRef]
22. Bar, N.; Bandyopadhyay, T.K.; Biswas, M.N.; Das, S.K. Prediction of pressure drop using artificial neural network for non-Newtonian liquid flow through piping components. *J. Pet. Sci. Eng.* **2010**, *71*, 187–194. [CrossRef]
23. Al-Wahaibi, T.; Mjalli, F.S. Prediction of Horizontal Oil-Water Flow Pressure Gradient Using Artificial Intelligence Techniques. *Chem. Eng. Commun.* **2013**, *201*, 209–224. [CrossRef]
24. Osman, E.-S.A.; Aggour, M.A. Artificial Neural Network Model for Accurate Prediction of Pressure Drop in Horizontal and Near-Horizontal-Multiphase Flow. *Pet. Sci. Technol.* **2002**, *20*, 1–15. [CrossRef]
25. Qinghua, W.; Honglan, Z.; Wei, L.; Junzheng, Y.; Xiaohong, W.; Yan, W. Experimental Study of Horizontal Gas-liquid Two-phase Flow in Two Medium-diameter Pipes and Prediction of Pressure Drop through BP Neural Networks. *Int. J. Fluid Mach. Syst.* **2018**, *11*, 255–264. [CrossRef]
26. Beale, M.H.; Hagan, M.T.; Demuth, H.B. *Neural Network Toolbox TM User's Guide*; MathWorks: Natick, MA, USA, 2018.

polymers

MDPI

Article

Penetration of Hydrogen into Polymer Electrolyte Membrane for Fuel Cells by Quantum and Molecular Dynamics Simulations

JinHyeok Cha *, Wooju Lee and Jihye Baek

Institute of Fundamental and Advanced Technology, Hyundai Motor Company, 37 Cheoldobangmulgwan-ro, Uiwang-si 16082, Gyeonggi-do, Korea; wooju@hyundai.com (W.L.); jh.baek@hyundai.com (J.B.)
* Correspondence: chajinhyeok@hyundai.com

Abstract: The advent of the Hydrogen Society created great interest around hydrogen-based energy a decade ago, with several types of vehicles based on hydrogen fuel cells already being produced in the automotive sector. For highly efficient fuel cell systems, the control of hydrogen inside a polymer-based electrolyte membrane is crucial. In this study, we investigated the molecular behavior of hydrogen inside a polymer-based proton-exchange membrane, using quantum and molecular dynamics simulations. In particular, this study focused on the structural difference of the pendent-like side chain polymer, resulting in the penetration ratio of hydrogen into the membrane deriving from the penetration depth of the membrane's thickness while keeping the simulation time constant. The results reveal that the penetration ratio of the polymer with a shorter side chain was higher than that with the longer side chain. This was justified via two perspectives; electrostatic and van der Waals molecular interactions, and the structural difference of the polymers resulting in the free volume and different behavior of the side chain. In conclusion, we found that a longer side chain is more trembling and acts as an obstruction, dominating the penetration of hydrogen inside the polymer membrane.

Keywords: polymer electrolyte membrane for fuel cell; molecular dynamics simulations; side chain; penetration

Citation: Cha, J.; Lee, W.; Baek, J. Penetration of Hydrogen into Polymer Electrolyte Membrane for Fuel Cells by Quantum and Molecular Dynamics Simulations. *Polymers* **2021**, *13*, 947. https://doi.org/10.3390/polym13060947

Academic Editors: Célio Bruno Pinto Fernandes, Salah Aldin Faroughi, Luís L. Ferrás and Alexandre M. Afonso

Received: 24 February 2021
Accepted: 17 March 2021
Published: 19 March 2021

1. Introduction

Hydrogen, as an energy resource, can make life significantly more eco-friendly. Although the hydrogen fuel cell was introduced by William Robert Grove 180 years ago [1], numerous potential applications have recently raised extreme engagement from many researchers, in various industrial sectors [2]. Automotive systems based on hydrogen fuel cells can already be found on the road, with examples such as "NEXO" produced by Hyundai Motor Company or "MIRAI" by Toyota representing the upcoming hydrogen-based automotive future.

A hydrogen fuel cell works based on electrochemistry, by passing hydrogen through the anode and oxygen through the cathode of the cell. More specifically, electrical power is generated via three simple steps: (i) the dissociation of hydrogen to the proton and electron; (ii) the conduction of electrons through electronic channels and protons through a proton-exchange membrane; and (iii) the synthesis of water by a proton, electron, and oxygen. A polymer electrolyte membrane for fuel cell (PEMFC) is commonly used in vehicles due to better operation at relatively low temperature, while other types of fuel cells (such as alkaline cell, phosphoric acid fuel cell, molten carbonate cell, direct methanol fuel cell and solid oxide cell) require a higher operation temperature [3,4].

A stack, one of the main components of PEMFCs, comprises of serially connected unit cells generating electricity. Each of them features a bipolar gas diffusion layer (GDL) and a membrane electrode assembly (MEA), which are directly responsible for the electrochemical reaction. Figure 1a shows a schematic of a proton exchange membrane in the fuel cell system. The proton exchange membrane is a key component of PEMFCs, acting excellently towards gas blocking and proton conduction. In particular, gas diffusion behavior inside MEA has a direct impact on the performance of the fuel cell, as the diffusion of hydrogen and oxygen at both the anode and cathode determines the efficiency of the fuel cell, hence shortened diffusion time induces faster proton transport and water conversion [5,6]. In addition, gas crossover through the PEMFC during long-term membrane operation causes the degradation of the MEA, due to the combustion reaction between H_2 and O_2 gases and the formation of reactive oxygen radicals (HO• and HOO• radicals) which attack membrane linkages [7–11].

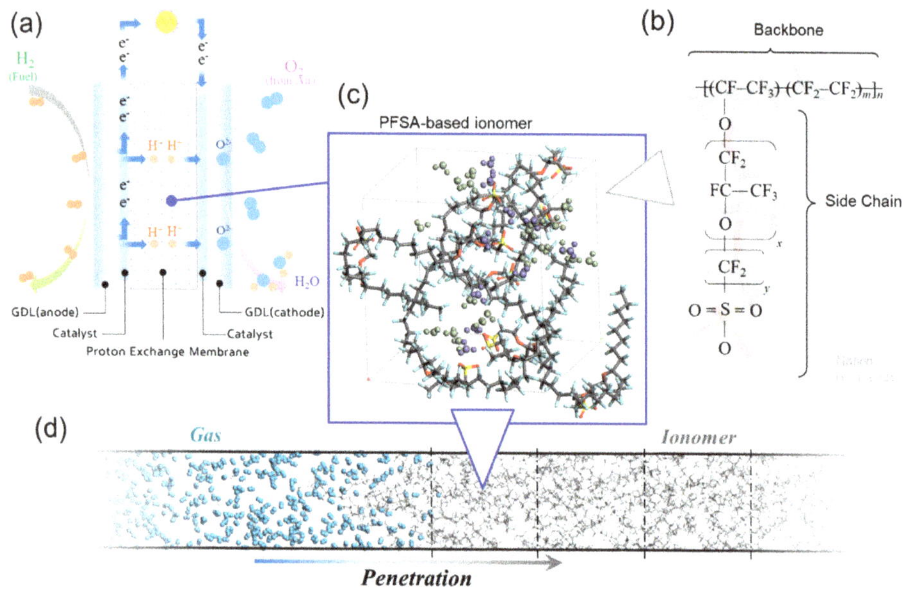

Figure 1. (**a**) Schematic of the operation of a fuel cell; (**b**) snapshot of the unit cell for the molecular dynamic (MD) simulation of the morphological effect of the side chain on gas penetration, containing H_3O^+ and H_2O with the hydration number of 3; (**c**) index x (backbone) and y (side chain) of the chemical structure of perfluorosulfonic acid determine the length of side chain; and (**d**) schematic of the MD simulation for hydrogen molecules' penetration into the ionomer region.

Gas is transported through carriers of polymeric surrounding, such as a polymer-based binder supporting catalysts at both electrodes, and a polymeric electrolyte membrane. Therefore, the interaction between the gas molecules and polymer needs to be clarified, in order to control the gas flow in and out. The transport of gas molecules in the PEM depends on the geometric properties of polymeric media as well as atomic interactions between gas and polymer molecules. The geometrical effects of polymeric media can be analyzed by using a theoretical model of diffusion in porous media. Several studies have been reported on the diffusion phenomena in the randomly distributed microstructure by employing fractal geometry theory [12,13], and those studies show a good agreement with the available experimental data and existing models reported in the literature. Even the theoretical model of diffusion in the porous media well predicts the effective transport properties of gas molecules, the effect of molecular interactions should be fully understood to design polymeric materials in molecular-level (e.g., chain length, electrostatic interaction, radius of gyration of polymer formed by different chain structures).

To explore the molecular behavior of gas (e.g., hydrogen at the anode and oxygen at the cathode) [14], numerous studies have employed molecular dynamic (MD) simulations and first-principles simulations [15–17]. MD simulation is a method of calculating and statistically processing material structures, thermodynamic properties, and reaction properties at atomic scales. It mathematically solves the kinetic trajectory of atoms and molecules based on Newton's equations of motion for the hundreds and hundreds of thousands of molecules that change over time by parametrizing the interaction potential between atoms. Kwon et al. evaluated the molecular behavior inside the polymeric membrane formed with a different equivalent weight (EW) of side chains at various hydrated levels by volume analysis that the lower EW of the ionomer exhibits higher diffusion coefficients of particles such as water, hydronium, and oxygen molecules [15]. Cha explored the morphological effect of the side chain on H_3O^+ transfer that short side chains induced more inter-chain cation movement than longer side chains [16]. Takeuchi et al. investigated that the local crystalline structure which inhibits the H_2 diffusion across the aligned polymer chains due to the void fraction in the structure, resulting in reducing H_2 permeability [17]. Jinnouchi et al. reported the transport properties of O_2 molecules inside Nafion ionomer using the density functional theory (DFT) calculation and MD simulations, showing that interfacial formation between Nafion and Pt significantly enhances O_2 permeation through the ionomer thin film [18]. Although all of these studies mainly used MD simulation to investigate the phenomena for molecular behavior in the polymeric membrane, however, there is still a lack of a more detailed understanding of the correlations between the structural characteristics of the polymer and the penetration of molecular behavior. To facilitate the research on polymer characteristics, it is necessary to focus on the interface between polymer and gas molecules, exploring the effect of properties of the polymer such as its type, structure, and ability for penetration by a specific gas.

In this study, we performed MD simulations focusing on hydrogen penetration into a polymer matrix. We employed a quantitative evaluation method, examining the effect of morphological differences of the polymer matrix on the penetration depth achieved. This study used two perspectives; molecular interaction (such as electrostatic and van der Waals interaction) and structural difference resulting in the free volume and behavior of polymer side chains. The results reveal that membranes based on polymer containing shorter side chains led to a higher penetration ratio than that of longer side chains. In addition, the increased trembling of the longer side chain compared to the shorter one acted as an obstruction, dominating the penetration ability of hydrogen inside the polymer.

2. Simulation Methodology

2.1. Models of Polymer Electrolyte Membranes and Hydrogen

A crucial factor of designing the MEA is that hydrogen from GDL should not penetrate into the polymer electrolyte membrane, resulting in the efficiency of the fuel cell system. Since the main purpose of this study is to evaluate the undesirable diffusion of undissociated hydrogen into the membrane, the dissociation process of hydrogen at the Pt particles was neglected. In other words, all of the hydrogen molecules penetrating the membrane were considered to be crossover. Based on this assumption, we only investigated the structural effects of the polymer on hydrogen penetration into the membrane.

We employed MD simulations to clarify the structural and kinetic properties of the polymer, which can calculate material structures, thermodynamic properties, and reaction properties at atomic scales. However, the MD simulation is hard to describe the varying hydration level and temperature of the real PEMFC system due to the large difference in time-scale between the simulation (ns) and operation (s). To simplify the calculation, we assumed that the hydration level and temperature were fixed. Several studies have reported selective penetration techniques under development, such as prevention strategies for the crossover of hydrogen from anode side, and the oxygen from cathode [14,19]. To investigate the effect of polymer structure on hydrogen penetration into the polymer electrolyte membrane, we employed perfluorosulfonic acid (PFSA)-based ionomer (commercially known as Nafion, most commonly used in fuel cells developed by DuPont Inc. more than 40 years ago) [20]. Figure 1b show the chemical structure of the polymer, including a backbone (–CF_2–) and pendent–like side chain (–O[$CF_2CF(CF_3)O]_x[CF_2]_ySO_3$–) groups (index x and y of Nafion is equal to 1 and 2, respectively). The equivalent weight of the Nafion employed in this study was 1.147.

To determine the structural difference of the side chain as longer and shorter, we varied "y" whilst keeping "x" constant. The structural length of the side chain is considered an impactful factor for the penetration of hydrogen at the surface of the polymer electrolyte membrane. The chain length ranged from 4.66 to 5.98 Å, and the number of atoms in a single chain was 582, 682, 982, and 1282, corresponding to y values of 1, 2, 5 and 8, respectively. The unit cell contained a single PFSA chain with 10 side chains, indicating 10 negatively charged sulfonate groups. Therefore, we added 10 positively charged H_3O^+ groups to the unit cell, to maintain the equivalent net charge constant. Moreover, each unit cell featured 20 H_2O molecules, as the hydration number (λ) was set to be 3, as shown in Figure 2c. The simulation system composed of 20 unit cells with adjusted density to about 1.7 g/cm^3, considering the hydration state [21]. The simulation system consisted of a rectangular cell with length ranging from 2.14 × 2.14 × 42.9 nm to 2.41 × 2.41 × 48.3 nm for the polymer membrane, and 2.30 × 2.30 × 6.33 nm for the hydrogen layer, resulting in a volume of 197.0–281.2 nm^3 and 33.5 nm^3, respectively. The density of the cell for hydrogen layer was kept constant at 0.05 g/cm^3, corresponding to ≈70 MPa, equivalent to the pressure in hydrogen storage tank.

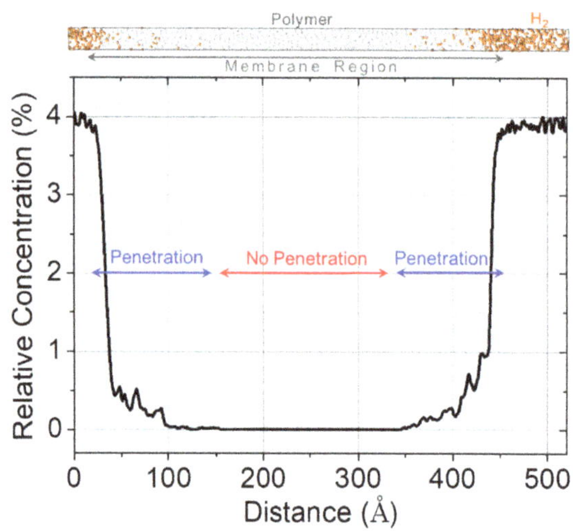

Figure 2. Relative concentration profile of hydrogen molecules. The simulation system is divided into penetrated and not penetrated regions. Their ratio is considered as a criterion for the comparison of penetration as a result of the side chain morphology.

2.2. Simulation Details

All MD simulations were based on the condensed-phase optimized molecular potentials for atomistic simulation studies (COMPASS) force field potential in the software package Materials Studio 2016 (BIOVIA Software Inc., San Diego, CA, USA), which is a general all-atom force field for the atomistic simulation of common organic molecules, inorganic small molecules, and polymers [22,23].

The Verlet velocity algorithm was used for the integration of the motion equations [24]. Simulations were performed at room temperature (298 K), and regulated using a Nose–Hoover–Langevin thermostat [25,26]. The NVT ensemble was employed to identify hydrogen molecular behavior with a 1 fs time step, where N, V and T correspond to the number of atoms, volume, and temperature of the simulation system, respectively. We carried out each simulation for 1 ns and repeated it 10–15 times for each different initial structure, excluding data from the initial 200 ps assuming unexpected molecular behavior.

Density functional theory (DFT) calculations were carried out to clarify the electrostatic interaction between hydrogen and the side chain using the Dmol^3 program, which is a theory for computing the shape of electrons within a molecule, and their energy based on quantum mechanics. This is one of the most widely used quantum mechanics calculations that allows to predict whether the molecule can exist in the world or not, and the form and properties of a particular molecule. Through the charge distribution of molecules calculated by DFT, it enabled accurately simulating the behavior of particles by representing the electrostatic interaction (Coulomb interaction) between molecules. The electronic exchange-correlation functional used was the Perdew–Burke–Emzerhog (PBE) functional with the generalized gradient approximation (GGA) [27], and the spin-polarized calculations were performed using a double numerical basis set with polarization functions (DNP). All electron relativistic effects were included for the treatment of core electrons in the models. The molecular binding energy of H_3O^+ or H_2 on the side chain was determined as follows:

$$E_{\text{binding energy}} = E_{\text{total energy}} - E_{\text{side chain}} - E_{\text{molecules}} \tag{1}$$

where $E_{\text{total energy}}$ is the total energy of the side chain with the bound molecules, and the $E_{\text{side chain}}$ and $E_{\text{molecules}}$ are the total energies of the side chain and the molecules, respectively. Note that in order to calculate the binding energy of H_2 on the side chain, the total energy includes the bonding state of H_3O^+ on side chain.

3. Results and Discussion

3.1. Quantitative Evaluation Model for the Effect of Side Chain on Penetration

There are unpredictably numerous polymeric structures for a given system, resulting from polymeric gyration, steric hindrance, and the free volume inside a system. However, it is practically impossible to explore the permeation of hydrogen into the polymer-based membrane for every possible structure. Thus, in this study, we focused on the structural effect of the side chain on hydrogen permeation, aiming to improve the design of polymer electrolyte membranes.

The concept of permeation implies that molecules pass through the membrane from the one side to the other. For most polymeric membranes, gas permeability is calculated by multiplying the diffusivity and solubility of the penetrant gas in the polymer [28]. However, it is very hard to observe the whole infiltration process from the one to the other side at once, because it requires an insurmountably long time and large ability to simulate it. MD simulation can describe a partial molecular behavior, for example on the membrane interface, focusing on the initial "penetration" of hydrogen into PFSA-based electrolyte membrane.

In this study, we investigated the instant interfacial phenomenon of molecular penetration from the hydrogen layer into the PFSA-based membrane using MD simulation, which can potentially provide physical and chemical information for the design of a hydrogen-controllable polymer for PEMFC fabrication. Figure 1d shows a snapshot of the MD simulation system, composed of a polymer region sufficiently long in the axial direction, and showing the combination of hydrogen and polymer layers during penetration. Then, we quantitatively evaluated the depth of hydrogen penetration into the polymer. The depth depends on the simulation time corresponding to solubility. The longer simulation time facilitates deeper hydrogen penetration into the polymer. However, the normalized depth ratio can overcome the dependence on constant simulation time, focusing on the structural effect of the polymer and its effect on hydrogen penetration.

Figure 2 shows the evaluation method employed to interpret results. First, we plotted the relative concentration profile of hydrogen in polymer over the axial direction for the entire simulation system. Then, we divided the polymer layer into two parts; the penetrated and the unpenetrated. The ratio of the length of membrane to penetration depth by hydrogen molecules ranged from 0 to 1, using normalized values as criteria for the comparison of penetration resulting from the morphology of the side chain. Whilst "0" meant that membrane was not at all penetrated, "1" implied full hydrogen penetration.

Figure 3 shows the effect of the side chain on hydrogen penetration into the polymer membrane. A shorter side chain allowed hydrogen to penetrate deeper inside the membrane compared to a longer chain. In particular, the PFSA-based polymer with the longest side chain in this study (eight repeating units) allowed for a 21.7% deeper hydrogen penetration compared to that with the shortest chain (one repeating unit). We devised three potential reasons to explain the obtained results. We assumed that the interaction caused from the electrostatic Coulomb force or van der Waals interaction between the hydrogen and each side chain resulted in hydrogen penetration. Structural differences of each side chain could also lead to a different free volume inside the system and accelerate or decelerate the molecular transport of hydrogen. Moreover, the different physical behavior of the side chain resulting from its length could facilitate or disturb hydrogen penetration inside the polymer membrane.

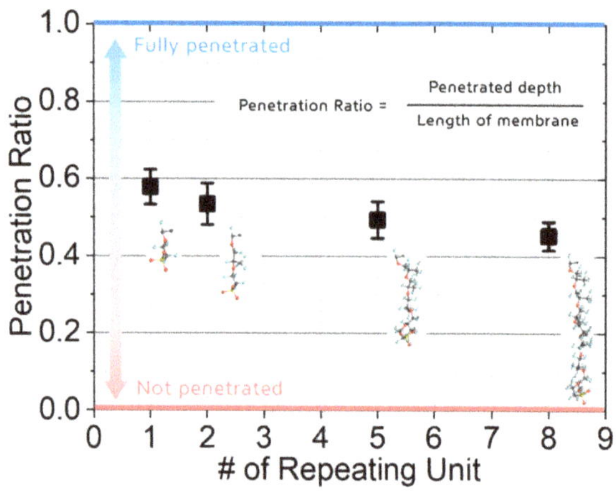

Figure 3. The effect of the side chain length on the penetration depth into polymeric membrane. The longer side chain (8 repeating units) causes hydrogen molecules to penetrate 21.7% deeper compared to the penetration achieved with the shortest side chain (1 repeating unit).

3.2. *Hydrogen Adsorption onto the Side Chain*

Hydrogen provided by the gas diffusion layer is dissociated with electron and proton with the aid of the catalyst. As described above, a membrane with selective permeability for proton and hydrogen is an essential factor for the fuel cell efficiency. Hydrogen interacts with Nafion electrostatically, and Nafion's polymeric structure contains a hydrophobic backbone ($-CF_2^-$) and a hydrophilic side chain having a negatively charged sulfonate group ($-SO_3^-$). This implies that the longer the length of a side chain is, the greater the difference in the charge distribution within the side chain will be. Therefore, we investigated the interaction of hydrogen with a side chain, having a negatively charged sulfonate group at the end, using DFT calculations. Figure 4 shows the comparison of the charge distribution for the side chains of various lengths, with or without H_3O^+ attached on the sulfonate group, corresponding to the lower and upper part of the figure respectively. The sulfonate group exhibits strong attractive interaction, leading to a hydrophilic network with H_3O^+. Negatively charged distribution along the side chain was neutralized by the attached H_3O^+ onto sulfonate group, irrespectively of the chain length. Consequently, we concluded that the electrostatic interaction between hydrogen and the side chain was weak enough to be considered negligible.

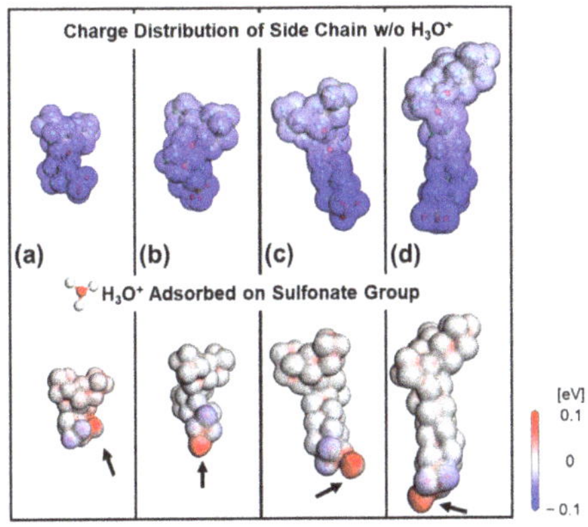

Figure 4. Charge distribution of the side chain before and after H_3O^+ is adsorbed on the negatively charged sulfonate group. H_3O^+adsorption leads to neutralization despite the chain length. The length index y corresponds to each of the (**a**)–(**d**) side chain is 1, 2, 5 and 8, respectively.

On the other hand, we considered the binding energy resulting from van der Waals forces driven by induced electrical interactions between two close atoms. The three positions selected to calculate binding energy are as shown in Figure 5a, close to the backbone, middle of the side chain, and around the adsorption of the sulfonated group and hydronium. The binding energy of the hydrogen molecules on side chain with adsorbed H_3O^+ ranged from −0.005 to −0.02 eV (Figure 5b–d), obviously referring to physical adsorption between them, with the distance ranging from 3.0 to 3.5 Å Therefore, we can safely conclude that van der Waals interaction has a negligible effect on hydrogen molecular behavior into the polymeric membrane, although the longer side chain offers a wider surface for physisorption.

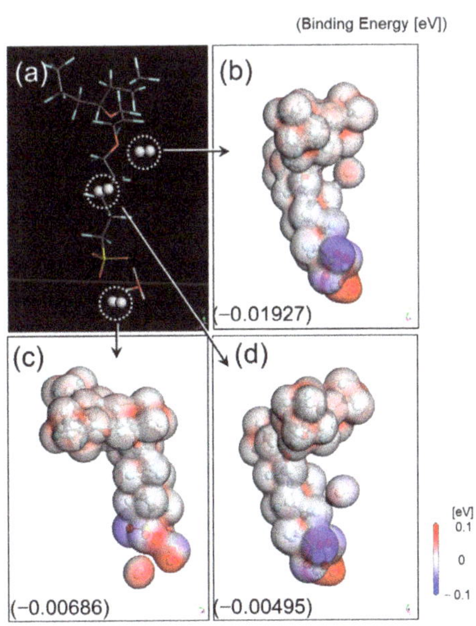

Figure 5. (**a**) Binding energy of hydrogen molecules on side chain with adsorbed H_3O^+; (**b**) possible positions for the adsorption of H_2 molecules; (**c**) charge distribution; and (**d**) binding energy.

3.3. Physical Obstruction against the Molecular Penetration

Hydrogen molecules are chemically stable without any reaction during the simulation, and DFT simulations also found only negligible chemical interactions through the binding energy calculations described in Section 3.2. Therefore, we focused on physical factors which can be divided into two categories: structural and kinetic. There are several representative polymer structures, such as linear, branched, cross-linked, and networked. Polymers containing side chain in particular, can be described by numerous complex morphologies. Thus, it is unfeasible to examine all these structures for their morphological influence, so we only varied the morphology of the side chain that causes hydrogen penetration, which is one of the most factors under the same composition. We found the effect of structure based on a single chain containing various side chain on hydrogen penetration. For that, we employed the radius of gyration (R_g), a basic concept for the feature of polymer structures, and free volume formed from it. For that, we employed the R_g, a basic concept for the feature of polymer structures, defined as the root-mean-square (RMS) distance of the collection of atoms in the molecule from their common center of mass, and calculated from the following Equation:

$$R_g{}^2 = \sum_{i=1}^{N} m_i s_i{}^2 / \sum_{i=1}^{N} m_i \quad (2)$$

where s_i, m_i and N denote the distance of atom i from the center of mass, the mass of atom i, and the total number of atoms, respectively. Figure 6a shows the dependence of R_g on the side chain length that the R_g of the longest longer side chain is 1. 5 times larger than that of the shortest. Since the difluoromethylene group tends to aggregate in water due to the hydrophobicity of the backbone, the longer side chain with a stronger hydrophilic group has higher R_g than shorter side chains. The increase in R_g resulting from the longer side chain led to the simulation system containing more free volume, since the system needed to adjust the volume in order to keep the density constant, as shown Figure 6b. The free volume measured for the longest side chain employed in this study was 1.5 times larger than that of the shortest. Although having more free volume could lead to increased hydrogen penetration, the simulation results contradicted that. The simulation was carried out for the instant phenomenon at the interface, which was not enough to thoroughly investigate the effect of free volume on hydrogen penetration, due to the low density (or pressure) inside the polymer layer where hydrogen behaves. For further clarification of the influence of free volume, the model should be built in a fully solvated state.

In addition to morphology, we also investigated the effect of the side chain on the polymer's kinetic behavior. More specifically, to find the kinetic behavior of the side chain, we calculated the molecular diffusivity, often called the diffusion coefficient (D), using Equation (3):

$$D = \frac{1}{6} \lim_{\Delta t \to \infty} \frac{dMSD}{d\Delta t} \quad (3)$$

where MSD, the mean square displacement of sulfur at the end of side chain, is given by

$$MSD \equiv \left\langle (x - x_0)^2 \right\rangle = \frac{1}{T} \sum_{t=1}^{T} (x(t) - x_0)^2 \quad (4)$$

where T is the average time and x_0 is the reference position of the particle. We found that an increased number of repeating units leads to the high diffusivity of the sulfur atom. Figure 6c shows the diffusivity of the sulfur atom at the end of side chains of various lengths, derived from the mean square displacement that increased with the number of repeating units, as shown in Figure 6d. Since the sulfur atoms are bonded at the end of the side chain, the diffusion of sulfur atoms presented in this study does not imply diffusion into other spaces, but rather the degree of movement within the space allowed by the side chain. Considering the analysis of diffusion coefficient, we believe that the more kinetic behavior of longer side chains mainly acts as a hindrance to hydration molecules penetration into the polymer membrane.

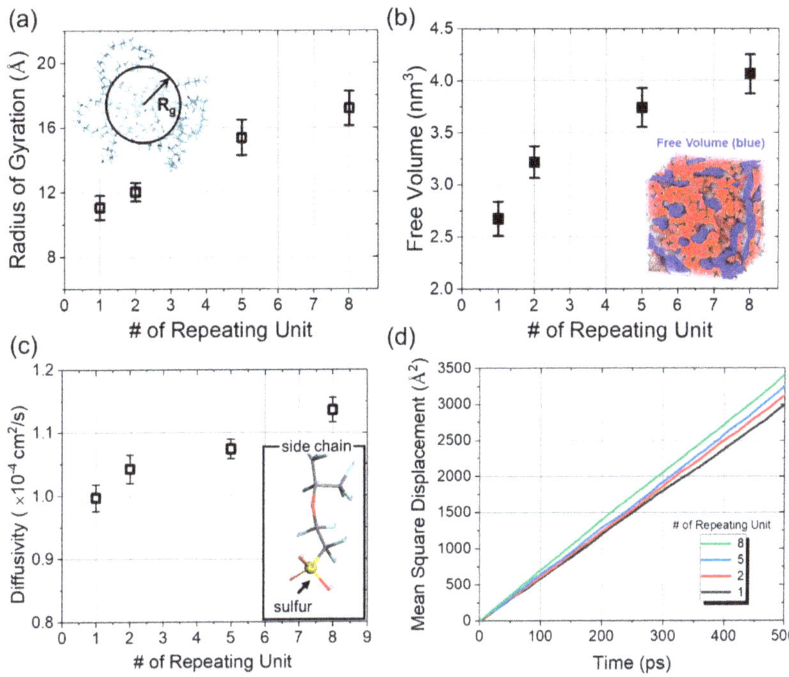

Figure 6. (**a**) Increase in the radius of gyration with the repeating unit number, which has a physical influence on the external molecular behavior; (**b**) the free volume with the repeating unit number for each simulation system, presenting the length of side chain; and (**c**) the increase in the diffusivity of the sulfur atom at the end of the side chain length, derived from (**d**) the mean square displacement increasing with the number of repeating units.

The operating temperature of PEMFC reached up to 65 °C decreases the activation loss of the catalyst and increases voltage loss (ohmic loss) and hydrogen crossover, and changes relatively to humidity to be lower than at room temperature. The MD simulation with the *NVT* ensemble employed in this study is unable to present the change in relative humidity, including the phase transition of water at every moment. Thus, we performed the simulations at room temperature where those changes do not occur. Since the temperature affects the behavior of hydrogen more than that of the polymer, we thus expect that higher temperature allows relatively more hydrogen penetration. In the near future, we carried out more simulations with various conditions to break through the limiting factors found in this study.

4. Conclusions

In summary, over the past decade, numerous studies have been reported on hydrogen-based energy systems, targeting industrial applications. Several types of vehicles based on hydrogen fuel cells have already been produced. For the fuel cell system to be efficient, it is crucial to control the hydrogen inside the polymer-based electrolyte membrane which transfers it, without its dissociation into protons and electrons, as it decreases the fuel efficiency. Thus, it is necessary to clarify the penetration mechanism of hydrogen on the surface of the polymer, and the effect of the polymer properties, such as the type, structure, and size on penetrations. In this study, we used molecular dynamics simulations to investigate the mechanism of hydrogen molecular behavior inside a polymer-based proton exchange membrane. Using a quantitative evaluation method and focusing on how deeply hydrogen penetrates into morphologically different polymers, this study was approached through two perspectives. The first was molecular interactions, such as electrostatic and van der Waals interactions, and the second was the structural difference, resulting in differences in free volume and the behavior of side chain. The results show that the polymer membrane containing a shorter side chain caused a higher penetration ratio than a longer side chain. In addition, the more trembling longer side chain acted as an obstruction, dominating the penetration of hydrogen inside the polymer. With the results obtained in this study as the beginning point, we further expanded the research to find the effect of the hydration level, different temperature, the addition of additives, and a hybrid membrane with a skeleton considering the improvement of durability, on hydrogen crossover, which contributes to estimating the optimal design factors for a polymer-based electrolyte membrane.

Author Contributions: Conceptualization, J.C., W.L. and J.B.; methodology, J.C. and W.L.; software, J.B.; validation, J.C. and J.B.; formal analysis, J.C. and J.B.; investigation, J.C.; data curation, J.C. and J.B.; writing—original draft preparation, J.C.; writing—review and editing, J.C.; visualization, J.C.; supervision, J.C.; project administration, J.C. All authors have read and agreed to the published version of the manuscript.

Funding: This research was supported by Hyundai Motor Company.

Institutional Review Board Statement: Not applicable.

Informed Consent Statement: Not applicable.

Data Availability Statement: The data presented in this study are available on request from the corresponding author.

Conflicts of Interest: The authors declare no conflict of interest.

References

1. Cameron, D.S. Fuel Cell Energy Generators. *Platin. Met. Rev.* **1978**, *22*, 38–46.
2. Wang, Y.; Chen, K.S.; Mishler, J.; Cho, S.C.; Adroher, X.C. A review of polymer electrolyte membrane fuel cells: Technology, applications, and needs on fundamental research. *Appl. Energy* **2011**, *88*, 981–1007. [CrossRef]
3. Mekhilef, S.; Saidur, R.; Safari, A. Comparative study of different fuel cell technologies. *Renew. Sustain. Energy Rev.* **2012**, *16*, 981–989. [CrossRef]
4. Smitha, B.; Sridhar, S.; Khan, A.A. Solid polymer electrolyte membranes for fuel cell applications—A review. *J. Memb. Sci.* **2005**, *259*, 10–26. [CrossRef]
5. Inaba, M.; Kinumoto, T.; Kiriake, M.; Umebayashi, R.; Tasaka, A.; Ogumi, Z. Gas crossover and membrane degradation in polymer electrolyte fuel cells. *Electrochim. Acta* **2006**, *51*, 5746–5753. [CrossRef]
6. Francia, C.; Ijeri, V.S.; Specchia, S.; Spinelli, P. Estimation of hydrogen crossover through Nafion® membranes in PEMFCs. *J. Power Sources* **2011**, *196*, 1833–1839. [CrossRef]
7. Wu, B.; Zhao, M.; Shi, W.; Liu, W.; Liu, J.; Xing, D.; Yao, Y.; Hou, Z.; Ming, P.; Gu, J.; et al. The degradation study of Nafion/PTFE composite membrane in PEM fuel cell under accelerated stress tests. *Int. J. Hydrogen Energy* **2014**, *39*, 14381–14390. [CrossRef]
8. Umeda, M.; Maruta, T.; Inoue, M.; Nakazawa, A. Cathode platinum degradation in membrane electrode assembly studied using a solid-state electrochemical cell. *J. Phys. Chem. C* **2008**, *112*, 18098–18103. [CrossRef]
9. Dubau, L.; Castanheira, L.; Maillard, F.; Chatenet, M.; Lottin, O.; Maranzana, G.; Dillet, J.; Lamibrac, A.; Perrin, J.C.; Moukheiber, E.; et al. A review of PEM fuel cell durability: Materials degradation, local heterogeneities of aging and possible mitigation strategies. *Wiley Interdiscip. Rev. Energy Environ.* **2014**, *3*, 540–560. [CrossRef]
10. Han, M.; Shul, Y.G.; Lee, H.; Shin, D.; Bae, B. Accelerated testing of polymer electrolyte membranes under open-circuit voltage conditions for durable proton exchange membrane fuel cells. *Int. J. Hydrogen Energy* **2017**, *42*, 30787–30791. [CrossRef]
11. Zhang, S.; Yuan, X.; Wang, H.; Mérida, W.; Zhu, H.; Shen, J.; Wu, S.; Zhang, J. A review of accelerated stress tests of MEA durability in PEM fuel cells. *Int. J. Hydrogen Energy* **2009**, *34*, 388–404. [CrossRef]
12. Liang, M.; Fu, C.; Xiao, B.; Luo, L.; Wang, Z. A fractal study for the effective electrolyte diffusion through charged porous media. *Int. J. Heat Mass Transf.* **2019**, *137*, 365–371. [CrossRef]
13. Xiao, B.; Wang, W.E.I.; Zhang, X.; Long, G.; Chen, H.; Cai, H.; Deng, L.I.N. A novel fractal model for relative permeability of gas diffusion layer in proton exchange membrane fuel cell with capillary pressure effect. *Fractals* **2019**, *27*, 1950012. [CrossRef]
14. Baik, K.D.; Hong, B.K.; Kim, M.S. Novel technique for measuring oxygen crossover through the membrane in polymer electrolyte membrane fuel cells. *Int. J. Hydrogen Energy* **2013**, *38*, 8927–8933. [CrossRef]

15. Kwon, S.H.; Kang, H.; Lee, J.H.; Shim, S.; Lee, J.; Lee, D.S.; Kim, C.M.; Lee, S.G. Investigating the influence of the side-chain pendants of perfluorosulfonic acid membranes in a PEMFC by molecular dynamics simulations. *Mater. Today Commun.* **2019**, *21*, 100625. [CrossRef]
16. Cha, J.H. Morphological effect of side chain on H_3O^+ transfer inside polymer electrolyte membranes across polymeric chain via molecular dynamics simulation. *Sci. Rep.* **2020**, *10*, 22014. [CrossRef] [PubMed]
17. Takeuchi, K.; Kuo, A.T.; Hirai, T.; Miyajima, T.; Urata, S.; Terazono, S.; Okazaki, S.; Shinoda, W. Hydrogen Permeation in Hydrated Perfluorosulfonic Acid Polymer Membranes: Effect of Polymer Crystallinity and Equivalent Weight. *J. Phys. Chem. C* **2019**, *123*, 20628–20638. [CrossRef]
18. Jinnouchi, R.; Kudo, K.; Kitano, N.; Morimoto, Y. Molecular Dynamics Simulations on O2 Permeation through Nafion Ionomer on Platinum Surface. *Electrochim. Acta* **2016**, *188*, 767–776. [CrossRef]
19. Lee, S.; Jang, W.; Kim, M.; Shin, J.E.; Park, H.B.; Jung, N.; Whang, D. Rational Design of Ultrathin Gas Barrier Layer via Reconstruction of Hexagonal Boron Nitride Nanoflakes to Enhance the Chemical Stability of Proton Exchange Membrane Fuel Cells. *Small* **2019**, *15*, 1–9. [CrossRef] [PubMed]
20. Grot, W.G. Perfluorinated ion exchange polymers and their use in research and industry. *Macromol. Symp.* **2011**, *82*, 161–172. [CrossRef]
21. Morohoshi, K.; Hayashi, T. Modeling and simulation for fuel cell polymer electrolyte membrane. *Polymers* **2013**, *5*, 56–76. [CrossRef]
22. Sun, H. COMPASS: An ab Initio Force-Field Optimized for Condensed-Phase Applicationss Overview with Details on Alkane and Benzene Compounds. *J. Phys. Chem. B* **1998**, *5647*, 7338–7364. [CrossRef]
23. Sun, H.; Ren, P.; Fried, J.R. The COMPASS and validation. *Comput. Theor. Polym. Sci.* **1998**, *8*, 229–246. [CrossRef]
24. Loup Verlet Computer "Experiments" on Classical Fluids. I. Thermodynamical Properties of Lennard-Jones Molecules. *Phys. Rev.* **1967**, *159*, 98–103. [CrossRef]
25. Samoletov, A.A.; Dettmann, C.P.; Chaplain, M.A.J. Thermostats for " Slow " Configurational Modes. *J. Stat. Phys.* **2007**, 1321–1336. [CrossRef]
26. Leimkuhler, B.; Noorizadeh, E.; Penrose, O. Comparing the Efficiencies of Stochastic Isothermal Molecular Dynamics Methods. *J. Stat. Phys.* **2011**, 921–942. [CrossRef]
27. Ernzerhof, M.; Gustavo, S. Assessment of the Perdew-Burke-Ernzerhof exchange-correlation functional. *J. Chem. Phys.* **1991**, *110*, 4173–4184. [CrossRef]
28. Wu, G.; Almquist, C.L.B.; Hwang, S.T. High gas permeability in open-structure membranes. *Korean J. Chem. Eng.* **2004**, *21*, 442–453. [CrossRef]